Graduate Texts in Physics

Graduate Texts in Physics

Graduate Texts in Physics publishes core learning/teaching material for graduate- and advanced-level undergraduate courses on topics of current and emerging fields within physics, both pure and applied. These textbooks serve students at the MS- or PhD-level and their instructors as comprehensive sources of principles, definitions, derivations, experiments and applications (as relevant) for their mastery and teaching, respectively. International in scope and relevance, the textbooks correspond to course syllabi sufficiently to serve as required reading. Their didactic style, comprehensiveness and coverage of fundamental material also make them suitable as introductions or references for scientists entering, or requiring timely knowledge of, a research field.

More information about this series at http://www.springer.com/series/8431

Timm Krüger • Halim Kusumaatmaja •
Alexandr Kuzmin • Orest Shardt • Goncalo Silva •
Erlend Magnus Viggen

The Lattice Boltzmann Method

Principles and Practice

Springer

Timm Krüger
School of Engineering
University of Edinburgh
Edinburgh, United Kingdom

Halim Kusumaatmaja
Department of Physics
Durham University
Durham, United Kingdom

Alexandr Kuzmin
Maya Heat Transfer Technologies
Westmount, Québec, Canada

Orest Shardt
Department of Mechanical and Aerospace Engineering
Princeton University
Princeton, NJ, USA

Goncalo Silva
IDMEC/IST
University of Lisbon
Lisbon, Portugal

Erlend Magnus Viggen
Acoustics Research Centre
SINTEF ICT
Trondheim, Norway

ISSN 1868-4513 ISSN 1868-4521 (electronic)
Graduate Texts in Physics
ISBN 978-3-319-83103-9 ISBN 978-3-319-44649-3 (eBook)
DOI 10.1007/978-3-319-44649-3

Printed on acid-free paper

This Springer imprint is published by Springer Nature
The registered company is Springer International Publishing AG Switzerland

Preface

Interest in the lattice Boltzmann method has been steadily increasing since it grew out of lattice gas models in the late 1980s. While both of these methods simulate the flow of liquids and gases by imitating the basic behaviour of a gas—molecules move forwards and are scattered as they collide with each other—the lattice Boltzmann method shed the major disadvantages of its predecessor while retaining its strengths. Furthermore, it gained a stronger theoretical grounding in the physical theory of gases. These days, researchers throughout the world are attracted to the lattice Boltzmann method for reasons such as its simplicity, its scalability on parallel computers, its extensibility, and the ease with which it can handle complex geometries.

We, the authors, are all young researchers who did our doctoral studies on the lattice Boltzmann method recently enough that we remember well how it was to learn about the method. We remember particularly well the aspects that were a little difficult to learn; some were not explained in the literature in as clear and straightforward a manner as they could have been, and some were not explained in sufficient detail. Some topics were not possible to find in a single place: as the lattice Boltzmann method is a young but rapidly growing field of research, most of the information on the method is spread across many, many articles that may follow different approaches and different conventions. Therefore, we have sought to write the book that the younger versions of ourselves would have loved to have had during our doctoral studies: an easily readable, practically oriented, theoretically solid, and thorough introduction to the lattice Boltzmann method.

As the title of this book says, we have attempted here to cover both the lattice Boltzmann method's *principles*, namely, its fundamental theory, and its *practice*, namely, how to apply it in practical simulations. We have made an effort to make the book as readable to beginners as possible: it does not expect much previous knowledge except university calculus, linear algebra, and basic physics, ensuring that it can be used by graduate students, PhD students, and researchers from a wide variety of scientific backgrounds. Of course, one textbook cannot cover everything, and for the lattice Boltzmann topics beyond the scope of this book, we refer to the literature.

The lattice Boltzmann method has become a vast research field in the past 25 years. We cannot possibly cover all important applications in this book. Examples of systems that are often simulated with the method but are not covered here in detail are turbulent flows, phase separation, flows in porous media, transonic and supersonic flows, non-Newtonian rheology, rarefied gas flows, micro- and nanofluidics, relativistic flows, magnetohydrodynamics, and electromagnetic wave propagation.

We believe that our book can teach you, the reader, the basics necessary to read and understand scientific articles on the lattice Boltzmann method, the ability to run practical and efficient lattice Boltzmann simulations, and the insights necessary to start contributing to research on the method.

How to Read This Book

Every textbook has its own style and idiosyncrasies, and we would like to make you aware of ours ahead of time.

The main text of this book is divided into four parts. First, Chaps. 1 and 2 provide background for the rest of the book. Second, Chaps. 3–7 cover the fundamentals of the lattice Boltzmann method for fluid flow simulations. Third, Chaps. 8–12 cover lattice Boltzmann extensions, improvements, and details. Fourth, Chap. 13 focuses on how the lattice Boltzmann method can be optimised and implemented efficiently on a variety of hardware platforms. Complete code examples accompany this book and can be found at https://github.com/lbm-principles-practice.

For those chapters where it is possible, we have concentrated the basic practical results of the chapter into an "in a nutshell" summary early in the chapter instead of giving a summary at the end. Together, the "in a nutshell" sections can be used as a crash course in the lattice Boltzmann method, allowing you to learn the basics necessary to get up and running with a basic LB code in very little time. Additionally, a special section before the first chapter answers questions frequently asked by beginners learning the lattice Boltzmann method.

Our book extensively uses index notation for vectors (e.g. u_α) and tensors (e.g. $\sigma_{\alpha\beta}$), where a Greek index represents any Cartesian coordinate (x, y, or z) and repetition of a Greek index in a term implies summation of that term for all possible values of that index. For readers with little background in fluid or solid mechanics, this notation is fully explained, with examples, in Appendix A.1.

The most important paragraphs in each chapter are highlighted, with a few keywords in bold. The purpose of this is twofold. First, it makes it easier to know which results are the most important. Second, it allows readers to quickly and easily pick out the most central concepts and results when skimming through a chapter by reading the highlighted paragraphs in more detail.

Instead of gathering exercises at the end of each chapter, we have integrated them throughout the text. This allows you to occasionally test your understanding as you read through the book and allows us to quite literally leave certain proofs as "an exercise to the reader".

Acknowledgements

We are grateful to a number of people for their help, big and small, throughout the process of writing this book.

We are indebted to a number of colleagues who have helped us to improve this book by reading and commenting on early versions of some of our chapters. These are Emmanouil Falagkaris, Jonas Latt, Eric Lorenz, Daniel Lycett-Brown, Arunn Sathasivam, Ulf Schiller, Andrey Ricardo da Silva, and Charles Zhou.

For various forms of help, including advice, support, discussions, and encouragement, we are grateful to Santosh Ansumali, Miguel Bernabeu, Matthew Blow, Paul Dellar, Alex Dupuis, Alejandro Garcia, Irina Ginzburg, Jens Harting, Oliver Henrich, Ilya Karlin, Ulf Kristiansen, Tony Ladd, Taehun Lee, Li-Shi Luo, Miller Mendoza, Rupert Nash, Chris Pooley, Tim Reis, Mauro Sbragaglia, Ciro Semprebon, Sauro Succi, Muhammad Subkhi Sadullah, and Alexander Wagner.

On a personal level, Alex wants to thank his family which allowed him to spend some family time on writing this book. Erlend wants to thank Joris Verschaeve for helping him get started with the LBM and his friends and family who worried about how much time he was spending on this book; it all worked out in the end. Halim wants to thank his family for the continuous and unwavering support and Julia Yeomans for introducing him to the wonder of the LBM. Goncalo thanks his family for their unlimited support, Alberto Gambaruto for introducing him to the LBM, and Viriato Semiao for the chance to start working in the field; special thanks to Irina Ginzburg for the opportunity to work with her and the countless discussions and teachings. Orest thanks his family, friends, and colleagues for their valuable support during his doctoral and postdoctoral studies. Timm thanks Aline for her understanding and support and Fathollah Varnik for introducing him to the LBM.

Finally, we would like to thank our editor Angela Lahee for her support and patience.

Edinburgh, UK	Timm Krüger
Durham, UK	Halim Kusumaatmaja
Brossard, QC, Canada	Alexandr Kuzmin
Princeton, NJ, USA	Orest Shardt
Lisbon, Portugal	Goncalo Silva
Trondheim, Norway	Erlend Magnus Viggen

The authors can be contacted at authors@lbmbook.com.

Contents

Part II Lattice Boltzmann Fundamentals

Acronyms

ADE	Advection-diffusion equation
BB	Bounce-back
BC	Boundary condition
BGK	Bhatnagar-Gross-Krook (see also SRT)
CBC	Characteristic boundary condition
CFD	Computational fluid dynamics
CPU	Central processing unit
D*d*Q*q*	*d*-dimensional set of *q* velocities
DPD	Dissipative particle dynamics
DSMC	Direct simulation Monte Carlo
ECC	Error-correcting code
FD(M)	Finite difference (method)
FE(M)	Finite element (method)
FV(M)	Finite volume (method)
GPU	Graphics processing unit
HP	Hermite polynomial
HPC	High-performance computing
IBB	Interpolated bounce-back
IBM	Immersed boundary method
LB(M/E)	Lattice Boltzmann (method/equation)
LBGK	Lattice BGK (i.e. LBM with BGK collisions)
LG(M/A)	Lattice gas (model/automaton)
MD	Molecular dynamics
MEA	Momentum exchange algorithm
Mlups	Million lattice updates per second
MPC	Multiparticle collision
MPI	Message-passing interface
MRT	Multiple relaxation time
NEBB	Non-equilibrium bounce-back (also called Zou-He)
NRBC	Nonreflecting boundary condition
NS(E)	Navier-Stokes (equations)

ODE	Ordinary differential equation
PDE	Partial differential equation
PML	Perfectly matched layer
PSM	Partially saturated method
RAM	Random access memory
SBB	Simple bounce-back
SPH	Smoothed-particle hydrodynamics
SRD	Stochastic rotation dynamics
SRT	Single relaxation time (see also BGK)
TRT	Two relaxation time
WSS	Wall shear stress

Frequently Asked Questions

Certain questions come up particularly frequently from people who are learning the lattice Boltzmann method. We have listed and answered many of these questions here, with references to the relevant book sections.

Getting Started

Q: How do I learn the basics of the LBM as quickly as possible?
A: We suggest referring to our summary sections, namely Sects. 3.2 and 3.3 for a general intro, Sect. 5.1 for an intro to boundary conditions, and Sect. 6.1 for an intro to forces.

Q: Why write vector quantities as, e.g. u_α instead of $\boldsymbol{u}$?
A: This index notation style is common in fluid mechanics due to its expressiveness; see Appendix A.1.

Q: How do I implement the LBM?
A: We cover this briefly and simply in Sect. 3.3 and cover it in more depth in Chap. 13.

Q: What is a "lattice", and how do I choose a good one?
A: Lattices, or velocity sets, are discussed in Sect. 3.4.7.

Q: Do you have some simple example code to help me get started?
A: We do, in Chap. 13 and at https://github.com/lbm-principles-practice.

Q: How do we convert between physical units and simulation "lattice" units?
A: This is explained in Sect. 7.1.

Q: How do I choose the simulation parameters?
A: Section 7.2 deals with this question.

Q: How do I implement a body force (density)?
A: We cover this in Chap. 6.

Q: How can I use the LBM to simulate steady incompressible flow?
A: You need to ensure that Ma^2 is small, or you can use the incompressible equilibrium covered in Sect. 4.3.2.

Q: How do I simulate heat diffusion and thermal flows?
A: LBM for heat flow is covered in Sect. 8.4.

Q: How can I model multiphase or multicomponent flows?
A: Chapter 9 covers multiphase and multicomponent flows.

Capabilities of the LBM

Q: When is LBM a good choice to solve the Navier-Stokes equation?
A: See Sect. 2.4 for a discussion.

Q: Is mass conserved in the LBM?
A: Mass is exactly conserved in the bulk fluid, but various types of boundary conditions may still not conserve mass (cf. Sect. 5.4.2).

Q: Since the Boltzmann equation describes gas dynamics, why can the LBM also be used to simulate liquids?
A: As shown in Sect. 4.1.4, the LBM behaves macroscopically like the Navier-Stokes equations, which describe the motion of both gases and liquids.

Q: What is Galilean invariance, and is it obeyed by the LBM?
A: Galilean invariance states that physical laws are the same in any inertial reference frame. An $O(u^3)$ error term (cf. Sect. 4.1) in the standard LBM, due to the minimal discretisation of velocity space (cf. Sect. 3.4), means that it is not obeyed in the LBM. However, with a good choice of the simulation's inertial frame, this is seldom an issue except for simulations with large flow velocity variations.

Boundary Conditions

Q: Where in the LBM algorithm are boundary conditions applied?
A: See Sect. 5.1.

Q: What is the difference between "fullway" and "halfway" bounce-back?
A: This is explained in Sect. 5.3.3. In this book, we almost exclusively consider halfway bounce-back due to its additional benefits.

Q: When using a no-slip boundary condition, where exactly in the system is no-slip enforced?
A: The location depends on the type of no-slip boundary condition employed: see Fig. 5.7 and Sect. 5.2.4.

Q: How can an open inflow or outflow boundary be simulated?
A: This is explained in Sect. 5.3.5.

Q: How are boundary conditions handled at 2D or 3D corners?
A: See Sects. 5.3.6 and 5.4.4.

Q: How can I implement curved boundaries instead of "staircase" boundaries?
A: This is covered in Chap. 11.

Q: What kind of boundary conditions is used for advection-diffusion LBM?
A: See Sect. 8.5.

Q: How do I compute the momentum exchange between fluid and walls?
A: See Sect. 5.4.3 for straight boundaries and Sect. 11.2.1 for more complex cases.

Pressure and Compressibility

Q: Why do I have sound waves in my simulations?
A: The LBM solves the *compressible* Navier-Stokes equations, which allow sound waves; see Sects. 12.1 and 12.3.

Q: Why is my pressure field not as accurate as the velocity field?
A: Sound waves generated in your system (cf. Sect. 12.3) may be reflected back into the system by its velocity or density boundary conditions. See Sect. 12.4 for more on this.

Q: Why is the speed of sound not exactly $1/\sqrt{3}$ in my simulation?
A: $1/\sqrt{3}$ is the "ideal" speed of sound in LB simulations. The actual sound speed in simulations is affected by viscosity and discretisation error. See Sect. 12.2.

Q: Assuming the simulated fluid is an ideal gas, what is its heat capacity ratio γ?
A: We explain in Sect. 1.1.3 that $\gamma = 1$ in the simulated isothermal fluid and that γ is rarely relevant in nonthermal simulations.

Advanced Questions

Q: How is the lattice Boltzmann equation derived?
A: We show the derivation in Sects. 3.4 and 3.5.

Q: What is the basis of the Boltzmann equation?
A: We explain this in Sect. 1.3.

Q: How can we prove that the lattice Boltzmann equation can be used to simulate the Navier-Stokes equations?
A: This can be shown through the Chapman-Enskog analysis covered in Sect. 4.1, and can furthermore be validated by simulations of concrete cases.

Q: How can I evaluate the stress tensor locally?
A: You can compute it from f_i^{neq} as shown in (3.6).

Q: What is the Hermite expansion?
A: We answer this in Sect. 3.4.

Q: What is the advantage of advanced collision operators?
A: We explain their benefits in Chap. 10.

Q: What other collision operators are available?
A: Other than the BGK operator covered in Chap. 3 and the MRT and TRT operators covered in Chap. 10, we have a short overview with references at the end of Sect. 10.1.

Q: My code seems to be slow. How can I accelerate it?
A: See Chap. 13 for advice on implementation and efficiency.

Q: How do I implement a parallelised code?
A: See Chap. 13 for implementation advice.

Q: There are many forcing schemes around. Which one should I take?
A: We compare different forcing schemes in Sect. 6.4.

Q: How can I increase the accuracy or stability of my simulations without significantly increasing the simulation time?
A: We have guidelines for increasing stability in Sect. 4.4.4 and for accuracy in Sect. 4.5.6. You can also use TRT or MRT collision operators instead of BGK; see Chap. 10.

Q: Why is the LBM equilibrium truncated to $O(u^2)$?
A: This is explained in Sect. 3.4.

Q: How can I get in touch with the book authors, e.g. to ask a question or to point out a mistake I found?
A: You can send an email to authors@lbmbook.com.

Part I
Background

Chapter 1
Basics of Hydrodynamics and Kinetic Theory

Abstract After reading this chapter, you will have a working understanding of the equations of fluid mechanics, which describe a fluid's behaviour through its conservation of mass and momentum. You will understand the basics of the kinetic theory on which the lattice Boltzmann method is founded. Additionally, you will have learned about how different descriptions of a fluid, such as the continuum fluid description and the mesoscopic kinetic description, are related.

While the lattice Boltzmann method (LBM) has found applications in fields as diverse as quantum mechanics and image processing, it has historically been and predominantly remains a computational fluid dynamics method. This is also the spirit of this book in which we largely develop and apply the LBM for solving fluid mechanics phenomena.

To facilitate discussions in subsequent chapters, we summarise in Sect. 1.1 the most basic theory of fluid dynamics. In particular, we will review the continuity, Navier-Stokes and energy equations which are direct consequences of conservation of mass, momentum and energy. However, fluid dynamics is a *continuum* description of fluids which treats them as continuous blobs of matter, ignoring the fact that matter is made up of individual molecules. Section 1.2 discusses various representations of a fluid, from the continuum level to the atomic level. Section 1.3 gives a basic introduction to *kinetic theory*, a finer description of a fluid where we track the evolution of its constituent molecules' *distributions* in coordinate and velocity space. The LBM springs from kinetic theory, making this description fundamental to this book.

1.1 Navier-Stokes and Continuum Theory

We give an overview of fluid dynamics, in particular the continuity equation (cf. Sect. 1.1.1), the Navier-Stokes equation (NSE, cf. Sect. 1.1.2) and the equation of state (cf. Sect. 1.1.3). This section is necessarily somewhat brief and cannot replace a proper fluid dynamics textbook, such as [1–4].

Throughout this book we utilise the index notation, using Greek indices to denote an arbitrary component of a vector or a tensor, e.g. $f_\alpha \in \{f_x, f_y, f_z\}$. Repeated indices

T. Krüger et al., *The Lattice Boltzmann Method*, Graduate Texts in Physics,
DOI 10.1007/978-3-319-44649-3_1

imply a summation, e.g. $a_\beta b_\beta = \sum_\beta a_\beta b_\beta = \boldsymbol{a}\cdot\boldsymbol{b}$. This style of notation is explained in more depth in Appendix A.1.

1.1.1 Continuity Equation

The field of fluid dynamics concerns itself with *macroscopic* phenomena of fluid motion. This implies that the fluid concept is a continuum one. Even when we speak about a fluid element, such a volume contains many molecules. This fluid element is small with respect to the system size, but is large in comparison to the size of each individual molecule and the typical distance between them. We will discuss the breakdown of this assumption in Sect. 1.2, but for most applications in fluid dynamics this is a very robust approximation.

Let us now consider a small fluid element with density ρ which occupies some stationary volume V_0. The mass of this fluid element is simply $\int_{V_0} \rho\, \mathrm{d}V$. If we consider the change of this mass per unit time, it must be due to fluid flow into or out of the volume element because fluid mass cannot be created or destroyed. Mathematically, this may be written as

$$\frac{\partial}{\partial t}\int_{V_0} \rho\, \mathrm{d}V = -\oint_{\partial V_0} \rho \boldsymbol{u} \cdot \mathrm{d}\boldsymbol{A} \tag{1.1}$$

where the closed area integral is taken over the boundary ∂V_0 of the volume element V_0, $\boldsymbol{u}$ is the fluid velocity, and we take the outward normal as the direction of $\mathrm{d}\boldsymbol{A}$. The surface integral on the right-hand side of (1.1) can be transformed into a volume integral using the divergence theorem to give

$$\int_{V_0} \frac{\partial \rho}{\partial t}\, \mathrm{d}V = -\int_{V_0} \nabla \cdot (\rho \boldsymbol{u})\, \mathrm{d}V. \tag{1.2}$$

This leads to

$$\frac{\partial \rho}{\partial t} + \nabla \cdot (\rho \boldsymbol{u}) = 0 \tag{1.3}$$

since the volume V_0 is stationary and arbitrary. Equation (1.3) is the **continuity equation** in fluid dynamics. It is a partial differential equation (PDE) reflecting the conservation of mass. The vector

$$\rho \boldsymbol{u} = \boldsymbol{j} \tag{1.4}$$

is called the **momentum density** or **mass flux density**.

In the literature, the continuity equation is also sometimes written in the forms

$$\frac{\partial \rho}{\partial t} + \boldsymbol{u} \cdot \nabla \rho + \rho \nabla \cdot \boldsymbol{u} = 0 \quad \text{or} \quad \frac{\mathrm{D}\rho}{\mathrm{D}t} + \rho \nabla \cdot \boldsymbol{u} = 0. \tag{1.5}$$

Here we have introduced the *material derivative*

$$\frac{\mathrm{D}}{\mathrm{D}t} = \frac{\partial}{\partial t} + \boldsymbol{u} \cdot \nabla \tag{1.6}$$

which denotes the rate of change as the fluid element moves about in space, rather than the rate of change $\partial/\partial t$ at a fixed point in space.

Exercise 1.1 Fluid conservation equations can be given in two main forms: *conservation form*, as in (1.3), or *material derivative form*, as in (1.5). Using the continuity equation, (1.3), show that for a general conserved quantity λ the two forms can be related as

$$\frac{\partial(\rho\lambda)}{\partial t} + \nabla \cdot (\rho \boldsymbol{u} \lambda) = \rho \frac{\mathrm{D}\lambda}{\mathrm{D}t}. \tag{1.7}$$

1.1.2 Navier-Stokes Equation

Similar to our analysis above, we can consider the change of momentum of a fluid element with density ρ and velocity $\boldsymbol{u}$, occupying a small volume V_0. For a simple ideal fluid, the change of net momentum can be due to (i) flow of momentum into or out of the fluid element, (ii) differences in pressure p and (iii) external body forces $\boldsymbol{F}$. Each of these contributions is written respectively on the right-hand side of the following momentum balance equation:

$$\frac{\mathrm{d}}{\mathrm{d}t} \int_{V_0} \rho \boldsymbol{u} \, \mathrm{d}V = -\oint_{\partial V_0} \rho \boldsymbol{u}\boldsymbol{u} \cdot \mathrm{d}\boldsymbol{A} - \oint_{\partial V_0} p \, \mathrm{d}\boldsymbol{A} + \int_{V_0} \boldsymbol{F} \, \mathrm{d}V. \tag{1.8}$$

Here, $\boldsymbol{uu}$ denotes the outer product with components $u_\alpha u_\beta$. Transforming the surface integrals into volume integrals using the divergence theorem, the above equation can be rewritten as

$$\int_{V_0} \frac{\partial(\rho \boldsymbol{u})}{\partial t} \mathrm{d}V = -\int_{V_0} \nabla \cdot (\rho \boldsymbol{u}\boldsymbol{u}) \, \mathrm{d}V - \int_{V_0} \nabla p \, \mathrm{d}V + \int_{V_0} \boldsymbol{F} \, \mathrm{d}V. \tag{1.9}$$

This leads to the **Euler equation**:

$$\frac{\partial(\rho \boldsymbol{u})}{\partial t} + \nabla \cdot (\rho \boldsymbol{u}\boldsymbol{u}) = -\nabla p + \boldsymbol{F}, \tag{1.10}$$

a PDE describing the conservation of momentum for an ideal fluid.

This momentum equation can be written in a more general form, called the **Cauchy momentum equation**:

$$\frac{\partial(\rho \boldsymbol{u})}{\partial t} + \nabla \cdot \boldsymbol{\Pi} = \boldsymbol{F}. \tag{1.11}$$

Here we have used the **momentum flux density tensor**, defined as

$$\Pi_{\alpha\beta} = \rho u_\alpha u_\beta - \sigma_{\alpha\beta}. \tag{1.12}$$

The term $\sigma_{\alpha\beta}$ is called the **stress tensor** and corresponds to the non-direct momentum transfer of the moving fluid. For simple fluids described by the Euler equation we find an isotropic stress $\sigma_{\alpha\beta} = -p\delta_{\alpha\beta}$; the stress tensor contains diagonal elements which are the same in all directions.

The momentum flux transfer in the Euler equation only includes momentum transfer which is reversible, either through the flow of mass or due to pressure forces which are conservative. For real fluids, we need to include a viscosity or internal friction term which causes dissipative and irreversible transfer of momentum from one fluid element to another, neighbouring element.

To establish the form of this *viscous stress tensor* $\boldsymbol{\sigma}'$, we first argue that such contribution must be zero when the flow is uniform, including rigid body translation and rotation. We may further argue that if the velocity gradients are small, then momentum transfer due to viscosity is well captured by terms which are proportional to the first derivatives of the velocity only [3]. A general tensor of rank two satisfying these two arguments is

$$\sigma'_{\alpha\beta} = \eta\left(\frac{\partial u_\alpha}{\partial x_\beta} + \frac{\partial u_\beta}{\partial x_\alpha}\right) + \zeta\delta_{\alpha\beta}\frac{\partial u_\gamma}{\partial x_\gamma} \tag{1.13}$$

where η and ζ are coefficients of viscosity. They are usually assumed to be isotropic and uniform, though this assumption will break down for more complex fluids.

The viscous stress tensor is often separated into a traceless *shear stress* and a *normal stress*:

$$\sigma'_{\alpha\beta} = \eta\left(\frac{\partial u_\alpha}{\partial x_\beta} + \frac{\partial u_\beta}{\partial x_\alpha} - \frac{2}{3}\delta_{\alpha\beta}\frac{\partial u_\gamma}{\partial x_\gamma}\right) + \eta_{\mathrm{B}}\delta_{\alpha\beta}\frac{\partial u_\gamma}{\partial x_\gamma}. \tag{1.14}$$

The first viscosity coefficient η is usually called the *shear viscosity* while the combination $\eta_B = 2\eta/3 + \zeta$ is normally called the *bulk viscosity*.

With the total stress tensor given as the sum of the pressure and viscosity contributions,

$$\sigma_{\alpha\beta} = \sigma'_{\alpha\beta} - p\delta_{\alpha\beta}, \tag{1.15}$$

we are now ready to write down the full momentum equation for a viscous fluid. Inserting (1.14) and (1.15) into (1.11), we obtain the **Navier-Stokes equation**

$$\frac{\partial(\rho u_\alpha)}{\partial t} + \frac{\partial(\rho u_\alpha u_\beta)}{\partial x_\beta} = -\frac{\partial p}{\partial x_\alpha} + \frac{\partial}{\partial x_\beta}\left[\eta\left(\frac{\partial u_\alpha}{\partial x_\beta} + \frac{\partial u_\beta}{\partial x_\alpha}\right) + \left(\eta_B - \frac{2\eta}{3}\right)\frac{\partial u_\gamma}{\partial x_\gamma}\delta_{\alpha\beta}\right] + F_\alpha. \tag{1.16}$$

Assuming the viscosities are constant, this can be simplified to give

$$\rho\frac{\mathrm{D}u_\alpha}{\mathrm{D}t} = -\frac{\partial p}{\partial x_\alpha} + \eta\frac{\partial^2 u_\alpha}{\partial x_\beta \partial x_\beta} + \left(\eta_B + \frac{\eta}{3}\right)\frac{\partial^2 u_\beta}{\partial x_\alpha \partial x_\beta} + F_\alpha. \tag{1.17}$$

The NSE can be simplified considerably if the flow may be regarded as incompressible with $\rho =$ const, so that the continuity equation, (1.3), reduces to $\nabla \cdot \boldsymbol{u} = 0$. In this case, we can write the NSE in its most common form, the **incompressible Navier-Stokes equation**

$$\rho\frac{\mathrm{D}\boldsymbol{u}}{\mathrm{D}t} = -\nabla p + \eta\Delta\boldsymbol{u} + \boldsymbol{F}. \tag{1.18}$$

Here, $\Delta = \nabla \cdot \nabla = \partial^2/(\partial x_\beta \partial x_\beta)$ is the *Laplace operator*.

Exercise 1.2 Let us consider the steady shear flow known as *Couette flow* where an incompressible fluid is sandwiched between two parallel plates as shown in Fig. 1.1(a). The separation between the two plates is d in the y-direction. The bottom plate is held fixed such that $\boldsymbol{u} = (0, 0, 0)^\top$ while the top plate moves with velocity $\boldsymbol{u} = (U, 0, 0)^\top$. There is no external force applied to the system. Starting from the

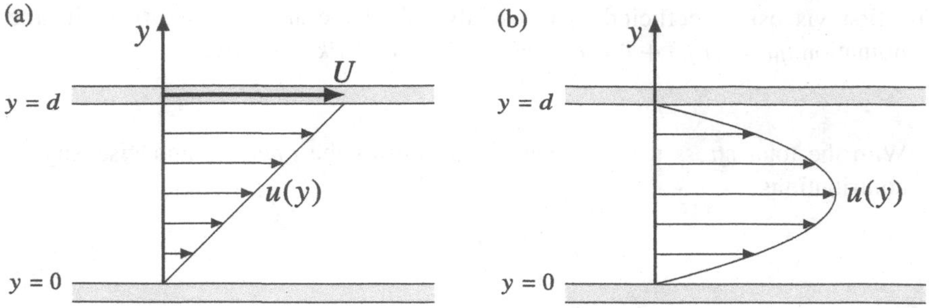

Fig. 1.1 Two fundamental steady-state flows between two plates. (**a**) Couette flow. (**b**) Poiseuille flow

incompressible NSE in (1.18), show that the velocity profile of the fluid is given by

$$u_x(y) = \frac{U}{d}y. \tag{1.19}$$

You should assume that there is no slip between the fluid and the parallel plates. In other words, the velocity of the fluid near the wall is equal to that of the wall.

Exercise 1.3 Let us now consider a steady, incompressible fluid flow commonly known as *Poiseuille flow*. The fluid is enclosed between two parallel plates and is moving in the x-direction, as shown in Fig. 1.1(b). The separation between the two plates is d along the y-axis. Poiseuille flow can in fact be driven by either (i) a constant pressure gradient or (ii) an external body force (such as gravity) in the x-direction. Assume the no-slip boundary condition between the fluid and the plates.

(a) For a pressure gradient driven flow, show that the fluid velocity profile is given by

$$u_x(y) = -\frac{1}{2\eta}\frac{\mathrm{d}p}{\mathrm{d}x}y(y-d). \tag{1.20}$$

(b) Derive the corresponding velocity profile when the flow is driven by an external body force.

1.1.3 Equations of State

At this point we have four equations that describe the behaviour of a fluid. The continuity equation, (1.3), describes the conservation of mass. The conservation of momentum is described by the Euler equation or the Navier-Stokes equation (one equation for each of the three spatial components) in (1.10) and (1.16), respectively.[1]

However, this system of equations is not closed. While we have five unknowns (density ρ, pressure p and the three velocity components u_x, u_y, u_z), we have only four equations to describe their evolution. Consequently, the system of equations is *unsolvable*, unless we can fix variables, e.g. by assuming the density to be constant.

We can add another equation to the system thanks to the *state principle* of equilibrium thermodynamics [2]. It relates the *state variables* that describe the local thermodynamic state of the fluid, such as the density ρ, the pressure p, the temperature T, the internal energy e, and the entropy s. We will defer a more detailed description of the temperature, the internal energy, and the entropy to Sect. 1.3. The state principle declares that any of these state variables can be related to any other two state variables through an *equation of state* [2].

The most famous such equation of state is the **ideal gas law**,

$$p = \rho R T. \tag{1.21}$$

It relates the pressure to the density and the temperature through the *specific gas constant* R, with units $[R] = \mathrm{J}/(\mathrm{kg\,K})$.[2]

Another equation of state for ideal gases expresses the pressure as a function of the density and the entropy, [2]

$$\frac{p}{p_0} = \left(\frac{\rho}{\rho_0}\right)^{\gamma} \mathrm{e}^{(s-s_0)/c_V}. \tag{1.22}$$

The constants p_0, ρ_0 and s_0 refer to values at some constant reference state. This equation makes use of the heat capacities at *constant volume* c_V and *constant pressure* c_p and their ratio γ, also known as the *adiabatic index*. These are defined generally as [2]

$$c_V = \left(\frac{\partial e}{\partial T}\right)_V, \quad c_p = \left(\frac{\partial(e + p/\rho)}{\partial T}\right)_p, \quad \gamma = \frac{c_p}{c_V}. \tag{1.23}$$

In an ideal gas, the two heat capacities are related as $c_p = c_V + R$.

[1] A fifth conservation equation for energy can also be derived, though we will only briefly address it later in Sect. 1.3.5 since it is less important both in fluid mechanics and in the LBM.

[2] The ideal gas law is expressed in many forms throughout science, often with quantities given in moles. Equation (1.21) is expressed using the state variables employed in fluid mechanics, the cost being that the specific gas constant R *varies between different gases*. Here, $R = k_B/m$, where k_B is Boltzmann's constant and m is the mass of the a gas molecule.

The attentive reader may have realised that any equation of state must introduce a *third* variable into the system of equations: for instance, (1.21) introduces the temperature T and (1.22) introduces the entropy s. Consequently, introducing the equation of state does not itself directly close the system of equations. The system can only be fully closed if an equation that describes the evolution of the third state variable is also derived from the aforementioned energy equation. However, the resulting system of equations is very cumbersome.

Instead, introducing the equation of state gives us more options to close the system of equations through suitable approximations. For example, most of acoustics is based on the assumption that s is approximately constant [2, 5]. This simplifies (1.22) to the *isentropic equation of state*

$$p = p_0 \left(\frac{\rho}{\rho_0} \right)^\gamma \tag{1.24}$$

and closes the system of equations.

In some cases we can also approximate the fluid as having a constant temperature $T \approx T_0$, which simplifies (1.21) to the **isothermal equation of state**

$$p = \rho R T_0 \tag{1.25}$$

that has a linear relationship between the pressure p and the density ρ.

Exercise 1.4 The isothermal equation of state is central to the LBM. Show that it is merely a special case of the isentropic equation of state, (1.24), with $\gamma = 1$.

Another approximation mentioned previously in Sect. 1.1.2 is the incompressible fluid of constant density $\rho = \rho_0$. In this case, an equation of state is not used since the incompressible continuity equation $\boldsymbol{\nabla} \cdot \boldsymbol{u} = 0$ and the incompressible NSE in (1.18) are by themselves sufficient. Together, these form a closed system of four equations for the four remaining variables p, u_x, u_y and u_z.

For small deviations from a reference state, nonlinear equations of state such as (1.22) may also be approximated by linearisation. For instance, using the total differential we may linearise any equation of state $p(\rho, s)$ as also illustrated in Fig. 1.2:

$$p = p_0 + p' \approx p_0 + \left(\frac{\partial p}{\partial \rho} \right)_s \rho' + \left(\frac{\partial p}{\partial s} \right)_\rho s'. \tag{1.26}$$

Here, the derivatives are evaluated about $p = p_0$, and the primed variables represent deviations from the reference state defined by p_0, ρ_0, and s_0.

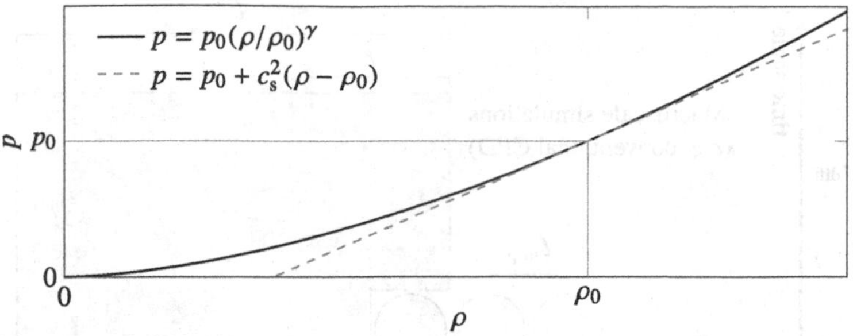

Fig. 1.2 Comparison of the isentropic equation of state in (1.24) and its linearised version in (1.27) for $\gamma = 5/3$

For the isentropic equation of state in (1.24), (1.26) is further simplified to

$$p \approx p_0 + c_s^2 \rho' \tag{1.27}$$

since the speed of sound c_s is in general given by the relation [2, 5]

$$c_s^2 = \left(\frac{\partial p}{\partial \rho}\right)_s . \tag{1.28}$$

In the case of the isentropic and isothermal equations of state in (1.24) and (1.25), we find that $c_s = \sqrt{\gamma R T_0}$ and $c_s = \sqrt{R T_0}$, respectively.

Note that the constant reference pressure p_0 is completely insignificant to the NSE as only the pressure gradient $\nabla p = \nabla(p_0 + p') = \nabla p'$ is present in the equation.[3] Therefore, the isothermal equation of state, which can be expressed as

$$p = c_s^2 \rho \quad \Longrightarrow p = c_s^2 \rho_0 + c_s^2 \rho', \tag{1.29}$$

can be used to model other equations of state in the linear regime if the entropy is nearly constant: as long as the speed of sound is matched, it does not matter if the reference pressure p_0 is different.

1.2 Relevant Scales

As discussed in the previous section, the mathematical descriptions of fluid dynamics rely on the continuum assumption where we operate at length and time scales sufficiently large that the atomistic picture can be averaged out. To formalise this discussion, let us start by considering the hierarchy of length scales associated with

[3]While p_0 is relevant to the energy equation that we will see later in Sect. 1.3.5, this equation is usually not taken into account in LB simulations.

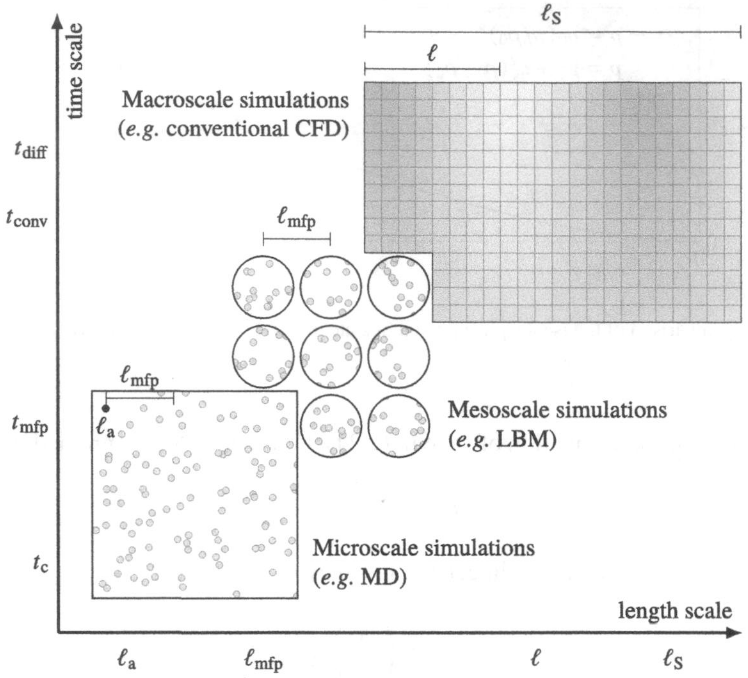

Fig. 1.3 The hierarchy of length and time scales in typical fluid dynamics problems. Depending on the level of details required, different simulation techniques are suitable

a typical fluid flow problem. If we stay within the classical mechanics picture, from small to large, we have (i) the size of the fluid atom or molecule ℓ_a, (ii) the mean free path (distance travelled between two successive collisions) ℓ_{mfp}, (iii) the typical scale for gradients in some macroscopic properties ℓ and (iv) the system size ℓ_S. The typical ordering of these length scales is $\ell_a \ll \ell_{mfp} \ll \ell \leq \ell_S$, as illustrated in Fig. 1.3.

In the context of fluids, one often refers to **microscopic**, **mesoscopic** and **macroscopic** descriptions as depicted in Fig. 1.3. In this book, "microscopic" denotes a molecular description and "macroscopic" a fully continuum picture with tangible quantities such as fluid velocity and density. Microscopic systems are therefore governed by Newton's dynamics, while the NSE is the governing equation for a fluid continuum. In between the microscopic and macroscopic description, however, is the "mesoscopic" description which does not track individual molecules. Rather, it tracks *distributions* or *representative collections* of molecules. Kinetic theory, which we will come back to in Sect. 1.3, is the mesoscopic fluid description on which the LBM is based.

Coupled to this hierarchy of length scales is the hierarchy of time scales. At very short times, we can define the collision time $t_c \sim \ell_a/v_T$, i.e. the duration of a collision event where $v_T = (k_B T/m)^{1/2}$ is the average thermal velocity of the molecules. Within Boltzmann's standard kinetic theory, we usually assume $t_c \to 0$, i.e. collisions happen instantaneously. Note that the thermal velocity v_T is different from the macroscopic fluid velocity, $u \ll v_T$.[4] Next we can define the mean flight time between two successive collisions, $t_{mfp} = \ell_{mfp}/v_T$. This is the time scale at which kinetic theory operates and where the system relaxes to local equilibrium through collision events. Local equilibrium, however, does not mean that the system is in global equilibrium. In fact, the opposite is often the case, and we are interested in studying these situations.

At longer time and larger length scales, there can exist hydrodynamic flow from one region of the fluid to another. Depending on whether we have advective (inertial regime) or diffusive (viscous regime) dynamics, the shortest (most relevant) time scale is either $t_{conv} \sim \ell/u$ or $t_{diff} \sim \ell^2/\nu$ where ν is the kinematic viscosity. The kinetic viscosity is related to the dynamic shear viscosity by $\eta = \rho\nu$. The ratio between these two hydrodynamic time scales is the well-known *Reynolds number*:

$$\mathrm{Re} = \frac{t_{diff}}{t_{conv}} = \frac{u\ell}{\nu}. \tag{1.30}$$

Both high and low Reynolds number flows are of interest. High Reynolds number flows, on the one hand, are usually dominated by turbulence and are relevant for vehicle aerodynamics, building designs, and many other applications. On the other hand, there is a surge of interest in low Reynolds number flows due to their importance in microfluidics and biophysics.

Another important macroscopic time scale is the acoustic time scale, $t_{sound} \sim \ell/c_s$, where c_s is the speed of sound in the fluid. This time scale determines how fast compression waves propagate in the fluid. When the acoustic time scale is fast in comparison to the advective time scale, the fluid behaves similarly to an incompressible fluid. Otherwise, the fluid compressibility is an important factor, which provides a number of additional physics such as shock waves. The *Mach number*

$$\mathrm{Ma} = \frac{t_{sound}}{t_{conv}} = \frac{u}{c_s} \tag{1.31}$$

defines the ratio between the acoustic and advective time scales. In practice, we can usually assume steady fluid flow with $\mathrm{Ma} \le 0.1$ to be incompressible.

We emphasise that there are a number of situations where the above-mentioned ordering of length and time scales is not satisfied. Examples include flows of rarified gases and nanofluidics. For the former, the mean free path becomes large enough so that it is comparable to the macroscopic length scale: $\ell_{mfp} \sim \ell$. On the other

[4] However, the thermal velocity v_T is of the order of the speed of sound c_s [2].

hand, the miniaturisation of fluidic devices makes the system size in nanofluidics comparable to the mean free path: $\ell_S \sim \ell_{mfp}$. A particularly useful parameter is therefore the *Knudsen number*

$$\mathrm{Kn} = \frac{\ell_{mfp}}{\ell} \tag{1.32}$$

that defines the ratio between the mean free path and the representative physical length scale. For Kn $\ll$ 1, the hydrodynamic picture (Navier-Stokes) is valid, whereas for Kn $\gtrsim$ 1, one has to go back to the kinetic theory description. As we shall see later in Sect. 1.3.5, the Knudsen number is in fact the (small) expansion parameter used in the Chapman-Enskog theory to derive the NSE from the Boltzmann equation. The Knudsen number is also closely related to the Mach and Reynolds numbers. It was first shown by von Kármán that

$$\mathrm{Kn} = \alpha \frac{\mathrm{Ma}}{\mathrm{Re}} \tag{1.33}$$

with α being a numerical constant. This relation is thus known as the *von Kármán relation*.

Dimensionless numbers such as the Reynolds, Knudsen and Mach numbers proliferate in the fluid mechanics literature. These numbers are in fact very useful. Primarily, we must appreciate that fluid flows which share the same dimensionless numbers provide the same physics upon a simple scaling by the typical length and velocity scales in the problem. This important statement is called the **law of similarity**.

To illustrate this, let us rewrite the incompressible NSE of (1.18) in its dimensionless form. We renormalise any length scale in the system by ℓ and velocity by the mean fluid velocity V, such that

$$\boldsymbol{u}^\star = \frac{\boldsymbol{u}}{V}, \quad p^\star = \frac{p}{\rho V^2}, \quad \boldsymbol{F}^\star = \frac{\boldsymbol{F}\ell}{\rho V^2}, \quad \frac{\partial}{\partial t^\star} = \frac{\ell}{V}\frac{\partial}{\partial t}, \quad \boldsymbol{\nabla}^\star = \ell \boldsymbol{\nabla} \tag{1.34}$$

and hence

$$\frac{\mathrm{d}\boldsymbol{u}^\star}{\mathrm{d}t^\star} = -\boldsymbol{\nabla}^\star p^\star + \frac{1}{\mathrm{Re}} \Delta^\star \boldsymbol{u}^\star + \boldsymbol{F}^\star. \tag{1.35}$$

As already mentioned before, the Reynolds number measures the relative importance of inertial to viscous terms in the NSE.

It is important to keep in mind that not all dimensionless numbers can be identical when comparing flows at smaller and larger length or time scales. Even if the

Reynolds number of two systems is the same, the Mach or Knudsen numbers are usually not. However, as long as the ordering of length and time scales is the same, their exact values often do not matter much. The key here is usually the separation of length and time scales. For example, if the Knudsen or Mach number is sufficiently small, their actual values are irrelevant for the hydrodynamic flows of interests where the Reynolds number is the key parameter. Therefore, usually it can be argued that all flows with the same Reynolds number are comparable to one another. In Chap. 7 we will get back to the non-dimensionalisation and how to take advantage of the law of similarity to convert parameters from the physical world to a simulation and back.

1.3 Kinetic Theory

Here we provide a concise summary of kinetic theory, the cornerstone of the LBM. Following the introduction in Sect. 1.3.1, we introduce the particle distribution function in Sect. 1.3.2. As we will see in later chapters, the assumption of a local equilibrium (cf. Sect. 1.3.3) is a crucial component of the LBM. Kinetic theory provides a kinetic description of gases; as such, molecular collisions play a central role. In Sect. 1.3.4 we discuss the collision operator in the Boltzmann equation. Conservation laws, such as mass and momentum conservation, follow from kinetic theory (cf. Sect. 1.3.5). Finally, we touch upon Boltzmann's $\mathcal{H}$-theorem in Sect. 1.3.6.

1.3.1 Introduction

As mentioned in Sect. 1.2, kinetic theory is a fluid description that lies between the *microscopic* scale where we track the motion of individual molecules and the *macroscopic* scale where we describe the fluid using more tangible quantities such as density, fluid velocity, and temperature. In the *mesoscopic* kinetic theory, we describe the *distribution* of particles in a gas, a quantity which evolves on timescales around the mean collision time t_{mfp}.

While kinetic theory can in principle be used to describe any fluid, it is most commonly applied to the simplest case of a dilute gas. There we can assume that the constituent molecules spend very little of their time actually *colliding* (i.e. $t_c \ll t_{\text{mfp}}$, using the terminology of Sect. 1.2). This is equivalent to assuming that the molecules almost always collide one-on-one, with three particles almost never simultaneously being involved in a collision. This assumption does not hold as well for dense gases where molecules are closer together and therefore spend more of their time colliding, and it does not hold at all for liquids where molecules are held close to each other by intermolecular attracting forces and thus constantly interact. While it is possible to formulate a kinetic theory of liquids [6], this is much more difficult than for dilute gases.

For simplicity, we will constrain our discussion in this section to the kinetic theory of *dilute monatomic* gases. Single atoms collide elastically, so that all translational energy is conserved in a collision. On the other hand, molecules consisting of several atoms have inner degrees of freedom; they may contain rotational and vibrational energy. Therefore, while total energy is always conserved in collisions, a collision between two such molecules may also be *inelastic* (i.e. translational energy becomes rotational or vibrational energy) or *superelastic* (i.e. rotational or vibrational energy becomes translational energy). In addition, molecular rotation and vibration must be treated quantum mechanically. The kinetic theory of polyatomic gases can be found in the literature [7–10]. However, the macroscopic behaviour of polyatomic and monatomic gases is largely similar.

Apart from the quantisation of rotational and vibrational energy in polyatomic gases, the kinetic theory of gases can be considered to be completely classical physics, as it is a statistical description of a large number of particles. As per Bohr's correspondence principle, the quantum behaviour of a system reduces to classical behaviour when the system becomes large enough.

1.3.2 The Distribution Function and Its Moments

The fundamental variable in kinetic theory is the particle **distribution function** $f(\boldsymbol{x}, \boldsymbol{\xi}, t)$. It can be seen as a generalisation of density ρ which also takes the microscopic *particle velocity* $\boldsymbol{\xi}$ into account. While $\rho(\boldsymbol{x}, t)$ represents the density of mass in *physical* space, $f(\boldsymbol{x}, \boldsymbol{\xi}, t)$ simultaneously represents the density of mass in both three-dimensional *physical* space *and* in three-dimensional *velocity* space. Therefore, f has the units

$$[f] = \mathrm{kg} \times \frac{1}{\mathrm{m}^3} \times \frac{1}{(\mathrm{m/s})^3} = \frac{\mathrm{kg\,s}^3}{\mathrm{m}^6}. \tag{1.36}$$

In other words, the distribution function $f(\boldsymbol{x}, \boldsymbol{\xi}, t)$ represents the density of particles with velocity $\boldsymbol{\xi} = (\xi_x, \xi_y, \xi_z)$ at position $\boldsymbol{x}$ and time t.

Example 1.1 To demonstrate how the distribution function extends the concept of density, let us consider a gas in a box of size $V = L_x \times L_y \times L_z$, as shown in Fig. 1.4. The total mass of the gas inside the box is of course given by the integral of the density over the box, $\int_V \rho\, \mathrm{d}^3x$. We can also calculate more specific things using the density: for instance, the mass in the *left half* of the box is $\int_{x=0}^{x=L_x/2} \rho\, \mathrm{d}^3x$. The distribution function f would let us find even more specific things: for example, the

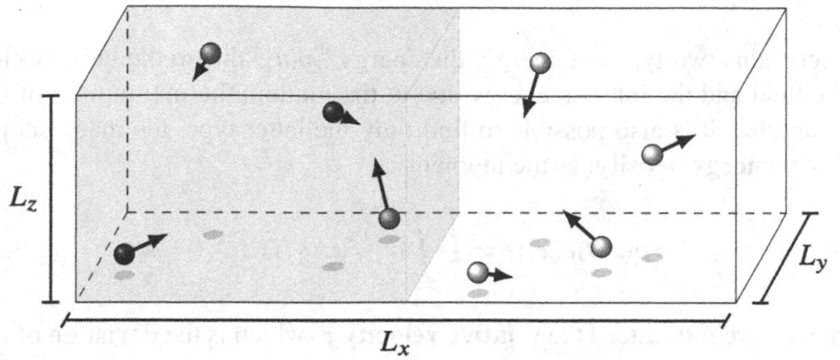

Fig. 1.4 Particles in a box. *Right-moving* particles in the *left half* of the box are marked as black. The total mass of such particles can be found from f as in (1.37)

mass of *right-moving* particles, i.e. particles with $\xi_x > 0$, in the *left half* of the box is

$$\int_{x=0}^{x=L_x/2} \int_{\xi_x>0} f \, \mathrm{d}^3\xi \, \mathrm{d}^3x. \tag{1.37}$$

The distribution function f is also connected to macroscopic variables like the density ρ and the fluid velocity u from its **moments**. These moments are integrals of f, weighted with some function of $\boldsymbol{\xi}$, over the entire velocity space. For instance, the macroscopic **mass density** can be found as the moment

$$\rho(\boldsymbol{x}, t) = \int f(\boldsymbol{x}, \boldsymbol{\xi}, t) \, \mathrm{d}^3\xi. \tag{1.38a}$$

By integrating over velocity space in this way, we are considering the contribution to the density of particles of *all possible velocities* at position $\boldsymbol{x}$ and time t.

We can also consider the particles' contribution $\boldsymbol{\xi} f$ to the momentum density. Again considering all possible velocities, we find the macroscopic **momentum density** as the moment

$$\rho(\boldsymbol{x}, t)\boldsymbol{u}(\boldsymbol{x}, t) = \int \boldsymbol{\xi} f(\boldsymbol{x}, \boldsymbol{\xi}, t) \, \mathrm{d}^3\xi. \tag{1.38b}$$

Similarly, we can find the macroscopic **total energy density** as the moment

$$\rho(\boldsymbol{x}, t)E(\boldsymbol{x}, t) = \frac{1}{2} \int |\boldsymbol{\xi}|^2 f(\boldsymbol{x}, \boldsymbol{\xi}, t) \, \mathrm{d}^3\xi. \tag{1.38c}$$

(continued)

This contains two types of energy; the energy $\frac{1}{2}\rho|\boldsymbol{u}|^2$ due to the bulk motion of the fluid and the internal energy due to the random thermal motion of the gas particles. It is also possible to find only the latter type, the macroscopic **internal energy density**, as the moment

$$\rho(\boldsymbol{x},t)e(\boldsymbol{x},t) = \frac{1}{2}\int |\boldsymbol{v}|^2 f(\boldsymbol{x},\boldsymbol{\xi},t)\,\mathrm{d}^3\xi. \tag{1.38d}$$

Here we have introduced the **relative velocity** $\boldsymbol{v}$, which is the deviation of the particle velocity from the local mean velocity:

$$\boldsymbol{v}(\boldsymbol{x},t) = \boldsymbol{\xi}(\boldsymbol{x},t) - \boldsymbol{u}(\boldsymbol{x},t). \tag{1.39}$$

These expressions for the fluid energy only consider the *translational* energy of the molecules, i.e. the energy due to the movement with their velocity $\boldsymbol{\xi}$. In the more difficult kinetic theory of polyatomic gases, the internal energy must include additional degrees of freedom, such as molecular vibrational and rotational energies.

Exercise 1.5 Consider a somewhat unrealistic spatially homogeneous gas where all particles are moving with the same velocity $\boldsymbol{u}$ so that the distribution function is $f(\boldsymbol{x},\boldsymbol{\xi},t) = \rho\,\delta(\boldsymbol{\xi} - \boldsymbol{u})$. Verify from its moments that its density is ρ and its momentum density is $\rho\boldsymbol{u}$. Additionally, find its moments of total energy density and internal energy density.

Exercise 1.6

(a) Show that the relative velocity moment of f is

$$\int \boldsymbol{v} f(\boldsymbol{x},\boldsymbol{\xi},t)\,\mathrm{d}^3\xi = 0. \tag{1.40a}$$

(b) Using this and the identity $|\boldsymbol{v}|^2 = |\boldsymbol{\xi}|^2 - 2(\boldsymbol{\xi}\cdot\boldsymbol{u}) + |\boldsymbol{u}|^2$, show that the total and internal energy densities at a given position and time are related as

$$\rho e = \rho E - \frac{1}{2}\rho|\boldsymbol{u}|^2. \tag{1.40b}$$

We can also find the pressure as a moment of the distribution function. There are several ways to do this. The most direct route, presented, e.g., in [11, 12], is to consider that particles impart momentum when bouncing off a surface. At higher particle velocities, more momentum is imparted, and more particles can bounce off in a given time. A closer analysis results in an expression for pressure as a moment of f.

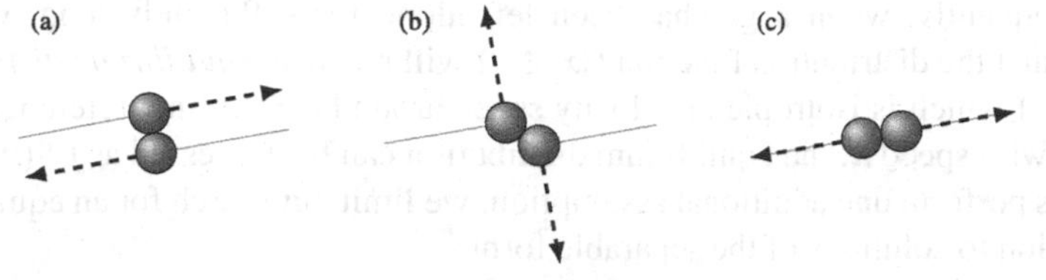

Fig. 1.5 Some collisions between hard spheres, with incoming paths shown in grey and outgoing paths in black. (**a**) Grazing collision. (**b**) Angled collision. (**c**) Head-on collision

A shortcut to the same expression for pressure can be made by using the equipartition theorem of classical statistical mechanics. This gives a specific internal energy density of $RT/2$ for each *degree of freedom* [2]. These degrees of freedom are typically molecular translation, vibration, and rotation. For monatomic gases, there is no inner molecular structure and there can be no vibration or rotation, leaving only the translational movement in the three spatial dimensions.

Thus, with three degrees of freedom, we can use the ideal gas law in (1.21) and find for an ideal monatomic gas that

$$\rho e = \frac{3}{2}\rho RT = \frac{3}{2}p. \tag{1.41}$$

Consequently, both the pressure and the temperature can be found proportional through the same moment as internal energy:

$$p = \rho RT = \frac{2}{3}\rho e = \frac{1}{3}\int |\boldsymbol{v}|^2 f(\boldsymbol{x}, \boldsymbol{\xi}, t)\, \mathrm{d}^3\xi. \tag{1.42}$$

Exercise 1.7 Show from (1.41) and the heat capacity definitions in (1.23) that the specific heat capacities and the heat capacity ratio of an ideal monatomic gas are

$$c_V = \frac{3}{2}R, \quad c_p = \frac{5}{2}R \quad \Longrightarrow \quad \gamma = \frac{5}{3}. \tag{1.43}$$

1.3.3 The Equilibrium Distribution Function

The outgoing directions of two elastically colliding hard spheres is highly sensitive to small variations in their initial relative positions, as illustrated in Fig. 1.5. This is not only true for hard spheres like pool balls that only really interact when they touch, but also for molecules that interact at a distance, e.g. *via* electromagnetic forces. Therefore, collisions tend to even out the angular distribution of particle velocities in a gas around the mean velocity $\boldsymbol{u}$.

Consequently, when a gas has been left alone for sufficiently long, we may assume that the distribution function $f(\boldsymbol{x}, \boldsymbol{\xi}, t)$ will reach an *equilibrium distribution* $f^{\mathrm{eq}}(\boldsymbol{x}, \boldsymbol{\xi}, t)$ which is isotropic in velocity space around $\boldsymbol{\xi} = \boldsymbol{u}$: in a reference frame moving with speed $\boldsymbol{u}$, the equilibrium distribution can be expressed as $f^{\mathrm{eq}}(\boldsymbol{x}, |\boldsymbol{v}|, t)$.

Let us perform one additional assumption: we limit our search for an equilibrium distribution to solutions of the separable form

$$f^{\mathrm{eq}}(|\boldsymbol{v}|^2) = f^{\mathrm{eq}}(v_x^2 + v_y^2 + v_z^2) = f^{\mathrm{eq}}_{\mathrm{1D}}(v_x^2) f^{\mathrm{eq}}_{\mathrm{1D}}(v_y^2) f^{\mathrm{eq}}_{\mathrm{1D}}(v_z^2). \tag{1.44a}$$

In other words, we assume that the 3D equilibrium distribution is the product of three 1D equilibrium distributions.

If we hold the magnitude of the velocity constant, i.e. with $|\boldsymbol{v}|^2 = v_x^2 + v_y^2 + v_z^2 =$ const, we find that

$$f^{\mathrm{eq}}(|\boldsymbol{v}|^2) = \mathrm{const} \quad \Longrightarrow \quad \ln f^{\mathrm{eq}}(v_x^2) + \ln f^{\mathrm{eq}}(v_y^2) + \ln f^{\mathrm{eq}}(v_z^2) = \mathrm{const}. \tag{1.44b}$$

This is fulfilled when each 1D equilibrium has a form like $\ln f^{\mathrm{eq}}_{\mathrm{1D}}(v_x^2) = a + b v_x^2$, with a and b being generic constants. Consequently,

$$\ln f^{\mathrm{eq}}_{\mathrm{1D}}(v_x) + \ln f^{\mathrm{eq}}_{\mathrm{1D}}(v_y) + \ln f^{\mathrm{eq}}_{\mathrm{1D}}(v_z) = 3a + b\left(v_x^2 + v_y^2 + v_z^2\right) = \mathrm{const}, \tag{1.44c}$$

and the full 3D equilibrium distribution is of the form

$$f^{\mathrm{eq}}(|\boldsymbol{v}|) = \mathrm{e}^{3a} \mathrm{e}^{b|\boldsymbol{v}^2|}. \tag{1.44d}$$

Since monatomic collisions conserve mass, momentum, and energy, the constants a and b can be found explicitly by demanding that f^{eq} has the same moments of density and energy as f.

Thus, the **equilibrium distribution** can be found to be

$$\begin{aligned} f^{\mathrm{eq}}(\boldsymbol{x}, |\boldsymbol{v}|, t) &= \rho \left(\frac{3}{4\pi e}\right)^{3/2} \mathrm{e}^{-3|\boldsymbol{v}|^2/(4e)} = \rho \left(\frac{\rho}{2\pi p}\right)^{3/2} \mathrm{e}^{-\rho|\boldsymbol{v}|^2/(2p)} \\ &= \rho \left(\frac{1}{2\pi RT}\right)^{3/2} \mathrm{e}^{-|\boldsymbol{v}|^2/(2RT)}. \end{aligned} \tag{1.45}$$

These different forms are related through (1.41).

This brief derivation follows the same lines as Maxwell's original derivation. The equilibrium distribution fulfills all the assumptions we have placed upon it, but we have not proven that it is *unique*. However, the same distribution can be found uniquely using more substantiated statistical mechanics [11], as done later by Boltzmann. In honor of these two, this equilibrium distribution is often called the *Maxwell-Boltzmann distribution*.

Exercise 1.8 Show that the moments in (1.38), applied to the equilibrium distribution in (1.45), result in a density ρ, a fluid velocity $\boldsymbol{u}$ and internal energy e. *Hint: Consider the symmetries of the integrands. If an integrand is spherically symmetric about* $\boldsymbol{v} = \boldsymbol{0}$, *the substitution* $\mathrm{d}^3\xi = 4\pi|\boldsymbol{v}|^2\,\mathrm{d}|\boldsymbol{v}|$ *can be performed.*

1.3.4 The Boltzmann Equation and the Collision Operator

Now we know what the distribution function $f(\boldsymbol{x}, \boldsymbol{\xi}, t)$ represents and what we can obtain from it. But how does it evolve? We will now find the equation that describes its evolution in time. For notational clarity, we will drop explicitly writing the dependence of f on $(\boldsymbol{x}, \boldsymbol{\xi}, t)$.

Since f is a function of position $\boldsymbol{x}$, particle velocity $\boldsymbol{\xi}$ and time t, its total derivative with respect to time t must be

$$\frac{\mathrm{d}f}{\mathrm{d}t} = \left(\frac{\partial f}{\partial t}\right)\frac{\mathrm{d}t}{\mathrm{d}t} + \left(\frac{\partial f}{\partial x_\beta}\right)\frac{\mathrm{d}x_\beta}{\mathrm{d}t} + \left(\frac{\partial f}{\partial \xi_\beta}\right)\frac{\mathrm{d}\xi_\beta}{\mathrm{d}t}. \tag{1.46}$$

Looking at each term on the right-hand side in order, we have $\mathrm{d}t/\mathrm{d}t = 1$, the particle velocity $\mathrm{d}x_\beta/\mathrm{d}t = \xi_\beta$, and from Newton's second law the specific body force $\mathrm{d}\xi_\beta/\mathrm{d}t = F_\beta/\rho$ which has the units of $[F/\rho] = \mathrm{N/kg}$.

Using the common notation $\Omega(f) = \mathrm{d}f/\mathrm{d}t$ for the total differential, we get the **Boltzmann equation**

$$\frac{\partial f}{\partial t} + \xi_\beta \frac{\partial f}{\partial x_\beta} + \frac{F_\beta}{\rho}\frac{\partial f}{\partial \xi_\beta} = \Omega(f). \tag{1.47}$$

This can be seen as a kind of advection equation: the first two terms represent the distribution function being advected with the velocity $\boldsymbol{\xi}$ of its particles. The third term represents forces affecting this velocity. On the right hand side, we have a source term, which represents the local redistribution of f due to collisions. Therefore, the source term $\Omega(f)$ is called the *collision operator*.

We know that collisions conserve the quantities of mass, momentum, and in our monatomic case, translational energy. These conservation constraints can be represented as moments of the collision operator, similarly to those in (1.38):

$$\text{mass conservation:} \quad \int \Omega(f)\, d^3\xi = 0, \tag{1.48a}$$

$$\text{momentum conservation:} \quad \int \boldsymbol{\xi}\,\Omega(f)\, d^3\xi = \mathbf{0}, \tag{1.48b}$$

$$\text{total energy conservation:} \quad \int |\boldsymbol{\xi}|^2\Omega(f)\, d^3\xi = 0, \tag{1.48c}$$

$$\text{internal energy conservation:} \quad \int |\boldsymbol{v}|^2\Omega(f)\, d^3\xi = 0. \tag{1.48d}$$

Exercise 1.9 Show using (1.39) that the total and internal energy conservation constraints are equivalent.

Boltzmann's original collision operator is of the form of a complicated and cumbersome double integral over velocity space. It considers all the possible outcomes of two-particle collisions for any choice of intermolecular forces. However, the collision operators used in the LBM are generally based on the much simpler **BGK collision operator** [13]:

$$\Omega(f) = -\frac{1}{\tau}\left(f - f^{\text{eq}}\right). \tag{1.49}$$

This operator, named after its inventors Bhatnagar, Gross and Krook, directly captures the relaxation of the distribution function towards the equilibrium distribution. The time constant τ, which determines the speed of this equilibration, is known as the **relaxation time**. The value of τ directly determines the transport coefficients such as viscosity and heat diffusivity, as we will show later in Sect. 4.1.

Any useful collision operator must both respect the conserved quantities as expressed in (1.48) and ensure that the distribution function f locally evolves towards its equilibrium f^{eq}. The BGK operator is the simplest possible collision operator given these constraints. However, it is not as exact as Boltzmann's original operator: the BGK operator predicts a *Prandtl number*, which indicates the ratio of viscosity and thermal conduction, of $\text{Pr} = 1$. Boltzmann's original operator correctly predicts $\text{Pr} \simeq 2/3$, a value also found in lab experiments on monatomic gases [10].

Exercise 1.10 For a force-free, spatially homogeneous case $f(\boldsymbol{x}, \boldsymbol{\xi}, t) \to f(\boldsymbol{\xi}, t)$, show that the BGK operator relaxes an initial distribution function $f(\boldsymbol{\xi}, t = 0)$

exponentially to the equilibrium distribution $f^{\mathrm{eq}}(\boldsymbol{\xi})$ as

$$f(\boldsymbol{\xi},t) = f^{\mathrm{eq}}(\boldsymbol{\xi}) + \left(f(\boldsymbol{\xi},0) - f^{\mathrm{eq}}(\boldsymbol{\xi})\right) \mathrm{e}^{-t/\tau}. \tag{1.50}$$

1.3.5 Macroscopic Conservation Equations

The macroscopic equations of fluid mechanics can actually be found directly from the Boltzmann equation, (1.46). We do this by taking the moments of the equation, i.e. by multiplying it with functions of $\boldsymbol{\xi}$ and integrating over velocity space.

For convenience, we introduce a general notation for the moments of f,

$$\begin{aligned} \Pi_0 &= \int f\,\mathrm{d}^3\xi = \rho, & \Pi_\alpha &= \int \xi_\alpha f\,\mathrm{d}^3\xi = \rho u_\alpha, \\ \Pi_{\alpha\beta} &= \int \xi_\alpha \xi_\beta f\,\mathrm{d}^3\xi, & \Pi_{\alpha\beta\gamma} &= \int \xi_\alpha \xi_\beta \xi_\gamma f\,\mathrm{d}^3\xi. \end{aligned} \tag{1.51}$$

The first two moments are already known as the moments for mass and momentum density. The second-order moment $\Pi_{\alpha\beta}$ will soon be shown to be the momentum flux tensor from (1.12). As we can see from their definitions, these moments are not altered if their indices are reordered.

To deal with the force terms, we need to know the moments of the force term, which we can find directly using multidimensional integration by parts as

$$\begin{aligned} \int \frac{\partial f}{\partial \xi_\beta}\,\mathrm{d}^3\xi &= 0, \\ \int \xi_\alpha \frac{\partial f}{\partial \xi_\beta}\,\mathrm{d}^3\xi &= -\int \frac{\partial \xi_\alpha}{\partial \xi_\beta} f\,\mathrm{d}^3\xi = -\rho\delta_{\alpha\beta}, \\ \int \xi_\alpha \xi_\alpha \frac{\partial f}{\partial \xi_\beta}\,\mathrm{d}^3\xi &= -\int \frac{\partial(\xi_\alpha\xi_\alpha)}{\partial \xi_\beta} f\,\mathrm{d}^3\xi = -2\rho u_\beta. \end{aligned} \tag{1.52}$$

1.3.5.1 Mass Conservation Equation

The simplest equation we can find from the Boltzmann equation is the *continuity equation* which describes the conservation of mass. Directly integrating the Boltzmann equation over velocity space, we find

$$\frac{\partial}{\partial t}\int f\,\mathrm{d}^3\xi + \frac{\partial}{\partial x_\beta}\int \xi_\beta f\,\mathrm{d}^3\xi + \frac{F_\beta}{\rho}\int \frac{\partial f}{\partial \xi_\beta}\,\mathrm{d}^3\xi = \int \Omega(f)\,\mathrm{d}^3\xi. \tag{1.53}$$

The integrals in each term can be resolved according to the moments in (1.38) and (1.52), in addition to the collision operator's mass conservation property in (1.48a). Thus, we find the *continuity equation* from (1.3) which describes mass conservation:

$$\frac{\partial \rho}{\partial t} + \frac{\partial(\rho u_\beta)}{\partial x_\beta} = 0. \tag{1.54}$$

Note that this equation only depends on the conserved moments ρ and ρu_α. It does not depend on the particular form of f, unlike the following conservation equations.

1.3.5.2 Momentum Conservation Equation

If we similarly take the first moment of the Boltzmann equation, i.e. we multiply by ξ_α before integrating over velocity space, we find

$$\frac{\partial(\rho u_\alpha)}{\partial t} + \frac{\partial \Pi_{\alpha\beta}}{\partial x_\beta} = F_\alpha. \tag{1.55}$$

The moment $\Pi_{\alpha\beta}$ defined in (1.51) is the *momentum flux tensor*, (1.12).

Exercise 1.11 By splitting the particle velocity as $\boldsymbol{\xi} = \boldsymbol{u} + \boldsymbol{v}$, show that the momentum flux tensor can be decomposed as

$$\Pi_{\alpha\beta} = \rho u_\alpha u_\beta + \int v_\alpha v_\beta f \, \mathrm{d}^3\xi. \tag{1.56}$$

Thus, the second moment of the Boltzmann equation becomes the *Cauchy momentum equation* previously seen in (1.11):

$$\frac{\partial(\rho u_\alpha)}{\partial t} + \frac{\partial(\rho u_\alpha u_\beta)}{\partial x_\beta} = \frac{\partial \sigma_{\alpha\beta}}{\partial x_\beta} + F_\alpha. \tag{1.57}$$

However, this equation is not closed as we do not know the *stress tensor*

$$\sigma_{\alpha\beta} = -\int v_\alpha v_\beta f \, \mathrm{d}^3\xi \tag{1.58}$$

explicitly. To approximate this stress tensor, we must somehow find an explicit approximation of the distribution function f. In Sect. 1.3.5.4 we will discuss how this may be done.

1.3.5.3 Energy Conservation Equation

Finally, we can find the energy equation from the trace of the second moment. In other words, we multiply by $\xi_\alpha \xi_\alpha$ before integrating over velocity space, resulting in

$$\frac{\partial(\rho E)}{\partial t} + \frac{1}{2}\frac{\partial \Pi_{\alpha\alpha\beta}}{\partial x_\beta} = F_\beta u_\beta. \tag{1.59}$$

We can simplify this in two steps. First, we can split the moment in the same way as for the momentum equation, giving the *total energy equation*

$$\frac{\partial(\rho E)}{\partial t} + \frac{\partial(\rho u_\beta E)}{\partial x_\beta} = \frac{\partial(u_\alpha \sigma_{\alpha\beta})}{\partial x_\beta} + F_\beta u_\beta - \frac{\partial q_\beta}{\partial x_\beta} \tag{1.60}$$

with the *heat flux* $\boldsymbol{q}$ given as the moment

$$q_\beta = \frac{1}{2}\int v_\alpha v_\alpha v_\beta f \, \mathrm{d}^3\xi. \tag{1.61}$$

Secondly, we can eliminate the bulk motion energy component $\frac{1}{2}\rho|\boldsymbol{u}|^2$ from the equation by subtracting (1.57) multiplied with u_α. The end result is the *internal energy equation*

$$\frac{\partial(\rho e)}{\partial t} + \frac{\partial(\rho u_\beta e)}{\partial x_\beta} = \sigma_{\alpha\beta}\frac{\partial u_\alpha}{\partial x_\beta} - \frac{\partial q_\beta}{\partial x_\beta}. \tag{1.62}$$

1.3.5.4 Discussion

Finding the macroscopic conservation equations from basic kinetic theory shows us that the mass equation is exact and invariable, while the momentum and energy equations depend on the stress tensor and the heat flux vector which themselves depend on the form of f.

At this point we do not know much about f except its value at equilibrium. It can be shown that approximating (1.58) and (1.61) by assuming $f \simeq f^{\mathrm{eq}}$ results in the Euler momentum equation from (1.10) and a simplified energy equation sometimes known as the *Euler energy equation*. Both are shown in (1.63).

Exercise 1.12 Assume that $f \simeq f^{\mathrm{eq}}$ for an ideal gas. From (1.58) and (1.61), show that the general momentum and internal energy conservation equations result in the Euler momentum and energy equations

$$\frac{\partial(\rho u_\alpha)}{\partial t} + \frac{\partial(\rho u_\alpha u_\beta)}{\partial x_\beta} = -\frac{\partial p}{\partial x_\alpha} + F_\alpha, \quad \frac{\partial(\rho e)}{\partial t} + \frac{\partial(\rho u_\beta e)}{\partial x_\beta} = -p\frac{\partial u_\beta}{\partial x_\beta}. \tag{1.63}$$

Both these Euler equations lack the viscous stress tensor $\boldsymbol{\sigma}'$ and the heat flux found in the Navier-Stokes-Fourier momentum and energy equations. We previously found the viscous stress in (1.14) while the heat flux is [2]

$$\boldsymbol{q} = -\kappa \nabla T, \tag{1.64}$$

κ being the fluid's thermal diffusivity.

The fact that the Euler equations are found for a particle distribution f at *equilibrium* indicates that the phenomena of viscous dissipation and heat diffusivity are connected to *non-equilibrium*, i.e. the deviation $f - f^{\text{eq}}$. How, then, can we find a more general form of f which takes this deviation into account?

The *Chapman-Enskog analysis* is an established method of connecting the kinetic and continuum pictures by finding non-equilibrium contributions to f. Its main idea is expressing f as a perturbation expansion about f^{eq}:

$$f = f^{\text{eq}} + \epsilon f^{(1)} + \epsilon^2 f^{(2)} + \dots . \tag{1.65}$$

The *smallness parameter* ϵ labels each term's order in the *Knudsen number* $\text{Kn} = \ell_{\text{mfp}}/\ell$, as defined in (1.32). For $\text{Kn} \to 0$, when the fluid is dominated by collisions, the particle distribution is approximately at equilibrium, and the fluid's behaviour is described by the Euler equation.

The perturbation of f combined with a non-dimensionalisation analysis lets us explicitly find the first-order perturbation $f^{(1)}$ from the macroscopic derivatives of the equilibrium distribution f^{eq} [12, 14]. Using this perturbation to approximate the stress tensor and heat flux moments in (1.58) and (1.61), we find the same stress tensor and heat flux as in (1.14) and (1.64), respectively. The resulting transport coefficients are simple functions of τ:

$$\eta = p\tau, \quad \eta_{\text{B}} = 0, \quad \kappa = \frac{5}{2} Rp\tau. \tag{1.66}$$

We will get back to the topic of the Chapman-Enskog analysis in Sect. 4.1.

What have we seen here? On a macroscopic scale, the **Boltzmann equation** describes the **macroscopic behaviour of a fluid**. To a zeroth-order approximation we have $f \approx f^{\text{eq}}$, giving the macroscopic equations of the *Euler model*. To a first-order approximation, $f \approx f^{\text{eq}} + \epsilon f^{(1)}$, we get the *Navier-Stokes-Fourier model* with its viscous stress and heat conduction.

It is also possible to go further; the second-order approximation $f \approx f^{\text{eq}} + \epsilon f^{(1)} + \epsilon^2 f^{(2)}$ results in the so-called *Burnett model* which in principle gives even more detailed and accurate equations for the motion of a fluid. In practice, however, the Burnett and Navier-Stokes-Fourier models are only distinguishable at high

Knudsen numbers [8], where the Burnett model predicts, e.g., ultrasonic sound propagation with a better agreement with experiments [15]. However, even higher-order approximations paradoxically give a *poorer* prediction of ultrasonic sound propagation [15]. A proposed reason for this strange result is that the Chapman-Enskog expansion is actually *asymptotic* [8], meaning that f *diverges* as more terms are added in its expansion. This also casts some doubt on the Burnett model. For the purpose of this book, we do not need to consider expansions beyond the first-order approximation $f \approx f^{\mathrm{eq}} + \epsilon f^{(1)}$.

1.3.6 Boltzmann's $\mathcal{H}$-Theorem

The thermodynamic property of *entropy* can also be related to the distribution function f. The *entropy density* is denoted by ρs, with the unit of $[\rho s] = \mathrm{J/kg\,m^3}$.

Boltzmann himself showed that the quantity

$$\mathcal{H} = \int f \ln f \, \mathrm{d}^3\xi \tag{1.67}$$

can only ever decrease and that it reaches its minimum value when the distribution function f reaches equilibrium.

We can see this directly from the Boltzmann equation in (1.46). By multiplying it with $(1 + \ln f)$, using the chain rule in reverse, and taking the zeroth moment of the resulting equation, we can find

$$\frac{\partial}{\partial t} \int f \ln f \, \mathrm{d}^3\xi + \frac{\partial}{\partial x_\alpha} \int \xi_\alpha f \ln f \, \mathrm{d}^3\xi = \int \ln f \, \Omega(f) \, \mathrm{d}^3\xi. \tag{1.68}$$

This equation is a balance equation for the quantity $\mathcal{H}$ and is found as a moment of the Boltzmann equation, similarly to the mass, momentum, and energy conservation equations in Sect. 1.3.5. Thus, the quantity $\int \xi_\alpha f \ln f \, \mathrm{d}^3\xi = \mathcal{H}_\alpha$ is the flux of the quantity $\mathcal{H}$ that we can split into an advective component $u_\alpha \int f \ln f \, \mathrm{d}^3\xi$ and a diffusive component $\int v_\alpha f \ln f \, \mathrm{d}^3\xi$.

For the BGK collision operator, the right-hand side of (1.68) can be found to be non-positive:

$$\begin{aligned}\int \ln f \,\Omega(f)\,\mathrm{d}^3\xi &= \int \ln\left(\frac{f}{f^{\mathrm{eq}}}\right)\Omega(f)\,\mathrm{d}^3\xi + \int \ln\left(f^{\mathrm{eq}}\right)\Omega(f)\,\mathrm{d}^3\xi \\ &= \frac{1}{\tau}\int \ln\left(\frac{f}{f^{\mathrm{eq}}}\right)\left(f^{\mathrm{eq}} - f\right)\mathrm{d}^3\xi \\ &= \frac{1}{\tau}\int f^{\mathrm{eq}} \ln\left(\frac{f}{f^{\mathrm{eq}}}\right)\left(1 - \frac{f}{f^{\mathrm{eq}}}\right)\mathrm{d}^3\xi \le 0.\end{aligned} \tag{1.69}$$

The $\ln(f^{\mathrm{eq}})\Omega(f)$ integral can be shown to disappear by inserting f^{eq} and using the conservation constraints of the collision operator in (1.48). The last inequality follows from the general inequality $\ln x(1 - x) \le 0$ for all $x > 0$. For $x = 1$, which corresponds to the equilibrium $f = f^{\mathrm{eq}}$, the inequality is exactly zero. This inequality can also be shown for Boltzmann's original collision operator and can be considered a necessary criterion for any collision operator in kinetic theory.

Consequently, (1.68) corresponds to the equation

$$\frac{\partial \mathcal{H}}{\partial t} + \frac{\partial \mathcal{H}_\alpha}{\partial x_\alpha} \le 0. \tag{1.70}$$

This shows us that $\mathcal{H}$ is not conserved in the system: it never increases, but instead it decreases, until the particle distribution reaches equilibrium. This is called the *Boltzmann $\mathcal{H}$-theorem*. It states that molecular collisions invariably drive the distribution function towards equilibrium.[5]

At first sight, this seems analogous to how the thermodynamic quantity of entropy always increases in a system unless the system has reached an equilibrium characterised by an entropy maximum. Indeed, for ideal gases $\mathcal{H}$ is actually **proportional to the entropy density** ρs [10, 16]:

$$\rho s = -R\mathcal{H}. \tag{1.71}$$

[5]A more expansive and rigorous explanation of the $\mathcal{H}$-theorem can be found elsewhere [8, 10].

References

1. G.K. Batchelor, *An Introduction to Fluid Dynamics* (Cambridge University Press, Cambridge, 2000)
2. P.A. Thompson, *Compressible-Fluid Dynamics* (McGraw-Hill, New York, 1972)
3. L.D. Landau, E.M. Lifshitz, *Fluid Mechanics* (Pergamon Press, Oxford, 1987)
4. W.P. Graebel, *Advanced Fluid Mechanics* (Academic Press, Burlington, 2007)
5. L.E. Kinsler, A.R. Frey, A.B. Coppens, J.V. Sanders, *Fundamentals of Acoustics*, 4th edn. (Wiley, New York, 2000)
6. M. Born, H.S. Green, Proc. R. Soc. A **188**(1012), 10 (1946)
7. C.S. Wang Chang, G. Uhlenbeck, Transport phenomena in polyatomic gases. Tech. Rep. M604-6, University of Michigan (1951)
8. S. Chapman, T.G. Cowling, *The Mathematical Theory of Non-uniform Gases*, 2nd edn. (Cambridge University Press, Cambridge, 1952)
9. T.F. Morse, Phys. Fluids **7**(2), 159 (1964)
10. C. Cercignani, *The Boltzmann Equation and Its Applications* (Springer, New York, 1988)
11. T.I. Gombosi, *Gaskinetic Theory* (Cambridge University Press, Cambridge, 1994)
12. D. Hänel, *Molekulare Gasdynamik* (Springer, New York, 2004)
13. P.L. Bhatnagar, E.P. Gross, M. Krook, Phys. Rev. E **94**(3), 511 (1954)
14. E.M. Viggen, The lattice Boltzmann method: Fundamentals and acoustics. Ph.D. thesis, Norwegian University of Science and Technology (NTNU), Trondheim (2014)
15. M. Greenspan, in *Physical Acoustics*, vol. IIA, ed. by W.P. Mason (Academic Press, San Diego, 1965), pp. 1–45
16. E.T. Jaynes, Am. J. Phys. **33**(5), 391 (1965)

Chapter 2
Numerical Methods for Fluids

Abstract After reading this chapter, you will have insight into a number of other fluid simulation methods and their advantages and disadvantages. These methods are divided into two categories. First, conventional numerical methods based on discretising the equations of fluid mechanics, such as finite difference, finite volume, and finite element methods. Second, methods that are based on microscopic, mesoscopic, or macroscopic particles, such as molecular dynamics, lattice gas models, and multi-particle collision dynamics. You will know where the particle-based lattice Boltzmann method fits in the landscape of fluid simulation methods, and you will have an understanding of the advantages and disadvantages of the lattice Boltzmann method compared to other methods.

While the equations of fluid mechanics described in Sect. 1.1 may *look* relatively simple, the behaviour of their solutions is so complex that analytical flow solutions are only available in certain limits and for a small number of geometries. In particular, the equations' non-linearity and the presence of boundary conditions of complex shape make it extremely difficult or even impossible to find analytical solutions. In most cases, we have to solve the equations numerically on a computer to find the flow field. The field of computational fluid dynamics (CFD) started soon after the advent of electronic computers, although numerical solution of difficult equations is a much older topic.[1]

At this point, a wide variety of methods for finding fluid flow solutions have been invented. Some of these methods are general-purpose methods, usable for any partial difference equation (PDE), which have been applied to fluids with minor adaptations. Other methods are more tailored for finding fluid flow solutions.

While the lattice Boltzmann method is the topic of this book, it is simply one of the many, many methods available today. Each of these methods has its own advantages and disadvantages, and the LB method is no exception. Therefore, this chapter briefly covers the most similar methods and relevant alternatives to LB, in order to give some perspective on where LB fits in the wider landscape of methods, and to give some idea of the cases for which LB can be better than other methods.

[1]Before electronic computers, numerical solutions were performed manually by people whose job title was "computer"!

T. Krüger et al., *The Lattice Boltzmann Method*, Graduate Texts in Physics,
DOI 10.1007/978-3-319-44649-3_2

One concept that we will be referring to often in this chapter is **order of accuracy**, which will be covered in more depth in Sect. 4.5.1. It is tied to *truncation errors*, i.e. the inherent errors when solving a PDE numerically. A numerical solution is always an approximation of the "true" solution, and typically deviates from it by truncation error terms proportional to resolution parameters like the time step Δt and spatial step Δx. For example, one particular numerical method could deviate from the "true" solution by terms $O(\Delta t^2) + O(\Delta x^4)$, so that as the resolution is made finer the deviation would decrease with the second power of Δt and the fourth power of Δx. This method is said to have a *second order* accuracy in time, and a *fourth order* accuracy in space. Section 2.1.1 shows where the truncation error comes from in the case of finite difference approximations.

Section 2.1 covers "conventional" Navier-Stokes solvers, i.e. "top-down" methods where the macroscopic fluid equations are directly discretised and solved by the aforementioned general-purpose methods. Section 2.2 covers particle-based methods, typically "bottom-up" methods based on microscopic or mesoscopic fluid descriptions. The LB method falls into the latter category. In Sect. 2.3 we summarise the two main categories of methods, and in Sect. 2.4 we explain where the lattice Boltzmann method fits in and give a brief overview of its advantages and disadvantages.

2.1 Conventional Navier-Stokes Solvers

Conventional numerical methods work by taking the equation (or coupled system of equations) of interest and directly solving them by a particular method of approximation. In the case of CFD, the basic equations to be solved are the continuity equation and the Navier-Stokes equation (or their incompressible counterparts). Additional equations, such as an energy equation and an equation of state, may augment these; the choice of such additional equations depends on the physics to be simulated and the approximations used.

The **derivatives** in these equations are always **discretised** in some form so that the equations may be solved approximately on a computer. One simple example is the **Euler approximation of a time derivative**. By definition, a variable's derivative is its slope over an infinitesimal interval Δt, and this can

(continued)

be approximated using a finite interval Δt as

$$\frac{\partial y(t)}{\partial t} = \lim_{\Delta t \to 0} \frac{y(t + \Delta t) - y(t)}{\Delta t} \approx \frac{y(t + \Delta t) - y(t)}{\Delta t}. \tag{2.1}$$

Unsurprisingly, the accuracy of this approximation increases as Δt is made smaller and thus closer to its infinitesimal ideal. Additionally, the accuracy depends on the solution itself; as a rule of thumb, a rapidly varying solution requires a smaller value of Δt to reach a good level of accuracy than a slowly varying solution does.

Example 2.1 The *forward Euler* method can be used to find a numerical solution to simple equations. Consider the equation $\partial y(t)/\partial t = -y(t)$, with $y(0) = 1$. If we did not know already that the answer is $y(t) = e^{-t}$ we might want to solve it step-by-step for discrete time steps $t_n = n\Delta t$ as

$$y_{n+1} = y_n + \Delta t \left.\frac{\partial y}{\partial t}\right|_{y=y_n} = (1 - \Delta t)y_n. \tag{2.2}$$

Here, y_n is the numerical approximation to $y(t_n)$. In this way, we would find $y_1 = (1 - \Delta t)y_0 = (1 - \Delta t)$, $y_2 = (1 - \Delta t)y_1$, and so forth. The resulting solutions for various values of Δt are shown in Fig. 2.1.

Exercise 2.1 Write a script implementing (2.2) from $t = 0$ to $t = 3$. Try out different values of Δt and show that the difference between the numerical solution $y_k|_{t_k=3}$ and the analytical solution $y(3) = e^{-3}$ varies linearly with Δt.

While the forward Euler method is the simplest and fastest method to step the solution forward in time, other methods such as the implicit backward Euler

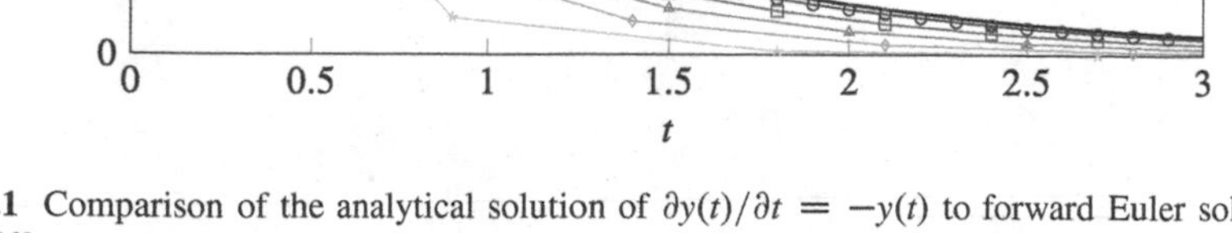

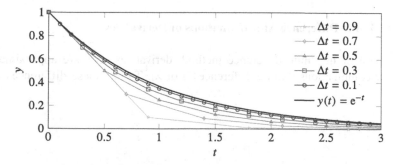

Fig. 2.1 Comparison of the analytical solution of $\partial y(t)/\partial t = -y(t)$ to forward Euler solutions with different values of Δt

method or Runge-Kutta methods beat it in stability and/or accuracy [1].[2] Typically, conventional methods for unsteady (i.e. time-dependent) CFD can use any of these methods in order to determine the solution at the next time step from the solution at the current time step.

However, these conventional CFD methods are distinguished by the approach they use to *discretise* the solution, i.e. how they use a finite set of numbers to represent the solution in continuous physical space. All these methods must represent the solution variables, such as fluid velocity $\boldsymbol{u}$ and pressure p, in such a way that their spatial derivatives can be found throughout the entire domain.

For many if not most conventional methods, this process of discretisation leads to matrix equations $\boldsymbol{Ax} = \boldsymbol{b}$, where $\boldsymbol{A}$ is a sparse matrix that relates the unknown discretised solution variables in the vector $\boldsymbol{x}$, and $\boldsymbol{b}$ represents the influence of boundary conditions and source terms. Solving such matrix equations by inverting $\boldsymbol{A}$ to find $\boldsymbol{x}$ is a linear algebra problem that lies at the heart of these methods, and finding efficient solution methods for such problems has been the topic of much research. Another common challenge of conventional incompressible Navier-Stokes solvers is to obtain a solution for the pressure Poisson equation.

In the following sections we will take a brief look at the basics of some of these methods, namely the finite difference, finite volume, and finite element methods. We will not cover the boundary-element method (BEM) [2], which is often used for creeping flows in complex geometries, or spectral methods for fluid dynamics [3].

2.1.1 Finite Difference Method

In the finite difference (FD) method, physical space is divided into a regular grid of nodes. In one dimension, these nodes are placed at the position $x_j = j\Delta x$. On each of these nodes, the solution variables are represented by a number; for a general quantity $\lambda(x)$, the exact solution $\lambda(x_j)$ is approximated by a discretised counterpart, denoted as λ_j.

2.1.1.1 Finite Difference Approximations of Derivatives

At the base of the finite difference method, derivatives of λ are approximated by linear combinations ("finite differences") of λ_j. To find these differences, we

[2]Stability and accuracy, especially in terms of the lattice Boltzmann method, are later covered in more detail in Sects. 4.4 and 4.5, respectively.

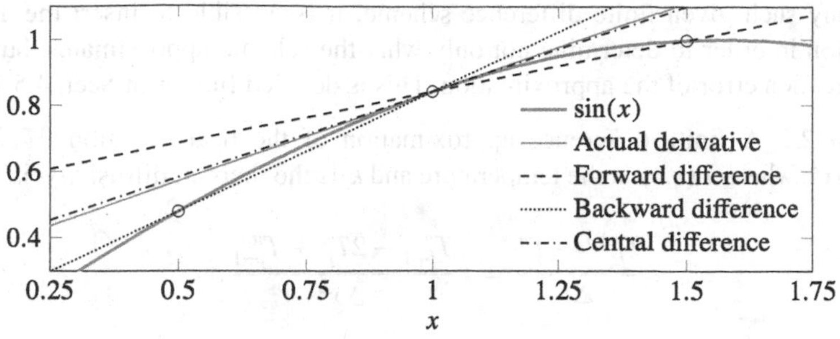

Fig. 2.2 Approximations of the derivative of sin(x) at $x = 1$, with $\Delta x = 0.5$

consider the Taylor series of $\lambda(x)$ about x_j:

$$\begin{aligned}\lambda(x_j + n\Delta x) &= \lambda(x_j) + (n\Delta x)\frac{\partial \lambda(x_j)}{\partial x} + \frac{(n\Delta x)^2}{2}\frac{\partial^2 \lambda(x_j)}{\partial x^2} + \ldots \\ &= \sum_{m=0}^{\infty} \frac{(n\Delta x)^m}{m!}\frac{\partial^m \lambda(x_j)}{\partial x^m}.\end{aligned} \tag{2.3}$$

From this we can find three simple approximations for the first-order derivative,

$$\left.\frac{\partial \lambda}{\partial x}\right|_{x_j} \approx \frac{\lambda_{j+1} - \lambda_j}{\Delta x}, \quad \left.\frac{\partial \lambda}{\partial x}\right|_{x_j} \approx \frac{\lambda_{j+1} - \lambda_{j-1}}{2\Delta x}, \quad \left.\frac{\partial \lambda}{\partial x}\right|_{x_j} \approx \frac{\lambda_j - \lambda_{j-1}}{\Delta x}. \tag{2.4}$$

These three approximations are called the *forward* difference,[3] the *central* difference, and the *backward* difference approximations, respectively.

Exercise 2.2 Prove that the four approximations in (2.4) are valid by letting e.g. $\lambda_{j+1} \longrightarrow \lambda(x_{j+1})$ and inserting (2.3), and show that the truncation error of the forward and backward difference approximations are $O(\Delta x)$ while that of the central difference approximation is $O(\Delta x^2)$.

The comparison in Fig. 2.2 indicates that central differences approximate the first derivative better, which is typically true and which can also be seen from its smaller $O(\Delta x^2)$ truncation error.

We can also find an approximation for the second-order derivative with a $O(\Delta x^2)$ truncation error:

$$\left.\frac{\partial^2 \lambda}{\partial x^2}\right|_{x_j} \approx \frac{\lambda_{j+1} - 2\lambda_j + \lambda_{j-1}}{\Delta x}. \tag{2.5}$$

[3]The forward difference approximation corresponds to the forward Euler approximation for time discretisation, shown in (2.1).

From any such given finite difference scheme, it is possible to insert the Taylor expansion in order to determine not only what the scheme approximates, but also the truncation error of the approximation. This is detailed further in Sect. 4.5.1.

Example 2.2 A finite difference approximation of the heat equation $\partial T/\partial t = \kappa \partial^2 T/\partial x^2$, where $T(x, t)$ is the temperature and κ is the thermal diffusivity, is

$$\frac{T_j^{n+1} - T_j^n}{\Delta t} = \kappa \frac{T_{j+1}^n - 2T_j^n + T_{j-1}^n}{\Delta x^2}. \tag{2.6}$$

Here the superscripts indicate the time step and the subscripts the spatial position, e.g. $T(x_j, t_n) \approx T_j^n$. We have used the forward Euler approximation from (2.1) to discretise the time derivative and (2.5) for the spatial second derivative. If we know the value of the solution at every point x_j at time t_n, along with the values at the edges of the system at all times, we can from these values determine the temperature at t_{n+1} for every point.

2.1.1.2 Finite Difference Methods for CFD

The finite difference method is simple in principle; just take a set of equations and replace the derivatives by finite difference approximations. However, this simple approach is often not sufficient in practice, and special techniques may be required for the set of equations in question. We will now touch on some problems and techniques of finding FD solutions of the Navier-Stokes equation, all of which are covered in more depth in the straightforward finite difference CFD book by Patankar [4].

We found above that the central difference scheme for first derivatives is typically more accurate than forward or backward schemes. However, in the advection term $\partial(\rho u_\alpha u_\beta)/\partial x_\beta$, information comes only from the opposite direction of the fluid flow, i.e. *upstream* or *upwind*.[4] Since the central difference scheme looks both upwind and downwind, it is possible to improve on it by using an *upwind scheme*, where either a forward or a backward scheme is used depending on the direction of fluid flow.

An issue requiring special treatment is the problem of *checkerboard instabilities*, where patterns of alternatingly high and low values emerge, patterns which in 2D are reminiscent of the black-and-white pattern on a checkerboard. A 1D example is shown in Fig. 2.3. In short, the reason behind this pattern is that a central difference scheme would report the first derivative as being zero, so that the rapidly varying field is felt as being uniform. Thus, there is nothing to stop the pattern from emerging.

A remedy to this problem is using a staggered grid as shown in Fig. 2.4, where different grids of nodes are used for different variables. These different grids are

[4]As a practical example, a deer can smell a hunter who is upwind of it, since the wind blows the hunter's scent towards the deer.

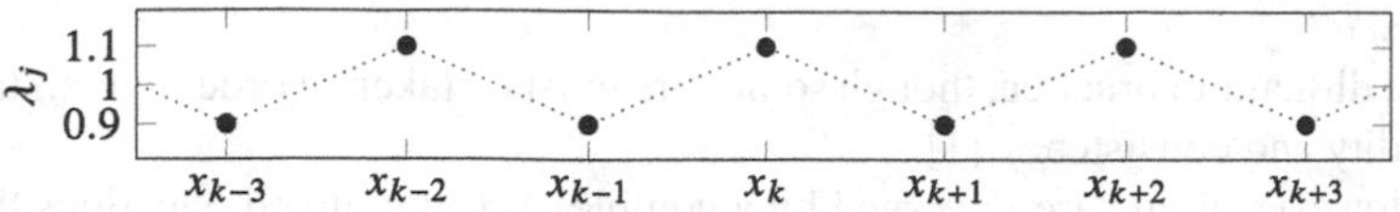

Fig. 2.3 A one-dimensional "checkerboard" field around x_k

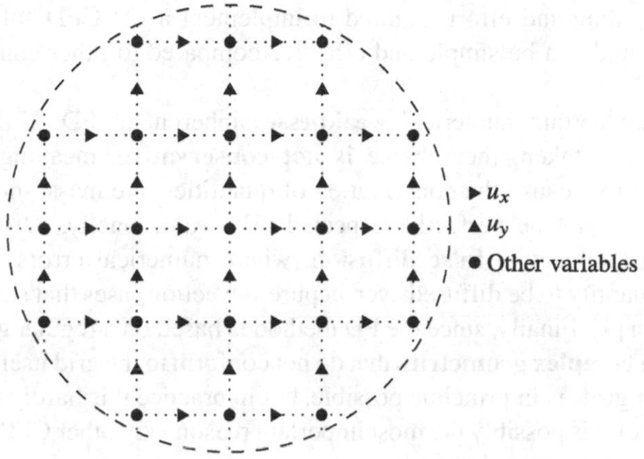

Fig. 2.4 A cutout of a staggered grid, where u_x and u_y are each stored in their own shifted grid of nodes

shifted relative to each other. Thus, when evaluating e.g. $\partial u_x/\partial x$ in one of the p nodes or $\partial p/\partial x$ in one of the u_x nodes, we can use a central difference scheme where only *adjacent* nodes are used, instead of having to skip a central node like the central difference scheme in (2.4) implies. Thus, a field like that shown in Fig. 2.3 is no longer felt as being uniform, and checkerboard instabilities cannot emerge.

The Navier-Stokes equation is nonlinear because of its advection term. Nonlinear equations are typically handled by iterating a series of "guesses" for the nonlinear quantity. This is additionally complicated by having to couple a simultaneous set of equations. In the classic FD algorithms for incompressible flow called SIMPLE and SIMPLER, guesses for the pressure field and the velocity field are coupled and successively iterated using equations tailored for the purpose. More information on these somewhat complex algorithms can be found elsewhere [4].

2.1.1.3 Advantages and Disadvantages

The crowning **advantage** of the finite difference method is that it is really quite **simple** in principle. For a number of simple equations it is not that much

(continued)

more difficult in practice, though some care must be taken in order to maintain stability and consistency [1].

However, fluids are governed by a complex set of coupled equations that contain several variables. Therefore, a number of special techniques need to be applied in order to use the FD method for CFD, which increases the amount of understanding and effort required to implement a FD CFD solver. Still, the FD method can be simple and effective compared to other conventional methods [5].

There are certain numerical **weaknesses** inherent to FD CFD. Unless special care is taken, the scheme is **not conservative**, meaning that the numerical errors cause the conservation of quantities like mass, momentum, and energy to not be perfectly respected [5]. Additionally, advective FD schemes are subject to **false diffusion**, where numerical errors cause the advected quantity to be diffused even in pure-advection cases that should have no diffusion [4]. Finally, since the FD method is based on a regular grid it **has issues with complex geometries** that do not conform to the grid itself [5]. (FD on irregular grids is in principle possible, but in practice it is hardly used [5].) The latter point is possibly the most important reason why other CFD methods have become more popular.

2.1.2 *Finite Volume Method*

In the finite volume (FV) method, space does not need to be divided into a *regular* grid. Instead, we subdivide the simulated volume V into many smaller *volumes* V_i, which may have different sizes and shapes to each other.[5] This allows for a better representation of complex geometries than e.g. the finite difference method, as illustrated in Fig. 2.5. In the middle of each finite volume V_i, there is a node where each solution variable $\lambda(x)$ is represented by its approximate average value $\bar{\lambda}_i$ within that volume.

2.1.2.1 Finite Volume Approximation of Conservation Equations

The FV method is not as general as the FD method which can in principle be used for any equation. Rather, the FV method is designed to solve *conservation equations*,

[5] We here use the term "volume" in a general sense, where a 2D volume is an area and a 1D volume is a line segment.

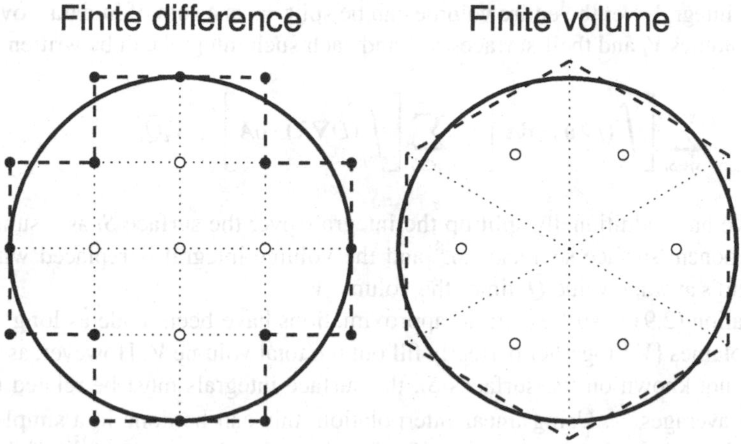

Fig. 2.5 Simple finite difference and finite volume discretisations of the volume inside a circular surface. The effective surface in each case is shown as *black dashed lines*, and interior nodes as *white circles*. To the right, the *dotted lines* show the finite volumes' interior edges

the type of equations which we typically find in e.g. fluid mechanics.[6] The FV method is *conservative* by design, which means that e.g. mass and momentum will always be conserved perfectly, unlike in the FD method.

To show the general principle of how FV approximates a conservation equation, we start with a steady advection-diffusion equation for a general quantity $\lambda(\boldsymbol{x}, t)$,

$$\nabla \cdot (\rho\lambda\boldsymbol{u}) = \nabla \cdot (D\nabla\lambda) + Q, \tag{2.7}$$

where the density ρ and the flow field $\boldsymbol{u}$ are assumed known, D is a diffusion coefficient for λ, and Q is a source term. By integrating this equation over the entire volume V and applying the divergence theorem, we get

$$\int_S (\rho\lambda\boldsymbol{u}) \cdot \mathrm{d}\boldsymbol{A} = \int_S (D\nabla\lambda) \cdot \mathrm{d}\boldsymbol{A} + \int_V Q\,\mathrm{d}V, \tag{2.8}$$

where S is the surface of the volume V and $\mathrm{d}\boldsymbol{A}$ is an infinitesimal surface normal element. The concept of the divergence theorem is as central to the FV method as it is for conservation equations in general: Sources and sinks of a quantity within a volume are balanced by that quantity's flux across the volume's boundaries.

[6] That is not to say that the FV method is limited to conservation equations; it can also be used to solve more general hyperbolic problems [6].

This integral over the entire volume can be split up as a sum of integrals over the finite volumes V_i and their surfaces S_i,[7] and each such integral can be written as

$$\sum_{s_j \in S_i} \left[\int_{s_j} (\rho \lambda \boldsymbol{u}) \cdot \mathrm{d}\boldsymbol{A} \right] = \sum_{s_j \in S_i} \left[\int_{s_j} (D \nabla \lambda) \cdot \mathrm{d}\boldsymbol{A} \right] + V_i \bar{Q}_i. \tag{2.9}$$

Here, we have additionally split up the integrals over the surface S_i as a sum over its component surface segments s_j,[8] and the volume integral is replaced with the integrand's average value $\bar{Q}_i$ times the volume V_i.

Equation (2.9) is still exact; no approximations have been made as long as the finite volumes $\{V_i\}$ together perfectly fill out the total volume V. However, as λ and $\nabla\lambda$ are not known on the surfaces S_j, the surface integrals must be related to the volume averages $\bar{\lambda}_i$. Using linear interpolation, this can be done in a simple way that leads to second-order accuracy [5]. Starting with the values of $\bar{\lambda}_i$ of the two volumes adjacent to the surface s_j, λ can be linearly interpolated between the two volumes' nodes so that each node point $\boldsymbol{x}_i$ has its corresponding volume's value of $\lambda(\boldsymbol{x}_i) = \bar{\lambda}_i$. At the point where the straight line between the two nodes crosses the surface s_j, we can find the linearly interpolated values of λ and $(\nabla\lambda) \cdot \mathrm{d}\boldsymbol{A}$. These values can then be applied to the entire surface in the surface integral.

Higher-order accuracy can be achieved by estimating the values of λ and $\nabla\lambda$ at more points on the surface, such as the surface edges which can be determined by interpolation from all the adjacent volumes [5]. Additionally, the interpolation of values on the surface may use node values from further-away volumes [5, 7].

2.1.2.2 Finite Volume Methods for CFD

While the basic formulations of finite volume and finite difference methods are different, CFD using FV methods bear many similarities to finite difference CFD, which is discussed in Sect. 2.1.1.2. For instance, for higher-order interpolation schemes, it is still generally a good idea to use more points in the upwind direction than in the downwind direction [5, 7]. Additionally, the iterative finite difference SIMPLE and SIMPLER schemes for CFD [4] and their descendants may also be adapted for finite volume simulations [7].

One difference is that the staggered grids generally used in FD CFD become too cumbersome to use for the irregular volumes typically used in FV CFD. While the issue of checkerboard instabilities is also present in the FV method for non-staggered grids, this is dealt with by the use of schemes that use more than two node values to approximate the first derivative at a point [7].

---

[7]For the internal surfaces between adjacent finite volumes, the surface integrals from the two volumes will cancel each other.

[8]In Fig. 2.5, S_i is the triangular surface around each volume, and s_j represents the straight-line faces of these triangles.

2.1.2.3 Advantages and Disadvantages

While finite volume methods are formulated differently to finite difference methods, the two methods are comparable in their relative simplicity. The FV method has some additional **advantages**, however. The control volume formulation makes it fundamentally **conservative**; e.g. mass and momentum will be conserved throughout the entire domain in a closed system. Additionally, the FV method is very appropriate for use with irregular grids, which means that **complex geometries can be captured well** (the grid is adapted to the geometry), and it is straightforward to "spend" more resolution on critical regions in the simulation by making the grid finer in these regions.

The **downside** of irregular grids is that making **appropriate grids for complex geometries** is itself a **fairly complex** problem; indeed, it is an entire field of study by itself. Additionally, **higher-order FV methods are not straightforward** to deal with, in particular in three dimensions and for irregular grids [5]. While FV is not as general a method as FD in terms of what equations it can solve, this is typically not an issue for the equations encountered in CFD.

2.1.3 Finite Element Methods

In finite element methods (FEM), PDEs are solved using an integral form known as the *weak form*, where the PDE itself is multiplied with a weight function $w(x)$ and integrated over the domain of interest. For example, the *Helmholtz equation* $\nabla^2\lambda + k^2\lambda = 0$ (a steady-state wave equation for wavenumber k, further explained in Sect. 12.1.4) in 1D becomes

$$\int w(x)\frac{\partial^2\lambda(x)}{\partial x^2}\,\mathrm{d}x + k^2\int w(x)\lambda(x)\,\mathrm{d}x = 0. \tag{2.10}$$

Generally, an unstructured grid can be used with FEM, with a discretised solution variable λ_i represented at each grid corner node $\boldsymbol{x}_i$. Between the grid corners, the variable $\lambda(\boldsymbol{x})$ is interpolated using *basis functions* $\phi_i(\boldsymbol{x})$ fulfilling certain conditions, i.e.

$$\lambda(x) \approx \sum_i \lambda_i\phi_i(x), \quad \text{for } \{\phi_i\} \text{ such that } \lambda(x_i) = \lambda_i, \quad \sum_i \phi_i(x) = 1 \tag{2.11}$$

in our 1D example. The simplest 1D basis functions are linear functions such that $\phi_i(x_i) = 1$, $\phi_i(x_{j\neq i}) = 0$, and are non-zero only in the interval (x_{i-1}, x_{i+1}). However, a large variety of basis functions that are not linear (e.g. quadratic and cubic ones) are also available, and the order of accuracy is typically tied to the order of the basis functions.

Usually, the basis functions themselves are chosen as weighting functions, $w(x) = \phi_i(x)$. This leads to a system of equations, one for each unknown value λ_i. Through the integrals, each value of λ_i in our 1D example is related with λ_{i-1} and λ_{i+1}, assuming linear basis functions.

The main advantage of FEM is that it is mathematically well-equipped for **unstructured grids** and for increasing the order of accuracy through **higher-order** basis functions (though these also require more unknowns λ_i). These grids can be dynamically altered to compensate for **moving geometry**, as in the case of simulating a car crash. One **disadvantage** of FEM is that, like FD methods, it is **not conservative** by default like FV methods are. Another disadvantage is its **complexity** compared to FD and FV methods. For instance, the integrals become tricky to solve on general unstructured grids. And as with FD and FV methods, solving the complex **Navier-Stokes** system of equations is **not straightforward**; see e.g. [8] for more on CFD with the FEM. The **checkerboard** instabilities described in Sect. 2.1.1.2 may appear here also unless special care is taken to deal with these [9].

2.2 Particle-Based Solvers

Particle-based solvers are not based on directly discretising the equations of fluid mechanics, and they thus take an approach distinctly different to that of the conventional solvers of the previous section. Instead, these methods represent the fluid using *particles*. Depending on the method, a particle may represent e.g. an atom, a molecule, a collection of molecules, or a portion of the macroscopic fluid. Thus, while conventional Navier-Stokes solvers take an entirely macroscopic view of a fluid, particle-based methods usually take a microscopic or mesoscopic view.

In this section we briefly present six different particle-based methods, ordered roughly from microscopic, via mesoscopic, to macroscopic. Methods that are related to or viable alternatives to the lattice Boltzmann method are described in more detail.

2.2.1 Molecular Dynamics

Molecular dynamics (MD) is at its heart a fundamentally simple microscopic method which tracks the position of particles that typically represent atoms or molecules. These particles interact through intermolecular forces $\boldsymbol{f}_{ij}(t)$ which are chosen to reproduce the actual physical forces as closely as possible.[9] Knowing the total force $f_i(t)$ on the ith particle from all other particles, we know its acceleration as per Newton's second law:

$$\frac{\mathrm{d}^2 x_i}{\mathrm{d}t^2} = \frac{f_i}{m_i} = \frac{1}{m_i} \sum_{j \neq i} f_{ij}. \tag{2.12}$$

The particle position x_i can then be updated numerically by integrating Newton's equation of motion. While there are many such *integrator* algorithms, a particularly simple and effective one is the Verlet algorithm [10],

$$x_i(t + \Delta t) = 2x_i(t) - x_i(t - \Delta t) + \frac{f_i(t)}{m_i} \Delta t^2. \tag{2.13}$$

This scheme uses the current and previous position of a particle to find its next position. The Verlet scheme can also be equivalently expressed to use the particle's velocity instead of its previous position [10].

Exercise 2.3 Using Taylor expansion as in (2.3), show that the truncation error of the Verlet algorithm in (2.13) is $O(\Delta t^4)$.

However, while MD is a great method for simulating microscale phenomena like chemical reactions, protein folding and phase changes, a numerical method that tracks individual molecules is far too detailed to be used for macroscopic phenomena—consider that a single gram of water contains over 10^{22} molecules. Therefore, MD is highly impractical as a Navier-Stokes solver, and more appropriate methods should be chosen for this application. More on MD and its applications can be found elsewhere [10–12].

2.2.2 Lattice Gas Models

Lattice gas models were first introduced in 1973 by Hardy, Pomeau, and de Pazzis as an extremely simple model of 2D gas dynamics [13]. Their particular model was subsequently named the HPP model after its inventors. In this model, fictitious particles exist on a square lattice where they stream forwards and collide in a manner that respects conservation of mass and momentum, much in the same way as molecules in a real gas. As the HPP lattice was square, each node had four neighbours and each particle had one of the four possible velocities c_i that would bring a particle to a neighbouring node in one time step.

[9] This straightforward force approach scales with the number N of particles as $O(N^2)$, though more efficient approaches that scale as $O(N)$ also exist [10].

However, it was not until 1986 that Frisch, Hasslacher, and Pomeau published a lattice gas model that could actually be used to simulate fluids [14]. Their model was also named after its inventors as the FHP model. The difference to the original HPP model is small, but significant: Instead of the square lattice and four velocities of the HPP model, the FHP model had a triangular lattice and six velocities $\boldsymbol{c}_i$. This change turned out to give the model sufficient lattice isotropy to perform fluid simulations [15–17].

Lattice gas models are especially interesting in the context of the lattice Boltzmann method, as **the LBM grew out of lattice gas models**. Indeed, early LB articles are difficult to read without knowledge of lattice gases, as these articles use much of the same formalism and methods.

2.2.2.1 Algorithm

Only up to one particle of a certain velocity can be present in a node at any time. Whether or not a particle of velocity $\boldsymbol{c}_i$ exists at the lattice node at $\boldsymbol{x}$ at time t is expressed by the *occupation number* $n_i(\boldsymbol{x}, t)$, where the index i refers to the velocity $\boldsymbol{c}_i$. This occupation number n_i is a Boolean variable with possible values of 0 and 1, representing the absence and presence of a particle, respectively.

This occupation number can be directly used to determine macroscopic observables: The mass density and momentum density in a node can be expressed as [16]

$$\rho(\boldsymbol{x}, t) = \frac{m}{v_0}\sum_i n_i(\boldsymbol{x}, t), \qquad \rho\boldsymbol{u}(\boldsymbol{x}, t) = \frac{m}{v_0}\sum_i \boldsymbol{c}_i n_i(\boldsymbol{x}, t), \tag{2.14}$$

respectively, where m is the mass of a particle and v_0 is the volume covered by the node.

There are two rules that determine the time evolution of a lattice gas. The first rule is *collision*, where particles that meet in a node may be redistributed in a way that conserves the mass and momentum in the node. Generally, collisions can be mathematically expressed as

$$n_i^\star(\boldsymbol{x}, t) = n_i(\boldsymbol{x}, t) + \Omega_i(\boldsymbol{x}, t), \tag{2.15}$$

where $n_i^\star$ is the post-collision occupation number and $\Omega_i \in \{-1, 0, 1\}$ is a collision operator that may redistribute particles in a node, based on all occupation numbers $\{n_i\}$ in that node [15]. Collisions must conserve mass ($\sum_i \Omega_i(\boldsymbol{x}, t) = 0$) and momentum ($\sum_i \boldsymbol{c}_i\Omega_i(\boldsymbol{x}, t) = \boldsymbol{0}$).

Which collisions may occur (i.e. the dependence of Ω_i on $\{n_i\}$) varies between different formulations of lattice gases, but in any case this is cumbersome to express mathematically [16, 17]. Rather, we represent graphically the two types of collisions

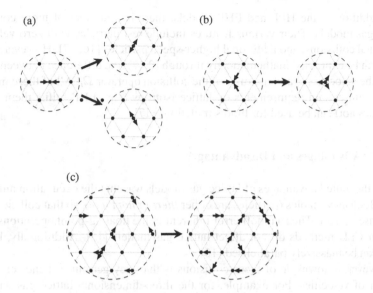

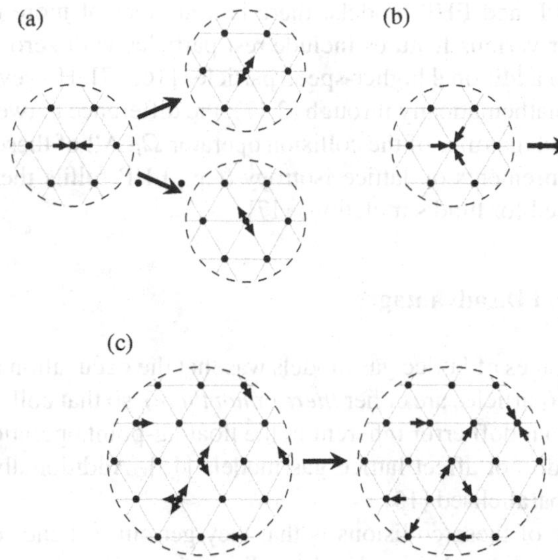

Fig. 2.6 Rules of the original FHP lattice gas model: collision and streaming. (**a**) Two-particle collision; the resolution is chosen randomly from the two options. (**b**) Three-particle collision. (**c**) Streaming

in the original FHP model [14] in Fig. 2.6. Figure 2.6a shows the two possible resolutions between head-on collisions of two particles, which are chosen randomly with equal probability. Figure 2.6b shows the resolution of a three-particle collision: When three particles meet with equal angles between each other, they are turned back to where they came from.

Exercise 2.4

(a) Show that the macroscopic quantities of (2.14) are preserved by these collisions.
(b) Show that there is another possible resolution for both the two-particle collision and the three-particle collision.

The second rule of a lattice gas is *streaming*: after collisions, particles move from their current node to a neighbouring node in their direction of velocity, as shown in Fig. 2.6c. The particle velocities $\boldsymbol{c}_i$ are such that particles move exactly from one node to another from one time step to the next. For the FHP model, which has six velocities of equal magnitude, we have $|\boldsymbol{c}_i| = \Delta x/\Delta t$, Δx being the distance between nodes and Δt being the time step. Thus, the streaming can be expressed mathematically as

$$n_i(\boldsymbol{x} + \boldsymbol{c}_i\Delta t) = n_i^\star(\boldsymbol{x}, t). \tag{2.16}$$

Both rules can be combined into a single equation:

$$n_i(\boldsymbol{x} + \boldsymbol{c}_i\Delta t, t + \Delta t) = n_i(\boldsymbol{x}, t) + \Omega_i(\boldsymbol{x}, t). \tag{2.17}$$

In addition to the HPP and FHP models, there is a number of more complex lattice gas models. Their various features include rest particles with zero velocity, additional collisions, and additional higher-speed particles [16, 17]. However, all of these can be expressed mathematically through (2.17); the difference between them lies in the velocities $\boldsymbol{c}_i$ and the rules of the collision operator Ω_i. All of these models which fulfil certain requirements on lattice isotropy (e.g., FHP fulfils them while HPP does not) can be used for fluid simulations [17].

2.2.2.2 Advantages and Disadvantages

One of the touted advantages of lattice gas models was that the occupation numbers n_i are Boolean variables (particles are either *there* or *not there*), so that collisions are in a sense perfect: The roundoff error inherent in the floating-point operations used in other CFD methods do not affect lattice gas models [15]. Additionally, lattice gases can be massively parallelised [15].

However, a downside of these collisions is that they get out of hand for larger number of velocities. For example, for the three-dimensional lattice gas with 24 velocities [15, 17], there are $2^{24} \approx 16.8 \times 10^6$ possible states in a node. The resolution of any collision in this model was typically determined by lookup in a huge table made by a dedicated program [15].

The FHP model additionally has problems with isotropy of the Navier-Stokes equations, which can only disappear in the limit of low Mach numbers, i.e. for a quasi-incompressible flow [15]. Additionally, lattice gases struggled to reach as high Reynolds numbers as comparable CFD methods [15].

The major issue with lattice gases, however, was **statistical noise**. Like real gases, lattice gases are teeming with activity at the microscopic level. Even for a gas at equilibrium, when we make a control volume smaller and smaller, the density (mass per volume) inside it will fluctuate more and more strongly with time: Molecules continually move in and out, and the law of large numbers applies less for smaller volumes. This is also the case with lattice gases, where the macroscopic values from (2.14) will fluctuate even for a lattice gas at equilibrium.

In one sense, it may be an advantage that lattice gases can qualitatively capture the thermal fluctuations of a real gas [16]. But if the goal is to simulate a macroscopic fluid, these fluctuations are a nuisance. For that reason, lattice gas simulations would typically report density and fluid velocity found through averaging in space and/or time (i.e. over several neighbouring nodes and/or several adjacent time steps), and even averaging over multiple ensembles (i.e. macroscopically similar but microscopically different realisations of the system) [16], though this could only reduce the problem and some noise would always remain.

The problem of statistical noise was more completely dealt with by the invention of the lattice Boltzmann method in the late 1980s [18–20]. This method was first introduced by tracking the occupation number's expectation value $f_i = \langle n_i \rangle$ rather than the occupation number itself, thus eliminating the statistical noise. This was the original method of deriving the LBM, and it was not fully understood how to derive it from the kinetic theory of gases presented in Sect. 1.3 until the mid-90s [21]. This more modern approach of derivation is the one that we will follow in Chap. 3.

2.2.3 Dissipative Particle Dynamics

Dissipative particle dynamics (DPD) is, like the LBM, a relatively new mesoscopic method for fluid flows. Originally proposed by Hoogerbrugge and Koelman [22] in 1992, it was later put on a proper statistical mechanical basis [23]. DPD can be considered a coarse-grained MD method that allows for the simulation of larger length and time scales than molecular dynamics (cf. Sect. 2.2.1) and avoids the lattice-related artefacts of lattice gases. Being a fully Lagrangian scheme without an underlying lattice, DPD is intrinsically Galilean invariant and isotropic.

In the following, we will summarise the essential ideas of DPD, as described in a recent review article by Liu et al. [24]. We will not cover smoothed dissipative particle dynamics (SDPD) [25] that is a special case of smoothed-particle hydrodynamics rather than an extension of DPD.

The basis of DPD are particles of mass m that represent clusters of molecules. These particles interact *via* three different forces: conservative (C), dissipative (D) and random (R). Unlike forces in MD, the conservative forces in DPD are soft, which allows larger time steps. The dissipative forces mimic viscous friction in the fluid, while random forces act as thermostat. All these forces describe additive pair interactions between particles (obeying Newton's third law), hence DPD conserves momentum. In fact, DPD is often referred to as a momentum-conserving thermostat for MD. The total force on particle i can be written as sum of all forces due to the presence of other particles j and external forces $\boldsymbol{f}^{\mathrm{ext}}$:

$$\boldsymbol{f}_i = \boldsymbol{f}^{\mathrm{ext}} + \sum_{j \neq i} \boldsymbol{f}_{ij} = \boldsymbol{f}^{\mathrm{ext}} + \sum_{j \neq i} \left(\boldsymbol{f}_{ij}^{\mathrm{C}} + \boldsymbol{f}_{ij}^{\mathrm{D}} + \boldsymbol{f}_{ij}^{\mathrm{R}} \right). \tag{2.18}$$

All interactions have a finite radial range with a cutoff radius r_{c}. Details of the radial dependence of the forces are discussed in [24]. Like in MD, a crucial aspect of the DPD algorithm is the time integration of the particle positions and velocities. Typical employed methods are a modified velocity-Verlet [26] or a leap-frog algorithm [27].

DPD obeys a fluctuation-dissipation theorem (if the radial weight functions are properly chosen) [23] and is particularly suited for hydrodynamics of complex fluids at the mesoscale with finite Knudsen number. Typical applications are suspended polymers or biological cells, but also multiphase flows in complex geometries. Solid boundaries are typically modelled as frozen DPD particles, while immersed soft structures (e.g. polymers) are often described by particles connected with elastic springs. Similarly to other mesoscopic methods, it is relatively easy to include additional physics, for instance the equation of state of multiphase fluids.

A disadvantage of DPD is that it contains a large number of parameters that have to be selected carefully. The choice of the radial weight functions is delicate and affects the emergent hydrodynamic behaviour. For example, to reach a realistically large viscosity, it is necessary to increase the cutoff distance r_c which in turn leads to more expensive simulations.

2.2.4 *Multi-particle Collision Dynamics*

In 1999, Malevanets and Kapral [28, 29] introduced the multi-particle collision (MPC) dynamics, which has since become a popular method in the soft matter community. The paradigm of MPC is to coarse-grain the physical system as much as possible while still resolving the essential features of the underlying problem.

Although MPC is nothing more than a modification of direct simulation Monte Carlo (DSMC, cf. Sect. 2.2.5) [30], we discuss both methods separately as they are normally used for completely different applications. In particular, MPC is commonly employed for systems with a small mean free path, while DSMC allows the simulation of rarefied gases with a large mean free path.

MPC is a method of choice for complex systems where both hydrodynamic interactions and thermal fluctuations are relevant. Due to its particle-based nature, it is relatively easy to implement coupled systems of solvent and solutes. Therefore, MPC is most suitable and often employed for the modelling of colloids, polymers, vesicles and biological cells in equilibrium and external flow fields. MPC particularly shows its strengths for systems with Reynolds and Péclet numbers between 0.1 and 10 and for applications where consistent thermodynamics is required and where the macroscopic transport coefficients (viscosity, thermal diffusivity, self-diffusion coefficient) have to be accurately known [31].

There exist also MPC extensions for non-ideal [32], multicomponent [33] and viscoelastic fluids [34]. We refer to [31, 35] for thorough reviews and to [36] for a recent overview.

2.2.4.1 Algorithm

The essential features of the MPC algorithm are: (i) alternating streaming and collision steps, (ii) local conservation of mass, momentum and, unlike standard LBM schemes, energy, (iii) isotropic discretisation. The last two properties ensure that MPC can be used as a viable Navier-Stokes solver.

The basic MPC setup comprises a large number of point-like particles with mass m. These particles can either be fluid or immersed (e.g. colloidal) particles. This feature allows a treatment of solvent and solutes on an equal footing. For example, immersed particles can be directly coupled by letting them participate in the collision and streaming steps [37]. This approach has been successfully employed in numerous colloid and polymer simulations (see [31] and references therein).

During propagation, space and velocity are continuous and the particles move along straight lines for one time step Δt:

$$\boldsymbol{x}_i(t + \Delta t) = \boldsymbol{x}_i(t) + \boldsymbol{c}_i(t)\Delta t, \tag{2.19}$$

where $\boldsymbol{x}_i$ and $\boldsymbol{c}_i$ are particle position and velocity. After propagation, particles collide. How the collision step looks like in detail depends on the chosen MPC algorithm. Generally, each particle-based algorithm with local mass and momentum conservation and an $\mathcal{H}$-theorem (analogous to that described in Sect. 1.3.6) is called a multi-particle collision algorithm.

One special case is the so-called stochastic rotation dynamics (SRD) algorithm. During collision, all particles are sorted into cells of a usually regular cubic lattice with lattice constant Δx. On average, there are N_c particles in each cell. The velocity $\boldsymbol{v}_i$ of each particle i in one cell is decomposed into the average cell velocity $\bar{\boldsymbol{v}}$ (as given by the average velocity of all particles in that cell) and the relative velocity $\delta \boldsymbol{v}_i$. The relative velocities are then rotated in space to give the post-collision velocities

$$\boldsymbol{v}_i^\star = \bar{\boldsymbol{v}} + \boldsymbol{R}\delta \boldsymbol{v}_i \tag{2.20}$$

where $\boldsymbol{R}$ is a suitable rotation matrix. In 2D, velocities are rotated by $\pm\alpha$ where α is a fixed angle and the sign is randomly chosen. In 3D, the rotation is defined by a fixed angle α and a random rotation axis. Rotations are the same for all particles in a given cell but statistically independent for different cells. Apart from this rotation, there is no direct interaction between particles. In particular, particles can penetrate each other, which makes a collision detection unnecessary. It can be shown that the resulting equilibrium velocity distribution is Maxwellian.

It should be noted that the originally proposed SRD algorithm [28, 29] violated Galilean invariance. This problem, which was particularly important for small time steps (i.e. a small mean free path), could be corrected by shifting the lattice by a random distance $d \in [-\Delta x/2, +\Delta x/2]$ before each collision step [38]. Furthermore, SRD does not generally conserve angular momentum; a problem that can be avoided as reviewed in [31].

Other collision models than SRD are available. For example, the Anderson thermostat (MPC-AT) [30, 39] is used to produce new particle velocities according to the canonical ensemble rather than merely rotating the existing velocity vectors in space. As noted earlier, DSMC is another MPC-like method that only differs in terms of the particle collisions.

It is also possible to implement a repulsion force between colloids and solvent particles [29] in order to keep the fluid outside the colloids. This, however, requires relatively large repulsion forces and therefore small time steps. Additional coupling approaches are reviewed in [31]. Slip [40] and no-slip boundary conditions [36] are available as well.

2.2.4.2 Advantages and Disadvantages

All MPC algorithms locally conserve mass and momentum[10] and have an $\mathcal{H}$-theorem which makes them unconditionally stable [28]. Due to its locality the MPC algorithm is straightforward to implement and to use, computationally efficient and easy to parallelise. MPC has been successfully ported to GPUs with a performance gain of up to two orders of magnitude [36]. But due to its strong artificial compressibility, MPC is not well suited for the simulation of Stokes flow ($\mathrm{Re} \to 0$) or compressible hydrodynamics [31].

Both hydrodynamics and thermal fluctuations are consistently taken into account. For example, interfacial fluctuations in binary fluids are accurately captured. The hydrodynamic interactions can be switched off [41], which makes it possible to study their relevance. However, it is recommended to use other methods like Langevin or Brownian Dynamics if hydrodynamics is not desired [31]. When hydrodynamics is included, it allows for larger time steps than in MD-like methods. Therefore longer time intervals can be simulated with MPC [31].

Compared to LBM, MPC naturally provides thermal fluctuations, which can be an advantage. Yet, for systems where those fluctuations are not required or even distracting, MPC is not an ideal numerical method. Conventional Navier-Stokes solvers or the LBM are more favorable in those situations [31].

As MPC is a particle-based method, immersed objects such as colloids or polymers can be implemented in a relatively straightforward fashion. This makes MPC particularly suitable for the simulation of soft matter systems. Additionally, the transport coefficients (viscosity, thermal diffusivity, self-diffusion coefficient) can be accurately predicted as function of the simulation parameters [31, 38, 42–44]. On the other hand, it is not a simple task to impose hydrodynamic boundary conditions, especially for the pressure. Furthermore, the discussions in [36] show that no-slip boundary conditions and forcing are not as well under control as for LBM. Multicomponent fluids and also surfactants can be simulated within the MPC framework [33]. However, the LBM seems to be more mature due to the larger number of works published.

[10]Most of them also conserve energy.

2.2.5 *Direct Simulation Monte Carlo*

The direct simulation Monte Carlo (DSMC) method was pioneered by Bird [45] in the 1960s. While its initial popularity was slowed down by limitations in computer technology, DSMC is currently considered a primary method to solve realistic problems in high Knudsen number flows. Typical applications range from spacecraft technology to microsystems. More details on DSMC can be found in dedicated books [46, 47] and review articles [48–50].

DSMC is a particle method based on kinetic theory, where flow solutions are obtained through the collisions of model particles. MPC (cf. Sect. 2.2.4) can be considered DSMC with a modified particle collision procedure [30]. Since DSMC is more than 30 years older than MPC and used for different applications, we provide a short independent overview of DSMC here.

Rather than seeking solutions of the governing mathematical model, e.g. the Boltzmann equation, DSMC incorporates the physics of the problem directly into the simulation procedure. Although this change in paradigm at first raised doubts on whether DSMC solutions were indeed solutions of the Boltzmann equation [51], modern studies have shown that the DSMC method is a sound physical simulation model capable of describing physical effects, even beyond the Boltzmann formulation [49, 50].

Fundamentally, DSMC simulations track a large number of statistically representative particles. While the position and velocity of particles is resolved deterministically during motion, particle collisions are approximated by probabilistic, phenomenological models. These models enforce conservation of mass, momentum and energy to machine accuracy.

The DSMC algorithm has four primary steps [49]:

1. Move particles, complying with the prescribed boundary conditions.
2. Index and cross-reference the particles. The particles are labelled inside the computational domain as a pre-requisite for the next two steps.
3. Simulate collisions. Pairs of particles are randomly selected to collide according to a given collisional model. (It is this probabilistic process of determining collisions that sets DSMC apart from deterministic simulation procedures such as MD.) This separates DSMC from MPC; the latter handling the collision of *all* particles in a cell simultaneously.
4. Sample the flow field. Within predefined cells, compute macroscopic quantities based on the positions and velocities of particles in the cell. This averaged data is typically not necessary for the rest of the simulation and is only used as output information.

DSMC is an explicit and time-marching technique. Depending on the sample size and the averaging procedure, it may produce statistically accurate results. The statistical error in a DSMC solution is inversely proportional to $\sqrt{N}$, N being the number of simulation particles [49]. On the other hand, the computational cost is proportional to N. The main drawback of DSMC is therefore the high computational demand of problems requiring a large N. This explains why DSMC is mostly used for dilute gases, where its accuracy/efficiency characteristics have no competitor. Nevertheless, the continuous improvement in computational resources is expected to expand the range of physical applications for DSMC simulations in the near future.

2.2.6 Smoothed-Particle Hydrodynamics

SPH was invented in the 1970s to deal with the particular challenges of 3D astrophysics. Since then it has been used in a large number of applications, in recent years also computer graphics where it can simulate convincing fluid flow relatively cheaply. Several books have been written on SPH, such as a mathematically rigorous introduction by Violeau [52] and a more practical introduction by Liu and Liu [53] on which the rest of our description will be based.

At the base of SPH is an interpolation scheme which uses point particles that influence their vicinity. For instance, any quantity λ at point $\boldsymbol{x}$ can be approximated as a sum over all particles, with each particle j positioned at $\boldsymbol{x}_j$:

$$\lambda(\boldsymbol{x}) = \sum_j \frac{m_j}{\rho_j} \lambda_j W(|\boldsymbol{x} - \boldsymbol{x}_j|, h_j). \tag{2.21}$$

Here, m_j is the particle's mass, ρ_j is the density at $\boldsymbol{x}_j$, λ_j is the particle's value of λ, and $W(|\boldsymbol{x} - \boldsymbol{x}_j|, h_j)$ is a kernel function with characteristic size h_j which defines the region of influence of particle j.[11] Thus, SPH particles can be seen as overlapping blobs, and the sum of these blobs at $\boldsymbol{x}$ determines $\lambda(\boldsymbol{x})$. For instance, the density ρ_i at particle i can be found by setting $\lambda(\boldsymbol{x}_i) = \rho(\boldsymbol{x}_i)$, so that

$$\rho_i = \sum_j \frac{m_j}{\rho_j} \rho_j W(|\boldsymbol{x}_i - \boldsymbol{x}_j|, h_j) = \sum_j m_j W(|\boldsymbol{x}_i - \boldsymbol{x}_j|, h_j). \tag{2.22}$$

The formulation of SPH and its adaptive resolution gives it a great advantage when dealing with large unbounded domains with huge density variations, such as in astrophysics. It can also deal with extreme problems with large deformations, such as explosions and high-velocity impacts, where more traditional methods may struggle. The particle formulation of SPH also allows for perfect conservation of mass and momentum.

On the downside, SPH has problems with accuracy, and it is not quite simple to deal with boundary conditions. Additionally, the formulation of SPH makes it

[11] While the kernel function can be e.g. Gaussian, it is advantageous to choose kernels that are zero for $|\boldsymbol{x} - \boldsymbol{x}_j| > h$, so that only particles in the vicinity of $\boldsymbol{x}$ need be included in the sum. Additionally, the fact that h_j can be particle-specific and varying allows adaptive resolution.

difficult to mathematically prove that the numerical method is consistent with the equations of the hydrodynamics that it is meant to simulate.

2.3 Summary

In this chapter, we have looked at various numerical methods for fluids in two main categories: *Conventional* methods, and *particle-based* methods.

Conventional methods are generic numerical methods, applied to solving the equations of fluid mechanics. These methods represent the fluid variables (such as velocity and pressure) as values at various points (nodes) throughout the domain. The interpretation of these node values varies. In the finite difference (FD) method, the idea of a continuous field is dropped in favour of a field defined only on a square grid of nodes. In the finite volume (FV) method, a node value represent the average of the fluid variables in a small volume around the node. In the finite element (FE) method, the continuous field is approximated by a kind of interpolation of the node values. These methods have in common that the node values are used to approximate the derivatives in the partial derivative equations in question.

While these general numerical methods are reasonably simple in principle (FE being somewhat more complex than FD or FV), they are complicated by the inherent difficulties of the equations of fluid mechanics. These represent a nonlinear, simultaneous system of equations where the solutions can behave in a very intricate way, especially in cases like turbulence or flows in complex geometries. Additionally, the pressure is implicit in the incompressible Navier-Stokes simulations. Such difficulties have caused the development of somewhat complex iterative algorithms such as SIMPLE and SIMPLER [4]. One troublesome issue that emerges is that of checkerboard instabilities, as described in Sect. 2.1.1.2, which can be dealt with by staggered grids or asymmetric schemes, both of which add complexity.

FD, FV, and FE simulations ultimately end up being expressed as matrix equations. Solving these equations efficiently is a pure linear algebra problem which is nevertheless important for these methods, and many different solution methods have been developed. Another complex mathematical problem that emerges when using irregular grids in FV and FE simulations is building this grid automatically for a given geometry, and this has also been a topic of extensive studies [54].

All in all, conventional methods have been thoroughly explored in the past decades, and are currently considered workhorse methods in CFD. Though it is possible to implement methods of higher order, in practice nearly all production flow solvers are second-order accurate [55].

Particle-based methods typically do not attempt to solve the equations of fluid mechanics directly, unlike conventional methods. Instead, they represent the fluid through particles, which themselves may represent atoms, molecules, collections or distributions of molecules, or portions of the macroscopic fluid. These methods are quite varied, and are often tailored to a particular problem.[12] It is therefore difficult to give a general summary of these methods as a whole.

However, it can be said that it can be difficult to relate the dynamics of some particle-based methods, such as smoothed-particle hydrodynamics, to a macroscopic description of the fluid. This makes it difficult to quantify generally the accuracy of these methods. Additionally, it must be said that the *microscopic* particle-based CFD methods are typically not appropriate for CFD. Even the lattice gas models described in Sect. 2.2.2, which were originally intended for flow simulations, had major problems with noise from fluctuations in the microscopic particle populations. For that reason, lattice gases were largely abandoned in favour of the very similar lattice Boltzmann method, which instead took a *mesoscopic* approach that eliminated this noise.

All in all, different solvers have different advantages and disadvantages, and different types of fluid simulations pose different demands on a solver. For that reason, it is generally agreed (e.g. [5, 15, 56, 57]) that there is **no one method** which **is generally superior** to all others.

2.4 Outlook: Why Lattice Boltzmann?

While we will not describe the lattice Boltzmann method in detail until Chap. 3, we will here compare it in general terms to the other methods of this chapter.

The lattice Boltzmann method (LBM) originally grew out of the lattice gas models described in Sect. 2.2.2. While lattice gases track the behaviour of concrete particles, the LBM instead tracks the *distribution* of such particles. It can be debated whether the LBM should be called a particle-based method when it only tracks particle distributions instead of the particles themselves, but it is clear that it has much in common with many of the methods described in Sect. 2.2.

The LBM has a strong physical basis, namely the Boltzmann equation described in Sect. 1.3.4. Well-established methods exist to link its dynamics to the macroscopic conservation equations of fluids.[13] It can thus be found that the "standard" LBM is a second-order accurate solver for the weakly compressible Navier-Stokes equation; this is detailed in Sect. 4.5.5. The "weak compressibility" refers to errors that become relevant as $\mathrm{Ma} \to 1$ (cf. Sect. 4.1).

---

[12]For instance, molecular dynamics is tailored to simulating phenomena on an atomic and molecular level, and smoothed-particle hydrodynamics was invented to deal with the largely empty domains of astrophysical CFD.

[13]The most common such method is covered in Sect. 4.1, with a number of alternative methods referenced in Sect. 4.2.5.

The LBM gains a major advantage from being based on the Boltzmann equation rather than the equations of fluid mechanics: It becomes much simpler to implement than conventional methods. However, there is a corresponding disadvantage: understanding and adapting the LBM typically requires some knowledge about the Boltzmann equation, in addition to knowledge about fluid mechanics.

In conventional methods, much of the complexity lies in determining derivative approximations non-locally from adjacent nodes. In particular, it is difficult to discretise the non-linear advection term $\boldsymbol{u} \cdot \nabla \boldsymbol{u}$. In contrast, the detail in the LBM lies in the particle description *within* the nodes themselves, causing that *"non-linearity is local, non-locality is linear"*[14]: interactions between nodes are entirely linear, while the method's non-linearity enters in a local collision process within each node. This property makes the LBM very amenable to high-performance computing on parallel architectures, including GPUs. Coupled with the method's simplicity, this means that parallelised LB simulations can be tailor-made for a particular case more quickly than simulations using a conventional method [15].

A number of publications have compared the LBM to other methods (e.g. [15, 56–59]). From these comparisons, some takeaway messages about the LBM's advantages (+) and disadvantages (−) can be found for a number of topics:

Simplicity and Efficiency

+ For solving the incompressible Navier-Stokes equation, the LBM is similar to pseudocompressible methods, which gain simplicity and scalability by allowing artificial compressibility [56].
+ Like pseudocompressible methods, the LBM does not involve the Poisson equation [56] which can be difficult to solve due to its non-locality.
+ The heaviest computations in the LBM are local, i.e. restricted to within nodes, further improving its amenability to parallelisation [15].
− LBM is memory-intensive. Propagating populations requires a large number of memory access events. As we will see in Sect. 13.3.2, these are a major bottleneck of LB computations.
− The LBM, being inherently time-dependent, is not particularly efficient for simulating steady flows [57].

Geometry

+ The LBM is well suited to simulating mass-conserving flows in complex geometries such as porous media [15, 56, 59].
+ Moving boundaries that conserve mass can be implemented particularly well in the LBM [56], making it an attractive method for soft matter simulations [59].

Multiphase and Multicomponent Flows

+ There is a wide range of multiphase and multicomponent methods available for the LBM [56].

[14] This concise description is attributed to Sauro Succi.

+ Coupled with the LBM's advantages in complex geometries, this means that it is well suited to simulating multiphase and multicomponent flows in complex geometries [15].
– As in other lattice-based methods, there are spurious currents near fluid-fluid interfaces (cf. Sect. 9.4.1).
– According to [56], no multiphase or multicomponent methods for the LBM have capitalised well on its kinetic origins, meaning that these methods are not very different from those in conventional CFD.
– The range of viscosities and densities are somewhat limited in multiphase and multicomponent simulations [56].

Thermal Effects

+ Thermal fluctuations, which originate on the microscale but are averaged out on the macroscale, can be incorporated into the LBM mesoscopically. We will not discuss fluctuations in this book. Instead, the interested reader should refer to [60–62] for ideal and to [63, 64] for non-ideal systems with thermal fluctuations.
– Energy-conserving (thermal) simulations are not straightforward in the LBM [15, 56]. We come back to this topic in Sect. 8.4.

Sound and Compressibility

+ As the LBM is a (weakly) compressible Navier-Stokes solver, it may be well-suited for simulating situations where sound and flow interact, such as aeroacoustic sound generation [65].
– The LBM is not appropriate for directly simulating long-range propagation of sound at realistic viscosities [56, 65].
– The LBM may not be appropriate for simulating strongly compressible (i.e. transonic and supersonic) flows [15, 56].

Other Points

+ The LBM is appropriate for simulating mesoscopic physics that are hard to describe macroscopically [15].

While the lattice Boltzmann method has many advantages, it is, like all other numerical methods for fluids, not well suited for all possible applications. However, the LBM is a relatively young method, and it is still evolving at a quick pace, meaning that the range of problems to which it can be applied well is still increasing.

References

1. R.J. LeVeque, *Finite Difference Methods for Ordinary and Partial Differential Equations: Steady State and Time Dependent Problems* (SIAM, Philadelphia, 2007)
2. C. Pozrikidis, *A Practical Guide to Boundary Element Methods with the Software Library BEMLIB* (CRC Press, Boca Raton, 2002)

3. C. Canuto, M.Y. Hussaini, A.M. Quarteroni, A. Thomas Jr, et al., *Spectral Methods in Fluid Dynamics* (Springer Science & Business Media, New York, 2012)
4. S.V. Patankar, *Numerical Heat Transfer and Fluid Flow* (Taylor & Francis, Washington, DC, 1980)
5. J.H. Ferziger, M. Peric, A. Leonard, *Computational Methods for Fluid Dynamics*, vol. 50, 3rd edn. (Springer, New York, 2002)
6. R.J. LeVeque, *Finite-Volume Methods for Hyperbolic Problems*. Cambridge Texts in Applied Mathematics (Cambridge University Press, Cambridge, 2004)
7. H.K. Versteeg, W. Malalasekera, *An Introduction to Computational Fluid Dynamics: The Finite Volume Method*, 2nd edn. (Pearson Education, Upper Saddle River, 2007)
8. O.C. Zienkiewicz, R.L. Taylor, P. Nithiarasu, *The Finite Element Method for Fluid Dynamics*, 7th edn. (Butterworth-Heinemann, Oxford, 2014)
9. C.A.J. Fletcher, *Computational Techniques for Fluid Dynamics*, vol. 2 (Springer, New York, 1988)
10. D. Frenkel, B. Smit, *Understanding Molecular Simulation: From Algorithms to Applications*, 2nd edn. Computational science series (Academic Press, San Diego, 2002)
11. M.P. Allen, D.J. Tildesley, *Computer Simulation of Liquids*. Oxford Science Publications (Clarendon Press, Oxford, 1989)
12. M. Karplus, J.A. McCammon, Nat. Struct. Biol. **9**(9), 646 (2002)
13. J. Hardy, Y. Pomeau, O. de Pazzis, J. Math. Phys. **14**(12), 1746 (1973)
14. U. Frisch, B. Hasslacher, Y. Pomeau, Phys. Rev. Lett. **56**(14), 1505 (1986)
15. S. Succi, *The Lattice Boltzmann Equation for Fluid Dynamics and Beyond* (Oxford University Press, Oxford, 2001)
16. J.P. Rivet, J.P. Boon, *Lattice Gas Hydrodynamics* (Cambridge University Press, Cambridge, 2001)
17. D.A. Wolf-Gladrow, *Lattice-Gas Cellular Automata and Lattice Boltzmann Models* (Springer, New York, 2005)
18. G.R. McNamara, G. Zanetti, Phys. Rev. Lett. **61**(20), 2332 (1988)
19. F.J. Higuera, S. Succi, R. Benzi, Europhys. Lett. **9**(4), 345 (1989)
20. F.J. Higuera, J. Jimenez, Europhys. Lett. **9**(7), 663 (1989)
21. X. He, L.S. Luo, Phys. Rev. E **56**(6), 6811 (1997)
22. P.J. Hoogerbrugge, J.M.V.A. Koelman, Europhys. Lett. **19**(3), 155 (1992)
23. P. Español, P. Warren, Europhys. Lett. **30**(4), 191 (1995)
24. M.B. Liu, G.R. Liu, L.W. Zhou, J.Z. Chang, Arch. Computat. Methods Eng. **22**(4), 529 (2015)
25. P. Español, M. Revenga, Phys. Rev. E **67**(2), 026705 (2003)
26. R.D. Groot, P.B. Warren, J. Chem. Phys. **107**(11), 4423 (1997)
27. I. Pagonabarraga, M.H.J. Hagen, D. Frenkel, Europhys. Lett. **42**(4), 377 (1998)
28. A. Malevanets, R. Kapral, J. Chem. Phys. **110**(17), 8605 (1999)
29. A. Malevanets, R. Kapral, J. Chem. Phys. **112**(16), 7260 (2000)
30. H. Noguchi, N. Kikuchi, G. Gompper, Europhys. Lett. **78**(1), 10005 (2007)
31. G. Gompper, T. Ihle, D.M. Kroll, R.G. Winkler, in *Advanced Computer Simulation Approaches for Soft Matter Sciences III*, Advances in Polymer Science (Springer, Berlin, Heidelberg, 2008), pp. 1–87
32. T. Ihle, E. Tüzel, D.M. Kroll, Europhys. Lett. **73**(5), 664 (2006)
33. E. Tüzel, G. Pan, T. Ihle, D.M. Kroll, Europhys. Lett. **80**(4), 40010 (2007)
34. Y.G. Tao, I.O. Götze, G. Gompper, J. Chem. Phys. **128**(14), 144902 (2008)
35. R. Kapral, in *Advances in Chemical Physics*, ed. by S.A. Rice (Wiley, New York, 2008), p. 89–146
36. E. Westphal, S.P. Singh, C.C. Huang, G. Gompper, R.G. Winkler, Comput. Phys. Commun. **185**(2), 495 (2014)
37. A. Malevanets, J.M. Yeomans, Europhys. Lett. **52**(2), 231 (2000)
38. T. Ihle, D.M. Kroll, Phys. Rev. E **67**(6), 066706 (2003)
39. E. Allahyarov, G. Gompper, Phys. Rev. E **66**(3), 036702 (2002)
40. J.K. Whitmer, E. Luijten, J. Phys. Condens. Matter **22**(10), 104106 (2010)

41. M. Ripoll, R.G. Winkler, G. Gompper, Eur. Phys. J. E **23**(4), 349 (2007)
42. N. Kikuchi, C.M. Pooley, J.F. Ryder, J.M. Yeomans, J. Chem. Phys. **119**(12), 6388 (2003)
43. T. Ihle, E. Tözel, D.M. Kroll, Phys. Rev. E **72**(4), 046707 (2005)
44. C.M. Pooley, J.M. Yeomans, J. Phys. Chem. B **109**(14), 6505 (2005)
45. G.A. Bird, Phys. Fluids **6**, 1518 (1963)
46. G.A. Bird, *Molecular Gas Dynamics and the Direct Simulation of Gas Flows* (Claredon, Oxford, 1994)
47. C. Shen, *Rarefied gas dynamics: Fundamentals, Simulations and Micro Flows* (Springer, New York, 2005)
48. G.A. Bird, Ann. Rev. Fluid Mech. **10**, 11 (1978)
49. E.S. Oran, C.K. Oh, B.Z. Cybyk, Ann. Rev. Fluid Mech. **30**, 403 (1998)
50. G.A. Bird, Comp. Math. Appl. **35**, 1 (1998)
51. W. Wagner, J. Stat. Phys. **66**, 1011 (1992)
52. D. Violeau, *Fluid Mechanics and the SPH Method: Theory and Applications*, 1st edn. (Oxford University Press, Oxford, 2012)
53. G.R. Liu, M.B. Liu, *Smoothed Particle Hydrodynamics: A Meshfree Particle Method* (World Scientific Publishing, Singapore, 2003)
54. P.J. Frey, P.L. George, *Mesh Generation: Application to Finite Elements* (Wiley, Hoboken, 2008)
55. Z. Wang, Prog. Aerosp. Sci. **43**(1–3), 1 (2007)
56. R.R. Nourgaliev, T.N. Dinh, T.G. Theofanous, D. Joseph, Int. J. Multiphas. Flow **29**(1), 117 (2003)
57. S. Geller, M. Krafczyk, J. Tölke, S. Turek, J. Hron, Comput. Fluids **35**(8-9), 888 (2006)
58. M. Yoshino, T. Inamuro, Int. J. Num. Meth. Fluids **43**(2), 183 (2003)
59. B. Dünweg, A.J.C. Ladd, in *Advances in Polymer Science* (Springer, Berlin, Heidelberg, 2008), pp. 1–78
60. A.J.C. Ladd, J. Fluid Mech. **271**, 285 (1994)
61. R. Adhikari, K. Stratford, M.E. Cates, A.J. Wagner, Europhys. Lett. **71**(3), 473 (2005)
62. B. Dünweg, U.D. Schiller, A.J.C. Ladd, Phys. Rev. E **76**(3), 036704 (2007)
63. M. Gross, M.E. Cates, F. Varnik, R. Adhikari, J. Stat. Mech. **2011**(03), P03030 (2011)
64. D. Belardinelli, M. Sbragaglia, L. Biferale, M. Gross, F. Varnik, Phys. Rev. E **91**(2), 023313 (2015)
65. E.M. Viggen, The lattice Boltzmann method: Fundamentals and acoustics. Ph.D. thesis, Norwegian University of Science and Technology (NTNU), Trondheim (2014)

Part II
Lattice Boltzmann Fundamentals

Chapter 3
The Lattice Boltzmann Equation

Abstract After reading this chapter, you will know the basics of the lattice Boltzmann method, how it can be used to simulate fluids, and how to implement it in code. You will have insight into the derivation of the lattice Boltzmann equation, having seen how the continuous Boltzmann equation is discretised in velocity space through Hermite series expansion, before being discretised in physical space and time through the method of characteristics. In particular, you will be familiar with the various simple sets of velocity vectors that are available, and how the discrete BGK collision model is applied.

In this chapter we provide a simple overview of the LBM. After presenting a general introduction in Sect. 3.1, we briefly outline the LBM without derivation in order to give an initial understanding (cf. Sect. 3.2). Section 3.3 contains general advice on implementing the algorithm on a computer. Together, these two sections should be sufficient to write simple LBM codes. Next, we present the derivation of the lattice Boltzmann equation by discretising the Boltzmann equation in two steps. First, in Sect. 3.4, we discretise velocity space by limiting the continuous particle velocity $\boldsymbol{\xi}$ to a discrete set of velocities $\{\boldsymbol{\xi}_i\}$. Secondly, we discretise physical space and time by *integrating along characteristics* in Sect. 3.5. The result of these two steps is the *lattice Boltzmann equation* (LBE). A thorough discussion of how the LBE is linked to fluid dynamics will be given in Chap. 4.

3.1 Introduction

The equations of fluid mechanics are notoriously difficult to solve in general. Analytical solutions can be found only for quite basic cases, such as the Couette or Poiseuille flows shown in Fig. 1.1. Situations with more complex geometries and boundary conditions must typically be solved using numerical methods. However, as we have seen in Chap. 2, the numerical methods used to solve the equations of fluid mechanics can be difficult both to implement and to parallelise.

The Boltzmann equation in Sect. 1.3 describes the dynamics of a gas on a mesoscopic scale. From Sect. 1.3.5 we also know that the Boltzmann equation leads

T. Krüger et al., *The Lattice Boltzmann Method*, Graduate Texts in Physics,
DOI 10.1007/978-3-319-44649-3_3

to the equations of fluid dynamics on the macroscale. Therefore, from a solution of the Boltzmann equation for a given case we can often find a solution to the NSE for the same case.[1]

The problem with this idea is that the Boltzmann equation is even more difficult to solve analytically than the NSE. Indeed, its fundamental variable, the distribution function $f(\boldsymbol{x}, \boldsymbol{\xi}, t)$, is a function of seven parameters: x, y, z, ξ_x, ξ_y, ξ_z and t. However, we can try a different approach; if we can solve the Boltzmann equation *numerically*, this may also indirectly give us a solution to the NSE.

The **numerical scheme for the Boltzmann equation** somewhat paradoxically turns out to be ***quite simple***, both **to implement and to parallelise**. The reason is that the force-free Boltzmann equation is a simple hyperbolic equation which essentially describes the advection of the distribution function f with the particle velocity $\boldsymbol{\xi}$. In addition, the source term $\Omega(f)$ depends only on the local value of f and not on its gradients.

Not only is the discretised Boltzmann equation simple to implement, it also has certain numerical advantages over conventional methods that directly discretise the equations of fluid mechanics, such as finite difference or finite volume methods. A major difficulty with these methods is discretising the advection term $(\boldsymbol{u} \cdot \nabla)\boldsymbol{u}$; complicated iterative numerical schemes with approximation errors are introduced to deal with this. Contrarily, the discretised Boltzmann equation takes a very different approach that results in exact advection.

In this chapter we will first discretise velocity space (cf. Sect. 3.4) and then space-time (cf. Sect. 3.5). Historically, the LBE was not found along these lines. It rather evolved out of the lattice gas automata described in Sect. 2.2.2 and many early articles on the LBM are written from this perspective. The connection between lattice gases and the LBE is described in detail, e.g., in [1].

Throughout this chapter we use the force-free form of the Boltzmann equation for the sake of simplicity. The inclusion of external forces in the LB scheme is covered in Chap. 6.

3.2 The Lattice Boltzmann Equation in a Nutshell

Before diving into the derivation of the LBE, we will first give a short summary. This serves two purposes: first, it is unnecessary to know all the details of the derivation

[1]Since the Boltzmann equation is more general, it also has solutions that *do not* correspond to Navier-Stokes solutions. The connections between these two equations will be further explored in Sect. 4.1.

in order to implement simple codes. Therefore, this section, along with Sect. 3.3, gives a quick introduction for readers who only wish to know the most relevant results. Secondly, readers who *are* interested in understanding the derivations will benefit from knowing the end results before going through the analysis.

3.2.1 Overview

The basic quantity of the LBM is the *discrete-velocity distribution function* $f_i(\boldsymbol{x}, t)$, often called the particle *populations*. Similar to the distribution function introduced in Sect. 1.3, it represents the density of particles with velocity $\boldsymbol{c}_i = (c_{ix}, c_{iy}, c_{iz})$ at position $\boldsymbol{x}$ and time t. Likewise, the mass density ρ and momentum density $\rho\boldsymbol{u}$ at $(\boldsymbol{x}, t)$ can be found through weighted sums known as *moments* of f_i:

$$\rho(\boldsymbol{x}, t) = \sum_i f_i(\boldsymbol{x}, t), \qquad \rho\boldsymbol{u}(\boldsymbol{x}, t) = \sum_i \boldsymbol{c}_i f_i(\boldsymbol{x}, t). \tag{3.1}$$

The major difference between f_i and the continuous distribution function f is that all of the argument variables of f_i are discrete. $\boldsymbol{c}_i$, to which the subscript i in f_i refers, is one of a small discrete set of velocities $\{\boldsymbol{c}_i\}$. The points $\boldsymbol{x}$ at which f_i is defined are positioned as a square lattice in space, with lattice spacing Δx. Additionally, f_i is defined only at certain times t, separated by a time step Δt.

The time step Δt and lattice spacing Δx respectively represent a time resolution and a space resolution in any set of units. One possible choice is SI units, where Δt is given in seconds and Δx in metres, and another possible choice would be Imperial units. The most common choice in the LB literature, however, is *lattice units*, a simple artificial set of units scaled such that $\Delta t = 1$ and $\Delta x = 1$. We can convert quantities between lattice units and physical units as easily as converting between two sets of physical units, such as SI and Imperial units. Alternatively, we can ensure that we simulate the same behaviour in two different systems of units by exploiting the *law of similarity* explained in Sect. 1.2. That way, we need only ensure that the relevant *dimensionless numbers*, such as the Reynolds number, are the same in both systems. We cover the topics of units and the law of similarity in more depth in Chap. 7.

The discrete velocities $\boldsymbol{c}_i$ require further explanation. Together with a corresponding set of weighting coefficients w_i which we will explain shortly, they form *velocity sets* $\{\boldsymbol{c}_i, w_i\}$. Different velocity sets are used for different purposes. These velocity sets are usually denoted by DdQq, where d is the number of spatial dimensions the velocity set covers and q is the set's number of velocities.

The most commonly used velocity sets to solve the Navier-Stokes equation are D1Q3, D2Q9, D3Q15, D3Q19 and D3Q27. They are shown in Fig. 3.3 and Fig. 3.4 and characterised in Table 3.1. The velocity components $\{c_{i\alpha}\}$ and weights $\{w_i\}$ are also collected in Tables 3.2, 3.3, 3.4, 3.5 and 3.6.

Typically, we want to use as few velocities as possible to minimise memory and computing requirements. However, there is a tradeoff between smaller velocity sets[2] (e.g. D3Q15) and higher accuracy (e.g. D3Q27). In 3D, the most commonly used velocity set is D3Q19.

We can also find a constant c_s in each velocity set as in (3.60). In the basic isothermal LBE, c_s determines the relation $p = c_s^2 \rho$ between pressure p and density ρ. Because of this relation, it can be shown that c_s represents the isothermal model's *speed of sound*.[3] In all the velocity sets mentioned above, this constant is $c_s^2 = (1/3)\Delta x^2/\Delta t^2$.

By discretising the Boltzmann equation in velocity space, physical space, and time, we find the *lattice Boltzmann equation*

$$f_i(\boldsymbol{x} + \boldsymbol{c}_i \Delta t, t + \Delta t) = f_i(\boldsymbol{x}, t) + \Omega_i(\boldsymbol{x}, t). \tag{3.2}$$

This expresses that particles $f_i(\boldsymbol{x}, t)$ move with velocity $\boldsymbol{c}_i$ to a neighbouring point $\boldsymbol{x} + \boldsymbol{c}_i \Delta t$ at the next time step $t + \Delta t$ as shown in Fig. 3.1.[4] At the same time, particles are affected by a collision operator Ω_i. This operator models particle collisions by redistributing particles among the populations f_i at each site.

While there are many different collision operators Ω_i available, the simplest one that can be used for Navier-Stokes simulations is the Bhatnagar-Gross-Krook (BGK) operator:

$$\Omega_i(f) = -\frac{f_i - f_i^{\mathrm{eq}}}{\tau} \Delta t. \tag{3.3}$$

It relaxes the populations towards an equilibrium f_i^{eq} at a rate determined by the relaxation time τ.[5]

This equilibrium is given by

$$f_i^{\mathrm{eq}}(\boldsymbol{x}, t) = w_i \rho \left(1 + \frac{\boldsymbol{u} \cdot \boldsymbol{c}_i}{c_s^2} + \frac{(\boldsymbol{u} \cdot \boldsymbol{c}_i)^2}{2c_s^4} - \frac{\boldsymbol{u} \cdot \boldsymbol{u}}{2c_s^2} \right) \tag{3.4}$$

with the weights w_i specific to the chosen velocity set. The equilibrium is such that its moments are the same as those of f_i, i.e. $\sum_i f_i^{\mathrm{eq}} = \sum_i f_i = \rho$ and $\sum_i \boldsymbol{c}_i f_i^{\mathrm{eq}} =$

[2]There is a limit to how small a velocity set can be, though, as it has to obey the requirements shown in (3.60) to be suitable for Navier-Stokes simulations. The smallest velocity set in 3D is D3Q13, but it has the disadvantage of being tricky to apply.

[3]This statement is not proven in this chapter, but rather in Sect. 12.1 on sound waves.

[4]Usually, the velocity sets $\{\boldsymbol{c}_i\}$ are chosen such that any spatial vector $\boldsymbol{c}_i \Delta t$ points from one lattice site to a neighbouring lattice site. This guarantees that the populations f_i always reach another lattice site during a time step Δt, rather than being trapped between them.

[5]Other collision operators are available which use additional relaxation times to achieve increased accuracy and stability (cf. Chap. 10).

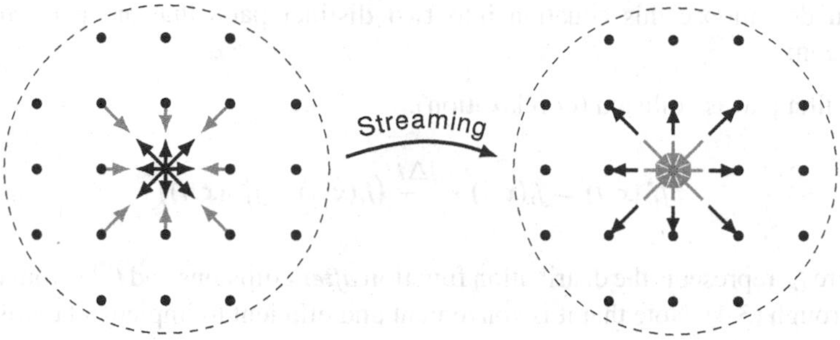

Fig. 3.1 Particles (*black*) streaming from the central node to its neighbours, from which particles (*grey*) are streamed back. To the left we see post-collision distributions $f_i^\star$ *before* streaming, and to the right we see pre-collision distributions f_i *after* streaming

$\sum_i \boldsymbol{c}_i f_i = \rho \boldsymbol{u}$.[6] The equilibrium f_i^{eq} depends on the local quantities density ρ and fluid velocity $\boldsymbol{u}$ only. These are calculated from the local values of f_i by (3.1), with the fluid velocity found as $\boldsymbol{u}(\boldsymbol{x}, t) = \rho \boldsymbol{u}(\boldsymbol{x}, t)/\rho(\boldsymbol{x}, t)$.

The link between the LBE and the NSE can be determined using the Chapman-Enskog analysis (cf. Sect. 4.1). Through this, we can show that the LBE results in macroscopic behaviour according to the NSE, with the kinematic shear viscosity given by the relaxation time τ as

$$\nu = c_s^2\left(\tau - \frac{\Delta t}{2}\right), \tag{3.5}$$

and the kinematic bulk viscosity given as $\nu_B = 2\nu/3$. Additionally, the viscous stress tensor can be calculated from f_i as

$$\sigma_{\alpha\beta} \approx -\left(1 - \frac{\Delta t}{2\tau}\right)\sum_i c_{i\alpha} c_{i\beta} f_i^{\text{neq}}, \tag{3.6}$$

where the *non-equilibrium* distribution $f_i^{\text{neq}} = f_i - f_i^{\text{eq}}$ is the deviation of f_i from equilibrium. (However, computing the stress tensor explicitly in this way is usually not a necessary step when performing simulations.)

3.2.2 The Time Step: Collision and Streaming

In full, the lattice BGK (LBGK) equation (i.e. the fully discretised Boltzmann equation with the BGK collision operator) reads

$$f_i(\boldsymbol{x} + \boldsymbol{c}_i \Delta t, t + \Delta t) = f_i(\boldsymbol{x}, t) - \frac{\Delta t}{\tau}\left(f_i(\boldsymbol{x}, t) - f_i^{\text{eq}}(\boldsymbol{x}, t)\right). \tag{3.7}$$

[6] As a consequence of this, mass and momentum are conserved in collisions. This conservation can be expressed mathematically as $\sum_i \Omega_i = 0$ and $\sum_i \boldsymbol{c}_i \Omega_i = \boldsymbol{0}$.

We can decompose this equation into two distinct parts that are performed in succession:

1. The first part is collision (or relaxation),

$$f_i^\star(\boldsymbol{x},t) = f_i(\boldsymbol{x},t) - \frac{\Delta t}{\tau}\left(f_i(\boldsymbol{x},t) - f_i^{\mathrm{eq}}(\boldsymbol{x},t)\right), \tag{3.8}$$

   where $f_i^\star$ represents the distribution function *after* collisions and f_i^{eq} is found from f_i through (3.4). Note that it is convenient and efficient to implement collision in the form

$$f_i^\star(\boldsymbol{x},t) = f_i(\boldsymbol{x},t)\left(1 - \frac{\Delta t}{\tau}\right) + f_i^{\mathrm{eq}}(\boldsymbol{x},t)\frac{\Delta t}{\tau}. \tag{3.9}$$

   This becomes particularly simple for $\tau/\Delta t = 1$, where $f_i^\star(\boldsymbol{x},t) = f_i^{\mathrm{eq}}(\boldsymbol{x},t)$.
2. The second part is streaming (or propagation),

$$f_i(\boldsymbol{x} + \boldsymbol{c}_i\Delta t, t + \Delta t) = f_i^\star(\boldsymbol{x},t), \tag{3.10}$$

   as shown in Fig. 3.1.

Overall, the **LBE concept** is straightforward. It **consists of two parts**: *collision* and *streaming*. The collision is simply an algebraic *local* operation. First, one calculates the density ρ and the macroscopic velocity $\boldsymbol{u}$ to find the equilibrium distributions f_i^{eq} as in (3.4) and the post-collision distribution $f_i^\star$ as in (3.9). After collision, we *stream* the resulting distribution $f_i^\star$ to neighbouring nodes as in (3.10). When these two operations are complete, one time step has elapsed, and the operations are repeated.

3.3 Implementation of the Lattice Boltzmann Method in a Nutshell

We cover here some details of the implementation of the LBE to achieve an efficient and working algorithm. After discussing initialisation in Sect. 3.3.1 and the time step algorithm in Sect. 3.3.2, we provide details about the underlying memory structure and coding hints in Sect. 3.3.3.

The case covered here is the simplest force-free LB algorithm in the absence of boundaries. This core LB algorithm can be extended by including boundary conditions (cf. Chap. 5) and forces (cf. Chap. 6). More in-depth implementation advice can be found in Chap. 13.

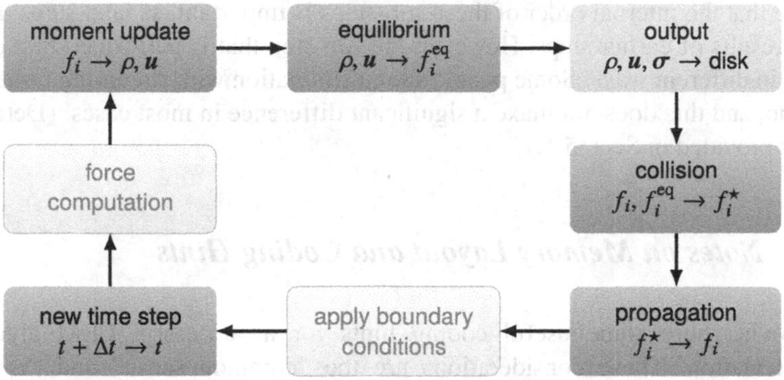

Fig. 3.2 An overview of one cycle of the LB algorithm. The *dark grey boxes* show sub-steps that are necessary for the evolution of the solution. The *light grey box* indicates the optional output step. The *pale boxes* show steps whose details are given in later chapters (namely, Chap. 5 for boundary conditions, and Chap. 6 for forces)

3.3.1 Initialisation

The simplest approach to initialising the populations at the start of a simulation is to set them to $f_i^{eq}(\boldsymbol{x}, t = 0) = f_i^{eq}(\rho(\boldsymbol{x}, t = 0), \boldsymbol{u}(\boldsymbol{x}, t = 0))$ *via* (3.4). Often, the values $\rho(\boldsymbol{x}, t = 0) = 1$ and $\boldsymbol{u}(\boldsymbol{x}, t = 0) = \boldsymbol{0}$ are used. More details about different initialisation schemes are given in Sect. 5.5.

3.3.2 Time Step Algorithm

Overall, the core LBM algorithm consists of a cyclic sequence of substeps, with each cycle corresponding to one time step. These substeps are also visualised in Fig. 3.2:

1. Compute the macroscopic moments $\rho(\boldsymbol{x}, t)$ and $\boldsymbol{u}(\boldsymbol{x}, t)$ from $f_i(\boldsymbol{x}, t)$ *via* (3.1).
2. Obtain the equilibrium distribution $f_i^{eq}(\boldsymbol{x}, t)$ from (3.4).[7]
3. If desired, write the macroscopic fields $\rho(\boldsymbol{x}, t)$, $\boldsymbol{u}(\boldsymbol{x}, t)$ and/or $\boldsymbol{\sigma}(\boldsymbol{x}, t)$ to the hard disk for visualisation or post-processing. The viscous stress tensor $\boldsymbol{\sigma}$ can be computed from (3.6).
4. Perform collision (relaxation) as shown in (3.9).
5. Perform streaming (propagation) *via* (3.10).
6. Increase the time step, setting t to $t + \Delta t$, and go back to step 1 until the last time step or convergence has been reached.

[7] For those who want ready-to-program expressions, the unrolled equilibrium functions for D1Q3 and D2Q9 are shown in (3.64) and (3.65).

Note that the internal order of these sub-steps is important, as later steps depend on the results of earlier steps. However, the sub-step that is performed first can be chosen in different ways. Some people start a simulation with streaming rather than collision, and this does not make a significant difference in most cases. (Details on this are provided in Sect. 5.5.)

3.3.3 *Notes on Memory Layout and Coding Hints*

We provide here some useful coding hints for a successful LBM algorithm implementation. These considerations are the "common sense" and "common practice" implementations, and Chap. 13 contains more sophisticated explanations and advanced coding guidelines.

3.3.3.1 Initialisation

We have to allocate memory for the macroscopic fields ρ and $\boldsymbol{u}$ and the populations f_i. Usually, taking 2D as an example, the macroscopic fields are allocated in two-dimensional arrays $\rho[N_x][N_y]$, $u_x[N_x][N_y]$ and $u_y[N_x][N_y]$ and the populations are in a three-dimensional array $f[N_x][N_y][q]$. Here, N_x and N_y are the number of lattice nodes in x- and y-directions and q is the number of velocities. Other memory layouts are discussed in Chap. 13. Depending on the programming language and structure of the loops that update the simulation domain, it may be necessary to exchange the order of N_x, N_y, and q in the array dimensions to improve the speed at which memory is read and written.

3.3.3.2 Streaming

The streaming step has to be implemented in a way that ensures that the streamed populations do not overwrite memory that still contains unstreamed populations. In other words, as we sweep through the domain, we cannot overwrite data that we will need to use later. There are three common ways to efficiently implement streaming:

- Run through memory in "opposite streaming direction" and use a small temporary buffer for a single population. For example, for the direction $(0, 1)$, at each node we read the population for this direction from below, which is the direction opposite to the direction that is being streamed, and save it to the current node. The difficulty with this method is that we need to sweep through memory in a different direction for each discrete velocity, which is a common source of bugs. However, some programming languages provide convenient functions to do this, for example `circshift` in Matlab and Octave, `numpy.roll` in Python, and `CSHIFT` in Fortran.

- Allocate memory for two sets of populations, f_i^{old} and f_i^{new}, in order to store the populations for two consecutive time steps. During streaming, read data from $f_i^{\text{old}}(\boldsymbol{x})$ and write to $f_i^{\text{new}}(\boldsymbol{x} + \boldsymbol{c}_i\Delta t)$. Swap f_i^{old} and f_i^{new} after each time step so that the new populations at the end of each time step become the old populations for the next step. In this way, one does not have to care about how to stream populations without incorrectly overwriting data. The resulting code is significantly easier but requires twice the memory.[8]
- Avoid streaming altogether, and perform a combined streaming and collision step in which the necessary populations are read from adjacent nodes when they are needed to compute macroscopic variables and do the collision calculations at each node. In this case as in the previous, a second set of populations is used to store the result of the combined streaming and collision step. This approach is described in greater detail in Chap. 13.

3.3.3.3 Updating Macroscopic Variables

As seen in (3.1), the local density ρ and velocity $\boldsymbol{u}$ are (weighted) sums of the populations f_i. It is advisable to unroll these summations, which means writing out the full expressions rather than using loops. Most velocity components are zero, and we do not want to waste CPU time by summing zeros. For example, for the D1Q3 velocity set we have:

$$\rho = f_0 + f_1 + f_2, \quad u_x = \frac{f_1 - f_2}{\rho}. \tag{3.11}$$

For the D2Q9 velocity set as defined in Table 3.3, we would implement:

$$\begin{aligned} \rho &= f_0 + f_1 + f_2 + f_3 + f_4 + f_5 + f_6 + f_7 + f_8, \\ u_x &= \left[(f_1 + f_5 + f_8) - (f_3 + f_6 + f_7)\right]/\rho, \\ u_y &= \left[(f_2 + f_5 + f_6) - (f_4 + f_7 + f_8)\right]/\rho. \end{aligned} \tag{3.12}$$

3.3.3.4 Equilibrium

As for the macroscopic variable calculations, it is possible to accelerate the equilibrium computation in (3.4). Instead of computing all q distributions f_i^{eq} within a single for-loop over i, we should write separate expressions for each of the q distributions.

[8]For example, for a moderate simulation domain of 100 × 100 × 100 lattice sites and the D3Q19 velocity set, one copy of the populations f_i requires about 145 MiB of memory, where we assume double precision (8 bytes per variable).

It is also strongly recommended to replace the inverse powers of the speed of sound, c_s^{-2} and c_s^{-4}, with new variables. For the standard lattices, the numerical values happen to be 3 and 9. However, it is better practice to keep distinct variables for those expressions as this will make it easier to switch to different velocity sets (with $c_s^2 \neq (1/3)\Delta x^2/\Delta t^2$) if necessary. In any case, numerical divisions are expensive, and we can implement the equilibrium distributions without a single division.

3.3.3.5 Collision

From (3.9) we see that the terms $(1 - \Delta t/\tau)$ and $\Delta t/\tau$ have to be computed for each time step, lattice site, and velocity direction. By defining two additional constant variables, e.g. $\omega = \Delta t/\tau$ and $\omega' = 1 - \omega$, which are computed only once, an expensive numerical division can be removed from the collision operation:

$$f_i^\star(\boldsymbol{x}, t) = \omega' f_i(\boldsymbol{x}, t) + \omega f_i^{\text{eq}}(\boldsymbol{x}, t). \tag{3.13}$$

This saves otherwise wasted CPU time. As detailed in Chap. 13, optimisations such as these can be performed automatically by the compiler if appropriate compilation options are selected. Running simulation codes compiled with no automatic optimisation is not recommended, and at least a basic level of optimisation should always be used. When compiled with no optimisation, the code in Chap. 13 took more than five times longer than when it was compiled with the highest level of optimisation.

3.4 Discretisation in Velocity Space

In Sect. 3.2 we gave an overview of LBM, now it is time to thoroughly derive the lattice Boltzmann equation from the Boltzmann equation. As a starting point we will first derive its velocity discretisation. One problem mentioned before is that the particle distribution function $f(\boldsymbol{x}, \boldsymbol{\xi}, t)$ spans the seven-dimensional space defined by the coordinates x, y, z, ξ_x, ξ_y, ξ_z and t. Solving equations in this high-dimensional space is usually computationally expensive and requires large-scale computers and programming efforts.

This apprehension, however, is often not justified. As we found in Sect. 1.3.5, the *moments* of the Boltzmann equation give the correct equations for mass, momentum and energy conservation. Thus, much of the underlying physics is not relevant if we are only interested in getting the correct *macroscopic* behaviour.

For example, the moments are nothing else than weighted integrals of f in velocity space. It is obvious that there is a vast number of different functions whose integrals are identical to those of f. As we will see shortly, there are approaches to simplify the continuous Boltzmann equation without sacrificing the macroscopic (i.e. moment-based) behaviour. The discretisation of velocity space allows us to reduce the continuous 3D velocity space to a small number of discrete velocities without compromising the validity of the macroscopic equations.

The velocity discretisation can be performed using the Mach number expansion [2] or the Hermite series expansion [3]. Both approaches give the same form of the equilibrium on the Navier-Stokes level. Although the Mach number expansion approach is simpler, we will follow the Hermite series approach as it provides a strong mathematical basis. Aside from delivering a variety of suitable discrete velocity sets, it can also correctly restore equations beyond the Navier-Stokes-Fourier level, i.e. Burnett-type equations [3].

Despite the rather heavy mathematical background of this section, the main idea is not difficult:

We will see that a **simplified equilibrium** f^{eq} and a **discrete velocity space** are **sufficient to obtain the correct macroscopic conservation laws**.

In contrast to the unknown distribution function f, the *equilibrium* distribution function f^{eq} is a known function of exponential form. f^{eq} can consequently be expressed through the exponential *weight function* (or *generating function*) of Hermite polynomials. Additionally, the mass and momentum moments can be represented as integrals of f^{eq} multiplied with Hermite polynomials.

These features let us apply two clever techniques in succession. First, we can express f^{eq} in a reduced form through a truncated sum of Hermite polynomials, while retaining the correct mass and momentum moment integrals. Secondly, the moment integrals are then of a form which lets us evaluate them exactly as a *discrete sum* over the polynomial integrand evaluated at specific points $\boldsymbol{\xi}_i$ (*abscissae*). Thus, f^{eq} becomes discrete rather than continuous in velocity space. These techniques can also be applied to the particle distribution f itself.

Here we deal with the discretisation in velocity space. Later, in Sect. 3.5, we will perform the space-time discretisation.

3.4.1 Non-dimensionalisation

Before we start with the Hermite series expansion, we *non-dimensionalise* the governing equations to simplify the subsequent steps. (Non-dimensionalisation is also useful when implementing computational models and for relating physical laws in form of mathematical equations to the real world, as discussed in Chap. 7.)

Let us recall the Boltzmann equation in continuous velocity space:

$$\frac{\partial f}{\partial t} + \xi_\alpha \frac{\partial f}{\partial x_\alpha} + \frac{f_\alpha}{\rho} \frac{\partial f}{\partial \xi_\alpha} = \Omega(f) \tag{3.14}$$

where $\Omega(f)$ is the collision operator. Its specific form is not important here.[9]

The Boltzmann equation describes the evolution of the distribution function $f(\boldsymbol{x}, \boldsymbol{\xi}, t)$, i.e. the density of particles with velocity $\boldsymbol{\xi}$ at position $\boldsymbol{x}$ and time t. In a force-free, homogeneous and steady situation, the left-hand-side of (3.14) vanishes, and the solution of the Boltzmann equation becomes the equilibrium distribution function f^{eq}. As we found in Sect. 1.3.3, f^{eq} can be written in terms of the macroscopic quantities of density ρ, fluid velocity $\boldsymbol{u}$ and temperature T as

$$f^{\text{eq}}(\rho, \boldsymbol{u}, T, \boldsymbol{\xi}) = \frac{\rho}{(2\pi RT)^{d/2}} \mathrm{e}^{-(\boldsymbol{\xi}-\boldsymbol{u})^2/(2RT)} \tag{3.15}$$

where d is the number of spatial dimensions and the gas constant $R = k_B/m$ is given by the Boltzmann constant k_B and the particle mass m.

Physical phenomena occur on certain space and time scales. For example, the wavelengths of tsunamis and electromagnetic waves in the visible spectrum are of the order of hundreds of kilometers and nanometers, respectively, and their propagation speeds are a few hundred meters per second and the speed of light. One can therefore classify these phenomena by identifying their characteristic scales. As covered in more detail in Sect. 1.2, we can analyse the properties of a fluid in terms of its characteristic length ℓ, velocity V and density ρ_0. A characteristic time scale is then given by $t_0 = \ell/V$.

Using stars to denote non-dimensionalised quantities, we first introduce the non-dimensional derivatives:

$$\frac{\partial}{\partial t^\star} = \frac{\ell}{V}\frac{\partial}{\partial t}, \quad \frac{\partial}{\partial x^\star} = \ell \frac{\partial}{\partial x}, \quad \frac{\partial}{\partial \xi^\star} = V\frac{\partial}{\partial \xi}. \tag{3.16}$$

This leads to the non-dimensional continuous Boltzmann equation:

$$\frac{\partial f^\star}{\partial t^\star} + \xi^\star_\alpha \frac{\partial f^\star}{\partial x^\star_\alpha} + \frac{F^\star_\alpha}{\rho^\star} \frac{\partial f^\star}{\partial \xi^\star_\alpha} = \Omega^\star(f^\star) \tag{3.17}$$

where $f^\star = fV^d/\rho_0$, $\boldsymbol{F}^\star = \boldsymbol{F}\ell/(\rho_0 V^2)$, $\rho^\star = \rho/\rho_0$ and $\Omega^\star = \Omega \ell V^2/\rho_0$. The non-dimensional equilibrium distribution function reads

$$f^{\text{eq}\star} = \frac{\rho^\star}{(2\pi\theta^\star)^{d/2}} \mathrm{e}^{-(\boldsymbol{\xi}^\star-\boldsymbol{u}^\star)^2/(2\theta^\star)} \tag{3.18}$$

with the non-dimensional temperature $\theta^\star = RT/V^2$.

[9] As we will see later in Chap. 10, the collision operator can have different forms all of which locally conserve the moments (mass, momentum and energy) and, thus, yield the correct macroscopic behaviour.

Exercise 3.1 Check that the above expressions are correct by showing that all quantities with a star are non-dimensional.

We will hereafter omit the symbol $\star$ indicating non-dimensionalisation in order to keep the notation compact. Note that when the Boltzmann equation is referred to from now on in this chapter, the non-dimensional version is implied unless otherwise specified.

We will perform the Hermite series expansion in the force-free case ($\boldsymbol{f} = \boldsymbol{0}$) to reduce the level of complexity. The effect of forces on the Hermite series expansion will be covered in Sect. 6.3.1.

The **non-dimensional, continuous and force-free Boltzmann equation** is

$$\frac{\partial f}{\partial t} + \xi_\alpha \frac{\partial f}{\partial x_\alpha} = \Omega(f). \tag{3.19}$$

The **non-dimensional equilibrium distribution** reads

$$f^{\mathrm{eq}}(\rho, \boldsymbol{u}, \theta, \boldsymbol{\xi}) = \frac{\rho}{(2\pi\theta)^{d/2}} \mathrm{e}^{-(\boldsymbol{\xi}-\boldsymbol{u})^2/(2\theta)}. \tag{3.20}$$

3.4.2 Conservation Laws

We have already discussed conservation laws in Sect. 1.3.4. We saw that the collision operator conserves certain moments of the distribution function. This conservation implies that the moments of the equilibrium distribution function f^{eq} and the particle distribution function f coincide:

$$\begin{aligned}
\int f(\boldsymbol{x}, \boldsymbol{\xi}, t)\, \mathrm{d}^3\xi &= \int f^{\mathrm{eq}}\left(\rho, \boldsymbol{u}, \theta, \boldsymbol{\xi}\right) \mathrm{d}^3\xi &&= \rho(\boldsymbol{x}, t),\\
\int f(\boldsymbol{x}, \boldsymbol{\xi}, t)\boldsymbol{\xi}\, \mathrm{d}^3\xi &= \int f^{\mathrm{eq}}\left(\rho, \boldsymbol{u}, \theta, \boldsymbol{\xi}\right) \boldsymbol{\xi}\, \mathrm{d}^3\xi &&= \rho\boldsymbol{u}(\boldsymbol{x}, t),\\
\int f(\boldsymbol{x}, \boldsymbol{\xi}, t)\frac{|\boldsymbol{\xi}|^2}{2}\, \mathrm{d}^3\xi &= \int f^{\mathrm{eq}}\left(\rho, \boldsymbol{u}, \theta, \boldsymbol{\xi}\right) \frac{|\boldsymbol{\xi}|^2}{2}\, \mathrm{d}^3\xi &&= \rho E(\boldsymbol{x}, t),\\
\int f(\boldsymbol{x}, \boldsymbol{\xi}, t)\frac{|\boldsymbol{\xi}-\boldsymbol{u}|^2}{2}\, \mathrm{d}^3\xi &= \int f^{\mathrm{eq}}\left(\rho, \boldsymbol{u}, \theta, \boldsymbol{\xi}\right) \frac{|\boldsymbol{\xi}-\boldsymbol{u}|^2}{2}\, \mathrm{d}^3\xi &&= \rho e(\boldsymbol{x}, t).
\end{aligned} \tag{3.21}$$

The dependence on space and time in f^{eq} enters only through $\rho(\boldsymbol{x}, t)$, $\boldsymbol{u}(\boldsymbol{x}, t)$ and $\theta(\boldsymbol{x}, t)$.

All conserved quantities on the right-hand side of (3.21) can be obtained as integrals of f or f^{eq} in velocity space. The basic idea of the Hermite series expansion is to turn the continuous integrals into discrete sums evaluated at certain points in velocity space (i.e. for specific values of $\boldsymbol{\xi}$). We will now discuss the properties of Hermite polynomials that form the basis of the Hermite series expansion.

3.4.3 *Hermite Polynomials*

Among the infinite number of different functions and polynomials, one particularly interesting set of polynomials used for the discretisation of integrals are the *Hermite polynomials* (HPs). They naturally appear in the quantum-mechanical wave functions for harmonic potentials. Before we make use of their integral discretisation properties, we will first characterise the HPs and give some examples.

The derivation of the LBE (and also working with the NSE) requires some knowledge of tensor notation. Readers without or with only little experience with tensors should take some time to learn the basics of tensor calculus [4, 5].

3.4.3.1 Definition and Construction of Hermite Polynomials

1D HPs are polynomials which can be obtained from the *weight function* (also called the *generating function*

$$\omega(x) = \frac{1}{\sqrt{2\pi}} \mathrm{e}^{-x^2/2}. \tag{3.22}$$

This weight function $\omega(x)$ allows us to construct the 1D HP of n-th order as

$$H^{(n)}(x) = (-1)^n \frac{1}{\omega(x)} \frac{\mathrm{d}^n}{\mathrm{d}x^n} \omega(x) \tag{3.23}$$

where $n \geq 0$ is an integer. The first six of these polynomials are

$$\begin{aligned} H^{(0)}(x) &= 1, & H^{(1)}(x) &= x, \\ H^{(2)}(x) &= x^2 - 1, & H^{(3)}(x) &= x^3 - 3x, \\ H^{(4)}(x) &= x^4 - 6x^2 + 3, & H^{(5)}(x) &= x^5 - 10x^3 + 15x. \end{aligned} \tag{3.24}$$

Exercise 3.2 Show that (3.22) and (3.23) lead to the HPs in (3.24).

We can extend the HP definition to d spatial dimensions [3, 6]:

$$\boldsymbol{H}^{(n)}(\boldsymbol{x}) = (-1)^n \frac{1}{\omega(\boldsymbol{x})} \boldsymbol{\nabla}^{(n)} \omega(\boldsymbol{x}), \quad \omega(\boldsymbol{x}) = \frac{1}{(2\pi)^{d/2}} \mathrm{e}^{-\boldsymbol{x}^2/2}. \tag{3.25}$$

This notation requires several comments. Both $\boldsymbol{H}^{(n)}$ and $\boldsymbol{\nabla}^{(n)}$ are tensors of rank n, i.e. we can represent $\boldsymbol{H}^{(n)}$ and $\boldsymbol{\nabla}^{(n)}$ by their d^n components $H^{(n)}_{\alpha_1\ldots\alpha_n}$ and $\nabla^{(n)}_{\alpha_1\ldots\alpha_n}$, respectively, where $\{\alpha_1, \ldots, \alpha_n\}$ are n indices running from 1 to d each. $\nabla^{(n)}_{\alpha_1\ldots\alpha_n}$ is a short notation for n consecutive spatial derivatives:

$$\nabla^{(n)}_{\alpha_1\ldots\alpha_n} = \frac{\partial}{\partial x_{\alpha_1}} \cdots \frac{\partial}{\partial x_{\alpha_n}}. \tag{3.26}$$

Derivatives are symmetric upon permutation of the indices if we assume that derivatives commute,[10] e.g. $\nabla^{(3)}_{xxy} = \nabla^{(3)}_{xyx} = \nabla^{(3)}_{yxx}$. We are mostly interested in the cases $d = 2$ or $d = 3$ where we write $\alpha_1, \ldots, \alpha_n \in \{x, y\}$ or $\{x, y, z\}$, respectively.

Example 3.1 To make this clearer, we explicitly write down the 2D ($d = 2$) HPs up to second order ($n = 0, 1, 2$). Using

$$\nabla^{(2)}_{xx} = \frac{\partial}{\partial x}\frac{\partial}{\partial x}, \quad \nabla^{(2)}_{xy} = \frac{\partial}{\partial x}\frac{\partial}{\partial y}, \quad \nabla^{(2)}_{yx} = \frac{\partial}{\partial y}\frac{\partial}{\partial x}, \quad \nabla^{(2)}_{yy} = \frac{\partial}{\partial y}\frac{\partial}{\partial y} \tag{3.27}$$

we find

$$H^{(0)} = 1 \tag{3.28}$$

for $n = 0$,

$$\begin{aligned} H^{(1)}_x &= -\frac{1}{\mathrm{e}^{-(x^2+y^2)/2}} \partial_x \mathrm{e}^{-(x^2+y^2)/2} = x, \\ H^{(1)}_y &= -\frac{1}{\mathrm{e}^{-(x^2+y^2)/2}} \partial_y \mathrm{e}^{-(x^2+y^2)/2} = y \end{aligned} \tag{3.29}$$

for $n = 1$ and

$$\begin{aligned} H^{(2)}_{xx} &= \frac{1}{\mathrm{e}^{-(x^2+y^2)/2}} \partial_x \partial_x \mathrm{e}^{-(x^2+y^2)/2} = x^2 - 1, \\ H^{(2)}_{xy} = H^{(2)}_{yx} &= \frac{1}{\mathrm{e}^{-(x^2+y^2)/2}} \partial_x \partial_y \mathrm{e}^{-(x^2+y^2)/2} = xy, \\ H^{(2)}_{yy} &= \frac{1}{\mathrm{e}^{-(x^2+y^2)/2}} \partial_y \partial_y \mathrm{e}^{-(x^2+y^2)/2} = y^2 - 1 \end{aligned} \tag{3.30}$$

for $n = 2$.

[10] This is the case for sufficiently smooth functions.

Exercise 3.3 Construct the eight third-order HPs $H^{(3)}_{\alpha_1\alpha_2\alpha_3}$ for $d = 2$.

3.4.3.2 Orthogonality and Series Expansion

Let us now turn our attention to the mathematical properties of the HPs that we will require for the Hermite series expansion.

One of the nice features of the Hermite polynomials is their orthogonality. In 1D, HPs are *orthogonal* with respect to $\omega(x)$:

$$\int_{-\infty}^{\infty} \omega(x) H^{(n)}(x) H^{(m)}(x)\, \mathrm{d}x = n!\delta^{(2)}_{nm} \tag{3.31}$$

where $\delta^{(2)}_{nm}$ is the usual Kronecker delta.

The orthogonality of the HPs can be generalised to d dimensions:

$$\int \omega(\boldsymbol{x}) H^{(n)}_{\boldsymbol{\alpha}}(\boldsymbol{x}) H^{(m)}_{\boldsymbol{\beta}}(\boldsymbol{x})\, \mathrm{d}^d x = \prod_{i=1}^{d} n_i!\, \delta^{(2)}_{nm} \delta^{(n+m)}_{\boldsymbol{\alpha\beta}}. \tag{3.32}$$

Here, $\delta^{(n+m)}_{\boldsymbol{\alpha\beta}}$ is a generalised Kronecker symbol which is 1 only if $\boldsymbol{\alpha} = (\alpha_1, \dots, \alpha_n)$ is a permutation of $\boldsymbol{\beta} = (\beta_1, \dots, \beta_m)$ and 0 otherwise. For example, (x, x, z, y) is a permutation of (y, x, z, x) but not of (x, y, x, y). n_x, n_y and n_z are the numbers of occurrences of x, y and z in $\boldsymbol{\alpha}$. For instance, for $\boldsymbol{\alpha} = (x, x, y)$ one gets $n_x = 2$, $n_y = 1$ and $n_z = 0$. For $d = 3$ in particular, (3.32) reads

$$\int \omega(\boldsymbol{x}) H^{(n)}_{\boldsymbol{\alpha}}(\boldsymbol{x}) H^{(m)}_{\boldsymbol{\beta}}(\boldsymbol{x})\, \mathrm{d}^3 x = n_x!\, n_y!\, n_z!\, \delta^{(2)}_{mn} \delta^{(n+m)}_{\boldsymbol{\alpha\beta}}. \tag{3.33}$$

Exercise 3.4 Show that (3.32) reduces to (3.31) for $d = 1$.

Example 3.2 We consider some concrete examples of (3.33). The following integrals vanish because the indices $\boldsymbol{\alpha}$ are no permutations of $\boldsymbol{\beta}$, although the order of both HPs is identical ($m = n$):

$$\begin{aligned} \int \omega(\boldsymbol{x}) H^{(1)}_{x}(\boldsymbol{x}) H^{(1)}_{y}(\boldsymbol{x})\, \mathrm{d}^3 x = 0, \\ \int \omega(\boldsymbol{x}) H^{(2)}_{xy}(\boldsymbol{x}) H^{(2)}_{xx}(\boldsymbol{x})\, \mathrm{d}^3 x = 0. \end{aligned} \tag{3.34}$$

The integral

$$\int \omega(\boldsymbol{x}) H^{(1)}_{x}(\boldsymbol{x}) H^{(2)}_{xy}(\boldsymbol{x})\, \mathrm{d}^3 x = 0 \tag{3.35}$$

vanishes because the orders are not identical: $m \neq n$. The remaining integrals

$$\begin{aligned}
&\int \omega(\boldsymbol{x}) H_x^{(1)}(\boldsymbol{x}) H_x^{(1)}(\boldsymbol{x})\, \mathrm{d}^3x = 1! = 1, \\
&\int \omega(\boldsymbol{x}) H_{xxx}^{(3)}(\boldsymbol{x}) H_{xxx}^{(3)}(\boldsymbol{x})\, \mathrm{d}^3x = 3! = 6, \\
&\int \omega(\boldsymbol{x}) H_{xxy}^{(3)}(\boldsymbol{x}) H_{xyx}^{(3)}(\boldsymbol{x})\, \mathrm{d}^3x = 2!\,1! = 2
\end{aligned} \tag{3.36}$$

do not vanish since $m = n$ and $\boldsymbol{\alpha}$ is a permutation of $\boldsymbol{\beta}$.

The 1D HPs form a complete basis in $\mathbb{R}$, i.e. any sufficiently well-behaved[11] continuous function $f(x) \in \mathbb{R}$ can be represented as a series of HPs:

$$f(x) = \omega(x) \sum_{n=0}^{\infty} \frac{1}{n!} a^{(n)} H^{(n)}(x), \quad a^{(n)} = \int f(x) H^{(n)}(x)\, \mathrm{d}x. \tag{3.37}$$

Again, this can be extended to d dimensions:

$$f(\boldsymbol{x}) = \omega(\boldsymbol{x}) \sum_{n=0}^{\infty} \frac{1}{n!} \boldsymbol{a}^{(n)} \cdot \boldsymbol{H}^{(n)}(\boldsymbol{x}), \quad \boldsymbol{a}^{(n)} = \int f(\boldsymbol{x}) \boldsymbol{H}^{(n)}(\boldsymbol{x})\, \mathrm{d}^d x. \tag{3.38}$$

The expansion coefficients $\boldsymbol{a}^{(n)}$ are also tensors of rank n, and the dot product $\boldsymbol{a}^{(n)} \cdot \boldsymbol{H}^{(n)}$ is defined as the full contraction $a^{(n)}_{\alpha_1\ldots\alpha_n} H^{(n)}_{\alpha_1\ldots\alpha_n}$, i.e. the summation over all possible indices.

We now have the necessary apparatus available to apply the Hermite series expansion to the equilibrium distribution function. This will eventually lead to the desired velocity discretisation.

3.4.4 *Hermite Series Expansion of the Equilibrium Distribution*

Let us now apply the Hermite series expansion in (3.38) to the equilibrium distribution function f^{eq} in $\boldsymbol{\xi}$-space:

$$\begin{aligned}
f^{\mathrm{eq}}(\rho, \boldsymbol{u}, \theta, \boldsymbol{\xi}) &= \omega(\boldsymbol{\xi}) \sum_{n=0}^{\infty} \frac{1}{n!} \boldsymbol{a}^{(n),\mathrm{eq}}(\rho, \boldsymbol{u}, \theta) \cdot \boldsymbol{H}^{(n)}(\boldsymbol{\xi}), \\
\boldsymbol{a}^{(n),\mathrm{eq}}(\rho, \boldsymbol{u}, \theta) &= \int f^{\mathrm{eq}}(\rho, \boldsymbol{u}, \theta, \boldsymbol{\xi})\, \boldsymbol{H}^{(n)}(\boldsymbol{\xi})\, \mathrm{d}^d \xi.
\end{aligned} \tag{3.39}$$

[11] The goal of this book is to give practical aspects of the derivation and usage of the LBM rather than a rigorous mathematical theory for some assumptions used. However, for interested readers we recommend [7] for a rigorous proof.

The **equilibrium distribution function** $f^{\text{eq}}(\boldsymbol{\xi})$ **has the same form as the weight function** $\omega(\boldsymbol{\xi})$ of the HPs in (3.25):

$$f^{\text{eq}}(\rho, \boldsymbol{u}, \theta, \boldsymbol{\xi}) = \frac{\rho}{(2\pi\theta)^{d/2}} \mathrm{e}^{-(\boldsymbol{\xi}-\boldsymbol{u})^2/(2\theta)} = \frac{\rho}{\theta^{d/2}} \omega\left(\frac{\boldsymbol{\xi}-\boldsymbol{u}}{\sqrt{\theta}}\right). \tag{3.40}$$

We can use this crucial relation to calculate the series coefficients:

$$\boldsymbol{a}^{(n),\text{eq}} = \frac{\rho}{\theta^{d/2}} \int \omega\left(\frac{\boldsymbol{\xi}-\boldsymbol{u}}{\sqrt{\theta}}\right) \boldsymbol{H}^{(n)}(\boldsymbol{\xi})\, \mathrm{d}^d\xi. \tag{3.41}$$

The substitution $\boldsymbol{\eta} = (\boldsymbol{\xi} - \boldsymbol{u})/\sqrt{\theta}$ yields

$$\boldsymbol{a}^{(n),\text{eq}} = \rho \int \omega(\boldsymbol{\eta}) \boldsymbol{H}^{(n)}\left(\sqrt{\theta}\boldsymbol{\eta} + \boldsymbol{u}\right) \mathrm{d}^d\eta. \tag{3.42}$$

These integrals can be directly computed, for example with the help of mathematical software packages:

$$\begin{aligned}
a^{(0),\text{eq}} &= \rho, \\
a_\alpha^{(1),\text{eq}} &= \rho u_\alpha, \\
a_{\alpha\beta}^{(2),\text{eq}} &= \rho\left(u_\alpha u_\beta + (\theta - 1)\delta_{\alpha\beta}\right), \\
a_{\alpha\beta\gamma}^{(3),\text{eq}} &= \rho\left[u_\alpha u_\beta u_\gamma + (\theta - 1)\left(\delta_{\alpha\beta}u_\gamma + \delta_{\beta\gamma}u_\alpha + \delta_{\gamma\alpha}u_\beta\right)\right].
\end{aligned} \tag{3.43}$$

A close look at (3.43) reveals that the coefficients in the Hermite series expansion of the equilibrium distribution function f^{eq} are related to the conserved moments; the first three coefficients are connected to the density, momentum and energy. At the same time, it turns out that the conserved quantities can also be represented by the Hermite series expansion coefficients of the d-dimensional particle distribution function f:

$$\begin{aligned}
a^{(0),\text{eq}} &= \int f^{\text{eq}}\, \mathrm{d}^d\xi &= \rho &= \int f\, \mathrm{d}^d\xi &= a^{(0)}, \\
a_\alpha^{(1),\text{eq}} &= \int f^{\text{eq}}\xi_\alpha\, \mathrm{d}^d\xi &= \rho u_\alpha &= \int f\xi_\alpha\, \mathrm{d}^d\xi &= a_\alpha^{(1)}, \\
\frac{a_{\alpha\alpha}^{(2),\text{eq}} + \rho d}{2} &= \int f^{\text{eq}}\frac{|\boldsymbol{\xi}|^2}{2}\, \mathrm{d}^d\xi &= \rho E &= \int f\frac{|\boldsymbol{\xi}|^2}{2}\, \mathrm{d}^d\xi &= \frac{a_{\alpha\alpha}^{(2)} + \rho d}{2}.
\end{aligned} \tag{3.44}$$

This is one of the reasons why the Hermite series expansion is so useful for the Boltzmann equation; the series coefficients are directly connected to the conserved moments or even coincide with them. Only the first three expansion coefficients ($n = 0, 1, 2$) are required to fulfill the conservation laws and represent the macroscopic equations, although some authors have indicated that the inclusion of higher expansion terms can improve the numerical stability and accuracy (cf. Chap. 10) [8–11].

To reproduce the relevant physics, i.e. to satisfy the conservation laws on the *macroscopic* level, **one does not need to consider the full mesoscopic equilibrium and particle distribution functions**. Instead, the **first three terms of the Hermite series expansion are sufficient** to recover the macroscopic laws for hydrodynamics. This allows for a significant reduction of numerical effort.

Limiting the expansion to the N-th order, the equilibrium and particle distribution functions can be represented as

$$
\begin{aligned}
f^{\mathrm{eq}}(\boldsymbol{\xi}) &\approx \omega(\boldsymbol{\xi}) \sum_{n=0}^{N} \frac{1}{n!} \boldsymbol{a}^{(n),\mathrm{eq}} \cdot \boldsymbol{H}^{(n)}(\boldsymbol{\xi}), \\
f(\boldsymbol{\xi}) &\approx \omega(\boldsymbol{\xi}) \sum_{n=0}^{N} \frac{1}{n!} \boldsymbol{a}^{(n)} \cdot \boldsymbol{H}^{(n)}(\boldsymbol{\xi})
\end{aligned}
\tag{3.45}
$$

where we have only denoted the $\boldsymbol{\xi}$-dependence.

We can now explicitly write down the equilibrium distribution function approximation up to the third moment, i.e. up to the second order in $\boldsymbol{\xi}$ ($N = 2$):

$$
\begin{aligned}
f^{\mathrm{eq}}(\rho, \boldsymbol{u}, \theta, \boldsymbol{\xi}) &\approx \omega(\boldsymbol{\xi})\rho \left[1 + \xi_\alpha u_\alpha + \left(u_\alpha u_\beta + (\theta - 1)\delta_{\alpha\beta}\right)\left(\xi_\alpha \xi_\beta - \delta_{\alpha\beta}\right)\right] \\
&= \omega(\boldsymbol{\xi})\rho Q(\boldsymbol{u}, \theta, \boldsymbol{\xi})
\end{aligned}
\tag{3.46}
$$

where $Q(\boldsymbol{u}, \theta, \boldsymbol{\xi})$ is a multi-dimensional polynomial in $\boldsymbol{\xi}$.

Exercise 3.5 Show the validity of (3.46).

At this point we should mention the Mach number expansion. If one expands the equilibrium distribution function in (3.20) up to second order in $\boldsymbol{u}$, then one will also obtain (3.46). However, at the next order (corresponding to energy conservation), the Hermite series and the Mach number expansions give different results. Due to the orthogonality of HPs, the Hermite series expansion does not mix the lower-order moments related to the NSE with the higher-order moments related to the energy equation and beyond. This is not the case for the Mach number expansion [12]. Thus, the Hermite series expansion is generally preferable.

Moreover, we will now see that the Hermite series expansion readily provides discrete velocity sets. This means we can finally discretise velocity space by replacing the continuous $\boldsymbol{\xi}$ by a set of suitable discrete velocities $\{\boldsymbol{\xi}_i\}$.

3.4.5 Discretisation of the Equilibrium Distribution Function

We have seen that the Hermite series expansion is a suitable expansion method since the equilibrium distribution function $f^{\mathrm{eq}}(\boldsymbol{\xi})$ has the same form as the Hermite weight function $\omega(\boldsymbol{\xi})$. But there is also another compelling reason to use Hermite polynomials: as mentioned earlier, it is possible to calculate integrals of certain functions by taking integral function values at a small number of discrete points, the so-called *abscissae*. We will only cover the basics of this technique here, leaving the details for Appendix A.4.

Let us take a 1D polynomial $P^{(N)}(x)$ of order N as example. The integral $\int \omega(x)P^{(N)}(x)\,\mathrm{d}x$ can be calculated exactly by considering integral function values in certain points x_i. This rule is called the *Gauss-Hermite quadrature rule*:

$$\int_{-\infty}^{+\infty} \omega(x)P^{(N)}(x)\,\mathrm{d}x = \sum_{i=1}^{n} w_i P^{(N)}(x_i). \tag{3.47}$$

Here, the n values x_i are the roots of the HP $H^{(n)}(x)$, i.e. $H^{(n)}(x_i) = 0$, and $N \leq 2n - 1$. This means, to exactly integrate a polynomial of order N, one requires at least $n = (N+1)/2$ abscissae x_i and associated weights w_i. An expression for the w_i is given in (A.32). Higher-order polynomials obviously require a larger number of abscissae and therefore higher-order HPs.

The generalisation to d dimensions reads

$$\int \omega(\boldsymbol{x})P^{(N)}(\boldsymbol{x})\,\mathrm{d}^d x = \sum_{i=1}^{n} w_i P^{(N)}(\boldsymbol{x}_i) \tag{3.48}$$

where each of the components of the multidimensional point $\boldsymbol{x}_i$ (i.e. $x_{i\alpha}$ with $\alpha = 1, \ldots, d$) is a root of the one-dimensional Hermite polynomial: $H^{(n)}(x_{i\alpha}) = 0$ (cf. Appendix A.4 for more details).

We can use the Gauss-Hermite quadrature rule to calculate moments and coefficients of the Hermite series expansion. Let us take a closer look at the definition of the coefficients for the equilibrium function in (3.42):

$$\boldsymbol{a}^{(n),\mathrm{eq}} = \rho \int \omega(\boldsymbol{\eta})\boldsymbol{H}^{(n)}\left(\sqrt{\theta}\boldsymbol{\eta} + \boldsymbol{u}\right)\mathrm{d}^d\eta = \rho \int \omega(\boldsymbol{\eta})\boldsymbol{P}^{(n)}(\boldsymbol{\eta})\,\mathrm{d}^d\eta. \tag{3.49}$$

Since the tensor-valued HP $\boldsymbol{H}^{(n)}(\boldsymbol{\eta})$ is clearly a polynomial of order n, we can write it as $\boldsymbol{P}^{(n)}(\boldsymbol{\eta})$. Obviously, we can apply the Gauss-Hermite quadrature rule to the integral on the right-hand side. It takes an additional effort to calculate the polynomial $\boldsymbol{P}^{(n)}(\boldsymbol{\eta})$ from the HP $\boldsymbol{H}^{(n)}(\sqrt{\theta}\boldsymbol{\eta}+\boldsymbol{u})$. Instead, we follow a simpler route and use the truncated series expansion of the equilibrium distribution function f^{eq} that yields the same macroscopic moments.

From (3.39) and (3.46) we first obtain

$$\boldsymbol{a}^{(n),\text{eq}} = \int f^{\text{eq}}(\boldsymbol{\xi})\boldsymbol{H}^{(n)}(\boldsymbol{\xi})\,\text{d}^d\xi = \rho\int \omega(\boldsymbol{\xi})Q(\boldsymbol{\xi})\boldsymbol{H}^{(n)}(\boldsymbol{\xi})\,\text{d}^d\xi \tag{3.50}$$

where we have again only denoted the $\boldsymbol{\xi}$-dependence. Now, we take the composed polynomial $\boldsymbol{R}(\boldsymbol{\xi}) = Q(\boldsymbol{\xi})\boldsymbol{H}^{(n)}(\boldsymbol{\xi})$ and apply the Gauss-Hermite quadrature rule:

$$\boldsymbol{a}^{(n),\text{eq}} = \rho\int \omega(\boldsymbol{\xi})\boldsymbol{R}(\boldsymbol{\xi})\,\text{d}^d\xi = \rho\sum_{i=1}^{n} w_i\boldsymbol{R}(\boldsymbol{\xi}_i) = \rho\sum_{i=1}^{n} w_iQ(\boldsymbol{\xi}_i)\boldsymbol{H}^{(n)}(\boldsymbol{\xi}_i). \tag{3.51}$$

This is the *discretised* Hermite series expansion with n being the required number of abscissae.

Let us turn our attention to the abscissae $\boldsymbol{\xi}_i$ now. To fulfill all relevant conservation laws, we have to ensure that the polynomial with the highest occurring degree can be correctly integrated. This highest order polynomial in (3.44) is related to the energy and is connected to the second-order HP $\boldsymbol{H}^{(2)}(\boldsymbol{\xi})$. If we further assume that the polynomial $Q(\boldsymbol{\xi})$ is of second order as in (3.46), the polynomial $\boldsymbol{R}(\boldsymbol{\xi}) = Q(\boldsymbol{\xi})\boldsymbol{H}^{(2)}(\boldsymbol{\xi})$ is of fourth order ($N = 4$). We know that we need $n \geq (N+1)/2$ and therefore at least $n = 3$ to calculate the moments exactly. The abscissae are therefore given by the roots of $H^{(3)}(\xi_{i\alpha})$. The actual choice of the abscissae and details of the velocity set construction are presented in Appendix A.4. We will also discuss the LB velocity sets in Sect. 3.4.7.

We have now successfully passed the "heavy" mathematical part of this section. There are only a few minor things left before we have completed the velocity discretisation.

Let us define the n quantities $f_i^{\text{eq}}(\boldsymbol{x},t) = w_i\rho(\boldsymbol{x},t)Q(\boldsymbol{u}(\boldsymbol{x},t),\theta(\boldsymbol{x},t),\boldsymbol{\xi}_i)$ as the equilibrium distribution function related to the velocity (direction) $\boldsymbol{\xi}_i$. Instead of having a continuous function $f^{\text{eq}}(\boldsymbol{\xi})$, we only consider a finite set of quantities $f_i^{\text{eq}} = f^{\text{eq}}(\boldsymbol{\xi}_i)$ (cf. (3.46)):

$$f_i^{\text{eq}} = w_i\rho\left[1 + \xi_{i\alpha}u_\alpha + \frac{1}{2}\left(u_\alpha u_\beta + (\theta-1)\delta_{\alpha\beta}\right)\left(\xi_{i\alpha}\xi_{i\beta} - \delta_{\alpha\beta}\right)\right]. \tag{3.52}$$

The discrete set $\{f^{\text{eq}}(\boldsymbol{\xi}_i)\}$ satisfies the **same conservation laws for the first three moments** (density, momentum, energy) as the continuous equilibrium function $f^{\text{eq}}(\boldsymbol{\xi})$.

We have not yet discretised space and time. This means that $f_i^{\text{eq}}(\boldsymbol{x},t)$ is still defined at every point in space and time, with the dependence on $\boldsymbol{x}$ and t entering through the *continuous* moments $\rho(\boldsymbol{x}, t)$, $\boldsymbol{u}(\boldsymbol{x}, t)$ and $\theta(\boldsymbol{x}, t)$. In Sect. 3.5 we will discuss the discretisation in space and time that is necessary to obtain the final form of the LBE suitable for computer simulations.

We can further simplify (3.52). The first simplification is the *isothermal assumption*. From the Boltzmann equation we can obtain a whole hierarchy of equations including the continuity equation, the Navier-Stokes equation and the energy equation [13, 14]. In this book, the main focus is on hydrodynamic applications with the explicit use of energy conservation. The energy equation can alternatively be simulated as a separate equation as discussed in Chap. 8. The isothermal assumption implies $\theta = 1$; it removes the temperature from the equilibrium distribution in (3.52).

Many abscissae $\boldsymbol{\xi}_i$ reported in Appendix A.4 contain unhandy factors of $\sqrt{3}$. A natural second simplification is to introduce a new particle velocity:

$$\boldsymbol{c}_i = \frac{\boldsymbol{\xi}_i}{\sqrt{3}}. \tag{3.53}$$

This way, we obtain velocity sets with integer abscissae as further discussed in Sect. 3.4.7.

The final form of the **discrete equilibrium distribution** reads

$$f_i^{\text{eq}} = w_i \rho \left(1 + \frac{c_{i\alpha} u_\alpha}{c_s^2} + \frac{u_\alpha u_\beta \left(c_{i\alpha} c_{i\beta} - c_s^2 \delta_{\alpha\beta}\right)}{2c_s^4}\right) \tag{3.54}$$

where c_s is the *speed of sound*. It will later appear in the equation of state as proportionality factor between pressure and density, cf. Sect. 4.1. Why c_s is called the speed of *sound* is shown later in Chap. 12.

Equation (3.54) is one of the most important equations in this book. It represents the discretised equilibrium distribution used in most LB simulations. The specific velocity sets, characterised by the numerical values of the abscissae $\{\boldsymbol{c}_i\}$, weights $\{w_i\}$ and the speed of sound c_s are discussed in Sect. 3.4.7.

3.4.6 *Discretisation of the Particle Distribution Function*

We have already shown the Hermite series expansion of the distribution function $f(\boldsymbol{\xi})$ in (3.45):

$$f(\boldsymbol{x}, \boldsymbol{\xi}, t) \approx \omega(\boldsymbol{\xi}) \sum_{n=0}^{N} \frac{1}{n!} \boldsymbol{a}^{(n)}(\boldsymbol{x}, t) \cdot \boldsymbol{H}^{(n)}(\boldsymbol{\xi}). \tag{3.55}$$

The first two coefficients of the expansion are the same as for the equilibrium distribution which is a consequence of the conservation laws.[12] The discrete velocities $\{\boldsymbol{c}_i\}$ are chosen such that those moments are conserved. Therefore, if we discretise the distribution function f in the same way as f^{eq}, it is guaranteed that the conservation laws of (3.44) are still satisfied.

Without repeating calculations that we have already performed for the equilibrium distribution function, one can discretise f as

$$f_i(\boldsymbol{x},t) = \frac{w_i}{\omega(\boldsymbol{c}_i)} f(\boldsymbol{x},\boldsymbol{c}_i,t) \tag{3.56}$$

where $\omega(\boldsymbol{c}_i)$ is added to satisfy the Gauss-Hermite rule:

$$\begin{aligned} a^{(n)}(\boldsymbol{x},t) &= \int f(\boldsymbol{x},\boldsymbol{c},t)\boldsymbol{H}^{(n)}(\boldsymbol{c})\,\mathrm{d}^d c = \int \frac{\omega(\boldsymbol{c})}{\omega(\boldsymbol{c})} f(\boldsymbol{x},\boldsymbol{c},t)\boldsymbol{H}^{(n)}(\boldsymbol{c})\,\mathrm{d}^d c \\ &\approx \sum_{i=1}^{q} \frac{w_i}{\omega(\boldsymbol{c}_i)} f(\boldsymbol{x},\boldsymbol{c}_i,t)\boldsymbol{H}^{(n)}(\boldsymbol{c}_i) = \sum_{i=1}^{q} f_i(\boldsymbol{x},t)\boldsymbol{H}^{(n)}(\boldsymbol{c}_i). \end{aligned} \tag{3.57}$$

We now have q functions $f_i(\boldsymbol{x},t)$. Similarly to $f_i^{\mathrm{eq}}(\boldsymbol{x},t)$, each is related to one discrete velocity $\boldsymbol{c}_i$, but all are still continuous in space and time.

We can finally write down the **discrete-velocity Boltzmann equation**:

$$\partial_t f_i + c_{i\alpha}\partial_\alpha f_i = \Omega(f_i), \quad i = 1,\ldots,q. \tag{3.58}$$

The macroscopic moments (density and momentum) are computed from the *finite sums*

$$\begin{aligned} \rho &= \sum_i f_i = \sum_i f_i^{\mathrm{eq}}, \\ \rho\boldsymbol{u} &= \sum_i f_i \boldsymbol{c}_i = \sum_i f_i^{\mathrm{eq}} \boldsymbol{c}_i, \end{aligned} \tag{3.59}$$

rather than from integrals of $f(\boldsymbol{\xi})$ or $f^{\mathrm{eq}}(\boldsymbol{\xi})$ in velocity space.

To finalise the velocity discretisation, we still have to discuss the available velocity sets and their properties.

[12] In the isothermal case, only density and momentum are considered.

3.4.7 *Velocity Sets*

We have seen how velocity space can be discretised. This naturally leads to the question which discrete velocity set $\{\boldsymbol{c}_i\}$ to choose. On the one hand, an appropriate set has to be sufficiently well-resolved to allow for consistent solutions of the Navier-Stokes or even Navier-Stokes-Fourier equations. On the other hand, the numerical cost of the algorithm scales with the number of velocities. It is therefore an important task to find a set with a minimum number of velocities, yet the ability to capture the desired physics. For a detailed discussion of the history and properties of velocity sets we refer to Chaps. 3 and 5 in the book by Wolf-Gladrow [1].

3.4.7.1 General Comments and Definitions

We name each velocity set by its number d of spatial dimensions and the number q of discrete velocities using the notation DdQq [15]. Two famous examples are D2Q9 (9 discrete velocities in 2D) and D3Q19 (19 discrete velocities in 3D) which we will shortly discuss more thoroughly. A velocity set for the LB algorithm is fully defined by two sets of quantities: the velocities $\{\boldsymbol{c}_i\}$ and the corresponding weights $\{w_i\}$. Another important quantity that can be derived from these two sets is the speed of sound c_s.

The **numbering of the velocities in a set** is not consistently handled throughout the literature. For instance, some authors choose the index i to run from 0 to $q-1$, others from 1 to q. Neither are there any fixed rules about how the different velocities are ordered. This means that one has to be careful when working from different articles, as they may use different orders. It is always recommended to sketch the chosen velocity set and clearly indicate the applied numbering scheme in a coordinate system. For this reason, the most common LB velocity sets are illustrated and their velocity vectors are numbered below.

Most velocity sets have one *rest velocity* with zero magnitude that represents stationary particles. This velocity is often assigned the index $i = 0$: $\boldsymbol{c}_0 = \boldsymbol{0}$. In this book, we will count from 0 to $q-1$ if the set has a rest velocity (with 0 indicating the rest velocity) and from 1 to q if there is no rest velocity. For example, D3Q19 has one rest velocity ($i = 0$) and 18 non-rest velocities ($i = 1, \ldots, 18$).

3.4.7.2 Construction and Requirements of Velocity Sets

There are various approaches to construct LB velocity sets. The first is based on the Gauss-Hermite quadrature rule described above. This leads to the D1Q3, D2Q7,

D2Q9, D3Q15, D3Q19 and D3Q27 velocity sets (cf. Appendix A.4 and [2, 3]). Additionally, D2Q9 and D3Q27 can also be constructed as tensor products of D1Q3. Another possibility is to construct a $(d-1)$-dimensional velocity set *via lattice projection* from a known velocity set in d dimensions. We will briefly discuss this later on.

Yet another possibility is to find general conditions a velocity set has to obey. Apart from the conservation of mass and momentum, a paramount requirement is the rotational *isotropy* of the lattice [16]. It depends on the underlying physics what a "sufficiently isotropic lattice" means. In most cases, LB is used to solve the NSE for which one requires all moments of the weight w_i up to the fifth order to be isotropic (recall fifth order integration via Hermite polynomials). This leads to the following conditions [17–19]:

$$\begin{aligned}
\sum_i w_i &= 1, \\
\sum_i w_i c_{i\alpha} &= 0, \\
\sum_i w_i c_{i\alpha} c_{i\beta} &= c_s^2 \delta_{\alpha\beta}, \\
\sum_i w_i c_{i\alpha} c_{i\beta} c_{i\gamma} &= 0, \\
\sum_i w_i c_{i\alpha} c_{i\beta} c_{i\gamma} c_{i\mu} &= c_s^4 (\delta_{\alpha\beta}\delta_{\gamma\mu} + \delta_{\alpha\gamma}\delta_{\beta\mu} + \delta_{\alpha\mu}\delta_{\beta\gamma}), \\
\sum_i w_i c_{i\alpha} c_{i\beta} c_{i\gamma} c_{i\mu} c_{i\nu} &= 0.
\end{aligned} \tag{3.60}$$

Additionally, all weights w_i have to be non-negative. Any velocity set which fails to satisfy these conditions is not suitable for LB as a Navier-Stokes solver.

As a general simple approach, one may specify a new velocity set $\{\boldsymbol{c}_i\}$ and use (3.60) as conditional equation for the unknown weights w_i and the speed of sound c_s. On the one hand, if only few velocities are chosen, not all conditions may be satisfied at the same time. For example, the D2Q5 lattice can only satisfy the first four equations in (3.60). On the other hand, the equation system may be overdetermined if more velocities are introduced. For example, for D3Q27 the weights w_i and the sound speed c_s are not uniquely determined and can vary [20]. These free parameters can be used to optimise certain properties, e.g. stability.[13]

LB may also be used to simulate advection-diffusion problems for which a lower level of isotropy is sufficient. In this case, only the first four equations in (3.60)

[13] Another example of the higher versatility of the D3Q27 velocity set is its ability to reproduce the Navier-Stokes dynamics with higher-order Galilean invariance, through an off-lattice implementation [21].

have to be fulfilled. If one wants to solve the Navier-Stokes-Fourier system of equations, which includes the energy equation, larger velocity sets are required since higher moments have to be resolved. This leads to so-called *extended lattices*. The discussion of the advection-diffusion equation, energy transfer and their commonly used lattices is put off until Chap. 8. In the following, we focus on the lattices used to simulate the NSE.

We will see in Sect. 3.5 that the LBE is spatially and temporary discretised on a lattice with lattice constant Δx and time step Δt. It is extremely convenient to use velocity sets where all velocities (in this context also called *lattice vectors*) $\boldsymbol{c}_i$ directly connect lattice sites, rather than ending up somewhere between them. We will therefore only consider velocity sets for which all components of the vectors $\boldsymbol{c}_i$ are integer multiples of $\Delta x/\Delta t$. Usually, Δx and Δt are chosen as being 1 in simulations as discussed in Chap. 7. Thus, the velocity vector components $c_{i\alpha}$ are usually integers. Still, it is possible to construct velocity sets without this restriction [3, 21].

3.4.7.3 Common Velocity Sets for Hydrodynamics

Figure 3.3 shows the common D1Q3 and D2Q9 sets for 1D and 2D simulations of hydrodynamics. The most popular sets for 3D are D3Q15 and D3Q19 as shown in Fig. 3.4, along with the less frequently used D3Q27 velocity set. The velocities and weights of those velocity sets are summarised in Table 3.1, and given explicitly in Tables 3.2–3.6.

Exercise 3.6 Show that all velocity sets in Table 3.1 obey the conditions in (3.60).

It is important to consider which of the three 3D schemes (D3Q15, D3Q19 or D3Q27) is most suitable in a given situation. First of all, D3Q15 is more computationally efficient than D3Q19 which in turn is more efficient than D3Q27. All of these lattices allow to recover hydrodynamics to leading order. However, the functional form of the truncation errors is different. Indeed, numerical errors are typically less significant and the stability is usually better for larger velocity sets.

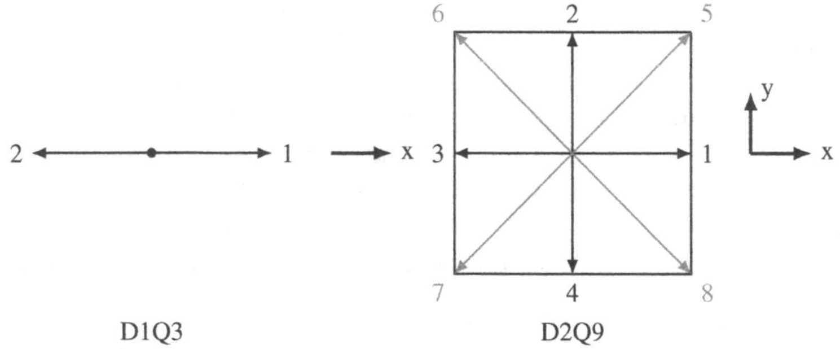

Fig. 3.3 D1Q3 and D2Q9 velocity sets. The square denoted by solid lines has an edge length $2\Delta x$. Velocities with length $|\boldsymbol{c}_i| = 1$ and $\sqrt{2}$ are shown in black and grey, respectively. Rest velocity vectors $\boldsymbol{c}_0 = \boldsymbol{0}$ are not shown. See Tables 3.1, 3.2 and 3.3 for more details

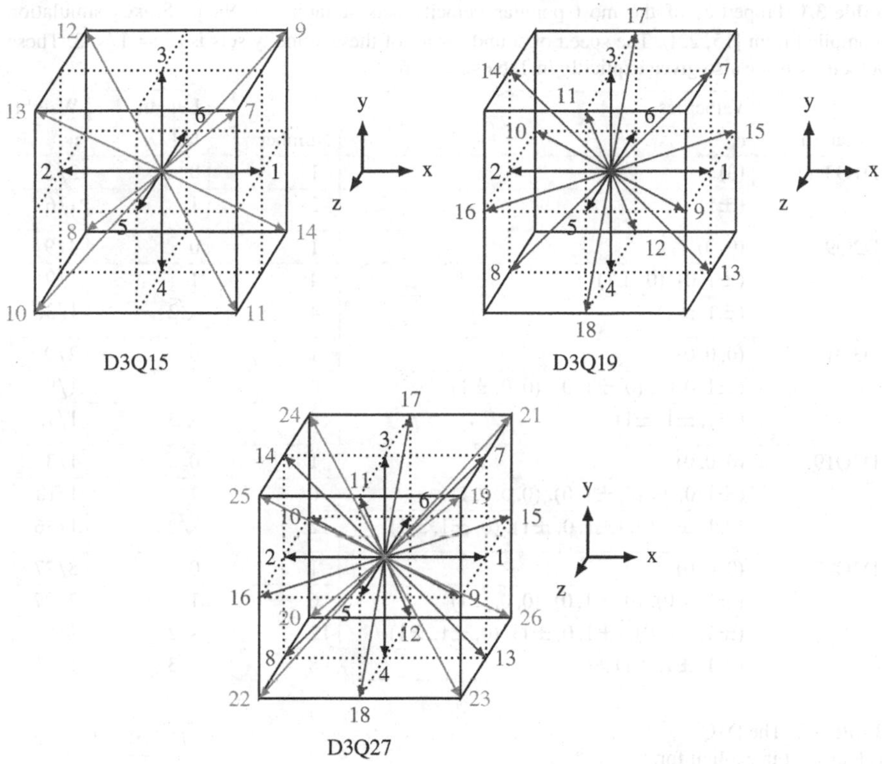

Fig. 3.4 D3Q15, D3Q19 and D3Q27 velocity sets. The cube denoted by solid lines has edge length $2\Delta x$. Velocities with length $|\boldsymbol{c}_i| = 1, \sqrt{2}, \sqrt{3}$ are shown in black, darker grey and lighter grey, respectively. Rest velocity vectors $\boldsymbol{c}_0 = \boldsymbol{0}$ are not shown. Note that D3Q15 has no $\sqrt{2}$-velocities and D3Q19 has no $\sqrt{3}$-velocities. See Tables 3.1, 3.4, 3.5 and 3.6 for more details

D3Q27 was disregarded for a long time as it was not considered superior to D3Q19 and requires 40% more memory and computing power. Recently, it was shown [18] that some truncation terms, the (non-linear) momentum advection corrections, are not rotationally invariant in D3Q15 and D3Q19, in contrast to D3Q27. This lack of isotropy may lead to problems whenever non-linear phenomena play an important role, e.g. in the simulation of high Reynolds number flows [23–25]. Therefore, D3Q27 is probably the best choice for turbulence modelling, but D3Q19 is usually a good compromise for laminar flows.

As a sidenote, there also exists a D3Q13 velocity set which only works within the framework of multi-relaxation-time LB (cf. Chap. 10) [26]. D3Q13 is an example of a lattice exhibiting the *checkerboard instability* [1] previously covered in Sect. 2.1.1. D3Q13 is the minimal velocity set to simulate the NSE in 3D and can be very efficiently implemented on GPUs [27].

We now turn our attention to the projection of velocity sets to lower-dimensional spaces.

Table 3.1 Properties of the most popular velocity sets suitable for Navier-Stokes simulation (compiled from [15, 22]). The speed of sound for all of these velocity sets is $c_s = 1/\sqrt{3}$. These velocity sets are also given explicitly in Tables 3.2–3.6

Notation	Velocities c_i	Number	Length $\lvert c_i \rvert$	Weight w_i
D1Q3	(0)	1	0	2/3
	(± 1)	2	1	1/6
D2Q9	$(0, 0)$	1	0	4/9
	$(\pm 1, 0), (0, \pm 1)$	4	1	1/9
	$(\pm 1, \pm 1)$	4	$\sqrt{2}$	1/36
D3Q15	$(0, 0, 0)$	1	0	2/9
	$(\pm 1, 0, 0), (0, \pm 1, 0), (0, 0, \pm 1)$	6	1	1/9
	$(\pm 1, \pm 1, \pm 1)$	8	$\sqrt{3}$	1/72
D3Q19	$(0, 0, 0)$	1	0	1/3
	$(\pm 1, 0, 0), (0, \pm 1, 0), (0, 0, \pm 1)$	6	1	1/18
	$(\pm 1, \pm 1, 0), (\pm 1, 0, \pm 1), (0, \pm 1, \pm 1)$	12	$\sqrt{2}$	1/36
D3Q27	$(0, 0, 0)$	1	0	8/27
	$(\pm 1, 0, 0), (0, \pm 1, 0), (0, 0, \pm 1)$	6	1	2/27
	$(\pm 1, \pm 1, 0), (\pm 1, 0, \pm 1), (0, \pm 1, \pm 1)$	12	$\sqrt{2}$	1/54
	$(\pm 1, \pm 1, \pm 1)$	8	$\sqrt{3}$	1/216

Table 3.2 The D1Q3 velocity set in explicit form

i	0	1	2
w_i	$\frac{2}{3}$	$\frac{1}{6}$	$\frac{1}{6}$
c_{ix}	0	+1	−1

Table 3.3 The D2Q9 velocity set in explicit form

i	0	1	2	3	4	5	6	7	8
w_i	$\frac{4}{9}$	$\frac{1}{9}$	$\frac{1}{9}$	$\frac{1}{9}$	$\frac{1}{9}$	$\frac{1}{36}$	$\frac{1}{36}$	$\frac{1}{36}$	$\frac{1}{36}$
c_{ix}	0	+1	0	−1	0	+1	−1	−1	+1
c_{iy}	0	0	+1	0	−1	+1	+1	−1	−1

Table 3.4 The D3Q15 velocity set in explicit form

i	0	1	2	3	4	5	6	7	8	9	10	11	12	13	14
w_i	$\frac{2}{9}$	$\frac{1}{9}$	$\frac{1}{9}$	$\frac{1}{9}$	$\frac{1}{9}$	$\frac{1}{9}$	$\frac{1}{9}$	$\frac{1}{72}$	$\frac{1}{72}$	$\frac{1}{72}$	$\frac{1}{72}$	$\frac{1}{72}$	$\frac{1}{72}$	$\frac{1}{72}$	$\frac{1}{72}$
c_{ix}	0	+1	−1	0	0	0	0	+1	−1	+1	−1	+1	−1	−1	+1
c_{iy}	0	0	0	+1	−1	0	0	+1	−1	+1	−1	−1	+1	+1	−1
c_{iz}	0	0	0	0	0	+1	−1	+1	−1	−1	+1	+1	−1	+1	−1

Table 3.5 The D3Q19 velocity set in explicit form

i	0	1	2	3	4	5	6	7	8	9	10	11	12	13	14	15	16	17	18
w_i	$\frac{1}{3}$	$\frac{1}{18}$	$\frac{1}{18}$	$\frac{1}{18}$	$\frac{1}{18}$	$\frac{1}{18}$	$\frac{1}{18}$	$\frac{1}{36}$	$\frac{1}{36}$	$\frac{1}{36}$	$\frac{1}{36}$	$\frac{1}{36}$	$\frac{1}{36}$	$\frac{1}{36}$	$\frac{1}{36}$	$\frac{1}{36}$	$\frac{1}{36}$	$\frac{1}{36}$	$\frac{1}{36}$
c_{ix}	0	+1	−1	0	0	0	0	+1	−1	+1	−1	0	0	+1	−1	+1	−1	0	0
c_{iy}	0	0	0	+1	−1	0	0	+1	−1	0	0	+1	−1	−1	+1	0	0	+1	−1
c_{iz}	0	0	0	0	0	+1	−1	0	0	+1	−1	+1	−1	0	0	−1	+1	−1	+1

3.4.7.4 Velocity Set Relations

Historically, the 2D lattice gas models described in Sect. 2.2.2 used a hexagonal D2Q6 velocity set [1]. The initial paradigm was that all velocities $\boldsymbol{c}_i$ should have the same magnitude and weight (i.e. single-speed velocity sets with $w_i = 1/q$ for all i) and therefore point at the surface of a single sphere in velocity space. This ruled out the existence of rest velocities.

While the D1Q2 and D2Q6 single-speed sets were fairly simple to determine, it was surprisingly difficult to find such a single-speed velocity set in 3D [1]. The solution to this problem was to construct a D4Q24 single-speed velocity set in four dimensions, with 24 velocities consisting of all the various spatial permutations of $(\pm 1, \pm 1, 0, 0)$. This 4D velocity set was then *projected* down to 3D:

$$\begin{pmatrix} \pm 1 \\ 0 \\ 0 \\ \pm 1 \end{pmatrix} \rightarrow \begin{pmatrix} \pm 1 \\ 0 \\ 0 \end{pmatrix}, \quad \begin{pmatrix} 0 \\ \pm 1 \\ 0 \\ \pm 1 \end{pmatrix} \rightarrow \begin{pmatrix} 0 \\ \pm 1 \\ 0 \end{pmatrix}, \quad \begin{pmatrix} 0 \\ 0 \\ \pm 1 \\ \pm 1 \end{pmatrix} \rightarrow \begin{pmatrix} 0 \\ 0 \\ \pm 1 \end{pmatrix},$$
$$\begin{pmatrix} \pm 1 \\ \pm 1 \\ 0 \\ 0 \end{pmatrix} \rightarrow \begin{pmatrix} \pm 1 \\ \pm 1 \\ 0 \end{pmatrix}, \quad \begin{pmatrix} \pm 1 \\ 0 \\ \pm 1 \\ 0 \end{pmatrix} \rightarrow \begin{pmatrix} \pm 1 \\ 0 \\ \pm 1 \end{pmatrix}, \quad \begin{pmatrix} 0 \\ \pm 1 \\ \pm 1 \\ 0 \end{pmatrix} \rightarrow \begin{pmatrix} 0 \\ \pm 1 \\ \pm 1 \end{pmatrix}. \tag{3.61}$$

Thus, the single-speed D4Q24 velocity set leads to a *multi-speed* D3Q18 velocity set, with velocities of length 1 and $\sqrt{2}$.

Note an important difference between the upper and the lower rows of (3.61). In the lower row, the projections are one-to-one: each resulting 3D velocity vector corresponds to only one 4D velocity vector. In the upper row, the projections are *degenerate*: each 3D vector here corresponds to *two different* 4D velocity vectors. This implies that the resulting multi-speed 3D velocity set has nonequal weights w_i such that the shorter velocity vectors in the upper row have twice the weight of the longer vectors in the lower row. This is in contrast to the single-speed D2Q6 and D4Q24 velocity sets, where all the velocities are weighted equally.

Table 3.6 The D3Q27 velocity set in explicit form

i	0	1	2	3	4	5	6	7	8	9	10	11	12	13	14	15	16	17	18	19	20	21	22	23	24	25	26
w_i	$\frac{8}{27}$	$\frac{2}{27}$	$\frac{2}{27}$	$\frac{2}{27}$	$\frac{2}{27}$	$\frac{2}{27}$	$\frac{2}{27}$	$\frac{1}{54}$	$\frac{1}{54}$	$\frac{1}{54}$	$\frac{1}{54}$	$\frac{1}{54}$	$\frac{1}{54}$	$\frac{1}{54}$	$\frac{1}{54}$	$\frac{1}{54}$	$\frac{1}{54}$	$\frac{1}{54}$	$\frac{1}{54}$	$\frac{1}{216}$	$\frac{1}{216}$	$\frac{1}{216}$	$\frac{1}{216}$	$\frac{1}{216}$	$\frac{1}{216}$	$\frac{1}{216}$	$\frac{1}{216}$
c_{ix}	0	+1	−1	0	0	0	0	+1	−1	+1	−1	0	0	+1	−1	+1	−1	0	0	+1	−1	+1	−1	+1	−1	−1	+1
c_{iy}	0	0	0	+1	−1	0	0	+1	−1	0	0	+1	−1	−1	+1	0	0	+1	−1	+1	−1	+1	−1	−1	+1	+1	−1
c_{iz}	0	0	0	0	0	+1	−1	0	0	+1	−1	+1	−1	0	0	−1	+1	−1	+1	+1	−1	−1	+1	+1	−1	+1	−1

In fact, all the velocity sets listed in Table 3.1 are related through such projections: D2Q9 is the two-dimensional projection of the D3Q15, D3Q19, and D3Q27 velocity sets, while D1Q3 is the one-dimensional projection of all of these.

Example 3.3 D2Q9 can be obtained from D3Q19 by projecting the latter onto the x-y-plane:

$$\begin{aligned} &\begin{pmatrix}0\\0\\0\end{pmatrix}, \begin{pmatrix}0\\0\\\pm1\end{pmatrix} \rightarrow \begin{pmatrix}0\\0\end{pmatrix}, && \begin{pmatrix}\pm1\\0\\0\end{pmatrix}, \begin{pmatrix}\pm1\\0\\\pm1\end{pmatrix} \rightarrow \begin{pmatrix}\pm1\\0\end{pmatrix}, \\ &\begin{pmatrix}0\\\pm1\\0\end{pmatrix}, \begin{pmatrix}0\\\pm1\\\pm1\end{pmatrix} \rightarrow \begin{pmatrix}0\\\pm1\end{pmatrix}, && \begin{pmatrix}\pm1\\\pm1\\0\end{pmatrix} \rightarrow \begin{pmatrix}\pm1\\\pm1\end{pmatrix}. \end{aligned} \tag{3.62}$$

The D2Q9 weights can also be obtained from this projection. For instance, the above shows us that the D2Q9 zero-velocity vector $\boldsymbol{c}_0$ is a projection of three different D3Q19 velocity vectors, whose total weight $1/3 + 1/18 + 1/18 = 4/9$ is the D2Q9 rest velocity weight. It is similarly easy to obtain the other D2Q9 weights from the D3Q19 weights.

Example 3.4 Similarly, D1Q3 can be obtained by projecting D2Q9 onto the x axis:

$$\begin{pmatrix}0\\0\end{pmatrix}, \begin{pmatrix}0\\\pm1\end{pmatrix} \rightarrow (0), \qquad \begin{pmatrix}\pm1\\0\end{pmatrix}, \begin{pmatrix}\pm1\\\pm1\end{pmatrix} \rightarrow (\pm1). \tag{3.63}$$

The D1Q3 rest velocity then obtains the weight $4/9 + 1/9 + 1/9 = 2/3$, while the non-rest velocities each obtain weights of $1/9 + 1/36 + 1/36 = 1/6$.

Exercise 3.7 Show that the D3Q27 velocity set becomes the D2Q9 and the D1Q3 velocity sets when projected down to two and one dimensions, respectively.

We have already mentioned that D3Q27 is a 3D generalisation of D1Q3. In addition, D3Q15 and D3Q19 can be obtained from D3Q27 *via* so-called *pruning* [3, 28], where some of the particle velocities are discarded in order to create a smaller velocity set. In turn, D3Q27 can be understood as a *superposition* of D3Q15 weighted by $1/3$ and D3Q19 weighted by $2/3$, as can be seen from Table 3.1. The relations between these velocity sets are shown in Fig. 3.5.

The fact that lower-dimensional velocity sets may be projections of higher-dimensional ones means that simulations can be simplified in cases where the simulated problem is invariant along one or more axes. For instance, a plane sound wave propagating along the x-axis is invariant in the y- and z-directions. The macroscopic results of such a simulation will be the same whether it is simulated with the D1Q3 velocity set or with the two- and three-dimensional velocity sets discussed above combined with periodic y- and z-boundary conditions [19].

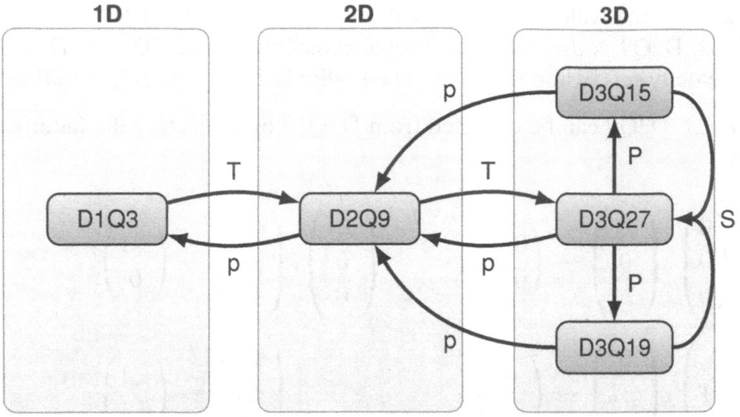

Fig. 3.5 Mutual relations of common LB velocity sets. Lower-dimensional velocity sets can be obtained from higher-dimensional ones *via* projection (p). D2Q9 and D3Q27 are tensor (T) products of D1Q3. D3Q15 and D3Q19 can be obtained from D3Q27 *via* pruning (P). D3Q27 can in turn be considered a superposition (S) of D3Q15 and D3Q19

3.4.7.5 Equilibrium Distributions

The **discrete equilibrium distribution function** in (3.54) is sufficient to recover Navier-Stokes behaviour and is valid for any of the above-mentioned velocity sets (D1Q3, D2Q9, D3Q15, D3Q19 and D3Q27) for which the speed of sound is $c_s = 1/\sqrt{3}$. Different equilibria may be used for other purposes; we will come back to this in Sect. 4.3 and Chap. 8.

In order to implement (3.54) numerically, we can write a for- or do-loop (which is less susceptible to bugs, but usually slower). We can also unroll the q discretised equilibrium distributions (which is normally more computationally efficient but more cumbersome to implement). For new LB users' convenience, we present, in the examples that follow, the explicit equilibrium distribution functions for D1Q3 and D2Q9 lattices that can be directly used in simulations.

Example 3.5 For the D1Q3 velocity set in Table 3.2, the equilibrium distribution reads (with $\boldsymbol{u} = (u)$ and $c_s^2 = 1/3$):

$$\begin{gathered} f_0^{\text{eq}} = \frac{\rho}{3}\left(2 - 3u^2\right), \\ f_1^{\text{eq}} = \frac{\rho}{6}\left[1 + 3\left(u + u^2\right)\right], \quad f_2^{\text{eq}} = \frac{\rho}{6}\left[1 - 3\left(u - u^2\right)\right]. \end{gathered} \tag{3.64}$$

Example 3.6 The D2Q9 equilibrium distribution for the chosen velocity vectors in Table 3.3 is given by (using $\boldsymbol{u} = (u_x, u_y)^\top$, $\boldsymbol{u}^2 = u_x^2 + u_y^2$ and $c_s^2 = 1/3$):

$$\begin{aligned}
f_0^{\mathrm{eq}} &= \frac{2\rho}{9}\left(2 - 3\boldsymbol{u}^2\right), \\
f_1^{\mathrm{eq}} &= \frac{\rho}{18}\left(2 + 6u_x + 9u_x^2 - 3\boldsymbol{u}^2\right), \quad f_5^{\mathrm{eq}} = \frac{\rho}{36}\left[1 + 3(u_x + u_y) + 9u_x u_y + 3\boldsymbol{u}^2\right], \\
f_2^{\mathrm{eq}} &= \frac{\rho}{18}\left(2 + 6u_y + 9u_y^2 - 3\boldsymbol{u}^2\right), \quad f_6^{\mathrm{eq}} = \frac{\rho}{36}\left[1 - 3(u_x - u_y) - 9u_x u_y + 3\boldsymbol{u}^2\right], \\
f_3^{\mathrm{eq}} &= \frac{\rho}{18}\left(2 - 6u_x + 9u_x^2 - 3\boldsymbol{u}^2\right), \quad f_7^{\mathrm{eq}} = \frac{\rho}{36}\left[1 - 3(u_x + u_y) + 9u_x u_y + 3\boldsymbol{u}^2\right], \\
f_4^{\mathrm{eq}} &= \frac{\rho}{18}\left(2 - 6u_y + 9u_y^2 - 3\boldsymbol{u}^2\right), \quad f_8^{\mathrm{eq}} = \frac{\rho}{36}\left[1 + 3(u_x - u_y) - 9u_x u_y + 3\boldsymbol{u}^2\right].
\end{aligned} \tag{3.65}$$

Exercise 3.8 Write down the equilibria for the 3D lattices.

3.4.7.6 Macroscopic Moments

We have already stated in (3.59) how to compute the moments in discretised velocity space. Using the general equilibrium distribution in (3.54) together with the isotropy conditions in (3.60), which hold for all the velocity sets we have discussed here (i.e. D1Q3, D2Q9, D3Q15, D3Q19 and D3Q27), we can find the equilibrium moments explicitly:

$$\begin{aligned}
\Pi^{\mathrm{eq}} &= \sum_i f_i^{\mathrm{eq}} = \rho, \\
\Pi_\alpha^{\mathrm{eq}} &= \sum_i f_i^{\mathrm{eq}} c_{i\alpha} = \rho u_\alpha, \\
\Pi_{\alpha\beta}^{\mathrm{eq}} &= \sum_i f_i^{\mathrm{eq}} c_{i\alpha} c_{i\beta} = \rho c_s^2 \delta_{\alpha\beta} + \rho u_\alpha u_\beta, \\
\Pi_{\alpha\beta\gamma}^{\mathrm{eq}} &= \sum_i f_i^{\mathrm{eq}} c_{i\alpha} c_{i\beta} c_{i\gamma} = \rho c_s^2 \left(u_\alpha \delta_{\beta\gamma} + u_\beta \delta_{\alpha\gamma} + u_\gamma \delta_{\alpha\beta}\right).
\end{aligned} \tag{3.66}$$

Exercise 3.9 Show the validity of these relations by explicitly calculating these moments of (3.54) using the conditions in (3.60).

One can already see some similarities with the NSE. For example, the NSE contains a term $\partial_\beta(p\delta_{\alpha\beta} + \rho u_\alpha u_\beta)$. Comparing this with the second-order moment of the discrete-velocity Boltzmann equation, we can *guess* (and we will *prove* it later in Sect. 4.1) that the equation of state for the LBE is $p = c_s^2 \rho$.

3.5 Discretisation in Space and Time

So far we have only performed the discretisation of velocity space. This section is dedicated to the final step towards the LBE: the discretisation of space and time.

As we discussed in Sect. 2.1, the space discretisation of some conventional CFD methods, such as finite volume or finite element methods, is arbitrary to some extent. Each volume or element can have many possible shapes, such as triangles, quads, tetrahedra, pyramids or hexes. This is not the case for the "classical" LBM. Though there does exist LBM discretisations on unstructured grids [29, 30], and local grid refinement of the LBM on structured grids is possible ([31] gives an overview of this), the most common form of space discretisation is a uniform and structured grid. As we will see, this also implies a strong coupling of the spatial and temporal discretisations in the LBM.

Overall, the original LB algorithm assumes that populations f_i move with velocity $\boldsymbol{c}_i$ from one lattice site to another. After one time step Δt, each population should exactly reach a neighbouring site. This is guaranteed if (i) the underlying spatial lattice is uniform and regular with lattice constant Δx and (ii) the velocity components are integer multiples of $\Delta x/\Delta t$, i.e. $c_{i\alpha} = n\Delta x/\Delta t$.

We have already mentioned in Sect. 3.4.7 that all common velocity sets obey this condition. Thus, we can expect that the populations starting at a lattice site at $\boldsymbol{x}$ end up at another lattice site at $\boldsymbol{x} + \boldsymbol{c}_i\Delta t$, i.e. that the populations do not "get stuck" between lattice sites.[14]

In the following, we will present how the discrete-velocity Boltzmann equation can be further discretised in physical space and time.

3.5.1 *Method of Characteristics*

Let us recall from (3.58) the non-dimensional force-free discrete-velocity Boltzmann equation with a general collision operator Ω_i that conserves density and momentum:

$$\partial_t f_i + c_{i\alpha}\partial_\alpha f_i = \Omega_i \tag{3.67}$$

where $f_i(\boldsymbol{x}, t) = f(\boldsymbol{x}, \boldsymbol{c}_i, t)$ is the particle distribution function discretised in velocity space. We also call f_i the population of particles moving in direction $\boldsymbol{c}_i$. Note that the form of the collision operator is not yet specified. We rather assume that it somehow depends on the discretised populations $\{f_i\}$ and the equilibrium populations $\{f_i^{\text{eq}}\}$. Also note that the equilibrium populations depend on the macroscopic quantities such as density and velocity which can be explicitly found through the moments

[14] This is not a hard-and-fast rule, though; it is possible to work with cases where $\boldsymbol{x} + \boldsymbol{c}_i\Delta t$ falls between lattice sites [3, 21].

of the populations $\{f_i\}$. Thus, we can assume that the collision operator Ω_i can be determined fully through the discretised populations $\{f_i\}$.

Equation (3.67) can be classified as a first-order hyperbolic partial differential equation (PDE). Each velocity $\boldsymbol{c}_i$ is a known constant. There exist a number of techniques to solve equations like (3.67). One particularly powerful approach to tackle such PDEs is the so-called *method of characteristics* (or *method of trajectories*).

This method exploits the existence of trajectories known as *characteristics* in the space of a PDE's independent variables, i.e. $\boldsymbol{x}$ and t for (3.67), which lets us simplify the PDE. What does this mean physically? Let us consider the hyperbolic equation

$$\frac{\partial g}{\partial t} + \boldsymbol{a} \cdot \nabla g = 0 \tag{3.68}$$

with a constant vector $\boldsymbol{a}$. This equation describes the advection of the quantity g at a velocity given by the vector $\boldsymbol{a}$. One can therefore simplify the solution of the PDE by defining a trajectory $\boldsymbol{x} = \boldsymbol{x}_0 + \boldsymbol{a}t$ or $\boldsymbol{x} - \boldsymbol{a}t = \boldsymbol{x}_0$ where $\boldsymbol{x}_0$ is an arbitrary constant.

Exercise 3.10 Show using the chain rule that any function $g = g(\boldsymbol{x} - \boldsymbol{a}t)$ solves (3.68).

In the method of characteristics, one can generally parametrise a PDE's independent variables in such a way that the PDE can be re-expressed as an ordinary differential equation (ODE). We can write the solution of (3.67) in the form $f_i = f_i(\boldsymbol{x}(\zeta), t(\zeta))$, where ζ parametrises a trajectory in space. We provide more details of the mathematical treatment in Appendix A.5.

By converting the left-hand side of (3.67) into a total derivative with respect to the parameter ζ, we find that the PDE becomes an *ordinary* differential equation

$$\frac{\mathrm{d}f_i}{\mathrm{d}\zeta} = \left(\frac{\partial f_i}{\partial t}\right)\frac{\mathrm{d}t}{\mathrm{d}\zeta} + \left(\frac{\partial f_i}{\partial x_\alpha}\right)\frac{\mathrm{d}x_\alpha}{\mathrm{d}\zeta} = \Omega_i\left(\boldsymbol{x}(\zeta), t(\zeta)\right). \tag{3.69}$$

For (3.69) to be equal to (3.67), we must have

$$\frac{\mathrm{d}t}{\mathrm{d}\zeta} = 1, \quad \frac{\mathrm{d}x_\alpha}{\mathrm{d}\zeta} = c_{i\alpha}. \tag{3.70}$$

Exercise 3.11 Show from (3.70) that the solution f_i follows a trajectory given by $\boldsymbol{x} = \boldsymbol{x}_0 + \boldsymbol{c}_i t$ where $\boldsymbol{x}_0$ is an arbitrary constant.

Now we want to integrate both sides of (3.69) along the trajectory, but we have to specify initial conditions first. Let us take a look at the trajectory passing the point $(\boldsymbol{x}_0, t_0)$, choosing $t(\zeta = 0) = t_0$ and $\boldsymbol{x}(\zeta = 0) = \boldsymbol{x}_0$. The integration from $\zeta = 0$ to $\zeta = \Delta t$ of (3.69) then yields

$$f_i(\boldsymbol{x}_0 + \boldsymbol{c}_i\Delta t, t_0 + \Delta t) - f_i(\boldsymbol{x}_0, t_0) = \int_0^{\Delta t} \Omega_i(\boldsymbol{x}_0 + \boldsymbol{c}_i\zeta, t_0 + \zeta)\,\mathrm{d}\zeta. \tag{3.71}$$

By the fundamental theorem of calculus, the integration of the left-hand side is exact. Note that the point $(\boldsymbol{x}_0, t_0)$ is arbitrary so that we can more generally write

$$f_i(\boldsymbol{x} + \boldsymbol{c}_i\Delta t, t + \Delta t) - f_i(\boldsymbol{x}, t) = \int_0^{\Delta t} \Omega_i(\boldsymbol{x} + \boldsymbol{c}_i\zeta, t + \zeta)\, \mathrm{d}\zeta. \tag{3.72}$$

However, the right-hand side is not as simple to determine as the left-hand side. We will show in the next sections how the integral may be *approximated*. A more rigorous treatment is available in Appendix A.5.

In (3.72) we can already see the discretisation pattern introduced above: during the time step Δt, the population $f_i(\boldsymbol{x}, t)$ moves from $\boldsymbol{x}$ to $\boldsymbol{x} + \boldsymbol{c}_i\Delta t$, giving $f_i(\boldsymbol{x} + \boldsymbol{c}_i\Delta t, t + \Delta t)$. This supports our aforementioned "naive" idea of space and time discretisation, where populations exactly reach neighbouring lattice sites after one time step, given that the lattice is uniform and regular and the chosen velocity set is such that $c_{i\alpha} = n\Delta x/\Delta t$.

3.5.2 *First- and Second-Order Discretisation*

There are several ways to approximate the right-hand side of (3.72). The most common general space-time integration methods include the *Crank-Nicolson* [32] and the *Runge-Kutta* schemes [33]. Although these methods, among others, in most cases allow a more accurate integration, the "classical" LBE employs the simple explicit *forward Euler* scheme previously seen in Sect. 2.1.

There are good reasons for this. Runge-Kutta-type schemes require tracking the populations f_i at several points in time (and for this reason they are known as *multi-step* schemes). This is memory-intensive, especially for D3Q27. The Crank-Nicolson time-space discretisation of the LBE [11, 32] leads to the original LBE after introducing new variables as we will see below (and in more detail in Appendix A.5). Other implicit discretisations of (3.72) lead to a linear system of equations to be solved. It is needless to say that this is computationally demanding, especially for D3Q27 with its 27 populations for each node, and is less attractive than it is for the four variables (i.e. pressure and three velocity components) of the incompressible Navier-Stokes equation in e.g. a finite volume solver.

One interesting alternative space-time discretisation of the discrete-velocity Boltzmann equation is the finite volume formulations of the LBE [29, 30, 34].[15] The finite-volume formulations are more flexible in terms of generating meshes that fit complex geometries [34], especially if one wants to use local grid refinement which is difficult to computationally implement for uniform grids [35]. However, explicit finite volume formulations [36, 37] (using forward Euler or Runge-Kutta formulations) are inferior in terms of stability and/or computational efficiency.

[15] We have described the general principles of such finite volume methods in Sect. 2.1.2.

Implicit formulations lead to a system of equations with high computational overhead. We will not discuss non-"classical" discretisations of the Boltzmann equation in more detail and instead refer to the literature indicated above.

The **beauty, yet weakness, of the LBE lies in its explicitness and uniform grid**. The explicit discretisation of velocities, space and time allows for a relatively easy setup of complex boundary conditions, e.g. for multiphase flows in porous media. Also, the uniform grid allows for effective parallelisation [27]. However, this comes with a price: there are some stability restrictions on the lattice constant Δx and the time step Δt.

We will now take a closer look at first-order and second-order discretisations of the right-hand side of (3.72).

3.5.2.1 First-Order Discretisation

The first-order discretisation, also denoted *rectangular discretisation*, approximates the collision operator integral by just one point:

$$f_i(\boldsymbol{x} + \boldsymbol{c}_i\Delta t, t + \Delta t) - f_i(\boldsymbol{x}, t) = \Delta t \Omega_i(\boldsymbol{x}, t). \tag{3.73}$$

The scheme in (3.73) is fully explicit and the most used for LB simulations. In its form

$$f_i(\boldsymbol{x} + \boldsymbol{c}_i\Delta t, t + \Delta t) = f_i(\boldsymbol{x}, t) + \Delta t \Omega_i(\boldsymbol{x}, t) \tag{3.74}$$

it is called **the *lattice Boltzmann equation* (LBE)**.

However, (3.74) alone is still not of much use since the collision operator Ω_i has not yet been explicitly specified. We will get back to it in Sect. 3.5.3. Yet, we could generally expect that the explicit scheme in (3.74) is of first-order accuracy in time as the right-hand side is determined by a first-order approximation. However, as we will see below, the second-order discretisation of the integral leads to the same form of the LBE. Thus, the LBE in (3.74) is actually second-order accurate in time.

3.5.2.2 Second-Order Discretisation

We obtain a more accurate approximation of the right-hand side of (3.72) *via* the *trapezoidal rule*:[16]

$$f_i(\boldsymbol{x} + \boldsymbol{c}_i\Delta t, t + \Delta t) - f_i(\boldsymbol{x}, t) = \Delta t \frac{\Omega_i(\boldsymbol{x}, t) + \Omega_i(\boldsymbol{x} + \boldsymbol{c}_i\Delta t, t + \Delta t)}{2}. \quad (3.75)$$

This is a second-order accurate discretisation.

Equation (3.75) is implicit since $\Omega_i(\boldsymbol{x} + \boldsymbol{c}_i\Delta t, t + \Delta t)$ depends on f_i at $t + \Delta t$. However, it is possible to transform this equation into an explicit form as detailed in Appendix A.5. This transformation introduces a change of variables, $f_i \to \bar{f}_i$ and results in [11, 38]

$$\bar{f}_i(\boldsymbol{x} + \boldsymbol{c}_i\Delta t, t + \Delta t) = \bar{f}_i(\boldsymbol{x}, t) + \Delta t \bar{\Omega}_i(\boldsymbol{x}, t) \quad (3.76)$$

with a slightly redefined collision operator.

Equation (3.76) is of the same form as (3.74), but with f_i changed to the transformed variable $\bar{f}_i$. Ideally, we would like to eliminate the untransformed variable f_i from (3.76) so that we can solve this equation for $\bar{f}_i$ without having to determine f_i. In fact, all common collision operators can be re-expressed with $\bar{f}_i$ instead of f_i. An important reason for this is that $\sum_i \bar{f}_i = \sum_i f_i = \rho$ and $\sum_i \boldsymbol{c}_i \bar{f}_i = \sum_i \boldsymbol{c}_i f_i = \rho \boldsymbol{u}$, as shown in Appendix A.5.

Therefore, we have found the somewhat surprising result that the second-order discretisation in (3.76) is of the same form as the first-order discretisation in (3.74), which indicates that both discretisations are actually second-order accurate [38, 39]. The second-order accuracy of the LBE in both forms (i.e. (3.74) and (3.76)) can also be proven by other methods [32].

As already mentioned above, neither method is very useful without a properly defined collision operator. We will now investigate the so-called BGK collision operator, the simplest and most widely used collision operator for the LBE. In Chap. 10 we will introduce and thoroughly discuss other, more elaborate operators.

3.5.3 BGK Collision Operator

We mentioned in Sect. 1.3.4 that the collision operator of the original Boltzmann equation considers all possible outcomes of binary collisions and has a rather

[16] Applying the trapezoidal rule when integrating along characteristics is not in general the best approach. As a simple example, applying the trapezoidal rule to the equation $\mathrm{d}f/\mathrm{d}\zeta = -f^3$ gives us $f(\zeta_0 + \Delta\zeta) - f(\zeta_0) = -[f(\zeta_0)^3 + f(\zeta_0 + \Delta\zeta)^3]\Delta t/2$, which would have to be solved implicitly, instead of the exact explicit result $f(\zeta_0 + \Delta\zeta) = 1/\sqrt{2\Delta t + 1/f(\zeta_0)^2}$. A more general second-order LB discretisation is shown in Appendix A.5, though in this case the end result is the same as here.

complicated and cumbersome mathematical form. This collision operator is only suitable for gas simulations, as it only accounts for binary collisions between molecules. However, due to the significantly larger densities, molecules in liquids can undergo more complicated interactions involving three and more particles. So one could naively assume that more complicated integrals accounting for all these possible interactions are required to characterise the collisions in liquids. Fortunately, this is not necessary.

As we discussed in Sect. 3.4, one does not need to know all underlying microscopic information to recover the macroscopic equations. This important observation can be used to simplify the collision operator *significantly*. In particular, one can get rid of any complicated integrals. The first step is to *approximate* the collision operator and write it in terms of the known variables, the populations f_i and the equilibrium populations f_i^{eq}. The simplest non-trivial functional form is a linear relation, so we assume that Ω_i should contain both f_i and f_i^{eq} only linearly.

Let us take a closer look at the form $\Omega_i \propto (f_i - f_i^{\text{eq}})$. Note that this linear form conserves mass and momentum, as we require for the Navier-Stokes behaviour:

$$\begin{aligned} \sum_i \Omega_i &\propto \sum_i (f_i - f_i^{\text{eq}}) = 0, \\ \sum_i \Omega_i \boldsymbol{c}_i &\propto \sum_i \left(f_i \boldsymbol{c}_i - f_i^{\text{eq}} \boldsymbol{c}_i\right) = \boldsymbol{0}. \end{aligned} \tag{3.77}$$

The most important property of **collision operators** is **mass and momentum conservation**. One can easily construct simple collision operators by writing down linear functions of $\{f_i\}$ and $\{f_i^{\text{eq}}\}$.

We now adapt from (1.49) the *Bhatnagar-Gross-Krook* (*BGK*) collision operator [40]:

$$\Omega_i = -\frac{f_i - f_i^{\text{eq}}}{\tau}. \tag{3.78}$$

What does (3.78) mean physically? It can be interpreted as the tendency of the population f_i to approach its equilibrium state f_i^{eq} after a time τ. This process is also called *relaxation towards equilibrium*, and τ is therefore denoted the *relaxation time*.

Exercise 3.12 Show that the solution of

$$\frac{\mathrm{d}f_i}{\mathrm{d}t} = -\frac{f_i - f_i^{\text{eq}}}{\tau} \tag{3.79}$$

leads to an exponential decay, $(f_i - f_i^{\text{eq}}) \propto \exp(-t/\tau)$ if we assume that f_i^{eq} is constant.

Substituting the BGK collision operator from (3.78) into the first-order approximation of the collision operator integral in (3.74) gives the **lattice Boltzmann equation with BGK collision operator**, also sometimes called the *lattice BGK (LBGK)* equation:

$$f_i(\boldsymbol{x}+\boldsymbol{c}_i\Delta t, t+\Delta t) = f_i(\boldsymbol{x},t) - \frac{\Delta t}{\tau}\left(f_i(\boldsymbol{x},t) - f_i^{\mathrm{eq}}(\boldsymbol{x},t)\right). \tag{3.80}$$

In Appendix A.5 we explain that substituting the BGK collision operator into the second-order accurate LBE leads to

$$\bar{f}_i(\boldsymbol{x}+\boldsymbol{c}_i\Delta t, t+\Delta t) = \bar{f}_i(\boldsymbol{x},t) - \frac{\Delta t}{\bar{\tau}}\left(\bar{f}_i(\boldsymbol{x},t) - f_i^{\mathrm{eq}}(\boldsymbol{x},t)\right) \tag{3.81}$$

with a redefined relaxation time $\bar{\tau} = \tau + \Delta t/2$. As (3.80) and (3.81) have the same form and f_i^{eq} can be constructed from $\bar{f}_i$ in the exact same way as from f_i, there is no practical difference between using the first- or second-order approximation of the collision integral. This is an unexpected result and a specific property of the LBE. The equivalence of the first- and second-order discretisations is one of several proofs of the second-order time accuracy of the LBE [32].

As we will see in Sect. 4.1, the very crude approximation of the original Boltzmann collision operator by the BGK operator works astonishingly well in most cases. In particular, we will show that the LBE with this simple BGK collision operator is able to reproduce the continuity and the Navier-Stokes equations. This is one of the main reasons why the LBM has become so popular.

Note that the BGK collision operator is not the only possible collision operator. For example, there exist two-relaxation-times (TRT) and multi-relaxation-times (MRT) collision operators that utilise more than just a single relaxation time. (The BGK operator is also often called a *single-relaxation-time (SRT)* collision operator.) These extended collision operators allow avoiding or mitigating some limitations of the BGK collision operator, such as stability and accuracy issues. We will get back to this more advanced topic in Chap. 10.

3.5.3.1 Under-, Full and Over-Relaxation

The lattice BGK equation, being discrete in time and space, differs from the continuous BGK equation in one major respect. While the latter always evolves f_i *towards* f_i^{eq} (see Exercise 3.12), the lattice BGK equation can also evolve f_i *immediately to* f_i^{eq} or even *past* f_i^{eq}. To see why this is so, we briefly look at the discrete analogue of the spatially homogeneous continuous BGK equation in (3.79). From (3.80), this is

$$f_i(t+\Delta t) = \left(1 - \frac{\Delta t}{\tau}\right) f_i(t) + \frac{\Delta t}{\tau} f_i^{\mathrm{eq}}. \tag{3.82}$$

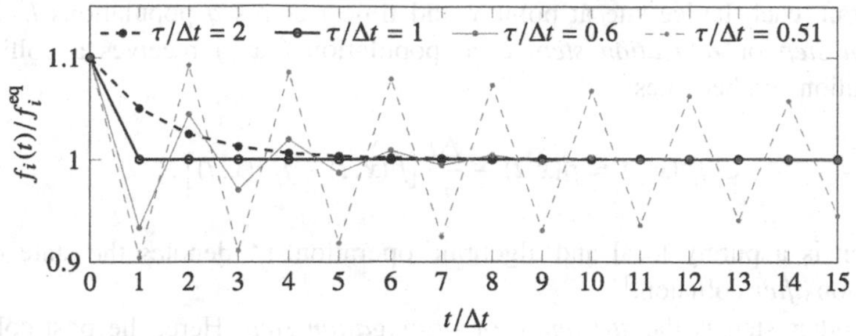

Fig. 3.6 Simple example of under-, full, and over-relaxation for the spatially homogeneous lattice BGK equation in (3.82) with an initial condition $f_i(0)/f_i^{\text{eq}} = 1.1$ and constant f_i^{eq}

Depending on the choice of $\tau/\Delta t$, we find that f_i relaxes in one of three different ways:

- **Under-relaxation** for $\tau/\Delta t > 1$, where f_i decays *exponentially towards* f_i^{eq} like in the continuous-time BGK equation.
- **Full relaxation** for $\tau/\Delta t = 1$, where f_i decays *directly to* f_i^{eq}.
- **Over-relaxation** for $1/2 < \tau/\Delta t < 1$, where f_i *oscillates around* f_i^{eq} with an exponentially decreasing amplitude.

These cases are illustrated in Fig. 3.6. A fourth, *unstable*, case is $\tau/\Delta t < 1/2$, where f_i oscillates around f_i^{eq} with an exponentially *increasing* amplitude. Consequently, $\tau/\Delta t \geq 1/2$ is a necessary condition for stability. We cover other stability conditions in Sect. 4.4.

Finally, we have successfully discretised the Boltzmann equation in velocity space, physical space and time. We have replaced the complicated collision operator by the simple BGK collision operator. Before starting to write LB simulations on a computer, it is helpful to understand the concept of separating (3.80) into streaming (or propagation) and collision (or relaxation) substeps.

3.5.4 Streaming and Collision

By having a close look at (3.80) we can identify two separate parts. One comes from the integration along characteristics, $f_i(\boldsymbol{x} + \boldsymbol{c}_i\Delta t, t + \Delta t) - f_i(\boldsymbol{x}, t)$. The other comes from the local collision operator, $-\Delta t[f_i(\boldsymbol{x}, t) - f_i^{\text{eq}}(\boldsymbol{x}, t)]/\tau$. We can therefore logically separate the LBGK equation into distinct streaming (or propagation) and collision steps.

Overall, each lattice site at point $\boldsymbol{x}$ and time t stores q populations f_i. In the *collision step* or *relaxation step*, each population $f_i(\boldsymbol{x},t)$ receives a collisional contribution and becomes

$$f_i^\star(\boldsymbol{x},t) = f_i(\boldsymbol{x},t) - \frac{\Delta t}{\tau}\left[f_i(\boldsymbol{x},t) - f_i^{\mathrm{eq}}(\boldsymbol{x},t)\right]. \tag{3.83}$$

Collision is a purely local and algebraic operation. $f_i^\star$ denotes the state of the population *after* collision.[17]

The other step is the *streaming* or *propagation step*. Here, the post-collision populations $f_i^\star(\boldsymbol{x},t)$ just stream along their associated direction $\boldsymbol{c}_i$ to reach a neighbouring lattice site where they become $f_i(\boldsymbol{x}+\boldsymbol{c}_i\Delta t, t+\Delta t)$:

$$f_i(\boldsymbol{x}+\boldsymbol{c}_i\Delta t, t+\Delta t) = f_i^\star(\boldsymbol{x},t). \tag{3.84}$$

This is a non-local operation. Practically, one has to copy the memory content of $f_i^\star(\boldsymbol{x},t)$ to the lattice site located at $\boldsymbol{x}+\boldsymbol{c}_i\Delta t$ and overwrite its old information. (One has to be careful not to overwrite populations which are still required.) One common strategy is to use two sets of populations, one for reading data, the other for writing data (see Chap. 13).

In summary, the implementation of the LBGK equation consists of two main substeps, **collision and streaming**:

$$\begin{aligned} f_i^\star(\boldsymbol{x},t) &= f_i(\boldsymbol{x},t) - \frac{\Delta t}{\tau}\left[f_i(\boldsymbol{x},t) - f_i^{\mathrm{eq}}(\boldsymbol{x},t)\right] && \text{(collision)},\\ f_i(\boldsymbol{x}+\boldsymbol{c}_i\Delta t, t+\Delta t) &= f_i^\star(\boldsymbol{x},t) && \text{(streaming)}. \end{aligned} \tag{3.85}$$

Now we have derived everything required to write a first LB simulation code, except boundary conditions and forces. The most important results of this chapter

[17]Note that is more convenient for code implementation to write (3.83) in the form

$$f_i^\star(\boldsymbol{x},t) = \left(1 - \frac{\Delta t}{\tau}\right) f_i(\boldsymbol{x},t) + \frac{\Delta t}{\tau} f_i^{\mathrm{eq}}(\boldsymbol{x},t).$$

The specific choice $\tau = \Delta t$ (which is quite common in LB simulations) leads to the extremely efficient collision rule

$$f_i^\star(\boldsymbol{x},t) = f_i^{\mathrm{eq}}(\boldsymbol{x},t),$$

i.e. the populations directly go to their equilibrium and forget about their previous state. We provide more details about efficient implementations of (3.83) in Chap. 13.

were already collected in Sect. 3.2, and simple implementation hints were covered in Sect. 3.3. In the next chapter, we will show that the LBGK equation actually simulates the NSE and consider accuracy and stability.

References

1. D.A. Wolf-Gladrow, *Lattice-Gas Cellular Automata and Lattice Boltzmann Models* (Springer, New York, 2005)
2. X. He, L.S. Luo, Phys. Rev. E **56**(6), 6811 (1997)
3. X. Shan, X.F. Yuan, H. Chen, J. Fluid Mech. **550**, 413 (2006)
4. K. Dullemond, K. Peeters, Introduction to Tensor Calculus. http://www.ita.uni-heidelberg.de/~dullemond/lectures/tensor/tensor.pdf (1991–2010)
5. J. Simmonds, *A Brief on Tensor Analysis* (Springer, New York, 1994)
6. H. Grad, Commun. Pure Appl. Math. **2**(4), 325 (1949)
7. N. Wiener, *The Fourier Integral and Certain of Its Applications* (Cambridge University Press, Cambridge, 1933)
8. D. d'Humières, I. Ginzburg, M. Krafczyk, P. Lallemand, L.S. Luo, Phil. Trans. R. Soc. Lond. A **360**, 437 (2002)
9. A. Kuzmin, A. Mohamad, S. Succi, Int. J. Mod. Phys. C **19**(6), 875 (2008)
10. I. Ginzburg, F. Verhaeghe, D. d'Humières, Commun. Comput. Phys. **3**, 427 (2008)
11. P. Dellar, Phys. Rev. E **64**(3) (2001)
12. Z. Guo, C. Zheng, B. Shi, T. Zhao, Phys. Rev. E **75**(036704), 1 (2007)
13. G. Uhlenbeck, G. Ford, *Lectures in Statistical Mechanics*. Lectures in applied mathematics (American Mathematical Society, Providence, 1974)
14. S. Chapman, T.G. Cowling, *The Mathematical Theory of Non-uniform Gases*, 2nd edn. (Cambridge University Press, Cambridge, 1952)
15. Y.H. Qian, D. d'Humières, P. Lallemand, Europhys. Lett. **17**(6), 479 (1992)
16. U. Frisch, B. Hasslacher, Y. Pomeau, Phys. Rev. Lett. **56**(14), 1505 (1986)
17. J. Latt, Hydrodynamic limit of lattice Boltzmann equations. Ph.D. thesis, University of Geneva (2007)
18. G. Silva, V. Semiao, J. Comput. Phys. **269**, 259 (2014)
19. E.M. Viggen, The lattice Boltzmann method: Fundamentals and acoustics. Ph.D. thesis, Norwegian University of Science and Technology (NTNU), Trondheim (2014)
20. I. Ginzburg, D. d'Humières, A. Kuzmin, J. Stat. Phys. **139**, 1090 (2010)
21. W.P. Yudistiawan, S.K. Kwak, D.V. Patil, S. Ansumali, Phys. Rev. E **82**(4), 046701 (2010)
22. S. Succi, *The Lattice Boltzmann Equation for Fluid Dynamics and Beyond* (Oxford University Press, Oxford, 2001)
23. A.T. White, C.K. Chong, J. Comput. Phys. **230**(16), 6367 (2011)
24. S.K. Kang, Y.A. Hassan, J. Comput. Phys. **232**(1), 100 (2013)
25. K. Suga, Y. Kuwata, K. Takashima, R. Chikasue, Comput. Math. Appl. **69**(6), 518 (2015)
26. D. d'Humières, M. Bouzidi, P. Lallemand, Phys. Rev. E **63**(6), 066702 (2001)
27. J. Tölke M. Krafczyk, Int. J. Comp. Fluid Dyn. **22**(7), 443 (2008)
28. I. Karlin, P. Asinari, Physica A **389**(8), 1530 (2010)
29. S. Ubertini, S. Succi, Prog. Comput. Fluid Dyn. **5**(1/2), 85 (2005)
30. M.K. Misztal, A. Hernandez-Garcia, R. Matin, H.O. Sørensen, J. Mathiesen, J. Comput. Phys. **297**, 316 (2015)
31. D. Lagrava, Revisiting grid refinement algorithms for the lattice Boltzmann method. Ph.D. thesis, University of Geneva (2012)
32. S. Ubertini, P. Asinari, S. Succi, Phys. Rev. E **81**(1), 016311 (2010)

33. J. Hoffmann, *Numerical Methods for Engineers and Scientists* (McGraw-Hill, New York, 1992)
34. H. Xi, G. Peng, S.H. Chou, Phys. Rev. E **59**(5), 6202 (1999)
35. O. Filippova, D. Hanel, J. Comput. Phys. **147**, 219 (1998)
36. S. Ubertini, S. Succi, Commun. Comput. Phys **3**, 342 (2008)
37. S. Ubertini, S. Succi, G. Bella, Phil. Trans. R. Soc. Lond. A **362**, 1763 (2004)
38. X. He, S. Chen, G.D. Doolen, J. Comput. Phys. **146**(1), 282 (1998)
39. J.D. Sterling, S. Chen, J. Comput. Phys. **123**(1), 196 (1996)
40. P.L. Bhatnagar, E.P. Gross, M. Krook, Phys. Rev. **94**, 511 (1954)

Chapter 4
Analysis of the Lattice Boltzmann Equation

Abstract After reading this chapter, you will be familiar with many in-depth aspects of the lattice Boltzmann method. You will have a detailed understanding of how the Chapman-Enskog analysis can be used to determine how the lattice Boltzmann equation and its variations behave on the macroscopic Navier-Stokes level. You will know a number of such variations that result in different macroscopic behaviour from the standard lattice Boltzmann equation. Necessary and sufficient conditions that serve as stability guidelines for lattice Boltzmann simulations will be known to you, along with how to improve the stability of a given simulation. You will also have insight into the accuracy of both general simulations and lattice Boltzmann simulations. For the latter, you will understand what the sources of inaccuracy are, and how they may be reduced or nullified.

This chapter is oriented towards the theory of the LBE. A thorough understanding is *not* required for performing LB simulations.

In Chap. 3 we derived the discrete numerical method known as the lattice Boltzmann equation (LBE) from the Boltzmann equation, the latter being a fundamental equation from the field of kinetic theory. The motivation for doing so is that we wish to use the LBE to simulate the behaviour of fluids. However, the LBM is a different approach to fluid simulation than standard Navier-Stokes solvers: while the latter start with a set of fluid conservation equations and discretise them, the LBM discretises another equation, which means that it is less obvious that it works as a method for simulating fluids.

We will shed some light on these topics by analysing the LBE in further depth. In Sect. 4.1, we apply a method known as the Chapman-Enskog analysis to determine the connection between the LBE and the macroscopic equations of fluid mechanics. Section 4.2 discusses additional important aspects of this analysis. Section 4.3 shows how important the choice of the equilibrium f_i^{eq} is to the resulting macroscopic behaviour, and discusses a few specific alternatives including the widely-used equilibrium for incompressible flow. Section 4.4 covers the requirements for stability of the LBE, i.e. the range of simulation parameters within which the LBE will be safe from errors that grow exponentially. Section 4.5 examines factors, among others time and space discretisation errors, that separate

T. Krüger et al., *The Lattice Boltzmann Method*, Graduate Texts in Physics,
DOI 10.1007/978-3-319-44649-3_4

LB solutions from the "true" physical solution, and discusses how these factors may be minimised.

4.1 The Chapman-Enskog Analysis

Now that we have found the LBE, we need to show that it *actually can be used* to simulate the behaviour of fluids. While we previously looked at the macroscopic behaviour of the undiscretised Boltzmann equation in Sect. 1.3.5 and found that it behaves according to the continuity equation and a general Cauchy momentum equation with an unknown stress tensor, we have not yet seen that the latter specifically corresponds to the NSE. We will show this correspondence for the LBE using the most common method: the *Chapman-Enskog analysis*. Other methods are also available and will be touched upon in Sect. 4.2.5.

The analysis is named after Sydney Chapman (1888–1970) and David Enskog (1884–1947), two mathematical physicists from the UK and Sweden, respectively. In 1917, both independently developed similar methods of finding macroscopic equations from the Boltzmann equation with Boltzmann's original collision operator. In his book on the kinetic theory of gases, Chapman later combined the two approaches into what is now known as the Chapman-Enskog analysis [1].

4.1.1 The Perturbation Expansion

In Sect. 1.3.5 we saw that the assumption $f \simeq f^{\text{eq}}$ results in the Euler momentum equation. Therefore, any macroscopic behaviour beyond the Euler equation must be connected to the *non-equilibrium* part of f, i.e. $f^{\text{neq}} = f - f^{\text{eq}}$. This is also true for a discretised velocity space: the general momentum equation in Sect. 1.3.5 only depends on moments of f which are also equal to the corresponding moments of f_i. Throughout the rest of this section we shall assume that velocity space is discretised.

It is not obvious how to determine this non-equilibrium part, however. This is where the Chapman-Enskog analysis comes in. At its heart is a perturbation expansion of f_i around the equilibrium distribution f_i^{eq} with the Knudsen number Kn as the expansion parameter. Using the label ϵ^n to indicate terms of order Kn^n,[1] the expansion is

$$f_i = f_i^{\text{eq}} + \epsilon f_i^{(1)} + \epsilon^2 f_i^{(2)} + \dots . \tag{4.1}$$

[1]For instance, the relative order of ϵ shows us immediately in (4.1) that $f_i^{(2)}/f_i^{\text{eq}} = O(\text{Kn}^2)$. However, in the literature it is often stated that $\epsilon = \text{Kn}$, unlike here where we treat it as a mere label. This is another possible approach to the expansion, where the Knudsen number is separated from the higher-order terms so that e.g. $f_i^{(2)}/f_i^{\text{eq}} = O(1)$ while $\epsilon^2 f_i^{(2)}/f_i^{\text{eq}} = O(\text{Kn}^2)$. Which approach to use is simply a matter of taste, and the following equations in this section are the same in both cases.

(Throughout the literature, the equilibrium distribution f_i^{eq} is often written as $f_i^{(0)}$, giving a fully consistent notation in the expansion.)

Introducing the smallness label ϵ lets us more easily group the terms according to their relative order in the Knudsen number. Central to this perturbation analysis is the concept that *in the perturbed equation, each order in* Kn *forms a semi-independent equation by itself.* As mentioned, the lowest-order terms in Kn give us the Euler momentum equation. Consequently, the higher-order terms may be seen as correction terms, analogously to how the viscous stress tensor in the NSE may be seen as a correction term to the Euler equation. The perturbation must be performed in such a way that the equations at different orders in Kn still retain some tie to each other so that the higher-order correction terms are connected to the lower-order equations.

In perturbation analyses, the perturbation terms at the two lowest orders together often result in a sufficiently accurate description of the system. We therefore make the **ansatz** that **only the two lowest orders in** Kn **are required to find the NSE**. Under this ansatz, we do not need to look closely at higher-order components of f_i than f_i^{eq} and $f_i^{(1)}$.

The derivation will be based on the LBE with the BGK collision operator,

$$f_i(\boldsymbol{x} + \boldsymbol{c}_i\Delta t, t + \Delta t) - f_i(\boldsymbol{x}, t) = -\frac{\Delta t}{\tau}\left(f_i(\boldsymbol{x}, t) - f^{\mathrm{eq}}(\boldsymbol{x}, t)\right). \tag{4.2}$$

While we could also perform the Chapman-Enskog analysis for a more general collision operator, we will use this BGK operator for its simplicity. (See Appendix A.2.3 for the analysis of the multi-relaxation-time collision operator).

Like any collision operator, the BGK operator must conserve mass and momentum. As per (3.77), this conservation can be expressed as

$$\sum_i f_i^{\mathrm{neq}} = 0, \quad \sum_i \boldsymbol{c}_i f_i^{\mathrm{neq}} = \boldsymbol{0}. \tag{4.3}$$

We can write this for the expanded f_i in (4.1). In the literature, these are often called the *solvability conditions*. These can be further strengthened by the assumption that they hold individually at each order [1], i.e.

$$\sum_i f_i^{(n)} = 0 \quad \text{and} \quad \sum_i \boldsymbol{c}_i f_i^{(n)} = \boldsymbol{0} \quad \text{for all } n \geq 1. \tag{4.4}$$

While this assumption simplifies the following derivation considerably, it can be done without and it is not always made in the literature, e.g. [2–4].

4.1.2 Taylor Expansion, Perturbation, and Separation

When we transformed the discrete-velocity Boltzmann equation in (3.58) into the LBE by discretising time and space, we used a first-order integral approximation and showed that it is indeed second-order accurate upon a redefinition of the populations f_i. By Taylor expanding the LBE, which results in

$$\Delta t\,(\partial_t + c_{i\alpha}\partial_\alpha)f_i + \frac{\Delta t^2}{2}(\partial_t + c_{i\alpha}\partial_\alpha)^2 f_i + O\left(\Delta t^3\right) = -\frac{\Delta t}{\tau}f_i^{\mathrm{neq}}, \tag{4.5}$$

we obtain an equation which is continuous in time and space but which retains the discretisation error of the LBE. Indeed, apart from the higher-order derivative terms

$$\sum_{n=2}^{\infty}\frac{\Delta t^n}{n!}(\partial_t + c_{i\alpha}\partial_\alpha)^n f_i \tag{4.6}$$

on the left-hand side, the Taylor expanded LBE is identical to the discrete-velocity Boltzmann equation in (3.58) from which we found the LBE.

In the following, we will neglect the terms with third-order derivatives or higher. The short explanation for this is that these terms tend to be very small and do not significantly affect the macroscopic behaviour. The longer justification, detailed in Appendix A.2.1, is that $\Delta t^n\,(\partial_t + c_{i\alpha}\partial_\alpha)^n f_i$ scales with $O(\mathrm{Kn}^n)$, and the terms at third order and higher can thus be neglected according to our ansatz that we only need the two lowest orders in the Knudsen number to find the NSE.

This assumes that the changes in f_i are slow, occurring only on a macroscopic scale. If e.g. numerical errors cause rapid changes in f_i, this assumption is no longer valid and the macroscopic equations that result from the Chapman-Enskog analysis are no longer suitable descriptions of the LBE's macroscopic behaviour.

We can get rid of the second-order derivative terms in (4.5) by subtracting $(\Delta t/2)(\partial_t + c_{i\alpha}\partial_\alpha)$ applied to the equation itself. The resulting equation, which we will be basing the rest of this analysis on, is

$$\Delta t\,(\partial_t + c_{i\alpha}\partial_\alpha)f_i = -\frac{\Delta t}{\tau}f_i^{\mathrm{neq}} + \Delta t\,(\partial_t + c_{i\alpha}\partial_\alpha)\frac{\Delta t}{2\tau}f_i^{\mathrm{neq}}. \tag{4.7}$$

Here we have neglected the $O(\Delta t^3)$ terms. The f^{neq} derivative terms on the right-hand side are the only non-negligible remnants of the discretisation error.[2]

[2]Using the second-order time and space discretisation described in Sect. 3.5, these terms would have been cancelled by terms from the Taylor expansion of the discretised BGK operator. At the two lowest orders in Kn, the discrete-velocity Boltzmann equation would thus be captured without error [5].

Before expanding f_i in (4.7) we make another ansatz: *it is also necessary to expand the time derivative into terms spanning several orders in* Kn.[3] Similarly labelling the spatial derivative without expanding it, the time and space derivatives become

$$\Delta t \partial_t f_i = \Delta t \left(\epsilon \partial_t^{(1)} f_i + \epsilon^2 \partial_t^{(2)} f_i + \dots \right), \quad \Delta t c_{i\alpha} \partial_\alpha f_i = \Delta t \left(\epsilon c_{i\alpha} \partial_\alpha^{(1)} \right) f_i. \tag{4.8}$$

These different components of ∂_t at various orders in Kn should not themselves be considered time derivatives [1]. Instead, they are terms at different orders in Kn that, when summed together, are equal to the time derivative. Such derivative expansions, often called *multiple-scale expansions*, are also used in general perturbation theory to deal with expansions that otherwise result in terms which grow without bound at one order but is cancelled by similar terms at higher orders [6]. We will get back to the interpretation of this expansion in Sect. 4.2.2.

If we apply both the expansion of f_i from (4.1) and the derivative expansion from (4.8) to (4.7) and separate the equation into terms of different order in Kn, we find

$$O(\epsilon): \qquad \left(\partial_t^{(1)} + c_{i\alpha} \partial_\alpha^{(1)} \right) f_i^{\text{eq}} = -\frac{1}{\tau} f_i^{(1)}, \tag{4.9a}$$

$$O(\epsilon^2): \qquad \partial_t^{(2)} f_i^{\text{eq}} + \left(\partial_t^{(1)} + c_{i\alpha} \partial_\alpha^{(1)} \right) \left(1 - \frac{\Delta t}{2\tau} \right) f_i^{(1)} = -\frac{1}{\tau} f_i^{(2)} \tag{4.9b}$$

for the two lowest orders in Kn.[4]

4.1.3 Moments and Recombination

Taking the zeroth to second moments of (4.9a) (i.e. multiplying by 1, $c_{i\alpha}$ and $c_{i\alpha} c_{i\beta}$, respectively, and then summing over i), we can find the $O(\epsilon)$ moment equations:

$$\partial_t^{(1)} \rho + \partial_\gamma^{(1)} (\rho u_\gamma) = 0, \tag{4.10a}$$

$$\partial_t^{(1)} (\rho u_\alpha) + \partial_\beta^{(1)} \Pi_{\alpha\beta}^{\text{eq}} = 0, \tag{4.10b}$$

[3] Given the assumptions made in this derivation, this expansion should be done even in the steady-state where $\partial_t f_i = 0$. However, it is possible to make fewer assumptions so that the time derivative expansion is not necessary. This, and the steady-state case in general, is discussed in more detail in Sect. 4.2.3.

[4] The parenthesis $(1 - \Delta t / 2\tau)$ in (4.9b) is the only remnant of the space and time discretisation error terms on the right-hand side of (4.7). If we had based this analysis directly on the discrete-velocity Boltzmann equation in (3.58), where there is no such discretisation error, the $-\Delta t / 2\tau$ term would not have been present here.

$$\partial_t^{(1)} \Pi_{\alpha\beta}^{\text{eq}} + \partial_\gamma^{(1)} \Pi_{\alpha\beta\gamma}^{\text{eq}} = -\frac{1}{\tau} \Pi_{\alpha\beta}^{(1)}. \tag{4.10c}$$

They contain the moments

$$\Pi_{\alpha\beta}^{\text{eq}} = \sum_i c_{i\alpha} c_{i\beta} f_i^{\text{eq}} = \rho u_\alpha u_\beta + \rho c_s^2 \delta_{\alpha\beta}, \tag{4.11a}$$

$$\Pi_{\alpha\beta\gamma}^{\text{eq}} = \sum_i c_{i\alpha} c_{i\beta} c_{i\gamma} f_i^{\text{eq}} = \rho c_s^2 \left(u_\alpha \delta_{\beta\gamma} + u_\beta \delta_{\alpha\gamma} + u_\gamma \delta_{\alpha\beta}\right), \tag{4.11b}$$

$$\Pi_{\alpha\beta}^{(1)} = \sum_i c_{i\alpha} c_{i\beta} f_i^{(1)}. \tag{4.11c}$$

The first two of these moments are known from (3.66), while the third is unknown as of yet. Note that the third moment $\Pi_{\alpha\beta\gamma}^{\text{eq}}$ is lacking a term $\rho u_\alpha u_\beta u_\gamma$ as f_i^{eq} only contains terms of up to $O(u^2)$. If we take the moment equations in (4.10) and reverse the expansions from (4.8) under the assumption that $\partial_t f_i \approx \epsilon \partial_t^{(1)} f_i$ (i.e. neglecting the contributions of higher-order terms in Kn), the first equation is the continuity equation while the second is the Euler equation. We will return to the third equation shortly.

Similarly taking the zeroth and first moments of (4.9b), we can find the $O(\text{Kn}^2)$ moment equations:

$$\partial_t^{(2)} \rho = 0, \tag{4.12a}$$

$$\partial_t^{(2)} (\rho u_\alpha) + \partial_\beta^{(1)} \left(1 - \frac{\Delta t}{2\tau}\right) \Pi_{\alpha\beta}^{(1)} = 0. \tag{4.12b}$$

These equations can both be interpreted as $O(\epsilon^2)$ corrections to the $O(\epsilon)$ equations above. The continuity equation is exact already at $O(\epsilon)$, so we might have expected that the $O(\epsilon^2)$ correction is zero. However, the $O(\epsilon^2)$ correction to the Euler momentum equation is non-zero, though given by the as of yet unknown moment $\Pi_{\alpha\beta}^{(1)}$.

Assembling the mass and momentum equations from their $O(\epsilon)$ and $O(\epsilon^2)$ component equations in (4.10) and (4.12), respectively, we find

$$\left(\epsilon \partial_t^{(1)} + \epsilon^2 \partial_t^{(2)}\right) \rho + \epsilon \partial_\gamma^{(1)} (\rho u_\gamma) = 0, \tag{4.13a}$$

$$\left(\epsilon \partial_t^{(1)} + \epsilon^2 \partial_t^{(2)}\right) (\rho u_\alpha) + \epsilon \partial_\beta^{(1)} \Pi_{\alpha\beta}^{\text{eq}} = -\epsilon^2 \partial_\beta^{(1)} \left(1 - \frac{\Delta t}{2\tau}\right) \Pi_{\alpha\beta}^{(1)}. \tag{4.13b}$$

Reversing the derivative expansions of (4.8), these equations become the continuity equation and a momentum conservation equation with an as of yet unknown viscous stress tensor

$$\sigma'_{\alpha\beta} = -\left(1 - \frac{\Delta t}{2\tau}\right)\Pi^{(1)}_{\alpha\beta}. \tag{4.14}$$

Note that the recombination of the two different orders in ϵ is done through the expanded time derivative. Indeed, without expanding the time derivative across multiple orders in Kn as in (4.8), (4.12b) would be missing its $\partial_t^{(2)}(\rho u_\alpha)$ term and would consequently be mistakenly predicting $\partial_\beta^{(1)}\Pi^{(1)}_{\alpha\beta} = 0$.[5]

The final piece of the puzzle is finding an explicit expression for the perturbation moment $\Pi^{(1)}_{\alpha\beta}$. This can be directly found from derivatives of the equilibrium moments using (4.10c). Doing so requires some amount of algebra which we have relegated to Appendix A.2.2. For the isothermal equation of state and the equilibrium distribution f_i^{eq} expanded only to $O(u^2)$, the end result is

$$\Pi^{(1)}_{\alpha\beta} = -\rho c_s^2\tau\left(\partial_\beta^{(1)}u_\alpha + \partial_\alpha^{(1)}u_\beta\right) + \tau\partial_\gamma^{(1)}\left(\rho u_\alpha u_\beta u_\gamma\right). \tag{4.15}$$

As we do not at any point in the derivation assume that $\partial_t\tau = 0$ or $\partial_\alpha\tau = 0$, this holds even if τ is a function of space and time. Of the two terms on the right-hand side, the first corresponds to a Navier-Stokes viscous stress tensor, while the second is an error term stemming from the lack of a correct $O(u^3)$ term in the equilibrium distribution f_i^{eq}.

This error term is negligible in most cases. From a closer look at the magnitudes of the two terms, we find that the $O(u^3)$ error term can be neglected if $u^2 \ll c_s^2$, which is equivalent to the condition $\mathrm{Ma}^2 \ll 1$ for the Mach number $\mathrm{Ma} = u/c_s$ [7].

For this reason, it is often stated in the literature (e.g. [8, 9]) that the LBM is only valid for *weakly compressible* phenomena, as opposed to the *strongly compressible* phenomena which occur for transonic and supersonic flow when Ma goes towards unity and beyond [10].[6]

4.1.4 Macroscopic Equations

We now have all the pieces to determine the macroscopic equations simulated by the LBE.

[5] As we will explain in Sect. 4.2.3, it is possible to work without the time derivative expansion. However, this invalidates some simplifying assumptions taken in the current derivation.

[6] Sound propagation can generally be considered a weakly compressible phenomenon; sound waves cause the pressure p, the density ρ, and the fluid velocity $\boldsymbol{u}$ to all fluctuate, though these fluctuations are weak enough that $\mathrm{Ma} = u/c_s \ll 1$ for all sounds within the range of normal human experience [11].

Inserting (4.15) (neglecting the $O(u^3)$ error term) into (4.13) and reversing the derivative expansion from (4.8), we finally find that the LBE solves the **continuity equation** as in (1.3) **and the NSE** as in (1.16):

$$\partial_t \rho + \partial_\gamma \left(\rho u_\gamma\right) = 0, \tag{4.16a}$$

$$\partial_t(\rho u_\alpha) + \partial_\beta(\rho u_\alpha u_\beta) = -\partial_\alpha p + \partial_\beta \left[\eta \left(\partial_\beta u_\alpha + \partial_\alpha u_\beta\right)\right] \tag{4.16b}$$

with

$$p = \rho c_s^2, \quad \eta = \rho c_s^2 \left(\tau - \frac{\Delta t}{2}\right), \quad \eta_B = \frac{2}{3}\eta. \tag{4.17}$$

The bulk viscosity η_B is not itself directly visible in (4.16b) but follows from comparison with (1.16).

In monatomic kinetic theory, the bulk viscosity η_B is normally found to be zero, while here we have it to be of the order of the shear viscosity η. This difference is caused by the use of the isothermal equation of state [12] which is fundamentally incompatible with the monatomic assumption.[7]

From (4.17) we can find that $\tau/\Delta t \geq 1/2$ is a necessary condition for stability, as $\tau/\Delta t < 1/2$ would lead to the macroscopically unstable situation of negative viscosity. The same condition was found in Sect. 3.5.3 from the argument that $\tau/\Delta t < 1/2$ would lead to a divergent under-relaxation.

We could also have found the same macroscopic equations as above using more general collision operators than the BGK operator. The only time that τ comes into play in the above derivation is as the relaxation time of the moment $\Pi_{\alpha\beta}$ in (4.10c). In the more general multiple-relaxation-time (MRT) collision operators, which we will take a closer look at in Chap. 10, each moment can relax to equilibrium at a different rate. Good choices of these various relaxation times will increase the stability and accuracy of LB simulations.[8]

[7] In fact, by imposing in the analysis the isentropic equation of state $p/p_0 = (\rho/\rho_0)^\gamma$ previously described in Sect. 1.1.3, we would find a bulk viscosity $\eta_B = \eta(5/3 - \gamma)$ [5]. In the monatomic limit $\gamma = 5/3$ we find $\eta_B = 0$, while in the isothermal limit $\gamma = 1$ we find $\eta_B = 2\eta/3$. We describe how to impose other equations of state in Sect. 4.3.3.

[8] It is also possible to alter the shear and bulk viscosity separately through a more complex relaxation of the $\Pi_{\alpha\beta}$ moment (cf. Chap. 10).

4.2 Discussion of the Chapman-Enskog Analysis

Despite its long history, the Chapman-Enskog analysis is still subject to debate and is generally difficult to digest and understand in its fundamental meaning [13]. Therefore, we take a brief look at some implications of the Chapman-Enskog analysis (velocity sets in Sect. 4.2.1, time scales in Sect. 4.2.2, stationary flows in Sect. 4.2.3 and the explicit distribution perturbation in Sect. 4.2.4) and its alternatives (cf. Sect. 4.2.5).

As this section discusses the details of Sect. 4.1, it is necessarily somewhat more complex. However, first-time readers may safely skip this entire section as it is not necessary for a basic understanding of the Chapman-Enskog analysis.

4.2.1 Dependence of Velocity Moments

The Hermite polynomial approach taken in Sect. 3.4 yields velocity sets $\boldsymbol{c}_i$ and equilibrium functions f_i^{eq} so that the zeroth- to second-order moments Π^{eq}, Π_α^{eq}, and $\Pi_{\alpha\beta}^{\text{eq}}$ of f_i^{eq} equal those of the continuous-velocity distribution f^{eq}.

However, the third-order moment $\Pi_{\alpha\beta\gamma}^{\text{eq}}$ is not fully correct, leading to the $\rho u_\alpha u_\beta u_\gamma$ error term in (4.15) that further leads to a similar error term in the macroscopic momentum equation. It is not difficult to see why it *cannot be* for any of the velocity sets given in Table 3.1: in all of these velocity sets, we find that $c_{i\alpha} \in \{-1, 0, +1\}\frac{\Delta x}{\Delta t}$ so that $c_{i\alpha}^3 = c_{i\alpha}(\frac{\Delta x}{\Delta t})^2$. Consequently,

$$\Pi_{xxx}^{\text{eq}} = \sum_i c_{ix}^3 f_i^{\text{eq}} = \left(\frac{\Delta x}{\Delta t}\right)^2 \sum_i c_{ix} f_i^{\text{eq}} = \left(\frac{\Delta x}{\Delta t}\right)^2 \Pi_x^{\text{eq}}, \tag{4.18}$$

i.e. the third-order *xxx*-, *yyy*-, and *zzz*-moments are proportional to the first-order *x*-, *y*-, and *z*-moments, respectively.

Exercise 4.1 From the general forms of Π_α^{eq} and $\Pi_{\alpha\beta\gamma}^{\text{eq}}$ in (3.66), show (4.18).

Other higher-order moments are also equal to lower-order ones, though the specific equalities vary between velocity sets. For the D2Q9 velocity set, we can find from the fact that $c_{i\alpha}^3 = c_{i\alpha}(\frac{\Delta x}{\Delta t})^2$ that there are only nine independent moments, as shown in Fig. 4.1.[9]

Exercise 4.2 Show the dependencies given in Fig. 4.1 by equations like (4.18).

Equation (4.18) shows us that at least some of the third-order moments are equal to first-order moments for all the velocity sets that we have examined previously. Consequently, it is not possible for these velocity sets to correctly

[9] In general, for a velocity set with q velocities there are q different independent moments, as we shall see in Chap. 10.

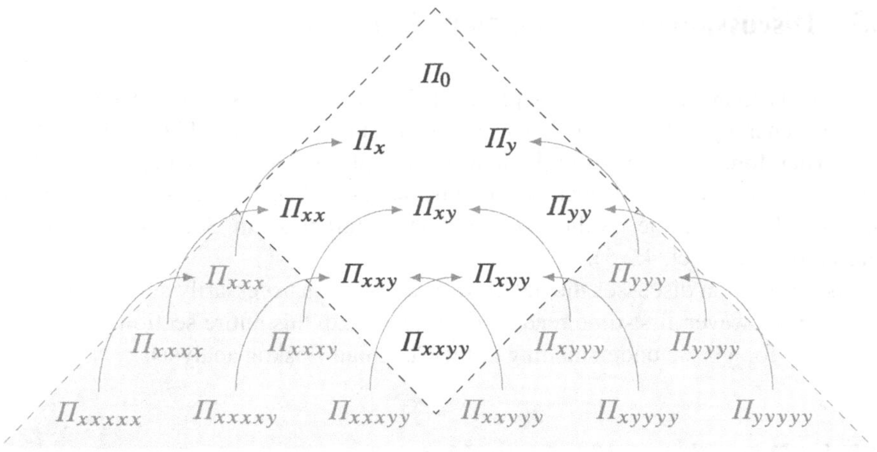

Fig. 4.1 Dependency map of D2Q9 velocity moments. Higher-order moments (*dark on light grey*) depend on nine lower-order independent moments (*black on white*)

recover the $\rho u_\alpha u_\beta u_\gamma$ term in $\Pi^{\text{eq}}_{\alpha\beta\gamma}$ and thus avoid the $O(u^3)$ error term in (4.15) which carries over to the macroscopic stress tensor. Additionally, these velocity sets are insufficient for some of the thermal models which we will touch upon in Sect. 8.4, as these require independent third-order moments. In cases where these issues cannot be safely ignored, extended velocity sets with more velocities and higher-order equilibrium distributions must be used.[10] These can be found using the same Hermite approach as in Sect. 3.4.

Exercise 4.3 Consider a rotated and rescaled D2Q9 velocity set consisting of the zero-velocity vector $(0, 0)$, four short velocity vectors $(\pm 1/2, \pm 1/2)\frac{\Delta x}{\Delta t}$ and four long velocity vectors $(\pm 1, 0)\frac{\Delta x}{\Delta t}$, $(0, \pm 1)\frac{\Delta x}{\Delta t}$. Show that the nine moments Π, Π_x, Π_y, Π_{xx}, Π_{xy}, Π_{yy}, Π_{xxx}, Π_{yyy}, and Π_{xxyy} are independent of each other. Show also that $\Pi_{xxy} = \Pi_{xyy} = \frac{1}{2}\frac{\Delta x}{\Delta t}\Pi_{xy}$.

4.2.2 *The Time Scale Interpretation*

In the Chapman-Enskog analysis presented in Sect. 4.1, the time derivative ∂_t was expanded into terms $\epsilon^n \partial_t^{(n)}$ at different orders of smallness in (4.8). This decomposition is sometimes interpreted as a decomposition into different *time scales*, i.e. "clocks ticking at different speeds". This interpretation may in some

[10] Alternatively, it has also been proposed to add correction terms to the LBE to counteract these errors [14].

cases lead to false conclusions. Whenever the link between different orders in $f_i^{(n)}$ is broken by the strengthened solvability conditions in (4.4), it is not difficult to show that the time scale interpretation leads to such a false conclusion.

From this interpretation, one would expect for steady-state that when the time derivative $\partial_t g$ of a function g vanishes, all "time scale" derivatives $\partial_t^{(n)} g$ can be removed as well. After all, "steady-state" means that nothing can change on *any* time scale.

Let us look at a steady Poiseuille flow. On the one hand, we know macroscopically that a Poiseuille flow obeys

$$\nabla p = \nabla \cdot \boldsymbol{\sigma}, \tag{4.19}$$

i.e. the flow is driven by a pressure gradient which is exactly balanced by the divergence of the viscous stress tensor $\boldsymbol{\sigma}$. On the other hand, (4.10) and (4.12) describe relations for the momentum on the ϵ- and ϵ^2-scales:

$$\partial_t^{(1)}(\rho u_\alpha) = -\partial_\beta^{(1)} \Pi_{\alpha\beta}^{\text{eq}}, \qquad \partial_t^{(2)}(\rho u_\alpha) = -\partial_\beta^{(1)} \left(1 - \frac{\Delta t}{2\tau}\right) \Pi_{\alpha\beta}^{(1)}. \tag{4.20}$$

Obviously, $\partial_t(\rho u_\alpha) = 0$ holds for steady Poiseuille flow. From the time scale interpretation, $\partial_t^{(1)}(\rho u_\alpha)$ and $\partial_t^{(2)}(\rho u_\alpha)$ must also vanish independently so that (4.20) becomes

$$\partial_\beta^{(1)} \Pi_{\alpha\beta}^{\text{eq}} = 0, \qquad \partial_\beta^{(1)} \left(1 - \frac{\Delta t}{2\tau}\right) \Pi_{\alpha\beta}^{(1)} = 0. \tag{4.21}$$

From (4.11a) and (4.14) we know that

$$\Pi_{\alpha\beta}^{\text{eq}} = \rho u_\alpha u_\beta + p\delta_{\alpha\beta}, \qquad \left(1 - \frac{\Delta t}{2\tau}\right) \Pi_{\alpha\beta}^{(1)} = -\sigma_{\alpha\beta}, \tag{4.22}$$

and thus (4.21) states

$$\partial_\alpha^{(1)} p = 0, \qquad \partial_\beta^{(1)} \sigma_{\alpha\beta} = 0, \tag{4.23}$$

i.e. no flow at all!

In other words: given (4.20), Poiseuille flow would be impossible if $\partial_t^{(1)}(\rho u_\alpha)$ and $\partial_t^{(2)}(\rho u_\alpha)$ vanished independently. It is therefore misleading to say that time itself is decomposed into different scales. It is rather the term $\partial_t g$ which is decomposed into contributions from different orders in the perturbation expansion, contributions which are themselves not time derivatives [1]. Only the *sum* of these contributions vanishes, i.e.

$$\partial_t(\rho u_\alpha) = 0 \implies \partial_t^{(1)}(\rho u_\alpha) + \partial_t^{(2)}(\rho u_\alpha) = 0 \implies \partial_\alpha^{(1)} p = \partial_\beta^{(1)} \sigma_{\alpha\beta}, \tag{4.24}$$

neglecting terms at $O(\epsilon^3)$ and higher orders.

In this section we have seen that setting $\partial_t^{(n)} f_i = 0$ leads to false conclusions in the Chapman-Enskog analysis. It must be pointed out that the reason that this does not work is the simplifying (but not, strictly speaking, necessary) assumption that leads from (4.3) to (4.4). Without this assumption we would not get zeros on the right-hand side of (4.21). We will discuss this assumption further at the end of Sect. 4.2.3.

4.2.3 Chapman-Enskog Analysis for Steady Flow

The Chapman-Enskog analysis performed in Sect. 4.1 depends on the expansion of the time derivative into components, each component at a different order in ϵ. The expanded time derivative gives us a unique way to reconnect the equations at different orders in ϵ after having performed the perturbation expansion.

For a time-invariant case with $\partial_t f_i = 0$, this approach is still valid; there is nothing wrong with keeping the time derivative and expanding it even though it is equal to zero. (Indeed, we saw in Sect. 4.2.2 that setting it to zero before expanding it may lead to false conclusions.) However, if we take the final time-dependent macroscopic equations from the Chapman-Enskog analysis in Sect. 4.1 and set the time derivative to zero, the resulting steady-state macroscopic equations are somewhat misleading. As we will soon see, the time-invariant stress tensor can be expressed differently to the time-variant one.

In a time-invariant case, the expanded time derivative is

$$\Delta t \partial_t f_i = \Delta t \left(\epsilon \partial_t^{(1)} f_i + \epsilon^2 \partial_t^{(2)} f_i + \dots \right) = 0, \quad \Longrightarrow \quad \epsilon \partial_t^{(1)} f_i = O(\epsilon^2). \tag{4.25}$$

Thus, $\partial_t^{(1)} f_i$ increases by one order in smallness in this case. Consequently, the mass and momentum moment equations in (4.10) become

$$\partial_\gamma^{(1)} (\rho u_\gamma) = -\partial_t^{(1)} \rho = O(\epsilon), \tag{4.26a}$$

$$\partial_\beta^{(1)} \left(\rho c_s^2 \delta_{\alpha\beta} + \rho u_\alpha u_\beta \right) = -\partial_t^{(1)} (\rho u_\alpha) = O(\epsilon). \tag{4.26b}$$

Therefore, the left-hand sides in these equations are one order higher in ϵ than they would be in a time-variant case. Similarly, the second-order moment equation in (4.10c), which gives us the viscous stress tensor through (4.14), becomes

$$\Pi_{\alpha\beta}^{(1)} = -\tau \left(\partial_t^{(1)} \Pi_{\alpha\beta}^{\text{eq}} + \partial_\gamma^{(1)} \Pi_{\alpha\beta\gamma}^{\text{eq}} \right) = -\tau \partial_\gamma^{(1)} \Pi_{\alpha\beta\gamma}^{\text{eq}} + O(\epsilon). \tag{4.27}$$

By applying (A.14) and (4.26a), this becomes

$$\Pi_{\alpha\beta}^{(1)} = -\tau c_s^2 \left[\partial_\beta^{(1)} (\rho u_\alpha) + \partial_\alpha^{(1)} (\rho u_\beta) \right] + O(\epsilon). \tag{4.28}$$

Finally, we can insert the above equations into (4.13) to find the steady macroscopic equations exactly predicted by the Chapman-Enskog analysis:

$$\partial_\gamma(\rho u_\gamma) = 0, \tag{4.29a}$$

$$\partial_\beta(\rho u_\alpha u_\beta) = -\partial_\alpha p + \partial_\beta \nu \left[\partial_\beta(\rho u_\alpha) + \partial_\alpha(\rho u_\beta)\right] \tag{4.29b}$$

with pressure $p = \rho c_s^2$, kinematic shear viscosity $\nu = c_s^2(\tau - \Delta t/2)$ and kinematic bulk viscosity $\nu_B = 2\nu/3$. This steady momentum equation lacks the $O(u^3)$ error term of the unsteady momentum equation. Instead, its stress tensor contains the gradients

$$\partial_\beta(\rho u_\alpha) + \partial_\alpha(\rho u_\beta) = \rho\left(\partial_\beta u_\alpha + \partial_\alpha u_\beta\right) + \left(u_\alpha \partial_\beta + u_\beta \partial_\alpha\right)\rho \tag{4.30}$$

instead of the gradients $\rho(\partial_\beta u_\alpha + \partial_\alpha u_\beta)$ found in the correct NSE. However, since $\partial_\alpha \rho = O(u^2)$ for steady flow at Ma $\ll$ 1 [10], the error in this macroscopic momentum equation remains at $O(u^3)$. (For a particular LBE variant designed for incompressible flow, this error disappears so that its momentum equation is error-free at steady-state. We describe this variant in Sect. 4.3.2.)

Exercise 4.4 Show that it is also possible to find the steady perturbation moment in (4.28) directly from the unsteady perturbation moment in (4.15). *Hint: you will need to use (A.13) along with (4.26).*

We must point out that this approach to the steady Chapman-Enskog analysis is not the only possible one. It is also possible to perform a steady analysis if we assume time invariance initially by setting $\partial_t f_i = 0$ *a priori* without expanding it [4, 15, 16]. However, without the expanded time derivative, we must link the equations at different order in ϵ in a different way. This can be done through the solvability conditions in (4.3) which can be expanded as [2–4]

$$\begin{aligned} \sum_i f_i^{\text{neq}} &= \sum_i \left(\epsilon f_i^{(1)} + \epsilon^2 f_i^{(2)} + \ldots\right) = 0, \\ \sum_i c_i f_i^{\text{neq}} &= \sum_i c_i \left(\epsilon f_i^{(1)} + \epsilon^2 f_i^{(2)} + \ldots\right) = \mathbf{0}. \end{aligned} \tag{4.31}$$

This approach is incompatible with the additional simplifying assumption made in (4.4) which breaks the link between orders by assuming that the solvability conditions hold individually at each order. Regardless, this approach results in the same steady macroscopic equations as the approach taken above without any loss of generality [4, 15, 16].

4.2.4 The Explicit Distribution Perturbation

In the Chapman-Enskog analysis in Sect. 4.1, we took the moments of (4.9) to find macroscopic equations given by moments of f_i and their derivatives. The only unknown in these equations, $\Pi^{(1)}_{\alpha\beta} = \sum_i c_{i\alpha} c_{i\beta} f_i^{(1)}$, was directly found through (4.10c), using the known moments $\Pi^{\text{eq}}_{\alpha\beta}$ and $\Pi^{\text{eq}}_{\alpha\beta\gamma}$ as detailed in Appendix A.2.2.

However, it is also possible to find the unknown $\Pi^{(1)}_{\alpha\beta}$ by a slightly more complicated two-step approach. First, we find the distribution perturbation $f_i^{(1)}$ through (4.9a) as

$$f_i^{(1)} = \left(\partial_t^{(1)} + c_{i\alpha}\partial_\alpha^{(1)}\right) f_i^{\text{eq}}, \tag{4.32}$$

and then we obtain $\Pi^{(1)}_{\alpha\beta}$ directly through the second moment of $f_i^{(1)}$.

To find $f_i^{(1)}$ explicitly, we can apply the time and space derivatives directly to the equilibrium distribution, (3.54). This distribution can for this purpose be expressed in the more convenient form

$$f_i^{\text{eq}} = w_i \rho \left(1 + \frac{c_{i\alpha} u_\alpha}{c_\text{s}^2} + \frac{Q_{i\alpha\beta} u_\alpha u_\beta}{2c_\text{s}^4}\right), \tag{4.33}$$

using the velocity tensor $Q_{i\alpha\beta} = c_{i\alpha} c_{i\beta} - c_\text{s}^2 \delta_{\alpha\beta}$.

Exercise 4.5 Show that the time derivative of f_i^{eq} can be expressed solely through spatial derivatives as

$$\begin{aligned} \partial_t^{(1)} f_i^{\text{eq}} = -w_i \Big[& \partial_\alpha^{(1)} (\rho u_\alpha) + \frac{c_{i\alpha}}{c_\text{s}^2} \partial_\beta^{(1)} (\rho u_\alpha u_\beta) + c_{i\alpha} \partial_\alpha^{(1)} \rho \\ & + \frac{Q_{i\alpha\beta}}{c_\text{s}^2} u_\alpha \partial_\beta^{(1)} \rho + \frac{Q_{i\alpha\beta}}{2c_\text{s}^4} \partial_\gamma^{(1)} (\rho u_\alpha u_\beta u_\gamma) \Big]. \end{aligned} \tag{4.34}$$

Hint: Use (4.10) and (A.13). Since $Q_{i\alpha\beta}$ is symmetric in α and β, $Q_{i\alpha\beta} A_{\alpha\beta} = Q_{i\alpha\beta} A_{\beta\alpha}$ for any tensor $A_{\alpha\beta}$.

After some algebra, we can find the **first-order distribution perturbation** explicitly as

$$\begin{aligned} f_i^{(1)} = -\frac{\tau w_i}{c_\text{s}^2} \Big[& Q_{i\alpha\beta} \rho \partial_\beta^{(1)} u_\alpha - c_{i\alpha} \partial_\beta^{(1)} (\rho u_\alpha u_\beta) \\ & + \frac{Q_{i\alpha\beta}}{2c_\text{s}^2} c_{i\gamma} \partial_\gamma^{(1)} (\rho u_\alpha u_\beta) - \frac{Q_{i\alpha\beta}}{2c_\text{s}^2} \partial_\gamma^{(1)} (\rho u_\alpha u_\beta u_\gamma) \Big], \end{aligned} \tag{4.35}$$

(continued)

where the last term is the $O(u^3)$ error term discussed in Sect. 4.1.3 and Sect. 4.2.1. The same result (neglecting the error term) can be found in [9, 17]. This result is also useful for understanding the role of the non-equilibrium part of f_i; we see that it is given largely by gradients of the macroscopic fluid velocity $\boldsymbol{u}$.

Exercise 4.6 Derive (4.35) from (4.32), (4.33), and (4.34).

Exercise 4.7

a) Take the second moment of (4.35), and show that this gives the same expression for $\Pi^{(1)}_{\alpha\beta} = \sum_i c_{i\alpha} c_{i\beta} f^{(1)}_i$ as in (4.15). *Hint: Use the isotropy conditions in (3.60).*
b) Show that (4.35) is consistent with the strengthened solvability conditions of (4.4) by showing that $\sum_i f^{(1)}_i = 0$ and $\sum_i c_{i\alpha} f^{(1)}_i = 0$.

4.2.5 Alternative Multi-scale Methods

While the Chapman-Enskog analysis is the classical multi-scale tool to link the LBE and its resulting macroscopic equations [1], alternative mathematical techniques exist. Since their detailed analysis is far outside the scope of this book, this section only provides a brief list of the most relevant examples, with further references for interested readers.

Similarly to the Chapman-Enskog expansion, most of these alternative multi-scale methods have their origins in kinetic theory and were later adapted to the LB field. Below, we illustrate three approaches that have followed this route.

The *asymptotic expansion* technique was pioneered by Sone [18] in the study of solutions of the Boltzmann equation at small Knudsen and finite Reynolds numbers. Such a technique diverts from the classical Chapman-Enskog analysis in two main aspects. First, the macroscopic solutions are formally expanded in a series of small Knudsen number, similarly to the mesoscopic variables. Secondly, it restricts the analysis of the dynamical processes to the diffusive scaling, i.e. $\Delta t \propto \Delta x^2 \sim \epsilon^2$ (where ϵ is some smallness parameter used in the expansion), based on the argument that the target solution belongs to the incompressible NSE. The asymptotic analysis was introduced in the LB field by [19], and later it experienced further developments in [20, 21]. After the Chapman-Enskog analysis, this technique is possibly the one with the broadest acceptance in the LB field.

The *Hermite expansion series* technique was introduced by Grad [22] as a way to approximate the solution of the Boltzmann equation in terms of a finite set of Hermite polynomials. Such a representation provides a convenient closure for describing the fluxes in the macroscopic laws. The idea is that the truncation order of the Hermite expansion of the mesoscopic variable determines the range of validity of the resulting macroscopic fluid system. This technique differs from the Chapman-Enskog analysis as it recovers the governing macroscopic dynamics in a non-perturbative way. The Hermite expansion series was introduced in the LB field by [23] and subsequently further developed by [24, 25].

The *Maxwellian iteration* technique was proposed by Ikenberry and Truesdell [26] as a systematic procedure to find a closure for the fluxes in the governing macroscopic flow equations. The idea is to work with a hierarchy of moments of the mesoscopic variables, which enables completing the missing information from the lower-order moments using higher-order ones. In order to assess whether the higher-order terms in this hierarchy should be neglected or retained, this process is supplemented by an order of magnitude analysis. Therefore, this technique employs elements from both the Grad expansion method and the Chapman-Enskog analysis. In the LB field its use was proposed by [27–29], exploiting the fact that it applies rather naturally to LB models working with the MRT collision operator.

The inspection of the macroscopic behaviour of LB models is also possible through more numerically oriented procedures. Even though they may introduce a higher level of abstraction, they have the advantage of providing access to the macroscopic information directly by Taylor expanding the discrete numerical system, without resorting to any results from kinetic theory. There are three main approaches. First, the *equivalent equation method* [30] where the Taylor expansion method is applied directly to the LBE. Secondly, the method proposed by [31] that, while following the same Taylor expansion procedure, applies it to a *recursive representation* of the LBE. Thirdly, the approach proposed by [16] where the LBE is first written in *recurrence form*, i.e. in terms of finite-difference stencils, with the macroscopic information recovered order by order by Taylor expanding the finite-difference operators.

The number of existing LB multi-scale approaches has led to debates about their equivalence, e.g. [32, 33]. The work [34] contributed to shed light on this issue by showing that the Chapman-Enskog analysis is nothing but a particular example of a general expansion procedure which also encompasses many other multi-scale methods such as the asymptotic expansion technique. As general conclusion, this study showed that distinctive multi-scale methods offer identical consistency information up to the same expansion order.

4.3 Alternative Equilibrium Models

In Chap. 3, we derived the discrete equilibrium distribution f_i^{eq} *via* an expansion of the continuous Maxwell-Boltzmann distribution f^{eq} in Hermite polynomials. We have also mentioned the simpler but less physically illuminating method of Taylor expanding f^{eq} to find f_i^{eq}.

However, a more heuristic approach can be taken [35–37]: instead of deriving f_i^{eq} directly from f^{eq}, a more general equilibrium distribution is chosen by an ansatz as, e.g.

$$f_i^{\text{eq}} = w_i\rho\left(1 + a_1 c_{i\alpha}u_\alpha + a_2 c_{i\alpha}c_{i\beta}u_\alpha u_\beta - a_3 u_\alpha u_\alpha\right). \tag{4.36}$$

Here, a_1, a_2, and a_3 are constants that may be kept throughout the Chapman-Enskog analysis. The resulting macroscopic equations will then be functions of these constants which may subsequently be chosen in order to get the desired macroscopic equations.

These macroscopic equations are not limited to the compressible flow case we have looked at previously in this chapter; in fact, this approach can be used to derive LBEs to solve a much more general class of PDEs. Indeed, this goes to show that a great deal of the physics of the LBE is determined by the choice of the equilibrium distribution f_i^{eq}. As a simple example for this, the subtle change of imposing a given flow field $\boldsymbol{u}$ in the basic LB equilibrium of (3.54) instead of taking $\boldsymbol{u}$ from the populations f_i, results in macroscopic behaviour according to the advection-diffusion equation.[11] We will cover this more in depth in Sect. 8.3.2.

However, it is not always necessary to derive alternative equilibria through this elaborate process. In some cases, the standard equilibrium in (3.54) may be altered directly in order to derive alternative models.

In this section, we will describe in some depth a few different models where the equilibrium distribution is changed in order to modify the macroscopic behaviour. We will then refer to other such models in the literature.

4.3.1 Linear Fluid Flow

The research field of acoustics is based almost entirely on linearised versions of the conservation equations [11] where we assume that the fluid quantities are small variations about a constant rest state,

$$\begin{aligned} \rho(\boldsymbol{x},t) &= \rho_0 + \rho'(\boldsymbol{x},t), \\ p(\boldsymbol{x},t) &= p_0 + p'(\boldsymbol{x},t), \\ \boldsymbol{u}(\boldsymbol{x},t) &= \boldsymbol{0} + \boldsymbol{u}'(\boldsymbol{x},t). \end{aligned} \tag{4.37}$$

The rest state constants are labelled with subscripted zeros while the primed variables represent deviations from this rest state. Terms which are nonlinear in

[11] Breaking the link between f_i and $\boldsymbol{u}$ in this way also breaks the conservation of momentum in collisions, as $\sum_i \boldsymbol{c}_i f_i \neq \sum_i \boldsymbol{c}_i f_i^{\text{eq}}$. Thus, this advection-diffusion model is an example of an alternative LBE where the physical meaning of f_i has changed.

these deviations are neglected.[12] This approximation corresponds, e.g., to creeping flows where viscosity dominates over advection, or to cases of linear (i.e. low-Ma) sound propagation.

By similarly neglecting nonlinear deviations from equilibrium in (3.54), we find a **linearised equilibrium distribution**

$$f_i^{\mathrm{eq}} = w_i \left(\rho + \rho_0 \frac{c_{i\alpha} u_\alpha}{c_s^2} \right). \tag{4.38}$$

Here, ρ_0 is the rest state density.

From this equilibrium distribution and the isotropy conditions in (3.60), we can directly find the equilibrium moments

$$\begin{aligned} \Pi^{\mathrm{eq}} = \rho, \qquad \Pi_\alpha^{\mathrm{eq}} = \rho_0 u_\alpha, \qquad \Pi_{\alpha\beta}^{\mathrm{eq}} = \rho c_s^2 \delta_{\alpha\beta}, \\ \Pi_{\alpha\beta\gamma}^{\mathrm{eq}} = \rho_0 c_s^2 \left(u_\alpha \delta_{\beta\gamma} + u_\beta \delta_{\alpha\gamma} + u_\gamma \delta_{\alpha\beta} \right). \end{aligned} \tag{4.39}$$

Exercise 4.8 Confirm these moments by recalculating them. *Hint: Use (3.60).*

Performing the Chapman-Enskog analysis as in Sect. 4.1 with this simplified equilibrium distribution results in the macroscopic mass and momentum conservation equations

$$\begin{aligned} \partial_t \rho + \rho_0 \partial_\alpha u_\alpha &= 0, \\ \rho_0 \partial_t u_\alpha &= -\partial_\alpha p + \eta \partial_\beta \left(\partial_\beta u_\alpha + \partial_\alpha u_\beta \right) \end{aligned} \tag{4.40}$$

with pressure $p = \rho c_s^2$ and dynamic shear viscosity $\eta = \rho_0 c_s^2 (\tau - \Delta t/2)$. By comparison with (4.16), we find that these are perfectly linearised versions of the continuity and Navier-Stokes equations that we get with the normal nonlinear equilibrium distribution f_i^{eq}. Interestingly, since this linear model drops any nonlinearity, it is *not* affected by the nonlinear $O(u^3)$ error terms in the momentum equation, unlike the standard model we have looked at previously.

Exercise 4.9 Find these macroscopic equations through a Chapman-Enskog analysis of the LBE with linearised equilibrium. (The linearised equilibrium makes the analysis significantly simpler!)

[12] The advection term in Navier-Stokes is one such nonlinear term. By neglecting it, we break the Galilean invariance of the fluid model. To avoid this issue, we could instead linearise the velocity around a non-zero rest state velocity as necessary.

In the case of steady flow, these linearised equations become the equations of *Stokes flow*, i.e. of viscosity-dominated steady flow at Re $\ll$ 1:

$$\begin{aligned} \partial_\alpha u_\alpha &= 0, \\ \eta \partial_\beta \left(\partial_\beta u_\alpha + \partial_\alpha u_\beta\right) - \partial_\alpha p &= 0. \end{aligned} \tag{4.41}$$

This linearised equilibrium has been used to ensure full linearity in LB acoustics simulations. This way, sound waves may be simulated using the complex exponential form in which a plane wave would be represented as varying with, e.g., $e^{i(\omega t - kx)}$ [5, 38]. We will get back to this technique in Sect. 12.1.2.

With the linearised equilibrium the LBE is fully linear, and it is therefore also possible to split f_i into a rest state $F_i^{\text{eq}} = \rho_0 w_i$ and a fluctuation $f_i' = f_i - F_i^{\text{eq}}$, each of which is a valid solution to the linear LBE. In fact, it is possible to implement an LBE that only tracks the fluctuation f_i', with a corresponding equilibrium $f_i'^{\text{eq}} = w_i(\rho' + \rho_0 u_\alpha c_\alpha / c_s^2)$. We can see such an LBE in (12.24) as part of the LBE linearisation analysis in Sect. 12.2.

4.3.2 Incompressible Flow

In the linearised model described in Sect. 4.3.1, we assumed that the density ρ, the pressure p and the fluid velocity $\boldsymbol{u}$ deviated only very little from a constant rest state. This assumption does not hold for flows at larger Reynolds numbers.

The deviation of *density* from a rest state may be smaller than that of the *velocity*. Indeed, it is known for steady flow at Ma $\ll$ 1 that $\rho'/\rho_0 = O(\text{Ma}^2)$ [10].[13] In this limit, we may derive an **LBE that approximates incompressibility** by neglecting terms in f_i^{eq} that are above $O(\text{Ma}^2)$:

$$f_i^{\text{eq}} = w_i \rho + w_i \rho_0 \left(\frac{c_{i\alpha} u_\alpha}{c_s^2} + \frac{u_\alpha u_\beta \left(c_{i\alpha} c_{i\beta} - c_s^2 \delta_{\alpha\beta}\right)}{2 c_s^4} \right). \tag{4.42}$$

[13]Note that this holds only for *steady flow* and not for cases where the flow field is time-dependent. For instance, we have for plane sound waves that $|\rho'/\rho_0| = O(\text{Ma})$ [11].

Such equilibria were proposed in the literature early in the history of the LBM [39, 40]. We can find the moments of this particular equilibrium using (3.60):

$$\begin{aligned} \Pi^{\mathrm{eq}} &= \rho, \qquad \Pi^{\mathrm{eq}}_{\alpha} = \rho_0 u_\alpha, \qquad \Pi^{\mathrm{eq}}_{\alpha\beta} = \rho c_s^2 \delta_{\alpha\beta} + \rho_0 u_\alpha u_\beta, \\ \Pi^{\mathrm{eq}}_{\alpha\beta\gamma} &= \rho_0 c_s^2 \left(u_\alpha \delta_{\beta\gamma} + u_\beta \delta_{\alpha\gamma} + u_\gamma \delta_{\alpha\beta} \right). \end{aligned} \tag{4.43}$$

We can perform the Chapman-Enskog analysis as in Sect. 4.1 using this equilibrium in order to find the simulated macroscopic equations. The resulting stress tensor is

$$\sigma_{\alpha\beta} = \left(\tau - \frac{\Delta t}{2}\right) \Big[\rho_0 c_s^2 \left(\partial_\beta u_\alpha + \partial_\alpha u_\beta\right) - \rho_0 \left(u_\alpha \partial_\gamma \left(u_\beta u_\gamma\right) + u_\beta \partial_\gamma \left(u_\alpha u_\gamma\right)\right) \\ - c_s^2 \left(u_\alpha \partial_\beta \rho + u_\beta \partial_\alpha \rho \right) \Big]. \tag{4.44}$$

Of the three terms in the brackets, the latter two are error terms. The first of these error terms is of a similar nature as the $O(u^3)$ error term in (4.15). The second error term appears due to the different pressure in the second and third equilibrium moments: ρc_s^2 in $\Pi^{\mathrm{eq}}_{\alpha\beta}$ and $\rho_0 c_s^2$ in $\Pi^{\mathrm{eq}}_{\alpha\beta\gamma}$.

Neglecting these error terms, the Chapman-Enskog analysis results in macroscopic mass and momentum equations

$$\partial_t p + \rho_0 c_s^2 \partial_\gamma u_\gamma = 0, \tag{4.45a}$$

$$\rho_0 \left(\partial_t u_\alpha + \partial_\beta u_\alpha u_\beta\right) = -\partial_\alpha p + \eta \partial_\beta \left(\partial_\beta u_\alpha + \partial_\alpha u_\beta\right), \tag{4.45b}$$

with pressure $p = c_s^2 \rho$ and dynamic shear viscosity $\eta = \rho_0 c_s^2 (\tau - \Delta t/2)$.

For a steady flow case, we can use a similar analysis to Sect. 4.2.3 to show that the error terms in (4.44) disappear. Consequently, the macroscopic equations in (4.45) (without time derivatives) are exactly recovered at the Navier-Stokes level. (However, higher-order truncation error terms may still be present.)

Exercise 4.10 Apply the analysis in Sect. 4.2.3 to the incompressible LB model to show that the steady macroscopic equations (i.e. (4.45) without time derivatives) are exactly recovered at the Navier-Stokes level.

4.3.3 Alternative Equations of State

Previously in Chap. 3 we found that the simple square velocity sets lead to an isothermal equation of state, $p = c_s^2 \rho$, and that a more realistic thermal model requires extended velocity sets. However, alternative equilibria may also be applied in order to perform simulations with different equations of state, although, as we soon shall see, this approach has some weaknesses.[14] Different equations of state also allow for speeds of sound c that are different from the isothermal speed of sound c_s. In general, the speed of sound is determined by the equation of state $p(\rho, s)$ as $c = \sqrt{(\partial p/\partial \rho)_s}$, where s is entropy [10, 11].

From the Chapman-Enskog analysis in Sect. 4.1, the pressure in the macroscopic momentum equation depends on the equilibrium moments:

$$\Pi^{\text{eq}}_{\alpha\beta} = p\delta_{\alpha\beta} + \rho u_\alpha u_\beta, \tag{4.46a}$$

$$\Pi^{\text{eq}}_{\alpha\beta\gamma} = p\left(u_\alpha \delta_{\beta\gamma} + u_\beta \delta_{\alpha\gamma} + u_\gamma \delta_{\alpha\beta}\right). \tag{4.46b}$$

The first of these moments directly determines the pressure gradient in the momentum equation. In the Chapman-Enskog analysis, terms stemming from the third-order moment cancel pressure terms in the second-order moment.

However, no equilibrium distribution for the velocity sets given in Sect. 3.4.7 can give a freely chosen pressure p in both these equilibrium moments. From the discussion in Sect. 4.2, these velocity sets are *too small* for the third-order equilibrium moments $\Pi^{\text{eq}}_{\alpha\beta\gamma}$ to be independent of lower-order equilibrium moments. Without an extended velocity set, we can only have a third-order moment like (4.46b) but with $p \to \rho c_s^2$.

While we can find an equilibrium with a freely chosen pressure p in the second-order equilibrium moment, this leads to an inconsistent pressure between the second- and third-order moments: p in the former and ρc_s^2 in the latter. Consequently, the aforementioned cancellation does not happen for a number of terms in the Chapman-Enskog analysis of an LBE with such an equilibrium, and we recover a macroscopic momentum equation containing several additional error terms [46]:

$$\begin{aligned}
&\partial_t(\rho u_\alpha) + \partial_\beta(\rho u_\alpha u_\beta) \\
&\quad = -\partial_\alpha p + \partial_\beta \rho c_s^2 \left(\tau - \frac{\Delta t}{2}\right)\left[\partial_\beta u_\alpha + \partial_\alpha u_\beta + \left(1 - \frac{c^2}{c_s^2}\right)\delta_{\alpha\beta}\partial_\gamma u_\gamma\right] \\
&\qquad - \partial_\beta\left(\tau - \frac{\Delta t}{2}\right)\partial_\gamma(\rho u_\alpha u_\beta u_\gamma) \\
&\qquad + \partial_\beta\left(\tau - \frac{\Delta t}{2}\right)\left[\left(c_s^2 - c^2\right)\delta_{\alpha\beta} u_\gamma \partial_\gamma \rho + \left(u_\alpha \partial_\beta + u_\beta \partial_\alpha\right)\left(\rho c_s^2 - p\right)\right].
\end{aligned} \tag{4.47}$$

Here, c_s is the isothermal speed of sound and $c = \sqrt{(\partial p/\partial \rho)_s}$ is the actual speed of sound for a given pressure $p(\rho, s)$.

[14] Changing the equilibrium is not the only way to achieve different equations of state, however. Various body force schemes also exist [41–45].

The first two lines of this momentum equation is the same as the momentum equation (4.16b) found from the Chapman-Enskog analysis in Sect. 4.1, except with a bulk viscosity of $\eta_B = \eta(5/3 - c^2/c_s^2)$ instead of $\eta_B = 2\eta/3$. This bulk viscosity unphysically depends on the velocity set constant c_s^2, but it is possible to alter the bulk viscosity in LB simulations (cf. Chap. 10 and [12, 17, 47]). The third line of this momentum equation is the same error that also exists in the usual isothermal LBE, as seen in (4.15). The last line contains new error terms that appear due to the aforementioned inconsistent pressure between the second- and third-order equilibrium moments.

However, these new error terms all scale as $O(\text{Ma}\rho')$. For steady flow at Ma $\ll 1$ it is known that $\rho'/\rho_0 = O(\text{Ma}^2)$ [10], and for plane sound waves we have that $|\rho'/\rho_0| = O(\text{Ma})$ [11]. Consequently, these new error terms are, in the worst case, of $O(\text{Ma}^2)$ and can therefore be safely neglected along with the previous $O(\text{Ma}^3)$ error terms, as long as Ma $\ll 1$.

Up to this point, we have only described the constraints on the moments of the equilibrium distribution f_i^{eq}, but we have not looked at what this equilibrium distribution can look like. In fact, a number of such equilibria have been proposed in the literature, both for general and specific equations of state [35, 42, 48–50]. However, many of these are specific to a particular velocity set, and those which are not have not been validated for all the velocity sets described in Sect. 3.4.7. It is not given that these equilibria are stable or accurate in all of these velocity sets [50].

As an **example** of an **LB equilibrium** that allows for a **non-isothermal equation of state** for simple velocity sets, we can look at the equilibrium for **D1Q3** which is uniquely given by [5, 46]

$$f_0^{\text{eq}} = \rho - p - \rho uu, \quad f_{i\neq 0}^{\text{eq}} = \frac{1}{2}\left(p + \rho u c_i + \rho uu\right). \tag{4.48}$$

The pressure p can be freely chosen, and its relation to ρ determines the speed of sound as described above. However, this equilibrium is still subject to stability conditions as described in Sect. 4.4, and not all equations of state will therefore be stable.

Exercise 4.11 Show that the equilibrium distribution in (4.48) has zeroth- and first-order moments $\Pi^{\text{eq}} = \rho$ and Π_x^{eq} and that its second- and third-order moments Π_{xx}^{eq} and Π_{xxx}^{eq} are given by (4.46) with $p \to \rho c_s^2$ in the third-order moment.

4.3.4 Other Models

LBM has originally been used to solve the (weakly compressible) NSE. Therefore, the standard equilibrium in (3.54) follows from an expansion of the Maxwell-

Boltzmann distribution. However, researchers have realised that the form of the macroscopic equations that are solved by the LB scheme can be changed by choosing different and appropriate equilibrium distributions. We conclude this section by mentioning a few recent developments where LBM is employed for other purposes than standard fluid dynamics.

LBM is often applied to shallow water problems. This research goes back to the late 1990s [50–52] and is still ongoing [53]. As we will also point out in Sect. 5.5, LBM can be used as Poisson solver for the pressure if a velocity field is defined [54]. This method is often applied to find a consistent initial state for an LB simulation. The NSE can also be solved *via* LBM in the vorticity-streamfunction formulation [55]. A few years ago, Mendoza et al. suggested LBM modifications to solve mildly relativistic fluid flows [56] and Maxwell's equations in materials [57]. More recent applications are LBM for image processing [58], solving elliptic equations [59] and consistent modelling of compressible flows [60]. In Chap. 9 we will cover multi-phase and multi-component flows. The so-called free energy approach, covered in Sect. 9.2, can be implemented by redefining the local equilibrium distribution function.

4.4 Stability

We have previously discussed how the discrete LBE may be used to solve the continuous NSE. Like any other numerical scheme, the LBM brings along a number of issues. Those include the stability and accuracy of simulations. Since the very beginning of the LBM, the analysis of its numerical stability has attracted considerable interest.

To simulate physical systems, the relevant dimensionless quantities need to be matched (cf. Chap. 7). This involves the choice of simulation-specific parameters. This choice is not arbitrary, and it is an art to map the physics of a real system onto a simulation. It is relatively easy to fall into the trap of choosing parameters leading to numerical instability. Instability in an LB simulation refers to situations where the *errors* of populations, density and velocity are exponentially growing. One should distinguish those simulations from ones where *solutions* grow exponentially. Both eventually lead to "Not a Number" (NaN) values for the observables. Usually instabilities are attributed to truncation errors and an ill-posed time evolution.

We will not present an exhaustive mathematical description of stability analysis as can be found in [61–63]; we will rather show some practical guidelines to improve numerical stability. After discussing some general concepts in Sect. 4.4.1, we will cover the BGK collision operator (cf. Sect. 4.4.2), followed by advanced collision operators (cf. Sect. 4.4.3). We will finish the discussion with simple guidelines for improving stability in Sect. 4.4.4.

4.4.1 Stability Analysis

A common way to analyse stability is through a linear *von Neumann analysis* [64]. This analysis allows representing a variable we are solving for in the Fourier form (wave form), $A \exp(\mathrm{i}\boldsymbol{k} \cdot \boldsymbol{x} - \Omega t)$, with $\boldsymbol{x}$ and t being discretised spatial coordinates and time and $\boldsymbol{k}$ being a vector wavenumber. Substituting this expression into the discretised and linearised equation, one searches for the dispersion relation $\Omega(\boldsymbol{k})$. Simulations are stable when $|\exp(-\Omega \Delta t)| \leq 1$ for all possible wavenumbers $\boldsymbol{k}$.

In CFD applications one typically encounters the *Courant number* $C = |\boldsymbol{u}|\Delta t/\Delta x$ which is linked to stability [65]. The Courant number has a concrete meaning: it compares the speed $\Delta x/\Delta t$ at which information propagates in the model with the physical speed $|\boldsymbol{u}|$ at which the fluid field is advected. If $\Delta x/\Delta t < |\boldsymbol{u}|$ (i.e. $C > 1$), the simulation cannot propagate the physical solution quickly enough, and this tends to make the simulation unstable. Therefore, $C \leq 1$ is often necessary for stability.

Unfortunately, the stability analysis is more complicated for the LBM; the Courant number is not the only determining factor. In addition to the parameters Δx, Δt and $\boldsymbol{u}$, LB has additional degrees of freedom: one or more relaxation times. Moreover, it involves multiple equations, one for each direction $\boldsymbol{c}_i$. Thus, finding the dispersion relation $\Omega(\boldsymbol{k})$ requires inversion of a $q \times q$ matrix, with q being the number of populations in a DdQq model. Despite those difficulties, there has been some progress in applying the linear von Neumann analysis to LBM [13, 66], especially for advection-diffusion applications [67]. By analysis of the dispersion condition $\Omega(\boldsymbol{k})$ for all possible wavenumbers $\boldsymbol{k}$ it is possible to yield some restrictions for the velocity $\boldsymbol{u}$.

Overall, for the LBM with the BGK collision operator, the **stability map** $|\boldsymbol{u}_{\text{max}}|(\tau)$ tells us the maximum achievable velocity magnitude for a given value of the relaxation time τ before instability sets in.

For example, if one wants to achieve a high Reynolds number $\mathrm{Re} = |\boldsymbol{u}|N\Delta x/\nu$, then it is possible to either increase the macroscopic velocity magnitude $|\boldsymbol{u}|$ or grid number N, or decrease the viscosity ν. The increase of N is limited by constraints on the required memory and run time. Thus, the usual approach is to decrease the viscosity and/or increase the macroscopic velocity. We will soon see that we cannot decrease the viscosity arbitrarily as this would lead to instability. Also, we cannot increase the velocity without limit since the LBE recovers the NSE only in the low-Mach limit (cf. Chap. 3). Large velocities can also lead to instability as we will shortly discuss. Thus, for high Reynolds number flows, there is always a compromise between maximum velocity and minimum viscosity for which simulations remain stable.

One distinguishes between **sufficient** and **necessary stability conditions.** A *necessary* condition *must* be fulfilled in order to achieve stability. For example, we know from arguments in both Sect. 3.5.3 and Sect. 4.1.4 that $\tau/\Delta t \geq 1/2$ is a necessary stability condition for the BGK operator. However, if a single necessary condition for stability is met, the simulation is not automatically stable because other necessary conditions may still not be met. If a *sufficient* condition is met, the simulation is *always* stable. Sufficient conditions can never be *less* restrictive than necessary conditions; if a sufficient condition holds, all necessary conditions also hold. The least restrictive combination of necessary conditions is called **optimal**. Optimal conditions are thus both necessary and sufficient at the same time. Sufficient, necessary, and optimal conditions are sketched in Fig. 4.2.

How do we use these general concepts to describe the stability of LBGK, though? When sufficient and/or necessary conditions are mentioned, they are usually specified for a specific value of τ. The optimal stability condition represents the union of necessary conditions (i.e. the largest stable value of $|\boldsymbol{u}|$) for a *range* of τ values.

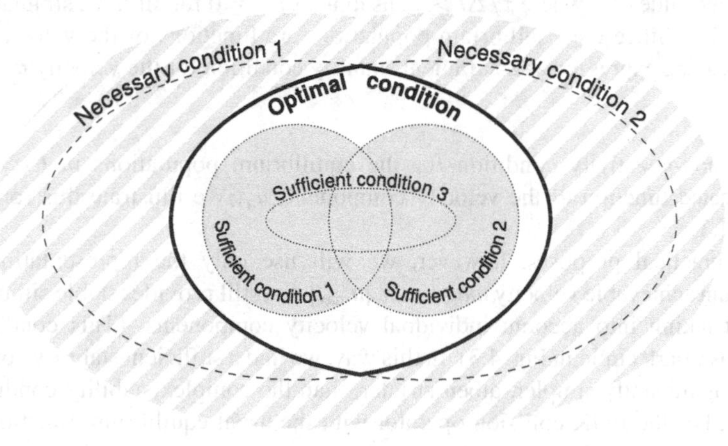

Fig. 4.2 Graphical representation of conditions, from most restrictive (inner) to least restrictive (outer). (Here, a condition is met if we are *inside* its region.) The region explicitly permitted by any of the given sufficient conditions is solid grey, and the region disallowed by any of the two necessary conditions is hatched. In this case, it is enough that these two necessary conditions are fulfilled, and their combination defines the optimal condition. (The optimal condition is the least restrictive sufficient condition possible; it is less restrictive than the given sufficient conditions)

For the LBM, necessary and sufficient stability conditions are normally obtained analytically in the bulk, far away from any boundaries. In most simulations, however, there are boundaries and round-off errors. Thus, the **analytically derived stability conditions** are often not very useful, but they **can act as guidelines**. It is always recommended to start with available analytical conditions and further improve stability for a given problem if that turns out to be necessary.

4.4.2 *BGK Stability*

We will now discuss sufficient and necessary stability conditions for the BGK collision operator. As mentioned, these conditions were obtained analytically for the bulk LBE without boundaries present, and can therefore only serve as guidelines.

For the **BGK** collision operator, a **sufficient stability condition** is the **non-negativity of all equilibrium populations** [63, 67, 68]. This condition holds for any value of τ where $\tau/\Delta t > \frac{1}{2}$: as long as $f_i^{\text{eq}} \geq 0$ for all i, the simulation is stable. Since the equilibrium populations are functions of the velocity $\boldsymbol{u}$, this can be expressed as a sufficient stability condition for the velocity $\boldsymbol{u}$.

The non-negativity condition for the equilibrium populations in (3.54) is a complicated function of the velocity components u_α. We illustrate this for D2Q9 in Fig. 4.3.

For practical purposes, however, we will use only the representation with maximum achievable velocity *magnitude* $|\boldsymbol{u}_{\max}|$ that still provides stable simulations without taking into account individual velocity components. (This condition is shown as a dark circle in Fig. 4.3). In this way, we find a sufficient stability condition that is significantly simpler, albeit stricter, than the complex stability condition in Fig. 4.3. For the BGK collision operator with the usual equilibrium function from (3.54), we can write it as

$$|\boldsymbol{u}_{\max}| < \begin{cases} \sqrt{\frac{2}{3}}\frac{\Delta x}{\Delta t} \approx 0.816\frac{\Delta x}{\Delta t} & \text{for D1Q3}, \\ \sqrt{\frac{1}{3}}\frac{\Delta x}{\Delta t} \approx 0.577\frac{\Delta x}{\Delta t} & \text{for D2Q9, D3Q15, D3Q19, D3Q27.} \end{cases} \tag{4.49}$$

In addition to the non-negativity of the equilibrium f_i^{eq}, some authors claim that also the non-negativity of the populations f_i is required to ensure stability [69]. Special

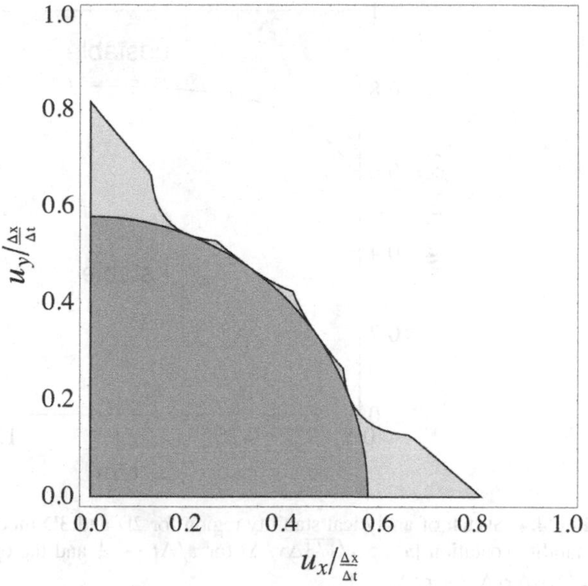

Fig. 4.3 Complex interplay of velocity components u_x and u_y that provide non-negative equilibrium populations in (3.54) for D2Q9. The *light grey area* denotes a sufficient stability condition that is independent of τ. The *darker area* represents a simplified but stricter sufficient condition that depends only on the velocity *magnitude*; $|\boldsymbol{u}| < |\boldsymbol{u}_{\max}|$

techniques to ensure that the populations remain positive have been proposed [70, 71].

For the **BGK** collision operator, an **optimal stability condition** for the range $\tau/\Delta t \geq 1$ is the **non-negativity of the rest equilibrium population** [67]: stability is guaranteed in the bulk if $f_0^{\text{eq}} > 0$ and $\tau/\Delta t \geq 1$. From this follows the velocity magnitude condition

$$|\boldsymbol{u}| < \sqrt{\frac{2}{3}}\frac{\Delta x}{\Delta t}. \tag{4.50}$$

For all other relaxation times, i.e. $\frac{1}{2} < \tau/\Delta t < 1$, the attained maximum velocity magnitude $|\boldsymbol{u}_{\max}|$ is a complicated function of τ [63]. In any case, the maximum velocity magnitude $|\boldsymbol{u}_{\max}|$ is bound between the values defined by two stability conditions: the sufficient condition for all $\tau/\Delta t > \frac{1}{2}$, and the optimal condition for $\tau/\Delta t \geq 1$. We illustrate this stability behaviour in Fig. 4.4.

Unfortunately, as we indicated earlier, those results were obtained analytically for the bulk LBE. In simulations with boundaries, it is not possible to have stable simulations for all values of $\tau/\Delta t > \frac{1}{2}$ for all velocity magnitudes $|\boldsymbol{u}| < |\boldsymbol{u}_{\max}|$ from (4.49). The stability condition typically deteriorates (i.e. $|\boldsymbol{u}_{\max}| \to 0$) with decreasing viscosity (i.e. $\tau/\Delta t \to \frac{1}{2}$), as shown in a number of publications. Some

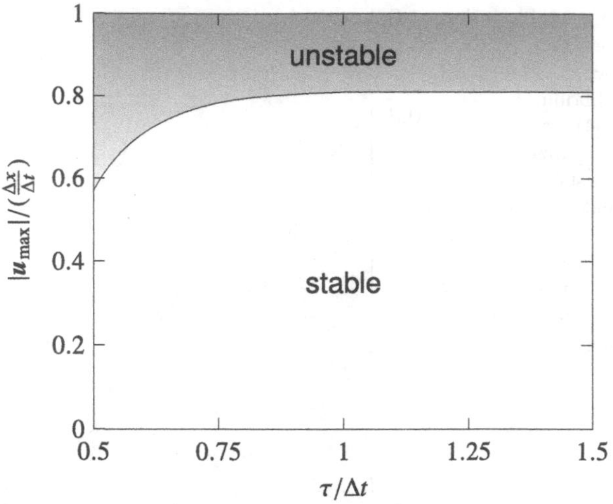

Fig. 4.4 Sketch of analytical stability region for 2D and 3D models. This includes the sufficient stability condition $|\boldsymbol{u}| < \sqrt{1/3}\Delta x/\Delta t$ for $\tau/\Delta t \to \frac{1}{2}$ and the optimal stability condition $|\boldsymbol{u}| < \sqrt{2/3}\Delta x/\Delta t$ for $\tau/\Delta t \geq 1$

of them show it through the non-negativity of populations f_i [69], some through the von Neumann analysis [62, 66].

It is important to emphasise that the sufficient stability boundaries for $\tau/\Delta t > \frac{1}{2}$ are not identical in these works. For example, in contrast to the analytical results presented above, the von Neumann stability analysis is sometimes performed numerically for certain wavenumbers $\boldsymbol{k}$. Thus, one should use results from publications [62, 66, 69] with caution. Overall, for general flows (including, for example, boundaries and numerical round-off errors), it is a safe assumption that the maximum achievable velocity magnitude $|\boldsymbol{u}_{\max}|$ approaches *zero* (rather than a finite value) for $\tau/\Delta t \to \frac{1}{2}$. For example, Niu et al. [62] investigated the dependence of $|\boldsymbol{u}_{\max}|$ on τ for D2Q9. They found a yet another linear relation for small τ,

$$|\boldsymbol{u}_{\max}|(\tau) = 8\left(\frac{\tau}{\Delta t} - \frac{1}{2}\right)\frac{\Delta x}{\Delta t} \quad \text{for} \quad \frac{\tau}{\Delta t} < 0.55, \tag{4.51}$$

and a constant maximum velocity, $|\boldsymbol{u}_{\max}|(\tau) = 0.4\Delta x/\Delta t$, for $\tau/\Delta t > 0.55$, see Fig. 4.5. Equation (4.51) is only valid for certain assumptions and may not hold for more general flow problems. However, we should emphasise that Fig. 4.5 represents a sketch of the stability picture in real simulations. (Note however that this picture is strongly dependent on the choice of boundary conditions [9].)

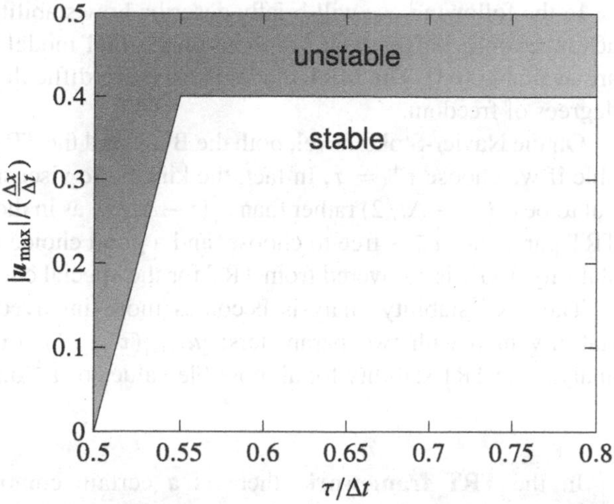

Fig. 4.5 Stability regions of $|\boldsymbol{u}_{\max}|/(\frac{\Delta x}{\Delta t})$ *vs.* $\tau/\Delta t$ (simplified version of Fig. 2 in [62])

As a **guideline to find stable parameters for small viscosities**, one should start with the sufficient stability condition for all $\tau/\Delta t > \frac{1}{2}$ in (4.49). If the simulations are unstable, then one needs to perform a few simulations with different values of τ and $\boldsymbol{u}$ to find an empirical relation $|\boldsymbol{u}_{\max}|(\tau)$ similar to that in Fig. 4.5 for the case at hand. This curve tends to be different for different flow problems.

There are other ways to improve the stability. For example, we can choose another collision operator, as discussed in more detail in Chap. 10. We will now briefly touch upon this.

4.4.3 *Stability for Advanced Collision Operators*

As we will explain in more detail in Chap. 10, BGK is only the simplest available collision operator with a single relaxation time τ. A possible extension is the *two-relaxation-time* (TRT) collision operator with two distinct relaxation times τ^+ and τ^-. An even more advanced model is the *multi-relaxation-time* (MRT) collision operator with q relaxation times τ_i for a DdQq velocity set.[15] We will show in Chap. 10 that the BGK model is a special case of the TRT model which in turn is a special case of the MRT model.

[15]Not all of these q relaxation times are independent, though.

In the following, we will briefly describe how stability can be improved with an advanced collision operator. We focus on the TRT model for which analytical results are available [63]. The MRT model is far more difficult to analyse due to its many degrees of freedom.

On the Navier-Stokes level, both the BGK and the TRT models are indistinguishable if we choose $\tau^+ = \tau$. In fact, the kinematic viscosity in the TRT model turns out to be $c_s^2(\tau^+ - \Delta t/2)$ rather than $c_s^2(\tau - \Delta t/2)$ as in the BGK case. The additional TRT parameter τ^- is free to choose, and a good choice can improve the numerical stability. BGK is recovered from TRT for the special choice $\tau^+ = \tau^- = \tau$.

The TRT stability analysis becomes more involved because now we have a stability map with two parameters: $|\boldsymbol{u}_{\max}|(\tau^+, \tau^-)$. Luckily, we do not need to analyse the TRT stability for all possible values of τ^+ and τ^-.

In the **TRT framework**, there is a certain combination of τ^+ and τ^- (historically called the *magic parameter* Λ) that governs the **stability and accuracy** of simulations [72]:

$$\Lambda = \left(\frac{\tau^+}{\Delta t} - \frac{1}{2}\right)\left(\frac{\tau^-}{\Delta t} - \frac{1}{2}\right). \tag{4.52}$$

A recommended choice is $\Lambda = 1/4$ [67]; this corresponds to $\tau/\Delta t = 1$ in the BGK case, and allows the same optimal stability from which the velocity condition in (4.50) follows. For any value of τ^+, one can always select the free parameter τ^- such that $\Lambda = \frac{1}{4}$. The advantage of the TRT model is therefore that the stability condition and the kinematic viscosity $\nu = c_s^2(\tau^+ - \Delta t/2)$ are *decoupled*. We can thus get optimal stability for any viscosity using TRT; this is clearly not possible with the BGK operator.

While the TRT collision operator with $\Lambda = 1/4$ is not a universal remedy that is sure to make simulations stable for any velocity magnitude, it can often improve stability, especially when the BGK relaxation time τ would be close to $\Delta t/2$.

4.4.4 Stability Guidelines

Finally, we present simple guidelines to improve the stability of LB simulations as also depicted in Fig. 4.6.

We assume that a given Reynolds number

$$\mathrm{Re} = \frac{|\boldsymbol{u}| N \Delta x}{c_s^2\left(\tau - \frac{\Delta t}{2}\right)} \tag{4.53}$$

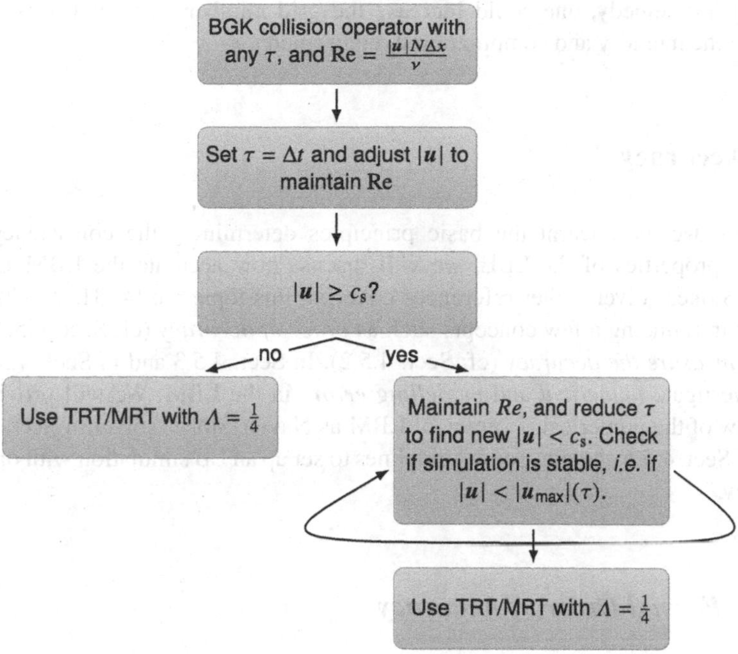

Fig. 4.6 Guidelines for improving simulation stability for a given Reynolds number

is to be simulated. Here, N is the number of lattice nodes (also called the grid number) along a characteristic length scale $\ell = N\Delta x$. We recommend keeping N fixed initially. This keeps the memory footprint and computational requirements under control.

As a rule of thumb, we always start with the BGK collision operator. If simulations are unstable with the current arbitrary τ, then the next step is to choose $\tau/\Delta t = 1$, the lowest τ value which allows optimal stability for the BGK collision operator, and adjust $|\boldsymbol{u}|$ to match the Reynolds number. After this adjustment, two cases are possible: $|\boldsymbol{u}| < c_s$ or $|\boldsymbol{u}| \geq c_s$.

If $|\boldsymbol{u}| < c_s$ (ensuring a sufficiently small Mach number) and the simulation is stable, then there is no need to further change the parameters. If simulations are unstable, then it is recommended to apply the TRT or MRT collision operator. Subsequently, the viscosity and velocity can be reduced while maintaining a constant Re and $\Lambda = 1/4$, the latter allowing for optimal stability.

If $|\boldsymbol{u}| \geq c_s$ for $\tau/\Delta t = 1$ then τ has to be decreased to reduce the velocity magnitude, while keeping Re unchanged. If the magnitude of the new scaled velocity $|\boldsymbol{u}|$ is larger than the obtained BGK collision operator stability curve, i.e. $|\boldsymbol{u}| > |\boldsymbol{u}_{\max}|(\tau)$ then one needs to further reduce $\boldsymbol{u}$ and τ (while keeping the Reynolds number) until the velocity magnitude is within the stability region. If the simulations are still unstable under these conditions, then one should use TRT or MRT models with $\Lambda = 1/4$ to allow optimal stability.

As a last remedy, one could increase the grid number N, though this would increase the memory and computational requirements.

4.5 Accuracy

Now that we have learnt the basic principles determining the consistency and stability properties of the LBE, we will discuss how accurate the LBM is as a Navier-Stokes solver. Other references covering this topic are [4, 31, 73–76]. We start by introducing a few concepts such as *order of accuracy* (cf. Sect. 4.5.1) and how to *measure the accuracy* (cf. Sect. 4.5.2). In Sect. 4.5.3 and in Sect. 4.5.4 we will investigate *numerical* and *modelling errors* in the LBM. We will provide an overview of the numerical accuracy of LBM as Navier-Stokes solver in Sect. 4.5.5. Finally, Sect. 4.5.6 contains some guidelines to set up an LB simulation with optimal accuracy.

4.5.1 Formal Order of Accuracy

Let us start by explaining what *accuracy order* means and how can it be formally determined. While we could discuss this subject in the framework of LBM, its theoretical analysis is a cumbersome task. As we have seen in Sect. 4.1, accessing the continuum description of the LBE requires carrying out a lengthy Chapman-Enskog analysis. In contrast, this kind of study is much more direct for finite difference schemes. Hence, in order to introduce concepts such as the formal order of accuracy and to show the relation between the accuracy order and the leading-order truncation errors, we will first investigate the finite difference method. The main conclusions of this analysis also hold for the LBE.

Consider the Couette flow depicted in Fig. 1.1(a) and assume it is impulsively started from rest. We are interested in the velocity field $u(y, t)$, given proper initial and boundary conditions (which are not relevant for our discussion here). In the bulk, the continuum formulation of this problem reads

$$\frac{\partial u}{\partial t} - \nu \frac{\partial^2 u}{\partial y^2} = 0 \tag{4.54}$$

with the kinematic viscosity ν.

Now, let us assume that we have discretised equation (4.54) using forward difference in time and centred second-order finite difference in space:

$$\frac{u_j^{n+1} - u_j^n}{\Delta t} - \nu \frac{u_{j+1}^n - 2u_j^n + u_{j-1}^n}{\Delta y^2} = 0. \tag{4.55}$$

Superscripts denote the time step and subscripts the spatial location.

The formal way to determine the accuracy of (4.55) as an approximation of (4.54) is to analyse the structure of the truncation errors in the discretised equation. We can obtain these errors in two steps. We first perform a Taylor expansion of the discrete solution around time step n and location j[16]:

$$u_j^{n+1} = u_j^n + \frac{\Delta t}{1!}\frac{\partial u}{\partial t}\bigg|_j^n + \frac{\Delta t^2}{2!}\frac{\partial^2 u}{\partial t^2}\bigg|_j^n + \frac{\Delta t^3}{3!}\frac{\partial^3 u}{\partial t^3}\bigg|_j^n + O(\Delta t^4),$$

$$u_{j+1}^n = u_j^n + \frac{\Delta y}{1!}\frac{\partial u}{\partial y}\bigg|_j^n + \frac{\Delta y^2}{2!}\frac{\partial^2 u}{\partial y^2}\bigg|_j^n + \frac{\Delta y^3}{3!}\frac{\partial^3 u}{\partial y^3}\bigg|_j^n + O(\Delta y^4),$$

$$u_{j-1}^n = u_j^n - \frac{\Delta y}{1!}\frac{\partial u}{\partial y}\bigg|_j^n + \frac{\Delta y^2}{2!}\frac{\partial^2 u}{\partial y^2}\bigg|_j^n - \frac{\Delta y^3}{3!}\frac{\partial^3 u}{\partial y^3}\bigg|_j^n + O(\Delta y^4).$$

We substitute these expressions into (4.55) and obtain, after some simplification,

$$\frac{\partial u}{\partial t} - \nu\frac{\partial^2 u}{\partial y^2} = -\frac{\Delta t}{2!}\frac{\partial^2 u}{\partial t^2} + \nu\frac{\Delta y^2}{4!}\frac{\partial^4 u}{\partial y^4} + O(\Delta t^2) + O(\Delta y^4). \tag{4.56}$$

Exercise 4.12 Show that the approximation of $\partial u/\partial t|_j^n$ using the centred difference formula[17] $(u_j^{n+1} - u_j^{n-1})/(2\Delta t)$ leads to a $O(\Delta t^2)$ leading truncation error.

Equation (4.56) reveals the form of the actual PDE solved by the discretisation scheme, namely (4.54) plus undesired truncation errors. Those errors are the difference between the discretised equation and the target continuum equation.

A **discretisation scheme** is *consistent* if its **truncation errors** tend to zero when Δt and Δy go to zero. The rate at which this happens establishes the *formal order of accuracy* of the discretisation scheme. (The scheme in the above example is first-order in time and second-order in space as the leading errors in (4.56) scale with Δt and Δy^2, respectively.) If the scheme is stable, then the order of the discretisation errors also dictates the *rate of convergence* of the numerical solution towards the target PDE solution. While consistency concerns with the *form of the governing equations*, convergence focuses on the *solution* itself.

[16]Here we assume that the solution is sufficiently smooth.

[17]This kind of approximation for the time derivative makes the algorithm effectively a three time level scheme, the so-called *leapfrog scheme*. Later in this section, we will see that the LBE employs a similar discrete form in its time evolution approximation.

4.5.2 Accuracy Measure

In simple situations, such as in Sect. 4.5.1, it is possible to evaluate the accuracy of a numerical scheme theoretically. However, this is often hardly feasible as simulations involve non-linear terms and/or are affected by other error sources (as discussed later in this section). Consequently, we need a more pragmatic approach rather than a theoretical measure.

Knowing the accuracy of a numerical simulation is important. It allows us, for example, to understand the sensitivity of our solution to the model parameters. We can also assess the quality of our code implementation, e.g., identify code mistakes (bugs) or other inconsistencies in the numerical algorithm or even in the model itself. Generally, code validation and verification are very important topics in CFD [77, 78].

A simple procedure to quantify the error of a numerical simulation consists in comparing it with a known analytical solution.[18] There are different approaches to do this [77–79]. In most cases in this book, we will use the so-called L_2 *error norm*. Given an analytically known quantity $q_{\text{a}}(\boldsymbol{x}, t)$, which is generally a function of space and time, and its numerical equivalent $q_{\text{n}}(\boldsymbol{x}, t)$, we define the L_2 error norm:

$$\epsilon_q(t) := \sqrt{\frac{\sum_{\boldsymbol{x}} \left(q_{\text{n}}(\boldsymbol{x}, t) - q_{\text{a}}(\boldsymbol{x}, t)\right)^2}{\sum_{\boldsymbol{x}} q_{\text{a}}^2(\boldsymbol{x}, t)}}. \tag{4.57}$$

The sum runs over the entire spatial domain where q is defined. An advantage of this definition is that local errors cannot cancel each other; the L_2 error is sensitive to any deviation from q_{a}.

The L_2 error can also be used as a criterion for convergence to steady flows. To do so, q_{a} and q_{n} are replaced by the numerical values of quantity q at a previous and the current time (or iteration) step, respectively. For example, we may define convergence by claiming that the L_2 deviation between the velocity field at subsequent time steps is below a threshold ϵ. For double precision arithmetic, typical chosen values are around $\epsilon = 10^{-7}$. However, we will revise this criterion in Sect. 4.5.3 when discussing the *iterative error*.

Some further remarks are necessary:

1. If $\boldsymbol{q}$ is a vector-valued quantity, e.g. the velocity, we replace $q_{\text{a}} \rightarrow \boldsymbol{q}_{\text{a}}$ and $q_{\text{n}} \rightarrow \boldsymbol{q}_{\text{n}}$, with e.g. $\boldsymbol{q}_{\text{a}}^2 = \boldsymbol{q}_{\text{a}} \cdot \boldsymbol{q}_{\text{a}}$.
2. If q_{a} is zero everywhere, the denominator in (4.57) vanishes and the error is not defined. In this case we need another suitable normalisation, such as a (non-zero) characteristic value of q.

---

[18] In case an analytical solution is not available, we can use the outcome of a finer mesh simulation and take it as reference solution.

Exercise 4.13 Show that the L_2 error ϵ_q is always non-negative, $\epsilon_q \geq 0$, and that equality only holds if $q_n(\boldsymbol{x}) = q_a(\boldsymbol{x})$ everywhere.

4.5.3 Numerical Errors

Approximating any system of *continuous* equations (with suitable boundary and initial conditions) by a system of *discrete* algebraic equations, we inevitably introduce errors. In the CFD community it is commonly accepted that there are three error sources: the round-off error, the iterative error and the discretisation error [77, 78]. We will take a look at all of them in the context of the LBM.

4.5.3.1 Round-Off Error

The *round-off error* results from computers having finite precision. While inherent to any digital computation, this error source can be mitigated by using more significant digits, e.g. by switching from single precision (with 6 to 9 decimal digits) to double precision (with 15 to 17 decimal digits). The importance of round-off error tends to increase with grid refinement and/or time step decrease.

The round-off error in the LBM grows as the ratio $u/c_s \propto \mathrm{Ma}$ becomes small, as explained in [73]. Of course we would like to keep Ma as low as possible to reduce compressibility errors, cf. Sect. 4.5.4. Hence, we are facing a conflict since loss of precision (due to the round-off error) and solution accuracy (due to compressibility errors) are difficult to control independently. Fortunately, in most practical applications with double precision, this problem turns out to be insignificant. The situation can be different for single precision computations, though.

Strategies to alleviate the round-off error in LB simulations have been considered. An early idea [73] consists of reformulating the LBE using alternative variables which, although mathematically equivalent to the original problem, reduce the round-off error. An alternative procedure to reduce round-off errors consists of replacing the standard equilibrium by the incompressible one [80]. This strategy, however, changes the mathematical form of the problem [40], replacing the weakly-compressible NSE by the incompressible NSE (represented in artificial compressibility form), cf. Sect. 4.3.2.

4.5.3.2 Iterative (Steady-State) Error

The iterative error results from the incomplete iterative convergence of the discrete equation solver. In explicit time-marching solvers, this error determines the accuracy up to which *steady-state* solutions are reproduced.

As we know, the LBM reaches steady state solutions through an explicit time-marching procedure. Hence, we can assume that such a (converged) solution may actually differ from the true steady state. This difference we call the *iterative error*. Even in cases where the LBM adopts other iterative procedures, such as time-implicit matrix formulations [81–83], we will still face a similar iterative error. In fact, the main benefit of these time-implicit techniques is an accelerated convergence, i.e. a shorter simulation time. However, these methods have never received broad acceptance, in large part due to their cumbersome algorithmic complexity. Possibly the best compromise to accelerate steady-state convergence while maintaining the algorithmic simplicity is the preconditioned formulation of the LBE [84–86].

One way to infer the impact of the iterative error is to quantify how much our solution changes between time steps (iterations), based on a suitable convergence criteria. It is common to evaluate the (absolute or relative) difference between successive iterations. Yet, this approach may lead to misleading conclusions, particularly when solutions display a slow convergence rate [87].

A more reliable alternative is based on evaluating the solution residual at each time step [87, 88]. The underlying idea is to measure the remainder (residual) after plugging the current numerical solution into the discretised equation we are solving. The residual tells us how far we are from the steady-state convergence. If the problem is well-posed, we should expect the residual to approach zero (up to the round-off error) as iterations converge.

In the LB community, the first strategy, based on the difference between successive iterations, remains the most popular approach. Further studies are required to understand the performance of both strategies to measure the steady-state convergence of LB simulations.

4.5.3.3 Discretisation Error

Recalling Sect. 4.5.1, the discretisation error results from approximating the continuous PDEs by a system of algebraic equations. This is the major error source separating our numerical solution from the exact solution of the underlying PDE. The discretisation error decreases with refining the grid and the time step.

For linear problems, the discretisation error can be directly related to the (spatial and temporal) truncation errors, as we did in Sect. 4.5.1. However, in more complex non-linear problems or indirect procedures to solve the hydrodynamic problem, such as LBM, the relationship between truncation and discretisation errors is not so evident. For this reason, the discretisation error is frequently estimated (cf. Sect. 4.5.2) rather than computed analytically. Providing certain conditions are fulfilled (as explained below), the rate of convergence estimates the discretisation error.

Both spatial and temporal truncation terms affect the discretisation error. Let us focus on the spatial discretisation error and assume it to be the dominating numerical error in ϵ_ϕ. Here, ϕ is the desired observable (which for this purpose is an arbitrary scalar field). Also for simplicity, instead of using the L_2 error norm, let us quantify ϵ_ϕ by simply taking the difference between the numerical estimate ϕ_i (where subscript i refers to the grid refinement level) and the exact analytical solution ϕ_0:

$$\epsilon_\phi = \phi_i - \phi_0 = \alpha \Delta x_i^p + O(\Delta x_i^{p+1}). \tag{4.58}$$

Δx_i is the spatial resolution for the chosen refinement level i, α is a constant to be determined, and p is the observed order of grid convergence. The key point here is the assumption that ϵ_ϕ follows a power-law relation for sufficiently fine meshes. In addition, $O(\Delta x_i^{p+1})$ stands for the higher-order error terms which we consider negligible here.

By determining the numerical solution ϕ_i for three or more mesh sizes, say $\Delta x_1 = \Delta x$, $\Delta x_2 = r\Delta x$ and $\Delta x_3 = r^2\Delta x$, we can estimate the convergence order p[19]:

$$p = \frac{\log\left(\frac{\phi_3 - \phi_2}{\phi_2 - \phi_1}\right)}{\log(r)}. \tag{4.59}$$

For solutions lying in the asymptotic range of convergence, i.e. for sufficiently low Δx and Δt so that the lowest-order terms in the truncation error dominate, the **order of accuracy** can be measured by its order of convergence according to (4.59).

The LBE is an $O(\epsilon^2)$ approximation of the NSE as shown in Sect. 4.1, where the second-order terms in the truncation error can be absorbed in the definition of the fluid viscosity. This leaves only third-order terms as the leading truncation error so that the method is effectively *second-order* accurate with respect to the NSE. As we have seen in Sect. 3.5, LBM is also second-order accurate in time. A way to identify this accuracy level is to measure the numerical convergence of LBM. Second-order convergence in space means that the error decreases quadratically with Δx when fixing the dimensionless ratio $\nu\Delta t/\Delta x^2$. Second-order convergence in time leads to a quadratical decrease of the error with Δt when fixing the spatial resolution Δx [73].[20]

While these convergence measures are valid estimators for the accuracy of LB solutions, we emphasise that the LB discretisation error displays a more complex structure. The reason is that, unlike in standard CFD procedures, the LBE is *not*

[19] Alternatively, if we know the exact solution ϕ_0, we need ϕ_i for only two mesh sizes. Other convergence estimators are reviewed in [77–79].

[20] Here we assume that other error sources are negligibly small. As we will discuss in Sect. 4.5.4, LBM may no longer support second-order time accuracy if the compressibility error is comparable to the discretisation error.

a direct discretisation of the NSE. In addition to the conventional dependence of truncation errors on Δx and Δt, truncation errors also depend on the LB relaxation parameter(s) [74, 75]. The specific relationship between the LB truncation errors and the relaxation parameters has been deduced for BGK [31] and for more advanced collision operators, such as TRT and MRT (cf. Chap. 10) [16, 76, 87, 89].

An interesting property of the LB discretisation is that, for **steady-state** solutions, the **spatial truncation errors** are solely determined by terms proportional to $(\tau - \Delta t/2)^2$ for the BGK collision operator (see Chap. 10 for advanced collision operators and their accuracy).

We can take advantage of this known functional form. In particular, we can select certain relaxation parameter values to tune the accuracy or stability (cf. Sect. 4.4). This is a distinctive feature of LBM and not available for standard CFD methods. Here we list possible improvements:

- The third-order spatial truncation error is proportional to $[(\tau/\Delta t - \frac{1}{2})^2 - \frac{1}{12}]$. According to the Chapman-Enskog analysis, this error appears at $O(\epsilon^3)$. It is the leading-order truncation of the Euler system of equations [31, 76, 87, 89]. The relaxation choice that cancels this error is $\tau = (1/\sqrt{12} + 1/2)\Delta t \approx 0.789\Delta t$ for BGK and is called the *optimal advection condition*.
- The fourth-order spatial truncation error is proportional to $[(\tau/\Delta t - \frac{1}{2})^2 - \frac{1}{6}]$. This truncation error appears at $O(\epsilon^4)$. It is the leading-order truncation of the viscous diffusion terms appearing in the NSE [31, 76, 87]. The relaxation choice that cancels this error is $\tau = (1/\sqrt{6} + 1/2)\Delta t \approx 0.908\Delta t$ for BGK and is called the *optimal diffusion condition*.
- The effect of the non-equilibrium f_i^{neq} on the evolution of f_i can be demonstrated [72, 76] to be proportional to $[(\tau/\Delta t - \frac{1}{2})^2 - \frac{1}{4}]$. Thus, $\tau = \Delta t$ removes this error. This choice makes the LBE equivalent to a central finite difference scheme [90]. According to the linear von Neumann analysis, this choice enables the *optimal stability condition* [63, 67] (cf. Sect. 4.4).

This list of discretisation coefficients is far from complete; more special cases can be found [87, 91]. Still, for most practical purposes, the items above are reasonable guidelines. The true impact of truncation corrections usually comes from several sources, possibly coupled in a complex way. As such, the choice of ideal relaxation value(s) does not have an immediate answer in general; it rather requires an educated guess.

The dependence of discretisation errors on the relaxation parameter(s) can have drawbacks. The **BGK** model can lead to an important **violation of physical behaviour** that has no parallel in standard CFD methods. The reason is that the relaxation time τ simultaneously controls the fluid viscosity and the discretisation errors. Solutions obtained with the BGK model generally exhibit τ-dependent and therefore **viscosity-dependent characteristics**. This contradicts to the fundamental physical requirement that hydrodynamic solutions are uniquely determined by their non-dimensional physical parameters.

We can overcome this disadvantage by choosing advanced collision operators, such as TRT or MRT (cf. Chap. 10). They allow us, at least for steady-state solutions, to keep the truncation errors independent of the fluid viscosity.

4.5.4 Modelling Errors

Sometimes a numerical scheme may not exactly reproduce the physics of interest, but rather approximate it in some sense. Still, such a choice can be totally justifiable, e.g. to keep the numerical procedure simple and/or efficient. In the end, we may have to deal with an approximation problem, i.e. a *modelling error*. The physical content of the LBE is essentially governed by the chosen lattice and the chosen form of the equilibrium distributions. These two features are the main sources of the LB modelling errors.

The velocity space discretisation is directly related to the chosen lattice (cf. Sect. 3.4). It defines which conservation laws (*via* the velocity moments) can be captured up to which accuracy level. It turns out that, for standard lattices (i.e. lattices whose isotropy is respected only up to the fifth order, cf. Sect. 3.4.7), the LBE cannot accurately capture energy transport [24, 92]. The LBE with standard lattices is an isothermal model (or, more rigourously, an athermal model [13, 93]).

Also the mass and momentum balance equations show limitations when simulated with standard lattices. As described in Sect. 4.2.1, these lattices are not sufficiently isotropic to accurately describe the third-order velocity moment. This leads to an $O(u^3)$ error term in the viscous stress tensor [7], (4.15). The LBE, therefore, does not exactly solve the desired NSE. While this cubic defect is negligible for slow flows, it may become important at high Mach numbers. The $O(u^3)$ error also affects the LBM by corrupting the Galilean invariance of the macroscopic equations [7].

Standard lattices introduce an $O(u^3)$ **error term** that **limits the LBM** to simulations of the **isothermal NSE in the weakly compressible regime.**

Another modelling error comes from the LB equilibrium. The form of the equilibrium dictates the physics simulated by the LBE (cf. Sect. 4.3). In fluid flow simulations, this equilibrium is usually given by (3.4) which reproduces the compressible NSE. But due to the $O(u^3)$ error, only the weakly compressible regime (small Ma) is accessible. Therefore, LBM is typically adopted as an explicit compressible scheme for the incompressible NSE [40, 80, 94].[21] In this case the equation reproduced by the LBM differs from the true incompressible NSE by the so-called *compressibility errors*. These are associated with gradients of the divergence of the density and velocity fields, and they typically scale with $O(\mathrm{Ma}^2)$ [73, 74].

A distinctive property of the compressibility error is that it is grid-independent. This means that, if we keep Ma constant while refining the grid, there will be a point where compressibility effects dominate the error and limit further convergence, even with arbitrarily fine meshes. In case that compressibility and discretisation errors are of the same order, convergence towards the *incompressible* NSE requires us to simultaneously decrease the Mach number ($\propto u/c_s$) and decrease the lattice constant ($\propto \Delta x/\ell$) where u and ℓ are macroscopic velocity and length scales, respectively.

Decreasing u/c_s and $\Delta x/\ell$ *with the same rate*, i.e. reducing compressibility and discretisation errors simultaneously, means that the time step has to scale like $\Delta t \propto \Delta x^2$. This means that decreasing the mesh spacing by a factor $r > 1$ requires a time step refined by a factor r^2 and leads to a reduction of the Mach number by a factor r. This has a number of consequences. First, increasing the grid resolution by a factor r leads to a total increase of computational requirements by a factor of $r^2 \cdot r^2 = r^4$ in 2D and $r^3 \cdot r^2 = r^5$ in 3D! The reason is that, in order to simulate the same physical time interval, the number of required time steps increases by a factor r^2. Secondly, the relation $\Delta t \propto \Delta x^2$ means that LBM becomes effectively only first-order accurate in time [20, 21, 31, 73, 74].

Exercise 4.14 Show that the condition $u/c_s \propto \Delta x/\ell$ leads to $\Delta t \propto \Delta x^2$.

In an attempt to get rid of the compressibility error altogether, it was suggested [39, 40] to replace the standard equilibrium in (3.4) by an incompressible variant, (4.42). This way, the incompressible NSE is approximated by an artificial compressibility system [40, 94, 95]. While for steady flows this system indeed fully removes the compressibility error, this "trick" is not completely effective in time-dependent flows [94, 96–98] (cf. Sect. 4.3.2).

[21] There are exceptions, though, e.g. the simulation of sound waves in Chap. 12.

In general, LB simulations for the incompressible NSE suffer from a $O(\mathrm{Ma}^2)$ compressibility error. When keeping this compressibility error equal to the discretisation error, this makes the **LB scheme effectively first-order accurate in time**.

4.5.5 *Lattice Boltzmann Accuracy*

It is possible to formulate the LBE in terms of finite difference stencils, called *recurrence equations* [72, 76]. Without going into details, it can be shown [76] that, in recurrence form, the momentum law reproduced by the LBE satisfies

$$\underbrace{\mathcal{T}(\boldsymbol{x},t)}_{\text{unsteady}} + \underbrace{\mathcal{A}(\boldsymbol{x},t)}_{\text{advection}} = \underbrace{-\mathcal{P}(\boldsymbol{x},t)}_{\text{pressure}} + \underbrace{\mathcal{D}(\boldsymbol{x},t)}_{\text{diffusion}} + \underbrace{O(\Delta x^2, p_s[\tau])}_{\text{spatial truncation}} + \underbrace{O(\Delta t^2, p_t[\tau])}_{\text{temporal truncation}}. \tag{4.60}$$

This generic form is nothing but a compact representation of (4.16b) at discrete level. Each term in (4.60) uses finite difference stencils rather than differential operators [76]. Let us analyse them individually.

1. The $\mathcal{T}$ operator is a three time level difference taken at time steps $(t - \Delta t, t, t + \Delta t)$. Its exact form depends on τ.[22]
2. The $\mathcal{A}$ and $\mathcal{P}$ operators are centred difference schemes for the advection and pressure terms, respectively.
3. The $\mathcal{D}$ operator denotes the Du Fort-Frankel approximation for the diffusion term [76, 99]. As known from [79], the consistency of this scheme requires $\Delta t \sim \Delta x^2$, a relation usually called *diffusive scaling* [20, 100].
4. The remaining two terms embody the truncation contributions coming from spatial and temporal terms, with an additional τ-dependent component in polynomial form. We denote the spatial and temporal contributions as $p_s[\tau]$ and $p_t[\tau]$, respectively. While the spatially varying polynomials $p_s[\tau]$ contain τ in form of $(\tau - \Delta t/2)^2$, the time dependent ones $p_t[\tau]$ feature a more complex τ-structure [16, 31, 76].[23]

[22] For those familiar with finite difference techniques, such an operator can be interpreted as a leapfrog scheme for the temporal derivative, yet not exactly identical due to the τ degree of freedom [76].

[23] This analysis is valid for the BGK model. For the TRT/MRT collision operators, the term $(\tau - \Delta t/2)^2$ should be interpreted as Λ for spatial truncations (cf. Chap. 10), while it is a combination of Λ and $(\Lambda^{\pm})^2$ terms for temporal truncations [76].

The **LBM is a numerical scheme approximating the NSE with second-order accuracy in space and time** (when τ is fixed). Yet, the continuum equations described by the LBE may not necessarily lead to the NSE. The difference is due to the modelling errors. The approximation is only successful at low velocities where $O(u^3)$ becomes negligible. Another modelling error is the compressibility error $O(\mathrm{Ma}^2)$ that further affects the LBE in case one is interested in the incompressible NSE. If this compressibility error and the discretisation error are simultaneously reduced, then the **time accuracy of the LBM reduces to first order**.

So far we have only addressed the behaviour of the LBE in the bulk. We have to keep in mind that the accuracy may be affected by other sources, such as boundary or initial conditions. We will address both topics in Chap. 5. If initial and boundary conditions are properly included, the LBM is a competitive Navier-Stokes solver. In fact, the LBM can even show better accuracy and stability characteristics than standard second-order Navier-Stokes solvers at identical grid resolutions. This supremacy mainly comes from the control over truncation errors in a grid-independent fashion, through the tuning of relaxation rates [4, 91, 101, 102].

4.5.6 Accuracy Guidelines

In a typical simulation scenario we want to optimise the accuracy while respecting the non-dimensional groups of the problem. The matching between the non-dimensional numbers of the physical problem, say the Reynolds number in hydrodynamics or the Péclet number in transport problems, and those in the LB scheme is the focus of Chap. 7. Here, our motivation is to discuss how to select values for the velocity u, the relaxation time τ through the viscosity, and the grid number N (cf. (4.53)) in order to reach the highest accuracy possible.

Increasing the grid number N obviously enhances accuracy. However, it also significantly impacts the run time and memory requirements of simulations, and it should therefore be kept as low as possible. Remember that a system size N means a memory overhead $\propto N^3$ in 3D. Hence, a judicious choice for the accuracy improvement should focus on the other two parameters, u and τ, while keeping N fixed. To match the Reynolds number in (4.53), we must adjust u and τ accordingly as they are not independent.

The velocity u does not strongly affect the accuracy, proving it is sufficiently small (cf. Sect. 4.5.4). Hence, the most important parameter for controlling the accuracy is the relaxation time τ.

It can be shown that, for steady-state problems [72], we can cancel the $O(\epsilon^3)$ or the $O(\epsilon^4)$ truncation errors with the following choice of τ:

$$\begin{aligned} O(\epsilon^3):\quad & \frac{\tau}{\Delta t} = \frac{1}{2} + \frac{1}{\sqrt{12}} \approx 0.789, \\ O(\epsilon^4):\quad & \frac{\tau}{\Delta t} = \frac{1}{2} + \frac{1}{\sqrt{6}} \approx 0.908. \end{aligned} \tag{4.61}$$

The case $\tau/\Delta t \approx 0.789$ is suitable for advection-dominated problems, while $\tau/\Delta t \approx 0.908$ is advantageous for diffusion-dominated problems. Other problems require the cancellation of other truncation errors. A more complete list of relevant values of τ can be found in [87, 91].

It turns out that for accurate and consistent LB simulations it is not recommendable to use the BGK collision operator. Results from BGK simulations are viscosity-dependent, which means they are *not* uniquely determined by the non-dimensional groups representing the physical problem. While for high grid refinement levels this defect may be kept within acceptable bounds (based on some subjective criteria), at coarser grids the effect of truncation terms is generally significant. A quick fix consists of making an educated guess for the value of τ and stick with it at all points of the domain. However, with this procedure we are not taking full advantage of the distinctive feature of LBM: the possibility to improve the accuracy of numerical solutions in a grid-independent fashion by tuning the relaxation parameters. Furthermore, in this way we are also compromising the computational efficiency of our simulation. The main reason is that a small value for τ is generally required to optimise accuracy, while a rapid steady-state convergence implies a large value of τ [72] (cf. Chap. 7).

The TRT/MRT collision operators (cf. Chap. 10) solve these problems. They enable the local prescription of the relaxation parameters without any change in viscosity. Taking the TRT collision operator as an example, its steady-state (or spatial truncation) errors depend on the combination of two relaxation times through the parameter $\Lambda = (\tau^+/\Delta t - 1/2)(\tau^-/\Delta t - 1/2)$. The relaxation parameter τ^+ is related to the viscosity in the NSE. The parameter τ^- is free. If one changes the viscosity (i.e. τ^+), then one should change τ^- accordingly to keep Λ constant. This is a very powerful tool as it allows performing accurate simulations even with $\tau^+/\Delta t \gg 1$; this would not be possible with BGK. In the TRT framework, (4.61) becomes

$$\begin{aligned} O(\epsilon^3):\quad & \Lambda = \frac{1}{12} \approx 0.083, \\ O(\epsilon^4):\quad & \Lambda = \frac{1}{6} \approx 0.166. \end{aligned} \tag{4.62}$$

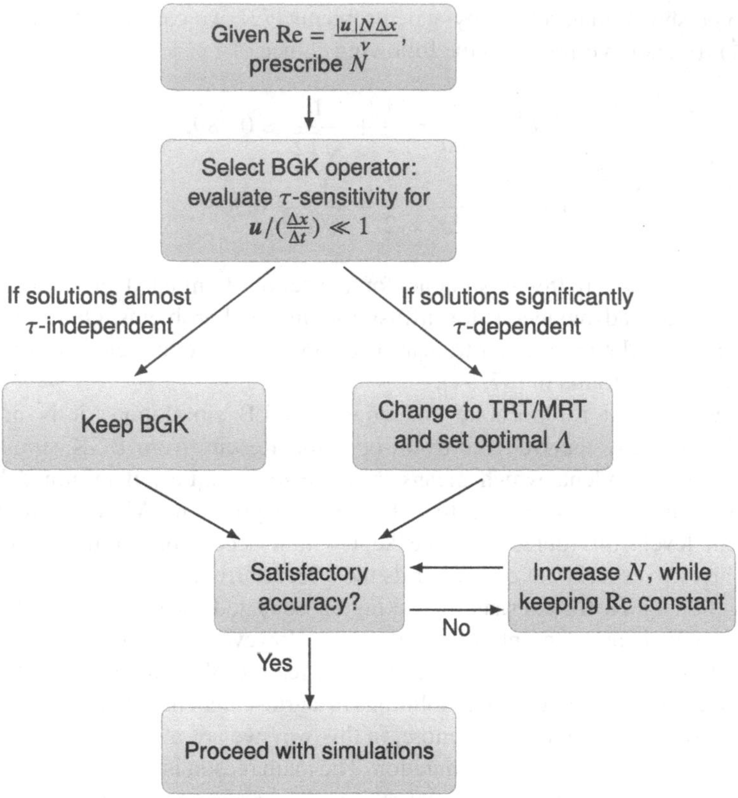

Fig. 4.7 Workflow for improving the accuracy of simulations that require a constant Reynolds number

We summarise the **workflow to maximise the simulation accuracy** for a given problem, also illustrated in Fig. 4.7.

Our first action should be to determine the relevant non-dimensional numbers, such as the Reynolds number. Then, based on some educated guess, we should select the grid number N. Its value should represent a compromise between the best spatial resolution possible and the available computational resources. Also, the choice of N should give us the possibility to vary τ while keeping u low to satisfy the low Ma requirement. The final parameter we should deal with is τ.

We should check whether solutions are too τ-sensitive. In case they significantly depend on τ, we may have to replace the BGK collision operator by the TRT or MRT operators. The next step is to find a suitable relaxation

(continued)

parameter, either τ or Λ, to optimise accuracy. For this task we can use our knowledge of the physical problem, e.g. if it is dominated by bulk phenomena, boundaries, advection or diffusion. Based on this, we can choose a proper value for τ or Λ from (4.61) or (4.62), respectively. A more meticulous adjustment on a trial and error basis is possible. In the end, if the attained accuracy is still unsatisfactory, we should take a larger grid number N and restart the process.

4.6 Summary

In this chapter we found the macroscopic conservation equations that the LBE reproduces when the grid size Δx and time step Δt tend to zero. The mass conservation equation is the continuity equation, while the momentum conservation equation is a weakly compressible NSE that corresponds to the incompressible NSE when Ma^2 goes to zero. (This "weak compressibility" comes from $O(u^3)$ error terms that are insignificant if $\mathrm{Ma}^2 \ll 1$.)

The Chapman-Enskog analysis in Sect. 4.1 establishes the connection between the "mesoscopic" LBE and the macroscopic mass and momentum equations. For other variants of the LBE, such as the alternative equilibrium models covered in Sect. 4.3, this analysis may also be used to determine the macroscopic equations that these variants imply. The Chapman-Enskog analysis is not the only possible approach here, though; we briefly touched on some alternatives in Sect. 4.2.5.

The picture of the LBE as a Navier-Stokes solver is complicated by the fact that the LBE is not a direct discretisation of the NSE, unlike many other fluid solvers. Instead, the LBE is a discretisation of the continuous Boltzmann equation from which the NSE can be restored.

The kinetic origin of the LBE affects its *consistency* as a Navier-Stokes solver. We saw in Sect. 4.1 that the NSE follows from the Boltzmann equation for *low* Knudsen numbers, which we can see as a condition on the changes in f_i being sufficiently small in time and space. Additionally, when velocity space is discretised, velocity-related $O(u^3)$ errors are introduced in the NSE due to the truncation of the equilibrium distribution. We discussed the necessity of this truncation in Sect. 4.2.1.

The *stability* of the LBE is not entirely simple, either. Stability analyses are often performed with the help of a linear von Neumann analysis, but applying this to the LBE is complicated by the fact that the LBE is a coupled *system* of equations; one equation for each velocity $\boldsymbol{c}_i$. Moreover, the equilibrium functions depend non-linearly on the populations f_i, which is an additional complication for stability analysis. However, we can still provide some stability *guidelines* for various relaxation parameters (cf. Sect. 4.4).

The *accuracy* study of the LBE is a difficult task to perform theoretically. Its analysis is only feasible in simple academic cases (cf. Sect. 4.5.1). In most practical situations, it is more convenient to estimate the accuracy through measurements (cf. Sect. 4.5.2). Still, some theoretical concepts, such as the range of errors which affect the LBM and their impact on the accuracy, should be understood in order to make good use of the LBE.

We saw in Sect. 4.5.3 that the *numerical errors* affecting the LBM come from three sources: the round-off error, the iterative error and the discretisation error. Typically, the first two have very limited impact compared to the discretisation error. The form of this error is controlled by the truncation terms, stemming from the discretisation of the continuous Boltzmann equation. In the LBM, the leading order spatial and temporal truncation terms decrease with Δx^2 and Δt^2. Additionally, these terms also depend on the relaxation parameter, e.g. τ in BGK. This extra degree of freedom allows controlling the LB discretisation error in a grid-independent fashion. If used judiciously, this can make LBM superior to standard second-order NSE solvers.

When used as Navier-Stokes solver, the LBM contains some *modelling errors* (cf. Sect. 4.5.4). The most common examples are the compressibility error and the cubic error. The compressibility error is what differentiates the LB solution from the true *incompressible* Navier-Stokes solution. However, as the compressibility error scales with $O(\mathrm{Ma}^2)$, it is negligible for sufficiently small Ma. The slow flow assumption in turn minimises the $O(u^3)$ error.

Considering its advantages and disadvantages, we can say that the LBM is a competitive second-order accurate Navier-Stokes solver due to its distinctive characteristics.

References

1. S. Chapman, T.G. Cowling, *The Mathematical Theory of Non-uniform Gases*, 2nd edn. (Cambridge University Press, Cambridge, 1952)
2. I. Ginzburg, F. Verhaeghe, D. d'Humières, Commun. Comput. Phys. **3**, 427 (2008)
3. Y.Q. Zu, S. He, Phys. Rev. E **87**, 043301 (2013)
4. G. Silva, V. Semiao, J. Comput. Phys. **269**, 259 (2014)
5. E.M. Viggen, The lattice Boltzmann method: Fundamentals and acoustics. Ph.D. thesis, Norwegian University of Science and Technology (NTNU), Trondheim (2014)
6. C.M. Bender, S.A. Orszag, *Advanced mathematical methods for scientists and engineers* (McGraw-Hill, New York, 1978)
7. Y.H. Qian, S.A. Orszag, Europhys. Lett. **21**(3), 255 (1993)
8. S. Geller, M. Krafczyk, J. Tölke, S. Turek, J. Hron, Comput. Fluids **35**(8-9), 888 (2006)
9. J. Latt, B. Chopard, O. Malaspinas, M. Deville, A. Michler, Phys. Rev. E **77**(5), 056703 (2008)
10. P.A. Thompson, *Compressible-Fluid Dynamics* (McGraw-Hill, New York, 1972)
11. L.E. Kinsler, A.R. Frey, A.B. Coppens, J.V. Sanders, *Fundamentals of Acoustics*, 4th edn. (Wiley, New York, 2000)
12. P. Dellar, Phys. Rev. E **64**(3) (2001)
13. P. Lallemand, L.S. Luo, Phys. Rev. E **61**(6), 6546 (2000)
14. N. Prasianakis, I. Karlin, Phys. Rev. E **76**(1) (2007)

15. I. Ginzburg, J. Stat. Phys. **126**, 157 (2007)
16. I. Ginzburg, Phys. Rev. E **77**, 066704 (2008)
17. J. Latt, Hydrodynamic limit of lattice Boltzmann equations. Ph.D. thesis, University of Geneva (2007)
18. Y. Sone, *Kinetic Theory and Fluid Dynamics* (Birkhäuser, Boston, 2002)
19. T. Inamuro, M. Yoshino, F. Ogino, Phys. Fluids **9**, 3535 (1997)
20. M. Junk, A. Klar, L.S. Luo, J. Comput. Phys. **210**, 676 (2005)
21. M. Junk, Z. Yang, J. Stat. Phys. **121**, 3 (2005)
22. H. Grad, Commun. Pure Appl. Maths **2**, 331 (1949)
23. X. Shan, H. X., Phys. Rev. Lett. **80**, 65 (1998)
24. X. Shan, X.F. Yuan, H. Chen, J. Fluid Mech. **550**, 413 (2006)
25. O. Malaspinas, P. Sagaut, J. Fluid Mech. **700**, 514 (2012)
26. E. Ikenberry, C. Truesdell, J. Ration. Mech. Anal. **5**, 1 (1956)
27. P. Asinari, T. Ohwada, Comp. Math. Appl. **58**, 841 (2009)
28. S. Bennett, P. Asinari, P.J. Dellar, Int. J. Num. Meth. Fluids **69**, 171 (2012)
29. W.A. Yong, W. Zhao, L.S. Luo, Phys. Rev. E **93**, 033310 (2016)
30. F. Dubois, Comp. Math. Appl. **55**, 1441 (2008)
31. D.J. Holdych, D.R. Noble, J.G. Georgiadis, R.O. Buckius, J. Comput. Phys. **193**(2), 595 (2004)
32. A.J. Wagner, Phys. Rev. E **74**, 056703 (2006)
33. D. Lycett-Brown, K.H. Luo, Phys. Rev. E **91**, 023305 (2015)
34. A. Caiazzo, M. Junk, M. Rheinländer, Comp. Math. Appl. **58**, 883 (2009)
35. B. Chopard, A. Dupuis, A. Masselot, P. Luthi, Adv. Complex Syst. **05**(02n03), 103 (2002)
36. D.A. Wolf-Gladrow, *Lattice-Gas Cellular Automata and Lattice Boltzmann Models* (Springer, New York, 2005)
37. B. Dünweg, A.J.C. Ladd, in *Advances in Polymer Science* (Springer, Berlin, Heidelberg, 2008), pp. 1–78
38. E.M. Viggen, Phys. Rev. E **87**(2) (2013)
39. Q. Zou, S. Hou, S. Chen, G.D. Doolen, J. Stat. Phys. **81**(1–2), 35 (1995)
40. X. He, L.S. Luo, J. Stat. Phys. **88**(3–4), 927 (1997)
41. X. Shan, H. Chen, Phys. Rev. E **47**(3), 1815 (1993)
42. X. Shan, H. Chen, Phys. Rev. E **49**(4), 2941 (1994)
43. H. Yu, K. Zhao, Phys. Rev. E **61**(4), 3867 (2000)
44. J.M. Buick, J.A. Cosgrove, J. Phys. A **39**(44), 13807 (2006)
45. A. Kupershtokh, D. Medvedev, D. Karpov, Comput. Math. Appl. **58**(5), 965 (2009)
46. E.M. Viggen, Phys. Rev. E **90**, 013310 (2014)
47. S. Bennett, A lattice Boltzmann model for diffusion of binary gas mixtures. Ph.D. thesis, University of Cambridge (2010)
48. F. Alexander, H. Chen, S. Chen, G. Doolen, Phys. Rev. A **46**(4), 1967 (1992)
49. B.J. Palmer, D.R. Rector, J. Comput. Phys. **161**(1), 1 (2000)
50. P.J. Dellar, Phys. Rev. E **65**(3) (2002)
51. R. Salmon, J. Mar. Res. **57**(3), 503 (1999)
52. J.G. Zhou, Comput. Method. Appl. M. **191**(32), 3527 (2002)
53. S. Li, P. Huang, J. Li, Int. J. Numer. Meth. Fl. **77**(8), 441 (2015)
54. R. Mei, L.S. Luo, P. Lallemand, D. d'Humières, Comput. Fluids **35**(8-9), 855 (2006)
55. S. Chen, J. Tölke, M. Krafczyk, Comput. Method. Appl. M. **198**(3-4), 367 (2008)
56. M. Mendoza, B.M. Boghosian, H.J. Herrmann, S. Succi, Phys. Rev. Lett. **105**(1), 014502 (2010)
57. M. Mendoza, J.D. Muñoz, Phys. Rev. E **82**(5), 056708 (2010)
58. J. Chen, Z. Chai, B. Shi, W. Zhang, Comput. Math. Appl. **68**(3), 257 (2014)
59. D.V. Patil, K.N. Premnath, S. Banerjee, J. Comput. Phys. **265**, 172 (2014)
60. K. Li, C. Zhong, Int. J. Numer. Meth. Fl. **77**(6), 334 (2015)
61. D.N. Siebert, L.A. Hegele, P.C. Philippi, Phys. Rev. E **77**(2), 026707 (2008)
62. X.D. Niu, C. Shu, Y.T. Chew, T.G. Wang, J. Stat. Phys. **117**(3–4), 665 (2004)

63. A. Kuzmin, I. Ginzburg, A. Mohamad, Comp. Math. Appl. **61**, 1090 (2011)
64. J. Hoffmann, *Numerical Methods for Engineers and Scientists* (McGraw-Hill, New York, 1992)
65. R.J. LeVeque, *Finite Difference Methods for Ordinary and Partial Differential Equations: Steady State and Time Dependent Problems* (SIAM, Philadelphia, 2007)
66. J.D. Sterling, S. Chen, J. Comput. Phys. **123**(1), 196 (1996)
67. I. Ginzburg, D. d'Humières, A. Kuzmin, J. Stat. Phys. **139**, 1090 (2010)
68. S. Suga, Int. J. Mod. Phys. C **20**, 633 (2009)
69. X. Aokui, Acta Mech. Sinica **18**(6), 603 (2002)
70. R. Brownlee, A. Gorban, J. Levesley, Physica A **387**, 385 (2008)
71. F. Tosi, S. Ubertini, S. Succi, H. Chen, I. Karlin, Math. Comp. Simul. **72**(2–6), 227 (2006)
72. D. d'Humières, I. Ginzburg, Comput. Math. Appl. **58**, 823 (2009)
73. P.A. Skordos, Phys. Rev. E **48**(6), 4823 (1993)
74. M. Reider, J. Sterling, Comput. Fluids **118**, 459 (1995)
75. R.S. Maier, Int. J. Mod. Phys. C **8**, 747 (1997)
76. I. Ginzburg, Commun. Comput. Phys. **11**, 1439 (2012)
77. P.J. Roache, *Verification and Validation in Computational Science and Engineering, 1998*, 1st edn. (Hermosa Publishers, New Mexico, 1998)
78. C.J. Roy, J. Comp. Phys. **205**, 131 (2005)
79. J.H. Ferziger, M. Peric, A. Leonard, *Computational Methods for Fluid Dynamics*, vol. 50, 3rd edn. (Springer, New York, 2002)
80. L.S. Luo, W. Lia, X. Chen, Y. Peng, W. Zhang, Phys. Rev. E **83**, 056710 (2011)
81. R. Verberg, A.J.C. Ladd, Phys. Rev. E **60**, 3366 (1999)
82. M. Bernaschi, S. Succi, H. Chen, J. Stat. Phys. **16**, 135 (2001)
83. J. Tölke, M. Krafczyk, E. Rank, J. Stat. Phys. **107**, 573 (2002)
84. O. Filippova, D. Hänel, J. Comp. Phys. **165**, 407 (2000)
85. Z. Guo, T.S. Zhao, Y. Shi, Phys. Rev. E **70**(6), 066706 (2004)
86. S. Izquierdo, N. Fueyo, J. Comput. Phys. **228**(17), 6479 (2009)
87. S. Khirevich, I. Ginzburg, U. Tallarek, J. Comp. Phys. **281**, 708 (2015)
88. L. Talon, D. Bauer, D. Gland, H. Auradou, I. Ginzburg, Water Resour. Res. **48**, W04526 (2012)
89. B. Servan-Camas, F. Tsai, Adv. Water Res. **31**, 1113 (2008)
90. R.G.M. van der Sman, Comput. Fluids **35**, 849 (2006)
91. I. Ginzburg, G. Silva, L. Talon, Phys. Rev. E **91**, 023307 (2015)
92. P.C. Philippi, L.A. Hegele, L.O.E. Santos, R. Surmas, Phys. Rev. E **73**, 056702 (2006)
93. P. Lallemand, L.S. Luo, Phys. Rev. E **68**, 1 (2003)
94. X. He, G.D. Doolen, T. Clark, J. Comp. Phys. **179**, 439 (2002)
95. A.J. Chorin, J. Comput. Phys **2**, 12 (1967)
96. P.J. Dellar, J. Comp. Phys. **190**, 351 (2003)
97. G. Hazi, C. Jimenez, Comput. Fluids **35**, 280–303 (2006)
98. G. Silva, V. Semiao, J. Fluid Mech. **698**, 282 (2012)
99. M.G. Ancona, J. Comp. Phys. **115**, 107 (1994)
100. S. Ubertini, P. Asinari, S. Succi, Phys. Rev. E **81**(1), 016311 (2010)
101. S. Marié, D. Ricot, P. Sagaut, J. Comput. Phys. **228**(4), 1056 (2009)
102. Y. Peng, W. Liao, L.S. Luo, L.P. Wang, Comput. Fluids **39**, 568 (2010)

Chapter 5
Boundary and Initial Conditions

Abstract After reading this chapter, you will be familiar with the basics of lattice Boltzmann boundary conditions. After also having read Chap. 3, you will be able to implement fluid flow problems with various types of grid-aligned boundaries, representing both no-slip and open surfaces. From the boundary condition theory explained in this chapter together with the theory given in Chap. 4, you will be familiar with the basic theoretical tools used to analyse numerical lattice Boltzmann solutions. Additionally, you will understand how the details of the initial state of a simulation can be important and you will know how to compute a good initial simulation state.

Traditional hydrodynamics theory is governed by a set of partial differential equations (PDEs) [1], describing the conservation of mass, momentum, and in some cases energy. For brevity, we shall refer to all these equations collectively as the Navier–Stokes equations (NSEs) in this chapter.

Recalling calculus, solutions of PDEs cannot be uniquely determined unless proper boundary and initial conditions are specified. Hence, the role of boundary and initial conditions within continuum fluid dynamics can be understood first of all as a mathematical necessity. Their purpose is to single out, from all admissible solutions of the NSEs, thc specific solution of the fluid flow problem being considered.

From a physical standpoint, the importance of boundary and initial conditions can be understood even more intuitively. Given that the general theoretical framework to model fluid flows (from the motion of air in the atmosphere to blood circulation in the human body) relies on the same set of equations, i.e. the NSEs, additional information must be given before a solution is possible. Otherwise, the formulation of the physical problem remains incomplete as the NSEs by themselves have no information in them about the particular flow one is interested in. The required information is contained in the boundary and initial conditions. Only after these conditions have been specified together with the NSEs one can consider the problem to be complete or, as stated in mathematics, well-posed.[1]

[1] Although no mathematical proof exists yet, this book takes for granted that solutions to the NSEs exist, are unique and vary smoothly with changes to initial conditions.

T. Krüger et al., *The Lattice Boltzmann Method*, Graduate Texts in Physics,
DOI 10.1007/978-3-319-44649-3_5

All things considered, it is not an exaggeration to say that the specification of boundary conditions (and also initial conditions for time-dependent problems) is one of the most important tasks in the setup of fluid flow problems. For that reason, this chapter covers this subject in detail in three parts. In the first part (i.e. Sect. 5.1 and 5.2), we review the role of boundary and initial conditions in the frame of continuum theories and standard numerical methods. This will facilitate the introduction of boundary conditions in the LBE, whose main concepts are discussed in the end of this first part. In the second part (i.e. Sect. 5.3), we explain how several LB boundary conditions work and how can they be implemented.[2] In part three (i.e. Sect. 5.4), we provide supplementary and more advanced material. Finally, we cover initial conditions and methods of initialising LB simulations in Sect. 5.5.

5.1 Boundary and Initial Conditions in LBM in a Nutshell

Before proceeding with the detailed exposition of boundary and initial conditions in LBM, let us briefly summarise them. Where do they appear in the LB algorithm? How do we implement them? What are their most relevant characteristics?

Initial and boundary conditions can be included in a typical LB algorithm as shown in Fig. 5.1, and as highlighted in bold in the following:

I. **LB initialisation**

   a. If the objective is a steady-state solution, it is sufficient to set initial populations to equilibrium $f_i(\boldsymbol{x}, t = 0) = f_i^{\text{eq}}\big(\rho(\boldsymbol{x}, t = 0), \boldsymbol{u}(\boldsymbol{x}, t = 0)\big)$. A typical choice for initial macroscopic fields is $\rho(\boldsymbol{x}, t = 0) = 1$ (simulation units) and $\boldsymbol{u}(\boldsymbol{x}, t = 0) = \boldsymbol{0}$.
   b. If the objective is a time-dependent solution with non-homogeneous initial conditions, the consistent initialisation of populations requires considering both equilibrium and non-equilibrium components: $f_i(\boldsymbol{x}, t = 0) = f_i^{\text{eq}}(\boldsymbol{x}, t = 0) + f_i^{\text{neq}}(\boldsymbol{x}, t = 0)$. For more details, see Sect. 5.5.

II. Main LB algorithm

   1. Compute the fluid density and velocity *via* (3.59).
   2. Compute the equilibrium populations $f_i^{\text{eq}}(\rho, \boldsymbol{u})$ from (3.54) to use in the BGK collision operator in (3.78).[3]
   3. If desired, output the macroscopic fields to the hard disk.
   4. Apply collision to find post-collision populations $f_i^{\star}$ according to (3.83).
   5. Propagate populations following (3.84).

[2]Chapter 11 extends this subject to the case of boundary conditions in complex geometries. This topic is further addressed in Sect. 12.4, where we discuss a special kind of boundary conditions designed to let sound waves exit the system smoothly.

[3]The extension to more advanced collision operators is discussed in Chap. 10.

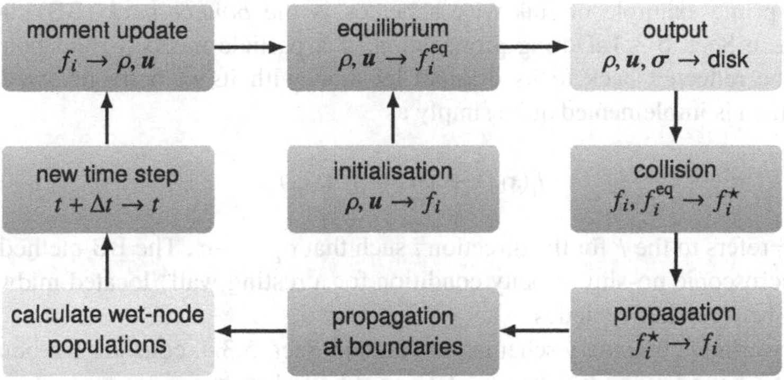

Fig. 5.1 An overview of one cycle of the LB algorithm, considering boundary conditions and the one-off computation of initial conditions (*centre*), but not considering forces. Optional sub-steps are shown in *light grey boxes*

6. **Propagate populations at boundaries according to the local boundary conditions.**
7. **If wet-node boundary conditions are used, compute their populations as necessary.**
8. Increment the time step and go back to step 1.

5.1.1 Boundary Conditions

The importance of boundary conditions should not be underestimated. Even though they apply to a small portion of the fluid domain, their influence may be felt everywhere in the flow solution (cf. the example in Sect. 5.2.3). Therefore, boundary conditions should be treated with great care. In this chapter we focus on *straight boundaries aligned with the lattice nodes*. More complex geometries are covered in Chap. 11.

In the LBE, the boundary conditions apply at boundary nodes $\boldsymbol{x}_\text{b}$ which are sites with at least one link to a solid and a fluid node. The formulation of LB boundary conditions is typically a non-trivial task. Rather than specifying the macroscopic variables of interest, such as ρ and $\boldsymbol{u}$, LB boundary conditions apply to the mesoscopic populations f_i, giving more degrees of freedom than the set of macroscopic variables. This gives rise to a non-uniqueness problem, evidenced by the "zoo" of LB boundary schemes available. The ensemble of LB boundary schemes can be divided into two big families: *link-wise* and *wet-node*. A distinguishing feature of LB boundary schemes is that, contrary to conventional numerical methods, their order of accuracy and exactness does not match. For example, boundary schemes of second-order accuracy do not accommodate exactly a parabolic solution. This source of confusion is explained in detail in Sect. 5.2.4.

The prime example of link-wise schemes is the *bounce-back* (BB) method, covered in Sect. 5.3.3. During propagation, if a particle meets a rigid boundary, it will be reflected back to its original location with its velocity reversed. This mechanism is implemented quite simply as

$$f_{\bar{i}}(\boldsymbol{x}_{\mathrm{b}}, t + \Delta t) = f_i^{\star}(\boldsymbol{x}_{\mathrm{b}}, t), \tag{5.1}$$

where $f_{\bar{i}}$ refers to the f for the direction $\bar{i}$ such that $c_{\bar{i}} = -c_i$. The BB method leads to a macroscopic no-slip velocity condition for a resting wall[4] located midway on the link between lattice nodes.

The wet-node boundary schemes, covered in Sect. 5.3.4, consider the boundary to lie *on* a boundary node. The populations at boundary nodes are reconstructed in an explicit way, i.e. using the macroscopic information from the actual boundary condition. The main challenge is that there are typically more unknown boundary populations than macroscopic conditions, which explains the large number of wet-node techniques developed to deal with this under-specified problem. Three examples of wet-node techniques covered in this book are:

- The *equilibrium scheme* (ES), which sets the boundary populations to an equilibrium prescribed by the desired density and velocity on the boundary, i.e. $f_i = f_i^{\mathrm{eq}}(\rho, \boldsymbol{u})$. It can be shown that this gives insufficient accuracy unless $\tau/\Delta t = 1$.
- The *non-equilibrium extrapolation method* (NEEM), which amends the equilibrium scheme by adding non-equilibrium information on the boundary gradients using values from nearby fluid nodes.
- The *non-equilibrium bounce-back method* (NEBB), which is an improvement of the first two. It directly operates on the non-equilibrium populations of the boundary node and leads to superior accuracy.

In Table 5.1 we summarise the main characteristics of the bounce-back and the above mentioned wet-node techniques.

The application of LB boundary conditions is not limited to the modelling of solid walls. They are also used for inflow/outflow conditions. This can be either through simple periodic flow conditions (cf. Sect. 5.3) or by more elaborate conditions where velocity or pressure are specified (cf. Sect. 5.3.5).

The prescription of boundary conditions at corners requires special attention, as explained in Sect. 5.3.6 and further elaborated on in Sect. 5.4.

5.1.2 Initial Conditions

Two popular approaches to prescribe consistent initial conditions for $f_i(\boldsymbol{x}, t = 0)$ are explained in Sect. 5.5: based on (i) an explicit Chapman-Enskog decomposition

[4] Moving boundaries are also possible and are discussed in Sect. 5.3.3.

Table 5.1 Comparison of link-wise and wet-node schemes for straight boundaries (abbreviations defined in main text). BB improves to 3rd-order accuracy and parabolic exactness if and only if $\tau/\Delta t = \sqrt{3/16} + 1/2 \approx 0.933$ for BGK. ES and NEEM improve to 3rd-order accuracy and parabolic exactness if and only if $\tau/\Delta t = 1$ for BGK

	Link-wise	Wet-node		
	BB	ES	NEEM	NEBB
Boundary location	Midway	On node	On node	On node
Accuracy	2nd-order	1st-order	2nd-order	3rd-order
Exactness	Linear	Constant	Linear	Parabolic
Stability	High	High	Moderate	Low
Mass conservation	Exact	Non-exact	Non-exact	Non-exact
Algorithm simplicity and extension to corners and 3D	Simple	Simple	Moderate	Complex

of the populations into equilibrium and non-equilibrium [2] and (ii) a modified LB scheme that is run before the actual simulation [3].

5.2 Fundamentals

The purpose of Sect. 5.2 is to review the basic principles of boundary and initial conditions. The presentation is divided in three parts. First, in Sect. 5.2.1, we discuss the role of boundary and initial conditions in the context of continuum fluid dynamics problems. Second, in Sect. 5.2.2 and Sect. 5.2.3, we extend this topic to the case of discrete numerical methods, taking as an example the standard finite difference technique. Third, in Sect. 5.2.4, we examine the notion of boundary conditions in the LBM. The first two parts may be skipped for users already familiar with fluid dynamics theory and conventional numerical methods. The third part, Sect. 5.2.4, is the key to understand the subsequent parts and we recommend its careful study before proceeding to the rest of this chapter.

5.2.1 Concepts in Continuum Fluid Dynamics

Whenever a flow problem is time-dependent, the specification of initial conditions in the entire domain is required.[5] That is, if $\boldsymbol{u}_0$ specifies the fluid velocity at some given instant t_0 (often conveniently chosen as $t_0 = 0$), then the time-dependent

[5]Exceptional cases of time-dependent problems where initial conditions have an immaterial role are discussed in Sect. 5.5.

solution of the NSE must respect the initial condition

$$\boldsymbol{u}(\boldsymbol{x}, t_0) = \boldsymbol{u}_0(\boldsymbol{x}). \tag{5.2}$$

When a physical process describes a variation of the solution along its spatial coordinates, the prescription of boundary conditions becomes imperative [4]. By definition, boundary conditions specify the behaviour of the PDE solution at the boundaries of the problem domain. If the solution is time-dependent, then the prescription of boundary conditions also needs to be set for all times.

Depending on which constraint they impose on the boundary, boundary conditions can be classified into three categories. All three are contained in the equation below, where φ denotes the solution of a generic PDE, $\boldsymbol{x}_\mathrm{B}$ the boundary location, and $\boldsymbol{n}$ the (outward) boundary normal:

$$b_1 \left.\frac{\partial \varphi}{\partial n}\right|_{(\boldsymbol{x}_\mathrm{B}, t)} + b_2 \varphi(\boldsymbol{x}_\mathrm{B}, t) = b_3. \tag{5.3}$$

- The *Dirichlet* condition is set by $b_1 = 0$ and $b_2 \neq 0$ in (5.3). This condition fixes thc value of φ on the boundary $\boldsymbol{x}_\mathrm{B}$ to b_3/b_2.
- The *Neumann* condition is given by $b_1 \neq 0$ and $b_2 = 0$. It fixes the flux of φ on the boundary $\boldsymbol{x}_\mathrm{B}$ to b_3/b_1.
- Finally, the *Robin* condition corresponds to $b_1 \neq 0$ and $b_2 \neq 0$. This condition entails a relation between the value and the flux of φ on the boundary $\boldsymbol{x}_\mathrm{B}$.

In hydrodynamics, the solution of the NSEs generally requires the prescription of boundary conditions for the fluid velocity and/or the stresses. These are normally Dirichlet and Neumann conditions for the fluid velocity.

The **Dirichlet condition** for the fluid velocity is

$$\boldsymbol{u}(\boldsymbol{x}_\mathrm{B}, t) = \boldsymbol{U}_\mathrm{B}(\boldsymbol{x}_\mathrm{B}, t) \tag{5.4}$$

where $\boldsymbol{U}_\mathrm{B}$ stands for the boundary velocity.

By denoting $\boldsymbol{n}$ and $\boldsymbol{t}$) as normal and tangential boundary vectors, the zero relative normal velocity, $[\boldsymbol{u} - \boldsymbol{U}_\mathrm{B}] \cdot \boldsymbol{n} = 0$, describes the impermeability of the material surface while the zero relative tangential velocity, $[\boldsymbol{u} - \boldsymbol{U}_\mathrm{B}] \cdot \boldsymbol{t} = 0$, is known as the no-slip velocity condition [5].

The **Neumann condition** for the fluid velocity reads

$$\boldsymbol{n} \cdot \boldsymbol{\sigma}(\boldsymbol{x}_\mathrm{B}, t) = \boldsymbol{T}_\mathrm{B}(\boldsymbol{x}_\mathrm{B}, t) \tag{5.5}$$

where $\boldsymbol{\sigma}$ denotes the sum of pressure and viscous contributions as defined in (1.15) and $\boldsymbol{T}_\mathrm{B}$ is the traction vector prescribed at boundary.

The continuity of the normal and tangential stresses is established, respectively, by $\boldsymbol{n} \cdot (\boldsymbol{\sigma} \cdot \boldsymbol{n}) = \boldsymbol{T}_{\mathrm{B}} \cdot \boldsymbol{n}$ and $\boldsymbol{n} \cdot (\boldsymbol{\sigma} \cdot \boldsymbol{t}) = \boldsymbol{T}_{\mathrm{B}} \cdot \boldsymbol{t}$. This specifies the mechanical balance at the boundary [5].

We can also take combinations of boundary conditions. For example, the entire boundary can be Dirichlet, or a part of it can be Dirichlet and the rest Neumann. Yet, one must be cautious when specifying the Neumann condition on the entire boundary. In this case, the problem does not have a unique solution [4] as, for instance, an arbitrary constant may be added to the solution without modifying it. Also, the same boundary may simultaneously be subject to different kinds of boundary conditions. For example, the interfacial boundary condition between two immiscible liquids involves both velocity and stresses, as will be discussed in Sect. 9.1.

5.2.2 Initial Conditions in Discrete Numerical Methods

When the NSEs are solved numerically, the *initial condition* is still specified by (5.2), but a more general step called *initialisation* is also necessary. Using these two terms indiscriminately may lead to confusion!

To differentiate between them, we should keep in mind that the initial conditions are set by the physics of the problem, i.e. they only apply to time-dependent problems. The initialisation procedure, on the other hand, is required even when the problem is steady because computations should not start with the memory filled with random values. Consequently, the initialisation step is always part of the numerical procedure, regardless of the time-dependency of the modelled problem.

Since this task involves no special efforts other than initialising appropriate arrays of data in the algorithm, we skip details on this topic here and instead refer the interested reader to the literature [6, 7]. However, given that the initialisation procedure does not extend in such a trivial way to LB implementations, we will return to this topic in Sect. 5.5 where we will discuss adequate strategies to initialise LB simulations.

5.2.3 Boundary Conditions in Discrete Numerical Methods

Although boundary conditions in numerical methods share the same fundamental goals of the continuous analytical case, they develop along conceptually different lines. In fact, while analytical boundary conditions apply as additional equations that select the solution of interest from an infinite family of admissible solutions, boundary conditions in numerical methods operate as part of the solution procedure. That is, numerical boundary conditions act in a dynamical manner, being part of the process responsible for the change of the state of the system towards the intended solution. For that reason, the task of specifying boundary conditions assumes an even greater importance in numerical methods. If they are not properly introduced in the numerical scheme, severe problems may arise during the evolution process. As a matter of fact, one of the most common causes for the divergence of numerical solutions comes from the incorrect implementation of boundary conditions.

Recalling Chap. 2, traditional numerical methods for solving the NSEs are usually designed to work directly on the equations governing the fluid flow problem. Taking the finite difference scheme (cf. Sect. 2.1.1) as an example, its discretisation procedure consists of replacing differential operators by appropriate finite difference operators. As a result, the original differential equations are converted into a system of algebraic equations from which fluid flow solutions may be obtained directly.

While constructing numerical boundary conditions as discretisations of their continuous counterparts, we have to be aware of an important guideline.

The **order of accuracy of discretised boundary conditions should never be inferior to that of the bulk solution.** Otherwise, the accuracy of the solution may degrade *everywhere*.

A naive argument which is often mistakenly employed by beginners is that the numerical accuracy of boundary conditions should be an issue of minor concern. The motivation behind this fallacy is that boundaries will only affect a small fraction of all grid points in the simulated domain. Unfortunately, this argument is incorrect! The accuracy of the boundary conditions necessarily interferes with the level of approximation reached in the bulk as the values at the bulk points depend on values at the boundary. In order to eradicate all doubts, let us illustrate this through an example.

Suppose we are interested in the finite difference solution of a 1D problem as depicted in Fig. 5.2. For example, the problem may be a thin film of an incompressible Newtonian fluid subject to an external acceleration a_x. Based on the time-independent NSEs, the description of the above problem reduces to

$$\nu \frac{\partial^2 u_x}{\partial y^2} = -a_x, \tag{5.6}$$

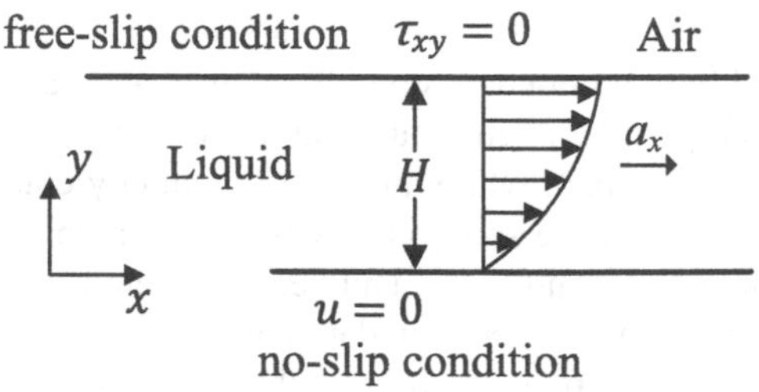

Fig. 5.2 Thin film of liquid flowing due to an external acceleration a_x. We assume that the ambient air is quiescent and the surface tension of the liquid is negligible

subject to the boundary conditions

$$u_x(y = 0) = 0, \quad \left.\frac{\partial u_x}{\partial y}\right|_{y=H} = 0. \tag{5.7}$$

Note that different kinds of conditions apply at each end of the fluid domain: a Dirichlet (no-slip velocity) condition at $y = 0$ and a Neumann (free-slip velocity) condition at $y = H$. The solution is the well-known parabolic Poiseuille profile:

$$u_x(y) = -\frac{a_x}{2\nu} y\,(y - 2H)\,. \tag{5.8}$$

The finite difference formulation of this problem replaces the derivatives in (5.6) by appropriate finite differences. In this way, the continuous target solution $u_x(y)$ is replaced by the values u_j ($j = 1, \ldots, N$) that approximate u_x at the discrete grid points y_j. For simplicity, the spatial discretisation adopted here uses a uniform mesh with spacing $\Delta y = H/(N-1)$. The grid points are placed at $y_j = (j-1)\Delta y$.

Approximating the differential operator in (5.6) by a second-order central scheme, its finite difference representation reads

$$\nu \frac{u_{j+1} - 2u_j + u_{j-1}}{\Delta y^2} = -a_x. \tag{5.9}$$

The discrete approximations of the boundary conditions are

$$u_1 = 0, \tag{5.10a}$$

$$\text{(F1)} \quad \left.\frac{\partial u_x}{\partial y}\right|_{y=H} \simeq \frac{u_N - u_{N-1}}{\Delta y} = 0, \tag{5.10b}$$

$$\text{(F2)} \quad \left.\frac{\partial u_x}{\partial y}\right|_{y=H} \simeq \frac{u_{N+1} - u_{N-1}}{2\Delta y} = 0. \tag{5.10c}$$

For comparison purposes, we discretise the spatial derivative in the Neumann condition in (5.7) with either formulation (F1) or (F2). Although both discretisations are consistent with the continuous problem, meaning that in either case (5.7) is recovered in the continuum limit $\Delta y \to 0$, they provide different approximations at the discrete level. The first approach (F1) uses a *first-order* one-sided difference scheme, while (F2) employs a *second-order* central difference approximation. As in the bulk case, the order of accuracy of the difference operators at the boundary is determined by the truncation error in the approximation of the differential operator (cf. Sect. 2.1.1 and Sect. 4.5.1).

Before proceeding further, it is instructive to explain (F2) in more detail. The approximation of the Neumann condition through a centred (second-order) scheme

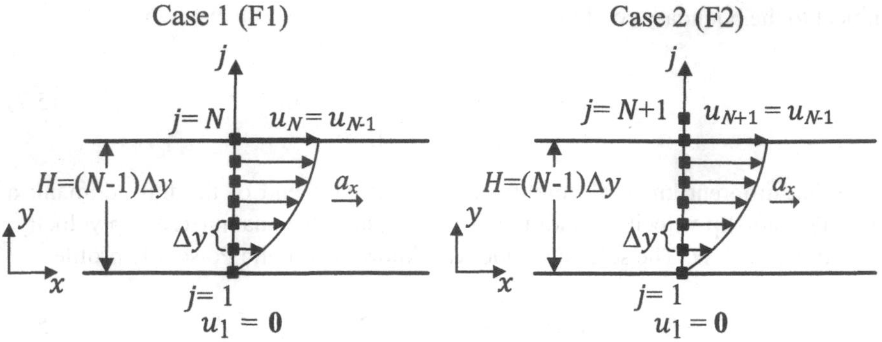

Fig. 5.3 Sketch of the finite difference formulation of the problem depicted in Fig. 5.2. Comparison of two approaches to specify the Neumann boundary condition at $y = H$ using the finite difference method

uses a "numerical trick" often employed in numerics. It consists in assuming the presence of an additional node placed beyond the boundary, where the Neumann condition is assigned.[6] In the above example, this boils down to the consideration of a virtual site at $j = N + 1$ as shown in Fig. 5.3. Due to the inclusion of such an additional node, one gets an extra equation describing the dynamics of the fluid at the boundary $j = N$. This equation, which is similar in form to that in the bulk, $\nu(u_{N+1} - 2u_N + u_{N-1})/\Delta y^2 = -a_x$, must however be corrected by the boundary condition: $u_{N+1} = u_{N-1}$. The resulting numerical approximation of the continuous problem at the boundary, $2\nu(-u_N + u_{N-1})/\Delta y^2 = -a_x$, thus attains the desired second-order accuracy.

Depending on whether we use (F1) or (F2) to approximate the Neumann condition, the finite difference solution of (5.9) takes one of two forms:

$$\text{(F1)} \quad u_j = -\frac{a_x}{2\nu} y_j \left(y_j - 2H + \Delta y\right), \tag{5.11a}$$

$$\text{(F2)} \quad u_j = -\frac{a_x}{2\nu} y_j \left(y_j - 2H\right). \tag{5.11b}$$

These solutions are illustrated in Fig. 5.4.

From the above solutions we can immediately conclude that (5.11b) reproduces *exactly* the analytical parabolic solution in (5.8). On the other hand, although (5.11a) also predicts a parabolic solution, its profile differs from the correct one due to an additional Δy term. This term originates from the lower accuracy of the boundary

[6]The understanding of (F2) is also relevant in the context of LB. By exploiting the similarity between LB and finite difference schemes, a procedure similar to (F2) was proposed as a way to prescribe boundary conditions in the LBM. The original suggestion from Chen and co-workers [8] consists of adding virtual lattice sites. More refined versions of this method were later developed based on an improved understanding of LB theory, e.g. [9–12].

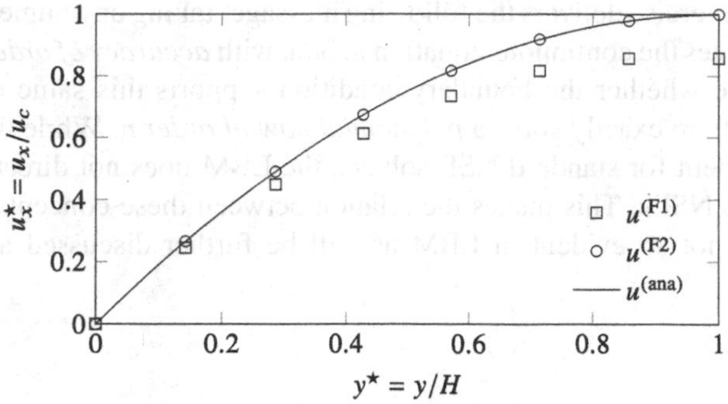

Fig. 5.4 Velocity solutions of (5.8), (5.11a) and (5.11b). The velocity is made non-dimensional with the centreline velocity $u_c = a_x H^2/(2\nu)$. The resolution is $N = 8$

condition in (F1), and it gives rise to an incorrect velocity slope at the boundary. The important observation is that, as (5.11a) shows, the mere choice of the discretisation of the boundary condition affects the solution u_j *everywhere*.

This example demonstrates that even if the discretisation is second-order accurate at all bulk points, a single point (in this case the Neumann boundary point) with a lower-order accuracy is sufficient to degrade the overall solution accuracy.

Example 5.1 Using as starting point the previous example, let us compute the accuracy (*via* the L2 error norm from Sect. 4.5.2) of both above schemes (F1) and (F2) and investigate: (i) How does the accuracy of each solution vary with the mesh resolution? (ii) What is the relation between this result and the accuracy of the numerical approximation?
Question (i): in case (F1), the numerical approximation becomes more accurate with finer resolution, displaying a first-order global improvement. In case (2), the numerical solution remains constant, i.e. it is mesh-independent.

Question (ii): the example involves approximations of both bulk and boundary descriptions. In the bulk, the viscous term is discretised adopting a second-order finite difference scheme. This approximation reads $\partial^2 u_x/\partial y^2 = (u_{j+1} - 2u_j + u_{j-1})/\Delta y^2 + E_{\mathrm{T}}$ where the truncation error is $E_{\mathrm{T}} = -(\Delta y^2/12)\partial^4 u_x/\partial y^4 - (\Delta y^4/360)\partial^6 u_x/\partial y^6 - O(\Delta y^6)$. Since we are searching for a parabolic solution $u_x(y)$, it follows that all derivatives higher than second-order are identically zero. This implies $E_{\mathrm{T}} = 0$, rendering the numerical approximation in the bulk exact. In the description of the Neumann boundary condition we have considered two cases. (F1): $\partial u_x/\partial y = (u_j - u_{j-1})/\Delta y + E_{\mathrm{T}}$ with $E_{\mathrm{T}} = -(\Delta y/2)\partial^2 u_x/\partial y^2 + (\Delta y^2/6)\partial^3 u_x/\partial y^3 - O(\Delta y^3)$; (F2): $\partial u_x/\partial y = (u_{j+1} - u_{j-1})/(2\Delta y) + E_{\mathrm{T}}$ with $E_{\mathrm{T}} = -(\Delta y^2/6)\partial^3 u_x/\partial y^3 - (\Delta y^4/120)\partial^5 u_x/\partial y^5 - O(\Delta y^6)$. It follows that (F1) yields $E_{\mathrm{T}} \neq 0$ because the term $O(\Delta y)$ does not vanish for a parabolic solution. On the contrary, (F2) keeps $E_{\mathrm{T}} = 0$, proving the exactness of the boundary approximation to describe the parabolic solution.

The above exercise delivers the following message: taking any numerical scheme that approximates the continuous equation in bulk with *accuracy of order n*, a simple test to evaluate whether the boundary condition supports this same order n is to assess its ability to exactly solve a *polynomial flow of order n*. While this condition may seem evident for standard NSE solvers, the LBM does not directly discretise the continuous NSEs. This makes the relation between these concepts in bulk and at boundaries not so evident in LBM as will be further discussed at the end of Sect. 5.2.4.

5.2.4 Boundary Conditions for LBM: Introductory Concepts

So far, we focussed on analytical and standard numerical procedures for boundary conditions in fluid flow problems. The rest of the chapter specifically addresses LB boundary conditions. This introductory section explains the following points:

- Which lattice sites should be subjected to boundary conditions?
- What differentiates boundary conditions in LBM from other more traditional numerical methods in fluid dynamics?
- How can boundary conditions be formulated for LBM?
- What determines the numerical accuracy of LB boundary conditions?

Throughout this chapter, we will concentrate on local techniques to implement boundary conditions. Such techniques are particularly suited to describe straight boundaries, aligned with the lattice directions.[7] We will discuss various techniques for curved boundaries in Chap. 11.

In this chapter we focus on 2D applications. Still, most LB boundary conditions covered here can be naturally extended to 3D. Section 5.4.4 discusses additional complexities that may arise in 3D problems.

5.2.4.1 Which Lattice Sites Should Be Subjected to Boundary Conditions?

The concept of nodes, or lattice sites, arises when mapping a continuous to a discrete domain. In this discrete space, a lattice site has to fit into one of three categories, as

[7] Although more complex boundary shapes and/or orientations can also be handled with the techniques discussed here, the resulting geometry will represent a "staircase" approximation of the true boundary. The implications are discussed in Sect. 11.1. Methods to treat smooth complex surfaces usually come at the price of increased complexity and/or the need for data from neighboring nodes, making them non-local schemes. While exceptions of local schemes for curved boundaries exist [13, 14], they are considerably more difficult to implement than comparable non-local techniques.

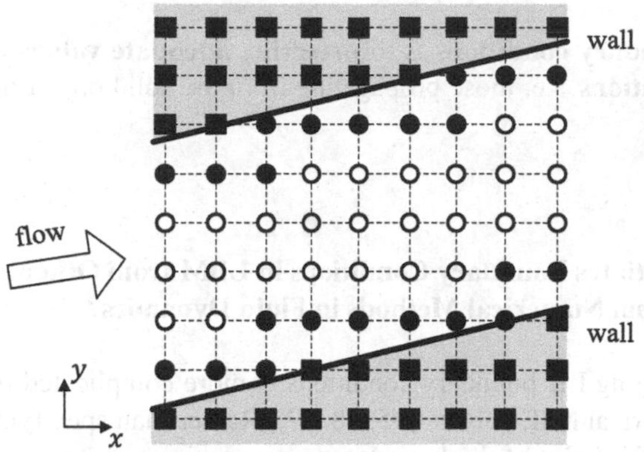

Fig. 5.5 Fluid, solid and boundary nodes in an inclined channel. The *grey-shaded domain* denotes the solid region

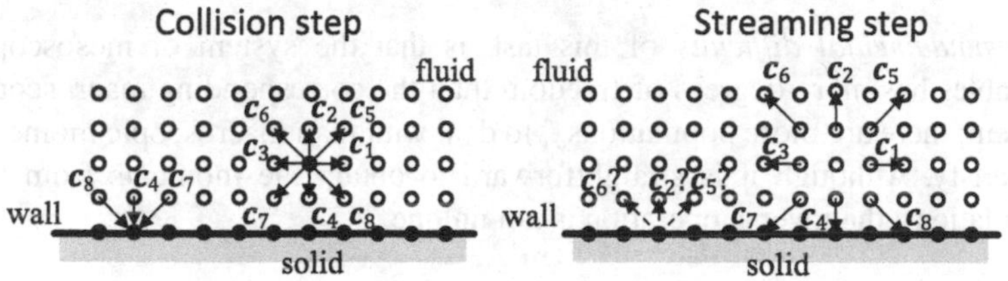

Fig. 5.6 Problem behind the prescription of boundary conditions in LBM: populations streaming from boundary to fluid nodes are unknown and have to be specified

illustrated in Fig. 5.5 [15–17]:

- **Fluid nodes** refer to sites where the LBE applies.
- **Solid nodes** are sites completely covered by the solid object where the LBE should not be solved.
- **Boundary nodes** link fluid and solid nodes; they require special dynamical rules to be discussed.

According to the above definitions, fluid nodes can be identified as nodes connected exclusively to other fluid or boundary nodes. Solid nodes are linked exclusively to other solid or boundary nodes. Boundary nodes have at least one link to a solid and a fluid node.

The *problem with boundary nodes* is illustrated in Fig. 5.6. During the streaming step, populations belonging to fluid nodes (denoted by solid arrows) will stream to neighbouring nodes. On the contrary, this behaviour is not possible for populations on boundary nodes pointing to the inner domain (denoted by dashed arrows): these incoming populations are not specified by the LBE; they must be determined by a different set of rules.

The **role of LB boundary conditions** is to **prescribe adequate values for the incoming populations**, i.e. those propagating from the solid object into the fluid region.

5.2.4.2 What Differentiates Boundary Conditions in LBM from Other more Traditional Numerical Methods in Fluid Dynamics?

It turns out that determining LB boundary conditions is more complicated than it is for conventional numerical NSE solvers [15, 18, 19]. Rather than specifying the macroscopic variables of interest at the boundary (e.g. velocity or pressure), here we must prescribe conditions for the *mesoscopic populations*.

The *fundamental difficulty* of this task is that the system of mesoscopic variables has more degrees of freedom than the corresponding macroscopic system; there are more populations f_i to deal with than macroscopic moments to satisfy. Although it is straightforward to obtain the moments from the populations, the inverse operation is not unique.

5.2.4.3 How Can Boundary Conditions Be Formulated for LBM?

Because of the aforementioned non-uniqueness, it is possible to develop distinct LB boundary conditions that attain "equally consistent" hydrodynamic behaviour [18–20].[8] This explains the "zoo" of approaches existing in this field, which is evidenced by over 160 works published until November 2015.[9]

Despite the large number of methods available, all **LB boundary conditions** for straight boundaries belong to one of **two groups**:

- the **link-wise** family where the boundary lies on *lattice links*,
- the **wet-node** family with the boundary located on *lattice nodes*.

[8]The same kind of "non-uniqueness" problem affects the specification of hydrodynamically consistent initial conditions in LBM (cf. Sect. 5.5).

[9]We obtained this number by performing a Web of Science search for articles with the words "lattice Boltzmann boundary condition" in the title.

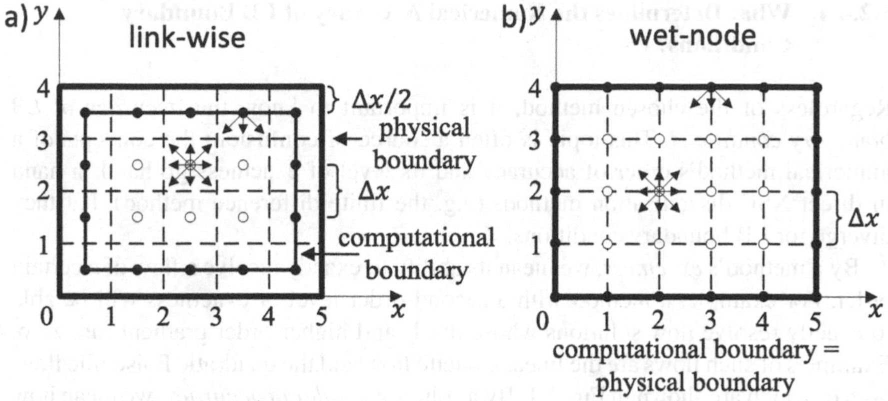

Fig. 5.7 Two discretisations of the same domain with **(a)** link-wise and **(b)** wet-node boundary conditions. Fluid nodes are illustrated as *open circles* (○), boundary nodes as *solid circles* (•)

Due to the possible two ways of locating the boundary node (i.e. the computational boundary) with the respect to the actual flow boundary (i.e. the physical boundary) two perspectives may be considered for the fluid domain discretisation in the LBM. On the one hand, with link-wise schemes the boundary node is shifted from the physical boundary, approximately midway between the solid and the boundary nodes [15, 21–24].[10] Hence, it is advantageous to consider the lattice nodes as locating at the centre of our computational cells. This way the surface of computational cells will coincide with the boarders of the physical domain, see Fig. 5.7a. On the other hand, in wet-node schemes the boundary nodes lie on the physical boundary [2, 18, 20, 25–29]. This time the lattice nodes shall be set on the vertices of our computational cells, in order to ensure computational cells coincide with the boarders of the physical domain, see Fig. 5.7b. While this distinction in the location of lattice nodes inside the computational cells is purely conceptual, recognising it proves useful in a number of situations. For example, it helps understanding the working principle behind each group of boundary schemes. Also, it makes more straightforward the numbering of the lattice nodes in the domain discretisation. Note that, depending on the approach chosen, a different number of lattice nodes is used if adopting the same number of computational cells in the domain discretisation; the two cases are illustrated in Fig. 5.7.

[10]We say "approximately" because in link-wise schemes the exact boundary location is not fixed. Rather, standard link-wise boundaries have a "second-order" dependence on the relaxation rates of the LB collision scheme. For example, with the BGK model this defect leads to a dependency of the no-slip wall location with the fluid viscosity [15, 21, 23]. We will discuss this issue in Sect. 5.3.3 using numerical examples and in Sect. 5.4.1 with a theoretical analyses. Such a "second-order" artefact is also disturbed by anisotropic effects, meaning the wall location will change according to the orientation of the boundary with respect to the lattice.

5.2.4.4 What Determines the Numerical Accuracy of LB Boundary Conditions?

Regardless of the chosen method, it is important to know the *accuracy of LB boundary conditions*. This topic is often a source of confusion: the concepts of a numerical method's order of accuracy and its level of exactness go hand in hand in direct NSE discretisation methods (e.g. the finite difference method), but they diverge for LB boundary conditions.

By a method's *exactness*, we mean its ability to exactly resolve a flow of a certain order. For example, a method with a second-order level of exactness will be able to exactly resolve flow solutions whose third- and higher-order gradients are zero. Examples of such flows are the linear Couette flow and the quadratic Poiseuille flow, both of which are shown in Fig. 1.1. By a solution's *order of accuracy*, we mean how its error scales with the resolution. For example, the error in a second-order accurate method is $O(\Delta x^2)$.

In the bulk, the LBE is said to be (spatially) second-order accurate (cf. Sect. 4.5). Moreover, since this Δx^2 truncation error is associated with the third-order spatial derivative, the second-order accuracy also means that LBM is exact for parabolic solutions, i.e. second-order exact. This conclusion is what we should expect from finite difference theory, and it has been demonstrated in LBM [23].

For LB boundaries, however, the connection between exactness and order of accuracy becomes more difficult to establish. The reason is that the boundary conditions must guarantee not only the correct value at the boundary, but also that the boundary is connected with the bulk solution in such a way that the solution's derivatives at the boundary are also correct.

This is naturally fulfilled by finite difference schemes (cf. Sect. 2.1.1). Boundary nodes with imposed solution values are connected to their neighbouring nodes through the same finite difference operators that approximate derivatives throughout the rest of the domain. This means we can exactly accommodate a parabolic solution on the boundary and throughout the domain if we use a second-order accurate finite difference scheme.

In LBM, on the other hand, the macroscopic boundary condition arises implicitly from the mesoscopic populations imposed at the boundaries. In order to inspect the explicit connection between the mesoscopic and macroscopic levels of description, we need to use multi-scale methods, such as the Chapman-Enskog analysis. However, at the boundary we have to perform this analysis in a different way than the bulk analysis in Sect. 4.1: at the boundary we need to determine the connection between the *mesoscopic populations* and the macroscopic picture, while in the bulk it is mainly the macroscopic conservation equations that we are after. We will later make use of the steady-state Chapman-Enskog approach already described in Sect. 4.2.3.

In the Chapman-Enskog expansion, the distribution function is decomposed as $f_i = f_i^{\text{eq}} + \epsilon f_i^{(1)} + \epsilon^2 f_i^{(2)} + O(\epsilon^3)$. In a steady-state situation with $\partial_t f_i = 0$, we can similarly to (4.9) show that the different orders are connected as

$$f_i^{(1)} = -\tau c_{i\alpha} \partial_\alpha^{(1)} f_i^{\mathrm{eq}}, \quad f_i^{(2)} = -\tau c_{i\alpha} \partial_\alpha^{(1)} \left(1 - \frac{\Delta t}{2\tau}\right) f_i^{(1)}. \tag{5.12}$$

Since the equilibrium f_i^{eq} depends on, for instance, the macroscopic fluid velocity $\boldsymbol{u}$, these equations show that $f_i^{(1)}$ will depend on $\nabla \boldsymbol{u}$ and $f_i^{(2)}$ on $\nabla^2 \boldsymbol{u}$.

Furthermore, we know from Sect. 4.1 that each expansion term's order in the label ϵ corresponds to its order in the Knudsen number $\mathrm{Kn} = \ell_{\mathrm{mfp}}/\ell$, with ℓ_{mfp} being the mean free path and ℓ being a characteristic macroscopic length. We can see this also directly from (5.12): as $\tau \sim \Delta t$, $c_{i\alpha} \sim \Delta x/\Delta t$, and $\partial_\alpha \propto 1/\ell$, we find that $f_i^{(1)}/f_i^{\mathrm{eq}} \propto \Delta x/\ell$ and $f_i^{(2)}/f_i^{\mathrm{eq}} \propto (\Delta x/\ell)^2$, i.e. that Δx takes the place of ℓ_{mfp} in the Knudsen number. Thus, for a constant ℓ, $f_i^{(n)}$ scales with Δx^n.

Resolving f_i at the boundary with **second-order accuracy** requires $f_i = f_i^{\mathrm{eq}} + \epsilon f_i^{(1)} + O(\Delta x^2)$. This is the boundary closure condition usually adopted in the LB literature, e.g. [18, 20, 28]. However, while this boundary accuracy complies with the accuracy order of the LBE in the bulk, its level of exactness is only of first order as it neglects $f_i^{(2)}$ that depends on $\nabla^2 \boldsymbol{u}$. In fact, the parabolic solution that is exactly captured in the bulk is not accommodated exactly at the boundary. Only the linear profile can be captured exactly, since we just account for the solution slope when including $f_i^{(1)}(\nabla \boldsymbol{u})$ alone.

In order to be second-order exact, so that parabolic solutions can be captured exactly, the boundary scheme needs to consider also the solution curvature by including the $f_i^{(2)}(\nabla^2 \boldsymbol{u})$ term in its closure condition. This is possible with **third-order accurate boundary schemes** $f_i = f_i^{\mathrm{eq}} + \epsilon f_i^{(1)} + \epsilon^2 f_i^{(2)} + O(\Delta x^3)$, as noted in a series of works by Ginzburg and co-workers [13, 15, 21, 30, 31].

The decision to choose a second- or third-order accurate boundary scheme is dictated by the exactness level we intend to reach versus the grid refinement level we are willing to invest. Obviously, an answer for this choice is problem-dependent and research in this field has been limited to a few specific cases. For example, studies in the context of porous media flows [15, 32, 33] have revealed that to reach the same level of precision offered by third-order accurate boundary schemes, it would be necessary to refine the mesh one order of magnitude more with second-order boundary schemes, and two orders of magnitude with first-order boundary schemes. (An example of the latter is when a bounce-back wall is not placed exactly midway on the lattice links, cf. Sect. 5.3.3.) However, more general conclusions still require a broader range of flow problems to be tested.

The above discussion forms the basis for the remainder of this chapter, in particular Sect. 5.3.3, Sect. 5.3.4 and Sect. 5.4.1.

5.3 Boundary Condition Methods

The second part of this chapter focuses on the boundary condition methods for the LBE. We will describe their working principles, how to implement them and what their advantages and disadvantages are, with particular focus on their accuracy.

In Sect. 5.3.1 we address periodic boundary conditions, which are then extended to including pressure variations in Sect. 5.3.2. Due to their importance, solid boundary conditions are described in two large subsections: Sect. 5.3.3 follows the bounce-back link-wise approach and Sect. 5.3.4 the wet-node approach. These two approaches are also applied to open boundary conditions, such as inlets and outlets, in Sect. 5.3.5. Finally, in Sect. 5.3.6, we elaborate on corners.

5.3.1 Periodic Boundary Conditions

Periodic boundary conditions arise from specific flow symmetry considerations and are intended to isolate a repeating flow pattern within a cyclic flow system. A common mistake performed by beginners is to confuse the periodicity of the geometry with the periodicity of the flow itself; as illustrated in Fig. 5.8 these two cases are not necessarily synonymous.

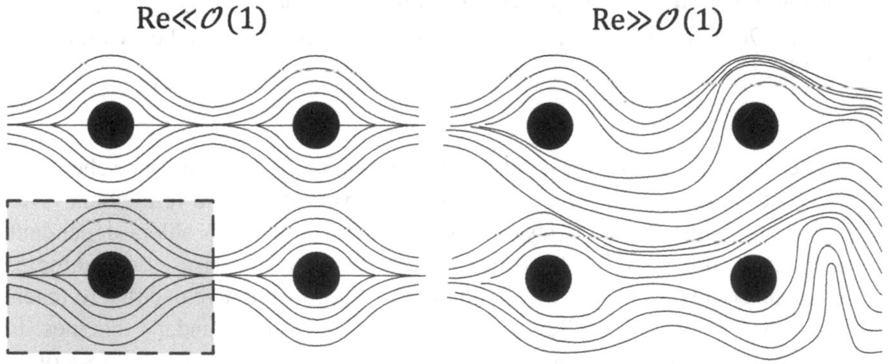

Fig. 5.8 Flow around a periodic array of cylinders. While the geometry is periodic, the periodicity of the flow solution varies depending on the Reynolds number Re. The *grey box* denotes the unit cell in the periodic flow pattern

Periodic boundary conditions apply only to **situations where the flow solution is periodic**, and they state that the **fluid leaving the domain on one side** will, instantaneously, **re-enter at the opposite side**. Consequently, periodic boundary conditions *conserve mass and momentum* at all times.

We note that, if the flow is periodic all over, it can only survive over time if an external source of momentum exists. Otherwise, and regardless its initial state, the flow will decay towards a state of homogeneous velocity (which can be non-zero), due to the action of viscosity.

In a fluid flow simulation, applying periodic boundary conditions is primarily justified by physical arguments, namely to isolate a unit cell in a repeating flow pattern. Furthermore, it may be used to enforce the fully-developed condition in a channel flow or to establish the appropriate conditions for simulating homogeneous isotropic turbulence.

Obviously, a fully periodic flow solution is unphysical since it would fill the entire universe. However, using periodic boundary conditions is often justified in situations where a *finite part* of the flow field can be approximated by a repeating pattern. For example, for not too large Reynolds numbers, the flow in a straight tube segment in a complex network of tubes can be considered invariant along the flow direction over a certain length. In this case it is appropriate to use periodic boundaries located in the straight tube. Another example is the simulation of turbulence that is often performed in periodic systems of size L although turbulent flow is never periodic. Obviously, this leads to finite size effects and flow structures larger than L cannot be captured. However, the assumption of periodicity on the scale L is good enough to capture the relevant physics on scales smaller than L.

Periodic boundaries also find other, less physically oriented, applications. For example, to **solve 2D flow problems with a pre-existing 3D code**, the easiest way is to adopt the **periodic flow condition along one of the Cartesian axis**. The system size along this axis should be as small as possible, ideally a single node. Still, it is more appropriate to use a lower-dimensional code for this purpose to reduce memory requirements and computational time.

For the NSEs, periodic flow conditions along a single dimension are applied as

$$\rho(\boldsymbol{x}, t) = \rho(\boldsymbol{x} + \boldsymbol{L}, t), \tag{5.13a}$$

$$\rho\boldsymbol{u}(\boldsymbol{x}, t) = \rho\boldsymbol{u}(\boldsymbol{x} + \boldsymbol{L}, t) \tag{5.13b}$$

where the vector $\boldsymbol{L}$ describes the periodicity direction and length of the flow pattern.

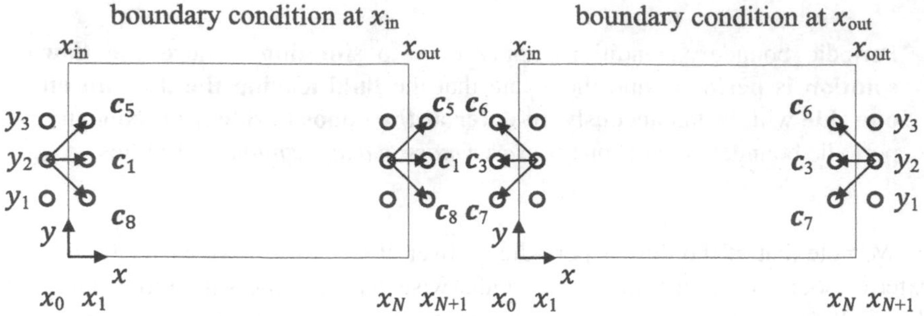

Fig. 5.9 Realisation of the periodic boundary condition (figure inspired by [37]). Layers of "virtual" nodes are added before and after the periodic boundaries, i.e. at $x_0 = x_1 - \Delta x$ and $x_{N+1} = x_N + \Delta x$, respectively

The periodic condition is straightforward in the LBM [34–36]. During propagation, the unknown incoming populations $f_i^\star$ on one side are given by those leaving the domain at the opposite side:

$$f_i^\star(\boldsymbol{x}, t) = f_i^\star(\boldsymbol{x} + \boldsymbol{L}, t). \tag{5.14}$$

For the 2D flow problem illustrated in Fig. 5.9, the periodic flow condition along the x-axis becomes

$$f_i^\star(\boldsymbol{x}, t) = f_i^\star(\boldsymbol{x} + \boldsymbol{L}, t) \quad \Longrightarrow \quad \begin{cases} f_1^\star(x_0, y_2, t) = f_1^\star(x_N, y_2, t) \\ f_5^\star(x_0, y_2, t) = f_5^\star(x_N, y_2, t) \\ f_8^\star(x_0, y_2, t) = f_8^\star(x_N, y_2, t) \end{cases}, \tag{5.15a}$$

$$f_{\bar{i}}^\star(\boldsymbol{x} + \boldsymbol{L}, t) = f_{\bar{i}}^\star(\boldsymbol{x}, t) \quad \Longrightarrow \quad \begin{cases} f_3^\star(x_{N+1}, y_2, t) = f_3^\star(x_1, y_2, t) \\ f_6^\star(x_{N+1}, y_2, t) = f_6^\star(x_1, y_2, t) \\ f_7^\star(x_{N+1}, y_2, t) = f_7^\star(x_1, y_2, t) \end{cases}. \tag{5.15b}$$

In the algorithm presented in (5.15), we have included an additional layer of "virtual" nodes[11] before and after the periodic boundaries at $x_0 = x_1 - \Delta x$ and at $x_{N+1} = x_N + \Delta x$, respectively. Before the streaming step, the populations $f_i^\star$ are copied into these nodes from the opposite periodic edge of the system according to (5.15). We call these nodes virtual because they are there for computational convenience rather than being part of the simulated physical system. The edges of the simulated system are at $x_{in} = (x_1 + x_0)/2 = x_1 - \Delta x/2$ and $x_{out} = (x_{N+1} + x_N)/2 = x_N + \Delta x/2$, giving a periodicity length of $L = x_{out} - x_{in} = N\Delta x$.

[11] The computational convenience of considering these extra layers of nodes in multithreading implementations of LB algorithms is given in Sect. 13.4.1.

But another algorithm can be followed where periodic conditions are implemented *without* virtual nodes. Here, we consider the opposing periodic edges of the flow domain as if they were attached together. In this case, periodic boundary conditions are implemented through a completion step in the streaming process. Post-streaming populations which enter the domain on one side are replaced by the post-collision populations, which leave the domain on the opposite side:

$$f_i(\boldsymbol{x}, t + \Delta t) = f_i^\star(\boldsymbol{x} + \boldsymbol{L} - \boldsymbol{c}_i \Delta t, t). \tag{5.16}$$

Applying (5.16) to the 2D flow case in Fig. 5.9 without virtual nodes, we find

$$f_i(\boldsymbol{x}, t + \Delta t) = f_i^\star(\boldsymbol{x} + \boldsymbol{L} - \boldsymbol{c}_i \Delta t, t) \quad \Longrightarrow \quad \begin{cases} f_1(x_1, y_2, t + \Delta t) = f_1^\star(x_N, y_2, t) \\ f_5(x_1, y_2, t + \Delta t) = f_5^\star(x_N, y_1, t) \\ f_8(x_1, y_2, t + \Delta t) = f_8^\star(x_N, y_3, t) \end{cases}, \tag{5.17a}$$

$$f_{\bar{i}}(\boldsymbol{x} + \boldsymbol{L} - \boldsymbol{c}_i \Delta t, t + \Delta t) = f_{\bar{i}}^\star(\boldsymbol{x}, t) \quad \Longrightarrow \quad \begin{cases} f_3(x_N, y_2, t + \Delta t) = f_3^\star(x_1, y_2, t) \\ f_6(x_N, y_2, t + \Delta t) = f_6^\star(x_1, y_1, t) \\ f_7(x_N, y_2, t + \Delta t) = f_7^\star(x_1, y_3, t) \end{cases}. \tag{5.17b}$$

Although (5.14) and (5.16) give perfectly identical results after streaming (in particular the location of the boundary is the same), it is generally easier to use virtual nodes. Both algorithms extend straightforwardly to 3D domains.

5.3.2 *Periodic Boundary Conditions with Pressure Variations*

While the standard periodic boundary conditions (cf. Sect. 5.3.1) are useful, there are flow problems with a periodic velocity field but a non-periodic pressure (or density) field. Since the fluid density and pressure are related according to $p = c_s^2 \rho$ in an isothermal fluid flow, it follows immediately that (5.13a) is no longer valid. Strategies exist to cope with this particular case of periodicity, called generalised periodic boundary conditions. If we now assume that the system is periodic with an additional prescribed density drop $\Delta\rho$ along $\boldsymbol{L}$, we can modify (5.13) into

$$\rho(\boldsymbol{x},t) = \rho(\boldsymbol{x}+\boldsymbol{L},t) + \Delta\rho, \tag{5.18a}$$

$$\rho\boldsymbol{u}(\boldsymbol{x},t) = \rho\boldsymbol{u}(\boldsymbol{x}+\boldsymbol{L},t) \tag{5.18b}$$

The momentum condition is kept unchanged.

However, this condition has a drawback: periodicity applies to fluid momentum, (5.18b). Hence, mass conservation and periodicity of velocity cannot be enforced at the same time. To overcome this limitation, the incompressible LB may be used [38, 39] (cf. Sect. 5.4.1). For incompressible flows, the generalised periodic boundary condition reads

$$p(\boldsymbol{x},t) = p(\boldsymbol{x}+\boldsymbol{L},t) + \Delta p, \tag{5.19a}$$

$$\boldsymbol{u}(\boldsymbol{x},t) = \boldsymbol{u}(\boldsymbol{x}+\boldsymbol{L},t) \tag{5.19b}$$

where Δp is a prescribed variation of the *pressure*.

While such generalised periodic boundary conditions have long been used in standard NSE solvers [40], they have only recently been developed for LBM [37, 41, 42]. We will now take a closer look at a simple and robust procedure proposed by Kim and Pitsch [37].

We start by considering the presence of a layer of virtual nodes at both ends of periodic boundaries. Following Fig. 5.9, we place these extra nodes at x_0 and x_{N+1}. Then, we decompose the populations on these virtual nodes into equilibrium f_i^{eq} and non-equilibrium f_i^{neq} parts. For the equilibrium part, we write

$$f_i^{\mathrm{eq}}(x_0,y,t) = f_i^{\mathrm{eq}}(p_{\mathrm{in}},\boldsymbol{u}_N), \tag{5.20a}$$

$$f_i^{\mathrm{eq}}(x_{N+1},y,t) = f_i^{\mathrm{eq}}(p_{\mathrm{out}},\boldsymbol{u}_1) \tag{5.20b}$$

where $\boldsymbol{u}_N$ and $\boldsymbol{u}_1$ denote the velocity at nodes x_N and x_1, respectively. The subscripts "in" and "out" denote the pressure values at the left and right physical boundaries, respectively. The individual values for p_{in} and p_{out} are assigned by the user to prescribe the intended pressure drop $\Delta p = p_{\mathrm{in}} - p_{\mathrm{out}}$ along the periodicity length $L = N\Delta x$. For the standard compressible LB equilibrium, the right-hand sides of (5.20) become $f_i^{\mathrm{eq}}(\rho_{\mathrm{in}},\boldsymbol{u}_N)$ and $f_i^{\mathrm{eq}}(\rho_{\mathrm{out}},\boldsymbol{u}_1)$.

The non-equilibrium part is copied from the corresponding image points inside the real domain:

$$f_i^{\mathrm{neq}}(x_0,y,t) = f_i^{\mathrm{neq}}(x_N,y,t), \tag{5.21a}$$

$$f_i^{\mathrm{neq}}(x_{N+1},y,t) = f_i^{\mathrm{neq}}(x_1,y,t). \tag{5.21b}$$

Non-equilibrium populations are determined *after* collision, i.e. $f_i^{\mathrm{neq}} = f_i^{\star} - f_i^{\mathrm{eq}}$.

Finally, we merge (5.20) and (5.21), i.e. $f_i^\star(x_0, y, t) = f_i^{\text{eq}}(x_0, y, t) + f_i^{\text{neq}}(x_0, y, t)$ and the same at x_{N+1}, and perform the streaming step.

Example 5.2 Taking the geometry depicted in Fig. 5.9, let us show how to implement the inlet/outlet boundary conditions for a streamwise invariant flow driven by a pressure difference Δp along the x-axis. The first task is to assign individual values for p_{in} and p_{out}. In incompressible flows, the absolute pressure value is defined up to an arbitrary constant. Thereby, we can set $p_{\text{out}} = 1$ (simulation units) and $p_{\text{in}} = p_{\text{out}} + \Delta p$. The generalised periodic boundary condition then prescribes the post-collision populations:

- Inlet boundary condition ($i \in \{1, 5, 8\}$):

$$f_i^\star(x_0, y, t) = f_i^{\text{eq}}(p_{\text{in}}, \boldsymbol{u}_N) + \left(f_i^\star(x_N, y, t) - f_i^{\text{eq}}(x_N, y, t)\right). \tag{5.22}$$

- Outlet boundary condition ($i \in \{3, 6, 7\}$):

$$f_i^\star(x_{N+1}, y, t) = f_i^{\text{eq}}(p_{\text{out}}, \boldsymbol{u}_1) + \left(f_i^\star(x_1, y, t) - f_i^{\text{eq}}(x_1, y, t)\right). \tag{5.23}$$

We could also apply (5.22) and (5.23) to all populations ($0 \le i \le 8$) without changing the final outcome. Obviously only those populations are relevant that propagate into the physical domain; the others do not play any role.

5.3.3 Solid Boundaries: Bounce-Back Approach

In hydrodynamics, the most common fluid-solid interface condition is the no-slip velocity boundary condition. Therefore, its correct implementation is crucial for modelling confined fluid flow phenomena and other problems involving solid boundaries.

The oldest LB boundary condition to model walls is the bounce-back method [21–23, 43–45]. Its concept was adopted from earlier lattice gas models (cf. Sect. 2.2.2 [43, 46–48]). Despite its age, it is still the most popular wall boundary scheme in the LB community, largely due to its simplicity of implementation.

Given the importance of the bounce-back scheme, this section is entirely devoted to it. First, we will introduce the basic bounce-back principle from a particle-based picture. After that, we will discuss different ways to realise the bounce-back rule in order to model (i) stationary and (ii) moving walls. Subsequently, we will point out the advantages and disadvantages of the bounce-back scheme. Finally, we will consolidate all previous theoretical elements by analysing two practical problems.

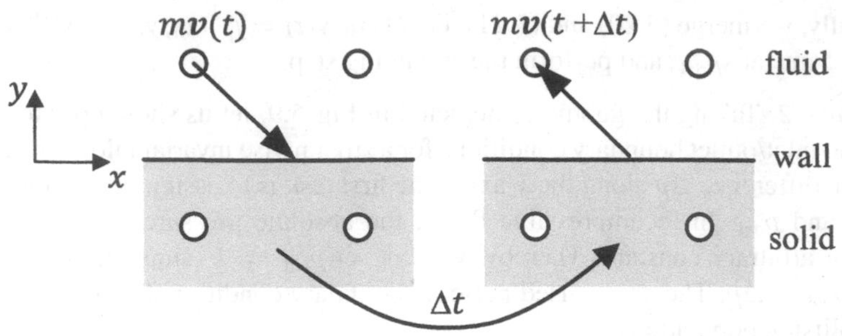

Fig. 5.10 Sketch of a moving particle with mass m and velocity $\boldsymbol{v}$ hitting a rigid wall. During the collision process both normal and tangential momentum components are reversed. The average particle momentum, before and after collision, is $\langle m\boldsymbol{v}\rangle = (m\boldsymbol{v}(t) + m\boldsymbol{v}(t + \Delta t))/2 = \boldsymbol{0}$ since $\boldsymbol{v}(t + \Delta t) = -\boldsymbol{v}(t)$

5.3.3.1 Principle of the Bounce-Back Method

The working principle of bounce-back boundaries is that **populations hitting a rigid wall** during propagation are **reflected back** to where they originally came from. This is illustrated in Fig. 5.10.

While it may not be immediately obvious why no-slip boundaries follow from this principle, we may understand this if we imagine populations as embodying fluid matter portions.[12] It follows that the *bounce-back of particles hitting a wall* implies *no flux across the boundary*, i.e. the wall is impermeable to the fluid. Similarly, the fact that *particles are bounced back* rather than bounced forward (i.e. specularly reflected) implies *no relative transverse motion between fluid and boundary*, i.e. the fluid does not slip on the wall. These two points illustrate how the bounce-back method on the population level mirrors the Dirichlet boundary condition for the macroscopic velocity at the wall.

The above particle-based explanation as shown in Fig. 5.10 should not be taken too literally. Rigorously explaining the macroscopic behaviour of the bounce-back rule (as any other mesoscopic model) requires the use of multi-scale expansion techniques, such as the Chapman-Enskog analysis. We will do this in Sect. 5.4.1.

[12] In an attempt to make this explanation of the bounce-back method more intuitive, we will often refer to the parameter f_i as "particles" instead of "particle distributions" or "populations", as it should be called more rigorously. We have to remember that LBM is *not* a true particle method, such as those presented in Chap. 2. Rather, LBM deals with discretised forms of continuous fields (cf. Chap. 3).

5.3.3.2 Fullway Versus Halfway Bounce-Back Method

The bounce-back method can be realised in two different ways:

1. In the first strategy, called *fullway bounce-back* [36], particles are considered to travel the complete link path from the boundary to the solid node, where the particle velocity is inverted in the next collision step; see Fig. 5.11a.
2. In the second strategy, called *halfway bounce-back* [22], that particles are considered to travel only half of the link distance so that the inversion of the particle velocity takes place during the streaming step; see Fig. 5.11b.

Both strategies introduce specific modifications of the LB algorithm. On the one hand, the fullway bounce-back method requires solid nodes where the populations are stored and then bounced back during the following collision step. On the other hand, solid nodes are not necessary for the halfway bounce-back method since the inversion of populations occurs during the streaming step. Algorithmically, we can say that the fullway bounce-back changes the collision step at solid nodes but leaves the usual streaming step unchanged, while the halfway bounce-back changes the streaming step but does not modify the collision step.

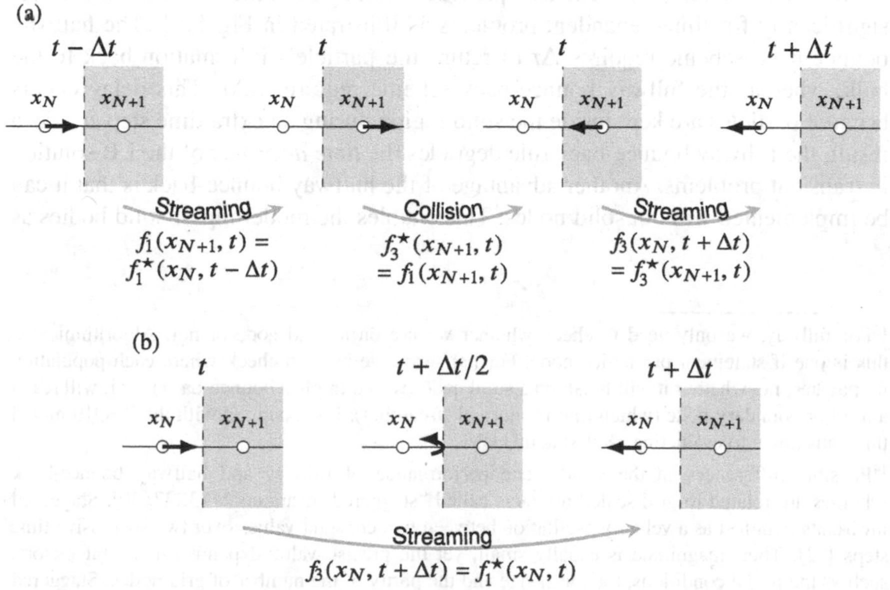

Fig. 5.11 Time evolution for **(a)** fullway bounce-back and **(b)** halfway bounce-back. The current time step is shown at the top of each action taken. In all pictures the *arrow* represents the particle's direction, the rightmost *grey shaded domain* is the solid region and the *dashed line* corresponds to the boundary

Despite what their names might suggest, **both the fullway and halfway** approaches assume that the **boundary is located approximately midway** between solid and boundary nodes, not *on* the solid nodes themselves.

Note, the midway location of the bounce-back wall is only *approximate*. Its exact placement depends on several factors that we will soon discuss. The important point to keep in mind is that the assumption of having the bounce-back boundary placed on a lattice node introduces a first-order error, even for straight boundaries. The setting of the wall location in the middle of a lattice link, i.e. between nodes, makes the method formally *second-order accurate*. This explains why the second interpretation is generally preferred and why the bounce-back rule is the classical example of a link-wise LB boundary condition.

The question that arises is: which strategy to implement? Fullway or halfway? There is no definitive answer. If simplicity is our main criterion, then fullway bounce-back wins. Here, the boundary treatment is independent of the direction of f_i and the execution time is shorter, cf. Chap. 13.[13] Yet, halfway bounce-back is more accurate for unsteady flows as explained below.

In steady state, both schemes provide similar outcomes.[14] But they differ significantly for time-dependent problems as illustrated in Fig. 5.11. The halfway bounce-back scheme requires Δt to return the particle's information back to the bulk, whereas the fullway bounce-back scheme requires $2\Delta t$. This delay occurs because particles are kept inside the solid region during an extra time step Δt. As a result, the fullway bounce-back rule degrades the *time accuracy* of the LB solution in transient problems. Another advantage of the halfway bounce-back is that it can be implemented without solid nodes. This enables the modelling of solid bodies as

[13]For fullway, we only need to check whether we are on a solid node or not. Algorithmically, this is one if-statement per lattice node. For half-way, we have to check where each population propagates, i.e. whether it will finish on a solid node (which implies bounce-back), or it will reach a fluid or boundary node (which implies normal streaming). For example, with the D3Q19 model this boils down to evaluating 18 if-statements!

[14]Possible differences in the steady-state performance of fullway and halfway bounce-back schemes are related to grid-scaled artefacts called "staggered invariants" [13, 22, 30]. Staggered invariants manifest as a velocity oscillation between two constant values over two successive time steps [22]. Their magnitude is usually small, yet the precise value depends on several factors, such as the initial conditions, the mesh size and the parity of the number of grid nodes. Staggered invariants are conserved by halfway bounce-back. For example, a typical channel flow along the x-axis, if initialised with $u_y(t = 0) = u_{y,0}$ and using an odd number of nodes along the y-axis, will conserve this constant transverse velocity $u_{y,0}$ throughout the channel width as a staggered invariant if halfway bounce-back is used [30]. However, with an even number of lattice nodes, this artefact vanishes. One way to correct this halfway bounce-back defect is by averaging the solution between two successive time steps as suggested in [22]. The fullway bounceback does not produce this artefacts since it delays the exchange of information between successive time steps. However, the conservation and stability properties of this scheme may be worse [30].

narrow as zero lattice widths, for example an infinitesimally thin plate. The above two points justify the general preference for the halfway implementation. For this reason, we will only consider halfway bounce-back in the remainder of this book, and we will omit the term "halfway" hereafter.

5.3.3.3 Resting Walls

Let us now discuss the case of a *resting wall.*

Populations leaving the boundary node $\boldsymbol{x}_\mathrm{b}$ at time t **meet the wall surface** at time $t+\frac{\Delta t}{2}$ where **they are reflected back** with a velocity $\boldsymbol{c}_{\bar{i}} = -\boldsymbol{c}_i$, arriving at time $t+\Delta t$ at the node $\boldsymbol{x}_\mathrm{b}$ from which they came. This is shown in Figs. 5.11b and 5.12. For these populations, the standard streaming step is replaced by

$$f_{\bar{i}}(\boldsymbol{x}_\mathrm{b}, t+\Delta t) = f_i^\star(\boldsymbol{x}_\mathrm{b}, t). \tag{5.24}$$

For the case depicted in Fig. 5.12, the implementation (5.24) reads

$$\begin{aligned} f_2(\boldsymbol{x}_\mathrm{b}, t+\Delta t) &= f_4^\star(\boldsymbol{x}_\mathrm{b}, t), \\ f_5(\boldsymbol{x}_\mathrm{b}, t+\Delta t) &= f_7^\star(\boldsymbol{x}_\mathrm{b}, t), \\ f_6(\boldsymbol{x}_\mathrm{b}, t+\Delta t) &= f_8^\star(\boldsymbol{x}_\mathrm{b}, t). \end{aligned} \tag{5.25}$$

Exercise 5.1 Write down (5.24) in the same form as (5.25) but for the case of a top wall.

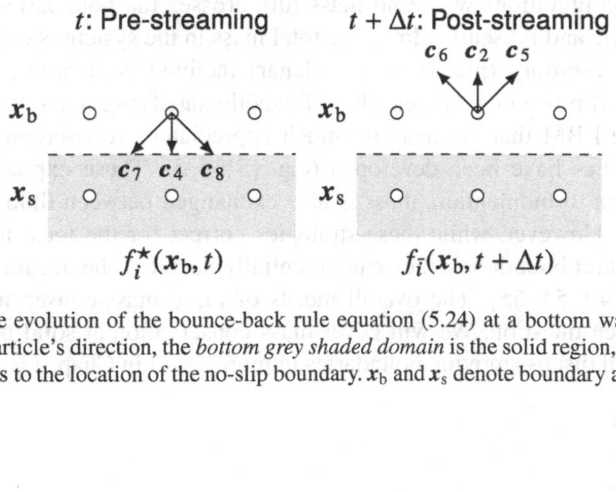

Fig. 5.12 Time evolution of the bounce-back rule equation (5.24) at a bottom wall. The *arrows* represent the particle's direction, the *bottom grey shaded domain* is the solid region, and the *dashed line* corresponds to the location of the no-slip boundary. $\boldsymbol{x}_\mathrm{b}$ and $\boldsymbol{x}_\mathrm{s}$ denote boundary and solid nodes, respectively

5.3.3.4 Moving Walls

The extension to *moving walls* is quite simple. It requires only a small correction to the standard bounce-back formula [22, 49]. Again, the role of this correction can be explained by resorting to our particle-based picture of the bounce-back dynamics. Since now the wall is not at rest, the bounced-back particles have to gain or lose a given amount of momentum after hitting the wall so that the outcome respects Galilean invariance [22, 50]. One way to show this is to transform to the rest frame of the wall, perform bounce-back there, and transform back to the initial frame.

The **bounce-back** formula for a **Dirichlet boundary condition** with a prescribed **wall velocity $\boldsymbol{u}_{\mathrm{w}}$** reads

$$f_{\bar{i}}(\boldsymbol{x}_{\mathrm{b}}, t + \Delta t) = f_i^{\star}(\boldsymbol{x}_{\mathrm{b}}, t) - 2w_i\rho_{\mathrm{w}}\frac{\boldsymbol{c}_i \cdot \boldsymbol{u}_{\mathrm{w}}}{c_{\mathrm{s}}^2} \tag{5.26}$$

where the subscript w indicates properties defined at the wall location $\boldsymbol{x}_{\mathrm{w}} = \boldsymbol{x}_{\mathrm{b}} + \frac{1}{2}\boldsymbol{c}_i\Delta t$.

Obviously, for a stationary boundary with $\boldsymbol{u}_{\mathrm{w}} = \boldsymbol{0}$, the correction vanishes and the above equation simplifies to (5.24).

When $\boldsymbol{u}_{\mathrm{w}} \neq \boldsymbol{0}$ an extra difficulty comes out: the local density value ρ_{w} at the wall, which may require some thought. If the incompressible LB model is used, there is no problem since the density is uniform throughout the flow domain. However, for the standard LB model, ρ varies with the pressure, and its value is not generally known at the wall. One solution is to estimate ρ_{w} as either the local fluid density $\rho(\boldsymbol{x}_{\mathrm{b}})$ or the system's average density $\langle\rho\rangle$. For steady flow, the difference $\rho(\boldsymbol{x}_{\mathrm{b}}) - \langle\rho\rangle$ is $O(\mathrm{Ma}^2)$ (where Ma is the Mach number) and therefore usually small [16].

In flow configurations where no mass flux crosses the boundaries, such as in parallel Couette and Poiseuille flows, the total mass in the system is conserved [13]. However, for arbitrary (planar or non-planar) inclined boundaries, none of the aforementioned procedures are capable of exactly satisfying mass conservation, a feature of the LBM that is otherwise much appreciated. To overcome this issue, several strategies have been developed (e.g. [51–53]). These explicitly consider that in addition to momentum, mass is also exchanged between fluid and moving solid regions. However, while these strategies correct for the local mass leakage across solid-fluid boundaries, they can potentially decrease the accuracy of the LB solution [13, 49, 54, 55]. The overall merits of these mass-conserving strategies remain an open question. We will cover mass conservation at solid boundaries in Sect. 5.4.2 and discuss moving boundaries in more detail in Chap. 11.

5.3.3.5 Advantages and Disadvantages

Besides its simplicity of implementation, the bounce-back boundary condition has other *advantages*:

1. It is a *stable* numerical scheme [15, 24], even when the bulk LB solution is brought close to the instability limit $\tau \to \Delta t/2$. This distinguishes bounce-back from other LB boundary conditions, particularly wet-node-based ones that are often a source of numerical instability.[15]
2. Since bounce-back is based on reflections, *mass conservation* is strictly guaranteed at resting boundaries. This is an important feature, especially in problems where the absolute mass is important. Exact mass conservation is a property often violated by wet-node techniques, where specific corrections need to be introduced [20, 28, 29].
3. It can be implemented straightforwardly for any number of spatial dimensions.

However, bounce-back also has *disadvantages*. The main points of criticism are:

1. The bounce-back rule can only approximate arbitrary surfaces through "staircase" shapes. In fact, the bounce-back condition only guarantees a higher than first-order accuracy if (i) the surface is aligned with the lattice and (ii) the wall cuts midway through the lattice links. Both these conditions are violated in general boundary configurations where the wall cuts the lattice links at varying distances. In this case, the bounce-back is first-order accurate [13, 15, 56]. We will cover more complex bounce-back methods avoiding the staircase approximation in Sect. 11.2.
2. The exact location of the no-slip boundary is *viscosity-dependent* when the bounce-back scheme is used together with the BGK collision model. This implies that the hydrodynamic solution will differ for the same grid at different viscosities, even though the governing physical parameters of the problem (e.g. Reynolds number) are fixed. This important violation of the physics of the problem is not found in standard discretisation methods of the NSEs. This defect may be solved by replacing the BGK model, which has only one free parameter τ, by a more complex collision model such as TRT [30, 57] or MRT [58–60] (cf. Chap. 10). We will explain the viscosity dependence through a Chapman-Enskog analysis in Sect. 5.4.1.

5.3.3.6 Numerical Evaluation of the Accuracy of the Bounce-Back Method

To conclude this section, we will clarify some aspects of the accuracy of the bounce-back method through the numerical solution of Couette and Poiseuille flows

[15] According to [20], instability is often triggered by the boundary condition rather than the bulk algorithm when LBM is running close to $\tau = \Delta t/2$.

(cf. Fig. 1.1). As their solutions display linear and parabolic spatial variations, respectively, they are ideal cases to investigate the first- and second-order analyses of the bounce-back rule. With this exercise we want to answer the following questions:

1. How is accuracy affected by assuming that the bounce-back wall lies on lattice nodes, rather than midway on the lattice links?
2. Why do we say that the location of the bounce-back no-slip boundary is *approximately*, instead of *exactly*, midway on the lattice links?

First-order analysis

Let us start with the first-order analysis for the linear Couette flow.

Using simulation units,[16] let us assume without loss of generality that the fluid density is $\rho = 1$ and the velocity is $u_\mathrm{w} = 0.1c$ at the top wall and zero at the bottom. The computational domain is resolved by 3 lattice nodes in length (an arbitrary choice as the flow is streamwise invariant) and by $N = 9$ lattice nodes along the channel height. We take the D2Q9 model. We initialise the system by setting $f_i(\boldsymbol{x}, t = 0) = f_i^{\mathrm{eq}}(\rho = 1, \boldsymbol{u} = \boldsymbol{0})$, cf. Sect. 5.5. The simulation is stopped when it reaches the steady-state criterion $L_2 \leq 10^{-10}$ for the fluid velocity (cf. Sect. 4.5.2). We use the BGK collision model with relaxation time $\tau = 0.9\Delta t$, but we will discuss other values of τ as well.

Figure 5.13 illustrates the unknown populations at the boundaries. Inlet and outlet regions are subject to periodic boundary conditions (cf. Sect. 5.3.1). We use the bounce-back method for the stationary bottom and the moving top wall.

- Bottom wall:

$$\begin{aligned} f_2(x, y_1, t + \Delta t) &= f_4^\star(x, y_1, t), \\ f_5(x, y_1, t + \Delta t) &= f_7^\star(x, y_1, t), \\ f_6(x, y_1, t + \Delta t) &= f_8^\star(x, y_1, t). \end{aligned} \tag{5.27}$$

- Top wall[17]:

$$\begin{aligned} f_4(x, y_N, t + \Delta t) &= f_2^\star(x, y_N, t), \\ f_7(x, y_N, t + \Delta t) &= f_5^\star(x, y_N, t) - \tfrac{1}{6c} u_\mathrm{w}, \\ f_8(x, y_N, t + \Delta t) &= f_6^\star(x, y_N, t) + \tfrac{1}{6c} u_\mathrm{w}. \end{aligned} \tag{5.28}$$

---

[16]Conversion between simulation and physical units will be discussed in Chap. 7.

[17]The numerical values for the correction terms come from $2w_i/c_\mathrm{s}^2 = 1/(6c^2)$, with $c_\mathrm{s}^2 = (1/3)c^2$ and using the D2Q9 lattice weights w_i given in Table 3.1.

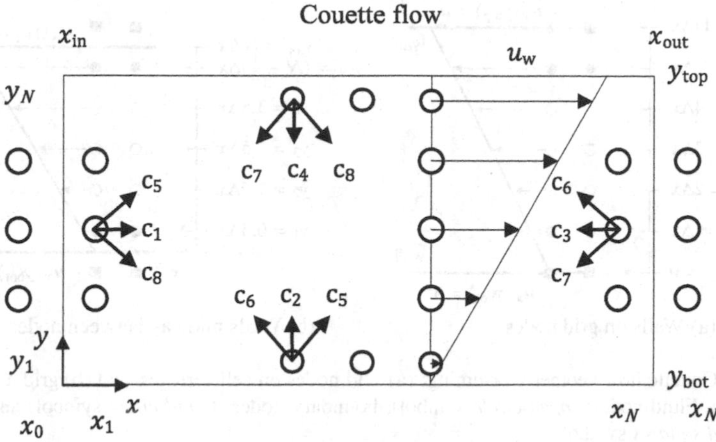

Fig. 5.13 Couette flow with unknown boundary populations annotated. The layer of wall nodes goes from x_0 to x_{N+1}, while the layer where inflow/outflow conditions are applied goes from y_1 to y_N. Corner nodes, e.g. (x_0, y_0), belong to wall layers

Let us now answer the first question and evaluate the effect of the bounce-back wall location on the accuracy of the LB solution. Let us recall Fig. 5.7 and assume the two possibilities for the location of grid nodes y_j ($j = 1, \dots, N$) with respect to the computational cells:

a) The grid nodes coincide with the cell vertexes, i.e. $y_j = j\Delta x$.
b) The grid nodes are placed at the cell centres, i.e. $y_j = (j - 0.5)\Delta x$.

To facilitate the analysis, we can assume the origin of the grid node system to coincide with the bottom wall. As such, the location of the bounce-back walls can be interpreted as follows:

a) In the first case, we assume bounce-back walls to be located on the solid nodes $y_0 = y_{bot} = 0$ and $y_{N+1} = y_{top} = (N + 1)\Delta x$. This means the height of the channel is $H = y_{top} - y_{bot} = (N + 1)\Delta x$, cf. Fig. 5.14a).
b) In the second case, the bounce-back walls are assumed to be located midway between boundary and (virtual) solid nodes. The boundary nodes are placed at $y_1 = 0.5\Delta x$ and $y_N = (N - 0.5)\Delta x$. The walls are located one-half lattice width outside these nodes, i.e. at $y_{bot} = 0$ and $y_{top} = N\Delta x$, respectively. This makes the channel height $H = y_{top} - y_{bot} = N\Delta x$, cf. Fig. 5.14b).

In order to evaluate the accuracy of the numerical solutions produced by each case, let us compare them with the exact solution:

$$u_x(y) = u_w \frac{y - y_{bot}}{y_{top} - y_{bot}}, \quad y_{bot} \le y \le y_{top} \tag{5.29}$$

where $y_{bot} = 0$ and $y_{top} = (N + 1)\Delta x$ for case a) and $y_{bot} = 0$ and $y_{top} = N\Delta x$ for case b).

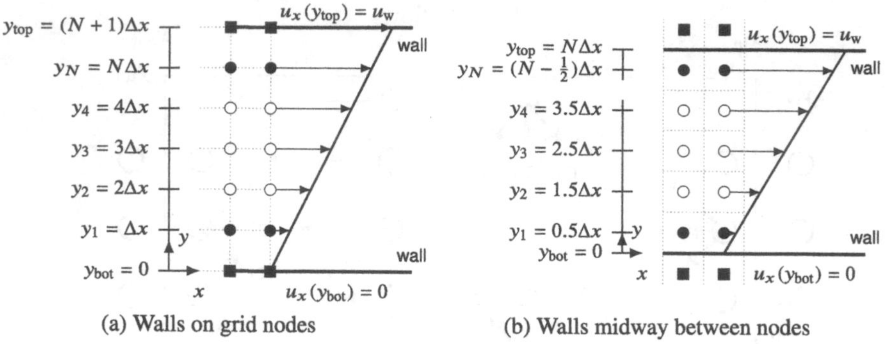

Fig. 5.14 Couette flow geometry assuming: (**a**) grid nodes on cell vertexes, and (**b**) grid nodes on cell centres. Fluid nodes: (*open circle* symbol), boundary nodes: (*solid circle* symbol) and solid node: (*filled squares* symbol)

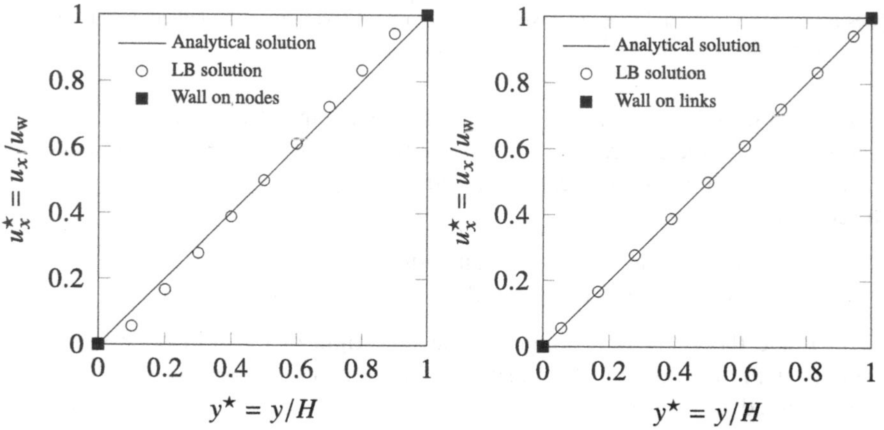

Fig. 5.15 Couette flow solution with BGK collision operator ($\tau/\Delta t = 0.9$) and bounce-back rule (*circles*) versus analytical solution (*line*). *Left panel*: wall is placed on lattice nodes; *right panel*: wall is placed midway between lattice nodes

Figure 5.15 shows the results. We see that case a) leads to a velocity solution different from the analytical one while in case b) the two solutions match exactly. This means that, when the boundary is located *midway between* grid nodes [44], the bounce-back method is (i) first-order exact and (ii) at least second-order accurate since both the boundary value and its first-order derivative are reproduced exactly. This leaves only an $O(\Delta x^2)$ truncation error according to the discussion at the end of Sect. 5.2.4.

Interestingly, this numerical test also shows that the Couette flow solution is independent of the choice of τ. Due to the linear nature of this problem we can establish that the bounce-back scheme is viscosity-independent at $O(\Delta x)$.

For **Couette flow** we observed that (i) **bounce-back** places the no-slip wall **exactly midway** between boundary and solid nodes and (ii) solutions are **viscosity-independent**. Since these conclusions are based on a linear solution, they only apply to **first-order** terms. This means that bounce-back, with the wall shifted by $\Delta x/2$ from boundary nodes, correctly prescribes the boundary value of the velocity and its first-order derivative.

Second-order analysis

The next question is whether the bounce-back method preserves the encouraging results observed for Couette flow also at the second order. To answer this question we proceed with an analysis of the pressure-driven parabolic Poiseuille flow.

We run this simulation with the same parameters as for the Couette problem, except for the boundary conditions. Now, the top and bottom walls are subject to a zero-velocity bounce-back condition while the inlet and outlet boundaries enforce a pressure-driven fully-developed flow condition using the generalised periodic boundary condition method (cf. Sect. 5.3.2).

Figure 5.16 illustrates the problem, including the missing populations at the boundaries. The bounce-back scheme at walls adopts (5.27) and (5.28) with $u_w = 0$. The inlet and outlet boundaries use the same formulas as in Example 5.2. Again, we

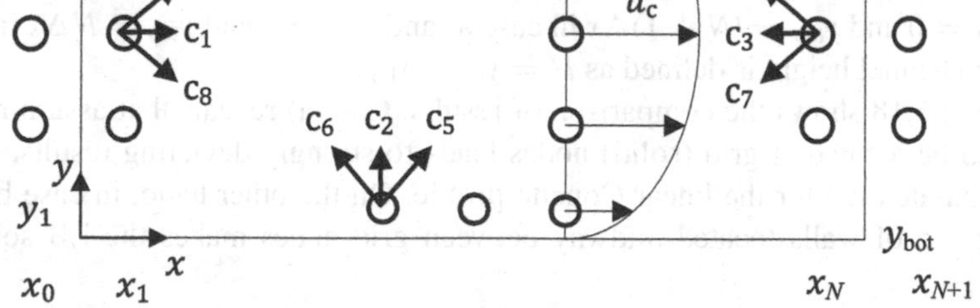

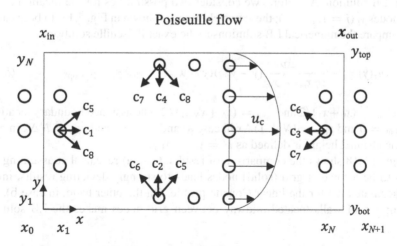

Fig. 5.16 Poiseuille flow with unknown boundary populations annotated. The layer of wall nodes goes from x_0 to x_{N+1}, while the layer where inflow/outflow conditions are applied goes from y_1 to y_N. Corner nodes, e.g. (x_0, y_0), belong to wall layers

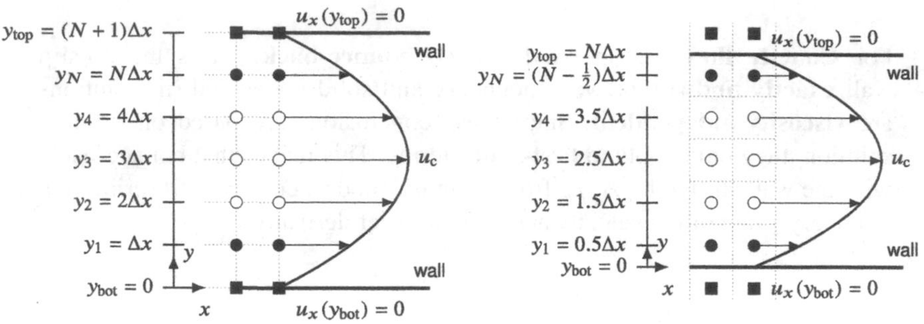

(a) Walls on grid nodes (b) Walls midway between nodes

Fig. 5.17 Poiseuille flow geometry assuming: (**a**) grid nodes on cell vertexes, and (**b**) grid nodes on cell centres. Fluid nodes: (*open circle* symbol), boundary nodes: (*solid circle* symbol) and solid nodes: (*filled squares* symbol)

set $p_{out} = 1$ (in simulation units) and compute the inlet pressure as $p_{in} = p_{out} + \Delta p$.[18] In order to fulfill the small-Ma requirement, we set the centreline velocity to $u_c = 0.1$ (simulation units) and relate it to the pressure difference Δp as

$$\frac{\Delta p}{x_{out} - x_{in}} = \frac{8\eta u_c}{(y_{top} - y_{bot})^2}.$$

Let us again look at the effect of the bounce-back wall location on the accuracy of the LB solution. As before, we consider two possibilities for the location of the grid nodes y_j ($j = 1, \ldots, N$); these two choices are shown in Fig. 5.17. In both cases, we compare the numerical LB solutions to the exact Poiseuille solution

$$u_x(y) = -\frac{1}{2\eta}\frac{\Delta p}{x_{out} - x_{in}}(y - y_{bot})(y - y_{top}), \quad y_{bot} \le y \le y_{top} \tag{5.30}$$

where $x_{in} = (x_0 + x_1)/2$ and $x_{out} = (x_N + x_{N+1})/2$. The assumed boundary locations are $y_{bot} = 0$ and $y_{top} = (N+1)\Delta x$ in case a) and $y_{bot} = 0$ and $y_{top} = N\Delta x$ in case b). The channel height is defined as $H = y_{top} - y_{bot}$.

Figure 5.18 shows the comparison of results. Case a) reveals that assuming the walls to be located at grid (solid) nodes leads to strongly deviating results, much less accurate than for the linear Couette profile. On the other hand, in case b), the assumption of walls located midway between grid nodes makes the LB solution

[18] Since this Poiseuille flow is driven by the pressure difference Δp, the incompressible LB model of Sect. 4.3.2 is appropriate. Recall that the standard LB equilibrium uses an isothermal equation of state where pressure and density relate linearly. Therefore, a pressure gradient inevitably leads to a gradient of density, which is incompatible with incompressible hydrodynamics. With the incompressible model this compressibility error can be avoided (at least in steady flows).

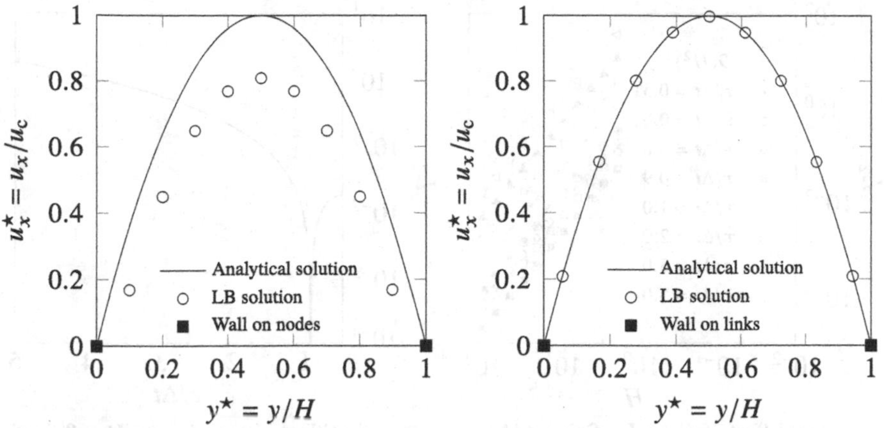

Fig. 5.18 Poiseuille flow solution with BGK collision operator ($\tau/\Delta t = 0.9$) and bounce-back rule (*circles*) versus analytical solution (*line*). *Left panel*: wall is placed at lattice nodes; *right panel*: wall is placed midway between lattice nodes

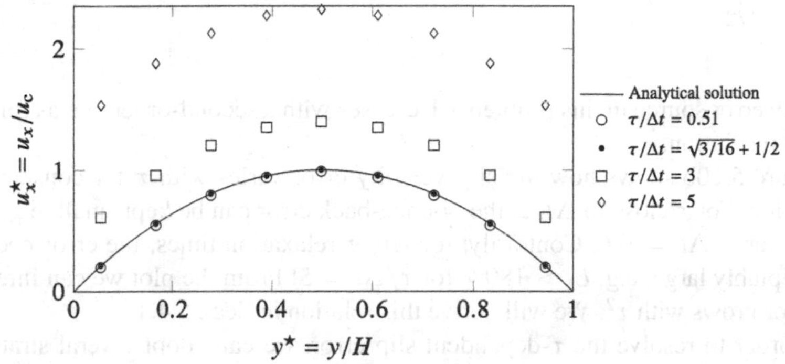

Fig. 5.19 Poiseuille flow solutions with bounce-back walls placed $\Delta x/2$ away from boundary nodes. Comparison between BGK profiles obtained with different values of τ and the analytical solution from (5.30)

almost coincide with the analytical parabolic profile (the deviations are essentially invisible in the plot).

Unlike the numerical Couette flow solution, the Poiseuille flow simulation is affected by the choice of τ. This is illustrated in Fig. 5.19. While the velocity profiles remain perfectly parabolic, we observe that the boundary velocity is clearly τ-dependent. We call this effect "numerical boundary slip".

Since this boundary slip artefact is not present in linear solutions, we can conclude that it stems from the $O(\Delta x^2)$ term in the bounce-back closure relation.[19] Indeed, Fig. 5.20a confirms that for fixed values of τ the slip velocity error (which is

[19]We will cover closure relations for boundary conditions in Sect. 5.4.1.

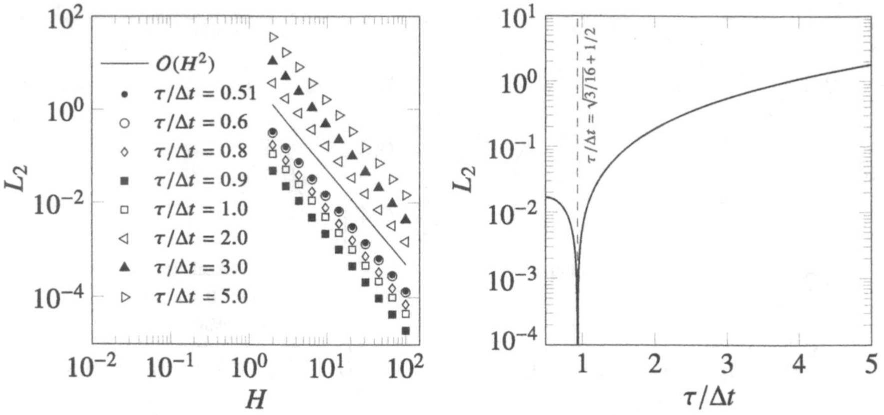

(a) Effect of H on L_2, fixing $\tau/\Delta t$. (b) Effect of τ on L_2, for $H = 9$ fixed.

Fig. 5.20 L_2 error of Poiseuille flow solutions with BGK model and bounce-back walls (assumed midway between boundary and solid nodes). Panel (**a**) shows that numerical solutions are second-order accurate with respect to the spatial resolution for all τ values (except $\tau/\Delta t = \sqrt{3/16} + 1/2$ which leads to third-order accuracy). Panel (**b**) shows the minimum of the error at $\tau/\Delta t = \sqrt{3/16} + 1/2$

the only error source in this problem) decreases with a second-order rate as function of grid resolution.

Figure 5.20b shows how the slip velocity error varies with τ for constant grid resolution. For τ close to $\Delta t/2$, the bounce-back error can be kept small, e.g. $L_2 \simeq 1.68\%$ for $\tau/\Delta t = 0.6$. Contrarily, for larger relaxation times, the error becomes unacceptably large, e.g. $L_2 > 180\%$ for $\tau/\Delta t = 5$! From the plot we can infer that the error grows with τ^2. We will derive this relation in Sect. 5.4.1.

In order to resolve the τ-dependent slip error, we can adopt several strategies. The most general approach is to replace the BGK collision model by the TRT or MRT model (cf. Chap. 10). In some particular cases where analytical solutions are accessible we can employ a viscosity calibration procedure [22, 31, 49, 61]. The idea consists in shifting the assumed wall position a certain distance away from the halfway location so that the no-slip velocity condition is fulfilled at the new wall location with the specific choice of τ. From Fig. 5.19 we see that this strategy would imply a wider channel whenever $\tau > \Delta t$.

Another approach to fix the numerical slip is to tune τ in such a way that the no-slip condition happens at the expected place. For example, if we want the no-slip condition *exactly* halfway between boundary and solid nodes, we have to set $\tau/\Delta t = \sqrt{3/16} + 1/2 \approx 0.93$. This way, the parabolic solution is exactly accommodated at the boundary and the bounce-back rule is turned into a *third-order accurate method* as per the discussion provided in Sect. 5.2.4. We can see the superiority of this choice of τ in Fig. 5.20b). We will explain the background of this optimum τ value in Sect. 5.4.1, see also more thorough discussions in [13, 15, 21, 23, 31, 33, 62].

For **Poiseuille flow** we observed that the bounce-back rule performs better with the walls located midway between boundary and solid nodes, just as for Couette flow. However, **the parabolic solution is generally not exact, due to an error depending on τ^2 and therefore on viscosity**. The slip error comes from the $O(\Delta x^2)$ term in the bounce-back closure relation. To avoid the unphysical viscosity-dependent slip, the safest procedure is to replace BGK by the TRT or MRT collision operators (cf. Sect. 10).

Still, no matter which collision model is adopted, the accuracy of the bounce-back rule depends on the relaxation parameters, and it can be further improved by carefully tuning them. That is to say, if with the BGK collision operator the formal third-order accuracy of the bounce-back rule in a straight midway wall is at $\tau/\Delta t = \sqrt{3/16} + 1/2 \approx 0.93$; with the TRT collision operator, the same condition holds for $\Lambda = 3/16$ which is also extensible to the MRT operator, cf. Sect. 10.

On top of the τ-dependent deficiency, the bounce-back setting of the no-slip condition is also anisotropic. For example, if the bounce-back wall is aligned with the diagonal lattice links (45° channel), the most accurate choice of τ is $\tau/\Delta t = \sqrt{3/8} + 1/2 \approx 1.11$ [15, 21, 33]. (In the TRT/MRT framework, this boils down to $\Lambda = 3/8$.) For arbitrary wall orientations the relaxation calibration of the bounce-back rule is no longer handy since the bounce-back rule becomes first-order accurate at best [13, 33], regardless the collision model employed. We will discuss alternative boundary conditions that are more suitable for arbitrary wall configurations in Chap. 11.

5.3.4 Solid Boundaries: Wet-Node Approach

Due to the viscosity dependence of bounce-back boundaries in combination with the BGK operator, another strategy was proposed for boundary conditions on straight walls: the so-called *wet-node* approach. Here, the boundary node is assumed to lie infinitesimally close to the actual boundary, but still inside the fluid domain. Thus, the boundary is effectively *on* the node, and the standard LB steps in the bulk (i.e. collision and streaming) also apply in the same way on the boundary nodes.

The idea of the wet-node approach is to assign suitable values for the unknown boundary populations such that the known and constructed populations reproduce the intended hydrodynamics at the boundary. The main challenge is that there are typically more unknown boundary populations than macroscopic conditions. This explains the large number of techniques developed to deal with this under-specified problem, e.g. [2, 9, 18, 20, 25–29]. Since covering all of them would be impossible in

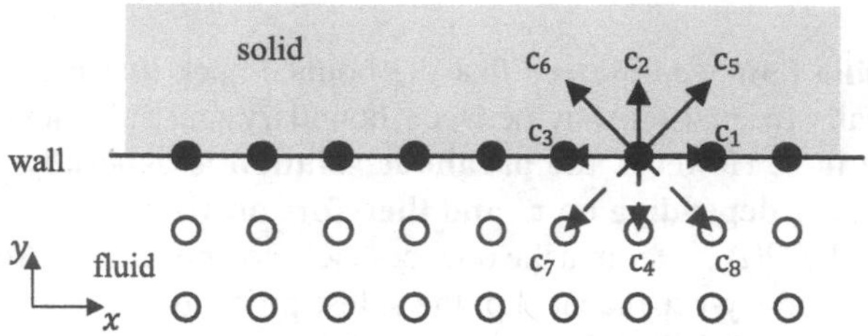

Fig. 5.21 Top wall coinciding with horizontal lattice links. The known boundary populations are represented by continuous vectors the unknown populations by dashed vectors

the context of this book, our focus is on three of the most popular approaches. These are the *equilibrium scheme* [23, 63], the *non-equilibrium extrapolation method* [9] and the *non-equilibrium bounce-back method* [27]. Although these approaches do not represent all wet-node techniques, they provide a good idea of the underlying philosophy.

5.3.4.1 Finding the Density on Boundaries

Before proceeding to the analysis of the wet-node approach itself, we will first introduce a general procedure to find the density (or pressure, for the incompressible model) on a straight boundary subject to a Dirichlet velocity condition [18, 27]. This procedure makes use of the flow continuity condition as a constraint to find the fluid density ρ from known fluid velocity $\boldsymbol{u}$ or mass flux $\rho\boldsymbol{u}$.[20]

Since the entire flow solution, boundaries included, must satisfy the continuity condition, it follows that one is also implicitly enforcing the wall density (or pressure) to a unique value while prescribing the wall velocity. However, the LBE does not work directly with the macroscopic fields. Consequently, the continuity condition does not arise naturally; rather it has to be enforced explicitly.

Fortunately, this is a rather simple task. The continuity of the fluid at the boundary is described by the well-known impermeable wall condition, i.e. zero relative normal velocity of the fluid. Hence, we just have to translate this condition to the populations f_i. For illustration purposes, let us see how to apply this procedure for a top wall (cf. Fig. 5.21). The extension to other wall orientations is straightforward, e.g. [18, 27].

[20]Although the LBM may reproduce the continuity condition in two different forms, depending on the equilibrium model adopted, i.e. $\partial_t \rho + \nabla \cdot (\rho \boldsymbol{u}) = 0$ for the standard compressible equilibrium or $c_s^{-2}\partial_t p + \rho_0 \nabla \cdot \boldsymbol{u} = 0$ for the incompressible equilibrium, the procedure described here remains applicable in both cases.

The impermeable wall condition is determined from the density and the vertical velocity component which in terms of populations is expressed as

$$\begin{aligned}\rho_{\mathrm{w}} &= \sum_i f_i = \underbrace{f_0 + f_1 + f_2 + f_3 + f_5 + f_6}_{\text{known}} + \underbrace{f_4 + f_7 + f_8}_{\text{unknown}}, \\ \rho_{\mathrm{w}} u_{\mathrm{w},y} &= \sum_i c_{iy} f_i = \underbrace{c\,(f_2 + f_5 + f_6)}_{\text{known}} - \underbrace{c\,(f_4 + f_7 + f_8)}_{\text{unknown}}\end{aligned} \tag{5.31}$$

where we have grouped the known and unknown populations (cf. Fig. 5.21) and introduced $c = \Delta x/\Delta t$. Combining the two equations above, we can determine ρ independently of the unknown populations as

$$\rho_{\mathrm{w}} = \frac{c}{c + u_{\mathrm{w},y}} \left[f_0 + f_1 + f_3 + 2\,(f_2 + f_5 + f_6) \right]. \tag{5.32}$$

Exercise 5.2 Show that the density on a bottom wall is given by

$$\rho_{\mathrm{w}} = \frac{c}{c - u_{\mathrm{w},y}} \left[f_0 + f_1 + f_3 + 2\,(f_4 + f_7 + f_8) \right]. \tag{5.33}$$

The major limitation of this procedure is that it only works on straight walls aligned with one of the coordinate axes. Still, the method is readily applicable to planar boundaries in either 2D or 3D problems.

5.3.4.2 Equilibrium Scheme

The equilibrium scheme [18, 23, 63] is possibly the simplest way to specify LB boundary conditions. It enforces the *equilibrium distribution* f_i^{eq} on the post-streaming boundary populations. The f_i^{eq} are readily available from the known macroscopic quantities at the boundary:

$$f_i(\boldsymbol{x}_{\mathrm{b}}, t) = f_i^{\mathrm{eq}} \left(\rho_{\mathrm{w}}, \boldsymbol{u}_{\mathrm{w}} \right). \tag{5.34}$$

The subscript b refers to the boundary node where wall properties (denoted by the subscript w) are imposed. Note that (5.34) applies to *all* populations f_i, instead of only the unknown ones.

Despite its attractive simplicity, the equilibrium scheme has a critical deficiency. While the basic principle of wet boundary nodes is that they have to be treated on an equal footing with bulk nodes, compatibility between boundary and bulk dynamics is limited to $\tau = \Delta t$ or $f_i^{\mathrm{neq}} = 0$. The latter condition is, however, of little interest since it leads to a trivial solution.[21] In the former condition of $\tau = \Delta t$, the LBM

---

[21] Recall that f_i^{neq} is related to the spatial derivatives of the flow field. Consequently, $f_i^{\mathrm{neq}} = 0$ corresponds to a spatially uniform solution.

becomes equivalent to a (second-order centred) finite difference (FD) scheme, so-called "finite Boltzmann scheme" [64]. Like in classical FD schemes, it becomes sufficient to prescribe the intended macroscopic solutions (i.e. $(\rho_\mathrm{w}, \boldsymbol{u}_\mathrm{w}) \to f_i^\mathrm{eq}$) on the boundary nodes. Only in this case do the LB solutions stay second-order accurate. Any other choice of τ breaks this equivalency between LBM and classical FD schemes [65, 66] and therefore degrades its accuracy to first order.

It is worth pointing out some of the advantages of the equilibrium scheme. One of them is its excellent stability. This is a remarkable feature considering the less stable behaviour usually exhibited by wet-node schemes. Another appealing advantage is that it can be easily applied to 2D and 3D problems.

Accuracy benchmark of the equilibrium scheme

We will now study the accuracy of the equilibrium scheme for a no-slip boundary condition by repeating the studies of the Couette and pressure-driven Poiseuille flows presented in Sect. 5.3.3. The description of the two flow problems is depicted in Fig. 5.22.

We make use of the same simulation parameters as employed in Sect. 5.3.3. The main difference is the implementation of the wall boundary conditions. We perform our analysis with the incompressible linear (Stokes) equilibrium as presented in Sect. 4.3.2 and further discussed in Sect. 5.3.2 and Sect. 5.3.3. Using the linear equilibrium simplifies the analysis without affecting numerical results since non-linear terms vanish in the problems studied here. The linear incompressible equilibrium populations on boundary nodes read

$$f_i^\mathrm{eq}(\boldsymbol{x}_\mathrm{b}, t) = w_i \left(\frac{p_\mathrm{w}}{c_\mathrm{s}^2} + \rho_0 \frac{\boldsymbol{c}_i \cdot \boldsymbol{u}_\mathrm{w}}{c_\mathrm{s}^2} \right) \tag{5.35}$$

where p_w and $\boldsymbol{u}_\mathrm{w}$ are known at the wall and ρ_0 denotes the constant density value.

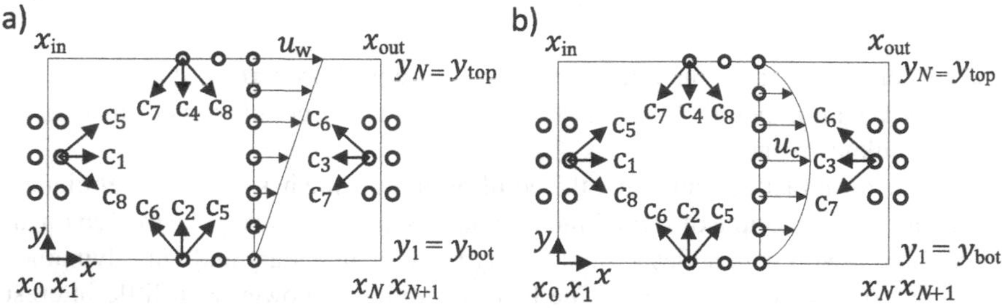

Fig. 5.22 Couette and Poiseuille flows showing the unknown boundary populations in the wet-node framework

We employ the following boundary conditions:

- Couette flow: $u_{w,x}(y_{bot}) = 0$ and $u_{w,x}(y_{top}) = 0.1c$
- Poiseuille flow: $u_{w,x}(y_{bot}) = u_{w,x}(y_{top}) = 0$

The transverse velocity u_y is always zero in both cases. The boundary pressure in (5.35) is found through the procedure explained above. However, since the velocity moments using the incompressible model are sightly different from the standard equilibrium case, it is worth re-writing them here. For example, for a boundary node located at the top wall (cf. Fig. 5.21), we have

$$\frac{p_w}{c_s^2} = \sum_i f_i = \underbrace{f_0 + f_1 + f_2 + f_3 + f_5 + f_6}_{\text{known}} + \underbrace{f_4 + f_7 + f_8}_{\text{unknown}}, \tag{5.36}$$

$$\rho_0 u_{w,y} = \sum_i c_{iy} f_i = \underbrace{c\,(f_2 + f_5 + f_6)}_{\text{known}} - \underbrace{c\,(f_4 + f_7 + f_8)}_{\text{unknown}}. \tag{5.37}$$

Combining them, we obtain

$$\frac{p_w}{c_s^2} = -\frac{\rho_0 u_{w,y}}{c} + \left[f_0 + f_1 + f_3 + 2\,(f_2 + f_5 + f_6)\right]. \tag{5.38}$$

Repeating this calculation for the bottom wall we obtain

$$\frac{p_w}{c_s^2} = \frac{\rho_0 u_{w,y}}{c} + \left[f_0 + f_1 + f_3 + 2\,(f_4 + f_7 + f_8)\right]. \tag{5.39}$$

We now have everything required to implement the equilibrium scheme. The results are shown in Fig. 5.23. Despite its simplicity, the lack of accuracy may compromise the usefulness of this boundary scheme. In fact, we can confirm that neither Couette nor Poiseuille flow solutions are accurately reproduced when $\tau \neq \Delta t$.[22]

As Fig. 5.23 illustrates, the equilibrium scheme is unable to capture the simple linear Couette profile for $\tau \neq \Delta t$. Due to the lack of non-equilibrium terms, the boundary scheme is unaware of the presence of velocity gradients near the wall. Obviously, this situation becomes worse for even steeper velocity gradients. Therefore, it is not a surprise that even larger discrepancies are observed for Poiseuille flow. Only for $\tau/\Delta t = 1$ the equilibrium scheme complies with the accuracy of LBM in bulk. In this case, both Couette and Poiseuille flow solutions are correct up to machine precision.

[22]Obviously, the use of other equilibrium distributions (e.g. the full standard equilibrium) will not lead to improvements as the problem identified here comes from neglecting the non-equilibrium part of the boundary populations.

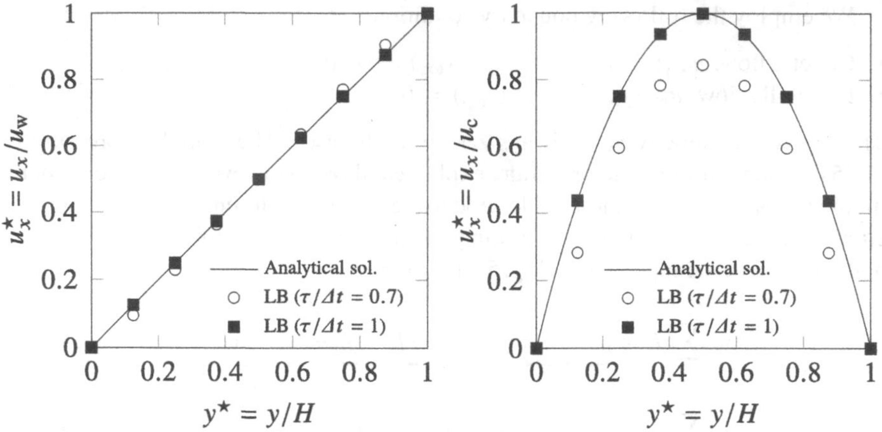

Fig. 5.23 LB solutions using the equilibrium boundary scheme for channel width $N = 9$. *Left plot*: Couette flow; *right plot*: Poiseuille flow. Results with $\tau/\Delta t = 0.7$ are inaccurate while $\tau/\Delta t = 1$ leads to machine-accurate solutions

5.3.4.3 Non-equilibrium Extrapolation Method

The previous example showed that, in general, the equilibrium scheme fails to be an accurate boundary method since it neglects the non-equilibrium part f_i^{neq} of the boundary populations f_i. As such, the next logical step of improvement is to include the non-equilibrium part. The new difficulty is how to find the non-equilibrium term for the boundary populations. As there is no unique answer to this question, several procedures have been proposed, resulting in a variety of wet-node techniques.

Perhaps the simplest way to determine the non-equilibrium part (although non-local) is to extrapolate its value from the fluid region where f_i^{neq} is known. This *non-equilibrium extrapolation* approach was originally proposed by Guo et al. [9]. It is applied to the post-streamed boundary populations according to

$$f_i(\boldsymbol{x}_\text{b}, t) = f_i^\text{eq}(\rho_\text{w}, \boldsymbol{u}_\text{w}) + \left(f_i(\boldsymbol{x}_\text{f}, t) - f_i^\text{eq}(\rho_\text{f}, \boldsymbol{u}_\text{f})\right) \tag{5.40}$$

where the non-equilibrium contribution comes from the fluid node $\boldsymbol{x}_\text{f}$ next to $\boldsymbol{x}_\text{b}$ along the boundary normal vector. Similarly to the equilibrium scheme, the non-equilibrium extrapolation method replaces *all* boundary populations, rather than only the unknown ones.

Equation (5.40) is consistent with the second-order accuracy of the LBE [9, 11]. However, in light of the discussion in Sect. 5.2.4, this is not sufficient to support the level of exactness of the LB solution in the bulk (parabolical exactness). We can easily understand this through the Chapman-Enskog analysis to unfold the content of f_i^{neq}. For this analysis it is sufficient to truncate the Chapman-Enskog expansion after the first order, i.e. $f_i^{\text{neq}} \simeq \epsilon f_i^{(1)}$ since failure at the first order already compromises the overall success.

Neglecting the $O(u^2)$ terms in (4.35), we have

$$f_i^{\text{neq}}(\boldsymbol{x}, t) \simeq -w_i \frac{\tau\rho}{c_s^2} Q_{i\alpha\beta} \partial_\beta u_\alpha \tag{5.41}$$

where $Q_{i\alpha\beta} = c_{i\alpha}c_{i\beta} - c_s^2\delta_{\alpha\beta}$. From (5.41) it follows that the zeroth order extrapolation of f_i^{neq} is equivalent to prescribing a *constant* $\nabla \boldsymbol{u}$. To capture a fluid velocity $\boldsymbol{u}$ that varies quadratically, we require at least a linearly correct approximation of $\nabla \boldsymbol{u}$ or, in the LB context, a first-order accurate extrapolation of f_i^{neq}. For that reason, the non-equilibrium extrapolation method in the form of (5.40) is unable to reproduce exact parabolic solutions, although it represents an improvement over the equilibrium scheme as we will see shortly.

Accuracy benchmark of the non-equilibrium extrapolation method

We now study the accuracy of the non-equilibrium extrapolation method by repeating the Couette and (pressure-driven) Poiseuille flow examples as carried out before (cf. Fig. 5.22). The boundary populations obey (5.40).

The non-equilibrium extrapolation method exactly solves the linear Couette flow, regardless of the value of τ. We could have expected this result since the first-order extrapolation of f_i^{neq} can accurately capture the constant velocity gradient of a linear profile. However, errors appear in the parabolic Poiseuille flow. By setting $f_i^{\text{neq}}(x, y_1) = f_i^{\text{neq}}(x, y_2)$ we are implicitly enforcing $\nabla \boldsymbol{u}|_{(x,y_1)} = \nabla \boldsymbol{u}|_{(x,y_2)}$, a condition that cannot be possibly satisfied in a parabolic solution. The effect of this inaccuracy is illustrated in Fig. 5.24; the error decreases with resolution at a second-

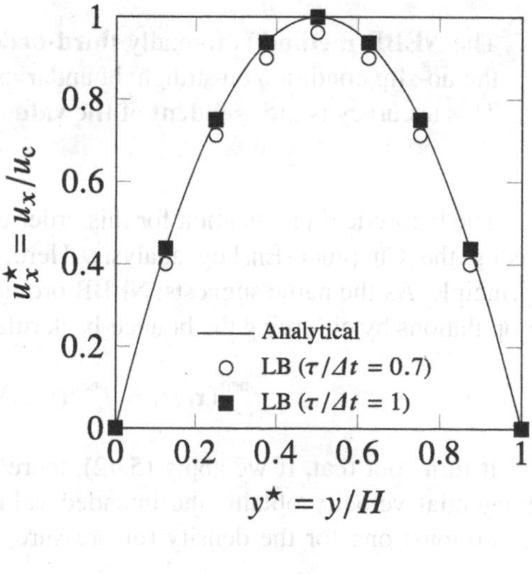

Fig. 5.24 LB solutions with the non-equilibrium extrapolation method for channel width $N = 9$. The result with $\tau/\Delta t = 0.7$ is inaccurate while $\tau/\Delta t = 1$ leads to machine-accurate solutions. Note the improvement compared to Fig. 5.23

order rate, as shown in [9]. Once again, this inaccuracy is absent with the relaxation time $\tau/\Delta t = 1$ (where the non-equilibrium does not contribute to solutions) and consequently the parabolic solution is described exactly.

This exercise confirms the necessity of supplying information about $f_i^{\text{neq}} = \epsilon f_i^{(1)}$ on the boundary node and that this is the necessary condition to achieve second order accuracy (recall Sect. 5.2.4). There are a number of different ways to reconstruct $f_i^{(1)}$ on boundary nodes [2, 18, 20, 27–29]. Since its hydrodynamic content is the velocity gradient on the boundary, cf. (5.41), one approach to construct f_i^{neq} is to directly evaluate $\nabla \boldsymbol{u}$ with finite differences [2, 18]. The downside of this procedure is that it requires information from neighbouring nodes, thereby losing locality as one of the main advantages of the LB algorithm. Furthermore, all methods addressing only $f_i^{(1)}$ are plagued by one key limitation: they are *not* enough accurate to accommodate the solution curvature on the boundary. This is evidenced by their overall inability to describe parabolic flow solutions.

5.3.4.4 Non-equilibrium Bounce-Back Method

In order to find f_i^{neq} locally and improve for the boundary scheme accuracy Zou and He [27] proposed an alternative method. This strategy has become quite popular and is often referred to by the names of its inventors (*Zou-He method*) or by its underlying principle (*non-equilibrium bounce-back method* or *NEBB method*).

The reason for the higher accuracy of the NEBB method comes from its ability to also capture the $f_i^{(2)}$ term, thus making it exact for parabolic solutions. This leads to an accuracy increase by one order, compared to standard second-order LB boundary schemes.

The **NEBB method** is formally **third-order accurate** in the prescription of the no-slip condition on straight boundaries coincident with the lattice nodes. This accuracy is **independent of the value of the relaxation parameters**.

The theoretical justification for this order of accuracy is provided in Sect. 5.4.1, using the Chapman-Enskog analysis. Here, let us explain the NEBB working principle. As the name suggests, NEBB provides values for the unknown boundary populations by enforcing the bounce-back rule for their non-equilibrium part:

$$f_{\bar{i}}^{\text{neq}}(\boldsymbol{x}_\text{b}, t) = f_i^{\text{neq}}(\boldsymbol{x}_\text{b}, t) \quad (\boldsymbol{c}_{\bar{i}} = -\boldsymbol{c}_i). \tag{5.42}$$

It turns out that, if we apply (5.42), there will be no way to guarantee that the tangential velocity obtains the intended value. So far we used two macroscopic conditions: one for the density (or pressure) and another for the normal velocity at the boundary, cf. (5.31). The condition for the tangential velocity at the boundary has not yet been linked with the prescription of the boundary populations.

To establish this link, we can add an extra term to (5.42) that includes this information by modifying only the tangential component of boundary populations. This term is called the *transverse momentum correction* N_t [67]. Its sign is determined by $\boldsymbol{t}$, a tangent unit vector along the wall. (By convention, we choose $\boldsymbol{t}$ to point along the positive direction of the Cartesian axis. Choosing the opposite convention would merely change the sign of N_t.) The modified NEBB rule reads

$$f_{\bar{i}}^{\text{neq}}(\boldsymbol{x}_{\text{b}}, t) = f_i^{\text{neq}}(\boldsymbol{x}_{\text{b}}, t) - \frac{\boldsymbol{t} \cdot \boldsymbol{c}_i}{|\boldsymbol{c}_i|} N_t. \tag{5.43}$$

In Example 5.3 we illustrate how to determine N_t from the tangential velocity.

Unlike the other two wet-node strategies previously described where the boundary scheme replaces *all* populations at the boundary, the NEBB method modifies *only the missing populations*. This has both advantages and disadvantages [18]. On the upside, NEBB does not touch the known populations, i.e. it makes full use of all information already available. According to [18, 20], this is believed to lead to improved accuracy. On the downside, it is difficult to generalise the scheme to different velocity sets, particularly to those with a large number of velocities. In fact, the NEBB is cumbersome to employ in 3D [27, 67, 68] (cf. Sect. 5.4.4). Furthermore, the boundary scheme itself may become a source of instability. It has been shown that the NEBB scheme introduces undesirable high wave number perturbations in the bulk solution [20, 69, 70], meaning that it cannot be applied to high Reynolds number flows [18, 20].

Example 5.3 Let us now apply the NEBB method to a Dirichlet velocity condition for a top wall as depicted in Fig. 5.21. For generality, assume a tangentially moving wall with fluid injection, i.e. $\boldsymbol{u}_{\text{w}} = (u_{\text{w},x}, u_{\text{w},y})^{\top}$. We solve the problem as follows:

1. The first step is to identify the missing populations, for a top wall f_4, f_7 and f_8.
2. We now have to determine the wall density independently of the unknown populations. For a top wall, the unknown density ρ_{w} is given by (5.32). As the wall velocity is also known, the equilibrium part f_i^{eq} of the boundary populations is now fully determined.
3. The next step is to express the unknown populations through the known populations and parameters, using (5.43):

$$\left.\begin{aligned} f_4^{\text{neq}} &= f_2^{\text{neq}} \\ f_7^{\text{neq}} &= f_5^{\text{neq}} - N_x \\ f_8^{\text{neq}} &= f_6^{\text{neq}} + N_x \end{aligned}\right\} \quad \Longrightarrow \quad \left\{\begin{aligned} f_4 &= f_2 + (f_4^{\text{eq}} - f_2^{\text{eq}}) \\ f_7 &= f_5 + (f_7^{\text{eq}} - f_5^{\text{eq}}) - N_x \\ f_8 &= f_6 + (f_8^{\text{eq}} - f_6^{\text{eq}}) + N_x \end{aligned}\right. . \tag{5.44}$$

Using the known equilibrium distributions from (5.35) we get

$$
\begin{aligned}
f_4 &= f_2 - \tfrac{2}{3c}\rho_w u_{w,y}, \\
f_7 &= f_5 - \tfrac{1}{6c}\rho_w (u_{w,x} + u_{w,y}) - N_x, \\
f_8 &= f_6 - \tfrac{1}{6c}\rho_w (-u_{w,x} + u_{w,y}) + N_x.
\end{aligned}
\tag{5.45}
$$

4. Now we compute N_x by resorting to the first-order velocity moment along the boundary tangential direction:

$$
\begin{aligned}
\rho_w u_{w,x} &= \sum_i c_{ix} f_i \\
&= c\,(f_1 + f_5 + f_8) - c\,(f_3 + f_6 + f_7) \\
&= c\,(f_1 - f_3) - c\,(f_7 - f_5) + c\,(f_8 - f_6) \\
&= c\,(f_1 - f_3) - \tfrac{1}{3}\rho_w u_{w,x} + 2N_x.
\end{aligned}
\tag{5.46}
$$

This gives

$$
N_x = -\frac{1}{2}(f_1 - f_3) + \frac{1}{3c}\rho_w u_{w,x}. \tag{5.47}
$$

5. Finally, we get closed-form solutions for all unknown populations, which are exactly those derived in [27]:

$$
\begin{aligned}
f_4 &= f_2 - \tfrac{2}{3c}\rho_w u_{w,y}, \\
f_7 &= f_5 + \tfrac{1}{2}(f_1 - f_3) - \tfrac{1}{2c}\rho_w u_{w,x} - \tfrac{1}{6c}\rho_w u_{w,y}, \\
f_8 &= f_6 - \tfrac{1}{2}(f_1 - f_3) + \tfrac{1}{2c}\rho_w u_{w,x} - \tfrac{1}{6c}\rho_w u_{w,y}.
\end{aligned}
\tag{5.48}
$$

Exercise 5.3 Show that the NEBB method for a Dirichlet velocity condition at a left wall results in

$$
\rho_w = \frac{c}{c - u_{w,x}}\left[f_0 + f_2 + f_4 + 2\,(f_3 + f_6 + f_7)\right], \tag{5.49}
$$

$$
f_1 = f_3 + \tfrac{2}{3c}\rho_w u_{w,x}, \tag{5.50}
$$

$$
f_5 = f_7 - \tfrac{1}{2}(f_2 - f_4) + \tfrac{1}{2c}\rho_w u_{w,y} + \tfrac{1}{6c}\rho_w u_{w,x}, \tag{5.51}
$$

$$
f_6 = f_8 + \tfrac{1}{2}(f_2 - f_4) - \tfrac{1}{2c}\rho_w u_{w,y} + \tfrac{1}{6c}\rho_w u_{w,x}. \tag{5.52}
$$

Example 5.4 We check the accuracy of the NEBB method by repeating the Couette and (pressure-driven) Poiseuille flow examples (cf. Fig. 5.22) as carried out previously. Here we obtain exact solutions up to machine accuracy in both cases (data not shown), independently of the choice of τ. This result confirms that the *NEBB method accommodates the bulk solution with third-order accuracy* for straight walls aligned with the lattice nodes.

In conclusion, the NEBB method has superior accuracy compared to the other wet-node schemes that we looked at. However, it is not the only available choice. For example, the methods proposed by Inamuro [26] or by Noble et al. [25] offer the same level of accuracy, for geometries with walls coinciding with lattice nodes. A more general wet-node technique was suggested by Ginzburg and d'Humières [13] in the early days of LB research. This wet-node boundary technique prescribes the wall velocity with third-order accuracy, for any boundary position and/or orientation. However, it has the disadvantage of being quite cumbersome to implement. Nevertheless, many subsequent schemes, e.g. [20, 28, 29, 71], have relied on ideas from [13] to develop simpler algorithms, but at the price of reducing the accuracy to second order.

Finally, we note that all wet-node methods, with the exception of [13], have a common limitation that is also shared by the simple bounce-back rule. When applied to walls not coinciding with lattice nodes, their accuracy reduces to first order. In this case, we have to use more advanced boundary schemes as covered in Chap. 11.

5.3.5 *Open Boundaries*

Due to computational constraints it is often necessary to truncate the simulation domain. Unfortunately, by cropping the physical domain, new boundary conditions have to be prescribed at places where the physical problem had originally no boundaries at all. We call these *open boundaries*.

Open boundaries consist of **inlets** or **outlets** where the flow either enters or leaves the computational domain and where we should impose, for example, **velocity or density profiles**. This is often non-trivial, and it can cause both physical and numerical difficulties.

Physically, the difficulty comes from the role of open boundaries: they are supposed to guarantee that what enters/leaves the computational domain is compatible with the bulk physics of the problem. However, this information is generally not accessible beforehand. Thus, we must rely on some level of *physical approximation*, e.g. by assuming fully developed flow conditions at inlet/outlet boundaries.

Numerically, open boundaries may also be complicated by a feature inherent to LBM: the presence of pressure waves, cf. Chap. 12. The "simple" enforcement of either velocity or density at boundaries will generally reflect these waves back into the computational domain. Since this is undesirable, special non-reflecting open boundary conditions may be required in some cases. These are covered in Sect. 12.4.

Open LB boundary conditions are still matter of active research, e.g. [72, 73]. While more advanced strategies exist (cf. e.g. Sect. 12.4), this section provides an introduction. We will discuss how to implement a Dirichlet boundary condition for an imposed velocity or pressure profile using (i) the link-wise bounce-back method [30, 49, 72] and (ii) the wet-node non-equilibrium bounce-back method [18, 27, 72]. Inlets and outlets are treated on an equal footing in both schemes.

5.3.5.1 Velocity Boundary Conditions: Bounce-Back Approach

The Dirichlet condition for fluid velocity at open boundaries is easily realised by the bounce-back technique [22, 49]. The formulation is the same as in (5.26), with $\boldsymbol{u}_\mathrm{w}$ determining the inlet and/or outlet boundary velocity. ($\boldsymbol{u}_\mathrm{w}$ may have both normal and tangential components). The bounce-back algorithm still places the inlet/outlet boundary, with its imposed velocity $\boldsymbol{u}_\mathrm{w}$, midway between the lattice nodes.

5.3.5.2 Pressure Boundary Conditions: Anti-bounce-Back Approach

The prescription of pressure at open boundaries can be performed with a technique similar to the bounce-back method, called the *anti-bounce-back* method [30, 31, 69]. In this method, the sign of the bounced-back populations is changed:

$$f_{\bar{i}}(\boldsymbol{x}_\mathrm{b}, t + \Delta t) = -f_i^\star(\boldsymbol{x}_\mathrm{b}, t) + 2w_i\rho_\mathrm{w}\left[1 + \frac{(\boldsymbol{c}_i \cdot \boldsymbol{u}_\mathrm{w})^2}{2c_\mathrm{s}^4} - \frac{u_\mathrm{w}^2}{2c_\mathrm{s}^2}\right]. \tag{5.53}$$

The notation in (5.53) is the same as in (5.26). As for the bounce-back method, the boundary is located $\Delta x/2$ outside the boundary node.

Equation (5.53) requires specification of $\boldsymbol{u}_\mathrm{w}$ which is not generally known. The problem is similar to finding ρ_w for the velocity Dirichlet correction in the standard bounce-back rule in (5.26). A possible way to estimate $\boldsymbol{u}_\mathrm{w}$ is by extrapolation. For example, following [69], we may estimate $\boldsymbol{u}_\mathrm{w} = \boldsymbol{u}(\boldsymbol{x}_\mathrm{b}) + \frac{1}{2}\left[\boldsymbol{u}(\boldsymbol{x}_\mathrm{b}) - \boldsymbol{u}(\boldsymbol{x}_{\mathrm{b}+1})\right]$, where $\boldsymbol{x}_\mathrm{b}$ and $\boldsymbol{x}_{\mathrm{b}+1}$ refer to the boundary node and the next interior node following the inward normal vector of the boundary.

Since only the square of the boundary velocity appears in (5.53), inaccuracies in the approximation of $\boldsymbol{u}_\mathrm{w}$ are $O(\mathrm{Ma}^2)$ and therefore usually small. Additionally, (5.53) can be further augmented with a correction term to eliminate second-order error terms, making it formally third-order accurate [30, 31, 69, 74].

5.3.5.3 Velocity Boundary Conditions: Wet-Node Approach

The wet-node formulation of velocity boundary conditions applies in the same manner for open boundaries as for walls. The only difference is that the velocity describes the mass flux entering or leaving the domain, rather than the motion of a solid wall. The methods discussed in Sect. 5.3.4 remain equally applicable here.

5.3.5.4 Pressure Boundary Conditions: Wet-Node Approach

In the wet-node approach, the algorithm for imposing pressure and velocity boundary conditions is the same. The only difference lies in the strategy to find the unknown macroscopic properties at the boundary. For instance, for a prescribed inlet velocity, the two equations in (5.31) are combined to find the unknown ρ_w which yields (5.32). Differently, for a prescribed inlet pressure, these two equations are combined to solve for the unknown wall normal velocity. Considering the top boundary depicted in Fig. 5.21 as an open boundary traversed by flow along the y-axis, then (5.31) solves for the boundary velocity:

$$u_{w,y} = -c + \frac{c}{\rho_w}\left[f_0 + f_1 + f_3 + 2(f_2 + f_5 + f_6)\right]. \tag{5.54}$$

The sign of $u_{w,y}$, i.e. inflow or outflow condition, is determined by the solution of (5.54). Consequently, this algorithm applies equally for inlets and outlets.

5.3.6 Corners

So far we limited the discussion on boundary conditions to straight surfaces. In both 2D and 3D domains, there are also other geometrical features that we have to consider.

In **2D domains** we have to deal with geometrical features where two straight surfaces intersect, called **corners**. In **3D domains**, we need to consider both **edges** and **corners**, i.e. places where two and three surfaces cross, respectively.

The focus of this section is on the treatment of 2D corners. The 3D case can be deduced from the principles provided here; we will get back to this in Sect. 5.4.4.

Usually the solution of the flow field should be smooth at corners.[23] While it is clear that corners are inherent to rectangular domains, they can sometimes be avoided by switching from velocity/pressure to periodic boundary conditions as shown in Fig. 5.25.

It is evident that the treatment of corners cannot be avoided in most practical flow geometries. The flow over a step, with concave and convex corners, is one such example, cf. Fig. 5.26a. Another example, shown in Fig. 5.26b, is a staircase approximation of an inclined wall.

[23]Here we will exclude cases where this requirement is violated. An example is the lid-driven cavity flow [62, 75], an often used benchmark problem with a velocity discontinuity at corners.

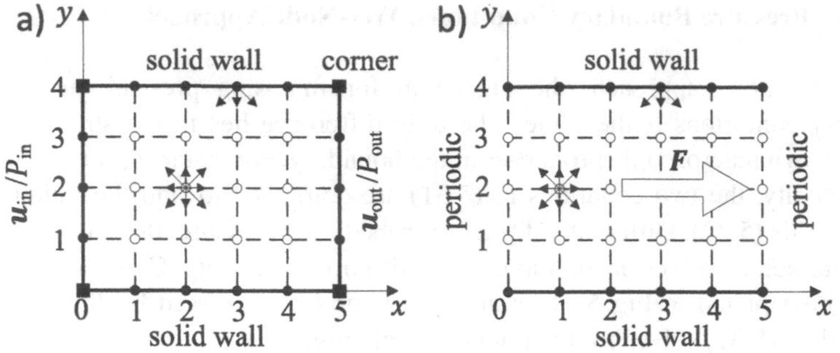

Fig. 5.25 Different implementations of a Poiseuille flow (Fluid nodes: *open circles* symbol; Boundary nodes: *solid circles* symbol; Corner nodes: *filled squares* symbol). (**a**) Velocity/pressure boundary conditions for the inlet and outlet and (**b**) periodic boundary conditions with a force density instead of a pressure gradient. While corners are necessary in (**a**), this is not the case in (**b**)

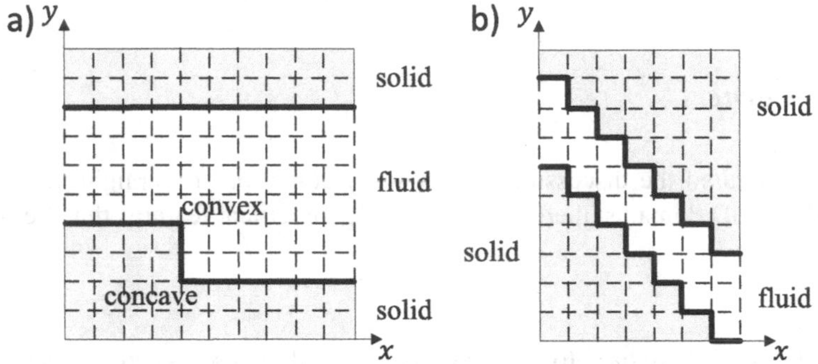

Fig. 5.26 Examples of flow geometries with corners: (**a**) channel with a sudden expansion, (**b**) discretisation of a diagonal channel

Even in problems where corners account for only a few points in the solution domain, like in Fig. 5.26a, the relevance of these points should not be underestimated. Recall from the discussion in Sect. 5.2.3 that one single point may contaminate the numerical solution everywhere [14]. Similarly to the case of planar boundaries, it turns out that the state of affairs in the modelling of corners is still not entirely settled.

One of the earliest systematic approaches to treat corners in LBM was proposed by Maier et al. [34]. Later, Zou and He [27] suggested another approach to specify corners, based on their NEBB method [27]. Yet, as this approach lacks generality in 3D implementations, Hecht and Harting [67] extended it to 3D (using D3Q19 as an example). As of today, numerous contributions have been made for both 2D and 3D problems that are compatible with different realisations of boundary condition. For example, [13, 14, 30] proposed corner approaches for link-wise methods, while [17, 20, 71] focused on wet-node boundary schemes.

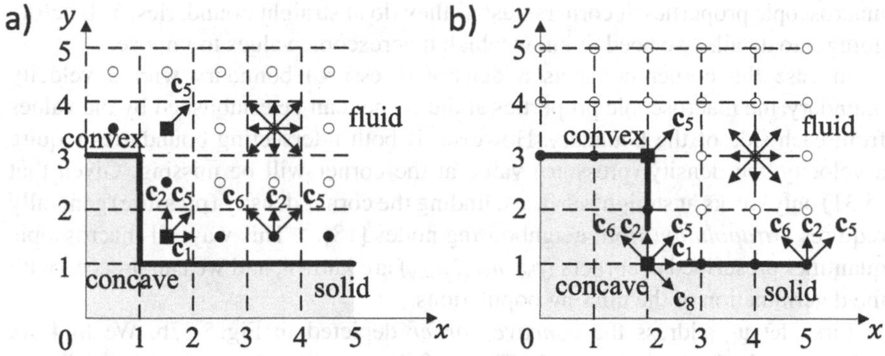

Fig. 5.27 Corner conditions for (**a**) link-wise (bounce-back) and (**b**) wet-node (NEBB) methods (Fluid nodes: *open circles* symbol; Boundary nodes: *solid circles* symbol; Corner nodes: *filled squares* symbol). Each case illustrates which unknown populations require specification at boundary and corner nodes

In the following we will limit our discussion to the implementation of corners for the bounce-back rule as a link-wise approach and for the NEBB method as a wet-node approach. As shown in Fig. 5.27, both methods have fundamental differences.

5.3.6.1 Corners and the Bounce-Back Rule

The bounce-back rule works in the same way for straight walls and corners, and it is the same for concave and convex cases: the unknown populations leaving corners are determined through the full reflection of the known incoming ones.

The first step is to identify the unknown corner populations (cf. Fig. 5.27a): f_1, f_2 and f_5 in the concave corner and solely f_5 in the convex corner. These populations are found through the standard bounce-back process in (5.24). All remaining populations passing through the corner are not subject to the bounce-back rule since they come from neighbouring fluid or boundary nodes.

From the application point of view, the *advantages and disadvantages of the bounce-back rule for corners* are similar to those at planar surfaces. The positive aspects remain the ease of implementation, strict conservation of mass and good stability characteristics (even close to $\tau = \Delta t/2$). On the other hand, the (possibly) lower accuracy, also featuring viscosity-dependent errors (if using the BGK model), remains the main disadvantages of the method.

5.3.6.2 Corners and the NEBB Method

In the wet-node approach, corners have to be treated in a special way. The big picture is that we have to ensure that known and unknown populations yield the desired macroscopic properties at corners, just as they do at straight boundaries. Yet, before going into details, we need to know which macroscopic values to impose.

In case the corner connects a density (pressure) boundary with a velocity boundary, the macroscopic properties at the corner can be established by the values from each side of the boundary. However, if both intersecting boundaries require a velocity, the density (pressure) value at the corner will be missing. Given that (5.31) only works at straight surfaces, finding the corner density (pressure) generally requires *extrapolation* from neighbouring nodes [18].[24] This way, all macroscopic quantities prescribed at corners (ρ_w, $u_{\mathrm{w},x}$, $u_{\mathrm{w},y}$) are known, and we can proceed with the determination of the missing populations.

First, let us address the *concave corner* depicted in Fig. 5.27b. We find six unknowns: f_0, f_1, f_2, f_5, f_6 and f_8. Three of them can be determined immediately by applying the non-equilibrium bounce-back rule:

$$\begin{aligned} f_1^\mathrm{neq} &= f_3^\mathrm{neq}, \\ f_2^\mathrm{neq} &= f_4^\mathrm{neq}, \\ f_5^\mathrm{neq} &= f_7^\mathrm{neq}. \end{aligned} \tag{5.55}$$

By substituting $f_i^\mathrm{neq} = f_i - f_i^\mathrm{eq}$ and using the standard LB equilibrium for f_i^eq, (5.55) becomes

$$\begin{aligned} f_1 &= f_3 + \tfrac{2}{3c}\rho_\mathrm{w} u_{\mathrm{w},x}, \\ f_2 &= f_4 + \tfrac{2}{3c}\rho_\mathrm{w} u_{\mathrm{w},y}, \\ f_5 &= f_7 + \tfrac{1}{6c}\rho_\mathrm{w}(u_{\mathrm{w},x} + u_{\mathrm{w},y}). \end{aligned} \tag{5.56}$$

However, the problem is not closed yet. There are still three unknown populations: f_0, f_6 and f_8. We can work out a solution by taking advantage of the fact that the number of unknown populations matches that of the conservation laws to satisfy:

$$\begin{aligned} \rho_\mathrm{w} &= \underline{f_0} + f_1 + f_2 + f_3 + f_4 + f_5 + \underline{f_6} + f_7 + \underline{f_8}, \\ \rho_\mathrm{w} u_{\mathrm{w},x} &= c(f_1 + f_5 + \underline{f_8}) - c(f_3 + \underline{f_6} + f_7), \\ \rho_\mathrm{w} u_{\mathrm{w},y} &= c(f_2 + f_5 + \underline{f_6}) - c(f_4 + f_7 + \underline{f_8}) \end{aligned} \tag{5.57}$$

where the underlined terms are unknown. In principle, the two non-rest populations, f_6 and f_8, can be determined from solving the second and third equations. Unfortunately, the system is non-invertible. Therefore, in order to make it solvable, we must introduce an additional constraint. A possible (but non-unique) constraint may

---

[24] A possible numerical approximation procedure is using the formula given in Sect. 5.3.5 to find $\boldsymbol{u}_\mathrm{w}$ in the anti-bounce-back approach, but now applied to ρ_w.

demand that f_6 and f_8 have equal magnitude. This choice leads to

$$\begin{aligned} f_6 &= \tfrac{1}{12c}\rho_w(u_{w,y} - u_{w,x}), \\ f_8 &= \tfrac{1}{12c}\rho_w(u_{w,x} - u_{w,y}). \end{aligned} \tag{5.58}$$

In the typical scenario of two intersecting walls at rest, (5.58) boils down to $f_6 = f_8 = 0$. Yet, the prescription of other values (even negative ones!) is also admissible. The reason is that, whatever their content is, the corner populations f_6 and f_8 do never propagate inside the fluid domain. As seen in Fig. 5.27b, these populations belong to links never pointing into the flow domain; these links are called *buried links* [34]. Thereby, buried populations never contaminate the bulk solution. The only demand on f_6 and f_8 is that they satisfy mass and momentum conservation in the corner, according to (5.57).

At last, we need to determine the rest population f_0. We can enforce the mass conservation at the corner site by using the first equation in (5.57):

$$f_0 = \rho_w - \sum_{i=1}^{8} f_i. \tag{5.59}$$

Altogether, the unknown populations in the concave corner are determined by (5.56), (5.58) and (5.59). It is interesting to note that, if the corner is at rest, i.e. $u_{w,x} = u_{w,y} = 0$, the present approach reduces to the (node) bounce-back rule, supplemented by (5.59).

Concerning the case of a convex corner in Fig. 5.27b, we identify only one missing population streaming out of the solid region: f_5. Consequently, if using the same procedure as for concave corners, the problem becomes over-specified; the no-slip condition cannot be satisfied at the convex corner node [34]. An alternative to overcome this problem is to employ the simple (node) bounce-back rule:

$$f_5 = f_7. \tag{5.60}$$

From this discussion we can conclude that the complexity of the wet-node approach for corners is its main drawback. This comes with a lack of flexibility in handling certain geometrical features such as convex corners. These problems are even worse in 3D where we have to deal with an even larger number of unknown populations [67], as we will discuss in Sect. 5.4.4. In addition to these problems, the NEBB method also suffers from other disadvantages, such as the violation of the exact mass conservation and its weak stability properties for $\tau \to \Delta t/2$, cf. Sect. 5.3.4. As advantage, this approach generally offers an accuracy superior to that of the bounce-back rule. Furthermore, the viscosity dependence that affects the bounce-back rule with BGK collision operator does not concern the wet-node approach.

5.3.7 Symmetry and Free-Slip Boundaries

Symmetry boundaries, where one half of the domain is the mirror image of the other, can be implemented in a simple way which is quite similar to the bounce-back method. Symmetry boundaries can be useful for reducing the time and memory requirements of simulations where such a symmetry plane exists. For instance a simulation of Poiseuille flow, which is symmetric, can be halved in size by simulating only half of the channel with a symmetry boundary at the channel centre as shown in Fig. 5.28.

First of all, let us consider what **mirror symmetry** implies. For a symmetry plane lying exactly midway on the links between two rows of nodes, the **populations on one side** of the boundary must **exactly mirror those on the other**. Concretely, for a symmetry plane lying at y, the populations on either side of the boundary are related as $f_i(x, y - \Delta x/2, t) = f_j(x, y + \Delta x/2, t)$, with i and j related so that only the normal velocity is reversed, i.e. $c_{j,\mathrm{n}} = -c_{i,\mathrm{n}}$. This is shown to the left in Fig. 5.29.

Now let us consider what happens after streaming in the main domain, i.e. the domain we want to simulate, and the mirror domain, i.e. the domain which we do not want to explicitly simulate. As shown to the right in Fig. 5.29, populations from a node $\boldsymbol{x}_\mathrm{b}$ in the main domain stream into the mirror domain, while mirrored populations from $\boldsymbol{x}_\mathrm{b}$'s mirror image stream to the main domain, to $\boldsymbol{x}_\mathrm{b}$ and its neighbours parallel to the symmetry plane. Thus, all populations entering the main domain from the mirror domain are exactly related to those going from the main domain to the mirror domain.

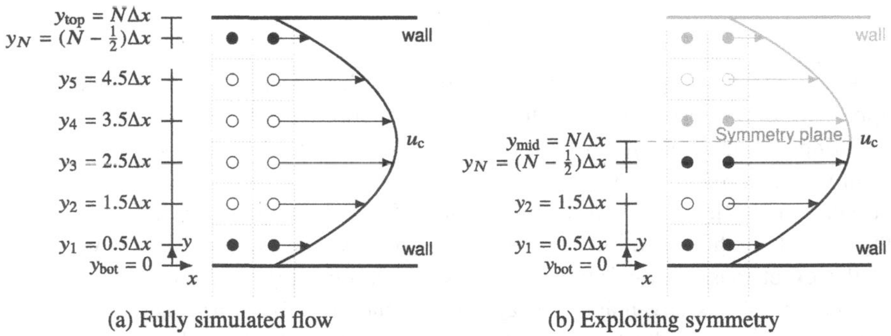

Fig. 5.28 Poiseuille flow geometry with bounce-back walls, showing fluid nodes (Fluid nodes: *open circles* symbol) and boundary nodes (Boundary nodes: *solid circles* symbol) (**a**) fully simulated with a system width of $N = 6$ nodes; (**b**) using a symmetry boundary, allowing a system width of $N = 3$ nodes

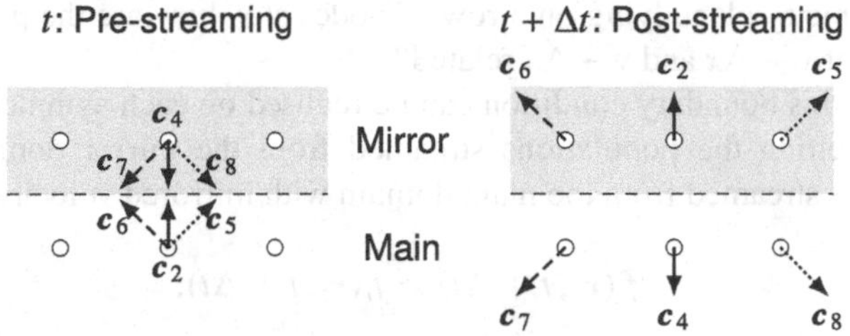

Fig. 5.29 Populations in a main and a mirrored domain, streaming across the boundary between the two domains. The *arrows* represent populations' directions, and populations that are identical to each other have identical arrow styles

We can therefore look at this **symmetry boundary** from a bounce-back-like perspective. **Populations leaving** the boundary node $\boldsymbol{x}_\mathrm{b}$ at time t **meet the symmetry surface** at time $t+\frac{\Delta t}{2}$ where **they are reflected specularly**, so that their resulting velocity $\boldsymbol{c}_j$ has its normal velocity component reversed from the incoming velocity $\boldsymbol{c}_i$, i.e. $c_{j,\mathrm{n}} = -c_{i,\mathrm{n}}$. The populations arrive at time $t + \Delta t$ at the node $\boldsymbol{x}_\mathrm{b}$ or one of its neighbours along the boundary as shown in Fig. 5.29. For these populations, the standard streaming step is replaced by

$$f_j(\boldsymbol{x}_\mathrm{b} + \boldsymbol{c}_{j,\mathrm{t}}\Delta t, t + \Delta t) = f_i^\star(\boldsymbol{x}_\mathrm{b}, t) \tag{5.61}$$

where $\boldsymbol{c}_{j,\mathrm{t}} = \boldsymbol{c}_{i,\mathrm{t}}$ is the *tangential* velocity of the populations, equalling $\boldsymbol{c}_i$ and $\boldsymbol{c}_j$ with their normal velocity set to zero.

A **free-slip boundary condition**, which enforces a zero normal fluid velocity $u_\mathrm{n} = 0$ but places no restrictions on the tangential fluid velocity $\boldsymbol{u}_\mathrm{t}$, can be **implemented in exactly the same way** as this symmetry boundary condition [35, 76, 77].

Exercise 5.4 For the D2Q9 case shown in Fig. 5.29, write down (5.61) explicitly for the populations that "bounce off" the boundary.

The free-slip condition gives an interesting insight into bounce-back schemes. The *normal* velocity mirroring of the reflected particles only results in a macroscopic condition of no *normal* velocity on the wall. Thus, it is the *tangential* velocity mirroring in the no-slip bounce-back scheme that additionally gives the condition of no *tangential* velocity on the wall.

Corners in the free-slip scheme are handled in the same way as corners in the bounce-back no-slip scheme, as described in Sect. 5.3.6.

Exercise 5.5 Consider another type of symmetry boundary condition where the symmetry plane lies exactly on a *line* of nodes instead of exactly midway on the links between two such lines of nodes as in the above method.

a) For a symmetry plane lying on a row of nodes at y, how are the populations on the nodes at $y + \Delta x$ and $y - \Delta x$ related?
b) Show that this boundary condition can be realised on each symmetry boundary node by setting the populations streamed from the mirror domain equal to populations streamed from the main domain with mirrored velocity, i.e. as

$$f_j(\boldsymbol{x}_\mathrm{b}, t + \Delta t) = f_i(\boldsymbol{x}_\mathrm{b}, t + \Delta t), \tag{5.62}$$

with $c_{j,\mathrm{n}} = -c_{i,\mathrm{n}}$ and $\boldsymbol{c}_{j,\mathrm{t}} = \boldsymbol{c}_{i,\mathrm{t}}$ as above.
c) Show that (5.62) imposes a free-slip wall condition, i.e. $u_\mathrm{n} = 0$.

5.4 Further Topics on Boundary Conditions

We present additional material on LB boundary conditions. Here we focus on theoretical aspects with proofs and details about results provided in the previous sections. This is of particular interest to those who wish to analyse and formulate new LB boundary schemes. In Sect. 5.4.1, we explain how to use theoretical tools, such as the Chapman-Enskog analysis, in order to evaluate the hydrodynamic characteristics of LB boundary conditions. We apply those results to both bounce-back and non-equilibrium bounce-back schemes. Then, in Sect. 5.4.2, we discuss the issue of mass conservation, followed by an analysis of the momentum exchange between fluid and walls in Sect. 5.4.3. Finally, in Sect. 5.4.4, we outline the main differences between 2D and 3D problems.

5.4.1 The Chapman-Enskog Analysis for Boundary Conditions

LB boundary conditions rely on specific closure rules. Those are different from the LB bulk rules of collision and propagation. Consequently, the results extracted in Sect. 4.1 for the macroscopic bulk behaviour cannot be expected to hold at boundaries. Still, the techniques introduced there are useful, and the present section shows how to use the Chapman-Enskog analysis to determine how LB boundary conditions behave macroscopically. Due to their conceptual differences, we will treat the bounce-back and the non-equilibrium bounce-back techniques separately.

5.4.1.1 Bounce-Back Method

As explained in Sect. 5.3.3, the macroscopic Dirichlet velocity condition described by the bounce-back rule in (5.26) applies at lattice links rather than at grid nodes.

However, the LBM, as a typical grid-based method, only provides solutions at the nodes. Hence, the understanding of what happens at links must be examined in terms of *continuation* behaviour. We can study this through a Taylor series expansion.

If we seek *third-order boundary accuracy* in order to reach the same level of exactness as the LBE in the bulk (cf. Sect. 5.2.4), then we shall use the *second-order Taylor expansion* to approximate the velocity $u_{\mathrm{w},x}$ at a wall point $\boldsymbol{x}_{\mathrm{w}}$ that is displaced by $\Delta x/2$ from a boundary node at $\boldsymbol{x}_{\mathrm{b}}$. As an example, we investigate a top wall so that $y_{\mathrm{w}} = y_{\mathrm{b}} + \Delta x/2$. For the tangential velocity component, we can write

$$u_x\big|_{y_\mathrm{w}} = u_x\big|_{y_\mathrm{b}} + \frac{\Delta x}{2}\,\partial_y u_x\big|_{y_\mathrm{b}} + \frac{1}{2}\left(\frac{\Delta x}{2}\right)^2 \partial_y\partial_y u_x\big|_{y_\mathrm{b}} + O\left(\Delta x^3\right). \tag{5.63}$$

For a **link-wise approach**, such as the bounce-back rule, to impose a wall velocity with formal **third-order accuracy** midway between nodes, its closure relation **must satisfy (5.63)** with $y_{\mathrm{w}} = y_{\mathrm{b}} + \Delta x/2$.

We can now perform a second-order Chapman-Enskog analysis to unfold the closure relation of the LB boundary scheme. If the solution is time-independent, $\partial_t f_i = 0$, the starting point is (5.12). For simplicity, let us focus on the linearised equilibrium from Sect. 4.3.1. In this case, (5.12) becomes

$$f_i^{\mathrm{eq}} = w_i\left(\rho + \rho_0\frac{c_{i\gamma}u_\gamma}{c_\mathrm{s}^2}\right), \tag{5.64a}$$

$$f_i^{(1)} = -\tau w_i c_{i\alpha}\partial_\alpha^{(1)}\left(\rho + \rho_0\frac{c_{i\gamma}u_\gamma}{c_\mathrm{s}^2}\right), \tag{5.64b}$$

$$f_i^{(2)} = \tau\left(\tau - \frac{\Delta t}{2}\right) w_i c_{i\alpha}c_{i\beta}\partial_\alpha^{(1)}\partial_\beta^{(1)}\left(\rho + \rho_0\frac{c_{i\gamma}u_\gamma}{c_\mathrm{s}^2}\right). \tag{5.64c}$$

The second-order Chapman-Enskog expansion of the bounce-back formula, (5.26), proceeds with the decomposition of its post-streaming and post-collision populations as follows:

$$f_{\bar{i}}^{\mathrm{eq}} + \epsilon f_{\bar{i}}^{(1)} + \epsilon^2 f_{\bar{i}}^{(2)} = f_i^{\mathrm{eq}} + \left(1 - \frac{\Delta t}{\tau}\right)\left(\epsilon f_i^{(1)} + \epsilon^2 f_i^{(2)}\right) - 2w_i\rho_0\frac{c_{i\gamma}u_{\mathrm{w},\gamma}}{c_\mathrm{s}^2} \tag{5.65}$$

with $u_{\mathrm{w},\gamma} = u_\gamma|_{\mathrm{w}}$. Then, by introducing (5.64) into (5.65) and performing some algebraic manipulations (using $p = c_\mathrm{s}^2\rho$ and $\partial_\alpha = \epsilon\partial_\alpha^{(1)}$), we obtain the steady-state macroscopic content of the bounce-back formula, (5.26), at node y_b:

$$\begin{aligned} c_{i\gamma}u_{w,\gamma} = c_{i\gamma}u_\gamma + \frac{\Delta t}{2}c_{i\alpha}\partial_\alpha(c_{i\gamma}u_\gamma) + \left(\tau - \frac{\Delta t}{2}\right)^2 c_{i\alpha}c_{i\beta}\partial_\alpha\partial_\beta(c_{i\gamma}u_\gamma) \\ - \left(\tau - \frac{\Delta t}{2}\right)\left[c_{i\alpha}\partial_\alpha\left(\frac{p}{\rho_0}\right) + \frac{\Delta t}{2}c_{i\alpha}c_{i\beta}\partial_\alpha\partial_\beta\left(\frac{p}{\rho_0}\right)\right]. \end{aligned} \tag{5.66}$$

To simplify the analysis, let us assume that this problem applies for a unidirectional horizontal flow aligned with the lattice so that $c_{i\gamma}u_\gamma = c_{ix}u_x$ and $c_{i\alpha}\partial_\alpha = c_{iy}\partial_y$. This way, only diagonal lattice links (i.e. those with $c_{ix}c_{iy} \neq 0$) play a role. In order to compare the resulting equation with (5.63), the pressure terms have to be re-expressed in the form of velocity corrections. This is possible by assuming that we are solving for a unidirectional pressure-driven flow, governed by $\partial_x(p/\rho_0) = c_s^2(\tau - \Delta t/2)\partial_y^2 u_x$ and $\partial_x^2(p/\rho_0) = 0$. Based on these arguments, we can re-express (5.66) to obtain the steady closure relation for the bounce-back rule on a top wall:

$$u_{w,x} = u_x\big|_{y_b} + \frac{\Delta x}{2}\partial_y u_x\big|_{y_b} + \underbrace{\frac{2c^2}{3}\left(\tau - \frac{\Delta t}{2}\right)^2}_{=\Delta x^2/8} \partial_y^2 u_x\Big|_{y_b}. \tag{5.67}$$

The equivalence between (5.67) and the target equation, (5.63), requires the coefficient of the second-order derivative term to be $\Delta x^2/8$. This is possible if and only if $(\tau - \Delta t/2)^2 = (3/16)\Delta t^2$ [15, 21].[25] Other τ values lead to a shift of the location of the effective boundary condition. This artefact can equivalently be interpreted as a velocity slip if we assume that the boundary remains halfway between lattice nodes. In any case, this *unphysical* dependence of the boundary condition on τ (and, therefore, on fluid viscosity) is a serious limitation of the bounce-back method combined with the BGK collision operator.

A solution for this problem is using an improved collision operator, such as TRT or MRT. In Chap. 10 we will discuss how those collision operators can be tuned to achieve a viscosity-independent boundary location.

5.4.1.2 Non-equilibrium Bounce-Back Method

Wet boundary nodes are designed to operate with the same dynamics as bulk nodes, and they should share the same macroscopic physics. We will explain how the hydrodynamic content of wet boundary nodes is determined, taking the NEBB method as an example. In addition, this exercise will helps us establishing its level of accuracy.

[25] Remember that $c_s^2 = c^2/3 = \Delta x^2/(3\Delta t^2)$.

Consider a bottom wall of a pressure-driven Poiseuille channel flow. The flow is horizontally aligned with the lattice links, and the wall coincides with the lattice nodes. In the bulk, the LBE solves this problem exactly: $\partial_x(p/\rho_0) = c_s^2(\tau - \Delta t/2)\partial_y^2 u_x$ [21, 23]. At boundaries, the NEBB method should reproduce this equation in the same way. Let us investigate through the second-order Chapman-Enskog analysis whether this is indeed the case.

For simplicity, let us focus on the diagonal link containing populations f_5 and f_7. Recalling Sect. 5.3.4, the NEBB scheme for these populations is

$$f_7^{\text{neq}} = f_5^{\text{neq}} - N_x. \tag{5.68}$$

When subjected to the second-order Chapman-Enskog analysis, (5.68) becomes

$$\epsilon f_7^{(1)} + \epsilon^2 f_7^{(2)} = \epsilon f_5^{(1)} + \epsilon^2 f_5^{(2)} - N_x. \tag{5.69}$$

Assuming, as in the previous example, that solutions are time-independent (cf. Sect. 4.2.3) and taking the linear equilibrium (cf. Sect. 4.3.1), the hydrodynamic content of $f_i^{(1)}$ and $f_i^{(2)}$ is given by (5.64b) and (5.64c), respectively. Pressure (density) and velocity at the wall are known from the prescribed macroscopic boundary condition. Furthermore, we know that, in this problem, the only non-zero gradients are $\partial_x\rho$, $\partial_y u_x$ and $\partial_y^2 u_x$. From this, the expressions for $f_{5,7}^{(1)}$ and $f_{5,7}^{(2)}$ can be greatly simplified. For example, f_5 becomes

$$\begin{aligned} f_5^{(1)} &= -\tau w_5 \left[c_{5x}\partial_x\rho_{\text{w}} + \frac{c_{5x}c_{5y}}{c_s^2}\rho_0\partial_y u_{\text{w},x} \right], \\ f_5^{(2)} &= \tau\left(\tau - \frac{\Delta t}{2}\right) w_5 \frac{c_{5x}c_{5y}^2}{c_s^2}\rho_0\partial_y^2 u_{\text{w},x}. \end{aligned} \tag{5.70}$$

Introducing the above equations for f_5 (and equivalently for f_7) into (5.69) and replacing the lattice-dependent parameters by numerical values for the D2Q9 lattice (cf. Table 3.1), we obtain

$$\frac{c}{18}\tau\partial_x\rho_{\text{w}} - \frac{c}{6}\tau\rho_0\left(\tau - \frac{\Delta t}{2}\right)\partial_y^2 u_{\text{w},x} + N_x = 0. \tag{5.71}$$

Now we have to determine the hydrodynamic form of N_x.

Example 5.5 Let us illustrate how to find the hydrodynamic content of the transverse momentum correction $N_x = -\frac{1}{2}(f_1 - f_3) + \frac{1}{3c}\rho_0 u_{\text{w},x}$. As usual, the analysis starts with the second-order Chapman-Enskog analysis:

$$N_x = -\frac{1}{2}\left(f_1^{\text{eq}} + \epsilon f_1^{(1)} + \epsilon^2 f_1^{(2)} - f_3^{\text{eq}} - \epsilon f_3^{(1)} - \epsilon^2 f_3^{(2)}\right) + \frac{1}{3c}\rho_0 u_{\text{w},x}. \tag{5.72}$$

With the assumption of steady unidirectional flow, using the linear equilibrium and introducing numerical values of the D2Q9 model, we obtain after some algebraic manipulations:

$$N_x = \frac{c}{9}\tau \partial_x \rho_{\mathrm{w}}. \tag{5.73}$$

By substituting the hydrodynamic content of N_x into (5.71) we find that the NEBB method correctly reproduces the desired macroscopic equation at the wall:

$$\partial_x p_{\mathrm{w}} - \frac{c^2}{3}\left(\tau - \frac{\Delta t}{2}\right)\rho_0 \partial_y^2 u_{\mathrm{w},x} = 0. \tag{5.74}$$

Exercise 5.6 Repeat the above analysis for the vertical link containing the populations f_1 and f_3. Demonstrate that its macroscopic content corresponds to the trivial solution $\partial_y \rho_{\mathrm{w}} = 0$, as initially assumed.

The **NEBB method** [27] exactly describes the pressure-driven (Poiseuille) channel flow on the boundary, thereby agreeing with the LBE in the bulk. In general flows, the NEBB method expresses the hydrodynamic solution with a third-order error, due to the $\mathcal{O}(\epsilon^3)$ terms disregarded in the second-order Chapman-Enskog analysis. Consequently, the NEBB method is a **third-order accurate boundary scheme** for **straight walls** coinciding with the lattice nodes.

5.4.2 *Mass Conservation at Solid Boundaries*

An important feature of LB boundary conditions is whether they conserve the total mass inside the system or not. Mass conservation is a crucial requirement for the modelling of many physical processes, such as compressible, multi-phase or multi-component flows.

The LB algorithm ensures that mass is conserved locally since collision leaves $\sum_i f_i = \rho$ invariant.[26] By inference, mass is also conserved globally in a periodic domain. However, there is no *a priori* guarantee that this property holds at boundary nodes which necessarily must behave differently than bulk fluid nodes, as explained in Sect. 5.2.4.

A mass conserving boundary node should guarantee the balance between the amount of mass carried by incoming and outgoing populations. Given that these

[26]Here, and throughout this discussion, we will assume no mass source is present.

boundary populations undergo different operations, depending on whether link-wise or wet-node methods are used, we will discuss both approaches separately.

5.4.2.1 Link-Wise Approach: Bounce-Back

The bounce-back method places the wall $\Delta x/2$ outside the boundary node. In order for the boundary node to hold its net mass constant it is necessary that what leaves the node towards the solid domain (after collision) is exactly balanced by what enters the node (after streaming), see (5.75) below.

The evolution described above is nothing but the particle reflection dynamics followed by the bounce-back rule, see Fig. 5.12. Thus, for straight walls, the bounce-back is always mass conserving; a property holding for walls at rest or with a prescribed tangential movement. Obviously mass is not conserved when a normal mass flux is prescribed, e.g. for normal velocity in the Dirichlet correction of the bounce-back formula, recall (5.26). In this case it is expected that the mass difference in the system is balanced by the mass injection at the wall.

However, two cases may exist where a *local* mass imbalance can occur. First, for boundary motions accompanied by the creation or destruction of fluid nodes to solid nodes and vice-versa [15, 52, 55, 61, 78]. A comparative study on refilling algorithms can be found in [79]. Second, for boundary shapes not aligning with the lattice nodes, where numerical errors due to the wall discretisation contaminate the local mass balance [13, 55].[27]

5.4.2.2 Wet-Node Approach: Non-equilibrium Bounce-Back

In the wet-node approach, the boundary node is considered to lie in the fluid domain, yet infinitesimally close to the solid region. This means that wet-node populations go through an extra step due to the boundary scheme, in addition to the stream-and-collide dynamics of the LB algorithm in the bulk. However, unlike in the bounce-back scheme where incoming boundary populations are found through simple reflection rules, these populations are *reconstructed* in wet-node approaches. This reconstruction does not automatically conserve the local mass [13], even for straight boundaries.

Mass conservation at a boundary node requires that the local mass m_{in} that is streamed in from neighbouring nodes equals the local mass m_{out} that is streamed out of the node after collision [20, 29, 71]. For a top wall as depicted in Fig. 5.6, the net mass Δm on the boundary node is

$$\Delta m = m_{\text{in}} - m_{\text{out}} = (f_2 + f_5 + f_6) - (f_4^\star + f_7^\star + f_8^\star). \tag{5.75}$$

[27]Depending on the definition of mass balance, as explained in [55], a third case may be considered: the presence of tangential density gradients along the wall surface. Here, a mass flux difference, proportional to the density gradient, may exist.

Let us examine under which circumstances $\Delta m \neq 0$ occurs. Considering a BGK collision $f_i^\star = (\Delta t/\tau)f_i^{\text{eq}} + (1 - \Delta t/\tau)f_i^{\text{neq}}$ with the linearised equilibrium f_i^{eq} and f^{neq} populations subject to the NEBB rule, we may re-express (5.75) as

$$\Delta m = \rho_{\text{w}} u_{\text{w},y} + \frac{\Delta t}{\tau}\left(f_4^{\text{neq}} + f_7^{\text{neq}} + f_8^{\text{neq}}\right). \tag{5.76}$$

Of course, like in bounce-back, a mass flux has to exist for either a normal boundary movement, or equivalently, for a boundary with vertical fluid injection. Both cases are represented by $\rho_{\text{w}} u_{\text{w},y} \neq 0$, and as expected lead to a physical change of mass at the node. However, even when this term is zero, we may still have $\Delta m \neq 0$ from the non-equilibrium contributions. This is a non-physical effect.

Once again we can use the Chapman-Enskog analysis to investigate the hydrodynamic content of the f_i^{neq} terms. For this analysis, the first-order decomposition $f_i^{\text{neq}} = \epsilon f_i^{(1)} + O(\epsilon^2)$ is sufficient, where $f_i^{(1)}$ is given in (5.64b). By using the numerical values for the D2Q9 velocity set, we can determine Δm up to $O(\epsilon^2)$ on a NEBB boundary node[28]:

$$\Delta m = \rho_{\text{w}} u_{\text{w},y} - \frac{\Delta t}{2}\partial_y(\rho_{\text{w}} u_{\text{w},y}) + \frac{\Delta t}{6}\partial_y \rho_{\text{w}}. \tag{5.77}$$

For steady flows moving parallel to straight walls, the NEBB method can guarantee mass conservation up to the numerical error of the scheme itself, i.e. $\Delta m \approx O(\epsilon^3)$. In some particular problems mass conservation may be achieved even exactly, by cancelling effects due to symmetry. However, in more complex geometries, the mass conservation condition may degrade with the same order as the NEBB numerical accuracy. In fact, whenever terms like $\partial_y(\rho_{\text{w}} u_{\text{w},y})$ become relevant, e.g. flow impinging an wall, the boundary will experience mass leakage [55]. In transient problems, the time-dependent solution has an analogous structure to (5.77), derived assuming steady-state. In this case, however, $\Delta m(t) \neq 0$ cannot be generally avoided due to the continuity condition $\partial_t \rho_{\text{w}} = -\partial_y(\rho_{\text{w}} u_{\text{w},y})$.

There exist wet-node methods where mass conservation is enforced exactly, e.g. [20, 29, 71]. However, their actual merits remain to be demonstrated in face of the theoretical criticisms put forward in [13, 55] which show that these methods may degrade the overall solution accuracy.

5.4.3 Momentum Exchange at Solid Boundaries

In cases such as wind or water flow around a ship or a bridge, one is interested in the force and torque acting on structures immersed in a fluid. Typical quantities of interest are the drag and lift coefficients.

[28]Note that we have used the no-slip condition $\partial_x u_x = \partial_x u_y = 0$ at the wall to simplify (5.77).

Generally, the total force $\boldsymbol{f}$ acting on a boundary area A can be written as the surface integral of the *traction* vector $\boldsymbol{t} = \boldsymbol{\sigma}_{\mathrm{w}} \cdot \hat{\boldsymbol{n}}$:

$$f_\alpha = \int \mathrm{d}A\, t_\alpha = \int \mathrm{d}A\, \sigma_{\mathrm{w},\alpha\beta}\hat{n}_\beta \tag{5.78}$$

where $\boldsymbol{\sigma}_{\mathrm{w}}$ is the stress tensor at the surface and $\hat{\boldsymbol{n}}$ the unit normal vector pointing from the surface into the fluid.

The key question is how to evaluate the integral in equation (5.78) in a computer simulation. The conventional approach is to compute the stress tensor $\boldsymbol{\sigma}$ at the surface, e.g. using finite differences, to approximate the integral in equation (5.78). This can introduce additional errors and requires tedious computations, in particular in 3D and for arbitrarily shaped surfaces [80].

Fortunately, the LBM allows direct access to the stress tensor, without those complicated extra efforts. In the bulk, the viscous shear stress follows directly from the second-order velocity moment of f_i^{neq} as shown in (4.14) [81, 82]. Also at boundaries nodes, LBM offers a direct way to evaluate (5.78). We know that wet boundary nodes also participate in collision and propagation. Therefore, we can compute the wall shear stress $\boldsymbol{\sigma}_{\mathrm{w}}$ on those nodes just as in the bulk, i.e. by computing the second-order velocity moment of f_i^{neq}.

However, for link-wise methods, such as bounce-back, the wall is shifted by $\Delta x/2$ away from the boundary nodes. In this case we cannot directly obtain the wall stress as in the bulk and a different approach is required.

5.4.3.1 Momentum Exchange in the Bounce-Back Method

Ladd [22, 61] suggested an approach to evaluate (5.78) for the bounce-back method. The key idea is that the LBM is a particle-based method where the populations f_i represent fluid elements with momentum $f_i \boldsymbol{c}_i$. Following Ladd's idea, we have to identify those populations that cross the boundary (both from the fluid into the solid and the other way around) and sum up all corresponding momentum contributions to obtain the momentum exchange at the wall. This procedure is called the *momentum exchange algorithm* (MEA). Note that the MEA provides only the traction vector $\boldsymbol{t}$ rather than the full stress tensor $\boldsymbol{\sigma}_{\mathrm{w}}$.

Consider a planar bottom wall along the x-axis in 2D as shown in Fig. 5.12. During propagation, the populations f_4, f_7 and f_8 stream from fluid to solid nodes. They hit the boundary half-way on their journey, are bounced back and continue their propagation as f_2, f_5, and f_6 towards their original nodes. Effectively it seems as if the populations f_4, f_7, and f_8 vanished into the boundary and f_2, f_5, and f_6 emerged from the wall.

What does this mean for the momentum exchange between fluid and wall? On the one hand, momentum is carried by f_4, f_7, and f_8 *to the wall,* but at the same time momentum is transported by f_2, f_5, and f_6 from the wall *to the fluid*. The difference is the net momentum transfer. Note that those populations which move parallel to the boundary are not relevant as they do not contribute to the momentum transfer between fluid and solid. The overall idea is very similar to the assessment of the local mass conservation at solid boundaries as explained in Sect. 5.4.2.

A population f_i moving from a boundary node at $\boldsymbol{x}_\text{b}$ to a solid node at $\boldsymbol{x}_\text{s} = \boldsymbol{x}_\text{b} + \boldsymbol{c}_i \Delta t$ is bounced back mid-way[29] at a wall location $\boldsymbol{x}_\text{w} = \frac{1}{2}(\boldsymbol{x}_\text{b} + \boldsymbol{x}_\text{s}) = \boldsymbol{x}_\text{b} + \frac{1}{2}\boldsymbol{c}_i \Delta t$. Those lattice links $\boldsymbol{c}_i$ connecting a fluid and a solid node are called *boundary links*. For each boundary link there are exactly two populations crossing the wall, one incoming population f_i^in (streaming from the fluid into the wall) and one outgoing population $f_{\bar{i}}^\text{out}$ (moving from the wall into the fluid). All populations moving along boundary links contribute to the momentum exchange. In order to keep track of the boundary links we denote them $\boldsymbol{x}_i^\text{w}$; this indicates that population f_i crosses the wall at location $\boldsymbol{x}_\text{w}$ in direction $\boldsymbol{c}_i$.

Let us pick out a single population, say f_4, and investigate its fate more closely. Just before bounce-back, the incoming population f_4^in carries the momentum $\boldsymbol{p}_4^\text{in} = f_4^\text{in}\boldsymbol{c}_4$, but right after, it becomes the outgoing population f_2^out with momentum $\boldsymbol{p}_2^\text{out} = f_2^\text{out}\boldsymbol{c}_2$. During bounce-back, the wall has to absorb the recoil such that the total momentum of the fluid and the wall is conserved. Bounce-back happens at location $\boldsymbol{x}_4^\text{w}$, so we can write the corresponding momentum exchange as $\Delta \boldsymbol{p}(\boldsymbol{x}_4^\text{w}) = \boldsymbol{p}_4^\text{in} - \boldsymbol{p}_2^\text{out}$.

Example 5.6 For a stationary wall, bounce-back dictates $f_2^\text{out} = f_4^\text{in}$. Hence, the momentum transfer due to the bounce-back of f_4 at $\boldsymbol{x}_4^\text{w}$ is $\Delta \boldsymbol{p}(\boldsymbol{x}_4^\text{w}) = \boldsymbol{p}_4^\text{in} - \boldsymbol{p}_2^\text{out} = f_4^\text{in}\boldsymbol{c}_4 - f_2^\text{out}\boldsymbol{c}_2 = 2f_4^\text{in}\boldsymbol{c}_4$ since $\boldsymbol{c}_2 = -\boldsymbol{c}_4$.

The same idea holds independently for all other populations which are bounced back anywhere at the boundary. The total momentum exchange between the fluid and the wall is the sum of all those contributions by all bounced back populations.[30] It is therefore straightforward to compute the momentum exchange for any surface which is described *via* simple bounce-back boundary conditions:

1. Identify all links between boundary and solid nodes with locations $\boldsymbol{x}_i^\text{w}$. If the boundary is stationary, the list can be constructed once and stored in memory for the whole simulation.[31]
2. At each time step (or each other interval when the momentum exchange is required), evaluate the incoming and bounced back populations f_i^in and $f_{\bar{i}}^\text{out}$ at each of the identified links.

---

[29]The MEA also works for curved boundaries as we will discuss in Sect. 11.2.

[30]Note that the number of identified links varies with the chosen lattice. For example, there will be more links for the same geometry when D3Q27 rather than D3Q15 is used. This does not affect the validity of the MEA, though.

[31]One possible way to implement this is to run over all solid nodes and identify all neighbouring boundary nodes [80, 83].

3. The total momentum exchange during one streaming step is the sum over all identified links and, due to $\boldsymbol{c}_{\bar{i}} = -\boldsymbol{c}_i$, can be written as

$$\Delta \boldsymbol{P} = \Delta x^3 \sum_{\boldsymbol{x}_i^{\mathrm{w}}} \Delta \boldsymbol{p}(\boldsymbol{x}_i^{\mathrm{w}}) = \Delta x^3 \sum_{\boldsymbol{x}_i^{\mathrm{w}}} \left(f_i^{\mathrm{in}} + f_{\bar{i}}^{\mathrm{out}} \right) \boldsymbol{c}_i. \tag{5.79}$$

The prefactor Δx^3 in 3D (or equivalently Δx^2 in 2D) ensures that the result is a momentum rather than a momentum density.

If we now assume that the momentum is exchanged smoothly during one time step Δt, we can easily find the force acting on the boundary:

$$\boldsymbol{f} = \frac{\Delta \boldsymbol{P}}{\Delta t}. \tag{5.80}$$

Similarly, the angular momentum exchange $\Delta \boldsymbol{L}$ and therefore the total torque $\boldsymbol{T} = \Delta \boldsymbol{L}/\Delta t$ acting on the wall can be computed. To do this, we have to replace $\Delta \boldsymbol{p}(\boldsymbol{x}_i^{\mathrm{w}})$ within the sum in (5.79) by $(\boldsymbol{x}_i^{\mathrm{w}} - \boldsymbol{x}_{\mathrm{ref}}) \times \Delta \boldsymbol{p}(\boldsymbol{x}_i^{\mathrm{w}})$ where $\boldsymbol{x}_{\mathrm{ref}}$ is a fixed reference point, e.g. the origin of the coordinate system.

Obviously the MEA is computationally much simpler than solving the integral in equation (5.78). In fact, neither the surface stress $\boldsymbol{\sigma}_{\mathrm{w}}$ nor the normal vector $\hat{\boldsymbol{n}}$ are required in the MEA. All necessary information is already contained in the populations participating in the bounce-back process.

5.4.3.2 Accuracy of the Momentum Exchange Method

Despite its simplicity, the MEA provides an accurate measure of the wall shear stress [22]. We can verify this through a Chapman-Enskog analysis.

Consider again a planar bottom wall modelled with the bounce-back method (cf. Fig. 5.12). Let us follow [23] and define Δp_x^{w} as the net tangential momentum (per time step and per unit area) transferred from fluid to solid across the wall, which is located midway between the solid node $\boldsymbol{x}_{\mathrm{s}}$ and the boundary node $\boldsymbol{x}_{\mathrm{b}}$. According to the momentum exchange principle, Δp_x^{w} can be computed as

$$\Delta p_x^{\mathrm{w}} = \frac{cV}{\Delta t A} \left[(f_5 - f_6)|_{\boldsymbol{x}_{\mathrm{b}}} - (f_8 - f_7)|_{\boldsymbol{x}_{\mathrm{s}}} \right] = c^2 \left[(f_5 - f_6) - (f_8^\star - f_7^\star) \right]\Big|_{\boldsymbol{x}_{\mathrm{b}}} \tag{5.81}$$

where in 3D (2D) we have $V = \Delta x^3$ ($V = \Delta x^2$) and $A = \Delta x^2$ ($A = \Delta x$). The evaluation of Δp_x^{w} using only local information at $\boldsymbol{x}_{\mathrm{b}}$ is possible by exploring the LB stream-and-collide dynamics.

As derived in Sect. 5.4.1, we know that the hydrodynamic content of f_i, up to the second order in steady state, is

$$\begin{aligned} f_i &= f_i^{\mathrm{eq}} + \epsilon f_i^{(1)} + \epsilon^2 f_i^{(2)} + O(\epsilon^3) \\ &= \left[1 - \epsilon\tau c_{i\alpha}\partial_\alpha^{(1)} + \epsilon^2\tau\left(\tau - \frac{\Delta t}{2}\right) c_{i\alpha}c_{i\beta}\partial_\alpha\partial_\beta\right] f_i^{\mathrm{eq}} + O(\epsilon^3). \end{aligned} \tag{5.82}$$

By introducing (5.82) into (5.81) and adopting the linear incompressible equilibrium (cf. Sect. 4.3.1) we get

$$\Delta p_x^{\mathrm{w}} = -c_s^2\left(\tau - \frac{\Delta t}{2}\right)\rho_0\,\partial_y u_x\Big|_{x_\mathrm{b}} + \frac{\Delta x}{2}c_s^2\left(\tau - \frac{\Delta t}{2}\right)\rho_0\,\partial_y^2 u_x\Big|_{x_\mathrm{b}}. \tag{5.83}$$

Equation (5.83) denotes the first-order Taylor expansion of the viscous stress $\sigma_{xy} = -\nu\rho_0\partial_y u_x$ exerted by the fluid on the wall located at $y_\mathrm{w} = y_\mathrm{b} - \Delta x/2$ [15]. We can rewrite this in the compact form $\sigma_{xy}|_{y_\mathrm{w}} = \sigma_{xy}|_{y_\mathrm{b}} - \frac{\Delta x}{2}\partial_y\sigma_{xy}|_{y_\mathrm{b}} + O(\Delta x^2)$. This proves the second-order accuracy of the MEA in computing σ_{xy} at the wall. This measure is, at the same time, exact for parabolic solutions, which may be surprising, considering that the bounce-back parabolic solution comes itself with a viscosity-dependent error. Such a result has been originally pointed out by [22] and is explained by the fact that for a linear shear stress solution the second-order derivative term, related to the viscosity-dependent error, is vanished.

5.4.4 Boundary Conditions in 3D

Up to this point we have only considered the application of boundary conditions in 2D domains. A natural question is: what changes in 3D? The purpose of this section is to summarise the major differences between 2D and 3D.

The first obvious difference is the geometry. While in 2D domains the prescription of boundary conditions is limited to edges and corners, in 3D we have to handle planes, edges and corners as illustrated in Fig. 5.30. Such a diversity of geometrical features evidently adds complexity to the treatment of boundary conditions.

The additional challenges in 3D are mostly related to the numerical implementation rather than the mathematical concept. Since 3D lattices typically contain more discrete velocity vectors than 2D lattices, we have to deal with more unknowns. This is summarised in Table 5.2.

But how does this added complexity affect the LB implementation of boundary conditions in practice? As discussed in Sect. 5.2.4, there are link-wise and wet-node procedures to treat boundary conditions. These two techniques are not equally simple to extend to 3D:

- Link-wise boundary methods specify the missing populations based on simple reflection rules, e.g. bounce-back. Therefore, they naturally extend to 3D.

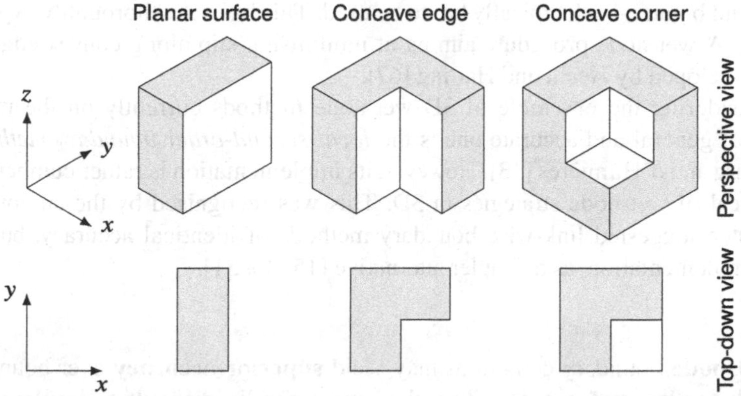

Fig. 5.30 Schematic representation of different geometrical features in 2D and 3D problems, inspired by [34]. White surfaces represent fluid-solid boundaries while grey surfaces represent cuts through the solid. The 3D plane degenerates to a 2D edge, and the two other 3D configurations degenerate to 2D corners

Table 5.2 Number of unknown populations for different lattices and boundary configurations

Configuration	D2Q9	D3Q15	D3Q19	D3Q27
Plane	–	5	5	9
Edge (concave)	3	8	9	15
Corner (concave)	5	10	12	17

- Wet node techniques are based on specific rules incorporating the consistency with bulk dynamics. Also, they are often designed to modify just the unknown boundary populations, as in, e.g., the NEBB method. Consequently, the added number of unknowns makes their extension to 3D non-trivial.[32]

A difficulty particular to wet-node boundaries is the handling of those populations that never stream into the fluid domain (buried links, cf. Sect. 5.3.6). In 3D, they appear in concave edges and corners. Strategies to deal with them can be found in [34, 67].

Convex boundary nodes, on the other hand, do *not* apply the same rules as in the concave case because fewer populations emanate from inside the wall. This naturally leads to an over-constrained problem in the prescription of the unknown populations. As the problem has too few unknowns, it follows that one cannot enforce the no-slip condition at convex boundary nodes using general wet-node

[32]Wet-node formulations that replace all populations are simpler to implement in 3D. Examples are the equilibrium scheme [23], the non-equilibrium extrapolation method [9], the finite-difference velocity gradient method [18] and the regularised method [18]. As they are based on reconstructing all populations, they are not sensitive to the number of populations. The downside of these approaches is that they either decrease the accuracy or increase the complexity of implementation, e.g. by making the scheme non-local [18].

rules, and bounce-back typically has to be used. This has been thoroughly explained in [34]. A wet-node procedure aiming at minimising slip along convex edges has been developed by Hecht and Harting [67].

Considering the ensemble of 3D wet-node methods currently on the market, the most general and accurate one is the *local second-order boundary method* by Ginzburg and d'Humières [13]. However, its implementation is rather cumbersome, as typical of wet-node strategies in 3D. This was recognised by the authors [13] who later suggested link-wise boundary methods of identical accuracy, but non-local implementation, as a simpler alternative [15, 30, 31].

Wet-node boundary conditions may yield **superior accuracy** over bounce-back **on flat surfaces**. However, they are generally **difficult to implement in 3D**, owing to the need to distinguish between populations according to their orientation to the wall and also the cumbersome treatment of edge and corner nodes. Contrarily, **link-wise** boundary schemes (in particular bounce-back) do not distinguish between populations and are **easy to implement**. In conclusion, in general 3D geometries, the link-wise boundary methods offer a better compromise between accuracy and ease of implementation compared to wet-node techniques.

Finally, we note that the difficulties in going from 2D to 3D are common to any problem where higher-order lattices are employed. Lattices with a larger number of velocities appear in a number of applications, like in thermal fluids and compressible hydrodynamics [84–86], rarefied gas flows [87, 88] or multiphase problems [89]. Despite their vast range of application, so far only few procedures have been devised to specify boundary conditions in such higher-order lattices, e.g. [90] in link-wise and [91] in wet-node approaches.

5.5 Initial Conditions

In Sect. 5.2.2 we have already explained the relevance of proper initial conditions for the NSE. Here, we address how an LB simulation can be initialised. The basic question is: given a known initial divergence-free velocity field $\boldsymbol{u}_0(\boldsymbol{x})$ for the incompressible NSE, how have the populations f_i to be chosen to recover a consistent initial macroscopic state? Skordos was the first who thoroughly discussed this issue in his seminal paper [2]. In the following, some general ideas about simulation initialisation (cf. Sect. 5.5.1) and the available LB initialisation schemes (cf. Sect. 5.5.2) are presented, followed by the decaying Taylor-Green vortex flow as a benchmark test (cf. Sect. 5.5.3).

5.5.1 Steady and Unsteady Situations

Before we discuss the available methods to initialise an LB simulation, it is worthwhile to take a look at typical simulation scenarios. Every LB simulation falls into exactly one of the following categories:

1. **Steady flows.** Since LBM is an inherently time-dependent method, it is generally not well suited for steady problems. Although one can of course use it to obtain steady solutions, it usually takes a larger number of iteration (i.e. time) steps compared to methods tailored for steady problems. In steady situations, e.g. flow through a porous medium with a constant pressure gradient, the initial state is normally not relevant for the final outcome. The reason is that all unphysical transients caused by the initial state decay after some time, and only the desired steady solution survives. In this context, there exist preconditioning techniques to accelerate convergence to steady state [74, 92].
2. **Unsteady flows after long times.** The LB algorithm is particularly powerful when it is applied to unsteady problems like suspension flows, flow instabilities or fluid mixing. In many cases, one is interested in the long-time behaviour of the system or the associated statistical properties such as suspension viscosity, particle diffusivity or the mass transfer coefficients in multi-component systems. It turns out that the exact choice of the initial conditions is often not relevant since unphysical transients decay and the system "finds" its proper state after some time. In other words: the statistical long-time behaviour of such systems is usually independent of the initial state.
3. **Time-periodic flows.** Some flows are time-periodic, for instance Womersley flow [93]. Fully converged time-periodic flows do not depend on the details of the initialisation, but in general it can take a long time until undesired transients have decayed. For Womersley flow, these transients can last for more than tens of oscillation periods (cf. Sect. 7.3.4).
4. **Initialisation-sensitive flows.** There are situations where the entire fate of a simulation depends on the initial state, for example turbulence or some benchmark tests [2] such as the decaying Taylor-Green vortex flow (cf. Sect. 5.5.3). For such systems, any error in the initial state propagates in time and can detrimentally affect the accuracy of the entire simulation.

5.5.2 Initial Conditions in LB Simulations

We first discuss the relationships between the populations f_i and the macroscopic fields (e.g. velocity and pressure) in the context of initial conditions. After thinking about the role of the order of collision and propagation, we discuss one particular initialisation scheme for LBM in more detail.

5.5.2.1 Role of Populations and Macroscopic Fields

Solving the unsteady NSE implies finding the velocity and the pressure fields $\boldsymbol{u}(\boldsymbol{x}, t)$ and $p(\boldsymbol{x}, t)$, respectively. This generally requires knowledge of the initial velocity and pressure fields $\boldsymbol{u}_0(\boldsymbol{x})$ and $p_0(\boldsymbol{x})$.

As shown in Sect. 4.1, the populations f_i can be decomposed into equilibrium and non-equilibrium parts. The equilibrium part, (3.54), can be easily constructed from the velocity and density (pressure) fields. However, in order to initialise an LB simulation properly, one also has to specify the strain rate tensor $\boldsymbol{S}_0$ with components $S_{0\alpha\beta}(\boldsymbol{x}) = \frac{1}{2}(\nabla_\alpha u_\beta + \nabla_\beta u_\alpha)|_{t=0}$ [94]. The first-order non-equilibrium populations $f_i^{(1)}$ are essentially proportional to the velocity gradients. Initialising populations by the equilibrium populations means that the initial state would not be second-order accurate. This is a direct consequence of the kinetic nature of the LB algorithm.

The initialisation of LB algorithms is thoroughly described in [3, 94] and, on a more mathematical basis, in [95, 96]. In the following, we will only report the most important points and refer to the literature where necessary.

The most common situation is that the initial solenoidal (divergence-free) velocity field $\boldsymbol{u}_0(\boldsymbol{x})$ is given, but neither the pressure $p_0(\boldsymbol{x})$ nor the strain rate $\boldsymbol{S}_0(\boldsymbol{x})$ are known. One could therefore be tempted to initialise the populations at equilibrium, i.e.

$$f_i(\boldsymbol{x}, t=0) = f_i^{\text{eq}}\left(\rho, \boldsymbol{u}_0(\boldsymbol{x})\right), \tag{5.84}$$

where ρ is some initial constant density. This will lead to an inconsistent state because the Poisson equation for the initial pressure,[33]

$$\Delta p_0 = -\rho\, \partial_\beta u_\alpha \partial_\alpha u_\beta \big|_{t=0}, \tag{5.85}$$

is not satisfied. In fact, Skordos [2] emphasised that an initialisation with a constant pressure is generally insufficient. Caiazzo [94] pointed out that a wrong pressure initialisation leads to undesired *initial layers* and numerical oscillations which potentially spoil the entire simulation [3]. In order to obtain a consistent initial pressure profile from the velocity, one has to solve the Poisson equation, (5.85). This can be done directly, or one may follow the iterative approach by Mei et al. [3] as summarised further below.

It is worth recalling that the pressure is not entirely determined by the velocity for weakly compressible schemes like most LB solvers; it is rather an independent field which requires its own initialisation [95]. However, since LB simulations are usually run in the limit of small Knudsen and Mach numbers, we assume that initialisations in accordance with the incompressible NSE are also good for the slightly compressible LBM.

[33] One obtains this equation by computing the divergence of the incompressible NSE.

If we assume for a moment that the initial pressure $p_0(\boldsymbol{x})$ is known, one can refine (5.84) as

$$f_i(\boldsymbol{x}, t=0) = f_i^{\text{eq}}\left(\rho_0(\boldsymbol{x}), \boldsymbol{u}_0(\boldsymbol{x})\right) \tag{5.86}$$

where

$$\rho_0(\boldsymbol{x}) = \bar{\rho} + \frac{p_0(\boldsymbol{x}) - \bar{p}}{c_s^2} \tag{5.87}$$

is the initial density profile with $\bar{\rho}$ being the average density and $\bar{p}$ an appropriate reference pressure. Equation (5.87) is nothing more than the equation of state of the standard LBE. The choice of $\bar{p}$ is arbitrary as it is a mere scaling factor in all equations. As will be discussed more thoroughly in Sect. 7.2.1, one usually chooses $\bar{\rho} = 1$ in lattice units.

But even if the pressure and therefore the density is set correctly, one still has to consider the non-equilibrium populations. For given velocity gradients and low fluid velocities, the BGK non-equilibrium can be approximated through (4.35) as

$$f_i^{\text{neq}} \simeq -w_i \frac{\tau\rho}{c_s^2} Q_{i\alpha\beta} \partial_\alpha u_\beta \tag{5.88}$$

where $Q_{i\alpha\beta} = c_{i\alpha}c_{i\beta} - c_s^2\delta_{\alpha\beta}$.

The velocity gradients can for example be computed from the velocity field $\boldsymbol{u}_0(\boldsymbol{x})$ analytically or through a finite difference scheme. The second approach is often employed since analytical expressions for the velocity are not always available.

Skordos [2] proposed an extended collision operator with finite-difference-based velocity gradients for initialisation. Holdych et al. [97] and van Leemput [95] reported consistent high-order initialisation schemes as generalisation of Skordos's approach. In comparison with finite difference approximation methods, the method by Mei et al. [3] does not only produce a consistent initial pressure, but also a consistent initial non-equilibrium field. Details of the latter approach are given below.

Generally, the populations should, if possible, be initialised according to

$$f_i(\boldsymbol{x}, t=0) = f_i^{\text{eq}}\left(\rho_0(\boldsymbol{x}), \boldsymbol{u}_0(\boldsymbol{x})\right) + f_i^{\text{neq}}\left(\rho_0(\boldsymbol{x}), \boldsymbol{S}_0(\boldsymbol{x})\right). \tag{5.89}$$

It turns out that several conclusions from Sect. 5.2.4 are also applicable here: if the populations $f_i^{(2)}$ are neglected during initialisation, the curvature $\nabla^2\boldsymbol{u}_0$ will not be correctly imposed and the resulting flow field will only be linearly exact and second-order accurate. In other words, to initialise a Poiseuille flow exactly, all orders up to $f_i^{(2)}$ have to be initialised correctly. More details are provided in [98].

5.5.2.2 Chicken or Egg? Order of Collision and Propagation

As mentioned before, in most situations a simulation is initially dominated by undesired transients until the physical solution dominates. In these cases it is not relevant whether to start a simulation with collision or propagation.

If, on the other hand, the transients are of interest, one has to start a simulation with the proper initialisation, followed by collision. The reason lies in the way the Boltzmann equation is discretised (cf. Chap. 3).

The LBE is the *explicit* discretisation of the continuous Boltzmann equation. This means that the equilibrium distribution, which is used for collision, is calculated using the known velocity and pressure fields. Only after this, propagation is performed. Initialising a simulation is basically the inverse of the velocity moments computation: instead of performing $f_i \to (\rho, \boldsymbol{u})$, one goes the other way around and executes $(\rho, \boldsymbol{u}) \to f_i$. Therefore, the equilibrium and non-equilibrium parts of the populations after initialisation assume a state compatible with the state after propagation, but before collision. Therefore, initialisation has to be followed by collision rather than propagation.[34]

We provide a simple example showing that the above argumentation is valid. Imagine a simulation with $\tau = \Delta t$. This means that, during collision, the non-equilibrium is set to zero everywhere and all populations relax to their equilibrium. The subsequent propagation step leads to the appearance of a new non-equilibrium if the flow field is not spatially homogeneous. In contrast, if the simulation is initialised with the correct non-equilibrium populations, a subsequent propagation will produce a new non-equilibrium which is inconsistent with a previous non-relaxed flow field. Therefore, one first has to perform collision. This example also shows that the deviatoric stress tensor (and all other moments) has to be computed after propagation rather than after collision. No matter how large the velocity gradients are, for $\tau = \Delta t$ they would always be zero after collision, which is obviously not correct.

5.5.2.3 Consistent Initialisation via a Modified LB Scheme

Mei et al. [3] proposed a modification of the incompressible[35] LB algorithm (cf. Sect. 4.3) to find a consistent initial state given a solenoidal velocity field $\boldsymbol{u}_0(\boldsymbol{x})$. The essential idea is to run the following algorithm until convergence has been obtained:

1. Initialise the populations f_i, e.g. with the initial velocity $\boldsymbol{u}_0(\boldsymbol{x})$ according to (5.84).
2. Compute the local density from $\rho(\boldsymbol{x}) = \sum_i f_i(\boldsymbol{x})$.

[34] A more accurate discussion of how to split the time step is found in recent works by Dellar [99] and Schiller [100].

[35] For this initialisation method, it is important to employ the incompressible LB algorithm. The standard equilibrium leads to large initial pressure errors.

3. Perform collision by using the modified incompressible equilibrium distributions $f_i^{\text{eq}}\left(\rho(\boldsymbol{x}), \boldsymbol{u}_0(\boldsymbol{x})\right)$, i.e. take the updated density field from step 2 but keep the initial velocity $\boldsymbol{u}_0(\boldsymbol{x})$ rather than recomputing it.
4. Propagate.
5. Go back to step 2 and iterate until the populations f_i (and the hydrodynamical fields) have converged to a user-defined degree. It is important to end this algorithm with propagation rather than collision; otherwise the non-equilibrium would be incorrect.

This algorithm does not obey momentum conservation (only density is conserved), it rather relaxes the velocity to its desired value at each point in space. The outcome is nearly independent of the choice of the relaxation parameter τ, although its choice affects the required number of iteration steps. It can be shown [3] that this procedure results in consistent initial populations, including the non-equilibrium part and therefore velocity gradients. For this reason, the resulting populations can be used as initial state for the actual LB simulation.

In fact, by modifying the LB algorithm as detailed above, one effectively solves the advection-diffusion equation for the density ρ as a passive scalar field and the Poisson equation for the pressure p in (5.85) [3]. This will become clearer in Sect. 8.3.

A disadvantage of this initialisation approach is that an initial force field is not taken into account, but other authors [94, 96] presented accelerated initialisation routines which also works in the presence of an initial force density. The major advantage is that this initialisation approach is much easier to implement than an additional Poisson solver, though depending on the parameters the solution of the advection-diffusion equation can be computationally demanding. Another advantage is that two outcomes are simultaneously available: the pressure field and consistent initial populations, including their non-equilibrium part.

We should mention a subtle detail of the above initialisation scheme. In the original paper [3], the authors used an MRT collision operator (cf. Chap. 10) with $\omega_{j_\alpha} = 1/\Delta t$, where ω_{j_α} is the relaxation frequency for the momentum density. Without going too much into detail here, the choice $\omega_{j_\alpha} = 1/\Delta t$ guarantees local momentum conservation so that the actual velocity field $\boldsymbol{u}(\boldsymbol{x}) \propto \sum_i f_i \boldsymbol{c}_i$ always equals its input value $\boldsymbol{u}_0(\boldsymbol{x})$ during the initialisation process. The same effect can be achieved with $\omega = 1/\tau = 1/\Delta t$ in the BGK model. For ω_{j_α} or $1/\tau$ different from $1/\Delta t$, momentum is not conserved and the velocity as computed from the populations does *not* match the specified velocity $\boldsymbol{u}_0$. We will see this in Sect. 5.5.3. However, the difference between the obtained velocity field and the input velocity field $\boldsymbol{u}_0(\boldsymbol{x})$ is of second order and therefore still compatible with a consistent initialisation of the LBM.

Recently, Huang et al. [96] proposed advanced initialisation schemes based on the asymptotic analysis of the LBM. We will not discuss these more sophisticated approaches here.

5.5.3 Example: Decaying Taylor-Green Vortex Flow

In order to show how relevant the correct initialisation of velocity, pressure and stress in a transient situation is, we simulate the decaying Taylor-Green vortex flow (cf. Sect. A.3 for its definition) with several different initialisation strategies following (5.89):

1. set velocity, pressure and stress analytically,
2. set velocity and pressure only (initialise with the equilibrium),
3. set velocity and stress only ($\rho_0 = 1$),
4. use the initialisation scheme by Mei et al. [3].

The simulation parameters are $N_x \times N_y = 96 \times 72$, $\tau = 0.8\Delta t$ ($\nu = 0.1\Delta x^2/\Delta t$), $\hat{u}_0 = 0.03\Delta x/\Delta t$, $\bar{\rho} = 1$ (in lattice units) and $p_0 = 0$. The standard equilibrium is used for the actual simulations while the incompressible equilibrium is employed for Mei's initialisation scheme.[36] The vortex decay time is $t_d \approx 840\Delta t$. We run the simulation for one decay time and show the L_2 error as defined in (4.57) as function of time for velocity $\boldsymbol{u}$, pressure p and the xx- and xy-components of the deviatoric stress tensor $\boldsymbol{\sigma}$. Furthermore, for Mei's initialisation scheme we use the same relaxation time τ as for the actual simulation (although in principle different relaxation times could be used), and the advection-diffusion scheme is terminated once the L_2 difference of the pressure profile at subsequent iteration steps falls below 10^{-10}. This convergence criterion is sufficient since an even lower threshold (we also tested 10^{-12}) did not result in more accurate simulations.

The results for the errors *after completed initialisation* are shown as function of time in Fig. 5.31. The major observation is that an incorrect pressure is much more problematic than an incorrect stress. Furthermore, Mei's initialisation scheme produces velocity and σ_{xx} errors nearly as small as the analytical initialisation while the pressure and σ_{xy} error is slightly smaller with Mei's approach. We see that the velocity error in the case of Mei's approach is initially not zero. The reason is that $\tau = 0.8\Delta t \neq \Delta t$ has been used. For $\tau = \Delta t$, the initial velocity error would be zero. Yet, the initial velocity error is of the same order as the typical velocity error and does not significantly affect the subsequent simulation results.

Disregarding the stress in the initialisation process increases the velocity error only during the first time steps. The pressure errors for initialised or non-initialised stress are virtually indistinguishable; even the stress errors behave similarly, except for very short times where they are 100%. In total, neglecting the initial stress does not lead to significant long-time deviations. The reason why the stress is much less important is that it represents a higher-order effect in the Chapman-Enskog expansion. An initially inconsistent stress is corrected after a few time steps without significant consequences for the flow. Note that, for $\tau = \Delta t$, a wrong (or missing)

[36]Using the incompressible equilibrium for the actual simulations does not result in significantly different results. This is no surprise since the incompressible model is only formally more accurate for *steady* flows.

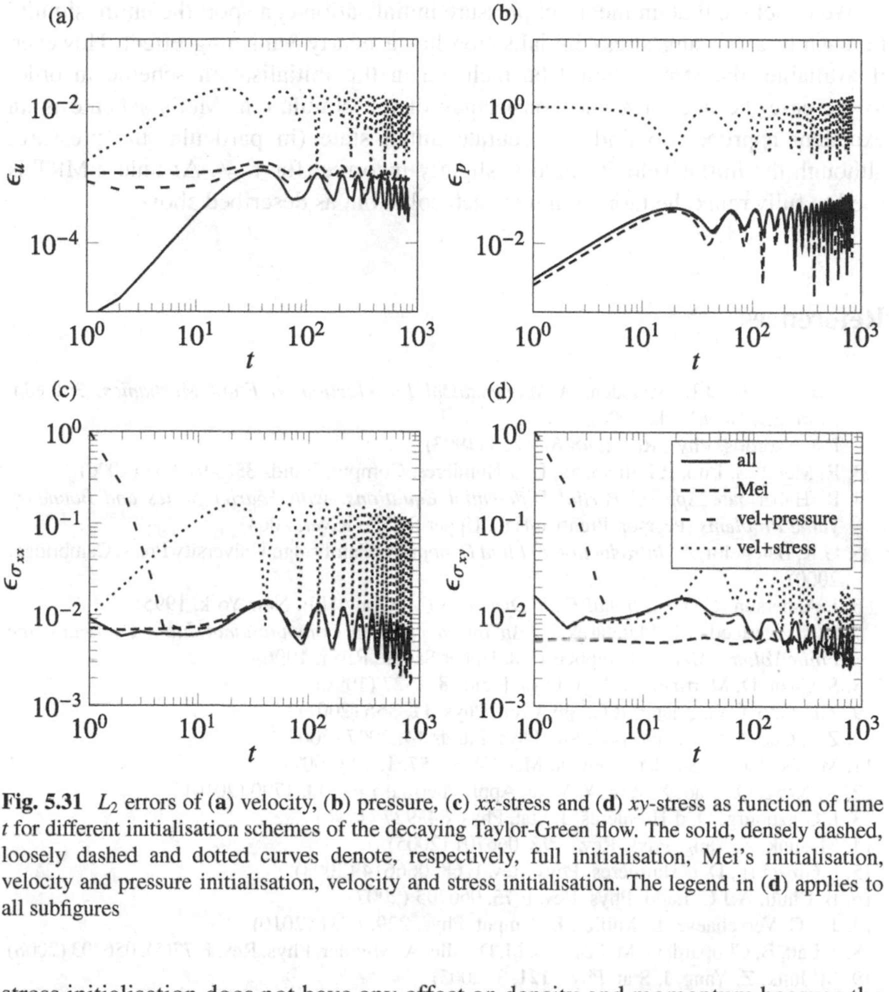

Fig. 5.31 L_2 errors of (**a**) velocity, (**b**) pressure, (**c**) *xx*-stress and (**d**) *xy*-stress as function of time t for different initialisation schemes of the decaying Taylor-Green flow. The solid, densely dashed, loosely dashed and dotted curves denote, respectively, full initialisation, Mei's initialisation, velocity and pressure initialisation, velocity and stress initialisation. The legend in (**d**) applies to all subfigures

stress initialisation does not have any effect on density and momentum because the simulation starts with collision and $\tau = \Delta t$ leads to the total extinction of any non-equilibrium distributions. This has been confirmed by simulations (data not shown here).

The situation is different when the pressure is initially ignored. The pressure error itself does never recover from the inconsistency and oscillates in the range 10–100%. The pressure error is nearly two orders of magnitude smaller when the pressure is properly initialised. The velocity error is less susceptible: without pressure initialisation it is around 1% whereas the full initialisation leads to velocity errors below 0.1%. Both stress errors behave qualitatively differently. The error of σ_{xx} strongly depends on the pressure initialisation, while the error of σ_{xy} does not. The explanation is that σ_{xx} errors, like pressure errors, are tightly related to compressibility artefacts while σ_{xy} errors are not.

We conclude that an incorrect pressure initialisation can spoil the entire simulation while a missing stress initialisation has a nearly vanishing effect. However, if available, the stress should be included in the initialisation scheme in order to increase the accuracy and consistency of the simulation. Mei's scheme is an excellent approach to find an accurate initial state (in particular the pressure) although the initial velocity field is slightly incorrect for $\tau \neq \Delta t$ unless MRT is used to fully relax the momentum in each collision, as described above.

References

1. A.J. Chorin, J.E. Marsden, *A Mathematical Introduction to Fluid Mechanics*, 3rd edn. (Springer, New York, 2000)
2. P.A. Skordos, Phys. Rev. E **48**(6), 4823 (1993)
3. R. Mei, L.S. Luo, P. Lallemand, D. d'Humières, Comput. Fluids **35**(8-9), 855 (2006)
4. R. Haberman, *Applied Partial Differential Equations: with Fourier Series and Boundary Value Problems* (Pearson Prentice Hall, Upper Saddle River, 2004)
5. G.K. Batchelor, *An Introduction to Fluid Dynamics* (Cambridge University Press, Cambridge, 2000)
6. J. Anderson, *Computational Fluid Dynamics* (McGraw-Hill, New York, 1995)
7. H.K. Versteed, M. Malalasekera, *An Introduction to Computational Fluid Dynamics, the Finite Volume Method* (Prentice-Hall, Upper Saddle River, 1996)
8. S. Chen, D. Martinez, R. Mei, Phys. Fluids **8**, 2527 (1996)
9. Z.L. Guo, C.G. Zheng, B.C. Shi, Chin. Phys. **11**, 366 (2002)
10. Z.L. Guo, C.G. Zheng, B.C. Shi, Phys. Fluids **14**, 2007 (2002)
11. M. Shankar, S. Sundar, Comput. Math. Appl. **57**, 1312 (2009)
12. X. Kang, Q. Liao, X. Zhu, Y. Yang, Appl. Thermal Eng. **30**, 1790 (2010)
13. I. Ginzbourg, D. d'Humières, J. Stat. Phys. **84**, 927 (1996)
14. M. Junk, Z. Yang, Phys. Rev. E **72**, 066701 (2005)
15. I. Ginzburg, D. d'Humières, Phys. Rev. E **68**, 066614 (2003)
16. B. Chun, A.J.C. Ladd, Phys. Rev. E **75**, 066705 (2007)
17. J.C.G. Verschaeve, B. Müller, J. Comput. Phys. **229**, 6781 (2010)
18. J. Latt, B. Chopard, O. Malaspinas, M. Deville, A. Michler, Phys. Rev. E **77**(5), 056703 (2008)
19. M. Junk, Z. Yang, J. Stat. Phys. **121**, 3 (2005)
20. J.C.G. Verschaeve, Phys. Rev. E **80**, 036703 (2009)
21. I. Ginzbourg, P.M. Adler, J. Phys. II France **4**(2), 191 (1994)
22. A.J.C. Ladd, J. Fluid Mech. **271**, 285 (1994)
23. X. He, Q. Zou, L.S. Luo, M. Dembo, J. Stat. Phys. **87**(1–2), 115 (1997)
24. M. Bouzidi, M. Firdaouss, P. Lallemand, Phys. Fluids **13**, 3452 (2001)
25. D.R. Noble, Chen, J.G. Georgiadis, R.O. Buckius, Phys. Fluids **7**, 203 (1995)
26. T. Inamuro, M. Yoshino, F. Ogino, Phys. Fluids **7**, 2928 (1995)
27. Q. Zou, X. He, Phys. Fluids **9**, 1591 (1997)
28. I. Halliday, L.A. Hammond, C.M. Care, A. Stevens, J. Phys. A Math. Gen. **35**, 157 (2002)
29. A.P. Hollis, I.H.H.M. Care, J. Phys. A Math. Gen. **39**, 10589 (2006)
30. I. Ginzburg, F. Verhaeghe, D. d'Humières, Commun. Comput. Phys. **3**, 427 (2008)
31. I. Ginzburg, F. Verhaeghe, D. d'Humières, Commun. Comput. Phys. **3**, 519 (2008)
32. C. Pan, L.S. Luo, C.T. Miller, Comput. Fluids **35**(8-9), 898 (2006)
33. S. Khirevich, I. Ginzburg, U. Tallarek, J. Comp. Phys. **281**, 708 (2015)
34. R.S. Maier, R.S. Bernard, D.W. Grunau, Phys. Fluids **8**, 1788 (1996)

35. S. Succi, *The Lattice Boltzmann Equation for Fluid Dynamics and Beyond* (Oxford University Press, Oxford, 2001)
36. M.C. Sukop, D.T. Thorne Jr., *Lattice Boltzmann Modeling: An Introduction for Geoscientists and Engineers* (Springer, New York, 2006)
37. S.H. Kim, H. Pitsch, Phys. Fluids **19**, 108101 (2007)
38. Q. Zou, S. Hou, S. Chen, G.D. Doolen, J. Stat. Phys. **81**, 35 (1995)
39. X. He, L.S. Luo, J. Stat. Phys. **88**, 927 (1997)
40. S.V. Patankar, C.H. Liu, E.M. Sparrow, ASME J. Heat Transfer **99**, 180 (1977)
41. J. Zhang, D.Y. Kwok, Phys. Rev. E **73**, 047702 (2006)
42. O. Gräser, A. Grimm, Phys. Rev. E **82**, 016702 (2010)
43. U. Frisch, B. Hasslacher, Y. Pomeau, Phys. Rev. Lett. **56**(14), 1505 (1986)
44. R. Cornubert, D. d'Humières, D. Levermore, Physica D **47**, 241 (1991)
45. D.P. Ziegler, J. Stat. Phys. **71**, 1171 (1993)
46. J. Hardy, Y. Pomeau, O. de Pazzis, J. Math. Phys. **14**(12), 1746 (1973)
47. J.P. Rivet, J.P. Boon, *Lattice Gas Hydrodynamics* (Cambridge University Press, Cambridge, 2001)
48. D.A. Wolf-Gladrow, *Lattice-Gas Cellular Automata and Lattice Boltzmann Models* (Springer, New York, 2005)
49. A.J.C. Ladd, R. Verberg, J. Stat. Phys. **104**(5–6), 1191 (2001)
50. A.J. Wagner, I. Pagonabarraga, J. Stat. Phys. **107**, 531 (2002)
51. C. Aidun, Y. Lu, J. Stat. Phys. **81**, 49 (1995)
52. C.K. Aidun, Y. Lu, E.J. Ding, J. Fluid Mech. **373**, 287 (1998)
53. S. Krithivasan, S. Wahal, S. Ansumali, Phys. Rev. E **89**, 033313 (2014)
54. N.Q. Nguyen, A.J.C. Ladd, Phys. Rev. E **66**(4), 046708 (2002)
55. X. Yin, G. Le, J. Zhang, Phys. Rev. E **86**(2), 026701 (2012)
56. O. Filippova, D. Hänel, J. Comput. Phys. **147**, 219 (1998)
57. I. Ginzburg, J. Stat. Phys. **126**, 157 (2007)
58. D. d'Humières, In Rarefied Gas Dynamics: Theory and Simulations, ed. B. Shizgal, D Weaver **159**, 450 (1992)
59. P. Lallemand, L.S. Luo, Phys. Rev. E **61**(6), 6546 (2000)
60. D. d'Humières, I. Ginzburg, M. Krafczyk, P. Lallemand, L.S. Luo, Phil. Trans. R. Soc. Lond. A **360**, 437 (2002)
61. A.J.C. Ladd, J. Fluid Mech. **271**, 311 (1994)
62. L.S. Luo, W. Lia, X. Chen, Y. Peng, W. Zhang, Phys. Rev. E **83**, 056710 (2011)
63. A.A. Mohamad, S. Succi, Eur. Phys. J. **171**, 213 (2009)
64. R.G.M. Van der Sman, Comput. Fluids **35**, 849 (2006)
65. D. d'Humières, I. Ginzburg, Comput. Math. Appl. **58**, 823 (2009)
66. I. Ginzburg, Commun. Comput. Phys. **11**, 1439 (2012)
67. M. Hecht, J. Harting, J. Stat. Mech. Theory Exp. **P**, 01018 (2010)
68. H. Chen, Y. Qiao, C. Liu, Y. Li, B. Zhu, Y. Shi, D. Sun, K. Zhang, W. Lin, Appl. Math. Model **36**, 2031 (2012)
69. S. Izquierdo, N. Fueyo, Phys. Rev. E **78**, 046707 (2008)
70. S. Izquierdo, P. Martinez-Lera, N. Fueyo, Comput. Math. Appl. **58**, 914 (2009)
71. A.P. Hollis, I. Halliday, C.M. Care, J. Comput. Phys. **227**, 8065 (2008)
72. S. Izquierdo, N. Fueyo, Phys. Rev. E **78**(4) (2008)
73. D. Heubes, A. Bartel, M. Ehrhardt, J. Comput. Appl. Math. **262**, 51 (2014)
74. L. Talon, D. Bauer, D. Gland, H. Auradou, I. Ginzburg, Water Resour. Res. **48**, W04526 (2012)
75. S. Hou, Q. Zou, S. Chen, G.D. Doolen, A.C. Cogley, J. Comput. Phys. **118**, 329 (1995)
76. A.R. da Silva, Numerical studies of aeroacoustic aspects of wind instruments. Ph.D. thesis, McGill University, Montreal (2008)
77. G. Falcucci, M. Aureli, S. Ubertini, M. Porfiri, Phil. Trans. R. Soc. A **369**, 2456 (2011).
78. P. Lallemand, L.S. Luo, J. Comput. Phys. **184**(2), 406 (2003)
79. S. Tao, J. Hu, Z. Guo, Comput. Fluids **133**, 1 (2016)

80. R. Mei, D. Yu, W. Shyy, L.S. Luo, Phys. Rev. E **65**(4), 041203 (2002)
81. T. Krüger, F. Varnik, D. Raabe, Phys. Rev. E **79**(4), 046704 (2009)
82. W.A. Yong, L.S. Luo, Phys. Rev. E **86**, 065701(R) (2012)
83. D. Yu, R. Mei, L.S. Luo, W. Shyy, Prog. Aerosp. Sci. **39**, 329 (2003)
84. Y. Chen, H. Ohashi, M. Akiyama, Phys. Rev. E **50**(4), 2776 (1994)
85. P.C. Philipi, L.A. Hegele, L.O.E. Santos, R. Surmas, Phys. Rev. E **73**, 056702 (2006)
86. A. Scagliarini, L. Biferale, M. Sbragaglia, K. Sugiyama, F. Toschi, Phys. Fluids **22**, 055101 (2010)
87. X. Shan, X.F. Yuan, H. Chen, J. Fluid Mech. **550**, 413 (2006)
88. S.H. Kim, H. Pitsch, I.D. Boyd, J. Comp. Phys. **227**, 8655 (2008)
89. C.E. Colosqui, M.E. Kavousanakis, A.G. Papathanasiou, I.G. Kevrekidis, Phys. Rev. E **87**, 013302 (2013)
90. J. Meng, Y. Zhang, J. Comp. Phys. **258**, 601 (2014)
91. O. Malaspinas, B. Chopard, J. Latt, Comput. Fluids **49**, 29 (2011)
92. Z. Guo, T.S. Zhao, Y. Shi, Phys. Rev. E **70**(6), 066706 (2004)
93. A.M.M. Artoli, A.G. Hoekstra, P.M.A. Sloot, Int. J. Mod. Phys. C **14**(6), 835 (2003)
94. A. Caiazzo, J. Stat. Phys. **121**(1–2), 37 (2005)
95. P. Van Leemput, M. Rheinlander, M. Junk, Comput. Math. Appl. **58**(5), 867 (2009)
96. J. Huang, H. Wu, W.A. Yong, Commun. Comput. Phys. **18**(02), 450 (2015)
97. D.J. Holdych, D.R. Noble, J.G. Georgiadis, R.O. Buckius, J. Comput. Phys. **193**(2), 595 (2004)
98. H. Xu, H. Luan, Y. He, W. Tao, Comput. Fluids **54**, 92 (2012)
99. P.J. Dellar, Comput. Math. Appl. **65**(2), 129 (2013)
100. U.D. Schiller, Comput. Phys. Commun. **185**(10), 2586 (2014)

Chapter 6
Forces

Abstract After reading this chapter, you will be able to add forces to lattice Boltzmann simulations while retaining their accuracy. You will know how a forcing scheme can be derived by including forces in the derivation of the lattice Boltzmann equation, though you will also know that there are a number of other forcing schemes available. You will understand how to investigate forcing models and their errors through the Chapman-Enskog analysis, and how initial and boundary conditions can be affected by the presence of forces.

Forces play an important role in many hydrodynamic problems (Sect. 6.1). Therefore, a proper discussion of force implementation in the LB algorithm is essential. Section 6.2 contains quick start instructions to implement an LB algorithm with forces. In Sect. 6.3 we show how to extend the force-free LBE derivation (i.e. the LBE derived in Chap. 3) to also reproduce a macroscopic body force at the hydrodynamic level. This derivation is based on the same discretisation steps (velocity followed by space-time) that are also used for the force-free LBE. Section 6.4 contains an overview of existing forcing schemes and a discussion of their differences and similarities. We will see that many of those schemes are equivalent if higher-order terms are neglected. In Sect. 6.5 we extend the Chapman-Enskog analysis to situations with forces to point out the detailed links between the LBE and the macroscopic PDEs it approximates. Furthermore, we investigate the errors associated with the selected forcing schemes. We analyse the influence of the forcing term on simulation initialisation and two types of boundary conditions in Sect. 6.6. In particular, we show how the bounce-back and the non-equilibrium bounce-back methods account for the presence of a force. Finally, in Sect. 6.7 we use a simple Poiseuille flow to demonstrate the previous theoretical elements in benchmark simulations.

T. Krüger et al., *The Lattice Boltzmann Method*, Graduate Texts in Physics,
DOI 10.1007/978-3-319-44649-3_6

6.1 Motivation and Background

Forces play a central role in many hydrodynamic problems. A prominent example is the gravitational acceleration $\boldsymbol{g}$ which can be cast into a force *density* $\boldsymbol{F}_g$ by multiplying it with the fluid density ρ:

$$\boldsymbol{F}_g = \rho \boldsymbol{g}. \tag{6.1}$$

In fact, in hydrodynamics we will mostly encounter force densities rather than forces since the momentum equation is a PDE for the momentum *density*. Forces are obtained by integrating surface stresses or bulk force densities. Mathematically, a force (density) is a momentum (density) source term, as can be seen from the Cauchy equation in (1.57).

Gravity leads to a number of effects which LBM can successfully simulate. If two fluids with different densities are mixed or if the temperature in a fluid is non-homogeneous, density gradients in the gravitational field lead to buoyancy effects and phenomena like the *Rayleigh-Bénard instability* [1] (cf. Sect. 8.4.1) or the *Rayleigh-Taylor instability* [2]. In the Rayleigh-Bénard instability, which is essential in studies of heat transfer, convection patterns develop when warmed fluid rises from a hot surface and falls after cooling. The Rayleigh-Taylor instability can occur when a layer of denser fluid descends as lower-density fluid below it rises. Gravity waves at a free water surface are another example [3].

Apart from gravity [4], there are several other physical problems where forces are important. Fluids in rotating reference frames are subject to radial and Coriolis forces [5–7]. Charged or magnetic particles immersed in a fluid exert forces on each other, and they may also be forced by external electromagnetic fields. This is particularly important for modelling the effects of external electric fields on regions of unbalanced charges (the electrical double layer, EDL) in electrolytes near a charged solid surface or liquid-liquid interface [8–11].

In incompressible flows, the driving mechanism of the pressure gradient field may be equivalently described by any *divergence-free* body force [12]. There exist cases where the problem physics specify pressure gradients, but where it is convenient to replace these with forces [12–14]. One reason for this is that the LB method may lose accuracy when solving pressure fields due to compressibility errors (cf. Sect. 4.5). While this change is possible in arbitrarily complex flow geometries, the task of finding an equivalent driving force field is only trivial in periodic flow configurations. This is often explored in LB simulations of porous media flows [14].

We will see in Chap. 9 that forces are also commonly used to model multi-phase or multi-component flows, although a mathematical description of these phenomena is usually based on the stress tensor. Furthermore, some algorithms for fluid-structure interactions, e.g. the immersed boundary method, rely on forces mimicking boundary conditions. We will discuss this in Sect. 11.4.

6.2 LBM with Forces in a Nutshell

We summarise the most important information about the implementation of forces in the bulk LBM and what a complete time step with forces looks like.

Assuming the BGK collision operator, we can write the order of operations in a single time step including forces, also illustrated in Fig. 6.1, in the following way:

1. Determine the force density $\boldsymbol{F}$ for the time step (e.g. gravity).
2. Compute the fluid density and velocity from

$$\rho = \sum_i f_i, \quad \boldsymbol{u} = \frac{1}{\rho}\sum_i f_i \boldsymbol{c}_i + \frac{\boldsymbol{F}\Delta t}{2\rho}. \tag{6.2}$$

3. Compute the equilibrium populations $f_i^{\text{eq}}(\rho, \boldsymbol{u})$ to construct the collision operator

$$\Omega_i = -\frac{1}{\tau}\left(f_i - f_i^{\text{eq}}\right). \tag{6.3}$$

4. If desired, output the macroscopic quantities. If required, the deviatoric stress is calculated as

$$\sigma_{\alpha\beta} \approx -\left(1 - \frac{\Delta t}{2\tau}\right)\sum_i f_i^{\text{neq}} c_{i\alpha} c_{i\beta} - \frac{\Delta t}{2}\left(1 - \frac{\Delta t}{2\tau}\right)\left(F_\alpha u_\beta + u_\alpha F_\beta\right). \tag{6.4}$$

5. Compute the source term

$$S_i = \left(1 - \frac{\Delta t}{2\tau}\right) w_i \left(\frac{c_{i\alpha}}{c_s^2} + \frac{\left(c_{i\alpha} c_{i\beta} - c_s^2 \delta_{\alpha\beta}\right) u_\beta}{c_s^4}\right) F_\alpha, \tag{6.5}$$

where the source S_i and forcing F_i terms are related as $S_i = (1 - \frac{1}{2\tau})F_i$

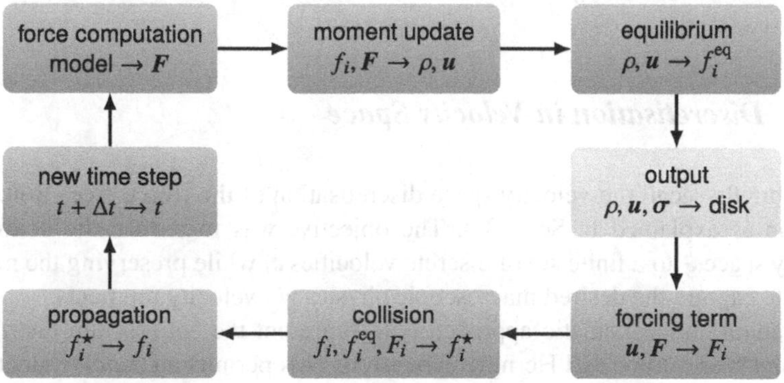

Fig. 6.1 An overview of one cycle of the LB algorithm, considering forces but not boundary conditions. The *light grey box* shows the optional output sub-step

6. Apply collision and source to find the post-collision populations:

$$f_i^\star = f_i + (\Omega_i + S_i)\Delta t. \tag{6.6}$$

7. Propagate populations.
8. Increment the time step and go back to step 1.

There are a few important remarks:

- The form of the force $\boldsymbol{F}$ depends on the underlying physics and is not itself given by the LB algorithm. Gravity is the simplest example.
- The velocity $\boldsymbol{u}$ in (6.2) contains the so-called *half-force correction*. This velocity $\boldsymbol{u}$ enters the equilibrium distributions and is also the macroscopic fluid velocity solving the Navier-Stokes equation. Using the *bare* velocity $\boldsymbol{u}^\star = \sum_i f_i \boldsymbol{c}_i/\rho$ would lead to first-order rather than second-order space-time accuracy (Sect. 6.3.2). The velocity $\boldsymbol{u}$ can be interpreted as the average velocity during the time step, i.e. the average of pre- and post-collision values.
- The forcing scheme presented here is based on a Hermite expansion (Sect. 6.3.1) and is the same as proposed by Guo et al. [15]. There are alternative ways to include forces, as discussed in Sect. 6.4.
- Any cyclic permutation of the above steps is permitted, as long as the simulation is properly initialised.

6.3 Discretisation

In Chap. 3 we have shown how to derive the LBE from the continuous Boltzmann equation in the absence of forces. Here we will revisit that derivation from Sect. 3.4 and Sect. 3.5, now highlighting the required steps to include forces. The two main steps are the discretisation in velocity space (Sect. 6.3.1) and the space-time discretisation (Sect. 6.3.2).

6.3.1 Discretisation in Velocity Space

Let us briefly recall the velocity space discretisation of the (force-free) Boltzmann equation as explained in Sect. 3.4. The objective was to reduce the continuous velocity space $\boldsymbol{\xi}$ to a finite set of discrete velocities $\boldsymbol{c}_i$ while preserving the model's ability to capture the desired macroscopic physics *via* velocity moments.

A natural and systematic approach is to represent the equilibrium distribution function f^{eq} as a truncated Hermite expansion. This permits an *exact* evaluation of macroscopic quantities (e.g. density and velocity) through a Gauss-Hermite quadrature. This procedure led to two important results: (i) a polynomial representation of f^{eq} in velocity space (cf. Sect. 3.4.5) and (ii) the description of particles' motion through a discrete velocity set (cf. Sect. 3.4.7). The question we aim to answer here is: *what is the equivalent polynomial representation in velocity space of the forcing term in the Boltzmann equation?* The following explanation is based on [16, 17].

Let us recall the continuous Boltzmann equation with a forcing term:

$$\frac{\partial f}{\partial t} + \xi_\alpha \frac{\partial f}{\partial x_\alpha} + \frac{F_\alpha}{\rho}\frac{\partial f}{\partial \xi_\alpha} = \Omega(f). \tag{6.7}$$

Our goal is to find the discrete velocity structure of the forcing term F_α which aligns with the velocity space discretisation of f^{eq} in Sect. 3.4.5. An evident problem is that F_α, contrarily to f^{eq}, does not appear as isolated term in (6.7). Rather, to deal with F_α we have to discretise the full term $\frac{F_\alpha}{\rho}\frac{\partial f}{\partial \xi_\alpha}$. Its discretisation in velocity space is simple if we keep the following two mathematical results in mind:

1. The Hermite series expansion of the distribution function $f(\boldsymbol{\xi})$ is

$$f(\boldsymbol{x},\boldsymbol{\xi},t) \approx \omega(\boldsymbol{\xi}) \sum_{n=0}^{N} \frac{1}{n!} \boldsymbol{a}^{(n)}(\boldsymbol{x},t)\cdot \boldsymbol{H}^{(n)}(\boldsymbol{\xi}). \tag{6.8}$$

2. The derivative property of Hermite polynomials reads

$$\omega(\boldsymbol{\xi})\boldsymbol{H}^{(n)} = (-1)^n \nabla_{\boldsymbol{\xi}}^n \omega(\boldsymbol{\xi}). \tag{6.9}$$

With their help we can rewrite the Hermite expansion of $f(\boldsymbol{\xi}_i)$ as follows:

$$f \approx \sum_{n=0}^{N} \frac{(-1)^n}{n!} \boldsymbol{a}^{(n)} \cdot \nabla_{\boldsymbol{\xi}}^n \omega. \tag{6.10}$$

This representation allows us to simplify the forcing contribution in (6.7):

$$\begin{aligned} \frac{\boldsymbol{F}}{\rho}\cdot\nabla_{\boldsymbol{\xi}} f &\approx \frac{\boldsymbol{F}}{\rho}\cdot \sum_{n=0}^{N} \frac{(-1)^n}{n!} \boldsymbol{a}^{(n)} \cdot \nabla_{\boldsymbol{\xi}}^{n+1}\omega \\ &\approx -\frac{\boldsymbol{F}}{\rho}\cdot \omega \sum_{n=1}^{N} \frac{1}{n!} n\boldsymbol{a}^{(n-1)} \cdot \boldsymbol{H}^{(n)}. \end{aligned} \tag{6.11}$$

The discretisation in velocity space can now be performed directly, by replacing the continuous $\boldsymbol{\xi}$ by a discrete set of $\boldsymbol{c}_i$. We rescale the velocities according to $\boldsymbol{c}_i = \boldsymbol{\xi}_i/\sqrt{3}$ and then renormalise the result by the lattice weights w_i. Recalling Sect. 3.4.5, this is similar to what we did in the construction of f^{eq}. Based on this procedure, the discrete form of the forcing term becomes:

$$F_i(\boldsymbol{x},t) = -\frac{w_i}{\omega(\boldsymbol{\xi})}\frac{\boldsymbol{F}}{\rho}\cdot\nabla_{\boldsymbol{\xi}} f\Big|_{\boldsymbol{\xi}\to\sqrt{3}\boldsymbol{c}_i}, \tag{6.12}$$

with the right-hand side given in (6.11). This way, we can write the discrete velocity Boltzmann equation with a forcing term similarly to (3.58):

$$\partial_t f_i + c_{i\alpha}\partial_\alpha f_i = \Omega_i + F_i, \quad i = 0, \dots, q-1. \tag{6.13}$$

The **truncation of the forcing term** up to second velocity order ($N = 2$), corresponding to the expansion of f^{eq}, reads

$$F_i = w_i \left(\frac{c_{i\alpha}}{c_s^2} + \frac{\left(c_{i\alpha}c_{i\beta} - c_s^2\delta_{\alpha\beta}\right) u_\beta}{c_s^4} \right) F_\alpha. \tag{6.14}$$

Its first three velocity moments are

$$\sum_i F_i = 0, \tag{6.15a}$$

$$\sum_i F_i c_{i\alpha} = F_\alpha, \tag{6.15b}$$

$$\sum_i F_i c_{i\alpha} c_{i\beta} = F_\alpha u_\beta + u_\alpha F_\beta. \tag{6.15c}$$

Exercise 6.1 Write down the explicit form of the forcing term F_i in (6.14) for the velocity sets D1Q3 (cf. Table 3.2) and D2Q9 (cf. Table 3.3). Compare the results obtained for F_i with the structure of f_i^{eq} expressed by (3.64) and (3.65), respectively.

The zeroth-order moment, (6.15a), denotes a mass source; it is zero in the present situation. The first-order moment, (6.15b), is a momentum source; it appears as a body force in the NSE. Finally, the second-order moment, (6.15c), is an energy source describing the power flux the body force exerts on the fluid [18].

The role of the second-order moment, (6.15c), is subtle. Its appearance, at first glance, may seem surprising as LBM is typically built upon an isothermal assumption. However, the (weakly) compressible regime reproduced by LBM with standard equilibrium still preserves a (weak) link to energy transport, although an isothermal one [19]. The purpose of (6.15c) is simply to remove the undesirable footprint left by this connection on the momentum equations. Otherwise, a spurious term, given by $F_\alpha u_\beta + u_\alpha F_\beta$, appears at the viscous stress level [15, 18, 20]. We explain this error source in more depth in Sect. 6.5.1.

On the other hand, in the incompressible regime the energy transport is totally decoupled from the momentum equation [21, 22]. As pointed out in Sect. 4.3.2, for steady problems the LBM with the incompressible equilibrium can reproduce the true incompressible NSE. Therefore, the cancelling of errors with link to compressibility is not required in this case, where we would expect the condition $\sum_i F_i c_{i\alpha} c_{i\beta} = 0$ instead. According to the force discretisation process above, this is equivalent to saying that F_i should be expanded only up to first order in velocity space:

$$F_i = w_i \frac{c_{i\alpha}}{c_s^2} F_\alpha. \tag{6.16}$$

This duality in the expansion order of F_i is explained in more detail in [5, 23].

6.3.2 Discretisation in Space and Time

We discussed the space-time discretisation of the force-free Boltzmann equation in Sect. 3.5. The idea was to replace the continuous space and time derivatives in the discrete-velocity Boltzmann equation, (6.13), by difference operators with discrete space and time steps (Δx and Δt). In the standard LBM, these discretisation steps are linked to the velocity space discretisation to ensure that populations f_i, travelling with discrete velocities $\boldsymbol{c}_i$, always reach neighbouring lattice sites within one time step Δt.

We seek a similar result in the presence of forces. The task consists of two parts [24–26]:

1. Advection, the left-hand side of (6.13), is identical to the force-free case (cf. Sect. 3.5). By applying the method of characteristics, i.e. defining $f_i = f_i(\boldsymbol{x}(\zeta), t(\zeta))$, where ζ parametrises a trajectory in space and time, the propagation step is *exact*, without any approximation:

$$\int_t^{t+\Delta t} \frac{\mathrm{d}f_i}{\mathrm{d}\zeta}\,\mathrm{d}\zeta = f_i(\boldsymbol{x} + \boldsymbol{c}_i\Delta t, t + \Delta t) - f_i(\boldsymbol{x}, t). \tag{6.17}$$

2. The only approximation appears in the treatment of the right-hand side of (6.13), collision, which now includes the forcing term F_i:

$$\int_t^{t+\Delta t} (\Omega_i + F_i)\,\mathrm{d}\zeta. \tag{6.18}$$

We can evaluate this integral in different ways [26]. We will now discuss two approximations, as already described for the force-free case in Sect. 3.5.

6.3.2.1 First-Order Integration

The least accurate procedure employs a rectangular discretisation. Here, the integral of collision and forcing terms is approximated by just one point:

$$\int_t^{t+\Delta t} (\Omega_i + F_i)\, \mathrm{d}\zeta = \left[\Omega_i(\boldsymbol{x}, t) + F_i(\boldsymbol{x}, t)\right] \Delta t + O(\Delta t^2). \tag{6.19}$$

Using this first-order approximation and the BGK collision operator, the LBE with a force assumes a form where all terms on the right-hand side are evaluated at $(\boldsymbol{x}, t)$:

$$f_i(\boldsymbol{x} + \boldsymbol{c}_i \Delta t, t + \Delta t) - f_i(\boldsymbol{x}, t) = -\frac{\Delta t}{\tau}\left(f_i - f_i^{\mathrm{eq}}\right) + F_i \Delta t. \tag{6.20}$$

Apart from the inclusion of $F_i \Delta t$ in (6.20), everything else is exactly as the unforced case in Chap. 3.

While being fully explicit, this scheme is only first-order accurate in time. In the absence of forces, this is not harmful since we can still obtain second-order accuracy providing the $\Delta t/2$ shift is considered in the viscosity-relaxation relation [26]: $\nu = c_s^2(\tau - \frac{\Delta t}{2})$ instead of $\nu = c_s^2 \tau$. The reason for this accuracy improvement is that both the "physical" viscous term and its leading-order error have the same functional form; the latter can be absorbed as a "physical" contribution by redefining the viscosity.

This "trick" does not work in the presence of forces, though. Hence, the first-order accuracy inevitably leads to macroscopic solutions corrupted by discrete lattice artefacts [4, 15]. We show their mathematical form in Sect. 6.5.1 and illustrate their quantitative effects in Sect. 6.7. We can eliminate these undesired artefacts by employing a second-order space-time discretisation.

6.3.2.2 Second-Order Integration

The trapezoidal discretisation is more accurate than the rectangular discretisation:

$$\int_t^{t+\Delta t} (\Omega_i + F_i)\, \mathrm{d}\zeta = \left(\frac{\Omega_i(\boldsymbol{x}, t) + \Omega_i(\boldsymbol{x} + \boldsymbol{c}_i \Delta t, t + \Delta t)}{2} + \frac{F_i(\boldsymbol{x}, t) + F_i(\boldsymbol{x} + \boldsymbol{c}_i \Delta t, t + \Delta t)}{2}\right) \Delta t + O(\Delta t^3). \tag{6.21}$$

However, we obtain second-order accuracy at the expense of a time-implicit scheme. Fortunately, this is not a problem since, as explained in Sect. 3.5, we can recover the explicit form by introducing a smart change of variables [19, 27]:

$$\bar{f}_i = f_i - \frac{(\Omega_i + F_i)\Delta t}{2}. \tag{6.22}$$

Using (6.22) and some simple algebra, the LBE for $\bar{f}_i$ takes the familiar form

$$\bar{f}_i(\boldsymbol{x}+\boldsymbol{c}_i\Delta t, t+\Delta t) - \bar{f}_i(\boldsymbol{x}, t) = \left[\Omega_i(\boldsymbol{x}, t) + F_i(\boldsymbol{x}, t)\right]\Delta t. \tag{6.23}$$

With the BGK collision operator this simplifies to

$$\bar{f}_i(\boldsymbol{x}+\boldsymbol{c}_i\Delta t, t+\Delta t) - \bar{f}_i(\boldsymbol{x}, t) = -\frac{\Delta t}{\tau+\Delta t/2}\left(\bar{f}_i - f_i^{\mathrm{eq}} - \tau F_i\right) \tag{6.24}$$

where, once again, all terms on the right-hand side are given at $(\boldsymbol{x}, t)$. The extension to other collision operators is straightforward (Sect. 10.5).

The **second-order accurate discretisation of the LBGK equation with forcing term** reads

$$\bar{f}_i(\boldsymbol{x}+\boldsymbol{c}_i\Delta t, t+\Delta t) - \bar{f}_i(\boldsymbol{x}, t) = -\frac{\Delta t}{\bar{\tau}}\left(\bar{f}_i - f_i^{\mathrm{eq}}\right) + \left(1 - \frac{\Delta t}{2\bar{\tau}}\right)F_i\Delta t \tag{6.25}$$

with a redefined relaxation parameter $\bar{\tau} = \tau + \Delta t/2$. Based on the new variable $\bar{f}_i$, the leading macroscopic moments are

$$\rho = \sum_i \bar{f}_i + \frac{\Delta t}{2}\sum_i F_i, \tag{6.26a}$$

$$\rho\boldsymbol{u} = \sum_i \bar{f}_i\boldsymbol{c}_i + \frac{\Delta t}{2}\sum_i F_i\boldsymbol{c}_{i\alpha}, \tag{6.26b}$$

$$\boldsymbol{\Pi} = \left(1 - \frac{\Delta t}{2\bar{\tau}}\right)\sum_i \bar{f}_i\boldsymbol{c}_i\boldsymbol{c}_i + \frac{\Delta t}{2\bar{\tau}}\sum_i f_i^{\mathrm{eq}}\boldsymbol{c}_i\boldsymbol{c}_i + \frac{\Delta t}{2\bar{\tau}}\left(1 - \frac{\Delta t}{2\bar{\tau}}\right)\sum_i F_i\boldsymbol{c}_i\boldsymbol{c}_i. \tag{6.26c}$$

In most cases, the notation of the redefined variables is dropped for convenience and f_i and τ are written instead of $\bar{f}_i$ and $\bar{\tau}$. The equilibrium populations f_i^{eq} have the same functional form as before. However, the velocity entering $f_i^{\mathrm{eq}}(\rho, \boldsymbol{u})$ is now given by (6.26b). The redefinition of the velocity in (6.26b) can be interpreted as averaging the velocity before and after forcing [29, 30]. If F_i is chosen to incorporate non-zero mass sources in addition to forces, the density entering $f_i^{\mathrm{eq}}(\rho, \boldsymbol{u})$ must also be redefined according to (6.26a) [28]. Sometimes, for convenience, the outcome from the space-time discretisation of the forcing term, as given in (6.25), is shortened to a source term S_i notation, with the two related as $S_i = (1 - \frac{1}{2\tau})F_i$.

Equation (6.26b) can lead to difficulties when $\boldsymbol{F}$ depends on $\boldsymbol{u}$, e.g. in Brinkman models [31–34] or Coriolis forces [5–7, 35]. Such a velocity-dependent force leads to an implicit form of (6.26b) in $\boldsymbol{u}$. For linear relations, $\boldsymbol{F} \propto \boldsymbol{u}$, and other analytically invertible dependencies we can easily solve (6.26b) for $\boldsymbol{u}$, e.g. [5, 31]. In more general cases, however, $\boldsymbol{u}$ has to be found numerically, e.g. [7, 35].

6.4 Alternative Forcing Schemes

In Sect. 6.3, we have shown how the forcing scheme can be constructed through a systematic procedure consistent with the overall LBE. However, there is a flood of articles about other LB forcing schemes.

This section aims at clarifying differences and similarities among some of the most popular forcing schemes. After recollecting important consequences of the presence of a force in Sect. 6.4.1, we show a few alternative forcing schemes in Sect. 6.4.2. We focus on results rather than on those lengthy calculations that can be found in the cited literature. The articles by Guo et al. [15] and Huang et al. [36] provide derivations and more detailed discussions. Also helpful in this context is the work by Ginzburg et al. [37] that discusses different, yet equivalent, ways of introducing the force in the LB equation.

6.4.1 General Observations

Based on the second-order velocity and space-time discretisations, the LBE with a force can be expressed as

$$f_i(\boldsymbol{x} + \boldsymbol{c}_i\Delta t, t + \Delta t) - f_i(\boldsymbol{x}, t) = \left[\Omega_i(\boldsymbol{x}, t) + S_i(\boldsymbol{x}, t)\right] \Delta t \tag{6.27}$$

where Ω_i is the BGK collision operator and $S_i = (1 - \frac{1}{2\tau})F_i$ denotes a source, with the forcing F_i given by (6.14). Guo et al. [15] derived the same result following an approach different from that in Sect. 6.3. Therefore, this scheme is often called *Guo forcing*.

It is important that the fluid velocity in the presence of a force is redefined to guarantee the second-order space-time accuracy (Sect. 6.3.2):

$$\boldsymbol{u} = \frac{1}{\rho}\sum_i \boldsymbol{c}_i f_i + \frac{\boldsymbol{F}\Delta t}{2\rho}. \tag{6.28}$$

This velocity also enters the equilibrium populations $f_i^{\text{eq}} = f_i^{\text{eq}}(\rho, \boldsymbol{u})$ and therefore the BGK collision operator $\Omega_i = -(f_i - f_i^{\text{eq}})/\tau$. Thus we can say that the fluid velocity in (6.28) and the equilibrium velocity $\boldsymbol{u}^{\text{eq}}$ (i.e. the velocity entering f_i^{eq}) are the same for Guo forcing.

The complexity in the LB literature is caused by the fact that there exist different force algorithms that decompose Ω_i and S_i differently but lead to essentially the same results on the Navier-Stokes level. To generalise the forcing method, let us write

$$\boldsymbol{u}^{\text{eq}} = \frac{1}{\rho}\sum_i f_i \boldsymbol{c}_i + A\frac{\boldsymbol{F}\Delta t}{\rho} \tag{6.29}$$

for the equilibrium velocity. A is a model-dependent parameter. For Guo forcing, we already know that $A = \frac{1}{2}$. Deviating from this value means that the collision operator Ω_i is modified. In turn, also the source term S_i has to be redefined to keep the sum $\Omega_i + S_i$ unchanged, at least to leading order.

Naively we can expect that we cannot distinguish forcing schemes macroscopically as long as the sum $\Omega_i + S_i$ is the same, no matter which individual forms Ω_i and S_i assume. In fact, there exist several forcing schemes for which $\Omega_i + S_i$ *nearly* has the same form as Guo forcing, only up to deviations of order F^2 or u^3. Therefore, all those methods can be considered equivalent as long as F and u are sufficiently small, which cannot always be guaranteed. Furthermore, there are other forcing schemes that result in different behaviour on the F and u^2 orders (or even worse); those methods are generally less accurate and should be avoided.

6.4.2 *Forcing Schemes*

Each different LB forcing scheme has a different set of expressions for A in (6.29) and the source term S_i. But not all of the proposed methods lead to acceptable hydrodynamic behaviour. In the following we collect a few selected forcing schemes that do recover the correct macroscopic behaviour. Table 6.1 provides a summary.

The **fluid velocity** needs to assume the form in (6.28), **independently** of the chosen forcing scheme. This is a pure consequence of the second-order time integration and not affected by details of the forcing scheme.

Table 6.1 Overview of accurate forcing schemes and how they modify the collision operator in (6.27), both directly and through the equilibrium velocity defined in (6.29). In any case the fluid velocity must obey (6.28) to ensure second-order time accuracy

Method	A	S_i
Guo et al. [15]	1/2	$\left(1-\frac{\Delta t}{2\tau}\right) w_i \left(\frac{\boldsymbol{c}_i-\boldsymbol{u}}{c_s^2}+\frac{(\boldsymbol{c}_i\cdot\boldsymbol{u})\boldsymbol{c}_i}{c_s^4}\right)\cdot\boldsymbol{F}$
Shan and Chen [38]	$\tau\Delta t$	0
He et al. [39]	1/2	$\left(1-\frac{\Delta t}{2\tau}\right)\frac{f_i^{\text{eq}}}{\rho}\frac{\boldsymbol{c}_i-\boldsymbol{u}}{c_s^2}\cdot\boldsymbol{F}$
Kupershtokh et al. [40]	0	$f_i^{\text{eq}}(\rho,\boldsymbol{u}^\star+\Delta\boldsymbol{u})-f_i^{\text{eq}}(\rho,\boldsymbol{u}^\star)$

All forcing schemes in Table 6.1 are equivalent up to terms of order F^2 or u^3 [36].[1] In the limit of small Mach number and small forces, all these methods yield basically the same results. The situation is different for multi-phase flows where forces in the vicinity of fluid-fluid interfaces can become large so that terms $\propto F^2$ (and also $\propto \nabla^2 \boldsymbol{F}$) are important.[2] We will not discuss the choice of the forcing scheme in the context of multi-phase flows here and refer to Sect. 9.3.2 and [36, 41] instead.

6.4.2.1 Guo et al. (2002)

This method is the same as derived in Sect. 6.3. Based on the Chapman-Enskog analysis, Guo et al. [15] performed a thorough analysis of the lattice effects in the presence of a force. In their article, which is an extension of previous work by Ladd and Verberg [20], the parameters assume the values $A = 1/2$ and $S_i = (1 - \frac{1}{2\tau})F_i$ with F_i as in (6.14). Guo et al. [15] showed that these choices remove undesired derivatives in the continuity and momentum equation due to time discretisation artefacts (cf. Sect. 6.3.2). In particular, $A = 0$ would lead to a term $\propto \nabla \cdot \boldsymbol{F}$ in the continuity equation and another term $\propto \nabla \cdot (\boldsymbol{uF} + \boldsymbol{Fu})$ in the momentum equation (cf. Sect. 6.5.2).

6.4.2.2 Shan and Chen (1993, 1994)

Shan and Chen [38] proposed $A = \tau/\Delta t$ and $S_i = 0$. Although their motivation was the simulation of multi-phase fluids (cf. Chap. 9), Shan and Chen's method is applicable to single-phase fluids as well.

6.4.2.3 He et al. (1998)

The essential idea of He et al. [39] was to approximate the forcing term in the kinetic equation by assuming a situation close to equilibrium:

$$\boldsymbol{F} \cdot \nabla_{\boldsymbol{c}} f \approx \boldsymbol{F} \cdot \nabla_{\boldsymbol{c}} f^{\text{eq}} = -\boldsymbol{F} \cdot \frac{\boldsymbol{c} - \boldsymbol{u}}{c_s^2} f^{\text{eq}}. \tag{6.30}$$

[1] Showing that these forcing schemes are equivalent to leading order is straightforward but involves lengthy calculations. We will not delve into details here and refer to [36] for a more qualitative analysis.

[2] Also in Brinkman and Coriolis force models, where $\boldsymbol{F} \propto \boldsymbol{u}$, the error term $\propto \nabla^2 \boldsymbol{F}$ is important [33, 34].

In the end this leads to $A = \frac{1}{2}$ and

$$S_i = \left(1 - \frac{\Delta t}{2\tau}\right) \frac{f_i^{\text{eq}}}{\rho} \frac{\boldsymbol{c}_i - \boldsymbol{u}}{c_s^2} \cdot \boldsymbol{F}. \tag{6.31}$$

6.4.2.4 Kupershtokh (2004)

Kupershtokh [42] proposed a simple forcing method based on kinetic theory, the so-called *exact difference method*. The idea is to include the force density $\boldsymbol{F}$ in such a way that it merely shifts f_i in velocity space. As a consequence, $A = 0$ and

$$S_i = f_i^{\text{eq}}(\rho, \boldsymbol{u}^\star + \Delta \boldsymbol{u}) - f_i^{\text{eq}}(\rho, \boldsymbol{u}^\star) \tag{6.32}$$

where $\boldsymbol{u}^\star = \sum_i f_i \boldsymbol{c}_i / \rho$ and $\Delta \boldsymbol{u} = \boldsymbol{F} \Delta t / \rho$. This essentially means that the equilibrium for a velocity $\boldsymbol{u}^\star$ is directly replaced by the equilibrium for a velocity $\boldsymbol{u}^\star + \Delta \boldsymbol{u}$. In particular, this scheme ensures that an equilibrium distribution remains in equilibrium upon the action of the force, independently of the chosen value of τ.

6.4.2.5 Other, Less Accurate Approaches

There exist several other forcing schemes in the LB literature. Guo et al. [15] reviewed a series of approaches [4, 16, 43, 44] and showed that all of them lead to certain unphysical terms in the continuity or momentum equations of the *weakly compressible* NSE. In other words, those forcing schemes have additional error terms which are more significant than u^3 or F^2.

However, the situation changes when modelling *steady incompressible* hydrodynamics. In this case, the most accurate forcing scheme is no longer Guo's [15], but the scheme proposed by Buick and Greated [4]. We will discuss the reason for this variation in Sect. 6.5.2; see also [5, 23].

Finally, we would like to emphasise that under some circumstances some of these models may still be appropriate choices, for example if the force density $\boldsymbol{F}$ is constant. We illustrate this case in Sect. 6.7 through a numerical example. Still, we strongly recommend to implement one of the generally more accurate models mentioned above since they are usually more accurate when boundary conditions are involved. We will demonstrate this analytically in Sect. 6.6 and numerically in Sect. 6.7.

Concluding, there exist **several different forcing schemes**. Many of these schemes (i.e. Guo, Shan-Chen, He, Kupershtokh) are equivalent up to higher-order terms (u^3 or F^2). Their differences are negligible as long as forces and their gradients are small (e.g. in the case of gravity). Other forcing schemes, however, lead to additional error terms on the Navier-Stokes level.

6.5 Chapman-Enskog and Error Analysis in the Presence of Forces

We look at the macroscopic behaviour of forces in the LBE. First we revisit the Chapman-Enskog analysis (cf. Sect. 4.1) and extend it to situations with forces (Sect. 6.5.1). Based on this analysis, we discuss the structure of errors created at hydrodynamic level due to incorrectly chosen LB force models (Sect. 6.5.2).

6.5.1 Chapman-Enskog Analysis with Forces

The Chapman-Enskog analysis (Sect. 4.1) reveals the consistency between the mesoscopic LBM and the macroscopic NSE. We will now extend the Chapman-Enskog analysis to situations with forces.

Historically, the Chapman-Enskog analysis applied to the forced LBM was pioneered in [45, 46]. Later, a number of authors [16, 20, 39, 42, 44] extended its formulation to include second-order terms, as given by (6.14). A subsequent improvement [4, 15] showed the necessity of correcting discrete lattice effects. These effects can be corrected in an *a priori* fashion through a systematic second-order discretisation of the LBE (Sect. 6.3.2) [24–26]. Even today, the study of a "clean" inclusion of forces in the LBE remains an active research topic involving, for example, perturbation (Chapman-Enskog) analyses [5, 23, 28, 29] or exact solutions of the LBE [32–34, 47].

The Chapman-Enskog analysis of the forced LBE is similar to the force-free case in Sect. 4.1. The difference is that now we are working with (6.25) as evolution equation, together with (6.26) for the velocity moments. Hence, the first question we need to answer is: what should be the expansion order of the forcing term F_i?

In order to be consistent with the remaining terms in the LBE, the **forcing term** must **scale** as $F_i = O(\epsilon)$ [4]. Therefore, we should at least have $F_i = \epsilon F_i^{(1)}$.

Considering $F_i = \epsilon F_i^{(1)}$, which is a valid assumption for most hydrodynamic problems,[3] the familiar steps from Sect. 4.1 lead to a hierarchy of ϵ-perturbed

[3]In certain cases, the forcing term requires a higher-order expansion. For example, for certain axisymmetric LB models [48, 49], the formal expansion of the forcing term is $F_i = \epsilon F_i^{(1)} + \epsilon^2 F_i^{(2)}$.

equations, similar to (4.9a) and (4.9b), now with a force term:

$$O(\epsilon): \qquad \left(\partial_t^{(1)} + c_{i\alpha}\partial_\alpha^{(1)}\right) f_i^{\text{eq}} - \left(1 - \frac{\Delta t}{2\tau}\right) F_i^{(1)} = -\frac{1}{\tau} f_i^{(1)}, \tag{6.33a}$$

$$O(\epsilon^2): \quad \partial_t^{(2)} f_i^{\text{eq}} + \left(\partial_t^{(1)} + c_{i\alpha}\partial_\alpha^{(1)}\right)\left(1 - \frac{\Delta t}{2\tau}\right)\left(f_i^{(1)} + \frac{\Delta t}{2} F_i^{(1)}\right) = -\frac{1}{\tau} f_i^{(2)}. \tag{6.33b}$$

In the presence of an external force, the hydrodynamic moments are no longer conserved. This leads to a redefinition of the solvability conditions for mass and momentum:

$$\sum_i f_i^{\text{neq}} = -\frac{\Delta t}{2} \sum_i F_i^{(1)}, \tag{6.34a}$$

$$\sum_i \boldsymbol{c}_i f_i^{\text{neq}} = -\frac{\Delta t}{2} \sum_i \boldsymbol{c}_i F_i^{(1)}. \tag{6.34b}$$

Likewise, the extension to "strengthened" order-by-order solvability conditions reads

$$\sum_i f_i^{(1)} = -\frac{\Delta t}{2} \sum_i F_i^{(1)} \qquad \text{and} \qquad \sum_i f_i^{(k)} = 0, \tag{6.35a}$$

$$\sum_i \boldsymbol{c}_i f_i^{(1)} = -\frac{\Delta t}{2} \sum_i \boldsymbol{c}_i F_i^{(1)} \qquad \text{and} \qquad \sum_i \boldsymbol{c}_i f_i^{(k)} = \boldsymbol{0} \tag{6.35b}$$

with $k \geq 2$, [5, 37], which results from $F_i^{(1)} \sim O(\epsilon)$, only affecting $f_i^{(1)}$ and not higher ϵ scales.

In order to proceed, we require the functional form of F_i. We continue with the specific form in (6.14) whose moments are given in (6.15). In particular, there are no mass sources, i.e. the right-hand sides in (6.34a) and (6.35a) vanish.

By taking the zeroth and first moments of (6.33a), we obtain at $O(\epsilon)$:

$$\partial_t^{(1)} \rho + \partial_\gamma^{(1)} (\rho u_\gamma) = 0, \tag{6.36a}$$

$$\partial_t^{(1)} (\rho u_\alpha) + \partial_\beta^{(1)} \Pi_{\alpha\beta}^{\text{eq}} = F_\alpha. \tag{6.36b}$$

Here, $\Pi^{\text{eq}}_{\alpha\beta} = \sum_i c_{i\alpha} c_{i\beta} f^{\text{eq}}_i = \rho u_\alpha u_\beta + \rho c_s^2 \delta_{\alpha\beta}$, according to (4.11a). Similarly, by taking the zeroth and first moments of (6.33b), we obtain at $O(\epsilon^2)$:

$$\partial^{(2)}_t \rho = 0, \tag{6.37a}$$

$$\partial^{(2)}_t (\rho u_\alpha) + \partial^{(1)}_\beta \left(1 - \frac{\Delta t}{2\tau}\right) \Pi^{(1)}_{\alpha\beta} = 0. \tag{6.37b}$$

By combining the mass and momentum equations in (6.36) and (6.37), respectively, we obtain

$$\left(\epsilon\partial^{(1)}_t + \epsilon^2\partial^{(2)}_t\right) \rho + \epsilon\partial^{(1)}_\gamma (\rho u_\gamma) = 0, \tag{6.38a}$$

$$\left(\epsilon\partial^{(1)}_t + \epsilon^2\partial^{(2)}_t\right) (\rho u_\alpha) + \epsilon\partial^{(1)}_\beta \Pi^{\text{eq}}_{\alpha\beta} = \epsilon F^{(1)}_\alpha - \epsilon^2\partial^{(1)}_\beta \left(1 - \frac{\Delta t}{2\tau}\right) \Pi^{(1)}_{\alpha\beta}. \tag{6.38b}$$

To close the moment system in (6.38), we require an expression of $\Pi^{(1)}_{\alpha\beta}$ in terms of known quantities. We can achieve this by taking the second moment of (6.33a),

$$\partial^{(1)}_t \Pi^{\text{eq}}_{\alpha\beta} + \partial^{(1)}_\gamma \Pi^{\text{eq}}_{\alpha\beta\gamma} - \left(1 - \frac{\Delta t}{2\tau}\right) \sum_i F^{(1)}_i c_{i\alpha} c_{i\beta} = -\frac{1}{\tau} \Pi^{(1)}_{\alpha\beta}. \tag{6.39}$$

Here we have used the identity

$$\Pi^{(1)}_{\alpha\beta} = \sum_i f^{(1)}_i c_{i\alpha} c_{i\beta} + \frac{\Delta t}{2} \sum_i F^{(1)}_i c_{i\alpha} c_{i\beta} \tag{6.40}$$

that can be deduced by applying the Chapman-Enskog decomposition to (6.26c).

$\Pi^{(1)}_{\alpha\beta}$ is the contribution responsible for the viscous stress at macroscopic level. Therefore, the role of $\sum_i F^{(1)}_i c_{i\alpha} c_{i\beta}$ is to remove spurious forcing terms possibly appearing in $\Pi^{(1)}_{\alpha\beta}$ so that its form is the same as for the force-free case (cf. (4.15)):

$$\Pi^{(1)}_{\alpha\beta} = -\rho c_s^2 \tau \left(\partial^{(1)}_\beta u_\alpha + \partial^{(1)}_\alpha u_\beta\right) + O(u^3). \tag{6.41}$$

Therefore, the viscous stress is still given by $\sigma_{\alpha\beta} = -\left(1 - \frac{\Delta t}{2\tau}\right) \Pi^{(1)}_{\alpha\beta}$, just as in the force-free case, (4.14).

Finally, we can re-assemble $\partial_t = \epsilon\partial_t^{(1)} + \epsilon^2\partial_t^{(2)}$ and use $\Pi_{\alpha\beta}^{\mathrm{eq}}$ and $\Pi_{\alpha\beta}^{(1)}$ to obtain from (6.38) the correct form (up to $O(u^3)$ error terms) of the unsteady NSE with forcing term:

$$\partial_t\rho + \partial_\gamma(\rho u_\gamma) = 0, \tag{6.42a}$$

$$\partial_t(\rho u_\alpha) + \partial_\beta\left(\rho u_\alpha u_\beta + \rho c_s^2\delta_{\alpha\beta}\right) = \partial_\beta\left[\eta\left(\partial_\beta u_\alpha + \partial_\alpha u_\beta\right)\right] + F_\alpha. \tag{6.42b}$$

As usual, the dynamic shear and bulk viscosities are $\eta = \rho c_s^2(\tau - \frac{\Delta t}{2})$ and $\eta_B = 2\eta/3$, respectively (cf. Sect. 4.1).

6.5.2 *Errors Caused by an Incorrect Force Model*

Now that we know how to perform the Chapman-Enskog analysis with forces, we can evaluate whether the selected forcing scheme introduces errors in the recovered hydrodynamic model. According to Sect. 6.3, the formulation of the force model comprises two steps: (i) velocity space discretisation and (ii) space-time discretisation. Each of these steps comes with different error sources in case we do not deal with them properly.

6.5.2.1 Discretisation of Velocity Space: The Issue of Unsteady and Steady Cases

We can recognise the impact of an incorrect velocity space discretisation by distinguishing between unsteady and steady phenomena.

In *unsteady* state, the term $\partial_t^{(1)}\Pi_{\alpha\beta}^{\mathrm{eq}}$ contains the contribution $F_\alpha u_\beta + u_\alpha F_\beta$ (see Exercise 6.2 below). This contribution can be exactly cancelled by $\sum_i F_i c_{i\alpha} c_{i\beta}$, providing the force term F_i is expanded up to the *second velocity order* as shown in (6.14) [15, 20]. This way, we can correctly recover the unsteady NSE with force, (6.42).

Exercise 6.2 Show that

$$\begin{aligned}\partial_t^{(1)}\Pi_{\alpha\beta}^{\mathrm{eq}} &= \partial_t^{(1)}\left(\rho u_\alpha u_\beta + \rho c_s^2\delta_{\alpha\beta}\right)\\ &= -\partial_\gamma^{(1)}\left(\rho u_\alpha u_\beta u_\gamma\right) - c_s^2\left(u_\alpha\partial_\beta^{(1)}\rho + u_\beta\partial_\alpha^{(1)}\rho\right)\\ &\quad - c_s^2\delta_{\alpha\beta}\partial_\gamma^{(1)}(\rho u_\gamma) + F_\alpha^{(1)}u_\beta + u_\alpha F_\beta^{(1)}.\end{aligned} \tag{6.43}$$

Hint: apply the procedure outlined in Appendix A.2.2, including a forcing term.

In *steady* state, however, the term $\partial_t^{(1)} \Pi_{\alpha\beta}^{\text{eq}}$ is immaterial (cf. Sect. 4.2.3). Hence, we could expect that the contribution $F_\alpha u_\beta + u_\alpha F_\beta$ no longer exists. This is not absolutely true, though.

On the one hand, when using the standard equilibrium, an identical term is retrieved due to the requirement that the shear stress depends on the gradients of the velocity $\boldsymbol{u}$ rather than on the gradients of the momentum $\rho\boldsymbol{u}$. Consequently, to cancel that term, $\sum_i F_i c_{i\alpha} c_{i\beta}$ is still required as a correction.

On the other hand, with the incompressible LB equilibrium, the steady incompressible NSE is recovered with no spurious terms as discussed in Sect. 4.3. Hence, unlike the previous cases, here we must set $\sum_i F_i c_{i\alpha} c_{i\beta} = 0$, i.e. F_i must be expanded only to the first velocity order. A second-order expansion of F_i would lead to an incorrect steady incompressible NSE affected by the divergence of $F_\alpha u_\beta + u_\alpha F_\beta$. We will illustrate this issue in Sect. 6.7 through numerical examples. A more detailed explanation of this subtle point can be found in [5, 23].

6.5.2.2 Discretisation of Space and Time: The Issue of Discrete Lattice Effects

We can understand the effect of an inaccurate space-time discretisation on the forcing term by repeating the Chapman-Enskog analysis, but this time with a first-order time integration scheme (cf. Sect. 6.3.1).

Let us assume a time-dependent process and a forcing term with second-order velocity discretisation, (6.14). It can be shown, see e.g. [4, 15], that the macroscopic equations reproduced in this case have the following incorrect form:

$$\begin{aligned} \partial_t \rho + \partial_\gamma(\rho u_\gamma) &= -\frac{\Delta t}{2} \partial_\gamma F_\gamma, \\ \partial_t(\rho u_\alpha) + \partial_\beta \left(\rho u_\alpha u_\beta + \rho c_s^2 \delta_{\alpha\beta}\right) &= \partial_\beta \left[\eta \left(\partial_\beta u_\alpha + \partial_\alpha u_\beta\right)\right] + F_\alpha \\ &\quad - \frac{\Delta t}{2} \left[\partial_t F_\alpha + \partial_\beta \left(u_\alpha F_\beta + F_\alpha u_\beta\right)\right]. \end{aligned} \tag{6.44}$$

The difference between (6.44) and the "true" NSE with a force, (6.42), lies in the added $O(\Delta t)$ error terms [4, 15]. They are called *discrete lattice artefacts* since they act on the same scale as the viscous term $\eta \sim O(\Delta t)$. Thereby, they corrupt the macroscopic equations below the truncation error $O(\Delta t^2)$. These discrete artefacts lead to inconsistencies in the macroscopic equations for *both* mass and momentum. Therefore discrete lattice artefacts are more problematical than an incorrect velocity space discretisation which "only" corrupts the momentum equation.

6.6 Boundary and Initial Conditions with Forces

So far we have limited the discussion about forces to the bulk solution. In Chap. 5 we have already discussed the topic of initial and boundary conditions, but without including the effect of forces. We will now point out the required modifications of initial (Sect. 6.6.1) and boundary conditions (Sect. 6.6.2) due to the presence of forces.

6.6.1 *Initial Conditions*

Initial conditions are necessary for time-dependent problems. But even steady flows must be subject to a proper initialisation. Otherwise, initial errors may be conserved during the simulation and contaminate the steady-state solution. In Sect. 5.5 we discussed two ways of initialising LB simulations. Let us revisit them and work out the necessary modifications when forces are present.

The simplest strategy is to initiate the populations with their *equilibrium* state, $f_i(\boldsymbol{x}, t = 0) = f_i^{\text{eq}}\left(\rho_0(\boldsymbol{x}), \boldsymbol{u}_0(\boldsymbol{x})\right)$, where ρ_0 and $\boldsymbol{u}_0$ refer to known initial density and velocity fields. We know from Sect. 6.3.2 that for problems with forces the macroscopic velocity is computed from $\rho\boldsymbol{u} = \sum_i f_i\boldsymbol{c}_i + \frac{\Delta t}{2}\boldsymbol{F}$. Therefore, to set an initial velocity $\boldsymbol{u}_0$ consistent with the force field, we take [50]

$$f_i(\boldsymbol{x}, t = 0) = f_i^{\text{eq}}\left(\rho_0(\boldsymbol{x}), \bar{\boldsymbol{u}}_0(\boldsymbol{x})\right), \quad \bar{\boldsymbol{u}}_0 = \boldsymbol{u}_0 - \frac{\boldsymbol{F}\Delta t}{2\rho_0}. \tag{6.45}$$

Obviously, for low-order forcing schemes, where the macroscopic velocity is computed from $\rho\boldsymbol{u} = \sum_i f_i\boldsymbol{c}_i$, the equilibrium initialisation is the same as in the force-free case, i.e. $\bar{\boldsymbol{u}}_0 = \boldsymbol{u}_0$.

As discussed in Sect. 5.5, a more accurate initialisation consists of adding the non-equilibrium populations f_i^{neq} to f_i^{eq}. Given that the leading order of f_i^{neq}, i.e. $f_i^{(1)}$, depends on $\boldsymbol{F}$, cf. (6.33a), the non-equilibrium term added to (6.45) must be redefined [5, 51]:

$$f_i^{\text{neq}} \approx -\frac{w_i\tau}{c_s^2}\rho Q_{i\alpha\beta}\partial_\alpha u_\beta - \frac{w_i\Delta t}{2c_s^2}\left(c_{i\alpha}F_\alpha + \frac{Q_{i\alpha\beta}}{2c_s^2}\left(u_\alpha F_\beta + F_\alpha u_\beta\right)\right) \tag{6.46}$$

where $Q_{i\alpha\beta} = c_{i\alpha}c_{i\beta} - c_s^2\delta_{\alpha\beta}$.

6.6.2 *Boundary Conditions*

Forces may also affect the operation of boundary conditions (cf. Sect. 5.2.4). We will discuss the consequences for both bounce-back and non-equilibrium bounce-back.

6.6.2.1 Bounce-Back

Although the principle of the bounce-back rule is not changed by the inclusion of forces, its accuracy does depend on the force implementation. If we do not work with the second-order space-time discretisation of the LBE, the macroscopic laws established by the bounce-back rule will be affected by discrete lattice artefacts. We demonstrate this issue by looking at a simple example: a hydrostatic equilibrium where a constant force (e.g. gravity) is balanced by a pressure gradient.

We choose the second-order space-time discretisation for the bulk dynamics (cf. (6.25)). Also, let us consider a time-independent process: $\partial_t f_i = 0$. Then, the Chapman-Enskog analysis yields up to $O(\epsilon)^4$:

$$f_i^{(1)} = -\tau c_{i\alpha}\partial_\alpha^{(1)} f_i^{\text{eq}} + \left(\tau - \frac{\Delta t}{2}\right) F_i^{(1)}. \tag{6.47}$$

Given that we are interested in the hydrostatic solution, i.e. $\boldsymbol{u} = \boldsymbol{0}$, the equilibrium reduces to $f_i^{\text{eq}} = w_i\rho$ and the forcing term to $F_i = w_i\boldsymbol{c}_i \cdot \boldsymbol{F}/c_s^2$. Inserting f_i^{eq} and F_i into (6.47), we get $f_i^{(1)} = -\tau w_i c_{i\alpha}\partial_\alpha^{(1)}\rho + (\tau - \Delta t/2)c_{i\alpha}F_\alpha/c_s^2$. The macroscopic behaviour of the populations for this hydrostatic problem is completely determined by $f_i = f_i^{\text{eq}} + \epsilon f_i^{(1)}$, without any approximation [4, 29].

The next step is transferring these results to the bounce-back formula applied at a resting wall, i.e. $f_{\bar{i}} = f_i^\star$ (cf. (5.24)). This way, one can describe the closure relation of the bounce-back rule in the form of a Chapman-Enskog decomposition:

$$f_{\bar{i}}^{\text{eq}} + \epsilon f_{\bar{i}}^{(1)} = f_i^{\text{eq}} + \left(1 - \frac{\Delta t}{\tau}\right)\epsilon f_i^{(1)} + \left(\tau - \frac{\Delta t}{2}\right)\Delta t \epsilon F_i^{(1)}. \tag{6.48}$$

After substituting the content of $f_i^{\text{eq}}, f_i^{(1)}$ and $F_i^{(1)}$ into (6.48) and undertaking some algebraic simplifications, we arrive at the hydrostatic solution established by the bounce-back rule at boundary node $\boldsymbol{x}_\text{b}$:

$$\left(\tau - \frac{\Delta t}{2}\right)\left(c_s^2\partial_\alpha\rho - F_\alpha\right)\Big|_{\boldsymbol{x}_\text{b}} = 0. \tag{6.49}$$

The first factor in (6.49) is positive due to the stability requirement $\tau > \frac{\Delta t}{2}$ (cf. Sect. 4.4) and can be cancelled. Hence, we conclude that the LBE with the bounce-back rule is *exact* for the hydrostatic pressure solution where we expect the balance $c_s^2\partial_\alpha\rho = F_\alpha$.

But does the correct hydrostatic balance also hold for a first-order space-time discretisation of the force? Based on the bulk analysis presented in Sect. 6.5.2, we might conclude that nothing changes because bulk errors have the form of force

[4] Equation (6.47) results from omitting the time derivatives in equation (6.33a) based on the Chapman-Enskog analysis for steady flows discussed in Sect. 4.2.3.

derivatives which in turn vanish for a constant body force. However, the closure in the bounce-back boundary conditions can retain discrete lattice artefacts even for a constant force, as shown in Exercise 6.3. More details can be found in [29, 34].

Exercise 6.3 Repeat the Chapman-Enskog analysis for a first-order time discretisation of the LBE. Show that the hydrostatic balance established by the bounce-back rule at boundary node $\boldsymbol{x}_\text{b}$ is then incorrectly predicted as

$$\left(\tau - \frac{\Delta t}{2}\right)\left(c_\text{s}^2 \partial_\alpha \rho - F_\alpha\right)\bigg|_{\boldsymbol{x}_\text{b}} = \frac{\Delta t}{2} F_\alpha(\boldsymbol{x}_\text{b}).$$

6.6.2.2 Non-equilibrium Bounce-Back

The fundamental principle of the wet-node technique is that boundary nodes follow the same rules as bulk nodes. Hence, to be consistent with the bulk, the algorithm for boundary nodes needs to be reformulated to account for the presence of a force as well. We demonstrate this for the non-equilibrium bounce-back (NEBB) method [52].

As we have seen in Sect. 5.3.4, wet boundary nodes must satisfy the macroscopic laws of bulk nodes through the velocity moments. Therefore, the first-order moment for the momentum is modified by the presence of a force when we use the second-order space-time discretisation in (6.26b). This leads to a number of changes in the NEBB algorithm.

Consider the top wall depicted in Fig. 5.21. As in Sect. 5.3.4, we will work in dimensional notation, which is noted by the presence of the particle velocity c that in lattice units is $c = 1$. The determination of the unknown wall density for the force-free case in (5.31) now changes to

$$\begin{aligned}
\rho_\text{w} &= \sum_i f_i = \underbrace{f_0 + f_1 + f_2 + f_3 + f_5 + f_6}_{\text{known}} + \underbrace{f_4 + f_7 + f_8}_{\text{unknown}}, \\
\rho_\text{w} u_y^\text{w} &= \sum_i f_i c_{iy} + \frac{\Delta t}{2} \sum_i F_i c_{iy} = \underbrace{c\,(f_2 + f_5 + f_6)}_{\text{known}} - \underbrace{c\,(f_4 + f_7 + f_8)}_{\text{unknown}} + \frac{F_y^\text{w} \Delta t}{2},
\end{aligned} \tag{6.50}$$

where index w refers to the macroscopic fluid properties evaluated at the wall, where wet boundary nodes lie. By combining these two equations we get

$$\rho_\text{w} = \frac{c}{c + u_y^\text{w}} \left(f_0 + f_1 + f_3 + 2\,(f_2 + f_5 + f_6) + \frac{F_y^\text{w} \Delta t}{2c} \right). \tag{6.51}$$

The unknown boundary populations still have to be determined by the bounce-back of their non-equilibrium components, i.e. (5.42). Yet, compared to the force-free case, now it is necessary to consider both tangential and normal momentum corrections N_α. The reason for that will become clear shortly. For now, let us consider the top wall in Fig. 5.21 and write the bounce-back of the non-equilibrium populations as[5]

$$\left.\begin{aligned} f_4^{\text{neq}} &= f_2^{\text{neq}} - N_y, \\ f_7^{\text{neq}} &= f_5^{\text{neq}} - N_x - N_y, \\ f_8^{\text{neq}} &= f_6^{\text{neq}} + N_x - N_y. \end{aligned}\right\} \implies \left\{\begin{aligned} f_4 &= f_2 + \left(f_4^{\text{eq}} - f_2^{\text{eq}}\right) - N_y, \\ f_7 &= f_5 + \left(f_7^{\text{eq}} - f_5^{\text{eq}}\right) - N_x - N_y, \\ f_8 &= f_6 + \left(f_8^{\text{eq}} - f_6^{\text{eq}}\right) + N_x - N_y. \end{aligned}\right. \tag{6.52}$$

Using the known equilibrium distributions, we get

$$\begin{aligned} f_4 &= f_2 - \frac{2\rho_{\text{w}} u_y^{\text{w}}}{3c} - N_y, \\ f_7 &= f_5 - \frac{\rho_{\text{w}}}{6c}\left(u_x^{\text{w}} + u_y^{\text{w}}\right) - N_x - N_y, \\ f_8 &= f_6 - \frac{\rho_{\text{w}}}{6c}\left(-u_x^{\text{w}} + u_y^{\text{w}}\right) + N_x - N_y. \end{aligned} \tag{6.53}$$

We compute N_x by resorting to the first-order velocity moment along the boundary tangential direction:

$$\begin{aligned} \rho_{\text{w}} u_x^{\text{w}} &= \sum_i f_i c_{ix} + \frac{\Delta t}{2} \sum_i F_i c_{ix} \\ &= c\left(f_1 + f_5 + f_8\right) - c\left(f_3 + f_6 + f_7\right) + \frac{F_x^{\text{w}} \Delta t}{2} \\ &= c\left(f_1 - f_3\right) - c\left(f_7 - f_5\right) + c\left(f_8 - f_6\right) + \frac{F_x^{\text{w}} \Delta t}{2} \\ &= c\left(f_1 - f_3\right) + \frac{\rho_{\text{w}} u_x^{\text{w}}}{3} + 2cN_x + \frac{F_x^{\text{w}} \Delta t}{2}. \end{aligned} \tag{6.54}$$

This gives

$$N_x = -\frac{1}{2}\left(f_1 - f_3\right) + \frac{\rho_{\text{w}} u_x^{\text{w}}}{3c} - \frac{F_x^{\text{w}} \Delta t}{4c}. \tag{6.55}$$

[5]The sign convention for the normal momentum correction is in line with the tangential case, cf. (5.43). If $\boldsymbol{n}$ and $\boldsymbol{t}$ denote the wall normal and the wall tangential vectors and if their positive sign coincides with the positive sign of the Cartesian axis, then the normal and tangential momentum corrections appear in the algorithm as $f_{\bar{i}}^{\text{neq}}(\boldsymbol{x}_{\text{B}}, t) = f_i^{\text{neq}}(\boldsymbol{x}_{\text{B}}, t) - (\boldsymbol{n} \cdot \boldsymbol{c}_i) N_n - (\boldsymbol{t} \cdot \boldsymbol{c}_i) N_t$.

Similarly, we compute N_y based on the first-order velocity moment along the boundary normal direction:

$$\begin{aligned}
\rho_\mathrm{w} u_y^\mathrm{w} &= \sum_i f_i c_{iy} + \frac{\Delta t}{2} \sum_i F_i c_{iy} \\
&= c\,(f_2 + f_5 + f_6) - c\,(f_4 + f_7 + f_8) + \frac{F_y^\mathrm{w} \Delta t}{2} \\
&= c\,(f_2 - f_4) - c\,(f_7 - f_5) + c\,(f_6 - f_8) + \frac{F_y^\mathrm{w} \Delta t}{2} \\
&= \frac{\rho_\mathrm{w} u_y^\mathrm{w}}{3} + 3cN_y + \frac{F_y^\mathrm{w} \Delta t}{2}.
\end{aligned} \tag{6.56}$$

We obtain

$$N_y = -\frac{F_y^\mathrm{w} \Delta t}{6c}. \tag{6.57}$$

Clearly, the normal momentum correction N_y is only relevant when forces are included.

In the end, the NEBB prescribes the unknown populations with forces, here for a top wall:

$$\begin{aligned}
f_4 &= f_2 - \frac{2\rho_\mathrm{w} u_y^\mathrm{w}}{3c} + \frac{F_y^\mathrm{w} \Delta t}{6c}, \\
f_7 &= f_5 + \frac{1}{2}(f_1 - f_3) - \frac{\rho_\mathrm{w} u_x^\mathrm{w}}{2c} - \frac{\rho_\mathrm{w} u_y^\mathrm{w}}{6c} + \frac{F_x^\mathrm{w} \Delta t}{4c} + \frac{F_y^\mathrm{w} \Delta t}{6c}, \\
f_8 &= f_6 - \frac{1}{2}(f_1 - f_3) + \frac{\rho_\mathrm{w} u_x^\mathrm{w}}{2c} - \frac{\rho_\mathrm{w} u_y^\mathrm{w}}{6c} - \frac{F_x^\mathrm{w} \Delta t}{4c} + \frac{F_y^\mathrm{w} \Delta t}{6c}.
\end{aligned} \tag{6.58}$$

The extension of (6.58) to other boundary orientations is straightforward (cf. Exercise 6.4).

The necessity of including force corrections in the NEBB method has been recognised in a number of works, e.g. [53–55]. These terms prevent the appearance of discrete lattice artefacts in the macroscopic laws of wet boundary nodes. However, those errors terms are proportional to $\boldsymbol{\nabla} \cdot \boldsymbol{F}$. Hence, they will only be macroscopically visible for *spatially varying* force fields. We will demonstrate this numerically in Sect. 6.7.

Exercise 6.4 Show that the Dirichlet velocity condition prescribed with the NEBB method at a left boundary takes the following form in the presence of a force $\boldsymbol{F} = (F_x, F_y)$:

$$\begin{aligned}
\rho &= \frac{c}{c - u_x^{\mathrm{w}}} \left(f_0 + f_2 + f_4 + 2\,(f_3 + f_6 + f_7) - \frac{F_x^{\mathrm{w}} \Delta t}{2c} \right), \\
f_1 &= f_3 + \frac{2\rho_{\mathrm{w}} u_x^{\mathrm{w}}}{3c} - \frac{F_x^{\mathrm{w}} \Delta t}{6c}, \\
f_5 &= f_7 - \frac{1}{2}\,(f_2 - f_4) + \frac{\rho_{\mathrm{w}} u_y^{\mathrm{w}}}{2c} + \frac{\rho_{\mathrm{w}} u_x^{\mathrm{w}}}{6c} - \frac{F_x^{\mathrm{w}} \Delta t}{6c} - \frac{F_y^{\mathrm{w}} \Delta t}{4c}, \\
f_8 &= f_6 + \frac{1}{2}\,(f_2 - f_4) - \frac{\rho_{\mathrm{w}} u_y^{\mathrm{w}}}{2c} + \frac{\rho_{\mathrm{w}} u_x^{\mathrm{w}}}{6c} - \frac{F_x^{\mathrm{w}} \Delta t}{6c} + \frac{F_y^{\mathrm{w}} \Delta t}{4c}.
\end{aligned} \tag{6.59}$$

6.7 Benchmark Problems

So far we have limited the discussion about forces in the LBE to theoretical analyses. While this helps us understanding basic features underlying LB forcing schemes, we have yet to see actual effects on LB simulations. The goal of this section, therefore, is to illustrate the true impact of the force inclusion, particularly when an incorrect force model is adopted. We will compare four possible forcing strategies (summarised in Table 6.2).

The alternative forcing schemes presented in Sect. 6.4 can be considered equivalent to scheme IV in Table 6.2. Although they behave differently at higher orders, these differences are not relevant for the examples that will follow.

6.7.1 Problem Description

We consider a 2D Poiseuille channel flow driven by a combined pressure gradient $\partial p/\partial x$ and body force F_x:

$$\rho\nu \frac{\partial u_x}{\partial y} = \frac{\partial p}{\partial x} - F_x. \tag{6.60}$$

Table 6.2 LB forcing schemes tested in Sect. 6.7. They have different velocity or space-time discretisation orders

Scheme	Velocity order	Space-time order	Examples of references
I	1st [(6.16)]	1st [(6.20)]	[43, 56–60]
II	2st [(6.14)]	1st [(6.20)]	[16, 18, 20, 44, 48, 61]
III	1st [(6.16)]	2st [(6.25)]	[4, 23, 37, 45, 46, 62]
IV	2st [(6.14)]	2st [(6.25)]	[15, 25, 28, 51, 63, 64]

The velocity solution is

$$u_x(y) = \frac{1}{2\rho\nu}\left(\frac{\partial p}{\partial x} - F_x\right)\left[y^2 - \left(\frac{H}{2}\right)^2\right] \tag{6.61}$$

where the no-slip condition ($u_x = 0$) holds at bottom/top walls ($y = \pm H/2$) as shown in Fig. 1.1b.

6.7.2 Numerical Procedure

For the bulk nodes we use the BGK collision operator with the incompressible equilibrium from Sect. 4.3.2. We will make some comments about the application of the standard (compressible) equilibrium later. We consider and individually discuss two different wall boundary schemes: the bounce-back and the non-equilibrium bounce-back (NEBB) methods.

The simulations are initialised by setting $f_i(\boldsymbol{x}, t = 0) = f_i^{\text{eq}}(\rho = 1, \boldsymbol{u} = \boldsymbol{0})$ as explained in Sect. 6.6.1; they are stopped when the velocity u_x reaches the steady-state criterion $L_2 \leq 10^{-10}$ between 100 consecutive time steps (cf. Sect. 4.5.2). The channel domain is discretised using $N_x \times N_y = 5 \times 5$ grid nodes. We evaluate the LB results for each of the four strategies presented in Table 6.2 and compare them with the analytical solution in (6.61) through the L_2 error norm.

6.7.3 Constant Force

Let us start by considering the simplest case: a purely force-driven Poiseuille flow ($\partial p/\partial x = 0$). We use periodic boundary conditions for the inlet and outlet (cf. Sect. 5.3.1) and the force magnitude is $F_x = 10^{-3}$ (in simulation units).

Since the force is uniform, any possible *bulk error* caused by an incorrect forcing scheme vanishes (cf. Sect. 6.5.2). However, boundaries can still lead to errors (cf. Sect. 6.6.2).

6.7.3.1 Bounce-Back

The errors for the LBGK model with bounce-back walls for several τ values are summarised in Table 6.3. The velocity discretisation of the force plays no role in this case. Differences exist in the space-time discretisation, though. While both strategies are able to reproduce the parabolic solution exactly, this happens at different values of τ. The reason is that spatial discretisation errors are cancelled for

Table 6.3 L_2 errors for Poiseuille flow with constant force and bounce-back at the walls (LBGK, grid resolution of $N_x \times N_y = 5 \times 5$). Results are identical for standard and incompressible equilibria

	ϵ_u [%]	
$\tau/\Delta t$	Schemes I and II	Schemes III and IV
0.6	5.91	5.18
0.8	5.04	2.85
$\sqrt{3/16} + 1/2 = 0.933$	3.16	2.04×10^{-12}
1.0	1.82	1.82
$\sqrt{13/64} + 5/8 = 1.076$	1.42×10^{-12}	4.20
1.2	3.72	8.83
1.4	11.60	18.17

specific values of τ, depending on the discretisation order of the force scheme [20, 43, 62, 65].

6.7.3.2 Non-equilibrium Bounce-Back

The NEBB method reproduces the dynamical rules of the bulk solution at boundary nodes. Consequently, for a constant force, the errors discussed in Sect. 6.5.2 vanish. This makes the NEBB method exact for the parabolic velocity solution in (6.61), regardless the forcing scheme employed.

6.7.4 Constant Force and Pressure Gradient

Let us now increase the complexity of the previous exercise by considering the simultaneous presence of a constant force and pressure gradient. In terms of implementation, the only modification concerns the inlet and outlet boundaries which are now modelled with pressure periodic boundary conditions (cf. Sect. 5.3.2). The relative fraction of the pressure gradient and the force density has no impact on the velocity solution of an incompressible flow, providing their combined effect is kept fixed. Without loss of generality, the overall magnitude is $\left(F_x - \partial p/\partial x\right) = 2 \times 10^{-3}$ (in simulation units), where we consider a 50/50 contribution from each term.

Similarly to the previous case, a constant force leads to vanishing force errors in the bulk, regardless the forcing strategy adopted (cf. Sect. 6.5.2). Yet, the closure relations at boundaries established by the bounce-back rule can differ, depending on the forcing scheme adopted.

Table 6.4 L_2 errors for Poiseuille flow with constant force and pressure gradient, bounce-back at the walls and pressure periodic conditions at inlet/outlet (LBGK, grid resolution of $N_x \times N_y = 5\times5$, incompressible equilibrium)

	ϵ_u (%)	
$\tau/\Delta t$	Schemes I and II	Schemes III and IV
0.6	5.55	5.18
0.8	3.94	2.85
$\sqrt{3/16}+1/2=0.933$	1.58	1.37×10^{-11}
1.0	5.78×10^{-13}	1.82
$\sqrt{13/64}+5/8=1.076$	2.10	4.20
1.2	6.27	8.83
1.4	14.89	18.17

6.7.4.1 Bounce-Back

Table 6.4 summaries the errors obtained with bounce-back. While the velocity discretisation order plays no role, the space-time discretisation is important.

In fact, we see that only the second-order space-time discretisation guarantees that the force-driven solution is unchanged when adding the pressure gradient. This follows from comparing Table 6.3 and Table 6.4: solutions are exactly equivalent for any τ value in that case. From a physical point of view, this result is expected since a constant force and a constant pressure gradient are equivalent in incompressible hydrodynamics.

However, this physical equivalence can be violated numerically when a less accurate space-time force discretisation is adopted. According to Table 6.4, the value where the solution becomes exact, $\tau = \Delta t$, now differs from the pure force-driven case where $\tau = (\sqrt{13/64}+5/8)\Delta t$ gives the exact solution.

6.7.4.2 Non-equilibrium Bounce-Back

Similarly to the purely force-driven case in Sect. 6.7.3, no force errors occur for a constant force. The explanation is the same as before.

6.7.5 Linear Force and Pressure Gradient

Finally, let us address the most interesting case in this exercise: the modelling of a spatially varying force. The force increases linearly along the streamwise direction, but the total contribution remains constant so that the overall magnitude remains locally $(F_x - \partial p/\partial x) = 2\times10^{-3}$ (in simulation units), with a 50/50 local contribution from each term. That means the slope of variation of each term is equal, but with different signs. More details about this test case are described in [23].

Compared to the last two cases, the key difference is that the body force is inhomogeneous now. Thus, the force bulk errors in Sect. 6.5.2 do not vanish any more. Both bulk and boundary errors can now interfere with the LB solution in case of an incorrect force implementation. This case allows us to identify the most accurate LB forcing scheme for *steady incompressible* flow problems.

6.7.5.1 Bounce-Back

According to Table 6.5, the first-order space-time discretisation never reaches the exact solution, regardless of the τ value. This is due to the non-vanishing bulk errors given by (6.44). However, also the velocity discretisation affects the bulk error. As outlined in Sect. 6.3 and Sect. 6.5.2, the correct modelling of steady incompressible hydrodynamics with a body force requires a *first-order* velocity discretisation of the forcing term. This is confirmed in Table 6.5 where only scheme III can reproduce the exact solution.

The reason for the exact solution only occurring for $\tau = (\sqrt{3/16} + 1/2)\Delta t$ is the τ-dependence of the bounce-back scheme. As explained in Sect. 5.3.3 and Sect. 5.4.1, only this value of τ locates the wall *exactly* halfway between nodes in a parabolic flow profile.

6.7.5.2 Non-equilibrium Bounce-Back

The NEBB method leads to essentially the same results as the bounce-back scheme; compare Table 6.5 and Table 6.6. Once again, a bulk error can corrupt the LB solutions if the velocity and space-time discretisations are not properly handled. To reproduce steady incompressible hydrodynamics exactly, scheme III is necessary (cf. Table 6.6). Here the exact solution is reproduced for any value of τ because: (i) scheme III leads to a velocity solution in bulk free from errors (cf. Sect. 6.5.2)

Table 6.5 L_2 errors for Poiseuille flow with linear force and pressure gradient, bounce-back at the walls and pressure periodic conditions at inlet/outlet (LBGK, grid resolution of $N_x \times N_y = 5 \times 5$, incompressible equilibrium)

	ϵ_u (%)			
$\tau/\Delta t$	Scheme I	Scheme II	Scheme III	Scheme IV
0.6	6.10	5.44	5.18	5.07
0.8	4.12	3.84	2.85	2.74
$\sqrt{3/16} + 1/2 = 0.933$	1.70	1.48	8.46×10^{-15}	0.12
1.0	0.51	0.51	1.82	1.94
$\sqrt{13/64} + 5/8 = 1.076$	2.00	2.21	5.47	4.32
1.2	6.19	6.39	8.83	8.96
1.4	14.81	15.03	18.17	18.32

Table 6.6 L_2 errors for Poiseuille flow with linear force and pressure gradient, non-equilibrium bounce-back at the walls and pressure periodic conditions at inlet/outlet (LBGK, grid resolution of $N_x \times N_y = 5 \times 5$, incompressible equilibrium)

	ϵ_u (%)			
$\tau/\Delta t$	Scheme I	Scheme II	Scheme III	Scheme IV
0.6	0.19	0.27	8.60×10^{-14}	0.05
0.8	0.42	0.41	4.41×10^{-15}	0.07
$\sqrt{3/16}+1/2$	0.63	0.63	1.24×10^{-14}	0.07
1.0	0.74	0.74	1.32×10^{-14}	0.07
$\sqrt{13/64}+5/8$	0.87	0.87	1.24×10^{-14}	0.07
1.2	1.08	1.08	2.39×10^{-14}	0.07
1.4	1.42	1.42	1.55×10^{-14}	0.07

and (ii) the NEBB scheme accommodates this solution at the wall in an exact and τ-independent way (cf. Sect. 6.6.2).

6.7.6 *Role of Compressibility*

The previous exercises used the incompressible equilibrium that allows for the exact description of steady incompressible flows (cf. Sect. 4.3.2). This explains why all test cases could reach an exact solution, providing the correct forcing scheme is chosen.

On the other hand, the standard equilibrium recovers the compressible NSE which approximates incompressible hydrodynamics in the limit of slow flows and small density (pressure) variations [4]. Associated with this are *compressibility errors*, as discussed in Sect. 4.5.4.

Compared to the discrete lattice artefacts, coming from the incorrect force modelling, and/or the velocity slip, created by the bounce-back boundary scheme, the compressibility errors typically have a secondary impact [4]. Still, they always contaminate the solutions. In this case, they preclude exact results even when the above error sources are corrected. This issue will be illustrated below, by repeating the previous exercises with the standard (compressible) equilibrium. As we shall see, although compressibility errors may obscure the clear identification of the force discretisation artefacts, the trends of the incompressible equilibrium remain. But this time, the lowest minimum in L_2, for a spatially varying force, is found in the forcing scheme with the second-order discretisation in velocity space (cf. Sect. 6.5.2), although differences are very small.

6.7.6.1 Constant Force

The first test case was the purely force-driven Poiseuille flow. Since no pressure variations occur in this setup, we have identical results for both the incompressible and the standard equilibria.

6.7.6.2 Constant Force and Pressure Gradient

The second test case considered the simultaneous presence of a constant force and pressure gradient. As pressure varies here, the LB solution now contains compressibility errors.

The effect of the velocity discretisation order remains negligible, although machine accuracy is never reached. This is in contrast to the incompressible case. Once again, the space-time discretisation has the largest effect, as shown in Fig. 6.2.

Using bounce-back boundaries (Fig. 6.2a), the first-order space-time discretisation features the L_2 minimum at $\tau = \Delta t$, while for the second-order discretisation it is at $\tau = (\sqrt{13/64} + 5/8)\Delta t$. This behaviour is similar to the incompressible case, Table 6.4, except that now the minimum does not correspond to the exact solution. The same kind of qualitative results occur when the NEBB method is used, yet without showing any clear minimum (cf. Fig. 6.2b). Obviously the NEBB has superior accuracy when the compressible equilibrium is used.

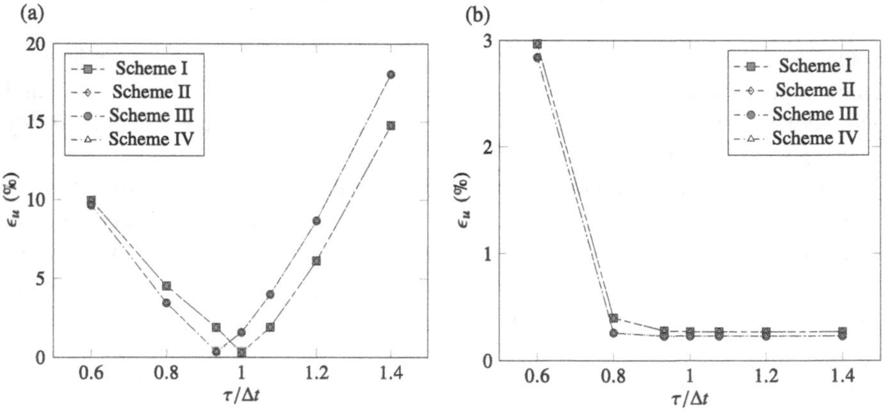

Fig. 6.2 L_2 errors for Poiseuille flow with constant force and pressure gradient, periodic conditions at inlet/outlet (LBGK, grid resolution of $N_x \times N_y = 5 \times 5$, compressible equilibrium). (**a**) Bounce-back. (**b**) Non-equilibrium bounce-back

6.7.6.3 Linear Force and Pressure Gradient

The third test case was a linearly varying body force and an according pressure gradient. In addition to the bulk errors caused by force artefacts, the pressure (density) variation also introduces compressibility errors.

While the order of the velocity discretisation has a slightly larger effect than in the previous problem, the LB solution is dominated by compressibility errors. In fact, using bounce-back, the second-order velocity discretisation is only more accurate for small values of τ (cf. Fig. 6.3a). With the NEBB method, the second-order velocity discretisation is more accurate for all values of τ (cf. Fig. 6.3b). Yet, the accuracy improvement due to a second-order velocity discretisation is marginal and not comparable to the incompressible case. Once again, the accuracy of the LB solution depends mostly on the space-time discretisation as shown in Fig. 6.3.

The conclusions are similar to those of the case with constant force and pressure gradient. The second-order space-time discretisation leads to minimum L_2 values. Still, due to the non-trivial interplay of force and compressibility errors, the second-order space-time discretisation does not perform better in the full range of τ, which is particularly noticeable for the bounce-back method. Generally, the NEBB method has smaller errors for the problem considered in this section.

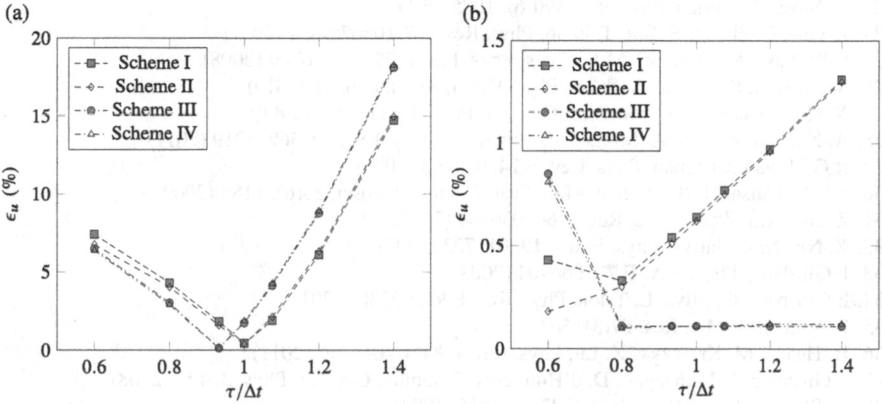

Fig. 6.3 L_2 errors for Poiseuille flow with linear force and pressure gradient, pressure periodic conditions at inlet/outlet (LBGK, grid resolution of $N_x \times N_y = 5 \times 5$, compressible equilibrium). (**a**) Bounce-back. (**b**) Non-equilibrium bounce-back

References

1. E. Bodenschatz, W. Pesch, G.Ahlers, Annu. Rev. Fluid Mech. **32**, 709 (2010)
2. S.I. Abarzhi, Phil. Trans. R. Soc. A **368**, 1809 (2010)
3. J. Lighthill, *Waves in Fluids*, 6th edn. (Cambridge University Press, Cambridge, 1979)
4. J.M. Buick, C.A. Greated, Phys. Rev. E **61**(5), 5307 (2000)
5. G. Silva, V. Semiao, J. Fluid Mech. **698**, 282 (2012)
6. P.J. Dellar, Comput. Math. Appl. **65**(2), 129 (2013)
7. R. Salmon, J. Mar. Res. **57**(3), 847 (1999)
8. J. Wang, M. Wang, Z. Li, J. Colloid Interf. Sci. **296**, 729 (2006)
9. M. Wang, Q. Kang, J. Comput. Phys. **229**, 728 (2010)
10. T.Y. Lin, C.L. Chen, Appl. Math. Model. **37**, 2816 (2013)
11. O. Shardt, S.K. Mitra, J.J. Derksen, Chem. Eng. J. **302**, 314 (2016)
12. S.H. Kim, H. Pitsch, Phys. Fluids **19**, 108101 (2007)
13. J. Zhang, D.Y. Kwok, Phys. Rev. E **73**, 047702 (2006)
14. L. Talon, D. Bauer, D. Gland, H. Auradou, I. Ginzburg, Water Resour. Res. **48**, W04526 (2012)
15. Z. Guo, C. Zheng, B. Shi, Phys. Rev. E **65**, 46308 (2002)
16. N.S. Martys, X. Shan, H. Chen, Phys. Rev. E **58**(5), 6855 (1998)
17. X. Shan, X.F. Yuan, H. Chen, J. Fluid Mech. **550**, 413 (2006)
18. L.S. Luo, Phys. Rev. E **62**(4), 4982 (2000)
19. P. Dellar, Phys. Rev. E **64**(3) (2001)
20. A.J.C. Ladd, R. Verberg, J. Stat. Phys. **104**(5–6), 1191 (2001)
21. S. Ansumali, I.V. Karlin, H.C. Öttinger, Phys. Rev. Lett. **94**, 080602 (2005)
22. J.R. Clausen, Phys. Rev. E **87**, 013309 (2013)
23. G. Silva, V. Semiao, Physica A **390**(6), 1085 (2011)
24. Z. Guo, C. Zheng, B. Shi, T. Zhao, Phys. Rev. E **75**(036704), 1 (2007)
25. R.W. Nash, R. Adhikari, M.E. Cates, Phys. Rev. E **77**(2), 026709 (2008)
26. S. Ubertini, P. Asinari, S. Succi, Phys. Rev. E **81**(1), 016311 (2010)
27. X. He, S. Chen, G.D. Doolen, J. Comput. Phys. **146**(1), 282 (1998)
28. A. Kuzmin, Z. Guo, A. Mohamad, Phil. Trans. Royal Soc. A **369**, 2219 (2011)
29. R.G.M. Van der Sman, Phys. Rev. E **74**, 026705 (2006)
30. S.D.C. Walsh, H. Burwinkle, M.O. Saar, Comput. Geosci. **35**(6), 1186 (2009)
31. Z. Guo, T.S. Zhao, Phys. Rev. E **66**, 036304 (2002)
32. X. Nie, N.S. Martys, Phys. Fluids **19**, 011702 (2007)
33. I. Ginzburg, Phys. Rev. E **77**, 066704 (2008)
34. I. Ginzburg, G. Silva, L. Talon, Phys. Rev. E **91**, 023307 (2015)
35. R. Salmon, J. Mar. Res. **57**(3), 503 (1999)
36. H. Huang, M. Krafczyk, X. Lu, Phys. Rev. E **84**(4), 046710 (2011)
37. I. Ginzburg, F. Verhaeghe, D. d'Humières, Commun. Comput. Phys. **3**, 427 (2008)
38. X. Shan, H. Chen, Phys. Rev. E **47**(3), 1815 (1993)
39. X. He, X. Shan, G. Doolen, Phys. Rev. E. Rapid Comm. **57**(1), 13 (1998)
40. A. Kupershtokh, D. Medvedev, D. Karpov, Comput. Math. Appl. **58**(5), 965 (2009)
41. D. Lycett-Brown, K.H. Luo, Phys. Rev. E **91**, 023305 (2015)
42. A. Kupershtokh, in *Proc. 5th International EHD Workshop, University of Poitiers, Poitiers, France* (2004), p. 241–246
43. X. He, Q. Zou, L.S. Luo, M. Dembo, J. Stat. Phys. **87**(1–2), 115 (1997)
44. L.S. Luo, Phys. Rev. Lett. **81**(8), 1618 (1998)
45. I. Ginzbourg, P.M. Adler, J. Phys. II France **4**(2), 191 (1994)
46. A.J.C. Ladd, J. Fluid Mech. **271**, 285 (1994)
47. Z. Guo, C. Zheng, B. Shi, Phys. Rev. E **83**, 036707 (2011)
48. I. Halliday, L.A. Hammond, C.M. Care, K. Good, A. Stevens, Phys. Rev. E **64**, 011208 (2001)
49. T. Reis, T.N. Phillips, Phys. Rev. E **75**, 056703 (2007)
50. I. Ginzburg, F. Verhaeghe, D. d'Humières, Commun. Comput. Phys. **3**, 519 (2008)

51. M. Gross, N. Moradi, G. Zikos, F. Varnik, Phys. Rev. E **83**(1), 017701 (2011)
52. Q. Zou, X. He, Phys. Fluids **9**, 1591 (1997)
53. A. D'Orazio, S. Succi, Future Generation Comput. Syst. **20**, 935 (2004)
54. A. Markus, G. Hazi, Phys. Rev. E **83**, 046705 (2011)
55. A. Karimipour, A.H. Nezhad, A. D'Orazio, E. Shirani, J. Theor. Appl. Mech. **51**, 447 (2013)
56. D.R. Noble, Chen, J.G. Georgiadis, R.O. Buckius, Phys. Fluids **7**, 203 (1995)
57. M. Bouzidi, M. Firdaouss, P. Lallemand, Phys. Fluids **13**, 3452 (2001)
58. D.A. Wolf-Gladrow, *Lattice-Gas Cellular Automata and Lattice Boltzmann Models* (Springer, New York, 2005)
59. M. Junk, A. Klar, L.S. Luo, J. Comput. Phys. **210**, 676 (2005)
60. T. Krüger, F. Varnik, D. Raabe, Phys. Rev. E **79**(4), 046704 (2009)
61. C.M. Pooley, K. Furtado, Phys. Rev. E **77**, 046702 (2008)
62. I. Ginzburg, D. d'Humières, Phys. Rev. E **68**, 066614 (2003)
63. K. Premnath, J. Abraham, J. Comput. Phys. **224**, 539 (2007)
64. J. Latt, Hydrodynamic limit of lattice Boltzmann equations. Ph.D. thesis, University of Geneva (2007)
65. M. Rohde, D. Kandhai, J.J. Derksen, H.E.A. Van den Akker, Phys. Rev. **67**, 066703 (2003)

Chapter 7
Non-dimensionalisation and Choice of Simulation Parameters

Abstract After reading this chapter, you will be familiar with how the "lattice units" usually used in simulations and articles can be related to physical units through unit conversion or through dimensionless numbers such as the Reynolds number. Additionally, you will be able to make good choices of simulation parameters and simulation resolution. As these are aspects of the lattice Boltzmann method that many beginners find puzzling, care is taken in this chapter to include a number of illustrative examples.

Being able to map the physical properties of a system to the lattice (and the results of a simulation back to a prediction about a physical system) is essential. LB simulations are mostly performed in "lattice units" where all physical parameters are represented by dimensionless numbers, as explained in Sect. 3.2.1. This requires converting any dimensional quantity into a non-dimensional or dimensionless *lattice* quantity (and *vice versa* to interpret the simulation results).

Unless the simulation program has a built-in functionality for performing unit conversion between physical and lattice units, the user is responsible for specifying simulation parameters in lattice units. Unit conversion is straightforward once the basic rules are clear (Sect. 7.1). Unfortunately, few references in the LB literature cover this topic coherently and rigorously.

LB simulations require not only a good knowledge of unit conversion. Due to intrinsic restrictions of the LB algorithm, it is crucial to balance the simulation parameters in such a way that a suitable compromise of accuracy, stability and efficiency is achieved (Sect. 7.2). We also provide numerous examples to demonstrate the non-dimensionalisation process (Sect. 7.3). Finally, a summary with the most important rules is given in Sect. 7.4.

7.1 Non-dimensionalisation

We introduce the underlying concepts and basic rules to non-dimensionalise physical parameters (Sect. 7.1.1). The *law of similarity* plays a crucial role (Sect. 7.1.2).

T. Krüger et al., *The Lattice Boltzmann Method*, Graduate Texts in Physics,
DOI 10.1007/978-3-319-44649-3_7

7.1.1 Unit Scales and Conversion Factors

In order to indicate a dimensional physical quantity, one requires a reference scale. For example, a length ℓ can be reported in multiples of a unit scale with length $\ell_0 = 1$ m. If a given length ℓ is ten times as long as the unit scale ℓ_0, we say that its length is 10 m: $\ell = 10\,\text{m} = 10\,\ell_0$. This convention is not unique. One may wish to express the length compared to a different unit scale ℓ'_0, e.g. $\ell'_0 = 1$ ft. How long is ℓ in terms of ℓ'_0? To answer this question we have to know how long ℓ_0 is compared to ℓ'_0. We know that 1 ft corresponds to 0.3048 m: $\ell'_0/\ell_0 = 0.3048\,\text{m/ft}$. Therefore, it is straightforward to report ℓ in feet, rather than metres:

$$\ell = 10\,\text{m} = 10\,\ell_0 = 10\,\ell'_0\frac{\ell_0}{\ell'_0} = 10\,\text{ft}\frac{1}{0.3048} \approx 32.81\,\text{ft} = 32.81\,\ell'_0. \tag{7.1}$$

Although the numerical values are not the same, the length ℓ is identical in both representations. This calculation seems to be trivial, but it is the key to understanding unit conversion and non-dimensionalisation.

Non-dimensionalisation is achieved by dividing a dimensional quantity by a chosen reference quantity of the same dimension. The result is a number which we will sometimes call the *lattice* value of the quantity, or the quantity's value in *lattice units*, when talking about the LBM. The reference quantity is called the *conversion factor*. Let us denote conversion factors by C and non-dimensionalised quantities by a star *,[1] then we can write for any length ℓ:

$$\ell^\star = \frac{\ell}{C_\ell}. \tag{7.2}$$

The conversion factor C_ℓ has to be chosen appropriately. How to do this for a given LB simulation we will discuss in Sect. 7.2; but let us first focus on the non-dimensionalisation itself.

Any mechanical quantity q has a dimension which is a combination of the dimensions of length ℓ, time t and mass m. In the following, let us only use SI units, i.e. metre (m) for length, second (s) for time and kilogramme (kg) for mass. The units of q, denoted as $[q]$, are therefore

$$[q] = [\ell]^{q_\ell}[t]^{q_t}[m]^{q_m}. \tag{7.3}$$

[1] Note that the star also denotes post-collision values. There is no danger of confusing both concepts in this chapter, though.

Table 7.1 Units of selected mechanical quantities

Quantity	Symbol	Unit	q_ℓ	q_t	q_m
Length	ℓ	m	1	0	0
Time	t	s	0	1	0
Mass	m	kg	0	0	1
Velocity	u	m/s	1	−1	0
Acceleration	a	m/s^2	1	−2	0
Force	f	kg m/s^2	1	−2	1
Force density	F	kg/(m^2 s^2)	−2	−2	1
Density	ρ	kg/m^3	−3	0	1

The exponents q_ℓ, q_t and q_m are numbers. Let us take the velocity u as an example. We know that the dimension of velocity is length over time:

$$[u] = \frac{\text{m}}{\text{s}} = [\ell]^1[t]^{-1} = [\ell]^{q_\ell}[t]^{q_t}[m]^{q_m}. \tag{7.4}$$

Therefore, we find $q_\ell = 1$, $q_t = -1$ and $q_m = 0$. It is a simple exercise to find the three numbers q_ℓ, q_t and q_m for any mechanical quantity q. Table 7.1 shows a few examples.

In practical situations, one does not only want to non-dimensionalise one quantity, but all of them. How can we achieve this in a consistent way?

Since three fundamental dimensions are sufficient to generate the dimension of any mechanical quantity, **one requires exactly three independent conversion factors** in order to define a unique non-dimensionalisation scheme.

Three conversion factors C_i $(i = 1, 2, 3)$ are independent if the relations

$$[C_i] = [C_j]^{a_i}[C_k]^{b_i} \quad (i \neq j \neq k \neq i) \tag{7.5}$$

have no solution for the numbers a_i and b_i. In other words: none of the dimensions of the three conversion factors must be a combination of the other two. For example, the conversion factors for length, time and velocity are dependent because $[C_u] = [C_\ell]/[C_t]$. But the conversion factors for velocity, force and time are independent. This argumentation is tightly related to the *Buckingham π theorem* [1].

Exercise 7.1 What are the exponents q_u, q_t, q_ρ for a force f if velocity u, time t and density ρ are used? (Answer: $q_\rho = 1$, $q_u = 4$, $q_t = 2$)

Let us call any set of three independent conversion factors *basic conversion factors*. The first step is to construct a system of three basic conversion factors. All other required conversion factors can then be easily constructed as we will see shortly.

It is completely arbitrary which basic conversion factors to choose, but in LB simulations one usually takes C_ℓ, C_t (or C_u) and C_ρ because length, time (or velocity) and density are natural quantities in any LB simulation. Using basic conversion factors for energy or force are possible but usually impractical.

Care must be taken when a 2D system is simulated. The dimension of density ρ is mass per volume, in 2D it is formally mass per area. But how can a 3D density be mapped to a 2D density? This dilemma can be circumvented by pretending that a simulated **2D system is nothing more than a 3D system with thickness of one lattice constant**, i.e. an $N_x \times N_y$ system is treated as an $N_x \times N_y \times 1$ system where N_x and N_y are the number of lattice sites along the x- and y-axes. This enables us to use 3D conversion factors for 2D problems without any restrictions.

7.1.2 Law of Similarity and Derived Conversion Factors

Let us pretend that we have a set of three basic conversion factors. How do we obtain the derived conversion factors for other mechanical quantities? First of all we note that physics is independent of units which are an arbitrary human construct. Ratios of physical phenomena are what matter: a pipe diameter of 1 m means that the diameter is the length of the path travelled by light in vacuum in 1/299 792 458 of a second, with time intervals defined through periods of radiation from a caesium atom. In particular, the physical outcome should not depend on whether we use dimensional or dimensionless quantities.

Related to this is the **law of similarity** in fluid dynamics which is nicely explained by Landau and Lifshitz [2]: two incompressible flow systems are dynamically similar if they have the same Reynolds number and geometry.

(continued)

The Reynolds number may be defined as

$$\mathrm{Re} = \frac{\ell U}{\nu} = \frac{\rho \ell U}{\eta} \tag{7.6}$$

where ℓ and U are typical length and velocity scales in the system and ρ, ν and η are the density, kinematic viscosity and dynamic viscosity of the fluid, respectively.

The law of similarity is, for instance, regularly used in the automotive or aircraft industries where models of cars or planes are tested in wind tunnels. Since the models are usually smaller than the real objects, one has to increase the flow velocity or decrease the viscosity in such a way that the Reynolds numbers in both systems match.[2] For example, if the model of a car is five times smaller than the car itself, the velocity of the air in the wind tunnel should be five times larger than in reality (given the same density and viscosity).

Another example is the flow in a T-junction. Consider two fluids: one with water-like viscosity and one with the same density but ten times more viscous. Both flow at the same average speed through a T-junction with circular pipes, but the more viscous fluid flows through pipes with ten times the diameter (and ten times the length of pipe before and after the junction). In this example, the Reynolds number is the same and the geometry is similar, and so we expect identical solutions (in non-dimensional space). Should we perform a simulation that matches the physical properties of the water flow or the more viscous flow? The answer is that the simulation should match the Reynolds number in a way that optimises the accuracy of the solution. The results of the simulation can then be applied to *both* physical systems.

This means that the Reynolds number must be identical in both unit systems (the physical and the lattice system):

$$\frac{\ell^\star U^\star}{\nu^\star} = \frac{\ell U}{\nu}. \tag{7.7}$$

Plugging in the definition of the conversion factors (e.g. $C_\ell = \ell/\ell^\star$) simply yields

$$\frac{C_\ell C_u}{C_\nu} = 1 \tag{7.8}$$

[2] Note that the definition of the Reynolds number is not unique: on the one hand, some people may choose ℓ for the length, others for the width of the considered system. On the other hand, U may be the average velocity or the maximum velocity—depending on the person defining the Reynolds number. This has to be kept in mind when comparing Reynolds numbers from different sources.

or, if for example C_ν shall be computed from C_ℓ and C_u:

$$C_\nu = C_\ell C_u. \tag{7.9}$$

Based on the dimensions of the conversion factors alone, one could have come to the conclusion that $C_\nu \propto C_\ell C_u$ must hold, although this is only a proportionality rather than an equality. The law of similarity for the Reynolds number, however, uniquely defines the relation of the conversion factors for viscosity, length and velocity.

This result can be generalised to all missing conversion factors. Any derived conversion factor C_q can be constructed directly by writing down a suitable combination of basic conversion factors of the form

$$C_q = C_1^{q_1} C_2^{q_2} C_3^{q_3} \tag{7.10}$$

without any additional numerical prefactor.[3] The problem reduces to finding a suitable set of numbers q_i, which is usually not a difficult task. The resulting expression for C_q is unique and guarantees that the conversion is consistent: the physics of the system, i.e. the characteristic dimensionless numbers are kept invariant, and the law of similarity is satisfied.

Example 7.1 For the given basic conversion factors C_ℓ, C_t and C_ρ, one can easily construct the conversion factor for pressure p. First we observe that $[p] = \mathrm{N/m^2} = (\mathrm{kg/m^3})\,\mathrm{m^2/s^2}$. Therefore one directly finds

$$C_p = \frac{C_\rho C_\ell^2}{C_t^2}. \tag{7.11}$$

We show some concrete examples of non-dimensionalisation procedures in Sect. 7.3.

Exercise 7.2 How does the pressure conversion factor look like if C_u rather than C_t is used? (Answer: $C_p = C_\rho C_u^2$).

Unit systems must not be mixed. This is a typical mistake made by most users at some point. Mixing unit systems causes inconsistencies in the definition of the conversion factors and a subsequent violation of the law of similarity. We strongly advise to clearly mark quantities in different unit systems (e.g. by adding * or subscripts like p for "physical" and l for "lattice").

(continued)

[3]This is true as long as we stick to base SI units. Mixing metres, kilometres and feet, for example, will generally lead to additional numerical prefactors. This is the beauty of using SI units!

Additionally, we recommend to perform all calculations in pure SI units without using prefixes like "milli" or "centi" which otherwise easily lead to confusion.

7.2 Parameter Selection

We discuss the relevance of parameters in the LBM in general (Sect. 7.2.1), the effect of accuracy, stability and efficiency (Sect. 7.2.2) and the strategies how to find simulation parameters for a given problem in particular (Sect. 7.2.3).

7.2.1 Parameters in the Lattice Boltzmann Method

Any LB simulation is characterised by a set of parameters:

- The lattice constant Δx is the distance between neighbouring lattice nodes in physical units, i.e. $[\Delta x] = \text{m}$.
- The physical length of a time step is denoted Δt, therefore $[\Delta t] = \text{s}$.
- The BGK relaxation parameter τ is often understood as a dimensionless quantity. Strictly speaking, τ is a relaxation time with $[\tau] = \text{s}$. This is how we understand τ in this book. Therefore, we write τ for the physical relaxation time and $\tau^\star$ for the dimensionless relaxation parameter.
- The dimensionless fluid density is $\rho^\star$. Its *average* value is usually set to unity, $\rho_0^\star = 1$.[4] This situation is slightly more complicated for multicomponent or multiphase simulations (Chap. 9), where the density can be different from unity.
- Another important parameter is the typical simulated velocity $U^\star$. It is usually not an input parameter but rather part of the simulation output. However, some boundary conditions require specification of $U^\star$ on the boundaries as inlet and outlet velocities. Additionally, it is desired to estimate the magnitude of $U^\star$ before the simulation is started in order to avoid unstable situations or very long computing times.

[4]Here it is important to remember that the physical density of an incompressible fluid is constant while the LB density can fluctuate. Therefore, we relate the physical density ρ to the average lattice density $\rho_0^\star$. The fluctuation of the LB density, $\rho'^\star$, is then related to the pressure, as shown in (7.16).

- In the standard LBM, the lattice speed of sound $c_s^\star$ is $\sqrt{1/3} \approx 0.577$. In order to operate in the quasi-incompressible limit, all simulated velocities have to be significantly smaller: $U^\star \ll c_s^\star$. In practice this means that the maximum value of $U^\star$ should be below 0.2.

The first step is to relate the physical parameters Δx, Δt, τ, ρ and U to their lattice counterparts $\Delta x^\star$, $\Delta t^\star$, $\tau^\star$, $\rho_0^\star$ and $U^\star$.

It is very common and recommended to set $\Delta x^\star = 1$, $\Delta t^\star = 1$ and $\rho_0^\star = 1$. This means that the **conversion factors for length, time and density** equal the dimensional values for the lattice constant, time step and density:

$$C_\ell = \Delta x, \quad C_t = \Delta t, \quad C_\rho = \rho. \tag{7.12}$$

It is extremely important to realise that Δx and $\Delta x^\star$ are the *same* quantity in *different* unit systems. The units defined by $\Delta x^\star = 1$ and $\Delta t^\star = 1$ are called **lattice units**.

Some authors prefer to set the total system size and the total simulation time, rather than the lattice constant and the time step, to unity. This is nothing more than a different but valid unit system. As a consequence, the conversion factors C_ℓ and C_t would be different.

Most non-dimensionalisation problems are caused by the fact that users confuse physical and non-dimensional quantities, i.e. Δx with $\Delta x^\star$, or Δt with $\Delta t^\star$. The parameters τ and $\tau^\star$ are connected through the conversion factor for Δt because τ has the dimension of time:

$$\tau = \tau^\star C_t = \tau^\star \Delta t. \tag{7.13}$$

The conversion factor for the velocity is $C_u = C_\ell / C_t = \Delta x / \Delta t$, and it is not independent. Thus, Δx, Δt and ρ form a complete set of basic conversion factors. Alternatively one may use Δx, C_u and ρ or other independent combinations. As we will see later, different choices of the set of conversion factors result in different strategies for non-dimensionalisation.

7.2.1.1 Viscosity

One of the important physical fluid properties is viscosity. From Sect. 4.1 we know that the kinematic lattice viscosity is related to the relaxation parameter according to $\nu^\star = c_s^{\star 2}(\tau^\star - 1/2)$. A typical problem is to relate the *dimensionless* relaxation parameter $\tau^\star$ to the *physical* kinematic viscosity ν since the latter is usually given by an experiment and the former has to be defined for a simulation. The conversion factor for ν is $C_\nu = C_\ell^2/C_t$ which directly follows from the dimension of ν: $[\nu] = \mathrm{m}^2/\mathrm{s}$.

The **kinematic viscosity** is related to the simulation parameters according to

$$\nu = c_s^{\star 2}\left(\tau^\star - \frac{1}{2}\right)\frac{\Delta x^2}{\Delta t}. \quad (7.14)$$

This is a consistency equation for the three simulation parameters $\tau^\star$, Δx and Δt, which means that these three parameters are not independent. Only two of them can be chosen freely.

We will soon consider the intrinsic limitations of the LB algorithm which further restrict the choice of parameters.

7.2.1.2 Pressure, Stress and Force

There are other additional quantities which are commonly encountered in LB simulations: pressure p, stress σ and force f. We know from Sect. 4.1 that the equation of state of the LB fluid is

$$p^\star = c_s^{\star 2}\rho^\star. \quad (7.15)$$

This is, however, not the entire truth. Only the pressure gradient ∇p rather than the pressure p by itself appears in the NSE. While the total pressure p *does* appear in the energy equation, this equation is not relevant for non-thermal LB models. The reference pressure is thus irrelevant; only pressure *changes* matter.

To connect with the physical pressure, one decomposes the LB density into its constant average $\rho_0^\star$ and deviation $\rho'^\star$ from the average:

$$\rho^\star = \rho_0^\star + \rho'^\star. \quad (7.16)$$

It is often wrongly assumed that the reference density $\rho_0^\star$ has to correspond to a physical reference pressure p_0, e.g. atmospheric pressure. Generally, the LB density can be converted to the physical pressure for non-thermal models as

$$p = p_0 + p' = p_0 + p'^\star C_p, \quad p'^\star = c_s^{\star 2}\rho'^\star, \quad (7.17)$$

where $C_p = C_\rho C_\ell^2/C_t^2 = C_\rho C_u^2$ is the conversion factor for pressure and p_0 is the physical reference pressure which can be freely specified by the user. Thus, the LB equation of state can model a number of realistic equations of state [3]; see also Sect. 1.1.3 for more details.

Example 7.2 A simulation with $\rho_0^\star = 1$ yields a density fluctuation $\rho'^\star = 0.03$ at a given point. The pressure conversion factor is known to be $C_p = 1.2 \cdot 10^3$ Pa. What is the physical pressure at that point? First we can compute the physical value of the pressure fluctuation from $p' = c_s^{\star 2} \rho'^\star C_p = \frac{1}{3} \cdot 0.03 \cdot 1.2 \cdot 10^3 \text{ Pa} = 12 \text{ Pa}$. However, since p_0 has not been specified, we do not know the absolute pressure in physical units. It may be wrong to compute it from $c_s^{\star 2} \rho_0^\star C_p$ which would give 400 Pa. Atmospheric pressure, for example is roughly 10^5 Pa and orders of magnitude larger than the "wrong" reference pressure of 400 Pa. This misconception often leads to confusion and incorrectly computed real-world pressure values.

The components of the **stress** tensor have the same dimension as a pressure, therefore the conversion factor $C_\rho C_\ell^2/C_t^2$ is always identical.

For example, we can obtain the lattice deviatoric stress tensor *via* (4.14) and then convert it to physical units by $\boldsymbol{\sigma} = \boldsymbol{\sigma}^\star C_\sigma$ with $C_\sigma = C_p$.

The procedure for forces is straightforward (cf. Table 7.1). A force which is often encountered in hydrodynamic situations is the drag or lift force acting on the surface of obstacles in the flow. It is obtained by computing the surface integral of the stress (cf. Sect. 5.4.3). The conversion factor for any force (no matter if body force or surface force) is $C_f = C_\rho C_\ell^4/C_t^2$. One has to be careful when talking about body forces, though: authors often write "body force" but actually mean "body force density" as in body force per volume. The conversion factor for a body force *density* F is obviously $C_F = C_f/C_\ell^3 = C_\rho C_\ell/C_t^2$.

Additional confusion is commonly caused by gravity. In physical terms, gravity g is an acceleration, $[g] = \text{m/s}^2$, not a force or force density. The gravitational force density F_g is given by $F_g = \rho g$. The conversion factor for gravity is C_ℓ/C_t^2. A precise language is therefore very helpful when talking about gravity (acceleration), force and force density.

The dimensions of pressure, stress and force density depend on the number of spatial dimensions. This has to be taken into account when 2D rather than 3D simulations are performed. It is always helpful to write down a table similar to Table 7.1 with all quantities of interest, their dimensions and chosen conversion factors.

7.2.2 *Accuracy, Stability and Efficiency*

Equation (7.14) tells us that there is an infinite number of possibilities to get the correct physical viscosity by balancing $\tau^\star$, Δx and Δt. The key question is: how does one choose these parameters? To answer this, we have to consider intrinsic limitations of the LB algorithm. In the end, one has to choose the parameters in such a way that simulation accuracy, stability and efficiency are reasonably considered. Here we will focus on the BGK collision operator. Note that some stability and accuracy problems of the BGK operator can be solved by using advanced collision operators (TRT or MRT, cf. Chap. 10) instead. We will see that there is no free lunch: increased accuracy and stability usually come at the expense of increased computing time. A similar, yet shorter discussion of this topic can be found in section 2.2 of [4].

7.2.2.1 Accuracy and Parameter Scaling

There are several error terms which affect the accuracy of an LB simulation (cf. Sect. 4.5):

- The spatial discretisation error scales like Δx^2 [5, 6].
- The time discretisation error scales like Δt^2 [5, 7, 8].
- The compressibility error for simulations in the incompressible limit is $\propto \mathrm{Ma}^2 \propto U^{\star 2}$ [5, 9]. Since $U^\star$ decreases with increasing $C_u = C_\ell/C_t = \Delta x/\Delta t$, this error scales like $\Delta t^2/\Delta x^2$.
- The BGK truncation error in space is proportional to $(\tau - 1/2)^2$ [10].

The user's task is to make sure that none of these error contributions plays too large a role. In fact, we observe that increasing the lattice resolution (decreasing Δx) alone does not necessarily reduce the error because the compressibility error will grow and dominate eventually. Reducing only the time step (decreasing Δt) does not decrease the spatial error. Thus, one needs to come up with certain relationships between Δx and Δt to control the error.

One particular relation between Δx and Δt is given by the *diffusive scaling*[5]: $\Delta t \propto \Delta x^2$. It guarantees that the leading order of the overall error scales like Δx^2. However, the LB algorithm then becomes effectively first-order accurate in time [6]. Additionally, since $U^{\star 2} \propto \Delta t^2/\Delta x^2 \propto \Delta x^2$ in the diffusive scaling, LB becomes second-order accurate in velocity.

[5]The expression "diffusive scaling" stems from the apparent similarity of $\Delta t \propto \Delta x^2$ and the diffusion equation, but there is no physical relation between both. Such a relation between the spatial and temporal scales is no special feature of the LB algorithm. It can also be found in typical time-explicit centred finite difference schemes, such as the DuFort-Frankel scheme.

The **diffusive scaling** leaves $\tau^\star$, and hence the non-dimensional viscosity $\nu^\star$, unchanged. This follows from (7.14). The diffusive scaling is the standard approach to test if an LB algorithm is second-order accurate: one performs a series of simulations, each with a finer resolution Δx than the previous. The overall velocity error should then decrease proportionally to Δx^2.

The *acoustic scaling* $\Delta t \propto \Delta x$ keeps the compressibility error unchanged. If one is only interested in incompressible situations, the speed of sound does not have any physical significance and any compressibility effects are undesired. The diffusive scaling is then the method of choice to reduce compressibility effects proportional to Δx^2. If, however, the speed of sound is a physically relevant parameter (as in compressible fluid dynamics and acoustics), the acoustic scaling must be chosen as it maintains the correct scaling of the speed of sound. Holdych et al. [6] emphasised that the numerical solution can only converge to the solution of the *incompressible* NSE when $\Delta t \propto \Delta x^\gamma$ with $\gamma > 1$ since the compressibility error remains constant for $\gamma = 1$.

We have also seen in Sect. 4.5 that the value of $\tau^\star$ affects the accuracy. $\tau^\star \gg 1$ should be avoided [6] because the error of the *bulk* BGK algorithm grows with $(\tau^\star - 1/2)^2$. Note that the presence of complex boundary conditions modelled with lower-order methods (such as simple bounce-back) leads to a non-trivial overall error dependence on $\tau^\star$. It is generally recommended to choose relaxation times around unity [6, 11, 12]. This is not feasible if the desired Reynolds number is large and the viscosity (and therefore $\tau^\star$) are small. We will get back to this point in Sect. 7.2.3.

The above discussion can only give a recommendation as to how Δx and Δt should be changed at the same time, e.g. for a grid convergence study, but it does not provide any information about the initial choice of Δx and Δt. We will discuss this in Sect. 7.2.3.

7.2.2.2 Stability

We have already discussed the stability of LB algorithms in Sect. 4.4. The essential results are that the relaxation parameter $\tau^\star$ should not be too close to $1/2$ and that the velocity $U^\star$ should not be larger than about 0.4 for $\tau^\star \geq 0.55$. Another result of Sect. 4.4 was that the achievable maximum velocity is decreasing when $\tau^\star$ approaches $1/2$. For $\tau^\star < 0.55$, we can approximate this relation by

$$\tau^\star > \frac{1}{2} + \alpha U^\star_{\mathrm{max}} \tag{7.18}$$

where α is a numerical constant which is of the order of $1/8$.

Example 7.3 On the one hand, if we choose $\tau^\star = 0.51$, the maximum velocity should be below 0.08. On the other hand, for an expected maximum velocity of 0.01, $\tau^\star$ could be as low as 0.50125.

The values reported above are only guidelines. The onset of instability also depends on the flow geometry and other factors. Simulations may therefore remain stable for smaller values of $\tau^\star$ or become unstable for larger values. In fact, instabilities are often triggered at boundaries rather than in the bulk [13].

Related to the stability considerations above is the *grid Reynolds number* Re_g that is defined by taking the lattice resolution Δx as length scale:

$$\mathrm{Re}_g = \frac{U^\star_{\max}\Delta x^\star}{\nu^\star} = \frac{U^\star_{\max}}{c_s^{\star 2}\left(\tau^\star - \frac{1}{2}\right)} \implies \tau^\star = \frac{1}{2} + \frac{U^\star_{\max}}{c_s^{\star 2}\mathrm{Re}_g}. \tag{7.19}$$

A comparison with (7.18) reveals that the **grid Reynolds number** Re_g should not be much larger than $O(10)$. The physical interpretation is that the lattice should always be sufficiently fine to resolve local vortices. In other words: the simulation usually remains stable as long as all relevant hydrodynamic length scales are resolved.

7.2.2.3 Efficiency

The *efficiency* of an LB simulation can have two meanings: performance and optimisation level of the simulation code on the one hand and required number of lattice sites and iterations for a given physical problem on the other hand. Here, we will only address the latter. We will address code optimisation in Chap. 13.

The total number of site updates required to complete a simulation is $N_s N_t$ where N_s is the total number of lattice sites and N_t is the required number of time steps. The memory requirements are proportional to N_s (LB is a quite memory-hungry method). It is clear that $N_s \propto 1/\Delta x^d$ in a situation with d spatial dimensions. Additionally, $N_t \propto 1/\Delta t$ holds independently of the chosen lattice. The finer space and time are resolved (i.e. the smaller Δx and Δt are), the more site updates are required. Assuming that the computing time for one lattice site and one time step is a fixed number (typical codes reach a few million site updates per second on modern desktop CPUs), the total required runtime T and memory M obey

$$T \propto \frac{1}{\Delta x^d \Delta t}, \quad M \propto \frac{1}{\Delta x^d}. \tag{7.20}$$

It is therefore important to choose Δt and especially Δx as large as possible to reduce the computational requirements. However, as we have seen before, a coarser resolution usually reduces the accuracy and brings the system closer to the stability limit.

7.2.3 *Strategies for Parameter Selection*

We have to consider the intrinsic limitations of the LB algorithm, as briefly discussed in the previous section, when choosing the simulation parameters for the BGK collision operator:

- If the spatial or temporal resolution shall be refined or coarsened, one should usually obey the diffusive scaling $\Delta t \propto \Delta x^2$.
- $\tau^\star$ should be close to unity if possible.
- For $\tau^\star$ close to $1/2$, $\tau^\star$ should obey (7.18) which is equivalent to the requirement of having a sufficiently small grid Reynolds number Re_g.
- In any case, the maximum velocity $U^\star_{\max}$ should not exceed 0.4 to maintain stability. If accurate results are desired, $U^\star_{\max}$ should even be smaller (below 0.1 or even below 0.03) due to the error caused by the truncation of the equilibrium distributions.
- Since the computing time and memory requirements increase very strongly with (powers of) the inverse of Δx, one should choose Δx as large as possible without violating the aforementioned limitations.

These intrinsic LB restrictions are also collected in Table 7.2.

7.2.3.1 Mapping of Dimensionless Physical Parameters

Any physical system can be characterised by dimensionless parameters like the Reynolds or Mach numbers. The first step before setting up a simulation is to identify these parameters and assess their relevance.

Inertia, for example, is relevant as long as Re is larger than order unity. Only flows with vanishing small Reynolds number (Stokes flow) do not depend on the actual value of Re: there is virtually no difference between the flow patterns for $\mathrm{Re} = 10^{-3}$ and $\mathrm{Re} = 10^{-6}$. The limit below which Re does not matter depends on

Table 7.2 Overview of intrinsic LB limitations and suggested remedies

Accuracy	(i) $\tau^\star$ not much larger than unity
	(ii) $U^\star_{\max} < 0.03$–0.1
	(iii) Δx sufficiently fine to resolve all flow features
	(iv) Δt sufficiently fine to reduce time discretisation artefacts
Stability	(i) $U^\star_{\max} < 0.4$
	(ii) $\tau^\star$ not too close to $1/2$, cf. (7.18)
	(iii) Sufficiently small grid Reynolds number, cf. (7.19)
Efficiency	Lattice constant Δx and time step Δt not unnecessarily small (memory requirements $\propto \Delta x^{-d}$, simulation runtime $\propto \Delta t^{-1}\Delta x^{-d}$ in d dimensions)
Interface width	Droplet/bubble radius significantly larger than interface width: $r^\star \gg d^\star = O(3)$

the flow details and the definition of Re. For many applications in fluid dynamics, Re is not small and, thus, it has to be properly mapped from the real world to the simulation, as per the law of similarity.[6]

LB is mostly used for the **simulation of incompressible fluids** where the Mach number is small. It is not necessary to map the exact value of Ma then; it is sufficient to guarantee that Ma is "small" in the simulation.

A lattice Mach number, which is defined as $\mathrm{Ma}^{(l)} = U^\star/c_s^\star$, is considered small if $\mathrm{Ma}^{(l)} < 0.3$. Larger Mach numbers lead to more significant compressibility errors. Note that many users try to match the lattice and real-world Mach numbers; this is not necessary in almost all situations where LBM is used.

Example 7.4 Let us investigate which effect a mapping of Re and Ma has on the simulation parameters. We have already found (7.14) that is based on the law of similarity for the Reynolds number. This equation poses one relation for the three parameters $\tau^\star$, Δx and Δt. If we now additionally claim that the lattice and physical Mach numbers shall be identical, we obtain

$$\mathrm{Ma}^{(l)} = \mathrm{Ma}^{(p)} \quad \Longrightarrow \quad \frac{U^\star}{c_s^\star} = \frac{U}{c_s} \quad \Longrightarrow \quad \frac{\Delta x}{\Delta t} = C_u = \frac{U}{U^\star} = \frac{c_s}{c_s^\star}. \tag{7.21}$$

Since $c_s^\star$ is known and c_s is defined by the physical system to be simulated, the ratio of Δx and Δt is fixed by the claim that the Mach numbers match (acoustic scaling). This leaves only one of the three parameters $\tau^\star$, Δx and Δt as an independent quantity.[7]

Luckily, in situations where compressibility effects are not desired, the Mach number mapping can be dropped. It is computationally more efficient to increase the lattice Mach number as much as possible, since the time step scales like $1/U^\star \propto 1/\mathrm{Ma}^{(l)}$.

Example 7.5 Typical wind speeds are a few metres per second while the speed of sound in air is about 330 m/s. The Mach number is then of the order of 0.01, and compressibility effects are not relevant. By choosing a simulation Mach number of $\mathrm{Ma}^{(l)} = 0.3$, while keeping the Reynolds number fixed, the simulation runs 30

[6] Some phenomena remain (inversely) proportional to Re, for example the drag coefficient. Thus, there is still a significant difference in the result of a simulation between $\mathrm{Re} = 10^{-3}$ and 10^{-6}, although inertia may be irrelevant in both.

[7] As we will cover in Chap. 12, treating c_s as physically relevant usually leads to small values of $\tau^\star$ or Δx and Δt, rendering LB an expensive scheme to simulate acoustic problems. The only way to decrease the lattice size is by reducing $\tau^\star$ which in turn can cause stability issues, though these can be alleviated, at least in part, by using other collision models than BGK.

times faster than a simulation with a mapping of the Mach number. Of course it is not automatically guaranteed that this simulation is sufficiently accurate and stable, but this example gives an idea about how useful it can be to increase the Mach number.

Starting from (7.6), the **relation of the simulation parameters in terms of Reynolds and Mach numbers** can be written in the useful form

$$\mathrm{Re}^{(\mathrm{l})} = \frac{\ell^\star U^\star}{\nu^\star} = \frac{\ell^\star U^\star}{c_s^{\star 2}\left(\tau^\star - \frac{1}{2}\right)} \quad \Longrightarrow \quad \frac{\mathrm{Re}^{(\mathrm{l})}}{\mathrm{Ma}^{(\mathrm{l})}} = \frac{\ell}{c_s^\star\left(\tau^\star - \frac{1}{2}\right)\Delta x}. \tag{7.22}$$

Here, $\ell^\star = \ell/\Delta x$ is a typical system length scale in lattice units.

In particular, we may choose $\ell^\star$ as the number of lattice sites along one axis of the system. The total number of lattice sites for the 3D space is therefore of the order of $\ell^{\star 3}$. For a target Reynolds number $\mathrm{Re}^{(\mathrm{l})}$ and a given system size ℓ, this equation allows to balance the simulation parameters in such a way that $\mathrm{Ma}^{(\mathrm{l})}$, $\tau^\star$ and Δx are under control. If all these parameters are set, one can find the time step from (7.14), which requires knowledge of the physical viscosity ν.

It should be noted that $\mathrm{Ma}^{(\mathrm{l})}/\mathrm{Re}^{(\mathrm{l})}$ is the lattice Knudsen number Kn. For $\tau^\star = O(1)$, Kn is basically $\Delta x/\ell$. Hydrodynamic behaviour is only expected for sufficiently small Knudsen numbers. This sets an upper bound for the lattice constant Δx.

7.2.3.2 Parameter Selection Strategies

Usually, the first step is to set the lattice density $\rho_0^\star$. It plays a special role in LB simulations because it is a pure scaling parameter which can be arbitrarily chosen, without any (significant) effect on accuracy, stability or efficiency.[8] The most obvious choice is $\rho_0^\star = 1$, as already mentioned. Other values are possible, but there is no good reason to deviate from unity. Changing $\rho_0^\star$ leads to a change of the conversion factor for the density ($\rho_0^\star \neq 1$ leads to $C_\rho \neq \rho$), and it will affect the lattice value of every quantity whose dimension contains mass (e.g. force, stress or energy).

A typical scenario is that there is a maximum lattice size $\ell^\star$ which can be handled by the computer. This suggests that the lattice constant Δx should be set next. The lattice Mach number (or the non-dimensional velocity $U^\star$) is then reasonably chosen, e.g. $\mathrm{Ma}^{(\mathrm{l})} = 0.1$ or $U^\star = 0.1$. For given system size $\ell^\star$, velocity $U^\star$ and Reynolds number, $\tau^\star$ can then be computed from (7.19). Now we have to

[8]Round-off errors may become important if $\rho_0^\star$ deviates too strongly from unity.

check *via* (7.18) whether the chosen values for $U^\star = \mathrm{Ma}^{(l)} c_s^\star$ and $\tau^\star$ provide stable simulations. If $\tau^\star$ is too small, (7.22) reveals that Δx should be decreased (increase $\ell^\star$) making simulations more expensive or $\mathrm{Ma}^{(l)}$ (increase $U^\star$) should be increased (less accurate, less stable). Once the parameters $U^\star$ and Δx are fixed, we can calculate Δt.

This scenario reveals some problems arising when large Reynolds numbers are simulated: it requires either large lattices, small relaxation parameters or large Mach numbers. It is possible only to a limited extent to reach large Reynolds numbers by increasing the Mach number and decreasing the relaxation parameter due to accuracy and stability issues. In the end, the only way is to increase the resolution which becomes progressively more computationally expensive.

Example 7.6 What is the maximum Reynolds number which can be achieved for a given lattice size? Assuming that we choose $\tau^\star$ near the stability limit, say $\tau^\star = 1/2 + U^\star/4$, we find

$$\mathrm{Re}^{(l)} = \frac{\ell^\star U^\star}{c_s^{\star 2}\left(\tau^\star - \frac{1}{2}\right)} = \frac{4\ell^\star}{c_s^{\star 2}}. \tag{7.23}$$

This shows that the achievable Reynolds number is limited by $O(10)\,\ell^\star$.

It is also possible to set the Mach number (or $U^\star$) and viscosity $\nu^\star$ (or $\tau^\star$) first. Here (7.18) has to be considered for stability reasons. Consequently, we can find the system size $\ell^\star$ and therefore the required lattice resolution Δx, and finally the time step Δt.

Another approach is to set the resolution Δx (or non-dimensional system size $\ell^\star$) and viscosity $\nu^\star$ first. Matching Re in physical and lattice systems gives the velocity $U^\star$ which needs to be less than 0.4 (or below 0.03–0.1 if high accuracy is required).

Overall, it is up to the user which strategy is most convenient. In any case, it has to be checked whether the accuracy, stability and efficiency conditions are reasonably satisfied. We will discuss several concrete examples in Sect. 7.3 to illustrate the parameter selection strategies.

7.2.3.3 Small Reynolds Numbers

We have seen above that it can be difficult to reach large Reynolds numbers in simulations. However, there is also an intrinsic limitation when it comes to small Reynolds numbers.

Equation (7.22) shows that a small Reynolds number can be reached by choosing a large Δx, a large relaxation parameter $\tau^\star$ or a small lattice Mach number $\mathrm{Ma}^{(l)}$. The resolution cannot be arbitrarily decreased; at some point the lattice domain is so small that the details of the flow are finer than the spacing between lattice nodes. It is not advisable to use $\tau^\star \gg 1$ as the numerical errors increase strongly with $\tau^\star$ (this can be avoided by using advanced collision operators such as TRT or MRT).

Therefore, the only unbounded way to reduce $\mathrm{Re}^{(l)}$ is to decrease $\mathrm{Ma}^{(l)}$ and therefore the flow velocity $U^\star$. This in turn means that the time step Δt becomes very small because $\Delta t = \Delta x / C_u$ and $C_u \propto 1/\mathrm{Ma}^{(l)}$ so that $\Delta t \propto \mathrm{Ma}^{(l)}$.

Example 7.7 How small can the Reynolds number be for a given resolution? A typical flow geometry has an extension of about 100 lattice constants, i.e. $\ell^\star = 100$. Assuming that we do not want to use $\tau^\star > 2$, (7.22) yields

$$\mathrm{Re}^{(l)} = U^\star \frac{100}{\frac{3}{2} c_s^{\star 2}}. \tag{7.24}$$

Reaching $\mathrm{Re}^{(l)} = 0.01$ therefore requires $U^\star \approx 10^{-4}$. The velocity is relatively small, and it takes a large number of time steps and therefore computing time to follow the development of the flow field. The situation becomes even worse when the spatial resolution is larger or smaller Reynolds numbers are required.

At this point it is advisable to consider a violation of the Reynolds number mapping and the law of similarity. As explained before, the flow field is in most cases not sensitive to the Reynolds number as long as it is sufficiently small. One of the examples where the Reynolds number is not important are capillary flows [14, 15] or microfluidic applications [16]. In these situations the time scale is not given by viscosity [17].

In some situations, when the Reynolds number is not important, we can use a **numerical Reynolds number** which is larger than the **physical Reynolds number** to accelerate the simulations: $\mathrm{Re}^{(l)} > \mathrm{Re}^{(p)}$. A word of warning: it is very tempting to speed up simulations by violating the law of similarity. One should always check if the simulation results are still valid, e.g. by benchmarking against analytical results or varying the numerical Reynolds number.

Concluding our achievements in this chapter so far, we have seen that LB is particularly useful for flow problems with intermediate Reynolds numbers, especially in the range between $\mathcal{O}(1)$ and $\mathcal{O}(100)$ [18].

7.3 Examples

The examples collected in this section shall underline the relevance of the parameter selection process for LB simulations. Poiseuille (Sects. 7.3.1, 7.3.2 and 7.3.3) and Womersley flow (Sect. 7.3.4) are covered in detail. Furthermore, we talk about surface tension in the presence of gravity (Sect. 7.3.5). It is recommended to read all examples in order of appearance since additional important concepts are introduced in each example.

7.3.1 Poiseuille Flow I

Consider a force-driven 2D Poiseuille flow. The physical parameters for channel diameter w, kinematic viscosity ν, density ρ and gravity g are $w = 10^{-3}\,\mathrm{m}$, $\nu = 10^{-6}\,\mathrm{m}^2/\mathrm{s}$, $\rho = 10^3\,\mathrm{kg/m}^3$ and $g = 10\,\mathrm{m/s}^2$. How should we choose the simulation parameters?

First of all we compute the expected centre velocity $\hat{u}$ of the flow. This is possible since the flow field is known analytically:

$$\hat{u} = \frac{gw^2}{8\nu} = 1.25\,\mathrm{m/s}. \tag{7.25}$$

We define the Reynolds number as

$$\mathrm{Re} = \frac{\hat{u}w}{\nu} = 1\,250, \tag{7.26}$$

but other definitions are possible, e.g. by taking the average velocity which is $\bar{u} = \hat{u}/2$ for 2D Poiseuille flow.

We may now set the resolution, e.g. $\Delta x = 5 \cdot 10^{-5}\,\mathrm{m}$, which corresponds to $w^\star = 20$. The lattice density is taken as $\rho_0^\star = 1$ and therefore $C_\rho = 10^3\,\mathrm{kg/m}^3$. Furthermore we choose $\tau^\star = 0.6$. The system is now fully determined (the user may start with a different set of initial values, though), and we can compute all dependent parameters.

Let us first obtain the conversion factor Δt by applying (7.14):

$$\Delta t = c_s^{\star 2}\left(\tau^\star - \frac{1}{2}\right)\frac{\Delta x^2}{\nu} = 8.33 \cdot 10^{-5}\,\mathrm{s}. \tag{7.27}$$

This means that 12 000 time steps are required for 1 s physical time. We can now compute the expected value of the lattice velocity.[9] We can do this by either using the law of similarity and the relation $\mathrm{Re}^{(\mathrm{p})} = \mathrm{Re}^{(\mathrm{l})} = \hat{u}^\star w^\star/\nu^\star$ or by computing the conversion factor for the velocity, $C_u = \Delta x/\Delta t = 0.6\,\mathrm{m/s}$. In either case we find $\hat{u}^\star = 2.08$ which is an invalid value because it is much larger than the speed of sound.

We have seen that the initially chosen parameters do not lead to a proper set of simulation parameters. This is very common, but one should not be discouraged. The invalidity of the simulation parameters could have been guessed before: the Reynolds number is nearly two orders of magnitude larger than the length scale $w^\star$ which is in conflict with the findings in (7.23). This means that we have to increase the spatial resolution. Let us try a finer resolution ($\Delta x = 1 \cdot 10^{-5}\,\mathrm{m}$ rather than $5 \cdot 10^{-5}\,\mathrm{m}$) and a reduced viscosity ($\tau^\star = 0.55$ instead of 0.6). The same

[9]If there is no analytical solution for the problem, one may run a test simulation and extract $\hat{u}^\star$.

calculation as before now yields $\hat{u}^\star = 0.208$, which may be acceptable for certain applications. The user may choose yet another parameter set; there is no general reason for sticking with the values provided here.

In order to run the simulation, the lattice value for the force density $F = \rho g$ has to be obtained. We first compute the conversion factor for g: $C_g = \Delta x/\Delta t^2 = 3.6 \cdot 10^6\,\text{m/s}^2$ where we have used the time conversion factor $\Delta t = 1.667 \cdot 10^{-6}\,\text{s}$. Therefore, we get $g^\star = 2.78 \cdot 10^{-6}$ and $F^\star = \rho_0^\star g^\star = 2.78 \cdot 10^{-6}$. The simulation can now be performed, and any results on the lattice can be mapped back to the physical system by using the known conversion factors.

We could have pursued alternative routes to find the simulation parameters, e.g.:

1. Choose $w^\star$ and $\hat{u}^\star$, which gives Δx and C_u.
2. Find the viscosity conversion factor $C_\nu = \Delta x^2/\Delta t = \Delta x C_u$ and therefore $\nu^\star$.
3. Compute $\tau^\star$ from $\nu^\star$.
4. Find the remaining conversion factors and check the validity of the parameters.

Another different approach is this:

1. Select $\hat{u}^\star$ and $\tau^\star$ and therefore C_u and $\nu^\star$.
2. Find the viscosity conversion factor C_ν.
3. Compute the lattice resolution $\Delta x = C_\nu/C_u$ and then $w^\star$ (if an integer value for $w^\star$ is required, the other parameters have to be slightly adapted).
4. Calculate all other required conversion factors and check the validity of the parameters.

Exercise 7.3 Find conversion factors for the above example by first selecting reasonable values for $\hat{u}^\star$ and $\tau^\star$.

If the **initially guessed simulation parameters** give invalid or otherwise unacceptable results, one has to modify one or two parameters (which are not necessarily the initially chosen parameters) while **updating the dependent parameters until the desired level of accuracy, stability and efficiency is obtained**.

The Reynolds number has to be kept invariant, which can be enforced by exploiting equation (7.22). This can be a frustrating process because one may find that the lattice becomes too large or the time step too small. However, this *a priori* analysis is necessary to assess whether the simulation is feasible at all. The alternative is to run a series of simulations blindly and adapt the parameters after each run. This very time consuming approach should be avoided.

After some trial and error the user will be able to make educated guesses for the initial parameter set based on the intrinsic limitations of the LB algorithm. It is absolutely justified to say that the parameter selection for LB simulations is an art which can be practised. We particularly recommend [17] for further reading.

7.3.2 *Poiseuille Flow II*

In the previous example, we obtained the simulation parameters for a Poiseuille flow after the physical set of parameters had been defined. Thanks to the law of similarity there is, however, another approach: one can set up simulations without any given dimensional parameter! For the Poiseuille flow it is absolutely sufficient to know the target Reynolds number.

Let us pursue this idea here. We assume that we only know the Reynolds number (e.g. Re $= 1250$ as in the previous example) but nothing else. We can then take advantage of (7.22) (where $U^\star$ has to be replaced by $\hat{u}^\star$) and start with an initial guess for any two independent parameters. Since we know that the Reynolds number is relatively large, we may want to run a well-resolved simulation with $w^\star = 100$. We may wish to limit the velocity to $\hat{u}^\star = 0.1$. From this we can directly compute $\tau^\star = 0.524$, which should yield a stable simulation (cf. (7.18)). The required force density $F^\star = \rho_0^\star g^\star$ can then be obtained from

$$\hat{u}^\star = \frac{g^\star w^{\star 2}}{8\nu^\star} \quad \Longrightarrow \quad g^\star = 6.4 \cdot 10^{-7} \tag{7.28}$$

after $\rho_0^\star$ has been chosen ($\rho_0^\star = 1$ in this case).

An interesting consequence is that we have all required simulation parameters, but not a single conversion factor. This is not a problem at all because the simulation is valid for all similar physical systems, i.e. for all systems with the same Reynolds number. The conversion factors can be obtained *a posteriori* by setting the physical scales *after* a successful simulation. For example, if w is known, one can compute Δx. The full set of conversion factors can only be obtained when three independent physical parameters are given (e.g. w, $\hat{u}$, ρ or F, ρ, w).

In physics it is often the case that **systems** are only **characterised by the relevant dimensionless parameters** without giving the scales like length or velocity themselves. This is sufficient to set up and run simulations, at least as long as *all* required dimensionless parameters are known.

This leads to the important question how many dimensionless parameters have to be known for a given physical system. We will give the answer in Sect. 7.3.4.

7.3.3 *Poiseuille Flow III*

It is very common to drive a Poiseuille flow by a pressure gradient rather than by gravity or a force density. The required boundary conditions for pressure-driven flow are provided in Sect. 5.3. Here, we address how to obtain the required pressure values at the inlet and outlet and which implications this brings along.

First we note that the pressure gradient $p' = \Delta p/\ell$ (where $\Delta p = p_{\text{in}} - p_{\text{out}} > 0$ is the pressure difference between inlet and outlet and ℓ is the length of the channel) and the force density F is simply $p' = F$. From (7.28) we immediately find

$$\hat{u}^\star = \frac{\Delta p^\star}{\rho_0^\star} \frac{w^{\star 2}}{8\ell^\star \nu^\star} \quad \Longrightarrow \quad \frac{\Delta \rho^\star}{\rho_0{}^\star} = \frac{8\ell^\star \nu^\star \hat{u}^\star}{c_s{}^{\star 2} w^{\star 2}} \tag{7.29}$$

for a 2D Poiseuille flow.[10] We have introduced the *average* density $\rho_0^\star$ to appreciate the fact that the density is not constant. In the second step, we replaced the pressure difference $\Delta p^\star$ by the density difference *via* $c_s^{\star 2}\Delta\rho^\star$. Equation (7.29) provides a direct expression for the required relative difference between the inlet and outlet *densities*.

In the incompressible limit, we require the **density difference** between any two points in the simulation, $\Delta\rho^\star$, to be **small compared to the average density** $\rho_0^\star$.

In particular, the right-hand-side of (7.29) has to be a small quantity. This is another restriction for an LB simulation and sets additional bounds, especially for the system length $\ell^\star$. Note that this problem does not exist for a force-driven flow.

Example 7.8 How long can we choose a channel for a typical set of simulation parameters ($\hat{u}^\star = 0.1$, $\nu^\star = 0.1$, $w^\star = 50$, $c_s^{\star 2} = 1/3$)? Assuming that we allow density variations of up to 5% ($\Delta\rho^\star/\rho_0{}^\star = 0.05$), the maximum channel length becomes $\ell^\star \approx 520$. This is only ten times more than the channel diameter. If longer channels are required, the user has to re-balance the simulation parameters on the right-hand-side of (7.29), keeping all relevant dimensionless parameters (here only the Reynolds number) invariant. In the end, this leads to an increased spatial resolution which allows larger ratios of $\ell^\star/w^\star$ but requires more expensive simulations.

How can we now choose the inlet and outlet densities in lattice units? If $\Delta\rho^\star$ and the average density $\rho_0^\star$ are known, we set

$$\rho_{\text{in}}^\star = \rho_0^\star + \frac{\Delta\rho^\star}{2} \quad \text{and} \quad \rho_{\text{out}}^\star = \rho_0^\star - \frac{\Delta\rho^\star}{2}. \tag{7.30}$$

Obviously, $\Delta\rho^\star = \rho_{\text{in}}^\star - \rho_{\text{out}}^\star$ is satisfied. In turn, the local pressure $p^\star$ can be obtained from the local density $\rho^\star$ according to $p^\star = p_0^\star + c_s^{\star 2}(\rho^\star - \rho_0^\star)$. The constant reference pressure $p_0^\star$ does not affect the simulation and is chosen by the user as discussed in (7.17).

[10] The numerical prefactor becomes 4 in a 3D Poiseuille flow with circular cross-section.

7.3.4 Womersley Flow

Womersley flow denotes a Poiseuille-like flow (channel width w, viscosity ν) with an oscillating pressure drop along the length ℓ of the channel: $\Delta p(t) = \Delta p_0 \cos \omega t$ with angular frequency ω. Since there exists an analytical solution for this unsteady flow, it is an ideal benchmark for Navier-Stokes solvers. Hence, Womersley flow has often been simulated with the LB algorithm (e.g. [19–21]). We will not show the analytical solution and refer the reader to the literature instead. Here we will rather consider the generic implications arising from simulating non-steady flows.

The Reynolds number for an oscillatory flow is usually defined through the velocity one would observe for $\omega = 0$: $\mathrm{Re} = \hat{u}_0 w/\nu$ where $\hat{u}_0$ is related to Δp_0 according to (7.29):

$$\hat{u}_0 = \frac{\Delta p_0 w^2}{8\ell \rho \nu} \tag{7.31}$$

for a 2D channel. Due to the presence of the frequency ω, which is obviously zero for simple Poiseuille flow, an additional dimensionless number is required to characterise the flow. There are different ways to construct such a number: we just have to write down a dimensionless combination containing ω and other suitable parameters. One possibility is $\omega w^2/\nu$. The *Womersley number* is defined as the square root

$$\alpha = \sqrt{\frac{\omega}{\nu}} w. \tag{7.32}$$

Before we continue with the discussion of unsteady flows in LB simulations, let us first address the issue of finding dimensionless numbers for a physical system.

The **Buckingham π theorem** [1] states that for Q independent quantities whose physical dimension can be constructed from D independent dimensions there are $N = Q - D$ independent dimensionless parameters.

For example, for a simple Poiseuille flow, there are $Q = 4$ independent quantities: channel width, flow velocity, viscosity and density. We know that any mechanical system is characterised by $D = 3$ independent dimensions: length, time and mass. This gives $N = 1$ parameter which is the Reynolds number. For each additional physical parameter, one dimensionless number is required. Womersley flow is characterised by $Q = 5$ quantities (the four Poiseuille parameters and the oscillation frequency) and therefore $N = 2$ dimensionless numbers.

Another very helpful way to look at this is the definition of characteristic time scales and their ratios. In a Poiseuille flow, there are two time scales: the advection time $t_a \sim w/\hat{u}$ and the diffusion or viscous time $t_\nu \sim w^2/\nu$. $N+1$ such time scales define N independent time scale ratios (e.g. $\mathrm{Re} \sim t_\nu/t_a$ for the Poiseuille flow). For Womersley flow, a third time scale is given by $t_\omega \sim 1/\omega$. A second suitable dimensionless number is therefore t_ω/t_ν or t_ω/t_a or appropriate combinations thereof (like $\alpha \sim \sqrt{t_\nu/t_\omega}$). Any additional dimensionless number would not be independent. For example, one can also define the *Strouhal number* $\mathrm{St} = \omega w/(2\pi\hat{u}_0) \sim t_a/t_\omega$. It can be expressed as a combination of Reynolds and Womersley numbers: $\alpha \sim \sqrt{\mathrm{Re}\,\mathrm{St}}$. Womersley flow is usually characterised by Re and α.

Let us now get back to the LBM. When setting up a Womersley flow (or any other unsteady flow) in an LB framework, one has to be aware of the subtleties of the LB algorithm. As LB is a local algorithm without a Poisson equation for the pressure, any information on the lattice propagates with a finite velocity comparable to the speed of sound.

> All **physical time scales** which shall be resolved *must* be **sufficiently long compared to the intrinsic sound wave (or acoustic) time scale** $t_s^\star \sim \ell^\star/c_s^\star$.

In other words, the acoustic time scale $t_s^\star$ on the one hand and the advection, diffusion and oscillation time scales ($t_a^\star$, $t_\nu^\star$ and $t_\omega^\star$) on the other hand have to be sufficiently separated (cf. section 3.1 in [22]) as long as one is interested in incompressible flows: $t_a^\star, t_\nu{}^\star, t_\omega{}^\star \gg t_s{}^\star$. This defines, for a given lattice size, a lower bound for the oscillation time scale and therefore an upper bound for the frequency (here we assume that $\ell^\star > w^\star$):

$$\omega^\star \ll \frac{c_s^\star}{\ell^\star}. \tag{7.33}$$

Any unsteady incompressible flow violating this condition is non-physical, and the results will not be a good approximation of the incompressible Navier-Stokes solution.

Having clarified this, we provide a suggested procedure to set up Womersley flow with known Re and α:

1. Select initial values for $\ell^\star$ and $w^\star$ which are compatible with the Reynolds number, e.g. $w^\star > 0.1\,\mathrm{Re}$ (stability).
2. Estimate the sound propagation time scale $t_s^\star \sim \ell^\star/c_s^\star$ and therefore the recommended maximum for $\omega^\star$ via $2\pi/\omega^\star = T^\star \gg t_s{}^\star$. It requires some trial and error to find a sufficient minimum ratio $T^\star/t_s{}^\star$, but one should at least use a factor of ten.
3. Balance $\omega^\star$ and $\nu^\star$ to match α via (7.32) by taking into account the law of similarity for the Womersley number: $\alpha^{(l)} = \alpha^{(p)}$. There is no unique way to

choose $\omega^\star$ and $\nu^\star$, but keep in mind already now that the relaxation parameter $\tau^\star$ should not be too close to $1/2$ or much larger than unity.

4. Estimate the velocity $\hat{u}_0^\star$ from Re using the selected values for $w^\star$ and $\nu^\star$.
5. Check the validity of $\hat{u}_0^\star$: if its value is not in the desired range (too large for reasonable stability or accuracy or too small for a feasible time step), the parameters have to be re-balanced while keeping Re and α invariant. This process is more complicated than for the Poiseuille flow because Re and α have to be considered simultaneously.
6. The required pressure difference $\Delta p_0^\star$ finally can be obtained from (7.31).

A parameter optimisation scheme for unsteady LB simulations is thoroughly discussed in [20].

Example 7.9 Let us consider a flow with $\text{Re} = 1000$ and $\alpha = 15$ which is typical for blood flow in the human aorta. First we choose $\rho_0^\star = 1$, $\ell^\star = 500$ and $w^\star = 100$ and find that $T^\star \gg 870$ should hold. We choose $T^\star = 8700$ and therefore $\omega^\star = 7.22 \cdot 10^{-4}$. Hence, we get $\nu^\star = \omega^\star w^{\star 2}/\alpha^2 = 0.0321$ and $\tau^\star = 0.596$ for a lattice with $c_s^{\star 2} = 1/3$. This is a reasonable value for the relaxation parameter. We find $\hat{u}_0^\star = \text{Re}\, \nu^\star / w^\star = 0.321$, though, which seems quite large. However, the maximum velocity actually observed in a Womersley flow is always smaller than $\hat{u}_0^\star$, especially for $\alpha > 1$. The reason is that the flow lags behind the pressure gradient and does not have enough time to develop fully. The required pressure difference follows from (7.31) and yields $\Delta p_0^\star = 4.12 \cdot 10^{-3}$ which is small compared to order unity. We can therefore assume that the present set of simulation parameters is suitable for a simulation of Womersley flow, but one may want to decrease the time step further to obtain a better separation of time scales.

LB simulations are usually subject to initial unphysical or at least undesired transients (cf. Sect. 5.5) that decay after some time. As a rule of thumb, the transient length in channel flow is given by the diffusion time scale $t_\nu = w^2/\nu$; for the above example we find $t_\nu^\star \approx 3.11 \cdot 10^5 \approx 36\, T^\star$! It may therefore be necessary to simulate about 40 full oscillations before the numerical solution converges. The actual transient length depends on the simulated flow and the chosen initial conditions. It is recommended to employ an on-the-fly convergence algorithm which aborts the simulation after a given convergence criterion has been satisfied.[11] Under the assumption that 40 oscillation periods are required, the simulation will run for 348 000 time steps. Further assuming a relatively slow D2Q9 code with 2 million site updates per second and taking the lattice size $\ell^\star \times w^\star = 500 \times 100$, the total simulation runtime will be 8 700 seconds or nearly three hours. A comparable 3D simulation would be roughly 100 times more expensive (assuming a circular cross-section with 100 lattice sites across the diameter), which is unfeasible for a serial code.

[11] One may check for temporal convergence by comparing the velocity profiles at times $t^\star$ and $t^\star - T^\star$ and applying a suitable error norm.

7.3.5 Surface Tension and Gravity

Let us now consider a droplet of a liquid in vapour (or the other way around: a vapour bubble in a liquid). The surface tension of the liquid-gas interface is γ. The droplet/bubble may be put on a flat substrate or into a narrow capillary. For the implementation of such a system, we refer to Chap. 9. Here, we focus on the unit conversion in presence of surface tension and gravity, which is independent of the underlying numerical model.

The density difference of the liquid and vapour phases is defined as $\Delta\rho = \rho_\mathrm{l} - \rho_\mathrm{v} > 0$. In the presence of the gravitational acceleration with magnitude g we can define the dimensionless *Bond number*

$$\mathrm{Bo} = \frac{\Delta\rho g r^2}{\gamma}. \tag{7.34}$$

It characterises the relative strength of gravity and surface tension effects where r is a length scale typical for the droplet/bubble. A common definition for r is the radius of a sphere with the same volume V as the droplet/bubble[12]:

$$r = \left(\frac{3V}{4\pi}\right)^{1/3}. \tag{7.35}$$

A small Bond number means that gravity is negligible and the droplet/bubble shape is dominated by the surface tension. This will result in a spherical shape of the droplet/bubble if it is not in contact with any wall or a shape like a spherical cap if it is attached to a wall. For larger Bond numbers, gravity is important, and the droplet/bubble is generally deformed. This is illustrated in Fig. 7.1.

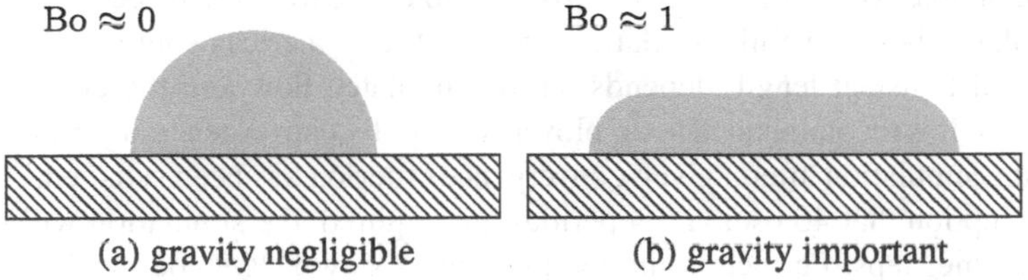

Fig. 7.1 Illustration of typical shapes of a droplet on a substrate (**a**) for small Bond number ($\mathrm{Bo} \ll 1$) with negligible gravity and (**b**) large Bond number ($\mathrm{Bo} \gg 1$) where gravity is important. In both cases the contact angle is 90°. More details about contact angles are given in Chap. 9

[12]This definition appreciates the fact that the droplet/bubble may be deformed and therefore have a more complex shape than a section of a sphere.

How do we find the simulation parameters in lattice units for a given Bond number? The first step is to set the lattice densities which are usually tightly related to the numerical model (cf. Chap. 9). Let us take the liquid density as reference and write

$$\rho_v^\star = \frac{1}{\lambda}\rho_l^\star =: \frac{1}{\lambda}\rho^\star \tag{7.36}$$

where $\lambda > 1$ is the achievable density ratio of the model. The next step is to choose the radius $r^\star$ properly. In lattice-based methods, interfaces between phases are always subject to a finite width of a few Δx ($d^\star = O(3)$) as we will detail in Chap. 9. We have to make sure that the radius $r^\star$ is significantly larger than this interface width, $r^\star \gg d^\star$, otherwise the simulated system will be dominated by undesired diffuse interface properties. Typical recommended thresholds are $r^\star \geq 8$, but one may obtain decent results for smaller resolutions. This is an important additional restriction in multiphase or multicomponent LB models.

Separating the known and unknown parameters, we get

$$\frac{\gamma^\star}{g^\star} \approx \frac{\rho^\star r^{\star 2}}{\mathrm{Bo}^{(l)}} \tag{7.37}$$

where we have approximated $\Delta\rho \approx \rho_l = \rho$, which is valid for $\lambda \gg 1$. Furthermore, we have taken advantage of the law of similarity in the form $\mathrm{Bo}^{(l)} = \mathrm{Bo}^{(p)}$. All quantities on the right-hand-side of (7.37) are known. This leaves us with one degree of freedom: it is up to the user to choose $\gamma^\star$ and $g^\star$ in such a way that (7.37) is satisfied. On the one hand, based on the underlying LB model, the surface tension may be restricted to a certain numerical range (similar to the restriction of $\tau^\star$). On the other hand, if $g^\star$ is chosen too big, the lattice acceleration and therefore velocity may be too large, and the simulation is less accurate or even unstable. Since the time step is effectively defined by $g^\star$ *via*

$$g = g^\star \frac{\Delta x}{\Delta t^2}, \qquad \Delta x = \frac{r}{r^\star}, \tag{7.38}$$

the user has to consider the total runtime of the simulation as well. Reducing $\gamma^\star$ and $g^\star$ will decrease the time step and result in a longer computing time.

We have not said anything about viscosity. The reason is that, so far, we have only taken advantage of the Bond number scaling. If for example the Reynolds number is a relevant parameter as well, i.e. if Re is not small, the simulation parameters have to be chosen simultaneously.

Example 7.10 Liquid glycerol with density $\rho_l = 1\,260\,\mathrm{kg/m^3}$ and kinematic viscosity $\nu = 8.49 \cdot 10^{-4}\,\mathrm{m^2/s}$ (dynamic viscosity $\eta = 1.07\,\mathrm{Pa\,s}$) is flowing in a vertical pipe of diameter $w = 0.015\,\mathrm{m}$ in the gravitational field $g = 9.81\,\mathrm{m/s^2}$. We want to simulate an air bubble in the liquid with radius $r = 4 \cdot 10^{-3}\,\mathrm{m}$ and large density contrast, $\lambda \gg 1$. The Reynolds number is defined according to the flow we would observe in the absence of the bubble (cf. Sect. 7.3.1):

$$\mathrm{Re} = \frac{\hat{u}w}{\nu} = \frac{gw^3}{4\nu^2} = 11.5. \tag{7.39}$$

The surface tension of glycerol in air at 20 °C is $\gamma = 6.34 \cdot 10^{-2}\,\mathrm{N/m}$ resulting in $\mathrm{Bo} = 3.12$. Note that the *confinement* $\chi := 2r/w = 0.533$ is also a relevant dimensionless parameter which is, however, easily mapped to the lattice as it is merely the ratio of two length scales. How should we select the simulation parameters? We start by setting $w^\star = 30$ and $r^\star = 8$ obeying the confinement scaling. The Reynolds number $\mathrm{Re}^{(\mathrm{p})} = \mathrm{Re}^{(\mathrm{l})}$ therefore restricts the ratio of gravity and viscosity:

$$\frac{g^\star}{\nu^{\star 2}} = \frac{4\,\mathrm{Re}}{w^{\star 3}} = 1.70 \cdot 10^{-3}. \tag{7.40}$$

We choose $\rho^\star = 1$ and find from the Bond number:

$$\frac{\gamma^\star}{g^\star} = \frac{\rho^\star r^{\star 2}}{\mathrm{Bo}^{(\mathrm{l})}} = 20.5. \tag{7.41}$$

We have one degree of freedom left because the choice of the three parameters $g^\star$, $\nu^\star$ and $\gamma^\star$ is restricted by two conditions. We may now set the surface tension first: $\gamma^\star = 0.06$. This gives $g^\star = 2.93 \cdot 10^{-3}$ and therefore $\nu^\star = 1.31$ and $\tau^\star = 4.44$. Additionally, we find the maximum velocity $\hat{u}^\star = 0.50$. These parameters are not acceptable, and the velocity and viscosity have to be reduced. Different strategies are available:

1. Keep $\gamma^\star$ fixed. This leads to the scalings $g^\star \propto r^{\star -2}$, $\nu^\star \propto r^{\star 1/2}$ and $\hat{u}^\star \propto r^{\star -1/2}$. Showing these relations is left as an exercise for the reader. We notice that we can either reduce the viscosity or the velocity by changing the resolution, but we cannot reduce both at the same time. It is therefore not possible to keep the surface tension unchanged.
2. Keep $r^\star$ fixed. Now we get $g^\star \propto \gamma^\star$, $\nu^\star \propto \gamma^{\star 1/2}$ and $\hat{u}^\star \propto \gamma^{\star 1/2}$. By reducing the surface tension we can therefore reduce the viscosity and the velocity simultaneously. But it depends on the selected numerical model if a decrease of surface tension is feasible at all.

The two above-mentioned approaches are usually combined to further balance the simulation parameters. One has to keep in mind, though, that every numerical method has intrinsic limitations which make certain combinations of dimensionless numbers (here: Re, Bo, χ) unfeasible if not impossible to access in a simulation.

7.4 Summary

We conclude this chapter by collecting the relevant results about unit conversion and parameter selection.

- Purely mechanical systems require exactly three conversion factors. The first step is to define three independent (basic) ones. All other conversion factors can be derived according to (7.10).
- Some dimensions (e.g. for pressure or density) are different in 2D and 3D. This problem can be circumvented by interpreting a 2D system as a 3D system with thickness Δx.
- The law of similarity (e.g. for the Reynolds, Womersley or Bond number) is the physical basis for consistent non-dimensionalisation: two systems with the same set of relevant dimensionless parameters are similar.
- Unit systems must not be mixed as this will lead to inconsistencies and wrong physical results.
- Typical basic conversion factors for LB simulations are those for length, time (or velocity) and density: C_ℓ, C_t (or C_u) and C_ρ. Other sets are possible but usually less practical.
- Relevant simulation parameters are the lattice spacing Δx, the time step Δt, the relaxation parameter $\tau^\star$, the average density $\rho_0^\star$ and the characteristic flow velocity $U^\star$ (a star * denotes non-dimensionalised parameters). They are not independent. Their relations are partially dictated by characteristic dimensionless numbers such as the Reynolds number. For example, the physical kinematic viscosity ν poses a condition for $\tau^\star$, Δx and Δt:

$$\nu = c_s^{\star 2}\left(\tau^\star - \frac{1}{2}\right)\frac{\Delta x^2}{\Delta t}. \tag{7.42}$$

- Choose simulations parameters in such a way that the intrinsic LB restrictions are considered (cf. Table 7.2).
- The LB algorithm is second-order accurate in space and first-order accurate in time when choosing the diffusive scaling $\Delta t \propto \Delta x^2$. This is the preferred scaling for resolution refinement if compressibility effects are unimportant (otherwise the acoustic scaling $\Delta t \propto \Delta x$ is the way to go).
- For incompressible simulations the Mach number is an unimportant parameter, as long as it is not too large. The correct scaling of the Mach number is therefore not necessary and should actually be avoided to increase the simulation efficiency. Another handy relation between the simulation parameters is given by

$$\frac{\mathrm{Re}^\star}{\mathrm{Ma}^\star} = \frac{\ell}{c_s^\star\left(\tau^\star - \frac{1}{2}\right)\Delta x} \tag{7.43}$$

where $\mathrm{Re}^\star = \ell^\star U^\star/\nu^\star$ and $\mathrm{Ma}^\star = U^\star/c_s^\star$.

- It is recommended to set the average lattice density $\rho_0^\star$ first. There is usually no good reason to deviate from $\rho_0^\star = 1$.
- LB simulations can be performed without specifying any conversion factors. The results can later be mapped to a physical system with the same dimensionless parameters. In particular, one can find all LB parameters from the dimensionless parameters without the necessity to specify absolute scales.
- Any LB simulation for incompressible flows has to be set up in a way such that the local density variation $\rho'^\star$ is small compared to $\rho_0^\star$.
- The number of independent simulation parameters is reduced by each imposed dimensionless parameter. Simulation parameters have to be chosen such that all relevant dimensionless numbers are correct, i.e. that all laws of similarity are satisfied. This is tightly related to the Buckingham π theorem and can also be understood from defining all relevant time scales and their ratios. For example, the Reynolds number is the ratio of the characteristic viscous and advection time scales. The Strouhal number is the ratio of the advection and the oscillation time scale of an oscillating flow.
- Any relevant time scale (e.g. advection, diffusion, oscillation) has to be significantly larger than the sound propagation time scale $t_s^\star \sim \ell^\star / c_s{}^\star$. Otherwise the physical properties of the system change faster than the simulation can adapt.

References

1. E. Buckingham, Phys. Rev. **4**, 345 (1914)
2. L.D. Landau, E.M. Lifshitz, *Fluid Mechanics* (Pergamon Press, Oxford, 1987)
3. E.M. Viggen, Phys. Rev. E **90**, 013310 (2014)
4. Y.T. Feng, K. Han, D.R.J. Owen, Int. J. Numer. Meth. Eng. **72**(9), 1111–1134 (2007)
5. P.A. Skordos, Phys. Rev. E **48**(6), 4823 (1993)
6. D.J. Holdych, D.R. Noble, J.G. Georgiadis, R.O. Buckius, J. Comput. Phys. **193**(2), 595 (2004)
7. S. Ubertini, P. Asinari, S. Succi, Phys. Rev. E **81**(1), 016311 (2010)
8. P.J. Dellar, Comput. Math. Appl. **65**(2), 129 (2013)
9. M. Reider, J. Sterling, Comput. Fluids **118**, 459 (1995)
10. D. d'Humières, I. Ginzburg, Comput. Math. Appl. **58**, 823 (2009)
11. T. Krüger, F. Varnik, D. Raabe, Phys. Rev. E **79**(4), 046704 (2009)
12. I. Ginzburg, D. d'Humières, A. Kuzmin, J. Stat. Phys. **139**, 1090 (2010)
13. J.C.G. Verschaeve, Phys. Rev. E **80**, 036703 (2009)
14. A. Kuzmin, M. Januszewski, D. Eskin, F. Mostowfi, J. Derksen, Chem. Eng. J. **171**, 646 (2011)
15. A. Kuzmin, M. Januszewski, D. Eskin, F. Mostowfi, J. Derksen, Chem. Eng. J. **178**, 306 (2011)
16. T. Krüger, D. Holmes, P.V. Coveney, Biomicrofluidics **8**(5), 054114 (2014)
17. M.E. Cates, J.C. Desplat, P. Stansell, A.J. Wagner, K. Stratford, R. Adhikari, I. Pagonabarraga, Philos. T. Roy. Soc. A **363**(1833), 1917 (2005)
18. S. Succi, *The Lattice Boltzmann Equation for Fluid Dynamics and Beyond* (Oxford University Press, Oxford, 2001)
19. X. He, G. Doolen, J. Comput. Phys. **134**, 306 (1997)
20. A.M.M. Artoli, A.G. Hoekstra, P.M.A. Sloot, Comput. Fluids **35**(2), 227 (2006)
21. R.W. Nash, H.B. Carver, M.O. Bernabeu, J. Hetherington, D. Groen, T. Krüger, P.V. Coveney, Phys. Rev. E **89**(2), 023303 (2014)
22. B. Dünweg, A.J.C. Ladd, in *Advances in Polymer Science* (Springer, Berlin, Heidelberg, 2008), pp. 1–78

Part III
Lattice Boltzmann Extensions, Improvements, and Details

Chapter 8
Lattice Boltzmann for Advection-Diffusion Problems

Abstract After reading this chapter, you will understand how the lattice Boltzmann equation can be adapted from flow problems to advection-diffusion problems with only small changes. These problems include thermal flows, and you will know how to simulate these as two interlinked lattice Boltzmann simulations, one for the flow and one for the thermal advection-diffusion. You will understand how advection-diffusion problems require different boundary conditions from flow problems, and how these boundary conditions may be implemented.

The LBM is not only used for fluid dynamics; it is also a powerful method to solve advection-diffusion problems. In fact, there is a growing interest in studying systems with coupled fluid dynamics and diffusion with LBM. First, we briefly summarise how LB advection-diffusion is implemented in Sect. 8.1. Then, we give a general overview of advection-diffusion problems in Sect. 8.2, and cover in more detail in Sect. 8.3 why and how the LBM can solve the advection-diffusion equation and which model extensions exist. Section 8.4 is dedicated to the important special case of thermal flows where the advection-diffusion equation is replaced by the heat or energy equation. Similarly to the Navier-Stokes equation, the advection-diffusion equation requires the specification of boundary conditions. We provide a short summary of simple diffusion boundary conditions for the LBM in Sect. 8.5. In Sect. 8.6 we demonstrate the suitability and accuracy of the LBM for advection-diffusion problems through a number of benchmark tests.

8.1 Lattice Boltzmann Advection-Diffusion in a Nutshell

Advection-diffusion problems are common in nature (cf. Sect. 8.2). They include mixing of and heat diffusion in fluids. The governing equation is the advection-diffusion equation (ADE, cf. (8.6)) for a scalar field C, which could be e.g. a concentration or temperature:

$$\frac{\partial C}{\partial t} + \nabla \cdot (C\boldsymbol{u}) = \nabla \cdot (D\nabla C) + q. \tag{8.1}$$

T. Krüger et al., *The Lattice Boltzmann Method*, Graduate Texts in Physics,
DOI 10.1007/978-3-319-44649-3_8

The left-hand side describes the advection of C in the presence of an external fluid velocity $\boldsymbol{u}$, while the right-hand side contains a diffusion term with diffusion coefficient D and a possible source term q.

The ADE and the Navier-Stokes equation have strong similarities (cf. Sect. 8.3). In fact, we can understand the NSE as an ADE for the fluid momentum density vector $\rho\boldsymbol{u}$. Therefore, the LBM is easily adapted to advection-diffusion problems. It turns out that the LBE

$$g_i(\boldsymbol{x} + \boldsymbol{c}_i\Delta t, t + \Delta t) - g_i(\boldsymbol{x}, t) = \Omega_i(\boldsymbol{x}, t) + Q_i(\boldsymbol{x}, t) \tag{8.2}$$

with

$$\Omega_i(\boldsymbol{x}, t) = -\frac{1}{\tau_g}\left(g_i(\boldsymbol{x}, t) - g_i^{\text{eq}}(\boldsymbol{x}, t)\right) \tag{8.3}$$

and suitable source terms Q_i solves the ADE for the concentration field $C = \sum_i g_i$. This collision operator results in a diffusion coefficient

$$D = c_s^2\left(\tau_g - \frac{\Delta t}{2}\right). \tag{8.4}$$

This is equivalent to the relation between kinematic viscosity ν and relaxation time τ for the standard LBM.

The main difference between LBM for Navier-Stokes and advection-diffusion problems is that in the latter there is only one conserved quantity: C. The velocity $\boldsymbol{u}$ is *not* obtained from g_i; it is rather imposed externally.

The equilibrium distribution typically assumes the form

$$g_i^{\text{eq}} = w_i C\left(1 + \frac{\boldsymbol{c}_i \cdot \boldsymbol{u}}{c_s^2} + \frac{(\boldsymbol{c}_i \cdot \boldsymbol{u})^2}{2c_s^4} - \frac{\boldsymbol{u} \cdot \boldsymbol{u}}{2c_s^2}\right). \tag{8.5}$$

It is also possible to employ a linear equilibrium given in (8.24), though this leads to undesired error terms (cf. Sect. 8.3.4). We can use the same lattices as for the standard LBM, but due to the less strict isotropy requirements for advection-diffusion problems, we can also use smaller and more efficient velocity sets (cf. Sect. 8.3.3).

One of the most important applications of the advection-diffusion model is thermal flows where the advection and diffusion of heat is coupled to the dynamics of the ambient fluid (cf. Sect. 8.4). In this case, the velocity $\boldsymbol{u}$ is provided by a Navier-Stokes solver, e.g. the LBM, whereas temperature acts back on the fluid. This feedback is often modelled as, but not always limited to, a temperature-density coupling. For example, the Boussinesq approximation assumes a temperature-dependent fluid density that can be modelled as a buoyancy force instead of changing the simulation density (cf. Sect. 8.4.1).

In thermal flows we have to distinguish between situations with and without energy conservation. If viscous heating and pressure work are relevant, an energy-conserving model is required (cf. Sect. 8.4.3). In many situations it is safe to assume that viscous dissipation does not lead to a large temperature increase, and energy conservation does not have to be satisfied (cf. Sect. 8.4.4).

The boundary conditions used for LB advection-diffusion are different to those used for Navier-Stokes LB. In particular, the simple bounce-back method is replaced by an anti-bounce-back method for Dirichlet boundary conditions. This is discussed in more detail in Sect. 8.5, followed by a number of benchmark tests in Sect. 8.6.

8.2 Advection-Diffusion Problems

Advection and diffusion are two common phenomena that we can observe in daily life and many hydrodynamic problems. Both are illustrated in Fig. 8.1. We have already seen that the Reynolds number in the NSE indicates the relative importance of advection and momentum diffusion as two important transport phenomena.

As an example of *advection*, oil spilled in a river is dragged along with the water, therefore moving downstream with the current. Clouds moving in the sky are another example: the clouds travel along with the wind in the atmosphere due to the drag force between both.

Diffusion is slightly less intuitive. At finite temperature, molecules show random motion, even if the average velocity in a medium is zero. A famous effect is the *Brownian motion* of dust particles on top of a water surface. The particles are constantly being randomly hit by water molecules, which leads to the characteristic Brownian trajectories. The mean square displacement of the dust particles increases linearly with time. Due to this kind of random molecular motion, two miscible fluids such as ethanol and water mix themselves when put into the same container. Both molecular species diffuse until the concentration of both ethanol and water is constant.

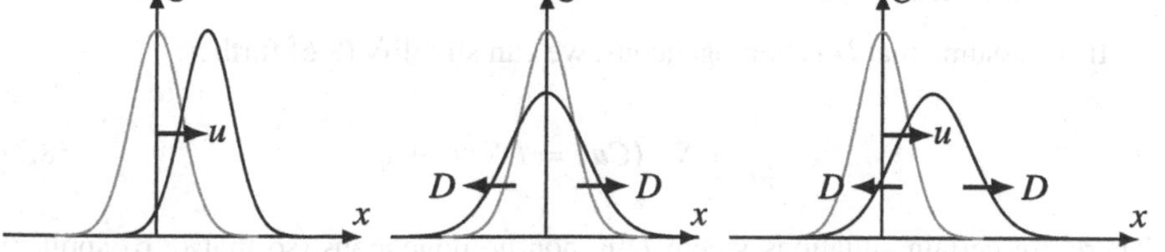

Fig. 8.1 Illustration of pure advection (*left*), pure diffusion (*middle*) and advection-diffusion (*right*). u and D are the advection velocity and diffusion coefficients, respectively. The *grey curve* is an initial concentration distribution $C(x, t = 0)$, the *black curve* shows the concentration at a later time

Thermal diffusion is relevant in many industrial applications, for instance to keep the temperature in a reactor as constant and homogeneous as possible. Local heat sources, e.g. due to chemical reactions, or heat sinks, e.g. caused by the presence of cold walls, cause a temperature gradient. Heat then diffuses from warmer to colder regions; therefore, heat diffusion tends to reduce temperature gradients.

Diffusion is often a slow process, in particular when the physical system is large. Since the mean square displacement grows linearly with time, diffusion is characterised by a $\sqrt{t}$ behaviour. Advection with a constant velocity u, however, scales linearly with time since the distance travelled at constant velocity is ut. It therefore often depends on the spatial scale whether advection or diffusion dominate. We will pick up this idea again shortly.

The **advection-diffusion equation** (ADE) or **convection-diffusion equation** for a scalar field C (e.g. the concentration of a chemical species or temperature) with isotropic diffusion coefficient D and source term q reads [1]

$$\frac{\partial C}{\partial t} + \nabla \cdot (C\boldsymbol{u}) = \nabla \cdot (D\nabla C) + q. \tag{8.6}$$

Equation (8.6) shows that there are three mechanisms that lead to a local change of C:

- *Advection* is caused be a prescribed advection velocity $\boldsymbol{u}$. This is typically the velocity of the ambient medium.
- The scalar field shows intrinsic *diffusion* according to the divergence of the term $\boldsymbol{j} = -D\nabla C$. This linear relation between the *diffusion flux* $\boldsymbol{j}$ and the gradient ∇C is called *Fick's first law*. We see that diffusion is generally driven by the gradient of C. The diffusion coefficient D is a material property that is normally temperature-dependent. Its physical unit is m^2/s, just like the kinematic viscosity.
- C may be locally produced or destroyed as indicated by the *source* term q. One possible mechanism is a chemical reaction that produces heat or consumes a given chemical species.

If we assume that D is homogeneous, we can simplify (8.6) further:

$$\frac{\partial C}{\partial t} + \nabla \cdot (C\boldsymbol{u}) = D\nabla^2 C + q. \tag{8.7}$$

There are certain situations where D is non-homogeneous (so that (8.6) applies) or non-isotropic (where D becomes a direction-dependent diffusion tensor $\boldsymbol{D}$). One example is systems with inhomogeneous temperature and therefore inhomogeneous diffusivity, and another is anisotropic diffusion in porous media. If not otherwise stated, we only consider the simplified ADE as in (8.7).

Equation (8.7) is general in the physical sense, as it describes not only a concentration C, but for example also the temperature T. In this case, one often writes the **heat equation**

$$\frac{\partial T}{\partial t} + \nabla \cdot (T\boldsymbol{u}) = \kappa \nabla^2 T + q \tag{8.8}$$

where κ is the **thermal diffusivity**.

To highlight the generality of the ADE, we continue writing C and keep in mind that C may be replaced by T (and D by κ) if we want to consider thermal problems.

Like the Reynolds number for the NSE, we can define a characteristic dimensionless number for advection-diffusion problems: the *Péclet number*. Introducing a characteristic velocity U, length L and time $T = L/U$, we can rewrite (8.7) without source term q as

$$\frac{\partial C}{\partial t^\star} + \nabla^\star \cdot (C\boldsymbol{u}^\star) = \frac{D}{LU} \nabla^{\star 2} C \tag{8.9}$$

where variables with a star $\star$ are dimensionless, e.g. $t^\star = t/T$ and $\nabla^\star = L\nabla$. The dimensionless prefactor on the right-hand side is the inverse Péclet number:

$$\mathrm{Pe} = \frac{LU}{D}. \tag{8.10}$$

Problems with large Péclet number are advection-dominated, i.e. the change of C due to the advection velocity $\boldsymbol{u}$ is more important than diffusive contributions. In the limit $\mathrm{Pe} \to 0$, for small or vanishing velocity or at small length scales, (8.7) is dominated by diffusion.

Thermal advection-diffusion problems often occur together with hydrodynamics. The ADE for temperature and the NSE for momentum are then coupled and have to be solved simultaneously, as we will discuss in more detail in Sect. 8.4. In this case, there is another relevant dimensionless number: the *Prandtl number* is the ratio of kinematic viscosity ν to thermal diffusivity κ:

$$\mathrm{Pr} = \frac{\nu}{\kappa}. \tag{8.11}$$

It is about 0.7 for air and 7 for water. If the diffusion of a substance in a viscous medium is considered instead of temperature diffusion, the *Schmidt number* is used:

$$\mathrm{Sc} = \frac{\nu}{D}. \tag{8.12}$$

There exist different numerical methods to solve advection-diffusion problems. The most commonly used is the finite difference method (Sect. 2.1.1). However, Wolf-Gladrow [2] showed that an LB-based diffusion solver can achieve higher diffusion coefficients than explicit finite-difference schemes since LBM permits larger time steps than finite-difference schemes with the same spatial resolution. In the remainder of this chapter, we will focus on the LBM as an advection-diffusion solver.

8.3 Lattice Boltzmann for Advection-Diffusion

In this section we show how the LBM can be used to simulate advection-diffusion problems. The governing equations are presented in Sect. 8.3.1. This requires a modified equilibrium distribution function (Sect. 8.3.2) and some comments about the lattice velocities (Sect. 8.3.3). The Chapman-Enskog analysis in Sect. 8.3.4 is similar to the analysis of the standard LBM for fluid dynamics. We conclude the section by mentioning a few extensions of the LB advection-diffusion model (Sect. 8.3.5). Particular usage of LB modelling of thermal flows will be covered in Sect. 8.4.

8.3.1 Similarities of Advection-Diffusion and Navier-Stokes

There are different approaches to construct an LB algorithm for advection-diffusion problems. A straightforward way is to start from the incompressible NSE and rewrite it as

$$\frac{\partial(\rho \boldsymbol{u})}{\partial t} + \nabla \cdot (\rho \boldsymbol{u}\boldsymbol{u} + p\boldsymbol{I}) = \nu \nabla^2 (\rho \boldsymbol{u}) + \boldsymbol{F}. \tag{8.13}$$

The similarity with the ADE in (8.7) becomes obvious when the following substitutions are performed:

$$\rho \boldsymbol{u} \to C, \quad \rho \boldsymbol{u}\boldsymbol{u} + p\boldsymbol{I} \to C\boldsymbol{u}, \quad \nu \to D, \quad \boldsymbol{F} \to q. \tag{8.14}$$

Instead of working with the momentum density vector $\rho \boldsymbol{u}$, we want to have the scalar C as the central observable. The kinematic viscosity ν is replaced by the diffusion coefficient D. There is no analogue for the fluid incompressibility in the advection-diffusion formalism. While the NSE involves two conserved quantities (density and momentum density), there is only one conserved quantity in the ADE (concentration).[1]

[1]Of course "conserved" quantities are only conserved in the absence of source terms.

Exercise 8.1 Show that the NSE is an ADE for the momentum density $\rho \boldsymbol{u}$. What is the meaning of the kinematic viscosity ν and the force density $\boldsymbol{F}$?

Since the functional forms of (8.7) and (8.13) are very similar, the next logical step is to use the same general LB algorithm as before:

$$g_i(\boldsymbol{x} + \boldsymbol{c}_i \Delta t, t + \Delta t) - g_i(\boldsymbol{x}, t) = \Omega_i(\boldsymbol{x}, t) + Q_i(\boldsymbol{x}, t). \tag{8.15}$$

Here, we write g_i to denote populations for the scalar C, Ω_i is the collision operator, and Q_i is responsible for the source term q.

The simplest collision model for advection-diffusion problems is the BGK operator:

$$\Omega_i(\boldsymbol{x}, t) = -\frac{1}{\tau_g} \left(g_i(\boldsymbol{x}, t) - g_i^{\text{eq}}(\boldsymbol{x}, t) \right). \tag{8.16}$$

We write τ_g to distinguish the relaxation time from τ in the standard LBM for the NSE. The application of the LBGK model to ADE problems goes back to works by Flekkoy [3] and Wolf-Gladrow [2] in the early 1990s.

It will be shown in Sect. 8.3.4 that the **diffusion coefficient** D in the BGK model is given by the relaxation time τ_g, similarly to the viscosity in the NSE:

$$D = c_s^2 \left(\tau_g - \frac{\Delta t}{2} \right). \tag{8.17}$$

The speed of sound c_s depends on the chosen lattice (Sect. 8.3.3).

Now we have to find out how to construct the equilibrium populations g_i^{eq} to recover the ADE, rather than the NSE, in the macroscopic limit.

8.3.2 *Equilibrium Distribution*

It would seem reasonable to assume that the zeroth and first moments of g_i obey

$$\sum_i g_i = \sum_i g_i^{\text{eq}}, \quad \sum_i g_i \boldsymbol{c}_i = \sum_i g_i^{\text{eq}} \boldsymbol{c}_i, \tag{8.18}$$

just as in the LBM for the NSE. However, *the second equality does not hold.* As we will see shortly, the momentum in the ADE is *not* conserved by collision.

To satisfy the substitutions in (8.14), we claim

$$\sum_i g_i^{\text{eq}} = C = \sum_i g_i, \tag{8.19a}$$

$$\sum_i g_i^{\text{eq}} \boldsymbol{c}_i = C\boldsymbol{u}. \tag{8.19b}$$

Let us construct an equilibrium in such a way that these moments are satisfied. The flow velocity $\boldsymbol{u}$ is an externally imposed field; it is not found from g_i.

The ADE is linear in C and $\boldsymbol{u}$, so we start with a *linear ansatz* for the equilibrium [3, 4]:

$$g_i^{\text{eq}} = w_i C(A + \boldsymbol{B}_i \cdot \boldsymbol{u}) \tag{8.20}$$

with an unknown scalar A and vectors $\boldsymbol{B}_i$. We have included the lattice weights w_i of the so far unspecified lattice. From (3.60), the weights' velocity moments up to second order are:

$$\sum_i w_i = 1, \quad \sum_i w_i c_{i\alpha} = 0, \quad \sum_i w_i c_{i\alpha} c_{i\beta} = c_s^2 \delta_{\alpha\beta}. \tag{8.21}$$

Inserting (8.20) into (8.19a) yields

$$\sum_i w_i C(A + \boldsymbol{B}_i \cdot \boldsymbol{u}) = C. \tag{8.22}$$

Since the right-hand side of this equation does no depend on $\boldsymbol{u}$, the vector $\boldsymbol{B}$ has to satisfy $\sum_i w_i B_{i\alpha} = 0$. We find $\sum_i w_i CA = C$ or, exploiting equation (8.21), $A = 1$. Combining (8.20) with (8.19b) gives

$$\sum_i w_i C u_\beta B_{i\beta} c_{i\alpha} = C u_\alpha. \tag{8.23}$$

From this and (8.21) we can infer $B_{i\alpha} = c_{i\alpha}/c_s^2$.

Thus, the simplest **equilibrium distribution** that leads to the **ADE** is the linear function

$$g_i^{\text{eq}} = w_i C \left(1 + \frac{\boldsymbol{c}_i \cdot \boldsymbol{u}}{c_s^2}\right) \tag{8.24}$$

(continued)

where the concentration C is obtained from

$$C = \sum_i g_i. \tag{8.25}$$

The velocity $\boldsymbol{u}$ is an external field that has to be provided. It could be (but does not have to be) the solution of the NSE. As the velocity $\boldsymbol{u}$ is not computed from the populations g_i, the ADE scheme does not conserve momentum.

The most commonly used lattices are the same as for fluid dynamics: D2Q9, D3Q15 or D3Q19, all with $c_s^2 = (1/3)\Delta x^2/\Delta t^2$; but reduced lattices are also possible (cf. Sect. 8.3.3).

We also could have obtained g_i^{eq} through a Hermite series expansion, as in Chap. 3. Since only the concentration C is conserved, the expansion should be at least *first* order in the Hermite polynomials, and therefore (8.24) is sufficient.

In general, the NSE and the ADE may have to be solved simultaneously. This is particularly important for thermal flows, such as evaporation or natural convection. We will discuss this in more detail in Sect. 8.4. For now we assume that the advection velocity $\boldsymbol{u}$ is given and not affected by the ADE.

Although the equilibrium is linear in C, it does not have to be linear in the velocity $\boldsymbol{u}$. An alternative quadratic equilibrium assumes the form [5, 6]

$$g_i^{\text{eq}} = w_i C\left(1 + \frac{\boldsymbol{c}_i \cdot \boldsymbol{u}}{c_s^2} + \frac{(\boldsymbol{c}_i \cdot \boldsymbol{u})^2}{2c_s^4} - \frac{\boldsymbol{u} \cdot \boldsymbol{u}}{2c_s^2}\right). \tag{8.26}$$

The choice of the equilibrium distribution has some subtle effects on the accuracy and convergence. We will briefly discuss this in Sect. 8.3.4.

Exercise 8.2 Show that the non-linear equilibrium in (8.26) and the linear equilibrium in (8.24) both fulfil the moments in (8.19). (This requires using the third-order lattice isotropy condition $\sum_i w_i c_{i\alpha} c_{i\beta} c_{i\gamma} = 0$ from (3.60).)

In many applications, **two LBEs** are **solved side by side**: one for the momentum and therefore the velocity $\boldsymbol{u}$, and the other for the field C. This is particularly straightforward if C is purely *passive*, i.e. if it is merely dragged along by the fluid, without affecting the NSE in return. This assumption is typically justified if C indicates a dilute chemical species where collisions between majority and minority components can be neglected [4]. If C is used for the temperature T, however, and the fluid density ρ is a function of T, then the NSE itself depends on the dynamics of C. A fully coupled system of equations has to be solved then. We will describe this in more detail in Sect. 8.4.

8.3.3 Lattice Vectors

The major difference between LBM for the NSE and the ADE is that the former requires velocity moments up to the second order, but the latter requires only the zeroth and first moments. We can therefore assume that we can get away with a lower-isotropy lattice when we are only interested in the ADE. Indeed it has been known for some time now that a square lattice is sufficient to resolve diffusion phenomena in 2D [7].

Although D2Q9 and D3Q15/D3Q19 are often employed for the ADE, it is sufficient to use D2Q5 and D3Q7 instead.[2] The D2Q5 lattice is defined by

$$\boldsymbol{c}_0 = \begin{pmatrix} 0 \\ 0 \end{pmatrix}, \quad \boldsymbol{c}_1 = \begin{pmatrix} 1 \\ 0 \end{pmatrix}, \quad \boldsymbol{c}_2 = \begin{pmatrix} 0 \\ 1 \end{pmatrix}, \quad \boldsymbol{c}_3 = \begin{pmatrix} -1 \\ 0 \end{pmatrix}, \quad \boldsymbol{c}_4 = \begin{pmatrix} 0 \\ -1 \end{pmatrix}. \tag{8.27}$$

Accordingly, D3Q7 has a rest velocity $\boldsymbol{c}_0$ and six velocities along the main lattice axes, such as $(1, 0, 0)^\top$ and $(0, 1, 0)^\top$. More details about those reduced lattices and their role in LBM for diffusion problems are given in [9–11].

It is even possible to drop the rest velocity and simulate diffusion on a D2Q4 lattice. According to Huang et al. [12], the accuracy can be comparable to D2Q5 and D2Q9, while the main difference is the numerical stability [10].

8.3.4 Chapman-Enskog Analysis

We will now discuss the relevant steps of the Chapman-Enskog analysis. The aim is to show that the BGK equation

$$g_i(\boldsymbol{x} + \boldsymbol{c}_i\Delta t, t + \Delta t) = g_i(\boldsymbol{x}, t) - \frac{\Delta t}{\tau_g}\left(g_i(\boldsymbol{x}, t) - g_i^{\text{eq}}(\boldsymbol{x}, t)\right) \tag{8.28}$$

indeed recovers the ADE. For simplicity, we show this using the linear ADE equilibrium of (8.24). We will also show how the relaxation time τ_g and the diffusion coefficient D are connected. Finally, we point out the existence of an error term that can reduce the convergence of the ADE to first order.

---

[2] Note that D2Q5 and D3Q7 are not sufficient for problems involving anisotropic diffusion with non-zero off-diagonal coefficients [8].

8.3.4.1 Analysis Procedure

The Chapman-Enskog analysis of (8.28) is a little simpler than the corresponding analysis of the standard LBE, shown in Sect. 4.1. The broad strokes are the same, and we refer to Sect. 4.1 for details on the parts that are identical.

One difference between the two derivations lies in the moments of g_i. The zeroth to second equilibrium moments are

$$\sum_i g_i^{\text{eq}} = C, \quad \sum_i c_{i\alpha} g_i^{\text{eq}} = C u_\alpha, \quad \sum_i c_{i\alpha} c_{i\beta} g_i^{\text{eq}} = c_s^2 C \delta_{\alpha\beta}, \tag{8.29}$$

where the two first equalities have been established earlier in this chapter, and the third follows from the ADE equilibrium in (8.24) and the isotropy conditions in (3.60). Since concentration is conserved in collisions, we have that $\sum_i g_i = \sum_i g_i^{\text{eq}} = C$. This leads us to assume a strengthened solvability condition analogous to (4.4), namely that $\sum_i g_i^{(n)} = 0$ for all $n \geq 1$. However, unlike momentum in the standard LBE, $\sum_i c_{i\alpha} g_i$ is *not* conserved in collisions, and we can therefore *not* assume that $\sum_i c_{i\alpha} g_i^{(n)} = 0$ for any $n \geq 1$.

The first part of the analysis proceeds exactly as in Sect. 4.1. The Taylor expansion of (8.28) leads to an analogue of (4.7), namely

$$\Delta t \left(\partial_t + c_{i\alpha}\partial_\alpha\right) g_i = -\frac{\Delta t}{\tau_g} g_i^{\text{neq}} + \Delta t \left(\partial_t + c_{i\alpha}\partial_\alpha\right) \frac{\Delta t}{2\tau_g} g_i^{\text{neq}}. \tag{8.30}$$

Expanding g_i and the derivatives in ϵ and separating the result into terms of different order, we find

$$O(\epsilon): \qquad \left(\partial_t^{(1)} + c_{i\alpha}\partial_\alpha^{(1)}\right) g_i^{\text{eq}} = -\frac{1}{\tau_g} g_i^{(1)}, \tag{8.31a}$$

$$O(\epsilon^2): \quad \partial_t^{(2)} g_i^{\text{eq}} + \left(\partial_t^{(1)} + c_{i\alpha}\partial_\alpha^{(1)}\right)\left(1 - \frac{\Delta t}{2\tau_g}\right) g_i^{(1)} = -\frac{1}{\tau_g} g_i^{(2)}. \tag{8.31b}$$

The zeroth and first moments of (8.31a) are

$$\partial_t^{(1)} C + \partial_\gamma^{(1)}(C u_\gamma) = 0, \tag{8.32a}$$

$$\partial_t^{(1)}(C u_\alpha) + c_s^2 \partial_\alpha C = -\frac{1}{\tau_g}\sum_i c_{i\alpha} g_i^{(1)}. \tag{8.32b}$$

As we discussed above, the right-hand side of (8.32b) does not disappear like it would in the analysis of the standard LBE where momentum is conserved. However, from (8.32b) we can find $\sum_i c_{i\alpha} g_i^{(1)}$ explicitly as

$$\sum_i c_{i\alpha} g_i^{(1)} = -\tau_g \left[\partial_t^{(1)}(Cu_\alpha) + c_s^2 \partial_\alpha C \right]. \tag{8.33}$$

The zeroth moment of (8.31b) is

$$\partial_t^{(2)} C + \partial_\gamma^{(1)} \left(1 - \frac{\Delta t}{2\tau_g}\right) \sum_i c_{i\gamma} g_i^{(1)} = 0. \tag{8.34}$$

Combining this equation with (8.32a), inserting for $\sum_i c_{i\gamma} g_i^{(1)}$ from (8.33), and reversing the derivative expansion, we finally find the ADE equation

$$\partial_t C + \partial_\gamma (Cu_\gamma) = \partial_\gamma \left(D \partial_\gamma C\right) + E, \tag{8.35a}$$

with a diffusion coefficient

$$D = c_s^2 \left(\tau_g - \frac{\Delta t}{2}\right). \tag{8.35b}$$

However, the equation also contains an error term

$$E = \partial_\gamma \left(\tau_g - \frac{\Delta t}{2}\right) \partial_t (Cu_\gamma), \tag{8.35c}$$

which cannot always be neglected.

It is possible to vary D locally by changing τ_g according to (8.35b) at each $\boldsymbol{x}$. Note, however, that τ_g should not be varied too much if the BGK collision operator is used. There exist model extensions that make larger variations of D possible, though. We will briefly summarise them in Sect. 8.3.5.

8.3.4.2 Error Term

We have seen that, up to $O(\Delta t^2)$ and $O(\epsilon^2)$, we encounter an error term E shown in (8.35c). This term interferes with the second-order convergence of the ADE based on the LBM.

Chopard et al. [13] showed that the error term can be rewritten as

$$E = -\frac{D}{c_s^2} u^2 \nabla^2 C \tag{8.36}$$

when the linear equilibrium, (8.24), is taken. As it has the same functional form as the governing diffusion term, we can write an effective velocity-dependent diffusion term $D(1 - u^2/c_s^2)\nabla^2 C$. This is clearly undesirable as we would expect accurate results only when $\mathrm{Ma}^2 \ll 1$.

For the quadratic equilibrium, (8.26), however, the error term assumes the form [13]

$$E = -\frac{D}{c_s^2}\partial_\alpha\left(\frac{C}{\rho_f}\partial_\alpha p_f\right) \tag{8.37}$$

if we assume that the external velocity field obeys the incompressible NSE. Here, p_f and ρ_f are the pressure and density of the ambient fluid. This error is velocity-independent.

We conclude that the **quadratic equilibrium** in (8.26) should be **preferred** over the linear equilibrium to avoid a velocity-dependent diffusivity.

Several improvements have been proposed to remove the error term in the first place. A common approach is to add an artificial source term that cancels the error [13, 14], but we will not discuss this further here.

8.3.5 Model Extensions

We point out a few of the various extensions for LB-based advection-diffusion modelling. The recent review paper by Karlin et al. [15] is a valuable source of further references. We will discuss thermal flows in more detail in Sect. 8.4.

8.3.5.1 Source Term

We have so far omitted a discussion of the source term q in (8.6). This term is important when reactions consume or produce chemical species (e.g. reactive transport in porous media [16], pattern formation [17, 18], the Keller-Segel chemotaxis model [19] or catalytic reactions [20]). Other systems that usually require a source term are thermal flows with heat sources (e.g. due to viscous dissipation) as further explained in Sect. 8.4.

To include a source q in the LB algorithm in (8.15), we define population sources

$$Q_i = w_i q. \tag{8.38}$$

Exercise 8.3 Show that $\sum_i Q_i = q$ and $\sum_i Q_i \boldsymbol{c}_i = \boldsymbol{0}$. This means that Q_i represents a concentration source but leaves the "momentum" $C\boldsymbol{u}$ unchanged.

Performing a Chapman-Enskog analysis with the additional terms Q_i, we find that the macroscopic ADE becomes [21]

$$\frac{\partial C}{\partial t} + \nabla \cdot (C\boldsymbol{u}) = \nabla \cdot (D\nabla C) + q - \frac{\Delta t}{2}\frac{\partial q}{\partial t}. \tag{8.39}$$

When we compare this result with (8.6), we see that there is obviously an unwanted term $(\Delta t/2)\partial q/\partial t$. This is a discrete lattice artefact, similar to that observed for forces in Sect. 6.3.

To remove this unphysical term, we can redefine the macroscopic moments according to [21]

$$C = \sum_i g_i + \frac{Q_i \Delta t}{2}, \quad Q_i = \left(1 - \frac{1}{2\tau_g}\right) w_i q. \tag{8.40}$$

This is completely analogous to the procedure in Sect. 6.3.2 to remove the discrete lattice artefact of the forcing scheme and reach second-order accuracy in time. A different approach is to add another term $\propto \partial C/\partial t$ to the evolution equation in (8.15) [22].

8.3.5.2 Advanced Physical Models

As mentioned earlier, the standard ADE involves a scalar diffusion coefficient D, but for more general applications it may be necessary to replace it by a diffusion tensor $\boldsymbol{D}$. This makes anisotropic and cross-diffusion possible: the former is caused by different diagonal elements of $\boldsymbol{D}$, the latter by non-zero off-diagonal elements of $\boldsymbol{D}$. Anisotropic diffusion is thoroughly discussed in various publications, such as [8, 23–26].

Like the standard LBM for fluid dynamics problems, LBM for the ADE can be extended to an axisymmetric formulation [27] or rectangular lattices [28]. Huang et al. [29] recently presented a multi-block approach for the thermal LBM. Li et al. [30] recently presented an advanced ADE solver for problems with variable coefficients.

Another important application is the simulation of *phase fields* for multi-phase or multi-component problems. We will get back to this in more depth in Sect. 9.2.

It is possible to modify the diffusion coefficient D not only through τ_g, but also through the choice of the equilibrium distribution [8, 10]. This allows better flexibility in terms of stability and range of achievable diffusion coefficients.

8.3.5.3 Stability Improvements and Advanced Collision Operators

The restrictions of the BGK collision operator, in particular stability (Sect. 4.4) and accuracy (Sect. 4.5), also apply to the ADE. This means that the relaxation time τ_g cannot be varied arbitrarily; values near $1/2$ lead to instability, values significantly larger than unity to reduced accuracy. It is therefore not possible to choose very small or large values for the Péclet, Prandtl and Schmidt numbers.

Suga [9, 24] performed stability analyses of LBM for the ADE. Perko and Patel [31] proposed a way to realise a large spatial variation of the diffusion coefficient within the BGK model. Yang et al. [32] improved the BGK stability by adding correction terms.

In recent years, advanced collision operators (particularly TRT and MRT, as detailed in Chap. 10) have become popular for ADE applications. Thorough analyses have been done of the TRT collision operator [10, 33, 34] and the more general MRT operator [25, 35, 36] applied to ADE problems.

8.4 Thermal Flows

The ADE model introduced in Sect. 8.3.1 is often used for situations where the field C does not affect the fluid dynamics. This means that on the one hand, the velocity $\boldsymbol{u}$ in (8.7) is given by a Navier-Stokes solver (which could be anything, e.g. LBM), an analytical solution or any otherwise defined flow field. On the other hand, since C does not enter the NSE, the fluid dynamics are not affected by the dynamics of C at all. In a one-way situation like this, C is called a *passive* field.

This simplification is often not applicable for thermal flows where the fluid dynamics can be strongly affected by the temperature field. Examples are evaporation and boiling and more general situations where the fluid density or viscosity change with temperature (cf. Sect. 1.1.3 for a discussion of the ideal gas law). Here we discuss the fully coupled situation where fluid dynamics and temperature affect each other.

We begin by introducing the Boussinesq approximation and briefly discussing the popular Rayleigh-Bénard convection in Sect. 8.4.1. In Sect. 8.4.2 we comment on the non-dimensionalisation in the presence of temperature. Section 8.4.3 deals with fully coupled thermal flows with energy conservation, i.e. viscous dissipation and pressure work act as heat source in the energy equation. In many cases, energy conservation is not required, and a simpler model can be used to model the coupled momentum and temperature equations (Sect. 8.4.4).

8.4.1 Boussinesq Approximation and Rayleigh-Bénard Convection

A relevant problem in nature and industry is the coupled dynamics of momentum and advection-diffusion in a thermal flow with temperature-dependent density. One of the best known problems is the *Rayleigh-Bénard convection*. This flow is well understood [37] and often used for code benchmarks, e.g. [4–6].

The geometrical setup for the Rayleigh-Bénard convection involves two parallel plates separated by a distance H. The bottom plate is kept at a higher temperature than the top plate (using, for instance, Dirichlet temperature boundary conditions as in Sect. 8.5.2 or [5]): $T_\mathrm{b} > T_\mathrm{t}$. At the same time, both plates are subject to the no-slip condition for the fluid momentum.

Now consider a fluid with a thermal expansion coefficient at constant pressure p:

$$\alpha = -\frac{\rho_0}{\rho^2}\left(\frac{\partial \rho}{\partial T}\right)_p . \tag{8.41}$$

The density depends on temperature, $\rho = \rho(T)$, and ρ_0 is the density for a reference temperature T_0: $\rho_0 = \rho(T_0)$. For a positive expansion coefficient α, the density decreases (i.e. the volume increases) with increasing temperature, just as we would expect it from most fluids, such as an ideal gas.

A typical but not necessary assumption is that the temperature and density changes are small so that we can linearise $\rho(T)$ about ρ_0 and T_0 and obtain

$$\rho(T) \approx \rho_0\left[1 - \alpha(T - T_0)\right]. \tag{8.42}$$

For larger temperature ranges it is necessary to use a more accurate description of $\rho(T)$, e.g. by introducing a temperature-dependent expansion coefficient $\alpha(T)$.

Let us turn our attention back to the Rayleigh-Bénard convection. In the presence of gravity with acceleration g, the fluid at the bottom plate will heat up, leading to a decrease of its density (if $\alpha > 0$) and therefore a buoyancy force. There exists a stationary solution to this problem with zero velocity everywhere. This state is called *conductive* because heat is only transported from the hot to the cold plate by conduction. However, if the system is perturbed (e.g. by adding small random momentum or temperature fluctuations) and the temperature gradient is sufficiently large, parts of the heated fluid will move upwards while colder fluid from the top will move down. This will eventually lead to convection.

The physics of the Rayleigh-Bénard convection is governed by the *Rayleigh number*

$$\mathrm{Ra} = \frac{g\alpha(T_\mathrm{b} - T_\mathrm{t})H^3}{\kappa\nu}, \tag{8.43}$$

the ratio of temperature-driven buoyancy and viscous friction forces. κ and ν are the thermal diffusivity and kinematic viscosity of the fluid, respectively. For $\mathrm{Ra} < 1708$, perturbations are dissipated by viscosity [38]. If Ra becomes larger, bouyancy can overcome dissipation, and convection sets in. Finding the critical Ra in simulations and comparing it with the above value from a stability analysis is a standard benchmark test.

We can model the advection and diffusion of the temperature field $T(\boldsymbol{x})$ with the LBM for the ADE as outlined in Sect. 8.3. The local advection velocity $\boldsymbol{u}(\boldsymbol{x})$ is then given by the fluid velocity that we can obtain from the LBM for the NSE.

This establishes the flow-to-temperature coupling. However, which effect does the temperature have on the flow?

Now we perform another approximation to couple the temperature back to the NSE. Above we have seen that small temperature changes lead to density fluctuations according to (8.42).

The **Boussinesq approximation** states that the effect of a small density change ($|\rho'|/\rho_0 = |\rho - \rho_0|/\rho_0 \ll 1$) creates a **buoyancy** force density

$$\boldsymbol{F}_b = (\rho(T) - \rho_0)\boldsymbol{g} = -\alpha\rho_0(T - T_0)\boldsymbol{g} \tag{8.44}$$

in the presence of a gravitational field with acceleration $\boldsymbol{g}$.

The buoyancy force vanishes for $\rho = \rho_0$, i.e. it describes only the contribution to the gravitational force that deviates from the "background" force $\rho_0\boldsymbol{g}$. The physical motivation of the Boussinesq approximation is that the effect of the density difference ρ' on inertia is much smaller than the buoyancy force $\rho'\boldsymbol{g}$. This justifies keeping the original density ρ_0 everywhere in the NSE, except for the buoyancy force that depends on the density *fluctuation* ρ'.

The advantage of using the Boussinesq approximation is that the temperature effect enters *only* through the body force density in (8.44). It is included like any other body force density in the NSE (cf. Chap. 6); the fluid density in the NSE itself is kept unchanged. This way we avoid fiddling around with the fluid density in the LBM, which could potentially lead to problems with mass conservation. Any occuring density fluctuation in the fluid can still be interpreted as pressure fluctuation as outlined in Sect. 7.3.3.

8.4.2 Non-dimensionalisation of the Temperature Field

Before we provide specific details about the LB-based modelling of thermal flows, we briefly discuss how to convert between physical and simulation units when temperature is important. In Chap. 7 we only considered mechanical quantities, such as velocity or pressure; temperature is a non-mechanical quantity.

The units of T and α are K and $1/\text{K}$, respectively. How can we now perform the unit conversion? It is important to realise that, for the Boussinesq approximation in Sect. 8.4.1, only the dimensionless combination $\alpha(T - T_0)$ is relevant and α and T are never converted independently. One should rather express all temperature-dependent terms as function of $\alpha(T-T_0)$. This has to be a small number to guarantee that the linear relation in (8.42) is justified. We therefore recommend to choose T_0 as the characteristic temperature of the system, for example its average temperature.

Having said this, the unit conversion does not pose any challenge. We only have to make sure that the Boussinesq approximation is valid. This requires an estimation of the typical magnitude of $\alpha(T - T_0)$ and consequently of the expected density variation ρ'.

Example 8.1 Let us consider water near $T_0 = 20\,°\mathrm{C}$ (293 K) where $\alpha = 2.1 \cdot 10^{-4}/\mathrm{K}$. We find that $\alpha(T - T_0)$ is small for all temperatures where water is in its liquid state. Note, however, that α varies significantly between the freezing and boiling points of water. Even if the Boussinesq approximation is valid, a non-linear relation $\rho(T)$ is required if the water temperature changes more than a few degrees.

8.4.3 *LBM for Thermal Flows with Energy Conservation*

In reality, fluids are always slightly compressible and dissipate kinetic energy due to viscous friction. This leads to compression work and viscous heating, and both mechanisms act as heat source or sink in the energy equation. In this case, we have to consider the compressible NSE

$$\partial_t(\rho u_\alpha) + \partial_\beta(\rho u_\alpha u_\beta) = -\partial_\alpha p + \partial_\beta \left[\eta \left(\partial_\beta u_\alpha + \partial_\alpha u_\beta \right) + \left(\eta_\mathrm{B} - \frac{2\eta}{3} \right) \delta_{\alpha\beta} \partial_\gamma u_\gamma \right] \tag{8.45}$$

and the energy equation

$$\partial_t \epsilon + \partial_\alpha(\epsilon u_\alpha) = \partial_\alpha(\kappa \partial_\alpha \epsilon) + \sigma_{\alpha\beta}\partial_\alpha u_\beta - p\partial_\alpha u_\alpha \tag{8.46}$$

which is the ADE for the internal energy density ϵ with two source terms. The first, $\sigma_{\alpha\beta}\partial_\alpha u_\beta$, reflects the heating due to viscous dissipation. The second source term, $-p\partial_\alpha u_\alpha$, is the pressure work that can be positive or negative, depending on whether the fluid is compressed or expanded. For (nearly) incompressible fluids we have $\partial_\alpha u_\alpha \approx 0$, and only the viscous heating is relevant.

To simultaneously solve the above equations with the LBM there are essentially two approaches that we will briefly present in the following.

8.4.3.1 Single-Population Model

From Sect. 1.3.5 we know that the moments of the distribution function $f(\boldsymbol{x}, \boldsymbol{\xi}, t)$ provide density, momentum density and energy density. Therefore, it has been suggested during the early years of LBM to use a *single set* of discrete populations f_i to recover the NSE *and* the energy equation [39, 40].

Recovering the correct energy equation requires third- and fourth-order velocity moments of the distribution function. We saw in Sect. 3.4.7 that the standard lattices that are used for the NSE (e.g. D2Q9 or D3Q19) do not have enough velocity components to correctly resolve velocity moments higher than second order. This means that we have to consider higher-order lattices, such as D3Q21 [41].[3]

The advantage of the single-population model is that the coupling between the NSE and the energy equation is automatically included; viscous dissipation and compression work emerge naturally and consistently from the mesoscopic model. Yet, single-population models in 3D turned out to be relatively unstable and suffer from a constant Prandtl number when using the BGK collision operator [42]. This is the main reason why they are not often employed today, and the two-population model (see below) has become more popular [15].

In section 5.5 of his book [43], Wolf-Gladrow provides a concise review of the single-population model and extensions to improve stability and allow a variable Prandtl number. See also [15] for a collection of more recent works improving the single-population model.

8.4.3.2 Two-Population Model

Another idea is to use two sets of populations, one for the momentum and one for the internal energy [6, 15, 44]. The main advantages of such an approach is the improved stability compared to the single-population model and that standard lattices (such as D2Q9 and D3Q19) are sufficient. However, if viscous heating and compression work are relevant and energy conservation is required, we have to include both as source terms in the energy equation.

He et al. [6] proposed a consistent two-population model with energy conservation. The governing mesoscopic equations for the temperature populations g_i are given by (8.15) with collision operator in (8.16) and the source term [6]

$$Q_i(\boldsymbol{x}, t) = -\left(1 - \frac{1}{2\tau_g}\right) f_i(\boldsymbol{x}, t) q_i(\boldsymbol{x}, t) \Delta t \tag{8.47}$$

with

$$q_i = (c_{i\alpha} - u_\alpha)\left[\frac{1}{\rho}(-\partial_\alpha p + \partial_\beta \sigma_{\alpha\beta}) + (c_{i\beta} - u_\beta)\partial_\alpha u_\beta\right]. \tag{8.48}$$

This specific form of the source term guarantees that the local heating in the energy equation is consistent with the dissipation and pressure work in the momentum

[3] D3Q21 is the D3Q15 lattice with six additional vectors $(\pm 2, 0, 0)^\top$, $(0, \pm 2, 0)^\top$ and $(0, 0, \pm 2)^\top$.

equation. Furthermore, the energy density becomes

$$\epsilon = \sum_i g_i - \frac{\Delta t}{2} \sum_i f_i q_i \tag{8.49}$$

to recover second-order time accuracy (cf. (8.40)). Note that the form of the equilibrium distribution function g_i^{eq} in [6] differs from those discussed in Sect. 8.3.2.

The original model [6] has two disadvantages: it is relatively complicated and lacks locality since the computation of the coupling terms q_i involves gradients. More recently Guo et al. [45] and Karlin et al. [15] reviewed and improved He's model [6] by replacing the conservation equation for the thermal energy by the conservation equation for the total energy. This subtle change of variables leads to a simplified algorithm with local coupling. Also the consideration of enthalpy has been proposed as an alternative simplification strategy [46].

8.4.4 *LBM for Thermal Flows Without Energy Conservation*

In many situations, the work from viscous dissipation and compression is so small that it does not significantly contribute to the heat balance. It is then sufficient to consider an ADE without heat source terms (unless there are other mechanisms that reduce or increase heat locally), together with the incompressible NSE. The governing equations (8.45) and (8.46) then become

$$\partial_t(\rho u_\alpha) + \partial_\beta(\rho u_\alpha u_\beta) = -\partial_\alpha p + \partial_\beta \eta \left(\partial_\beta u_\alpha + \partial_\alpha u_\beta\right) \tag{8.50}$$

and

$$\partial_t \epsilon + \partial_\alpha(\epsilon u_\alpha) = \partial_\alpha(\kappa \partial_\alpha \epsilon). \tag{8.51}$$

Note that the internal energy density and temperature are related by $\epsilon = dRT/2$ [47] (with d being the number of spatial dimensions and R the specific gas constant), and (8.51) can therefore be simply written as an ADE for temperature with the same transport coefficient κ. This means that we can use the simple algorithm from Sect. 8.3 to simulate the advection and diffusion of the temperature field.

Using the Boussinesq approximation from Sect. 8.4.1, we now explain the algorithm for the coupled dynamics of the fluid momentum and the temperature field (see also [48]):

1. Initialise the system with two sets of populations, f_i for the NSE and g_i for the ADE of the temperature field, (8.8). Specify gravity $\mathbf{g}$, the fluid viscosity η, the thermal conductivity κ and the thermal expansion coefficient α as desired.

Define a reference temperature T_0, taking Sect. 8.4.2 into account. If required, implement boundary conditions for the NSE and the ADE as specified in Chap. 5 and Sect. 8.5, respectively.

2. Using the fluid velocity $\boldsymbol{u}$ as external velocity, perform one time step of the LB algorithm for the populations g_i as detailed in Sect. 8.3. In particular, evaluate the temperature $T(\boldsymbol{x})$ from (8.25).
3. Compute the buoyancy force density $\boldsymbol{F}_\text{b}(\boldsymbol{x})$ from (8.44).
4. Perform one time step of the standard LB algorithm (cf. Chap. 3) for the populations f_i. Use the buoyancy force $\boldsymbol{F}_\text{b}(\boldsymbol{x})$ to drive the flow (cf. Chap. 6). Evaluate the new fluid velocity $\boldsymbol{u}$.
5. Go back to step 2 for the next time step.

8.5 Boundary Conditions

In this section we provide a concise presentation of the most common boundary conditions for the ADE. The boundary conditions are similar to those for the NSE in Chap. 5, but there is an important difference between the normal and tangential behaviour as outlined in Sect. 8.5.1. Here, we cover Dirichlet (Sect. 8.5.2) and Neumann (Sect. 8.5.3) boundary conditions. Robin boundary conditions, i.e. linear combinations of Dirichlet and Neumann conditions, are omitted here, but details can be found elsewhere [49]. A comparison and discussion of various boundary conditions for the ADE is available in [12, 49]. More recently, the immersed boundary method has been employed for advection-diffusion boundaries (see, e.g., [21]).

8.5.1 Normal and Tangential Conditions

Before we start with the presentation of boundary conditions, it is helpful to understand some underlying concepts better. Let $\boldsymbol{n}$ be the normal vector at a boundary point. We define $\boldsymbol{n}$ in such a way that it points into the fluid. If a field C cannot penetrate a boundary, then there is no flux $\boldsymbol{j}$ of C through the boundary and we call the boundary *impermeable* to C. This, however, is only a condition for the normal component of $\boldsymbol{j}$ and not its tangential component.

The **normal flux at a boundary,** $\boldsymbol{j} \cdot \boldsymbol{n}$, is related to a normal gradient of C, $\partial C/\partial n$:

$$D\frac{\partial C}{\partial n} = -\boldsymbol{j} \cdot \boldsymbol{n}. \tag{8.52}$$

(continued)

For an impermeable wall, $\boldsymbol{j} \cdot \boldsymbol{n} = 0$, and therefore $\partial C / \partial n = 0$. It is still possible that the value of C changes along the boundary, though. Unlike in fluid dynamics where usually both the relative normal *and* tangential velocities between fluid and boundary are zero, we can only impose the normal behaviour of C at a boundary. The **tangential flux** is a pure consequence of the gradient of C along the boundary.

Example 8.2 Imagine a quiescent fluid with an initially inhomogeneous concentration $C(\boldsymbol{x})$ in a container. This could be sugar in a cup of hot but unstirred coffee. According to the ADE in (8.7), diffusion fluxes will lead to a homogenisation of C eventually. This is also true in the direct vicinity of the wall, where the diffusion flux can only be tangential. The sugar can diffuse along the surface of the cup but can never penetrate it.

Fluxes are also often specified at inlets and outlets, leading to positive (inward) and negative (outward) fluxes, respectively. As we will show below, the normal and tangential conditions can be implemented through Dirichlet or Neumann boundary conditions.

8.5.2 Dirichlet Boundary Conditions

Dirichlet boundary conditions specify the value of a field at a wall, e.g. a concentration $C = C_\mathrm{w}$ or temperature $T = T_\mathrm{w}$. One example for a Dirichlet boundary condition in fluid dynamics is a pressure boundary condition. Therefore, the non-equilibrium bounce-back (Zou-He) scheme from Sect. 5.3.4 could be employed to simulate a constant concentration in the ADE. A disadvantage of this approach is the rather complicated implementation in 3D. We focus on two alternatives that are suitable for arbitrary geometries with staircase approximation: the *anti-bounce-back scheme* and *Inamuro's boundary condition*. We show a numerical example using both boundary conditions in Sect. 8.6.2.

8.5.2.1 Anti-Bounce-Back Scheme

The anti-bounce-back scheme [23] (cf. Sect. 5.3.5 for its hydrodynamic counterpart) is nearly as simple as the standard bounce-back method. Let $\boldsymbol{x}_\mathrm{b}$ be a fluid node close to a wall (i.e. a boundary node) and $\boldsymbol{x}_\mathrm{w}$ be the intersection of the wall and a lattice link $\boldsymbol{c}_i$ (cf. Fig. 8.2). For bounce-back-like schemes, the standard implementation relies on the assumption that the surface is located half-way between a boundary and a solid node, so we have $\boldsymbol{x}_\mathrm{w} = \boldsymbol{x}_\mathrm{b} + \frac{1}{2}\boldsymbol{c}_i \Delta t$. The anti-bounce-back algorithm

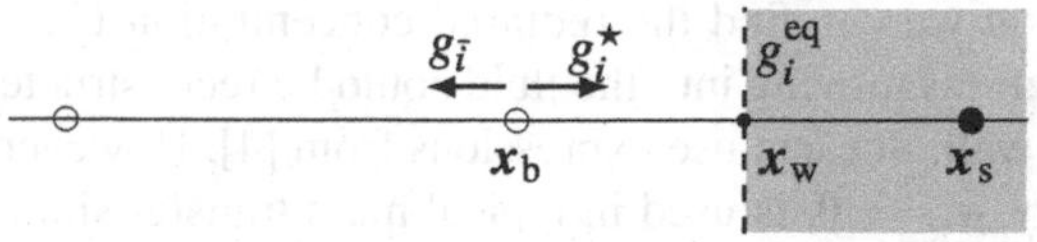

Fig. 8.2 Schematic representation of the anti-bounce-back rule. $\boldsymbol{x}_\text{b}$ and $\boldsymbol{x}_\text{s}$ denote a boundary and a solid node, respectively. The wall is located half-way between both nodes at $\boldsymbol{x}_\text{w}$

reads

$$g_{\bar{i}}(\boldsymbol{x}_\text{b}, t + \Delta t) = -g_i^\star(\boldsymbol{x}_\text{b}, t) + 2g_i^{\text{eq}}(\boldsymbol{x}_\text{w}, t + \Delta t) \tag{8.53}$$

where the superscript $\star$ denotes post-collision populations. In contrast to the usual bounce-back rule that recovers a zero-velocity condition, here we observe a minus sign. The anti-bounce-back algorithm recovers the zero-gradient condition for the concentration.

If the wall has zero velocity and the imposed wall concentation is C_w, then the anti-bounce-back condition simplifies to

$$g_{\bar{i}}(\boldsymbol{x}_\text{b}, t + \Delta t) = -g_i^\star(\boldsymbol{x}_\text{b}, t) + 2w_i C_\text{w}. \tag{8.54}$$

This method works with any arbitrary geometry.

8.5.2.2 Inamuro's Boundary Condition

Inamuro et al. [4, 50] proposed another boundary condition to impose a concentration value. For a wall with normal $\boldsymbol{n}$, unknown populations g_i moving into the fluid obey $\boldsymbol{c}_i \cdot \boldsymbol{n} > 0$. Inamuro et al. assumed that those pre-collision populations can be represented through the equilibrium populations with a yet unknown concentration C',

$$g_i = g_i^{\text{eq}}(C', \boldsymbol{u}_\text{w}), \tag{8.55}$$

where the equilibrium assumes the form of (8.24) or (8.26).

Inamuro's boundary condition is a wet-node model, i.e. the wall nodes are located just inside the fluid domain and participate in collision. This way we can find the unknown value C'. Right after propagation, we know all populations satisfying $\boldsymbol{c}_i \cdot \boldsymbol{n} \leq 0$ at the wall node. Assuming that the desired concentration at a wall node is C_w, we can write this concentration as the sum of known and unknown populations:

$$C_\text{w} = \sum_i g_i = \underbrace{\sum_{\boldsymbol{c}_i \cdot \boldsymbol{n} > 0} g_i}_{\text{unknown}} + \underbrace{\sum_{\boldsymbol{c}_i \cdot \boldsymbol{n} \leq 0} g_i}_{\text{known}} = \sum_{\boldsymbol{c}_i \cdot \boldsymbol{n} > 0} g_i^{\text{eq}}(C', \boldsymbol{u}_\text{w}) + \sum_{\boldsymbol{c}_i \cdot \boldsymbol{n} \leq 0} g_i. \tag{8.56}$$

From this equation we can find the required concentration C'. After C' has been found, the populations moving into the fluid could be reconstructed *via* (8.55).

For a moving wall, one can use expressions from [4]. However, usually the case with zero velocity, $\boldsymbol{u}_\mathrm{w} = \boldsymbol{0}$, is used in typical mass transfer simulations. Then, the result is

$$C' = \frac{C_\mathrm{w} - \sum_{\boldsymbol{c}_i \cdot \boldsymbol{n} \le 0} g_i}{\sum_{\boldsymbol{c}_i \cdot \boldsymbol{n} > 0} w_i}. \tag{8.57}$$

8.5.3 Neumann Boundary Conditions

Neumann boundary conditions specify a normal gradient, e.g. $\partial C/\partial n$, and therefore a normal flux $\boldsymbol{j} \cdot \boldsymbol{n}$ at a wall, according to (8.52). Here, two different procedures are commonly used: one is to directly impose the flux, the other is to transform the Neumann boundary condition into a Dirichlet condition.

8.5.3.1 Inamuro's Flux Boundary Condition

The species flux is computed from the first moment of g_i: $\boldsymbol{j} = \sum_i g_i \boldsymbol{c}_i$. For a given normal $\boldsymbol{n}$, the normal flux is

$$\boldsymbol{j} \cdot \boldsymbol{n} = \sum_i g_i \boldsymbol{c}_i \cdot \boldsymbol{n}. \tag{8.58}$$

We can now repeat Inamuro's approach for the Dirichlet boundary condition in Sect. 8.5.2, but this time for the flux $\boldsymbol{j}$ rather than the concentration C. We still assume that the unknown populations are computed from (8.55) with an unknown C' [50]:

$$\begin{aligned} \boldsymbol{j} \cdot \boldsymbol{n} = \sum_i g_i \boldsymbol{c}_i \cdot \boldsymbol{n} &= \underbrace{\sum_{\boldsymbol{c}_i \cdot \boldsymbol{n} > 0} g_i \boldsymbol{c}_i \cdot \boldsymbol{n}}_{\text{unknown}} + \underbrace{\sum_{\boldsymbol{c}_i \cdot \boldsymbol{n} \le 0} g_i \boldsymbol{c}_i \cdot \boldsymbol{n}}_{\text{known}} \\ &= \sum_{\boldsymbol{c}_i \cdot \boldsymbol{n} > 0} g_i^\mathrm{eq}(C', \boldsymbol{u}_\mathrm{w}) \boldsymbol{c}_i \cdot \boldsymbol{n} + \sum_{\boldsymbol{c}_i \cdot \boldsymbol{n} \le 0} g_i \boldsymbol{c}_i \cdot \boldsymbol{n}. \end{aligned} \tag{8.59}$$

If we again assume a resting wall, we get $g_i^\mathrm{eq} = w_i C'$ and finally

$$C' = \frac{\boldsymbol{j} \cdot \boldsymbol{n} - \sum_{\boldsymbol{c}_i \cdot \boldsymbol{n} \le 0} g_i \boldsymbol{c}_i \cdot \boldsymbol{n}}{\sum_{\boldsymbol{c}_i \cdot \boldsymbol{n} > 0} w_i \boldsymbol{c}_i \cdot \boldsymbol{n}}. \tag{8.60}$$

8.5.3.2 Transformation of Neumann into Dirichlet Boundary Condition

The second procedure is to turn the diffusion flux condition into a concentration condition. For example, for a known normal $\boldsymbol{n}$ the flux at the wall can be approximated by the desired wall concentration C_{w} at $\boldsymbol{x}_{\mathrm{w}}$ and the known concentration C_{b} at a boundary node located at $\boldsymbol{x}_{\mathrm{b}}$ along the normal direction:

$$\boldsymbol{j} \cdot \boldsymbol{n} = D\frac{\partial C}{\partial n} \approx D\frac{C_{\mathrm{b}} - C_{\mathrm{w}}}{|\boldsymbol{x}_{\mathrm{b}} - \boldsymbol{x}_{\mathrm{w}}|}. \tag{8.61}$$

Obviously this procedure becomes more complicated when $\boldsymbol{n}$ is not aligned with the lattice. We refer readers to [23] for imposing Dirichlet and Neumann boundary condition for more complex wall shapes.

We can now estimate the wall concentration:

$$C_{\mathrm{w}} \approx C_{\mathrm{b}} - \frac{\boldsymbol{j} \cdot \boldsymbol{n}|\boldsymbol{x}_{\mathrm{b}} - \boldsymbol{x}_{\mathrm{w}}|}{D}. \tag{8.62}$$

Let us say a few words about the accuracy of this approach. There are two aspects to consider: the implementation accuracy of the Dirichlet boundary condition itself and the approximation of the normal flux through the concentration derivative. The simple approximation in (8.61) will generally suffice for a first-order accurate boundary condition, and a more accurate gradient estimate at the boundary is necessary for a higher-order accuracy.

One widely used application of this scheme is the outflow boundary condition. It is often assumed that the concentration C does not change along the outflow direction at the outlet. This translates to a vanishing diffusion flux and $C_{\mathrm{w}} = C_{\mathrm{b}}$. We can easily implement this condition by evaluating C at the fluid plane just before the outlet and impose those values on the outlet plane. One approach is to copy the post-collision populations to the outlet plane before propagation. We demonstrate this approach in Sect. 8.6.3 and Sect. 8.6.4.

Another important note is that for impermeable walls one can use the simple bounce-back rule because of mass conservation. The non-zero tangential velocity is defined on boundary nodes, while bounce-back implies zero velocity on the middle plane between bounce-back nodes and boundary nodes. Thus, there is no contradiction in using bounce-back to mimic impermeable boundaries. In fact, we employ this approach to successfully simulate a symmetric flow in Sect. 8.6.4.

8.6 Benchmark Problems

We present a series of benchmark tests for the bulk ADE algorithm and its boundary conditions. The aim of this section is two-fold: to show analytical solutions for general benchmark tests and to demonstrate the performance of LB advection-diffusion. We will take a closer look at four different tests: advection-diffusion of a Gaussian hill (cf. Sect. 8.6.1), a more complicated example of diffusion from the walls of a two-dimensional cylinder (cf. Sect. 8.6.2), the concentration layer development in a uniform velocity field (cf. Sect. 8.6.3) and finally the concentration layer development in Poiseuille flow (cf. Sect. 8.6.4).

8.6.1 Advection-Diffusion of a Gaussian Hill

This example shows how a species with concentration C with an initial Gaussian profile develops in the presence of a uniform velocity field $\boldsymbol{u}$. We will observe both advection, due to the non-zero velocity, and diffusion, due to the inhomogeneous concentration.

The initial concentration profile is taken as Gaussian

$$C(\boldsymbol{x}, t = 0) = C_0 \exp\left(-\frac{(\boldsymbol{x} - \boldsymbol{x}_0)^2}{2\sigma_0^2}\right) \tag{8.63}$$

with width σ_0. In the presence of a homogeneous advection velocity $\boldsymbol{u}$, the time evolution has an analytical solution [8]:

$$C(\boldsymbol{x}, t) = \frac{\sigma_0^2}{\sigma_0^2 + \sigma_D^2} C_0 \exp\left(-\frac{(\boldsymbol{x} - \boldsymbol{x}_0 - \boldsymbol{u}t)^2}{2\left(\sigma_0^2 + \sigma_D^2\right)}\right), \quad \sigma_D = \sqrt{2Dt}. \tag{8.64}$$

In order to avoid the treatment of special boundary conditions, we choose periodic boundaries in a large 2D domain with $N_x \times N_y = 512 \times 512$ lattice nodes on a D2Q9 lattice. The initial centre of the Gaussian hill is at $\boldsymbol{x}_0 = (200, 200)^\top \Delta x$, and the initial width of the Gaussian hill is $\sigma_0 = 10\Delta x$. In our simulations, we use $C_0 = 1$ (in lattice units). We employ the BGK and the two-relaxation-time (TRT) collision operators (see Sect. 10.6).

We examine both the diffusion-dominated (small Péclet number) and the advection-dominated (large Péclet number) regimes. For Pe $= 0$ we choose the velocity $\boldsymbol{u} = \boldsymbol{0}$ and the diffusion coefficient $D = 1.5\Delta x^2/\Delta t$, i.e. $\tau_g = 5\Delta t$. Figure 8.3 shows the concentration contour plot and the concentration profile for $y = 200\Delta x$ at $t = 200\Delta t$. Note that we have intentionally defined a rather large relaxation time τ_g that leads to significant errors for the BGK collision operator. We see that the TRT collision operator is superior in terms of accuracy. Here, we have chosen the so-called *magic parameter* $\Lambda = 1/6$ that minimises the TRT error in a pure diffusion case, as detailed in Sect. 10.6. However, this choice of Λ does not cancel the error completely. We see a small deviation from the analytical curve, even for the TRT results.

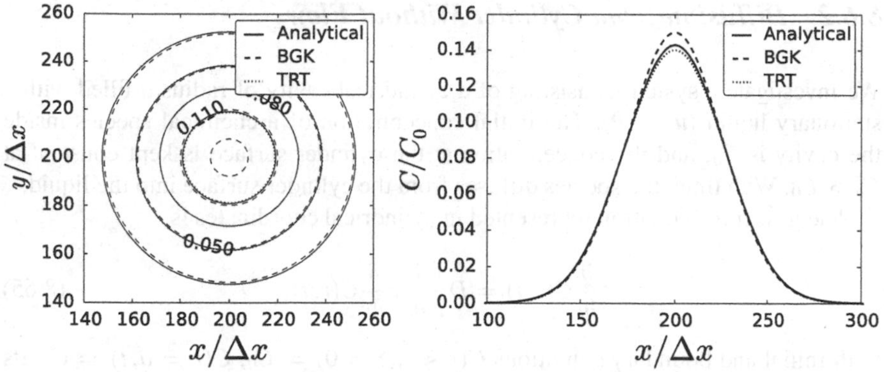

Fig. 8.3 Concentration contour plot (*left*) and concentration profile (*right*) of the Gaussian hill at $t = 200\Delta t$ in the pure diffusion case (zero advection velocity). The diffusion coefficient is $D = 1.5\Delta x^2/\Delta t$, i.e. $\tau_g = 5\Delta t$. The BGK collision operator results show deviations, while the TRT collision operator with $\Lambda = 1/6$ performs better

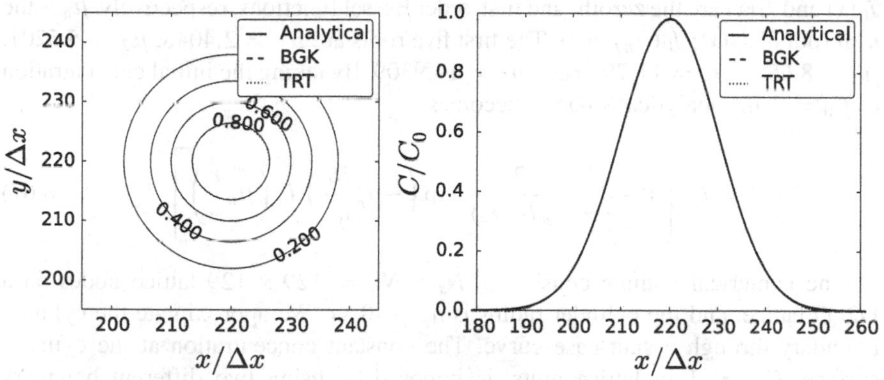

Fig. 8.4 Concentration contour plot (*left*) and concentration profile (*right*) of the Gaussian hill at $t = 200\Delta t$ in the advection-dominated case. The diffusion coefficient is $D = 0.0043\Delta x^2/\Delta t$, i.e. $\tau_g = 0.513\Delta t$, and the uniform velocity is $\boldsymbol{u} = (0.1, 0.1)^\top \Delta x/\Delta t$. The TRT results are obtained with the optimal advection parameter $\Lambda = 1/12$

Let us now investigate the advection-dominated regime with a small diffusion coefficient ($D = 0.0043\Delta x^2/\Delta t$, i.e. $\tau_g = 0.513\Delta t$) and an advection velocity $\boldsymbol{u} = (0.1, 0.1)^\top \Delta x/\Delta t$. As a consequence, we expect the Gaussian hill to essentially keep its initial shape while being advected. Figure 8.4 illustrates the results at $t = 200\Delta t$. Here, the differences between analytical, BGK and TRT collision operators are small. The TRT results are obtained with the magic parameter $\Lambda = 1/12$ for optimal advection accuracy (cf. Sect. 4.5 and Sect. 10.6).

8.6.2 Diffusion from Cylinder Without Flow

We investigate a system consisting of a cylindrical cavity of radius a filled with a stationary liquid ($\boldsymbol{u} = \boldsymbol{0}$). The initial concentration of a chemical species inside the cavity is C_0, and the concentration at the cylinder surface is kept constant at $C_c > C_0$. With time, the species diffuses from the cylinder surface into the liquid.

The governing equation represented in cylindrical coordinates is

$$\frac{\partial}{\partial t} C(r,t) = D \frac{1}{r} \frac{\partial}{\partial r} r \frac{\partial}{\partial r} C(r,t), \tag{8.65}$$

with initial and boundary conditions $C(r < a, t = 0) = C_0$, $C(r = a, t) = C_c$. Its analytical solution is [51]

$$\frac{C(r,t) - C_c}{C_0 - C_c} = \sum_{n=1}^{\infty} \frac{2}{\mu_n J_1(\mu_n)} \exp\left(-\mu_n^2 \frac{Dt}{a^2}\right) J_0\left(\mu_n \frac{r}{a}\right). \tag{8.66}$$

$J_0(x)$ and $J_1(x)$ are the zeroth- and first-order Bessel functions, respectively. μ_n is the n-th root of $J_0(x)$: $J_0(\mu_n) = 0$. The first five roots are $\mu_1 = 2.4048$, $\mu_2 = 5.5201$, $\mu_3 = 8.6537$, $\mu_4 = 11.7915$ and $\mu_5 = 14.9309$. By taking the initial concentration as $C_0 = 0$, the analytical solution becomes

$$C(r,t) = C_c \left[1 - \sum_{n=1}^{\infty} \frac{2}{\mu_n J_1(\mu_n)} \exp\left(-\mu_n^2 \frac{Dt}{a^2}\right) J_0\left(\mu_n \frac{r}{a}\right)\right]. \tag{8.67}$$

The numerical domain consists of $N_x \times N_y = 129 \times 129$ lattice nodes on a D2Q9 lattice, and the cylinder radius is $a = 40\Delta x$. We approximate the cylinder boundary through a stair-case curve. The constant concentration at the cylinder surface, $C_c = 1$ in lattice units, is imposed by using two different boundary conditions: anti-bounce-back and Inamuro's boundary condition. We use the BGK collision operator with $\tau_g = 0.516\Delta t$ and $D = 0.0052\Delta x^2/\Delta t$.

Figure 8.5 shows the comparison of both boundary conditions at two different times. The curves correspond to the concentration profile along the x-axis. We observe that the simulation captures the time evolution of the concentration field well. Both boundary conditions perform similarly, and we can see small deviations near the cylinder boundary ($r \to a$) caused by the stair-case nature of the boundary discretisation.

8.6.3 Diffusion from Plate in Uniform Flow

We now look at the concentration layer development near a semi-infinite plate as illustrated in Fig. 8.6. Here we assume a uniform fluid velocity u_0 along the x-axis and a constant concentration C_p at the surface of the plate.

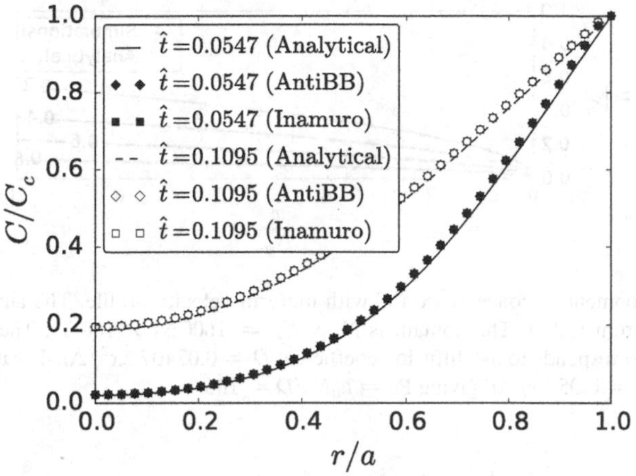

Fig. 8.5 Concentration profiles along the x-axis for the diffusion from a cylinder. The concentration at the cylinder surface is kept constant at $C_c = 1$ in lattice units. The diffusion coefficient is $D = 0.0052\Delta x^2/\Delta t$. Profiles are shown at two dimensionless times $\hat{t} = Dt/a^2$ for both investigated boundary conditions (anti-bounce-back and Inamuro's approach) and are compared with the analytical solution from (8.67)

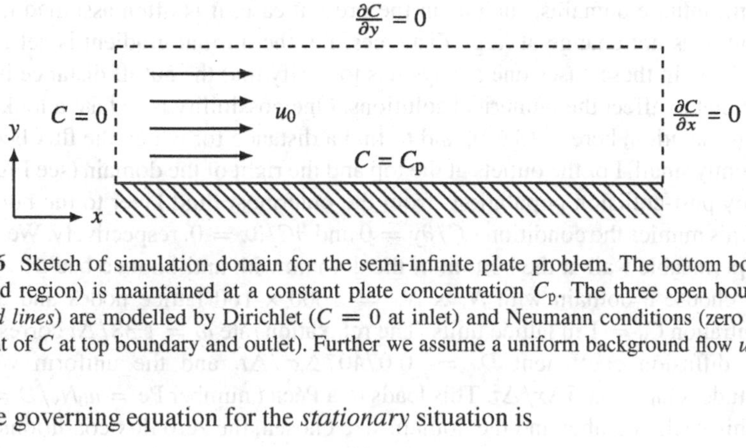

Fig. 8.6 Sketch of simulation domain for the semi-infinite plate problem. The bottom boundary (hatched region) is maintained at a constant plate concentration C_p. The three open boundaries (*dashed lines*) are modelled by Dirichlet ($C = 0$ at inlet) and Neumann conditions (zero normal gradient of C at top boundary and outlet). Further we assume a uniform background flow u_0

The governing equation for the *stationary* situation is

$$\frac{\partial C}{\partial x} u_0 = D \frac{\partial^2 C}{\partial y^2} \tag{8.68}$$

with the boundary conditions $C(x = 0, y) = 0$ and $C(x, y = 0) = C_p$. In (8.68) we have assumed that the diffusion of the species along the x-axis is negligible compared to its advection (high Péclet number), therefore omitting the term $D\partial^2 C/\partial x^2$. The analytical solution of (8.68) is [52]

$$C(x, y) = C_p \operatorname{erfc}\left(\frac{y}{\sqrt{4Dx/u_0}}\right) \tag{8.69}$$

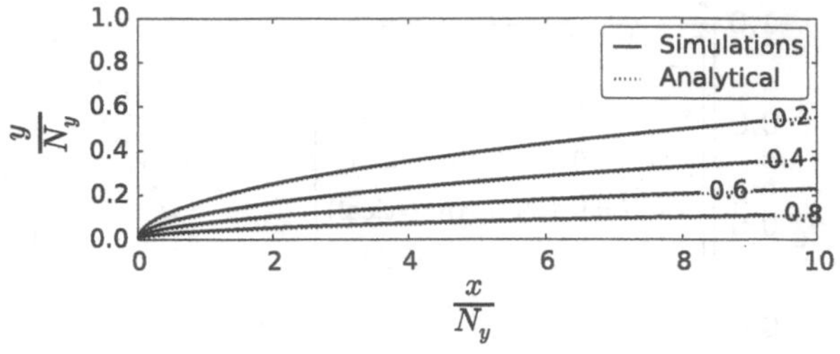

Fig. 8.7 Development of concentration C with uniform velocity profile. The analytical solution is constructed from (8.69). The domain is $N_x \times N_y = 1600\Delta x \times 160\Delta x$. The relaxation rate $\omega = 1.38/\Delta t$ corresponds to the diffusion coefficient $D = 0.07407\Delta x^2/\Delta t$. The uniform velocity magnitude is $u_0 = 0.05\Delta x/\Delta t$, giving $\mathrm{Pe} = u_0 N_y/D = 106$

where $\mathrm{erfc}(x)$ is the complementary error function:

$$\mathrm{erfc}(x) = 1 - \mathrm{erf}(x) = 1 - \frac{2}{\sqrt{\pi}} \int_0^x e^{-x'^2} \mathrm{d}x' = \frac{2}{\sqrt{\pi}} \int_x^\infty e^{-x'^2} \mathrm{d}x'. \tag{8.70}$$

Although this example looks simple, there is one interesting issue to discuss. For semi-infinite domains, such as in the present case, it is often assumed that the solution does not change at large distances, i.e. the normal gradient is set to zero somewhere. In these cases one always has to verify that the cutoff distance is large enough not to affect the numerical solutions. One possibility is to take a look at the expected solution, here in (8.69), and to find a distance for which the flux becomes sufficiently small. For the outlets at the top and the right of the domain (see Fig. 8.6), we copy post-collision populations from the outermost fluid layer to the boundary layer; this mimics the condition $\partial C/\partial y = 0$ and $\partial C/\partial x = 0$, respectively. We model the impermeable wall at the bottom of the domain with anti-bounce-back.

We choose a domain with $N_x \times N_y = 1600 \times 160$ lattice nodes and a plate concentration $C_\mathrm{p} = 1$ in lattice units. The relaxation rate $\omega = 1.38/\Delta t$ corresponds to the diffusion coefficient $D = 0.07407\Delta x^2/\Delta t$, and the uniform velocity magnitude is $u_0 = 0.05\Delta x/\Delta t$. This leads to a Péclet number $\mathrm{Pe} = u_0 N_y/D = 106$. With this Péclet number and the domain size chosen, the zero flux conditions at the top wall and at the outlet do not strongly affect the numerical solution, as evidenced in Fig. 8.7.

8.6.4 *Diffusion in Poiseuille Flow*

Finally we present a slightly more complicated example: the concentration layer development in a Poiseuille flow between two parallel plates with distance $2H$.

The governing stationary ADE for the concentration field C is

$$\frac{\partial C}{\partial x}u_x = D\frac{\partial^2 C}{\partial y^2} \tag{8.71}$$

with the known velocity profile

$$u_x(y) = u_0\left[1-\left(\frac{y}{H}\right)^2\right], \quad (-H \leq y \leq H) \tag{8.72}$$

and the concentration boundary conditions $C(y = \pm H) = C_\mathrm{p}$ at the surface of the plates. At the inlet we impose $C = 0$, and for the inlet we assume a zero-gradient condition: $\partial C/\partial x = 0$. As in Sect. 8.6.3, we neglect the term $D\partial^2 C/\partial x^2$, which implies a large Péclet number.

We note that the problem has a symmetry plane at $y = 0$ as indicated in Fig. 8.8. Thus, we use only one half of the channel ($0 \leq y \leq H$) and employ an additional symmetry condition: $\partial C/\partial y|_{y=0} = 0$.

The procedure to obtain the analytical solution for this problem is presented in detail in the appendix of [53]. The final solution is

$$C(x,y) = C_\mathrm{p}\left[1-\sum_m C_m \mathrm{e}^{-m^4 x/(H\mathrm{Pe})-m^2y^2/(2H^2)} {}_1F_1\left(-\frac{m^2}{4}+\frac{1}{4},\frac{1}{2},m^2\frac{y^2}{H^2}\right)\right] \tag{8.73}$$

where ${}_1F_1$ is the hypergeometric function, $\mathrm{Pe} = u_0H/D$ is the Péclet number and m is the root of the following expression:

$${}_1F_1\left(\frac{1}{4}-\frac{m^2}{4},\frac{1}{2},m^2\right) = 0. \tag{8.74}$$

The first ten roots of (8.74) are 1.2967, 2.3811, 3.1093, 3.6969, 4.2032, 4.6548, 5.0662, 5.4467, 5.8023 and 6.1373.

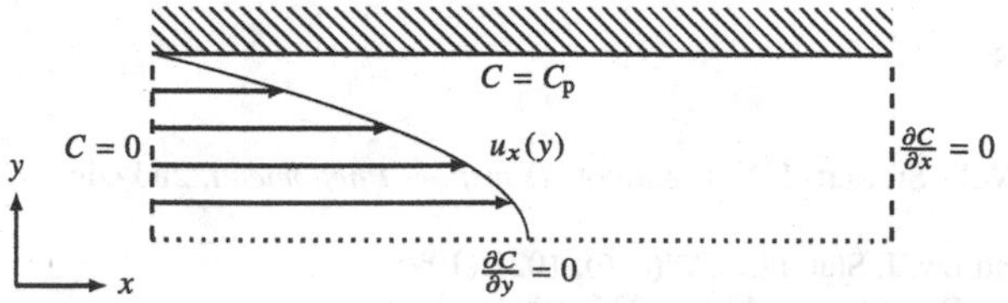

Fig. 8.8 Diffusion from a plate in Poiseuille flow with parabolic velocity profile $u_x(y)$. The imposed concentration at the plate (hatched pattern) is C_p. Due to symmetry (*dotted line*) only one half of the channel is simulated. Inlet and outlet (*dashed lines*) are open Dirichlet and Neumann boundaries, respectively

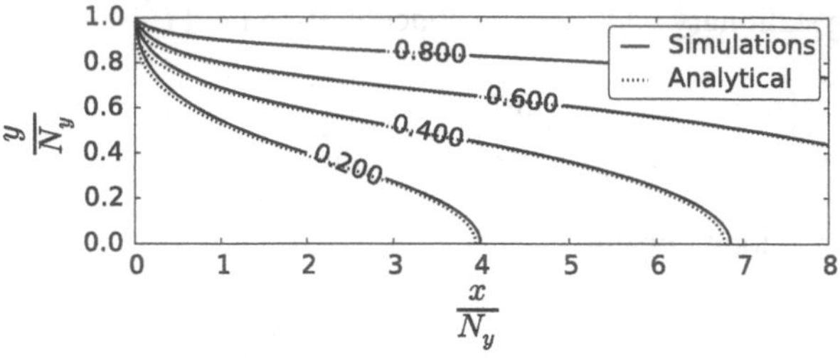

Fig. 8.9 Comparison of the analytical concentration contours and simulations with anti-bounce-back conditions for concentration at the inlet and the top wall. The simulations were done for the diffusion coefficient $D = 0.2857$ with a 160×1600 grid. The centreline velocity is $u_0 = 0.05$, and the Péclet number is 27.65

The coefficients C_m can be found through the integrals of the hypergeometric function:

$$C_m = -C_\mathrm{p} \frac{\int_0^1 (1-\xi^2) \mathrm{e}^{-m^2\xi^2/2} {}_1F_1\left(-\frac{m^2}{4}+\frac{1}{4}, \frac{1}{2}, m^2\xi^2\right) \mathrm{d}\xi}{\int_0^1 (1-\xi^2) \mathrm{e}^{-m^2\xi^2} {}_1F_1\left(-\frac{m^2}{4}+\frac{1}{4}, \frac{1}{2}, m^2\xi^2\right)^2 \mathrm{d}\xi}. \tag{8.75}$$

For $C_\mathrm{p} = 1$, the first ten coefficients, C_0–C_9, are $1.2008, -0.2991, 0.1608, -0.1074, 0.0796, -0.0627, 0.0515, -0.0435, 0.0375$ and -0.0329.

We performed a simulation on a D2Q9 lattice with $N_x \times N_y = 1600 \times 160$ nodes, diffusion coefficient $D = 0.2857\Delta x^2/\Delta t$ and centreline velocity $u_0 = 0.05\Delta x/\Delta t$. The Péclet number is $\mathrm{Pe} = 27.65$. We use bounce-back for the symmetry boundary condition and anti-bounce-back for the inlet and the plate surface (constant concentration). For the Neumann condition at the outlet we copied the concentration values of the last layer in the fluid domain to the outlet layer.

Figure 8.9 shows the comparison of the analytical and numerical results. For the analytical solution we used the first ten series terms (up to $m = 9$) as given above. We observe a good agreement.

References

1. R.B. Bird, W.E. Stewart, E.N. Lightfoot, *Transport Phenomena*, 2nd edn. (Wiley, New York, 1960)
2. D. Wolf-Gladrow, J. Stat. Phys. **79**(5–6), 1023 (1995)
3. E.G. Flekkoy, Phys. Rev. E **47**(6), 4247 (1993)
4. T. Inamuro, M. Yoshino, H. Inoue, R. Mizuno, F. Ogino, J. Comp. Phys. **179**(1), 201 (2002)
5. X. Shan, Phys. Rev. E **55**(3), 2780 (1997)
6. X. He, S. Chen, G.D. Doolen, J. Comput. Phys. **146**(1), 282 (1998)
7. T. Toffoli, N.H. Margolus, Physica D **45**(1-3), 229 (1990)
8. I. Ginzburg, Adv. Water Resour. **28**(11), 1171 (2005)
9. S. Suga, Int. J. Mod. Phys. C **17**(11), 1563 (2006)

10. I. Ginzburg, D. d'Humières, A. Kuzmin, J. Stat. Phys. **139**(6), 1090 (2010)
11. L. Li, C. Chen, R. Mei, J.F. Klausner, Phys. Rev. E **89**(4), 043308 (2014)
12. H.B. Huang, X.Y. Lu, M.C. Sukop, J. Phys. A: Math. Theor. **44**(5), 055001 (2011)
13. B. Chopard, J.L. Falcone, J. Latt, Eur. Phys. J. Spec. Top. **171**, 245 (2009)
14. Z. Chai, T.S. Zhao, Phys. Rev. E **87**(6), 063309 (2013)
15. I.V. Karlin, D. Sichau, S.S. Chikatamarla, Phys. Rev. E **88**(6), 063310 (2013)
16. Q. Kang, P.C. Lichtner, D. Zhang, J. Geophys. Res. **111**(B5), B05203 (2006)
17. S.G. Ayodele, F. Varnik, D. Raabe, Phys. Rev. E **83**(1), 016702 (2011)
18. J. Zhang, G. Yan, Comput. Math. Appl. **69**(3), 157 (2015)
19. X. Yang, B. Shi, Z. Chai, Comput. Math. Appl. **68**(12, Part A), 1653 (2014)
20. J. Kang, N.I. Prasianakis, J. Mantzaras, Phys. Rev. E **89**(6), 063310 (2014)
21. T. Seta, Phys. Rev. E **87**(6), 063304 (2013)
22. B. Shi, B. Deng, R. Du, X. Chen, Comput. Math. Appl. **55**(7), 1568 (2008)
23. I. Ginzburg, Adv. Water Resour. **28**(11), 1196 (2005)
24. S. Suga, Int. J. Mod. Phys. C **20**, 633 (2009)
25. H. Yoshida, M. Nagaoka, J. Comput. Phys. **229**(20), 7774 (2010)
26. I. Ginzburg, Commun. Comput. Phys. **11**, 1439 (2012)
27. A. Mohamad, *Lattice Boltzmann Method: Fundamentals and Engineering Applications with Computer Codes*, 1st edn. (Springer, New York, 2011)
28. R.G.M. Van der Sman, Phys. Rev. E **74**, 026705 (2006)
29. R. Huang, H. Wu, Phys. Rev. E **89**(4), 043303 (2014)
30. Q. Li, Z. Chai, B. Shi, Comput. Math. Appl. **70**(4), 548 (2015)
31. J. Perko, R.A. Patel, Phys. Rev. E **89**(5), 053309 (2014)
32. X. Yang, B. Shi, Z. Chai, Phys. Rev. E **90**(1), 013309 (2014)
33. B. Servan-Camas, F. Tsai, Adv. Water Resour. **31**, 1113 (2008)
34. I. Ginzburg, Adv. Water Resour. **51**, 381 (2013)
35. Z. Chai, T.S. Zhao, Phys. Rev. E **90**(1), 013305 (2014)
36. R. Huang, H. Wu, J. Comput. Phys. **274**, 50 (2014)
37. E.D. Siggia, Ann. Rev. Fluid Mech. **26**, 137 (1994)
38. W.H. Reid, D.L. Harris, Phys. Fluids **1**(2), 102 (1958)
39. F.J. Alexander, S. Chen, J.D. Sterling, Phys. Rev. E **47**(4), R2249 (1993)
40. Y. Chen, H. Ohashi, M. Akiyama, Phys. Rev. E **50**(4), 2776 (1994)
41. Y.H. Qian, J. Sci. Comput. **8**(3), 231 (1993)
42. G.R. McNamara, A.L. Garcia, B.J. Alder, J. Stat. Phys. **81**(1–2), 395 (1995)
43. D.A. Wolf-Gladrow, *Lattice-Gas Cellular Automata and Lattice Boltzmann Models* (Springer, New York, 2005)
44. Y. Peng, C. Shu, Y.T. Chew, Phys. Rev. E **68**(2), 026701 (2003)
45. Z. Guo, C. Zheng, B. Shi, T.S. Zhao, Phys. Rev. E **75**(3), 036704 (2007)
46. S. Chen, K.H. Luo, C. Zheng, J. Comput. Phys. **231**(24) (2012)
47. P.A. Thompson, *Compressible-Fluid Dynamics* (McGraw-Hill, New York, 1972)
48. Z. Guo, B. Shi, C. Zheng, Int. J. Numer. Meth. Fluids **39**(4), 325 (2002)
49. T. Zhang, B. Shi, Z. Guo, Z. Chai, J. Lu, Phys. Rev. E **85**(016701), 1 (2012)
50. M. Yoshino, T. Inamuro, Int. J. Num. Meth. Fluids **43**(2), 183 (2003)
51. A. Polyanin, A. Kutepov, A. Vyazmin, D. Kazenin, *Hydrodynamics, Mass and Heat Transfer in Chemical Engineering* (Taylor and Francis, London, 2002)
52. M. Ozisik, *Heat Conduction* (Wiley, New York, 1993)
53. A. Kuzmin, M. Januszewski, D. Eskin, F. Mostowfi, J. Derksen, Chem. Eng. J. **225**, 580 (2013)

Chapter 9
Multiphase and Multicomponent Flows

Abstract After reading this chapter, you will be able to expand lattice Boltzmann simulations by including non-ideal fluids, using either the free-energy or the Shan-Chen pseudopotential method. This will allow you to simulate fluids consisting of multiple phases (e.g. liquid water and water vapour) and multiple components (e.g. oil and water). You will also learn how the surface tension between fluid phases/components and the contact angle at solid surfaces can be varied and controlled.

We start by introducing the physical basis of multiphase and multicomponent flows in Sect. 9.1. In particular, we cover the concepts of the order parameter, surface tension, contact angle and thermodynamic consistency, and we discuss the differences between sharp and diffuse interface models. We then introduce and analyse two popular classes of LB multiphase and multicomponent models: the free-energy model in Sect. 9.2 and the pseudopotential (or Shan-Chen) model in Sect. 9.3. Section 9.2 and Sect. 9.3 can be read independently, but the prior study of Sect. 9.1 is strongly recommended. In Sect. 9.4 we will discuss limitations and extensions of both models, e.g. how to increase the range of physical parameters and how to improve accuracy and numerical stability. Finally, in Sect. 9.5, we provide a few example applications demonstrating the usefulness and suitability of LB multiphase and multicomponent methods.

There exist even more LB methods for multiphase and multicomponent problems. We cannot review all of them here. The most popular of those methods is the colour method [1–6]. Being rooted in lattice gas automata, the colour method is in fact the earliest multicomponent extension to the LBM. Other multiphase and multicomponent methods have also been suggested, e.g. [7–10].

Apart from reading this chapter, we recommend the study of a number of recent articles. Scarbolo et al. [11] developed a unified framework to analyse the similarities and differences of the free-energy and the Shan-Chen models. Chen et al. [12] performed a critical and topical review of the Shan-Chen method. Liu et al. [13] provide an extensive overview of the colour, the Shan-Chen and the free-energy models. The book by Huang, Sukop and Lu [14] is also dedicated to multiphase LB methods.

T. Krüger et al., *The Lattice Boltzmann Method*, Graduate Texts in Physics,
DOI 10.1007/978-3-319-44649-3_9

9.1 Introduction

One of the most popular applications of the LBM is simulating *multiphase* and *multicomponent* flows.

Multiphase and multicomponent flows refer to flows comprising two (or more) different fluids which differ by their physical properties, such as density, viscosity, conductivity *etc.* For single-component multiphase flows, the liquid and gas phases of the same substance are in coexistence. These two phases can interconvert from one to another: the gas can condense to form more liquid, and the liquid can evaporate. A typical example is liquid water and water vapour. Contrarily, multicomponent flows contain two (or more) different substances, for example water and oil. In a multicomponent flow, the substances do not interconvert. Instead, we have to account for the diffusion between these components.

Multiphase and multicomponent flows are important for a wide range of applications [15]. For example, emulsions are formed when one attempts to mix several immiscible liquids [16]. This is ubiquitously exploited in the food, pharmaceutical and personal care industries. Other examples include enhanced oil recovery [17], high-performance heat exchangers [18], polymer processing [19] and microfluidics [20].

In practice, the distinction between multiphase and a multicomponent flows can be quite blurry. Many flows are in fact a mixture between the two, where the liquid and gas phases can separately comprise of several components. One example is cooking: boiling water (liquid water and water vapour) with olive oil. In cases where there is no transfer of material between the different fluid domains and inertia plays a negligible role (e.g. low Reynolds number flow), equivalent results are obtained whether we are using a single-component multiphase or a multicomponent model [21], assuming the material parameters (e.g. viscosity, surface tension, *etc.*) used in the two types of simulations are equivalent. In this chapter, we will focus on two-phase and two-component flows, but not a mix between them.

To distinguish the two fluid phases in a multiphase flow, or similarly the two components in a multicomponent flow, we introduce a concept called the *order parameter*.[1] For multiphase flows, this order parameter is the fluid density. The gas and liquid phases are uniquely characterised by their values of density ρ_g and ρ_l.

For multicomponent flows, density is often not a suitable parameter. For example, the densities of water and oil are quite similar. Instead, a more effective order

[1]This concept can easily be extended to systems with more than two components by introducing more order parameters.

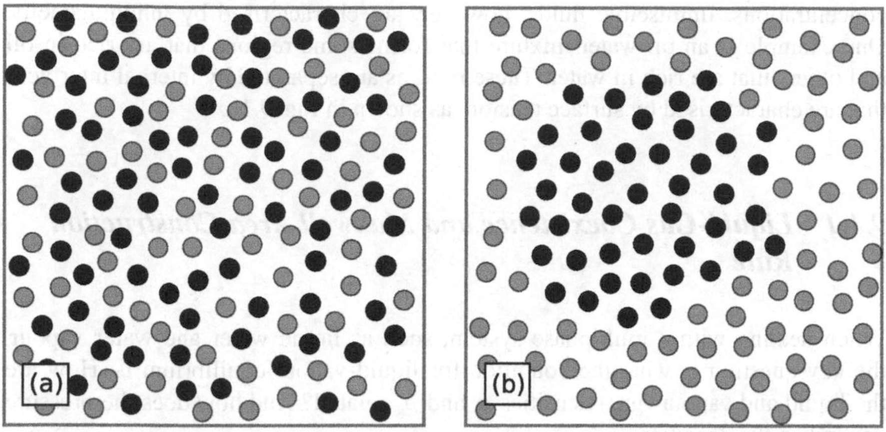

Fig. 9.1 Illustrations of binary fluid mixtures which are (**a**) miscible and (**b**) immiscible. In (**b**), the fluid particles separate into regions which are black-rich and grey-rich. At the interface, the black (grey) particles lose favourable interactions with black (grey) particles and gain less favourable interactions with grey (black) particles, resulting in an excess energy for forming the interface

parameter ϕ is given by

$$\phi = \frac{\rho^{(1)} - \rho^{(2)}}{\rho^{(1)} + \rho^{(2)}} \tag{9.1}$$

where $\rho^{(1)}(\boldsymbol{x}, t)$ and $\rho^{(2)}(\boldsymbol{x}, t)$ are the local densities of components 1 and 2. We denote the densities of the pure components $\rho_{\mathrm{b}}^{(1)}$ and $\rho_{\mathrm{b}}^{(2)}$. These values are also called *bulk densities*. The two *bulk phases* correspond to cases where (i) $\rho^{(1)} = \rho_{\mathrm{b}}^{(1)}$, $\rho^{(2)} = 0$ and (ii) $\rho^{(1)} = 0$, $\rho^{(2)} = \rho_{\mathrm{b}}^{(2)}$, respectively.[2] It is easy to verify that this leads to two distinct bulk values for the order parameter:

$$\phi = \begin{cases} +1 & \text{for component 1,} \\ -1 & \text{for component 2.} \end{cases} \tag{9.2}$$

Multicomponent fluids, i.e. systems comprising different fluids, can be *miscible* or *immiscible*. Miscible fluids can form a completely homogeneous mixture without internal interfaces, as illustrated in Fig. 9.1a. For example, ideal gases are always miscible. Water and ethanol are also miscible, at least over a wide range of

[2]In reality we cannot write $\rho^{(1)} = 0$ or $\rho^{(2)} = 0$ since the local density of a given component is never exactly zero. For example, in a water-oil mixture, one can always find a few water molecules in the oil-rich phase and the other way around. However, these minority densities are usually so small that we can neglect them here.

concentrations. Immiscible fluids, however, are characterised by inhomogeneity. One example is an oil-water mixture that forms some regions that are rich in oil and others that are rich in water. These regions are separated by internal interfaces that are characterised by surface tension, as shown in Fig. 9.1b.

9.1.1 Liquid-Gas Coexistence and Maxwell Area Construction Rule

When dealing with a multiphase system, such as liquid water and water vapour, the key question is what the condition for liquid-vapour equilibrium is. How are the liquid and vapour (gas) densities ρ_l and ρ_g related? And how does the pressure depend on the densities?

In nature we observe many situations with coexisting fluid phases or components. The physical requirement of having coexisting phases or components puts a constraint on the **equation of state**, which describes a complex interdependency between pressure p, molar volumes υ (alternatively: density $\rho \propto 1/\upsilon$) and order parameter ϕ for a given temperature T. (We introduced the concept of equations of state in Sect. 1.1.3.) The equation of state $p = p_b(\rho, \phi, T, \ldots)$ uniquely defines the bulk (i.e. the region not close to any interface) thermodynamic state of the multiphase and multicomponent system.

Let us now focus on one of the most commonly used equations of state, namely the *van der Waals equation* for a liquid-vapour system, shown in Fig. 9.2. We can see that the pressure-molar volume curve has a minimum and a maximum. Any equation of state that displays this property allows for two coexisting bulk fluids. In fact, the thermodynamic states between these two extrema are unstable. If the system is prepared at any of these intermediate states, it will spontaneously phase separate into liquid and gas domains [22]. For a given pressure p_0, the molar volumes of the liquid and gas phases are, respectively, υ_l and υ_g.

From Fig. 9.2, we see that there is a range of pressures for which two distinct molar volumes υ can be adopted for the same bulk pressure value p_b. To decide which exact pressure value the system will adopt (i.e. which p_0 the system will relax to) and, correspondingly, which values of molar volumes the liquid and gas phases will assume, we need to use the so-called Maxwell area construction rule [22].

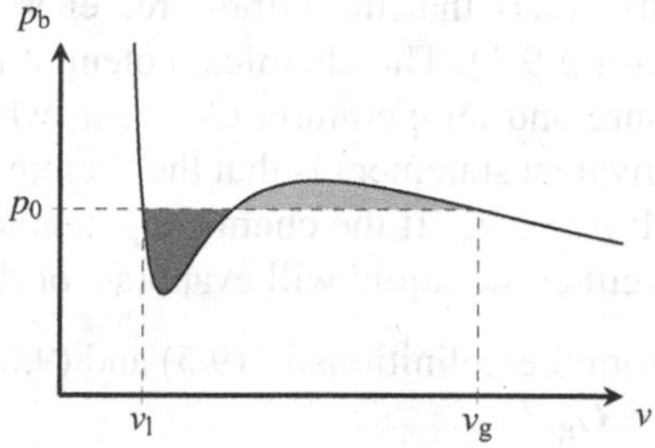

Fig. 9.2 Maxwell area construction rule for the van-der-Waals equation of state for a fixed temperature T. Note that the phase transition occurs at a pressure p_0 that equalises the areas below (*dark grey*) and above (*light grey*) the pressure curve

The **Maxwell area construction rule** postulates that, for a given temperature T, the liquid-gas coexistence happens at a pressure p_0 such that both shaded areas in Fig. 9.2 are identical:

$$\int_{v_g}^{v_l} (p_0 - p_b(v', T))\, dv' = 0. \tag{9.3}$$

The molar volumes of the gas and the liquid both satisfy

$$p_0 = p_b(v_g, T) = p_b(v_l, T). \tag{9.4}$$

In essence, the Maxwell area construction rule states that, at coexistence, the Gibbs free energy G, or equivalently the chemical potential μ, of the liquid and gas phases must be equal.[3] To see this, we note that the difference in the Helmholtz free energy F in both fluid phases is given by [22]

$$F_l - F_g = -\int_{v_g}^{v_l} p_b(v', T)\, dv' \tag{9.5}$$

and that the Gibbs free energy is [22]

$$G_{l/g} = F_{l/g} + p_0 v_{l/g}. \tag{9.6}$$

[3]For readers unfamiliar with Gibbs and Helmholtz free energies, their descriptions can be found in most textbooks on thermodynamics, e.g. [22]. Briefly, Gibbs free energy is usually used when the system is under constant pressure and temperature, while the Helmholtz free energy is taken when the system is under constant volume and temperature.

Equation (9.3) essentially states that the Gibbs free energy of the system obeys $G = G_l = G_g$ (see Exercise 9.1). The chemical potential is the molar Gibbs free energy at constant pressure and temperature, $G = \mu n$, where n is the number of moles. Therefore, an equivalent statement is that the chemical potential in the liquid and gas phases are equal: $\mu_l = \mu_g$. If the chemical potentials of the liquid and gas phases are not the same, either the liquid will evaporate or the gas will condense.

Exercise 9.1 Starting from the definitions in (9.5) and (9.6), show that (9.3) leads to the condition that $G_l = G_g$.

The molar volume υ is not a convenient parameter to use in LB simulations. Since it is proportional to the inverse of density, $\upsilon \propto 1/\rho$, the Maxwell area construction rule can easily be rewritten in terms of the density:

$$\int_{\rho_g}^{\rho_l} \left(p_0 - p_b(\rho', T)\right) \frac{d\rho'}{\rho'^2} = 0, \quad p_0 = p_b(\rho_g, T) = p_b(\rho_l, T). \tag{9.7}$$

Equation (9.7) provides three equations for the three unknowns p_0, ρ_g and ρ_l. Given the form of the equation of state $p_b(\rho, T)$ at a fixed temperature T, this system of equations can be uniquely solved.

Any model for the equation of state that satisfies (9.7) is *thermodynamically consistent*. Ideally all models should follow this requirement, which is the case for free-energy multiphase and multicomponent models (cf. Sect. 9.2). However, this is not necessarily true for the Shan-Chen model. In practice, this means that the recovered liquid and vapour densities do not exactly assume their expected values. We will elaborate on this issue in Sect. 9.3.

All these considerations are valid for the bulk, far away from any interface. Now we have to look closer at the effect interfaces have on the thermodynamic behaviour.

9.1.2 Surface Tension and Contact Angle

The richness of the multiphase and multicomponent flow behaviour comes, among others, from the interfaces formed between the bulk fluid phases. The presence of surface tension gives rise to complex viscoelastic behaviour, even though each phase/component in the flow itself may be a simple Newtonian fluid [23, 24].

A key concept for multiphase and multicomponent flows is the *surface tension* γ. It is the energy per unit area required to form the interface between the two fluid phases or components.[4] Therefore, surface tension is often given in Joule per square metre or, more commonly, Newton per metre.

[4]This definition is strictly valid only for simple liquids. More generally, the energy per unit area for stretching the interface is given by $\Gamma = \gamma + d\gamma/d\epsilon$ where ϵ is the strain. For simple liquids we have $d\gamma/d\epsilon = 0$ and $\Gamma = \gamma$.

Surface tension is caused by molecular interactions. Like molecules in a fluid typically attract each other. As illustrated in Fig. 9.1b, if such a molecule is in the bulk region, it will interact on average with z molecules of the same species, where z is the coordination number or the average number of neighbours. If this molecule is at an interface, it will lose interactions with approximately $z/2$ neighbours of like molecules. Furthermore, for multicomponent systems, the molecule will pick up less favourable interactions with molecules of a different species at the interface. This excess energy associated with an interface is usually positive, and it is a function of temperature. Thermodynamically, any physical system will prefer to minimise the amount of surface energy and therefore the total interface area.

If the volume bounded by the interface is not constrained, for example in the case of soap films, the shape of the interface will adopt one of the so-called *minimal surfaces*. Examples of well-known minimal surfaces include the plane (which is the trivial case), the catenoids, and the Schwarz triply periodic minimal surfaces [25]. For minimal surfaces, the surface is locally flat and there is no pressure jump across the interface. More precisely, the mean curvature of the surface is zero, although the Gaussian curvature may assume a non-zero value.[5]

For many multiphase and multicomponent flows, however, we find closed interfaces that enclose a certain volume, e.g. oil droplets in a pot of water or rain drops in air. In general, at mechanical equilibrium, the pressures on either side of these interfaces are different; the pressure at the inside is higher than the pressure outside. The pressure difference satisfies the so-called Laplace pressure [26].

Consider a droplet of one fluid (e.g. a liquid) suspended in another fluid (e.g. a gas). The **Laplace pressure** is

$$p_\mathrm{l} - p_\mathrm{g} = \gamma_\mathrm{lg}\left(\frac{1}{R_1} + \frac{1}{R_2}\right) \tag{9.8}$$

where R_1 and R_2 are the local curvature radii and γ_lg is the liquid-gas surface tension. This is illustrated in Fig. 9.3a.

Equation (9.8) has two important physical interpretations. First, it is a consequence of the force balance between the work done by the pressures on either side of the interface, and the energy penalty from changing the interfacial area. Secondly, it states that in equilibrium the mean curvature of the interface between the two fluid phases is constant. If this condition is not satisfied, it results in a force in the

[5] At a given point on a surface, we can define two radii of curvature, as shown in Fig. 9.3a. The mean curvature is simply defined as the average $(1/R_1 + 1/R_2)/2$, while the Gaussian curvature is the product $1/(R_1 R_2)$. Since one of the curvature radii can be negative and the other positive (e.g. a saddle surface), the mean curvature can vanish, even for a non-planar surface.

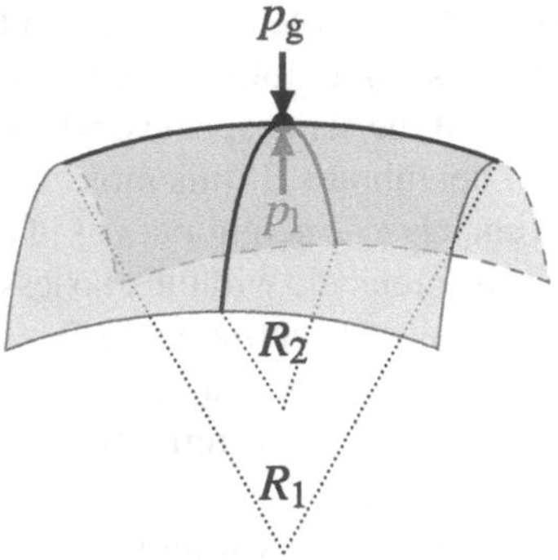

(a) $p_l - p_g = \gamma_{lg}\,(1/R_1 + 1/R_2)$

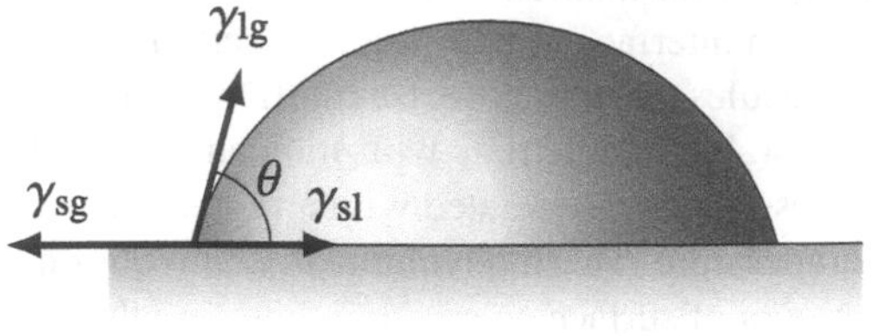

(b) $\gamma_{lg}\cos\theta = \gamma_{sg} - \gamma_{sl}$

Fig. 9.3 Schematic diagrams for (**a**) the Laplace pressure and (**b**) Young's contact angle. Each point at the interface can be characterised by two independent radii of curvature that can be positive or negative. In (**a**), the surface is convex and both radii are positive. The average curvature $(1/R_1 + 1/R_2)/2$ and the surface tension γ_{lg} are related to the pressure jump (Laplace pressure) $p_l - p_g$ across the interface. In (**b**), a liquid droplet is in contact with a surface and forms a contact angle θ. For this angle, all surface tension force components tangential to the surface are in mechanical equilibrium

hydrodynamic equations of motion, driving the system towards equilibrium. We will see later how this is accounted in the Navier-Stokes equation.

In many (if not most) situations, multicomponent and multiphase flows are also confined by solid surfaces, e.g. in porous media and microfluidics. The different fluid phases may have different affinities to these surfaces. This is usually quantified by a material property called the contact angle θ as shown in Fig. 9.3b.

For a droplet of one fluid (e.g. a liquid) surrounded by another fluid (e.g. a gas), we write

$$\cos\theta = \frac{\gamma_{sg} - \gamma_{sl}}{\gamma_{lg}} \tag{9.9}$$

for the **contact angle**. Here, γ_{sl}, γ_{sg} and γ_{lg} are, respectively, the solid-liquid, solid-gas and liquid-gas surface tensions. The contact angle is usually defined with respect to the liquid phase. We can understand the contact angle as a consequence of mechanical stability at the contact line where all three phases are in contact with each other [27, 28].

When $\theta < 90°$ in (9.9), the liquid phase preferably wets the solid surface. Such a surface is usually called a *hydrophilic*, or more generally a *lyophilic* surface. In contrast, a surface is called *hydrophobic* or *lyophobic* when $\theta > 90°$, i.e. when the gas phase has a favourable interaction with the solid. In the wetting literature, special terms are also reserved for $\theta = 0°$ (complete wetting), $\theta = 90°$ (neutral wetting) and $\theta > 160°$ (*superhydrophobic*).

The wetting properties can significantly affect the fluid flow near a solid boundary [29, 30]. Furthermore, it has been demonstrated that we can take advantage of surface patterning, both chemical and topographical, to control the motion of fluids [31, 32].

9.1.3 Sharp and Diffuse Interface Models

There are two different approaches to model multiphase and multicomponent flows: (i) sharp and (ii) diffuse interface models.

In the *sharp interface model*, the interface is a 2D boundary which is usually represented by a distinct computational mesh. The motion of this interface needs to be explicitly tracked, and we require a Navier-Stokes solver on either side of the boundary. Furthermore, the fluid velocity at the boundary must be continuous, and there is a stress jump normal to the interface corresponding to the Laplace pressure in (9.8). There are various publications describing in detail how such a sharp interface model can be efficiently implemented, which include volume-of-fluid [33], front-tracking [34], and immersed boundary [35] methods.

Contrarily, the models employed in the LB community usually belong to the *diffusive interface approach*. A typical 1D order parameter profile (density for multiphase flow; ϕ as defined in (9.2) for multicomponent flow) across a diffuse interface is shown in Fig. 9.4. Far from the interface, for $x \to \pm\infty$, the order parameter approaches the bulk values. The order parameter profile smoothly varies across the interface between the two bulk values.

The length scale that characterises the variation in the density profile across the interface is called the *interface width*. For a real physical system, this is usually of the order of nanometres. In the computational domain, the interface width is chosen to be several lattice spacings for the simulations to be stable. However, this does not necessarily mean that the grid spacing in multiphase and multicomponent simulations is assigned to several nanometres; this would limit the applicability of those simulations to nanoscale systems. Instead, we take advantage of the separation

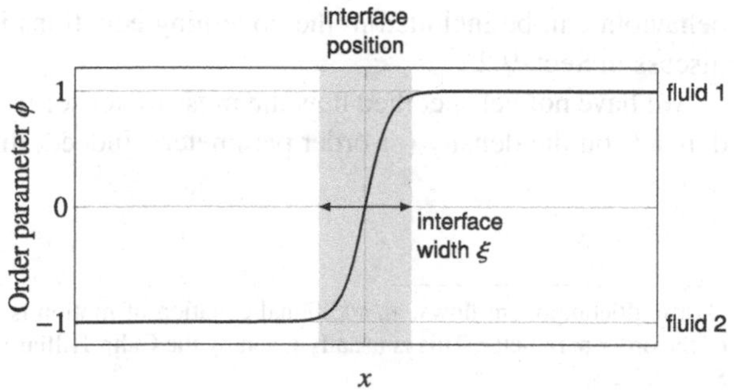

Fig. 9.4 A typical interface profile in the diffuse interface model. The order parameter varies smoothly across the interface to assume its bulk value on either side of the interface. For a multiphase system, the density varies from the gas density ρ_g to the liquid density ρ_l across the interface

of length scales. Ideally we work in a regime where the simulation results do not depend on the interface width. This can be achieved when the interface width is small enough, typically by an order of magnitude smaller than the first important length scale (e.g. the diameter of a droplet). It is often not necessary to truthfully represent the ratio between the physical length scale and the interface width, which makes diffuse-interface multiphase simulations possible in the first place.

The **key advantage of diffuse interface models** is that the motion of the interface need not be tracked explicitly. All fluid nodes can be treated on an equal footing, whether they are in the bulk of the fluid or at the interface. There is no need to introduce any additional mesh for the interface. Thus, diffuse interface models are convenient for studying problems with complex surface geometries.

The density (or order parameter) variation in diffuse interface models is smooth. This allows us to incorporate the description of surface tension into the bulk fluid equations of motion, more specifically in the description of the *pressure tensor* $\boldsymbol{P}$ that also varies smoothly across the interface.

In the definition of pressure as a tensor, $P_{\alpha\beta}$ corresponds to force per unit area in the β-direction on a surface pointing in the α-direction. For a homogeneous and isotropic fluid, which is the case we encountered up to this point, the pressure is the same in all directions. This isotropy means the pressure tensor is given by $P_{\alpha\beta} = p_{\mathrm{b}}\delta_{\alpha\beta}$, and we can treat pressure as a scalar.

When an interface is involved, isotropy is clearly broken: the directions normal and tangential to the interface do not behave in the same way. The fluid equations of motion for diffuse interfaces are given by none other than the continuity and Navier-Stokes equations with a modified pressure tensor[6]:

$$\partial_t\rho + \partial_\alpha(\rho u_\alpha) = 0, \tag{9.10}$$

$$\partial_t(\rho u_\alpha u_\beta) = -\partial_\beta P_{\alpha\beta} + \partial_\beta \eta(\partial_\beta u_\alpha + \partial_\alpha u_\beta). \tag{9.11}$$

It is important to note that the divergence of the pressure tensor, $-\partial_\beta P_{\alpha\beta}$, is equivalent to a body force density F_α. Therefore, the multiphase and multicomponent behaviour can be included in the governing equations in different ways as we will discuss in Sect. 9.2.

We have not yet specified how the pressure tensor looks like, in particular, how it depends on the density (or order parameter). Indeed, this is where the different LB

[6]For multicomponent flows, an additional equation of motion is needed to describe the evolution of the order parameter. This is usually given by the Cahn-Hilliard or Allen-Cahn equation, see e.g. Sect. 9.2.2.3.

models distinguish themselves. Broadly speaking, these models can be categorised into a bottom-up or a top-down approach.

In a **bottom-up approach**, the starting point is often kinetic theory, and some form of interactions are postulated between the fluids at the level of the Boltzmann equation. Similar to many other lattice- and particle-based simulation techniques, separation between different fluid phases and components can be induced by tuning the interaction potentials. The Shan-Chen method (cf. Sect. 9.3) is one famous example. In particular, the Shan-Chen model makes use of an additional body force density rather than a modified pressure tensor.

In a **top-down approach**, we start by writing down the free energy of the fluids (cf. Sect. 9.2). The form of the free energy functional should capture intended features of the thermodynamics of the system, e.g. phase separation and surface tension between different fluids. The corresponding chemical potential and pressure tensor can then subsequently be derived.

9.1.4 Surface Tension and Young-Laplace Test

While the detailed form of the pressure tensor is model specific, irrespective of the model, the pressure tensor must describe an equation of state that allows for phase coexistence between several fluid phases/components, and it must account for the surface tension. In Sect. 9.1.1 we discussed the van der Waals equation of state, one of the most popular equations of state for a multiphase flow. Other equations of states are possible and will be discussed later in this chapter.

The surface tension, in diffuse interface models, is typically introduced *via* a surface tension force given by [36, 37]

$$\boldsymbol{F} = \kappa\rho\boldsymbol{\nabla}\Delta\rho. \tag{9.12}$$

For a multicomponent flow, the same form applies except we replace the density ρ by the order parameter ϕ.

At this point, it is also useful to recognise that the relevant term in the Navier-Stokes equation is the divergence of the pressure tensor, not the pressure tensor itself. In this context, we can immediately show that

$$\begin{aligned}
\partial_\alpha p_\mathfrak{b} - F_\alpha &= \partial_\alpha p_\mathfrak{b} - \kappa\rho\partial_\alpha\partial_\gamma\partial_\gamma\rho = \partial_\alpha p_\mathfrak{b} - \kappa\partial_\alpha(\rho\partial_\gamma\partial_\gamma\rho) + \kappa(\partial_\alpha\rho)\partial_\gamma\partial_\gamma\rho \\
&= \partial_\alpha p_\mathfrak{b} - \kappa\partial_\alpha(\rho\partial_\gamma\partial_\gamma\rho) + \kappa\partial_\gamma\left((\partial_\alpha\rho)(\partial_\gamma\rho)\right) - \kappa(\partial_\gamma\rho)\partial_\alpha\partial_\gamma\rho \\
&= \partial_\alpha p_\mathfrak{b} - \kappa\partial_\alpha(\rho\partial_\gamma\partial_\gamma\rho) + \kappa\partial_\gamma\left((\partial_\alpha\rho)(\partial_\gamma\rho)\right) - \frac{\kappa}{2}\partial_\alpha\left((\partial_\gamma\rho)^2\right)
\end{aligned}$$

$$
\begin{aligned}
&= \partial_\beta \left[\left(p_b - \frac{\kappa}{2}(\partial_\gamma \rho)^2 - \kappa \rho \partial_\gamma \partial_\gamma \rho \right) \delta_{\alpha\beta} + \kappa (\partial_\alpha \rho)(\partial_\beta \rho) \right] \\
&= \partial_\beta P_{\alpha\beta}. \qquad (9.13)
\end{aligned}
$$

The term in the square bracket defines the pressure tensor $P_{\alpha\beta}$: it contains information about the equation of state and the fluid-fluid surface tension. We also note that it is the $\kappa(\partial_\alpha \rho)(\partial_\beta \rho)$ term that causes the pressure tensor to be anisotropic.

Given the pressure tensor $\boldsymbol{P}$, the surface tension can be computed. It is defined as the mismatch between the normal and transversal components of the pressure tensor, integrated across the interface in its normal direction [28, 38]:

$$
\gamma = \int_{-\infty}^{\infty} (P_\mathrm{n} - P_\mathrm{t})\, \mathrm{d}\hat{n} \qquad (9.14)
$$

where $\hat{\boldsymbol{n}}$ is a unit vector normal to the interface.

To clarify the notation, let us take an example where the interface is located at $x = 0$ and spans across the y-z plane. In such a case, $\hat{\boldsymbol{n}}$ is in the x-direction, $P_\mathrm{n} = P_{xx}$ and $P_\mathrm{t} = P_{yy} = P_{zz}$. Using the definition of the pressure tensor in (9.13), we can show that

$$
P_\mathrm{n} = P_{xx} = \left(p_b - \frac{\kappa}{2}(\partial_\gamma \rho)^2 - \kappa \rho \partial_\gamma \partial_\gamma \rho \right) + \kappa (\partial_x \rho)(\partial_x \rho), \qquad (9.15)
$$

$$
P_\mathrm{t} = P_{yy} = P_{zz} = \left(p_b - \frac{\kappa}{2}(\partial_\gamma \rho)^2 - \kappa \rho \partial_\gamma \partial_\gamma \rho \right), \qquad (9.16)
$$

$$
\gamma = \int_{-\infty}^{\infty} (P_\mathrm{n} - P_\mathrm{t})\, \mathrm{d}x = \kappa \int_{-\infty}^{\infty} \left(\frac{\mathrm{d}\rho}{\mathrm{d}x} \right)^2 \mathrm{d}x. \qquad (9.17)
$$

Note that we can also compute the interface profile $\rho(x)$ or $\phi(x)$, given the functional form of the pressure tensor. This is demonstrated in Appendix A.7 and Appendix A.8.

The surface tension as defined in (9.14) is not always straightforward to compute in simulations. There is, however, a simpler way to measure γ by exploiting the Laplace pressure relation in (9.8). In practice, this is usually achieved by simulating a spherical domain of fluid 1 with radius R, surrounded by fluid 2. Depending on the system of interest, this can be (i) a liquid droplet in a gas phase, (ii) a gas bubble in a liquid phase or (iii) a liquid droplet in another liquid phase. This procedure is called the *Young-Laplace test* or just *Laplace test*.

We have to distinguish between a spherical droplet/bubble in 3D and a circular droplet/bubble in 2D. While in 3D we have two principal curvature radii, there is only a single radius in 2D. Therefore, the pressure difference between the inside (phase/component 1) and the outside (phase/component 2) assumes the form

$$p^{(1)} - p^{(2)} = \begin{cases} \gamma/R & \text{(2D)}, \\ 2\gamma/R & \text{(3D)}. \end{cases} \tag{9.18}$$

By computing the pressure values at the centre of the droplet/bubble, i.e. $p^{(1)}$, and far away from the interface in the exterior phase/component, i.e. $p^{(2)}$, we can then obtain the surface tension γ.

For this test to be successful we have to be aware of several issues. First, it is crucial to use a sufficiently large droplet/bubble. If it is too small, the interior will be dominated by the shape of the diffuse interface and the measured pressure $p^{(1)}$ will not represent the correct value of the bulk pressure in the interior phase. A reasonable radius to start with is $R \sim 10\Delta x$, assuming the interface width is 2–4 grid nodes. Secondly, it is not obvious how to define the radius of a droplet/bubble with a diffuse interface. Many researchers define the interface as the surface where either the order parameter becomes zero or the density reaches the average of the bulk gas and liquid densities. It is good practice to run several simulations with different droplet radii to show that the measured curve $p^{(1)} - p^{(2)}$ *vs.* $1/R$ is linear, with the gradient of the curve being the surface tension γ in 2D and 2γ in 3D.

9.2 Free-Energy Lattice Boltzmann Model

In this section we will focus on the free-energy lattice Boltzmann models, covering both multiphase and multicomponent systems [39, 40]. The free-energy approach has a *top-down* philosophy: we start with a free-energy functional that contains the thermodynamics of the intended systems, and then other relevant physical quantities can be derived from this functional. Thus, an attractive feature of free-energy LB models is that, by design, they are always thermodynamically consistent.[7] This is in contrast to the Shan-Chen method covered in Sect. 9.3, where we begin by postulating interactions between the lattice fluids, and the thermodynamics of the multiphase and multicomponent systems emerge from these interactions.

So, what information is contained in the free energy functional? It prescribes the free energy that a given system has in a particular arrangement. For particle-based models, the energy depends on the position and orientation of the particles. For continuum models, such as the ones we have here, the situation is similar, except that the energy now depends on collective, coarse-grained variables. For example, density is a suitable collective variable for multiphase flows; the relative concentration (order parameter) is appropriate for multicomponent flows. In thermodynamic equilibrium,

[7] Thermodynamic consistency is defined in Sect. 9.1.1. See also Appendix A.7 where this is shown explicitly for the Landau multiphase model.

the free energy functional is minimised. If the system is out of equilibrium, e.g. due to external influence, the free energy is not minimised and there is a *thermodynamic force* driving the system towards equilibrium. In the context of the Navier-Stokes equation, such thermodynamic force can be equivalently represented as a body force or as gradient in the fluid pressure tensor.

For multiphase and multicomponent flows, the free energy functional usually consists of three terms:

$$\Psi = \int_V [\psi_b + \psi_g] \, dV + \int_A \psi_s \, dA \tag{9.19}$$

where ψ_b, ψ_g and ψ_s are functions of space and time. The first term, ψ_b, describes the bulk free energy. This term, most importantly, must lead to an equation of state that allows for the coexistence of several fluid phases and/or components. The equation of state for an isothermal ideal gas, $p_b = c_s^2 \rho$, does not have this capacity. A wide range of models have been proposed in the literature for the bulk free energy. The simplest models correspond to Landau free-energy models [41–43], which are essentially Taylor expansions in terms of the order parameters. These models are very popular due to their simplicity. However, more complex and realistic bulk free energies, such as the van der Waals [39, 40, 44] or Peng-Robinson models [45], may also be used.

The second term, ψ_g, is a gradient term which penalises any variation in the order parameter, be it the fluid density for multiphase flows or the relative concentration for multicomponent flows. This term captures the free energy of the interface between two fluid phases or components. Its form can be adjusted to handle surface tension and/or bending energy of the interface.

The last term, ψ_s, describes the interaction between the fluid and the surrounding solid. This term is required when the physics of wetting phenomena are relevant.

In the following, we will demonstrate how the equation of state, the pressure tensor and the chemical potential can be written down. We will show how suitable LB schemes can be devised to solve the hydrodynamic equations of motion for multiphase (cf. Sect. 9.2.1) and multicomponent flows (cf. Sect. 9.2.2).

9.2.1 *Liquid-Gas Model*

Here we will focus on the multiphase free-energy LBM. Following the *top-down* philosophy, we will start by describing the bulk thermodynamics of the multiphase fluid, followed by how the thermodynamics enter the fluid equations of motion, and subsequently the LB algorithm. For many applications of multiphase LBM, the wettability of the surface confining the fluid is also important. We will discuss how suitable surface thermodynamics can be introduced, and show that it enters our LB algorithm as a boundary condition.

9.2.1.1 Bulk Thermodynamics

We start by investigating the bulk properties far away from any solid boundaries, i.e. ψ_s does not play a role. The simplest multiphase model is a two-phase system where a liquid is in coexistence with its own vapour.

For pedagogical reasons, we will use a **model based on Landau theory** [41] so that we can derive all relevant quantities analytically:

$$\Psi = \int_V [\psi_b + \psi_g]\,dV, \tag{9.20}$$

$$\psi_b = p_c\left(\nu_\rho^2 - \beta\tau_w\right)^2 + \mu_0\rho - p_0, \tag{9.21}$$

$$\psi_g = \frac{\kappa}{2}(\nabla\rho)^2 \tag{9.22}$$

where we have defined the reduced density $\nu_\rho = (\rho - \rho_c)/\rho_c$ and the reduced temperature $\tau_w = (T_c - T)/T_c$. The subscript c indicates the critical point of the liquid-gas system, such that p_c, ρ_c and T_c describe the pressure, density and temperature at the critical point. β is a constant which can be tuned to control the liquid-gas density ratio, and κ is a constant which controls the magnitude of the surface tension. The constants μ_0 and p_0 are the reference chemical potential and pressure of the fluids. We will describe the relevance of these parameters in more detail below.

Other bulk free energy functionals, leading to more realistic equations of state, can be used instead for ψ_b. The machineries for deriving the pressure tensor, surface tension and LB schemes are the same as the ones we will show below for the Landau model. The form for the gradient free energy functional, ψ_g, is the one most commonly used to capture surface tension, but once again, it is not unique. Additional terms can be introduced, e.g. to account for the bending energy of the interface [46]. In standard LB methods, we also have total mass conservation such that (9.20) is subject to the constraint $\int_V \rho\,dV = \text{const.}$

The Landau free energy functional is written such that, below the critical temperature T_c (for positive τ_w), the system will favour two bulk solutions corresponding to $\nu_\rho^2 - \beta\tau_w = 0$. The positive branch $\nu_l = (\rho - \rho_c)/\rho_c = +\sqrt{\beta\tau_w}$ is the liquid state, while the negative branch $\nu_g = (\rho - \rho_c)/\rho_c = -\sqrt{\beta\tau_w}$ is the gas state. Above the critical temperature, the system cannot exhibit liquid-gas coexistence. Mathematically speaking, the solution for ν_ρ becomes imaginary above T_c in this model. ρ_c and p_c are the density and pressure at the critical point of the material, where the liquid and gas phases are indistinguishable. In this model, the liquid and gas densities can be varied by tuning the value of $\beta\tau_w$, as shown in Fig. 9.5.

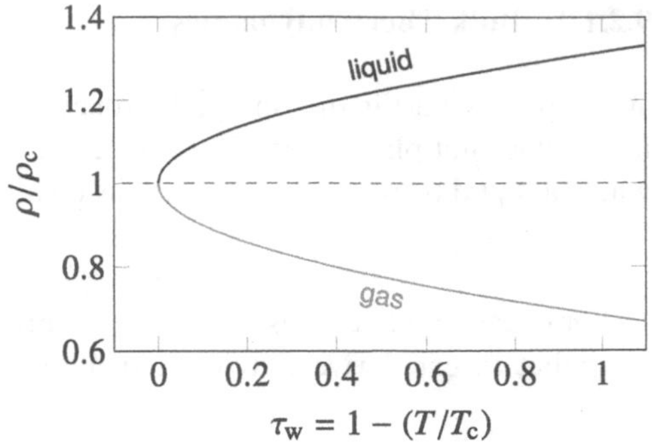

Fig. 9.5 The bulk liquid and gas densities for the Landau model given in (9.21) as a function of the reduced temperature τ_{w}. We have used $\beta = 0.1$ for this plot. Above the critical temperature T_{c} (i.e. for $\tau_{\mathrm{w}} < 0$) the liquid and gas phases are indistinguishable

The gradient term in (9.22) penalises changes in the density. This is key to the formulation of surface tension in this model. To appreciate this statement, let us first derive the *chemical potential*.

The chemical potential is defined as the free energy cost (gain) for adding (removing) materials to (from) the system. Mathematically, this is given by

$$\mu \equiv \frac{\delta(\psi_{\mathrm{b}} + \psi_{\mathrm{g}})}{\delta\rho} = \frac{4p_{\mathrm{c}}}{\rho_{\mathrm{c}}}\nu_\rho\left(\nu_\rho^2 - \beta\tau_{\mathrm{w}}\right) + \mu_0 - \kappa\Delta\rho. \tag{9.23}$$

In thermodynamic equilibrium, the chemical potential is constant everywhere in space. If it is not constant, there will be a free energy gain by transferring fluid material from one part of the system to another. In other words, there will be a *thermodynamic force*.

When the system is in one of the bulk free energy minimum solutions, either in the liquid or in the gas phase, then $\nu_\rho^2 - \beta\tau_{\mathrm{w}} = 0$ and the gradient term in (9.23) also vanishes. Therefore, we find $\mu = \mu_0$ in the liquid and gas bulk phases. Now, our statement that the chemical potential is constant everywhere in space includes the liquid-gas interface where the density varies. For simplicity, let us assume the interface is flat and is located at $x = 0$. The differential equation in (9.23), after setting $\mu = \mu_0$, thus reads

$$\frac{4p_{\mathrm{c}}}{\rho_{\mathrm{c}}}\nu_\rho\left(\nu_\rho^2 - \beta\tau_{\mathrm{w}}\right) - \kappa\rho_{\mathrm{c}}\Delta\nu_\rho = 0 \tag{9.24}$$

with boundary conditions $\nu_\rho = \pm\sqrt{\beta\tau_{\mathrm{w}}}$ (corresponding to the liquid and gas bulk densities) for $x = \pm\infty$.

Equation (9.24) has the following solution for the **liquid-gas interface profile**:

$$\nu_\rho = \sqrt{\beta\tau_{\mathrm{w}}}\tanh\left(\frac{x}{\sqrt{2}\xi}\right) \tag{9.25}$$

(continued)

where $\xi = \sqrt{\kappa\rho_c^2/(4\beta\tau_w p_c)}$ is defined as the interface width, as shown in Fig. 9.4. While ξ can take any value in the analytical model, to have a good numerical resolution in LBM, ξ is usually chosen to be a few lattice spacings.

Furthermore, the **surface tension of the liquid-gas interface** can then be calculated by integrating the free energy density across the interface:

$$\gamma_{\mathrm{lg}} = \int_{-\infty}^{\infty} \left[p_c \left(\nu_\rho^2 - \beta\tau_w \right)^2 + \frac{\kappa}{2}(\nabla\rho)^2 \right] \mathrm{d}x = \frac{4}{3}\sqrt{2\kappa p_c}(\beta\tau_w)^{3/2}\rho_c. \tag{9.26}$$

We can ignore the terms $\mu_0\rho$ and p_0 from the above integral because they only contribute to a constant in the free energy. An alternative derivation for the liquid-gas interfacial profile is also given in Appendix A.7 for the free-energy model.

9.2.1.2 Equations of Motion

In a multiphase model, the continuum equations of motion for the fluid are described by the continuity and Navier-Stokes equations (in the absence of external forces):

$$\partial_t\rho + \partial_\beta\left(\rho u_\beta\right) = 0, \tag{9.27}$$

$$\partial_t(\rho u_\alpha) + \partial_\beta\left(\rho u_\alpha u_\beta\right) = -\partial_\beta P_{\alpha\beta} + \partial_\beta\eta\left(\partial_\beta u_\alpha + \partial_\alpha u_\beta\right). \tag{9.28}$$

The thermodynamics of the multiphase system, including the description of the surface tension, enter the equations of motion through the pressure tensor $P_{\alpha\beta}$.

The **pressure tensor** $P_{\alpha\beta}$ can be derived (up to a constant contribution in space throughout the simulation domain) by requiring [47]

$$\partial_\beta P_{\alpha\beta} = \rho\partial_\alpha\mu. \tag{9.29}$$

This equation states that the presence of a thermodynamic force leads to a pressure tensor gradient for the fluids. This statement is general, not only for Landau models. For our specific Landau model, using the definition of the chemical potential μ in (9.23), it follows that

$$P_{\alpha\beta} = \left(p_b - \frac{\kappa}{2}(\partial_\gamma\rho)^2 - \kappa\rho\partial_\gamma\partial_\gamma\rho\right)\delta_{\alpha\beta} + \kappa(\partial_\alpha\rho)(\partial_\beta\rho), \tag{9.30}$$

$$p_b = p_c(\nu_\rho + 1)^2\left(3\nu_\rho^2 - 2\nu_\rho + 1 - 2\beta\tau_w\right). \tag{9.31}$$

(continued)

Equation (9.31) is the **equation of state** for this model. It relates the bulk pressure of the fluid to other thermodynamic quantities such as density and temperature.

Exercise 9.2 Starting from the equation of state in (9.31), show that the pressures at the bulk liquid and gas densities, $\nu_\rho = (\rho - \rho_c)/\rho_c = \pm\sqrt{\beta\tau_w}$, for the free-energy model in (9.20) are equal and satisfy

$$p_0 = p_c(1 - \beta\tau_w)^2. \tag{9.32}$$

Exercise 9.3 In (9.14) we defined the surface tension as the integral of the mismatch between the normal and transversal components of the pressure tensor. By evaluating the integral using (9.30), show that we can recover the same expression for the liquid-gas surface tension as in (9.26).

Exercise 9.4 An alternative approach to derive the pressure tensor is by exploiting equation (9.13) and the standard relation for the equation of state [40, 41]

$$p_b = \rho\partial_\rho\psi_b - \psi_b. \tag{9.33}$$

Verify that the pressure tensor derived using this approach is the same as in (9.30). Furthermore, by substituting the expressions for the pressure tensor and the chemical potential, show that (9.29) is satisfied.

9.2.1.3 The Lattice Boltzmann Algorithm

The thermodynamics of the multiphase flow is encoded in the modified pressure tensor. Therefore, the next step in free-energy LBM is to translate the description of the pressure tensor into the LB equation. This can be done either through a forcing term[8] F_i, a pressure tensor term G_i, or a mix between the two [48]:

$$\begin{aligned} f_i(\boldsymbol{x} + \boldsymbol{c}_i\Delta t, t + \Delta t) &= f_i(\boldsymbol{x}, t) - \frac{f_i(\boldsymbol{x}, t) - f_i^{\text{eq}}(\boldsymbol{x}, t)}{\tau}\Delta t \\ &\quad + \left(1 - \frac{\Delta t}{2\tau}\right) F_i(\boldsymbol{x}, t)\Delta t + G_i(\boldsymbol{x}, t)\Delta t. \end{aligned} \tag{9.34}$$

Here we will limit our discussion to the BGK collision operator. The extension to multiple relaxation times is straightforward.

[8] We note that our convention here follows that of Chap. 6. In the literature, sometimes the prefactor $\left(1 - \frac{1}{2\tau}\right)$ is included in the definition of F_i itself.

The properties of the forcing term F_i have been discussed in detail in Chap. 6. The moments of the pressure tensor term G_i have the following properties:

$$\begin{aligned}\sum_i G_i &= 0,\\ \sum_i G_i c_{i\alpha} &= 0,\\ \sum_i G_i c_{i\alpha} c_{i\beta} &= A_{\alpha\beta},\\ \sum_i G_i c_{i\alpha} c_{i\beta} c_{i\gamma} &= 0.\end{aligned} \tag{9.35}$$

It is important that only the second moment of the pressure tensor term is non-zero. It does not change the density or momentum of the fluid.

Pressure tensor approach: In the original free-energy LBM, the thermodynamics of the multiphase system is completely accounted for through a pressure tensor term, such that $A_{\alpha\beta} = P_{\alpha\beta} - c_s^2 \rho \delta_{\alpha\beta}$. This term corresponds to how the pressure tensor departs from the ideal gas scenario. The forcing term F_i is then set to zero, unless there is an additional external body force (e.g. gravity), which can be dealt with as in the usual way (cf. Chap. 6).

In the literature, it is further customary to absorb the **pressure tensor term** G_i into the the **equilibrium distribution** f_i^{eq}. Its resulting form for $i \neq 0$ is given by

$$\begin{aligned} f_i^{\mathrm{eq}} &= w_i \rho \left(1 + \frac{c_{i\alpha} u_\alpha}{c_s^2} + \frac{u_\alpha u_\beta \left(c_{i\alpha} c_{i\beta} - c_s^2 \delta_{\alpha\beta}\right)}{2 c_s^4}\right) \\ &+ \frac{w_i}{c_s^2}\left(p_{\mathrm{b}} - c_s^2 \rho - \kappa \rho \Delta \rho\right) + \kappa \sum_{\alpha,\beta} w_i^{\alpha\beta} (\partial_\alpha \rho)(\partial_\beta \rho). \end{aligned} \tag{9.36}$$

The corresponding expression for f_0^{eq} is given in (9.39). The first line in (9.36) is identical to that for standard single-phase and single-component equilibrium distribution functions. The second line accounts for surface tension (the gradient terms) and the deviation of the equation of state p_{b} from the ideal gas case, which must allow for a coexistence between the liquid and gas phases.

The density gradients can be computed using finite difference schemes. We strongly recommend to choose stencils which are at least second-order accurate. Inexactness in computing these derivatives is one of the main reasons for the appearance of spurious velocities that affect the accuracy of LBM, and in some cases its stability. Improving the stencil isotropy of the numerical derivatives has been shown to strongly reduce spurious velocities close to the liquid-gas interface [49–51].

The coefficients $w_i^{\alpha\beta}$ in (9.36) are chosen to minimise the spurious velocities at the interface. For example, following the work of Furtado and Pooley [51], for D3Q19 we can use

$$\begin{aligned}
w_{1,2}^{xx} = w_{3,4}^{yy} = w_{5,6}^{zz} &= \tfrac{5}{12},\\
w_{3-6}^{xx} = w_{1,2,5,6}^{yy} = w_{1-4}^{zz} &= -\tfrac{1}{3},\\
w_{1-6}^{xy} = w_{1-6}^{yz} = w_{1-6}^{zx} &= 0,\\
w_{7-10,13-16}^{xx} = w_{7-8,11-14,17-18}^{yy} = w_{9-12,15-18}^{zz} &= -\tfrac{1}{24},\\
w_{11,12,17,18}^{xx} = w_{9,10,15,16}^{yy} = w_{7,8,13,14}^{zz} &= \tfrac{1}{12},\\
w_{7,8}^{xy} = w_{11,12}^{yz} = w_{9,10}^{zx} &= \tfrac{1}{4},\\
w_{13,14}^{xy} = w_{17,18}^{yz} = w_{15,16}^{zx} &= -\tfrac{1}{4},\\
w_{9-12,15-18}^{xy} = w_{7-10,13-16}^{yz} = w_{7,8,11-14,17,18}^{zx} &= 0.
\end{aligned} \tag{9.37}$$

Similarly, for D2Q9 we have

$$\begin{aligned}
w_{1,2}^{xx} = w_{3,4}^{yy} &= \tfrac{1}{3},\\
w_{3,4}^{xx} = w_{1,2}^{yy} &= -\tfrac{1}{6},\\
w_{5-8}^{xx} = w_{5-8}^{yy} &= -\tfrac{1}{24},\\
w_{1-4}^{xy} &= 0,\\
w_{5-8}^{xy} &= \tfrac{1}{4}.
\end{aligned} \tag{9.38}$$

The expression for the equilibrium distribution function for $i = 0$ is quite lengthy, but in practice we can exploit conservation of mass to write

$$f_0^{\mathrm{eq}} = \rho - \sum_{i\neq 0} f_i^{\mathrm{eq}}. \tag{9.39}$$

Force approach: The thermodynamics of the multiphase system can be equivalently taken into account through a forcing term.

The appropriate **force density** due to the thermodynamics of a multiphase system is the divergence of the non-ideal terms in the pressure tensor

(continued)

$F_\alpha = -\partial_\beta(P_{\alpha\beta} - c_s^2\rho\delta_{\alpha\beta})$. Additional external forces (e.g. gravity) can also be added to the definition of the total force if they are present. We set $G_i = 0$ in (9.34) and the equilibrium distribution functions take identical forms as those for single-phase flow:

$$f_i^{eq} = w_i\rho\left(1 + \frac{c_{i\alpha}u_\alpha}{c_s^2} + \frac{u_\alpha u_\beta\left(c_{i\alpha}c_{i\beta} - c_s^2\delta_{\alpha\beta}\right)}{2c_s^4}\right), \tag{9.40}$$

$$F_i = w_i\left(\frac{\boldsymbol{c}_i - \boldsymbol{u}}{c_s^2} + \frac{(\boldsymbol{c}_i \cdot \boldsymbol{u})\,\boldsymbol{c}_i}{c_s^4}\right) \cdot \boldsymbol{F}, \tag{9.41}$$

$$\rho\boldsymbol{u} = \sum_i f_i\boldsymbol{c}_i + \frac{\boldsymbol{F}\Delta t}{2}. \tag{9.42}$$

Here we have used the method of Guo et al. [52] to implement the body force. More detailed explanations of the forcing term in LBM are discussed in Chap. 6.

At the level of numerical implementation, the forcing approach can be implemented in two different ways. First, in the so-called *pressure form* (not to be confused with the pressure tensor approach), the force density at every time step is computed as

$$F_\alpha = -\partial_\beta P_{\alpha\beta} + \partial_\alpha c_s^2\rho, \tag{9.43}$$

as we have written above. Secondly, in the *potential form*, the force density is

$$F_\alpha = -\rho\partial_\alpha\mu + \partial_\alpha c_s^2\rho. \tag{9.44}$$

Analytically the two forms are of course equivalent, cf. (9.29). However, upon discretisation, they are not exactly identical since the derivatives are usually approximated using finite difference schemes [53, 54]. Numerical evidence suggests that schemes which employ the potential form have lower spurious velocities [49, 55]. However, an important caveat is that the potential form is no longer written in a conservative form (i.e. as a divergence). This means momentum conservation is no longer satisfied exactly for the discretised potential form [53]. In his implementation, Wagner [55] also introduced a small amount of numerical viscosity to render the simulations stable.

As we shall see below, the pressure tensor approach as currently stated is inadequate for most applications because it does not satisfy Galilean invariance. The pressure tensor approach generally also produces higher spurious velocities [51, 55]. However, an advantage of using the pressure tensor approach is that we do not need to compute the third-order derivative in ρ, which is unavoidable in the force approach (potential form). Computing third-order derivatives require information from more neighbours, which affect parallelisation, and are more expensive to compute. In this context, to avoid computing third derivatives, a possible hybrid approach is to rewrite (9.29) as

$$\partial_\beta P_{\alpha\beta} = \rho \partial_\alpha \mu = \partial_\alpha(\mu\rho) - \mu \partial_\alpha \rho. \tag{9.45}$$

The first term can be absorbed in the equilibrium distribution, by defining a modified isotropic pressure $\tilde{p}_b = \mu\rho$, while the second term is introduced as a forcing term $F_\alpha = -\mu\partial_\alpha\rho$. The suitable equilibrium distribution function for $i > 0$ in the hybrid approach is then

$$f_i^{eq} = w_i\rho\left(\frac{\mu}{c_s^2} + \frac{c_{i\alpha}u_\alpha}{c_s^2} + \frac{u_\alpha u_\beta\left(c_{i\alpha}c_{i\beta} - c_s^2\delta_{\alpha\beta}\right)}{2c_s^4}\right). \tag{9.46}$$

9.2.1.4 Galilean Invariance

The original free-energy LB algorithm was shown to break Galilean invariance [48, 56, 57]. This is a serious limitation as real physical phenomena do not depend on the frame of reference. To appreciate this issue, let us consider a Chapman-Enskog analysis of (9.34). The analysis results in the continuity equation and the following momentum conservation equation [48]:

$$\begin{aligned}\partial_t(\rho u_\alpha) + \partial_\beta\left(\rho u_\alpha u_\beta\right) &= -\partial_\beta\left(A_{\alpha\beta} + c_s^2\rho\delta_{\alpha\beta}\right) + F_\alpha \\ &\quad + \partial_\beta\rho\nu\left(\partial_\beta u_\alpha + \partial_\alpha u_\beta + \partial_\gamma u_\gamma\delta_{\alpha\beta}\right) \\ &\quad - \nu\partial_\beta\left(u_\alpha\partial_\gamma A_{\beta\gamma} + u_\beta\partial_\gamma A_{\alpha\gamma} + \partial_\rho A_{\alpha\gamma}\partial_\gamma(\rho u_\gamma)\right) \\ &\quad + \nu\partial_\beta\partial_\gamma(\rho u_\alpha u_\beta u_\gamma).\end{aligned} \tag{9.47}$$

We remember that $A_{\alpha\beta} = P_{\alpha\beta} - c_s^2\rho\delta_{\alpha\beta}$ and $F_\alpha = 0$ for the pressure tensor approach, while $A_{\alpha\beta} = 0$ and $F_\alpha = -\partial_\beta(P_{\alpha\beta} - c_s^2\rho\delta_{\alpha\beta})$ for the forcing approach, as described above. The first two rows on the right-hand side of (9.47) correspond to the desired Navier-Stokes equation. The last term is an error term which is also present for standard LBM, cf. Sect. 4.1. This term is negligible for $\mathrm{Ma}^2 \ll 1$, which is usually the case for multiphase flow.

Thus, the problematic terms are those in the third row. They break Galilean invariance in the free-energy multiphase model. To restore it, we need to add appropriate correction terms. As an example, let us consider the pressure tensor approach where $A_{\alpha\beta} = P_{\alpha\beta} - c_s^2\rho\delta_{\alpha\beta}$. To restore Galilean invariance, one option is to introduce a body force

$$F_\alpha = \nu\partial_\beta\left(u_\alpha\partial_\gamma A_{\beta\gamma} + u_\beta\partial_\gamma A_{\alpha\gamma} + \partial_\rho A_{\alpha\gamma}\partial_\gamma(\rho u_\gamma)\right) \tag{9.48}$$

which cancels all the leading error terms in (9.47). Another option is to modify $A_{\alpha\beta}$ such that

$$A_{\alpha\beta} = P_{\alpha\beta} - c_s^2 \rho \delta_{\alpha\beta} - \nu \left(\partial_\alpha(\rho u_\beta) + \partial_\beta(\rho u_\alpha) + \partial_\gamma(\rho u_\gamma)\delta_{\alpha\beta} \right). \tag{9.49}$$

This approach cancels most of the error terms, leaving terms which are proportional to the second derivatives of the fluid pressure, which are usually small for systems close to equilibrium and for moderate Reynolds number.

9.2.1.5 A Practical Guide to Simulation Parameters

The Landau free-energy model is originally designed to describe physical multi-phase systems close to the critical point beyond which the liquid and gas phases are no longer distinguishable. However, in practice the Landau model has been exploited for liquid-gas systems far from the critical point. This assumption is valid for cases where the details of the equation of state are irrelevant for the problems at hand. For more realistic equations of state, the van der Waals [39, 40, 44] or Peng-Robinson models [45] should be used.

In the Landau model there are effectively four free parameters to tune the thermodynamics of the liquid-gas system: $\beta\tau_w$, κ, p_c and ρ_c. As a starting point, the following **parameters** are often chosen **for the multiphase Landau model**, all in lattice units: $\beta\tau_w = 0.03$, $\kappa = 0.004$, $p_c = 0.125$ and $\rho_c = 3.5$ [41, 58]. These parameters can be modified with considerations as detailed below.

- $\beta\tau_w$ is important for adjusting the liquid-gas density ratio, since $\rho_l/\rho_g = (1 + \sqrt{\beta\tau_w})/(1 - \sqrt{\beta\tau_w})$. Theoretically, $\beta\tau_w$ can be chosen such that the density ratio is ~ 1000, which is the case for most liquid-gas systems. In practice, however, this is not possible with the algorithms described thus far. The highest density ratio is limited to ~ 10. This is due to spurious velocities which act to destabilise the simulations. We will discuss this issue in Sect. 9.4.1 and Sect. 9.4.2, together with approaches that have been developed to reduce these spurious velocities, and hence achieve realistic density ratios.
- Combined with $\beta\tau_w$, we can choose ρ_c to tune the actual values of the liquid and gas densities, since $\rho_{l,g} = \rho_c(1 \pm \sqrt{\beta\tau_w})$.
- Given the choices for $\beta\tau_w$ and ρ_c, we can use κ and p_c to control two physically meaningful variables: the interface width $\xi = \sqrt{\kappa\rho_c^2/(4\beta\tau_w p_c)}$ and the surface tension $\gamma_{lg} = \frac{4}{3}\sqrt{2\kappa p_c}(\beta\tau_w)^{3/2}\rho_c$. In LBM we usually want the interface width ξ to be $\sim 2 - 3$ lattice spacings. Anything smaller will result in the interfacial profile being resolved very poorly. Larger ξ is in principle better, but this makes

the simulation very expensive, since any relevant physical length scale in the simulation should be at least an order of magnitude larger. Similar to the density ratio, the possible values of the surface tension are usually limited by spurious velocities. As a rule of thumb, the generated spurious velocity increases with the chosen surface tension. In the Landau model, the surface tension (in lattice units) is usually taken to be $\gamma_{\mathrm{lg}} \sim O(10^{-2})$ or less.

It is worth noting that one useful advantage of the Landau free-energy model is that we can compute all equilibrium quantities analytically. This allows us to initialise the desired liquid and gas domains with appropriate densities given by $\rho_{\mathrm{l,g}} = \rho_{\mathrm{c}}(1 \pm \sqrt{\beta\tau_{\mathrm{w}}})$. Furthermore, rather than a step density change, it is advisable to implement a smooth interface, following the tanh profile[9] given in (9.25), with an interface width $\xi = \sqrt{\kappa\rho_{\mathrm{c}}^2/(4\beta\tau_{\mathrm{w}}p_{\mathrm{c}})}$.

Exercise 9.5 Perform the Young-Laplace test as presented in Sect. 9.1.4. Prepare a liquid droplet of radius R surrounded by the gas phase in a periodic system. After the system has equilibriated, measure the bulk liquid and gas densities. Compute the corresponding bulk pressures from (9.31). You will notice that bulk liquid and gas densities deviate slightly from $\rho = \rho_c(1 \pm \sqrt{\beta\tau_{\mathrm{w}}})$. This is because the interface is no longer flat, and the deviation is necessary to account for the Laplace pressure. Make sure you compute these values far from the liquid-gas interface. Repeat this calculation for several values of droplet radius R and plot the pressure difference as function of $1/R$. Note the differences between 2D and 3D as pointed out in (9.18). You should reproduce a similar result to Fig. 9.10 as obtained with the Shan-Chen method.

9.2.1.6 Surface Thermodynamics

In multiphase flows we are often interested in cases where the fluids are in contact with solid surfaces. In general, the liquid and gas phases can interact differently with the surface, which results in the liquid phase wetting or dewetting the surface. The degree of affinity for the liquid and gas phases is often described by the contact angle, as defined in (9.9).

Similar to the bulk thermodynamics, we have different options to represent the **surface thermodynamics**. Following Cahn [59], we choose a surface energy term

$$\Psi_{\mathrm{s}} = \int_A \psi_s \, \mathrm{d}A = -\int_A h\rho_{\mathrm{s}} \, \mathrm{d}A \tag{9.50}$$

(continued)

[9] In fact, the tanh profile provides a good initial interfacial profile for most multiphase and multicomponent models.

where ρ_s is the value of the density at the surface and the integral is taken over the solid surface in our simulation domain. The parameter h is an effective interaction potential between the fluid and the solid surface, which we take to be constant. If h is positive, the fluid molecules interact favourably with the surface, and as such the liquid phase will be preferred over the gas phase close to the surface, as it lowers the free energy of the system more than the gas phase. If h is negative, the solid-fluid interaction is unfavourable, and the gas phase is preferably close to the solid surface. We will see later in (9.65) that the parameter h enters the LB equation as a boundary condition for the density gradient.

To derive an explicit relation between the variable h and the contact angle θ, we will use techniques from calculus of variation. First we will obtain the liquid and gas densities at the solid surface, followed by the solid-liquid and solid-gas surface tensions. Then, using Young's formula in (9.9), we can relate the surface tensions to the contact angle.

Let us start by computing the free energy changes upon variation in the fluid density in (9.19). The rationale behind this is that the contact angle is an equilibrium material parameter, and at equilibrium, the free energy functional is minimised. We find that

$$\begin{aligned}
\delta\Psi &= \int_V \left[\frac{4p_c}{\rho_c}\nu_\rho\left(\nu_\rho^2 - \beta\tau_w\right) + \mu_0\right]\delta\rho\,\mathrm{d}V + \int_V \kappa(\nabla\rho)\cdot(\nabla\delta\rho)\,\mathrm{d}V \\
&\quad - \int_A h\delta\rho_s\,\mathrm{d}A \\
&= \int_V \left[\frac{4p_c}{\rho_c}\nu_\rho\left(\nu_\rho^2 - \beta\tau_w\right) + \mu_0 - \kappa\Delta\rho\right]\delta\rho\,\mathrm{d}V + \int_V \kappa\nabla\cdot(\delta\rho\nabla\rho)\,\mathrm{d}V \\
&\quad - \int_A h\delta\rho_s\,\mathrm{d}A \\
&= \int_V \left[\frac{4p_c}{\rho_c}\nu_\rho\left(\nu_\rho^2 - \beta\tau_w\right) + \mu_0 - \kappa\Delta\rho\right]\delta\rho\,\mathrm{d}V \\
&\quad + \int_A \kappa\left[(\nabla\rho\cdot\hat{\boldsymbol{n}}) - h\right]\delta\rho_s\,\mathrm{d}A.
\end{aligned} \tag{9.51}$$

To derive (9.51), we have used the divergence theorem to convert one of the volume integrals into a surface integral. We have also used the convention that $\hat{\boldsymbol{n}}$ is the unit vector normal to the surface (pointing outward, not inward). Now the integrand for the volume integral is nothing but our definition for the chemical potential in (9.23). The new term in the presence of a solid surface is the surface integral. Setting

$\delta\Psi/\delta\rho_s = 0$, which is valid for an equilibrium solution, we obtain

$$\kappa(\nabla\rho \cdot \hat{\boldsymbol{n}}) = \kappa\nabla_\perp\rho = h. \tag{9.52}$$

This sets the gradient of the density normal to the solid surface.

To compute the liquid and gas densities at the solid surface, we will exploit the so-called *Noether theorem* [60]. In our context, this allows us to compute a quantity which is conserved across the spatial dimension at equilibrium:

$$\frac{\delta(\psi_b + \psi_g)}{\delta(\nabla\rho)} \cdot \nabla\rho - (\psi_b + \psi_g) + \mu_0\rho = \text{const} \tag{9.53}$$

and therefore

$$\frac{\kappa}{2}(\nabla\rho)^2 - p_c\left(\nu_\rho^2 - \beta\tau_w\right)^2 = \text{const} = 0. \tag{9.54}$$

For the last equality, we have used the fact that far from the interface (in the bulk), $\nu_\rho = \pm\sqrt{\beta\tau_w}$ and $\nabla\rho = \mathbf{0}$. Substituting (9.52) into (9.54), we find that the values of the density at the surface may take four possible values corresponding to

$$\nu_{\rho,s} = \frac{\rho_s - \rho_c}{\rho_c} = \pm\sqrt{\beta\tau_w \pm h\sqrt{\frac{1}{2\kappa p_c}}}. \tag{9.55}$$

To decide which solutions are physically admissible, let us consider the following argument. If $h > 0$, we argued above that the fluid molecules have favourable interactions with the solid surface. As such, we expect to have a local increase in the fluid density close to the surface. The relevant solutions for the liquid and gas phases are thus

$$\nu_{sl} = \sqrt{\beta\tau_w}\sqrt{1 + \frac{h}{\beta\tau_w\sqrt{2\kappa p_c}}} = \sqrt{\beta\tau_w}\sqrt{1 + \Omega}, \tag{9.56}$$

$$\nu_{sg} = -\sqrt{\beta\tau_w}\sqrt{1 - \frac{h}{\beta\tau_w\sqrt{2\kappa p_c}}} = -\sqrt{\beta\tau_w}\sqrt{1 - \Omega} \tag{9.57}$$

where we have defined $\Omega = h/(\beta\tau_w\sqrt{2\kappa p_c})$. It is also straightforward to see that for $h < 0$ the above solutions give us a local decrease in fluid density close to the surface.

The solid-liquid and solid-gas surface tensions can be calculated similarly to the liquid-gas surface tension in (9.26), except that now we also have to take into account the contributions from the surface energy term. For the solid-liquid surface tension, assuming a flat solid interface at $x = 0$, we find

$$\gamma_{sl} = -h\rho_{sl} + \int_0^\infty \left[p_c\left(\nu_\rho^2 - \beta\tau_w\right)^2 + \frac{\kappa}{2}(\nabla\rho)^2\right]\mathrm{d}x. \tag{9.58}$$

To evaluate this integral, we take advantage of Noether's theorem given in (9.54) and use a change of variables such that

$$\int_0^\infty \left[p_c \left(\nu_\rho^2 - \beta \tau_w \right)^2 + \frac{\kappa}{2} \left(\frac{d\rho}{dx} \right)^2 \right] dx = \int_{\nu_{sl}}^{\nu_l} \left[\rho_c \sqrt{2\kappa p_c} \left(\beta \tau_w - \nu_\rho^2 \right) \right] d\nu_\rho. \tag{9.59}$$

After some algebra, it is then possible to show that

$$\gamma_{sl} = -h\rho_c + \frac{\gamma_{lg}}{2} - \frac{\gamma_{lg}}{2}(1+\Omega)^{3/2} \tag{9.60}$$

with γ_{lg} given in (9.26).

The solid-gas surface tension can be derived in a similar way, and we obtain

$$\gamma_{sg} = -h\rho_c + \frac{\gamma_{lg}}{2} - \frac{\gamma_{lg}}{2}(1-\Omega)^{3/2}. \tag{9.61}$$

The contact angle follows from substituting the values of the surface tensions into Young's law in (9.9) to give

$$\cos\theta = \frac{\gamma_{sg} - \gamma_{sl}}{\gamma_{lg}} = \frac{(1+\Omega)^{3/2} - (1-\Omega)^{3/2}}{2}. \tag{9.62}$$

Equation (9.62) can be inverted to give a relation between the phenomenological parameter Ω and the equilibrium contact angle θ:

$$\Omega(\theta) = 2\,\mathrm{sgn}(\pi/2 - \theta)\sqrt{\cos(\alpha/3)[1 - \cos(\alpha/3)]} \tag{9.63}$$

where $\alpha(\theta) = \arccos(\sin^2\theta)$ and the function $\mathrm{sgn}(x)$ returns the sign of x. Figure 9.6 shows how Ω depends on the contact angle θ in (9.63).

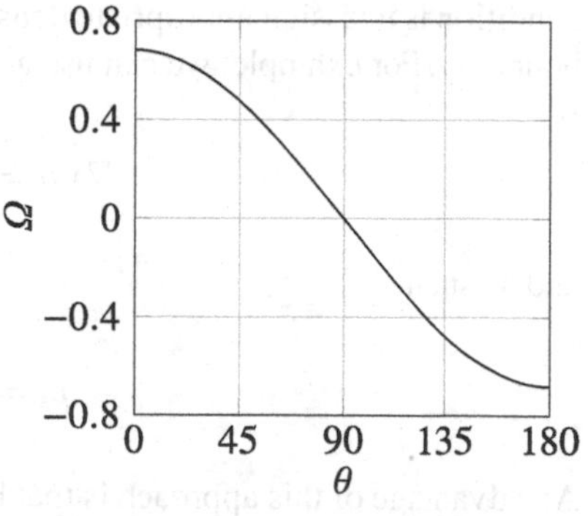

Fig. 9.6 The phenomenological parameter $\Omega = h/(\beta\tau_w\sqrt{2\kappa p_c})$ as a function of the contact angle θ, corresponding to (9.63)

9.2.1.7 Wetting Boundary Condition

At this point, we have elaborated on the surface thermodynamics required to introduce preferential wetting between the liquid and gas phases on the solid surface. When dealing with multiphase flow (similarly multicomponent flow) using free-energy LBM, in addition to standard no-slip boundary condition for the fluid velocity, we also need an additional wetting boundary condition.

To realise the **wetting condition in the LB equation**, we need **two key equations**. First, from (9.63) we have an analytical expression relating the desired contact angle θ to the parameter h which is an input in our model:

$$h = \beta\tau_{\mathrm{w}}\sqrt{2\kappa p_{\mathrm{c}}}\Omega = 2\beta\tau_{\mathrm{w}}\sqrt{2\kappa p_{\mathrm{c}}}\,\mathrm{sgn}(\pi/2-\theta)\sqrt{\cos(\alpha/3)[1-\cos(\alpha/3)]}. \tag{9.64}$$

Secondly, the parameter h enters the LB equation through the wetting boundary condition in (9.52), which sets the gradient of the density normal to the solid surface:

$$\nabla\rho\cdot\hat{\boldsymbol{n}} = \nabla_{\perp}\rho = \frac{h}{\kappa}. \tag{9.65}$$

The unit normal vector $\hat{\boldsymbol{n}}$ points outward, i.e. into the solid.

We note that the boundary condition in (9.65) specifies the normal gradient of the density at the solid surface. If we use the bounce-back approach to implement the no-slip boundary condition, then the position of the solid wall is usually halfway between two lattice nodes, with a fluid (subscript f below) and a solid (subscript s) node on each side of the wall. A common implementation of the wetting boundary condition is to assign appropriate density values to the solid nodes neighbouring the boundary. For example, we can use a standard finite difference scheme to write

$$\nabla_{\perp}\rho = \frac{\rho_{\mathrm{s}}-\rho_{\mathrm{f}}}{\Delta x} = \frac{h}{\kappa}, \tag{9.66}$$

and as such,

$$\rho_{\mathrm{s}} = \rho_{\mathrm{f}} + \frac{h}{\kappa}\Delta x. \tag{9.67}$$

An advantage of this approach is that higher order gradients can be calculated in the same way as in the bulk, since neighbouring solid nodes are assigned appropriate density values. For more complex geometries, for example surfaces which do not follow a lattice axis, the wetting boundary conditions can be implemented in a similar way. This typically gives a set of linear equations that must be solved

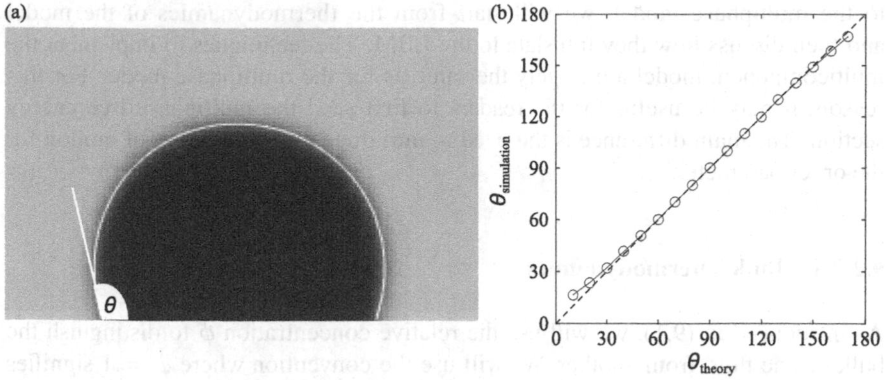

Fig. 9.7 (a) A typical simulation result for the contact angle test, as discussed in Exercise 9.6. The white line marks the contour where the local density assumes the value $(\rho_l + \rho_g)/2$. The contact angle is measured locally at the contact line where the liquid-gas interface meets the solid. **(b)** The comparison between the prescribed and the measured contact angles. Excellent agreement is obtained apart from very small and very large contact angles. This discrepancy is due to the finite width of the interface

simultaneously. Furthermore, if a solid node is surrounded by several fluid nodes (e.g. a corner), its density value can be defined in multiple ways via (9.67). In this case, we usually then take the average value.

Exercise 9.6 Implement the wetting boundary condition. Following (9.64), compute the suitable value of h for a given contact angle θ. This effectively sets the value of the density gradient normal to the solid surface, as discussed in (9.65). To set up the simulation, prepare solid nodes at the top and bottom of your simulation box, as shown in Fig. 9.7a. You may use periodic boundary conditions in the other directions. Then place a liquid droplet of radius R (use e.g. $R = 30\Delta x$) surrounded by the gas phase next to one of the solid planes, and let the system equilibrate. After the simulation reaches a steady state, measure the contact angle the droplet forms with the solid surface and compare the result with the prescribed contact angle. This is best done by fitting the interfacial profile of the droplet to the equation for a circle (in 2D) or a sphere (in 3D). The contact angle is the angle formed by the liquid-gas interface and the solid surface. Repeat the simulation for contact angles ranging from 0° to 180°. You should obtain excellent agreement to within 2–3° for $\theta = 20$–160°. Larger deviations are observed for very small and very large contact angles, as shown in Fig. 9.7b.

9.2.2 Binary Fluid Model

We will now move on to a multicomponent system. For simplicity, we will consider a binary fluid where the fluid is a mixture of two distinct fluid components. Similar to the multiphase model, we will start from the thermodynamics of the model and then discuss how they translate to the LBM. The techniques to implement the multicomponent model are largely the same as for the multiphase model. For this reason, it may be useful for the readers to first read the multiphase free energy section. The main difference is the need to introduce a new equation of motion for the order parameter.

9.2.2.1 Bulk Thermodynamics

As introduced in (9.1), we will use the relative concentration ϕ to distinguish the bulk of one fluid from another. We will use the convention where $\phi = 1$ signifies fluid 1 (e.g. water) and $\phi = -1$ fluid 2 (e.g. oil). The isosurface $\phi = 0$ then corresponds to the interface between the two fluids (e.g. oil-water interface). Given these conventions, we want to construct a free energy functional that has two minima at $\phi = \pm 1$ and provides an energy penalty which scales with the area of the fluid-fluid interfaces. The proportionality constant is the surface tension. We will start with the bulk properties far away from any solid boundary.

The **simplest model for a multicomponent system** which captures the physics mentioned above is given by the following **Landau free energy** [42, 61]:

$$\Psi = \int_V \left[\psi_\mathrm{b} + \psi_\mathrm{g}\right] \mathrm{d}V = \int_V \left[c_\mathrm{s}^2 \rho \ln \rho + \frac{A}{4}\left(\phi^2 - 1\right)^2 + \frac{\kappa}{2}(\nabla\phi)^2\right] \mathrm{d}V. \tag{9.68}$$

The first term in the bracket is the ideal gas free energy, and we will assume here that the density of the two fluids are the same. Otherwise we would need terms that couple the density ρ and the order parameter ϕ. For the multicomponent Landau model, the second term is key, and it is easy to see that it has two bulk minima at $\phi \pm 1$ for $A > 0$. When $A < 0$ the two fluids are miscible. Extensions to more than two fluid components have also been proposed, in particular for ternary systems [43, 62, 63]. The final term, the gradient term, accounts for surface tension.

Let us consider the fluid-fluid interface and derive an analytical expression for both the surface tension and the interface width. Taking the functional derivative of the free energy in (9.68) with respect to ϕ leads to an equation for the *chemical potential*:

$$\mu \equiv \frac{\delta(\psi_\mathrm{b} + \psi_\mathrm{g})}{\delta\phi} = -A\phi + A\phi^3 - \kappa\Delta\phi = \mathrm{const.} \tag{9.69}$$

We know that in equilibrium the chemical potential must be constant in space. Otherwise there would be a thermodynamic force density corresponding to $\boldsymbol{F} = -\phi \nabla \mu$. We can set the constant in the above equation to zero, $\mu = 0$, by taking the bulk behaviour of fluids 1 or 2, where $\phi = \pm 1$ and $\Delta \phi = 0$.

For simplicity, we shall now assume that the interface between the two fluids is flat and located at $x = 0$. The bulk behaviour at $x = \pm\infty$ is such that $\phi = \pm 1$, respectively.

Equation (9.69) allows an **interface solution** of the form

$$\phi = \tanh\left(\frac{x}{\sqrt{2}\xi}\right) \tag{9.70}$$

where $\xi = \sqrt{\kappa/A}$ is defined as the **interface width**. The **surface tension** of the interface between fluids 1 and 2 can then be calculated by integrating the free energy density across the interface. Using (9.70) for the order parameter profile, we obtain

$$\gamma_{12} = \int_{-\infty}^{\infty} \left[\frac{A}{4}\left(\phi^2 - 1\right)^2 + \frac{\kappa}{2}(\nabla \phi)^2\right] \mathrm{d}x = \sqrt{\frac{8\kappa A}{9}}. \tag{9.71}$$

9.2.2.2 Surface Thermodynamics

To account for the interactions between the fluids and the solid, here we can prescribe a surface energy contribution given by [59]

$$\Psi_\mathrm{s} = \int_A \psi_\mathrm{s} \, \mathrm{d}A = -\int_A h\phi_\mathrm{s} \, \mathrm{d}A \tag{9.72}$$

where ϕ_s is the value of the order parameter at the surface and the integral is taken over the system's solid surface. The readers will notice that this has the same form as for the liquid-gas model, except that ψ_s now depends on the order parameter ϕ, and not the density.

Minimisation of the total free energy functional with respect to ϕ at the solid boundary leads to a Neumann boundary condition in the gradient of order parameter ϕ:

$$\kappa(\nabla \phi \cdot \hat{\boldsymbol{n}}) = \kappa \nabla_\perp \phi = h. \tag{9.73}$$

The important consequence of this equation is that the contact angle of the surface can be implemented by setting the perpendicular derivative of the order parameter. In our convention, the normal unit vector is pointing outward, into the solid. We leave the derivation of (9.73) to the readers since the mathematical steps are identical to those for the liquid-gas model in Sect. 9.2.1.

The variable h can be used to tune the contact angle. For $h > 0$, fluid 1 interacts favourably with the solid (more wetting) compared to fluid 2. The opposite is true for $h < 0$. To derive an explicit relation between the variable h and the contact angle θ, we can exploit Noether's theorem [60]

$$\frac{\delta(\psi_{\mathrm{b}} + \psi_{\mathrm{g}})}{\delta(\boldsymbol{\nabla}\phi)} \cdot \boldsymbol{\nabla}\phi - (\psi_{\mathrm{b}} + \psi_{\mathrm{g}}) = \mathrm{const} \tag{9.74}$$

to show that

$$\frac{\kappa}{2}(\boldsymbol{\nabla}\phi)^2 - \frac{A}{4}\left(\phi^2 - 1\right)^2 = \mathrm{const} = 0. \tag{9.75}$$

In the last step, we have used the fact that far from the interface (in the bulk), $\phi = \pm 1$ and $\boldsymbol{\nabla}\phi = \mathbf{0}$.

Substituting (9.73) into (9.75), we find that the values of the order parameter at the surface may take four possible values: $\phi_{\mathrm{s}} = \pm(1 \pm (2h^2/\kappa A)^{1/2})^{1/2}$. To decide which solutions are physically admissible, let us consider the following arguments. For fluid 1, the bulk solution is given by $\phi = 1$. If $h > 0$, there is an energy gain for having higher concentration of fluid 1 at the solid surface such that the order parameter for fluid 1 at the solid surface $\phi_{\mathrm{s1}} > 1$. On the other hand, if $h < 0$, it is favourable to have $\phi_{\mathrm{s1}} < 1$. A solution satisfying this requirement is

$$\phi_{\mathrm{s1}} = +\sqrt{1 + \sqrt{\frac{2}{\kappa A}}h}. \tag{9.76}$$

Using similar arguments, we can conclude that for fluid 2, the order parameter at the solid surface is

$$\phi_{\mathrm{s2}} = -\sqrt{1 - \sqrt{\frac{2}{\kappa A}}h}. \tag{9.77}$$

The fluid-solid surface tensions can be calculated in a similar way to the fluid-fluid surface tension, except that now we also have to take into account the contributions from the surface energy term. For the tension between fluid 1 and the solid surface, assuming a flat interface at $x = 0$, we have

$$\gamma_{\mathrm{s1}} = -h\phi_{\mathrm{s1}} + \int_0^\infty \left[\frac{A}{4}\left(\phi^2 - 1\right)^2 + \frac{\kappa}{2}(\boldsymbol{\nabla}\phi)^2\right] \mathrm{d}x. \tag{9.78}$$

To evaluate this integral, we take advantage of Noether's theorem in (9.75) and introduce a change of variables:

$$\int_0^\infty \left[\frac{A}{4}\left(\phi^2 - 1\right)^2 + \frac{\kappa}{2}\left(\frac{\mathrm{d}\phi}{\mathrm{d}x}\right)^2\right] \mathrm{d}x = \int_{\phi_{\mathrm{s1}}}^1 \left[\sqrt{\frac{A\kappa}{2}}\left(1 - \phi^2\right)\right] \mathrm{d}\phi. \tag{9.79}$$

Using the definition of ϕ_{s1} given in (9.76), it is straightforward to show that

$$\gamma_{s1} = \frac{\gamma_{12}}{2}\left[1 - (1+\Omega)^{3/2}\right] \tag{9.80}$$

where $\Omega = h\sqrt{2/(\kappa A)}$ and $\gamma_{12} = \sqrt{8\kappa A/9}$. The surface tension between fluid 2 and the solid can be derived in a similar way, and we obtain

$$\gamma_{s2} = \frac{\gamma_{12}}{2}\left[1 - (1-\Omega)^{3/2}\right]. \tag{9.81}$$

The contact angle follows from substituting the values of the surface tensions into Young's law, (9.9), to give (with θ defined as the contact angle of fluid 1)

$$\cos\theta = \frac{\gamma_{s2} - \gamma_{s1}}{\gamma_{12}} = \frac{(1+\Omega)^{3/2} - (1-\Omega)^{3/2}}{2}. \tag{9.82}$$

Equation (9.82) can be inverted to give a relation between the phenomenological parameter Ω and the equilibrium contact angle θ:

$$\Omega = 2\,\mathrm{sgn}(\pi/2 - \theta)\sqrt{\cos(\alpha/3)\left[1 - \cos(\alpha/3)\right]} \tag{9.83}$$

where $\alpha(\theta) = \arccos(\sin^2\theta)$ and the function $\mathrm{sgn}(x)$ returns the sign of x. This is exactly the same equation as in (9.63), except for the definition of $\Omega = h\sqrt{2/(\kappa A)}$ in the multicomponent model.

9.2.2.3 Equations of Motion

Before we write down the LB equations for a binary fluid, let us review the corresponding continuum equations of motion. The fluid motion is described by the continuity and Navier-Stokes equations, as described in (9.27) and (9.28). The key additional physics due to the thermodynamics of a binary fluid is contained in the *pressure tensor*. The pressure tensor needs to satisfy the condition

$$\partial_\beta P_{\alpha\beta} = \rho\partial_\alpha\left[\frac{\delta(\psi_b + \psi_g)}{\delta\rho}\right] + \phi\partial_\alpha\left[\frac{\delta(\psi_b + \psi_g)}{\delta\phi}\right]. \tag{9.84}$$

This is a generalisation of (9.29), where in principle we now have two variables, the density ρ and the order parameter ϕ, which can vary in space. This equation can be simplified to

$$\partial_\beta P_{\alpha\beta} = \partial_\alpha(c_s^2\rho) + \phi\partial_\alpha\mu. \tag{9.85}$$

The first term has the same form as the hydrodynamic pressure for the standard lattice Boltzmann model. The thermodynamic of the multicomponent model is contained in the second term. We remind the reader that, in thermodynamic equilibrium, the chemical potential has to be the same everywhere. Any inhomogeneity leads to a body force proportional to the gradient of the chemical potential, driving the system to equilibrium. Using the definition of μ in (9.69), it follows that

$$P_{\alpha\beta} = \left(p_\mathrm{b} - \frac{\kappa}{2}(\partial_\gamma\phi)^2 - \kappa\phi\partial_\gamma\partial_\gamma\phi\right)\delta_{\alpha\beta} + \kappa(\partial_\alpha\phi)(\partial_\beta\phi), \tag{9.86}$$

$$p_\mathrm{b} = c_\mathrm{s}^2\rho + A\left(-\frac{1}{2}\phi^2 + \frac{3}{4}\phi^4\right). \tag{9.87}$$

Equation (9.87) is the *equation of state* for the binary fluid model. p_b can be interpreted as the bulk pressure far from the interface, where the gradient terms are zero. In this model, the value of ϕ usually deviates slightly from ± 1 in the bulk when there is a Laplace pressure difference between the two fluid domains.

The order parameter itself evolves through the **Cahn-Hilliard equation**

$$\frac{\partial\phi}{\partial t} + \nabla\cdot(\phi\boldsymbol{u}) = \nabla\cdot\left(M\nabla\mu\right). \tag{9.88}$$

This equation is also sometimes called the *interface-capturing equation* in multicomponent flows. The second term on the left-hand side is an advection term, where the order parameter moves along with the fluid. The diffusive term on the right-hand side accounts for the motion of the order parameter due to inhomogeneities in the chemical potential. In many cases, the mobility parameter M is taken to be constant, although in general it is a function of the fluid order parameter [64, 65]. The mobility parameter is also important in the context of contact line motion, as it controls the effective contact line slip length [66].

9.2.2.4 The Lattice Boltzmann Algorithm

We now describe an LB algorithm that solves (9.27), (9.28), and (9.88). For a binary fluid, we need to define two distribution functions, $f_i(\boldsymbol{x}, t)$ and $g_i(\boldsymbol{x}, t)$, corresponding to the density and relative concentration of the two fluids. The physical variables are related to the distribution functions by

$$\rho = \sum_i f_i, \quad \rho u_\alpha = \sum_i f_i c_{i\alpha} + \frac{F_\alpha \Delta t}{2}, \quad \phi = \sum_i g_i, \tag{9.89}$$

Here we have chosen to evolve both the density and the order parameter using LBM. This is not a requirement at all. It is possible to solve the continuity and Navier-Stokes equations using LBM (via the $f_i(\boldsymbol{x}, t)$ only), and solve the Cahn-Hilliard equation using a different method (e.g. finite difference).

The equations of motion for multicomponent flows are similar to those for advection-diffusion systems described in Chap. 8. As usual, the LB algorithm for multicomponent flows can be broken into two steps. For simplicity, we here use the BGK collision operator. Extensions to the MRT collision operator follow the same route as described in Chap. 10. The collision and propagation steps read

$$\begin{aligned} f_i^\star(\boldsymbol{x}, t) &= f_i(\boldsymbol{x}, t) - \frac{\Delta t}{\tau}(f_i(\boldsymbol{x}, t) - f^{\mathrm{eq}_i}(\boldsymbol{x}, t)) + \left(1 - \frac{\Delta t}{2\tau}\right) F_i(\boldsymbol{x}, t)\Delta t, \\ g_i^\star(\boldsymbol{x}, t) &= g_i(\boldsymbol{x}, t) - \frac{\Delta t}{\tau_\phi}\left(g_i(\boldsymbol{x}, t) - g_i^{\mathrm{eq}}(\boldsymbol{x}, t)\right). \end{aligned} \tag{9.90}$$

and

$$\begin{aligned} f_i(\boldsymbol{x} + \boldsymbol{c}_i\Delta t, t + \Delta t) &= f_i^\star(\boldsymbol{x}, t), \\ g_i(\boldsymbol{x} + \boldsymbol{c}_i\Delta t, t + \Delta t) &= g_i^\star(\boldsymbol{x}, t). \end{aligned} \tag{9.91}$$

f_i^{eq} and g_i^{eq} are local equilibrium distribution functions. The relaxation parameters τ and τ_ϕ are related to the transport coefficients ν and M in the hydrodynamic equations through

$$\nu = c_s^2\left(\tau - \frac{\Delta t}{2}\right), \tag{9.92}$$

$$M = \Gamma\left(\tau_\phi - \frac{\Delta t}{2}\right) \tag{9.93}$$

where Γ is a tunable parameter that appears in the equilibrium distribution as shown below. Since ν and M are positive quantities, the values of the relaxation times τ and τ_ϕ have to be larger than $\Delta t/2$.

Like for the one-component multiphase flows, the physics of surface tension can be implemented in two different ways. The first approach is to modify the equilibrium distribution functions f_i^{eq} to fully represent the pressure tensor. The suitable form of f_i^{eq} for $i \neq 0$ is given by

$$\begin{aligned} f_i^{\mathrm{eq}} = w_i\rho &\left(1 + \frac{c_{i\alpha}u_\alpha}{c_s^2} + \frac{u_\alpha u_\beta\left(c_{i\alpha}c_{i\beta} - c_s^2\delta_{\alpha\beta}\right)}{2c_s^4}\right) \\ &+ \frac{w_i}{c_s^2}\left(p_b - c_s^2\rho - \kappa\phi\Delta\phi\right) + \kappa\sum_{\alpha,\beta} w_i^{\alpha\beta}(\partial_\alpha\phi)(\partial_\beta\phi). \end{aligned} \tag{9.94}$$

The form of (9.94) is similar to that of the equivalent equation for the liquid-gas model in (9.36). The difference is that now p_b is given by (9.87), and the relevant gradient terms are for the order parameter ϕ rather than for the density ρ. The derivatives are usually approximated through finite difference schemes. The equilibrium distribution function for $i = 0$ can be obtained by exploiting conservation of mass, such that

$$f_0^{\mathrm{eq}} = \rho - \sum_{i \neq 0} f_i^{\mathrm{eq}}. \tag{9.95}$$

The second approach is to implement a forcing term. As already discussed in Sect. 9.2.1 for a multiphase fluid, this can be done by either using the pressure form, $F_\alpha = -\partial_\beta (P_{\alpha\beta} - c_s^2 \rho \delta_{\alpha\beta})$, or the potential form, $F_\alpha = -\phi \partial_\alpha \mu$. Remember that $\partial_\beta P_{\alpha\beta} = \partial_\alpha c_s^2 \rho + \phi \partial_\alpha \mu$, as given in (9.85). Since the derivatives of the order parameter (mostly obtained through a finite difference scheme) are only approximate, these two forms are not exactly identical numerically, therefore resulting in a loss of exact momentum conservation in the LB scheme.

The potential form requires the computation of third-order derivatives of the order parameter which are expensive to compute if we want to maintain accuracy. To alleviate this issue, a common mathematical trick is to rewrite the derivative of the pressure tensor as [13, 67]

$$\partial_\beta P_{\alpha\beta} = \partial_\alpha c_s^2 \rho + \phi \partial_\alpha \mu = \partial_\alpha \left(c_s^2 \rho + \phi \mu \right) - \mu \partial_\alpha \phi. \tag{9.96}$$

Thus, a possible hybrid approach is (i) to modify the equilibrium distribution functions to account for the bulk pressure term corresponding to $c_s^2 \rho + \phi \mu$ and (ii) to introduce a forcing term given by $F_\alpha = \mu \partial_\alpha \phi$. In this case, the suitable equilibrium distribution functions for $i \neq 0$ are

$$f_i^{\mathrm{eq}} = w_i \rho \left(1 + \frac{\phi \mu}{\rho c_s^2} + \frac{c_{i\alpha} u_\alpha}{c_s^2} + \frac{u_\alpha u_\beta \left(c_{i\alpha} c_{i\beta} - c_s^2 \delta_{\alpha\beta} \right)}{2 c_s^4} \right). \tag{9.97}$$

For the equilibrium distribution function g_i^{eq} of the order parameter, a comparison with (8.29) is appropriate. The key difference between the advection-diffusion equation in Chap. 8 and the Cahn-Hilliard equation is in the form of the diffusion term. For the latter, we have a term that is proportional to $\Delta \mu$, whereas for the former ΔC where C is the concentration. As such, the equilibrium distribution function g_i^{eq} has to obey

$$\begin{aligned} \sum_i g_i^{\mathrm{eq}} &= \phi, \\ \sum_i g_i^{\mathrm{eq}} c_{i\alpha} &= \phi u_\alpha, \\ \sum_i g_i^{\mathrm{eq}} c_{i\alpha} c_{i\beta} &= \Gamma \mu \delta_{\alpha\beta} + \phi u_\alpha u_\beta. \end{aligned} \tag{9.98}$$

The form of g_i^{eq} that satisfies these conditions is

$$g_i^{\text{eq}} = w_i \left(\frac{\Gamma\mu}{c_s^2} + \frac{\phi u_\alpha c_{i\alpha}}{c_s^2} + \frac{\phi u_\alpha u_\beta \left(c_{i\alpha} c_{i\beta} - c_s^2 \delta_{\alpha\beta}\right)}{2c_s^4} \right) \quad (i \neq 0) \tag{9.99}$$

and

$$g_0^{\text{eq}} = \phi - \sum_{i \neq 0} g_i^{\text{eq}}. \tag{9.100}$$

Compared to (8.26), the key difference is the term $\Gamma\mu/c_s^2$ in (9.99). For the Cahn-Hilliard equation, we have $\Gamma\mu$ rather than the concentration C in the advection-diffusion equation.

9.2.2.5 A Practical Guide to Simulation Parameters

The Landau free-energy model for a binary fluid is simple. There are only two free parameters: κ and A. They can be varied to tune the interface width $\xi = \sqrt{\kappa/A}$ and the surface tension $\gamma_{12} = \sqrt{(8\kappa A)/9}$. Similar to the liquid-gas model, the interface width is usually chosen to be $\sim 2-3$ lattice spacings, and the surface tension (in lattice units) is limited to $\gamma_{12} = O(10^{-2})$ or less due to the presence of spurious velocities.

For the binary model, we also have to choose the values of Γ and τ_ϕ which control the mobility parameter M in the Cahn-Hilliard equation via $M = \Gamma (\tau_\phi - \Delta t/2)$. A common practice is to set $\tau_\phi = \Delta t$ such that the distribution functions for $g_i(\boldsymbol{x}, t)$ are always relaxed to equilibrium, which simplifies the algorithm. The parameter Γ can be varied across a wide range of values, typically $\Gamma = 10^{-2} - 10$, while keeping the simulation stable.

Exercise 9.7 Repeat the Laplace pressure and contact angle benchmarks, as discussed in Exercises 9.5 and 9.6, for the binary fluid model.

9.3 Shan-Chen Pseudopotential Method

As we have discussed in Sect. 9.1, there are several ways to model multiphase or multicomponent flows within the LBM. For example, in Sect. 9.2 we described the commonly used free-energy method, a "top-down" approach. We started with a *macroscopic* concept, the free energy, and ended up with a force that can lead to phase separation.

Another way is to introduce a "bottom-up" approach by, e.g., postulating a *microscopic* interaction between fluid elements. This could be in the form of interaction potentials that eventually lead to the macroscopic separation of phases. Historically, this is how the *Shan-Chen (SC) model* was presented [68, 69].

The advantage of the SC approach is its intrinsic simplicity and mesoscopic nature. Surface tension is an emergent effect. This is akin to the LBM itself: LBM is based on simple mesoscopic rules with emergent transport coefficients, in particular the fluid viscosity. Additionally, for the multicomponent model, each of several different population sets directly represents a fluid component. This is a very intuitive approach to multicomponent physics, perhaps more so than the phase order parameters for free energy.

To keep this overview accessible also to an inexperienced audience, we omit technical details and refer to the literature instead. For example, for a deep explanation of the bottom-up approach, we encourage the reader to study [70]. We also recommend going through general SC review articles, such as [12, 13]. The earlier articles about the method, e.g. [68, 69, 71], are well worth reading.

We will start by explaining the general SC concepts in Sect. 9.3.1. Then we will distinguish between the two most important special cases: SC for a single-component multiphase system (e.g. liquid water and water vapour) in Sect. 9.3.2 and SC for a multicomponent system without phase change (e.g. a water-oil mixture) in Sect. 9.3.3. An overview of limitations and available extensions of the SC method (and the free-energy method), such as spurious currents and limited density ratio, will be discussed in Sect. 9.4.

9.3.1 General Considerations

In the following we will motivate the SC model and show fundamental concepts and equations. We will provide the basis for both the multiphase and the multicomponent cases that are covered in Sect. 9.3.2 and Sect. 9.3.3, respectively.

As a multiphase example, the coexistence of a liquid and a gas phase is caused by an attractive force between molecules in the liquid phase. The strength of this intermolecular force is tightly related to the boiling point and the vapour pressure of the liquid. For example, the dipolar molecules of water show a strong intermolecular interaction that leads to a relatively high boiling point of 100°C at normal pressure. Methane molecules, on the other hand, do not have a permanent dipole, and the attraction between CH_4 molecules is much weaker. As such, the boiling point of methane at -162°C is much lower compared to water.

Furthermore, different molecules in a multicomponent mixture (e.g. oil and water) interact differently with each other: the interaction between two water molecules is different from the interaction of two oil molecules or even between a water and an oil molecule.

These considerations raise the question of whether it is possible to simulate liquid-vapour or multicomponent systems by introducing a suitable local interaction force between fluid elements. In fact, this is exactly the underlying idea of the SC model. We will see shortly that the addition of a relatively simple interaction force defined at lattice nodes can be used to model both multiphase and multicomponent systems with or without surface tension. Of course, not every interaction force is suitable for this purpose.

Ideally, for multiphase systems, the force should have a thermodynamically consistent form, i.e. the values of the pressure and the equilibrium densities for a given temperature should be the same as those derived from thermodynamic principles using the Maxwell area construction rule (cf. Sect. 9.1). We will get back to this in Sect. 9.3.2.

Even if we find a suitable force in terms of thermodynamic consistency, it is not guaranteed that its discretised form allows for stable simulations. It is known that large surface tensions or large liquid-gas density ratios can lead to numerically "stiff" forces that cause negative LB populations and therefore instability [12].

In what follows we focus on a multiphase system for simplicity, but the results can be extended to multicomponent problems. In order to find a functional form for the interaction force, we have to consider its origin. We can assume that intermolecular forces act between pairs of molecules and are additive. Therefore, a higher density of molecules will lead to stronger forces. As a consequence, we expect that the magnitude of the interaction between fluid elements at $\boldsymbol{x}$ and $\tilde{\boldsymbol{x}} \neq \boldsymbol{x}$ is proportional to $\rho(\boldsymbol{x})\rho(\tilde{\boldsymbol{x}})$. Additionally, the interaction is a strong function of the distance between the fluid elements. We can thus introduce a kernel function $G(\boldsymbol{x}, \tilde{\boldsymbol{x}})$ that carries the information about the spatial dependency of the force. Also, the total force acting on a fluid element at $\boldsymbol{x}$ is the integral over all possible interaction sites $\tilde{\boldsymbol{x}}$. We can finally write the interaction force density at $\boldsymbol{x}$ as [68, 72]

$$\boldsymbol{F}^{\mathrm{SC}}(\boldsymbol{x}) = -\int (\tilde{\boldsymbol{x}} - \boldsymbol{x})\, G(\boldsymbol{x}, \tilde{\boldsymbol{x}}) \psi(\boldsymbol{x}) \psi(\tilde{\boldsymbol{x}})\, \mathrm{d}^3\tilde{x}. \tag{9.101}$$

Here we have replaced the density ρ by an effective density function ψ that is also called the *pseudopotential*. The prefix "pseudo" indicates that ψ represents an effective density, rather than the fluid density ρ.

The reason for using ψ rather than ρ is the possible numerical instability mentioned earlier. A widely accepted and often used form of the pseudopotential is

$$\psi(\rho) = \rho_0 \left[1 - \exp(-\rho/\rho_0)\right] \tag{9.102}$$

with a reference density ρ_0 that in simulation units is mostly set to unity. The pseudopotential in (9.102) is bounded between 0 and ρ_0 for any value of the density ρ. Therefore, the interaction force in (9.101) remains finite, even for large densities.

Exercise 9.8 By performing a Taylor expansion, show that $\psi(\rho) \approx \rho$ for $\rho \ll \rho_0$.

Another common form of the pseudopotential is simply the fluid density itself:

$$\psi(\rho) = \rho. \tag{9.103}$$

In this case there is no bound, and the interaction force in (9.101) can diverge for large ρ. There are even more functional forms in use throughout the literature (cf. Sect. 9.4).

The next step is the spatial discretisation of (9.101). This means that we want to restrict $\boldsymbol{x}$ and $\tilde{\boldsymbol{x}}$ to lattice nodes. Furthermore, we claim that the interaction force is short-ranged, i.e. fluid elements at $\boldsymbol{x}$ only interact with other fluid elements at $\tilde{\boldsymbol{x}}$ that are in the vicinity. Therefore, $G(\boldsymbol{x}, \tilde{\boldsymbol{x}}) = 0$ for sufficiently large $|\boldsymbol{x} - \tilde{\boldsymbol{x}}|$. Finally, $G(\boldsymbol{x}, \tilde{\boldsymbol{x}})$ should be isotropic and therefore a function of $|\boldsymbol{x} - \tilde{\boldsymbol{x}}|$ only.

There exist different discretisations for $G(\boldsymbol{x}, \tilde{\boldsymbol{x}})$ [72, 73], but the most common involves interactions between lattice nodes that are connected by one of the vectors $\boldsymbol{c}_i \Delta t$ [74, 75]:

$$G(\boldsymbol{x}, \tilde{\boldsymbol{x}}) = \begin{cases} w_i G & \text{for } \tilde{\boldsymbol{x}} = \boldsymbol{x} + \boldsymbol{c}_i \Delta t, \\ 0 & \text{otherwise.} \end{cases} \tag{9.104}$$

The simplest form of the **discretised Shan-Chen force for a single component** is represented through a sum of pseudopotential interactions with nearest lattice neighbours [68]:

$$\boldsymbol{F}^{\text{SC}}(\boldsymbol{x}) = -\psi(\boldsymbol{x}) G \sum_i w_i \psi(\boldsymbol{x} + \boldsymbol{c}_i \Delta t) \boldsymbol{c}_i \Delta t. \tag{9.105}$$

The sum runs over all velocities $\boldsymbol{c}_i$ of the underlying lattice (e.g. D2Q9 or D3Q19, as illustrated in Fig. 9.8) and the w_i are the usual lattice weights. The pseudopotential $\psi(\boldsymbol{x})$ is given by (9.102) or (9.103). The coefficient G is a simple scalar that controls the strength of the interaction. It is attractive for negative and repulsive for positive G. A more mathematical and thermodynamic rationale for this force will be provided in Sect. 9.3.2.

Exercise 9.9 The SC model violates *local momentum* conservation as the interaction force is not local. Show that the *global momentum* is conserved when the system is fully periodic. Use the fact that the interaction force between two fluid elements at $\boldsymbol{x}$ and $\tilde{\boldsymbol{x}}$ satisfies Newton's third law.

The SC model can be easily extended to systems with S fluid components. In this case, we label different components (e.g. water and oil) with the indices $1 \leq \sigma, \tilde{\sigma} \leq S$ and write

$$\boldsymbol{F}^{\text{SC}(\sigma)}(\boldsymbol{x}) = -\psi^{(\sigma)}(\boldsymbol{x}) \sum_{\tilde{\sigma}} G_{\sigma\tilde{\sigma}} \sum_i w_i \psi^{(\tilde{\sigma})}(\boldsymbol{x} + \boldsymbol{c}_i \Delta t) \boldsymbol{c}_i \Delta t \tag{9.106}$$

for the force density acting on component σ at location $\boldsymbol{x}$. The new sum runs over all S values of $\tilde{\sigma}$, including $\tilde{\sigma} = \sigma$. The coefficients $G_{\sigma\tilde{\sigma}} = G_{\tilde{\sigma}\sigma}$ with $\tilde{\sigma} \neq \sigma$

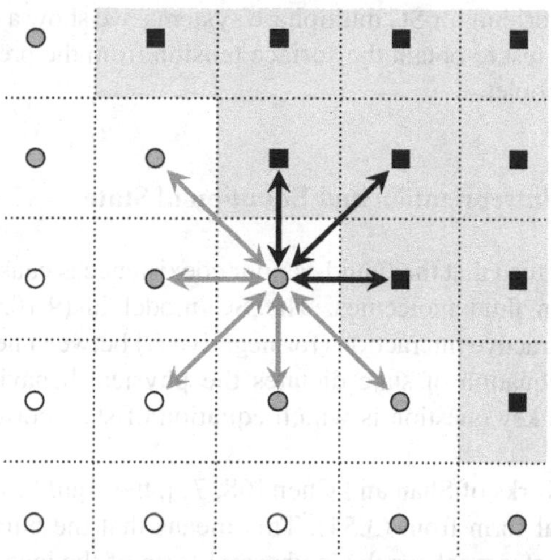

Fig. 9.8 The Shan-Chen force is often implemented along node pairs that are connected through one of the lattice vectors $c_i \Delta t$. In D2Q9, the central node interacts with its eight neighbours, as indicated by the *arrows*. These nodes can be in the fluid region (*circles*) or in the solid region (*squares*; cf. Sect. 9.3.2 and Sect. 9.3.3 for more details about the treatment of solid boundaries). To be consistent with the remainder of this book, we distinguish between fluid nodes (*white*) and boundary nodes (*grey*), the latter being neighbours of at least one solid node

denote the molecular *interactions between different fluid components*. Those are often repulsive. Each of the fluid components can potentially self-interact; this is captured by the coefficients $G_{\sigma\sigma}$. The SC model in (9.106) may thus be used to model a mixture of two (or more) fluids that could all exist in the liquid and gas phases, i.e. a multicomponent-multiphase system.

In Sect. 9.3.2 we will discuss in more detail how to use the SC model to simulate a multiphase system with a single component (i.e. $S = 1$ and $G = G_{11} \neq 0$), while in Sect. 9.3.3 we will discuss multicomponent systems without phase change (i.e. $S > 1$ and $G_{\sigma\sigma} = 0$). We will not consider multicomponent-multiphase problems in this book (i.e. problems with $S > 1$ *and* $G_{\sigma\sigma} \neq 0$). Those systems are investigated in [76] and reviewed in section 6 in [12].

9.3.2 Multiphase Model for Single Component

Here we deal with a single fluid component that can coexist in two phases, e.g. liquid water and water vapour. First we will investigate the physical content of the SC model in (9.105) and show how it can lead to phase separation and surface tension. We will then take a closer look at a planar interface and discuss the issue of thermodynamic consistency within the SC model. Later we will present how solid boundaries with given wetting properties can be simulated. After providing a summary of the algorithm for SC multiphase systems, we show a classical example: the Young-Laplace test to obtain the surface tension from the pressure jump across the interface of a droplet.

9.3.2.1 Physical Interpretation and Equation of State

In Sect. 9.3.1 we argued that the liquid-vapour coexistence is caused by an attractive interaction between fluid molecules. The SC model in (9.105) is one possible realisation of an attractive interaction (for negative G) between neighbouring lattice nodes. Since the equation of state dictates the physical behaviour of the liquid-vapour system, the key question is which equation of state corresponds to the SC force.

In the original works of Shan and Chen [68, 72], the equilibrium distribution f_i^{eq} is taken in its usual form from (3.54). This means that the bulk pressure, i.e. far away from any interfaces, obeys the isothermal form of the ideal equation of state, $p_{\text{b}} = c_s^2\rho$ (cf. Sect. 1.1.3). From thermodynamics we know that the ideal equation of state cannot invoke liquid-vapour coexistence. Therefore, the SC equation of state must include additional terms. In fact, the desired phase separation is linked to the SC force, as we will now investigate in more detail.

Let us first Taylor-expand the pseudopotential $\psi(\boldsymbol{x} + \boldsymbol{c}_i\Delta t)$ about $\boldsymbol{x}$:

$$\psi(\boldsymbol{x} + \boldsymbol{c}_i\Delta t) = \psi(\boldsymbol{x}) + c_{i\alpha}\Delta t\partial_\alpha\psi(\boldsymbol{x}) + \frac{1}{2}c_{i\alpha}c_{i\beta}\Delta t^2\partial_\alpha\partial_\beta\psi(\boldsymbol{x}) + \frac{1}{6}c_{i\alpha}c_{i\beta}c_{i\gamma}\Delta t^3\partial_\alpha\partial_\beta\partial_\gamma\psi(\boldsymbol{x}) + \ldots \tag{9.107}$$

Substituting this into (9.105) gives

$$\boldsymbol{F}^{\text{SC}}(\boldsymbol{x}) = -G\psi(\boldsymbol{x})\sum_i w_i\boldsymbol{c}_i\Delta t\Big(\psi(\boldsymbol{x}) + c_{i\alpha}\Delta t\partial_\alpha\psi(\boldsymbol{x}) + \frac{1}{2}c_{i\alpha}c_{i\beta}\Delta t^2\partial_\alpha\partial_\beta\psi(\boldsymbol{x}) + \ldots\Big). \tag{9.108}$$

Due to the symmetry of the velocity sets, shown in (3.60), the terms $\sum_i w_i\boldsymbol{c}_i$ and $\sum_i w_i\boldsymbol{c}_i c_{i\alpha}c_{i\beta}$ vanish.

Including expansion terms from (9.107) up to third order, the **continuum form of the Shan-Chen force** becomes [77]

$$\boldsymbol{F}^{\text{SC}}(\boldsymbol{x}) = -G\psi(\boldsymbol{x})\left(c_s^2\Delta t^2\nabla\psi(\boldsymbol{x}) + \frac{c_s^4\Delta t^4}{2}\nabla\Delta\psi(\boldsymbol{x})\right). \tag{9.109}$$

Exercise 9.10 Derive (9.109) by taking advantage of (3.60) for the moments of the weights w_i. Note that the result in (9.109) generally depends on the underlying lattice structure.

The first term on the right-hand side of (9.109) has the form of a gradient:

$$-c_s^2\Delta t^2 G\psi(\boldsymbol{x})\boldsymbol{\nabla}\psi(\boldsymbol{x}) = -\frac{c_s^2\Delta t^2 G}{2}\boldsymbol{\nabla}\psi^2(\boldsymbol{x}). \tag{9.110}$$

Therefore we can include it in the equation of state.

The **equation of state of the multiphase SC model** in (9.105) is

$$p_b(\rho) = c_s^2\rho + \frac{c_s^2\Delta t^2 G}{2}\psi^2(\rho). \tag{9.111}$$

The SC contribution leads to a non-ideal term that allows for the coexistence of a liquid and a vapour phase.

Exercise 9.11 Plot the bulk pressure $p_b(\rho)$ from (9.111) with the pseudopotential in (9.102). For the sake of simplicity, set $\rho_0 = 1$, $c_s^2 = 1/3$ and $\Delta t = 1$. Show that, if $G < -4$, there exist two distinct density values for a given pressure value, i.e. gas and liquid with respective densities ρ_g and ρ_l can coexist. What happens for $G \geq -4$?

The second term proportional to $G\psi\boldsymbol{\nabla}\Delta\psi$ in (9.109) looks nearly like the surface tension term $k\rho\boldsymbol{\nabla}\Delta\rho$ in (9.12). Obviously, we expect deviations when $\psi \neq \rho$.

One can show that the SC pressure tensor $\boldsymbol{P}^{SC}$, which is defined by $\boldsymbol{\nabla}\cdot\boldsymbol{P}^{SC} = \boldsymbol{\nabla}(c_s^2\rho) - \boldsymbol{F}^{SC}$ and has been introduced in Sect. 9.1, assumes the form (setting $\Delta t = 1$ for simplicity) [77]

$$P_{\alpha\beta}^{SC} = \left(c_s^2\rho + \frac{c_s^2 G}{2}\psi^2 + \frac{c_s^4 G}{4}(\boldsymbol{\nabla}\psi)^2 + \frac{c_s^4 G}{2}\psi\Delta\psi\right)\delta_{\alpha\beta} - \frac{c_s^4 G}{2}(\partial_\alpha\psi)(\partial_\beta\psi). \tag{9.112}$$

This pressure tensor differs from the thermodynamically consistent pressure tensor in (9.30). However, the resulting surface tension behaviour and the density profiles are acceptable for many practical purposes. Furthermore, there exist modifications of the SC model that allow for improved thermodynamic consistency (cf. Sect. 9.4).

The **Shan-Chen force** from (9.105) introduces two terms in the Navier-Stokes equation: one leads to a **non-ideal equation of state**, (9.111), the other acts

(continued)

like a **surface tension** from (9.12). The single parameter G, which appears in both terms, can be changed to control the phase separation. This is the reason why G is sometimes referred to as a **temperature-like parameter**.

9.3.2.2 Planar Interface and Thermodynamic Consistency

A suitable test for checking thermodynamical consistency is the planar interface between phases. For the free-energy liquid-gas model, the density profile across the interface satisfies the Maxwell area construction rule, cf. Appendix A.7. Yet, we have to investigate thermodynamic consistency in the context of the SC model. Below we collect the final results of the calculations, with algebraic details shown in Appendix A.8.

For the **Shan-Chen model**, there is an expression similar to the **Maxwell area construction rule** that allows to obtain the phase transition densities. The coexistence pressure is

$$p_0 = c_s^2 \rho_g + \frac{c_s^2 \Delta t^2 G}{2} \psi^2(\rho_g) = c_s^2 \rho_l + \frac{c_s^2 \Delta t^2 G}{2} \psi^2(\rho_l). \tag{9.113}$$

where the liquid and gas densities ρ_l and ρ_g obey

$$\int_{\rho_g}^{\rho_l} \left(p_0 - c_s^2 \tilde{\rho} - \frac{c_s^2 \Delta t^2 G}{2} \psi^2(\tilde{\rho}) \right) \frac{\psi'(\tilde{\rho})}{\psi^2(\tilde{\rho})} \mathrm{d}\tilde{\rho} = 0. \tag{9.114}$$

Instead of having the multiplier $1/\tilde{\rho}^2$ in the thermodynamically consistent model in (9.7), we now have an expression depending on the pseudopotential and its derivative, $\psi'(\rho)/\psi^2(\rho)$, with $\psi' = \mathrm{d}\psi/\mathrm{d}\rho$.

The obvious choice to satisfy the thermodynamic consistency is $\psi \propto \rho$. However, for large liquid-gas density ratios this leads to large gradients and eventually numerical instability. In Sect. 9.4.2 we show how to choose different equations of state. For the van der Waals equation of state, for example, the achievable liquid-gas density ratio with $\psi \propto \rho$ is around 10. However, reverting to the exponential pseudopotential in (9.102) for the same equation of state allows increasing the liquid-gas density ratio by an additional factor of 3–5.

Although the expressions for the Maxwell area construction rule and its SC equivalent are different, the phase separation densities ρ_l and ρ_g for the particular equation of state in (9.111) are *similar* to their thermodynamically consistent

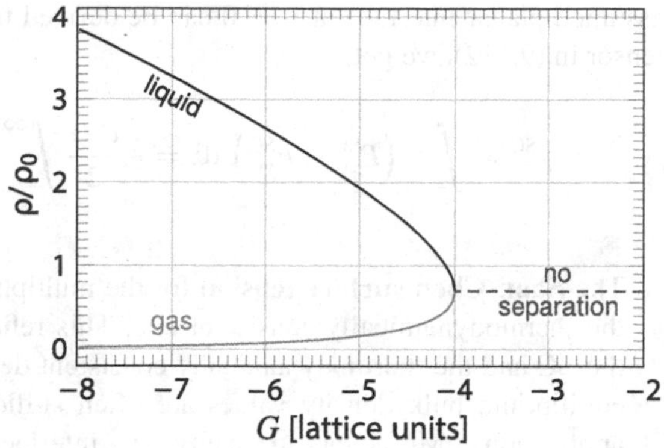

Fig. 9.9 Phase separation densities for (9.111) with the pseudopotential in (9.102) versus the temperature-like parameter G (in lattice units). The densities represent the solution of (9.114)

counterparts in many situations. More particular examples of possible equations of state are given in [78]. The Peng-Robinson and Carnahan-Sterling equations of state give practically the same gas and liquid density values while the results obtained from the van der Waals equation of state are significantly different.

In Fig. 9.9 we present the gas and liquid densities from (9.114) as function of the temperature-like parameter G. We do not present the thermodynamically consistent Maxwell-area reconstruction curve, as both curves would be indistinguishable in this figure.[10] We can see that phase separation occurs only for $G < -4$. The point at which the liquid and gas densities become equal ($G = -4$) is called *critical point*. For the equation of state in (9.111) with the pseudopotential in (9.102), the critical density is $\rho_{\mathrm{crit}}/\rho_0 = \ln 2$. Figure 9.9 is widely used to initialise simulations consistently, i.e. by setting the initial density as ρ_{l} and ρ_{g} in the liquid and gas phases, respectively. For example, for $G = -6$ the initial densities are $\rho_{\mathrm{g}} = 0.056$ and $\rho_{\mathrm{l}} = 2.659$ with the associated liquid-gas density ratio 47.

For the original SC model with the equation of state $p = c_s^2\rho + c_s^2\Delta t^2 G\psi^2(\rho)/2$, simulations can be stable for G as low as ≈ -7. This defines the achievable liquid-gas density ratio around 60–80 and the maximum surface tension of 0.1 (in lattice units) [79]. To reach higher liquid-gas density ratios, one needs to revert to extensions as covered in Sect. 9.4.

Exercise 9.12 Derive the values of the critical parameters (G and ρ_{crit}). Use the fact that for a phase transition the first and second derivatives of the equation of state $p_{\mathrm{b}}(\rho)$ with respect to density vanish at the critical point.

Finally, we can compute the SC liquid-gas surface tension at the planar interface and compare it with its thermodynamic consistent form. The surface tension for an

[10] Differences usually become visible at low densities when the density is plotted logarithmically.

assumed planar interface at $x = 0$ can be derived from (9.14). Using the pressure tensor in (9.112), we get

$$\gamma^{\mathrm{SC}} = \int_{-\infty}^{\infty} \left(P_{xx}^{\mathrm{SC}} - P_{yy}^{\mathrm{SC}}\right) \mathrm{d}x = -\frac{c_s^4 G}{2} \int_{-\infty}^{\infty} \left(\frac{\mathrm{d}\psi}{\mathrm{d}x}\right)^2 \mathrm{d}x. \tag{9.115}$$

The **Shan-Chen surface tension** for the multiphase model is different from the thermodynamically consistent one. This reflects the difference between the SC and the thermodynamically consistent density profiles. However, the equilibrium bulk density values are often sufficiently similar for practical applications. Moreover, in reality the interface width is extremely thin (nanometres), and computational tools such as LBM are not able to resolve them properly. The diffuse interface method is just a way to describe multiphase physics numerically. Researchers are mainly interested in macroscopic parameters, such as the bulk densities and surface tension, rather than the exact shape of the interface between phases.

It is not feasible to express the pseudopotential gradient in (9.115) analytically and derive an expression for the surface tension. Instead, the Young-Laplace test yields the surface tension without evaluating the integral. We show the Young-Laplace results for the multiphase model in Sect. 9.3.2.5.

9.3.2.3 Boundary Conditions and Contact Angle

In the fluid domain, the velocity and pressure boundary conditions for the multiphase SC model are the same as for the standard LBM (cf. Chap. 5). In addition to pressure and velocity boundary conditions, we also have to consider multiphase effects near the boundary.

As pointed out in Sect. 9.1 and Sect. 9.2, fluid interfaces in contact with a solid boundary will assume a certain contact angle. The easiest way to include solid nodes with a given wetting behaviour is the introduction of a SC-like interaction force between solid nodes and boundary nodes (i.e. fluid nodes near solid nodes) as illustrated in Fig. 9.8:

$$\boldsymbol{F}^{\mathrm{SC,s}}(\boldsymbol{x}) = -G\psi(\boldsymbol{x}) \sum_i^{\mathrm{solid}} w_i \psi(\rho_s) \boldsymbol{c}_i \Delta t. \tag{9.116}$$

This force acts on a fluid node at $\boldsymbol{x}$. The sum runs over all directions $\boldsymbol{c}_i$ for which $\boldsymbol{x}+\boldsymbol{c}_i\Delta t$ is a solid node. Solid nodes are assigned an effective density ρ_s. The contact angle at the solid boundary is indirectly controlled by the value of ρ_s.

The standard SC force in (9.105) continues to act between neighbouring fluid nodes. Therefore, we can write the total SC force at a fluid node at $\boldsymbol{x}$ as

$$\boldsymbol{F}^{\mathrm{SC}}(\boldsymbol{x}) = -G\psi(\boldsymbol{x}) \left(\sum_i^{\mathrm{fluid}} w_i \psi(\boldsymbol{x} + \boldsymbol{c}_i \Delta t) \boldsymbol{c}_i \Delta t + \sum_i^{\mathrm{solid}} w_i \psi(\rho_{\mathrm{s}}) \boldsymbol{c}_i \Delta t \right). \tag{9.117}$$

Solid and fluid nodes are treated on an equal footing; the only difference is that the density is prescribed at solid nodes. Therefore, the solid treatment is straightforward and easy to implement.

One needs to be attentive to avoid unnecessary condensation or evaporation near those boundaries. It is required to have densities consistent with the value for G from Fig. 9.9. For example, if the effective density is above the critical density ρ_{crit}, then the density in the fluid near this boundary will be driven towards the liquid density ρ_{l}, thus giving condensation (that may be undesired).

In order to achieve full wetting (contact angle 0°), we choose the bulk liquid density for the solid density: $\rho_{\mathrm{s}} = \rho_{\mathrm{l}}$. For a completely hydrophobic surface (contact angle 180°), we select $\rho_{\mathrm{s}} = \rho_{\mathrm{g}}$. All other contact angles can be realised by taking a solid density value between these two extremes.

For the free-energy model, the contact angle depends on the order parameter gradient (cf. (9.65)). A similar expression may be developed for SC involving the pseudopotential integrals [80]. However, it is difficult to apply in simulations due to its integral form.

9.3.2.4 Algorithm and Forcing Schemes

We briefly describe the relevant steps of the numerical algorithm for multiphase SC simulations. The choice of the forcing scheme is particularly important.

One should initialise the domain with gas and liquid densities obtained from (9.113) (or from Fig. 9.9) for a specific value of G. A common situation is a single liquid drop in vapour (or a single gas bubble in a liquid). It is recommended to implement an initially smooth interface. Starting with a density step change at the interface can cause instability for large liquid-gas density ratios. If the simulation involves solid boundaries with certain wetting properties, one may initialise a droplet as a spherical cap at a wall with a contact angle close to its expected value.

Another situation occurs if one is interested in phase separation, i.e. growth of liquid domains over time. To achieve his, the initial density is taken between ρ_{g} and ρ_{l}, and a small random fluctuation is imposed. For example, if the chosen average density of the system is $\bar{\rho}$, one could initialise the density with random values in the interval $\bar{\rho}(1 \pm 0.001)$, i.e. with a 0.1% perturbation. The tendency of the system to minimise its interface area will lead to an amplification of these perturbations and phase separation eventually.

In the following we will only consider the LBGK algorithm. For a short discussion of the MRT collision operator for multiphase and multicomponent models, see Sect. 9.4.

Each time step of the LB simulation after successful initialisation can be written as follows:

1. Find the fluid density $\rho = \sum_i f_i$ everywhere.
2. Calculate the SC force density $\boldsymbol{F}^{\text{SC}}$ from (9.105). For fluid sites interacting with a solid wall, apply (9.116) to satisfy the wetting condition. If additional forces, such as gravity, act on the fluid, sum up all force contributions.
3. Compute the equilibrium distributions f_i^{eq} as usual. This involves the fluid velocity

$$\boldsymbol{u} = \frac{1}{\rho}\left(\sum_i f_i \boldsymbol{c}_i + \frac{\boldsymbol{F}\Delta t}{2}\right) \tag{9.118}$$

where $\boldsymbol{F}$ includes all forces acting on the fluid. The velocity $\boldsymbol{u}$ is taken both as the equilibrium velocity $\boldsymbol{u}^{\text{eq}}$ and as the physical velocity that solves the Navier-Stokes equation.
4. Use Guo's approach (cf. Sect. 6.3) to include the force in the collision step. See below for additional comments.
5. Collide and stream as usual. Hydrodynamic boundary conditions, such as bounce-back, are included in the normal way (cf. Chap. 5).
6. Go back to step 1.

In fact, the only novelty in this algorithm is the calculation of the SC force from the density; everything else is the standard LBM with forces. To the LB algorithm, the SC forces behave as every other external force.

Historically, the force density in the original SC works (and also in many more recent publications) was implemented *via* a modification of the equilibrium velocity without additional forcing terms in the LBE:

$$\boldsymbol{u}^{\text{eq}} = \frac{1}{\rho}\left(\sum_i f_i \boldsymbol{c}_i + \tau \boldsymbol{F}\right). \tag{9.119}$$

Although this so-called Shan-Chen forcing approach, which is also discussed in Sect. 6.4, tends to be more stable, it leads to τ-dependent surface tension [81]. Therefore, we recommend to follow the algorithm as summarised above. See also [77] for a recent and careful discussion of forcing in the SC method.

Exercise 9.13 Perform the contact angle test similarly to that explained in Exercise 9.6. To control the contact angle in the SC model, vary the wall density ρ_{s} between ρ_{g} and ρ_{l}. In contrast to the free-energy model to which Exercise 9.6 applies, it is not possible to set the contact angle in the SC model directly. Measure the contact angle θ for different wall densities ρ_{s} and produce a diagram $\theta(\rho_{\text{s}})$. As we cannot compare this curve with theoretical predictions in the SC model, we use the curve $\theta(\rho_{\text{s}})$ as a constitutive law to find the appropriate wall density for a desired contact angle.

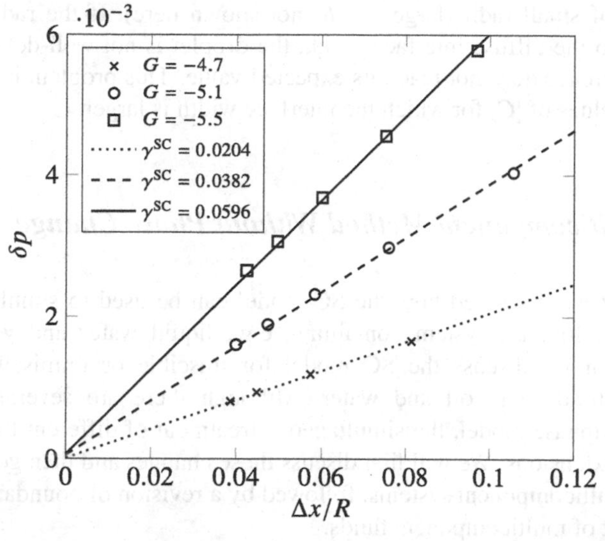

Fig. 9.10 Results of the Young-Laplace test for a liquid droplet in a gas phase. The pressure difference δp across the droplet interface is proportional to the inverse droplet radius R with the surface tension γ^{SC} as the proportionality factor. Simulations are shown for three different interaction parameters G. Pressure difference and surface tension are shown in lattice units. Simulations involve the standard SC model with the pseudopotential in (9.102) and Guo's forcing. The system size is $64\Delta x \times 64\Delta x$

9.3.2.5 Example: Young-Laplace Test

In Sect. 9.1.2 we have seen that in mechanical equilibrium the curved surface of a droplet leads to an increase of the interior pressure. For a 2D droplet, the pressure difference $\delta p = p_l - p_g$, the surface tension γ and the droplet radius R satisfy

$$\delta p = p_l - p_g = \frac{\gamma}{R}. \tag{9.120}$$

Instead of evaluating (9.115) to determine the SC surface tension γ^{SC}, we simulate a liquid droplet in the gas phase[11].

We define the droplet radius R as the radial position where the density profile reaches $(\rho_g + \rho_l)/2$. Due to the diffuse nature of the interface, other radius definitions are possible; this would lead to different interpretations of the surface tension. The pressure is computed from (9.111). We average the pressure over 4×4 grid nodes at the droplet centre and far away from the droplet surface, respectively.

It is common to simulate different droplet radii for fixed G to show that the measured pressure difference is indeed proportional to the inverse radius. The surface tension is then obtained from a linear fit. The results for different interaction parameters G are shown in Fig. 9.10. Note that the Young-Laplace test fails to work

[11]We could also simulate a gas bubble in a liquid.

in the limit of small radii (large $\Delta x/R$, not shown here). If the radius becomes comparable to the diffuse interface width, the droplet is not well-defined, and the pressure difference does not reach its expected value. This problem is more severe for smaller values of $|G|$ for which the interface width is larger.

9.3.3 Multicomponent Method Without Phase Change

In Sect. 9.3.2 we discussed how the SC model can be used to simulate a single-component multiphase system containing, e.g., liquid water and water vapour. In this section we discuss the SC model for miscible or immiscible mixtures of different fluids, e.g. oil and water. Although there are several similarities with the multiphase model, the simultaneous treatment of different fluids requires algorithmic extensions. We will first discuss these changes and then get back to the physics of multicomponent systems, followed by a revision of boundary conditions in the context of multicomponent fluids.

This section is mostly based on early works about the SC approach for multicomponent systems [68, 69, 71] with newer developments mentioned in passing. Shan [82] demonstrated how to derive this model from continuum kinetic theory. For recent review articles covering several extensions of the LBM for multicomponent flows we refer to [12, 13]. Also see Sect. 9.4 for a more detailed discussion of the limitations and extensions of the SC multicomponent method.

9.3.3.1 Shan-Chen Force and Algorithmic Implications

Let us take another look at the general SC interaction force for S components in (9.106). If we are only interested in the mixture of S ideal fluids, we can assume that each fluid component σ does not interact with itself. Therefore, the SC force only includes interactions between different fluid components, $\sigma \neq \tilde{\sigma}$:

$$\boldsymbol{F}^{\mathrm{SC}(\sigma)}(\boldsymbol{x}) = -\psi^{(\sigma)}(\boldsymbol{x}) \sum_{\tilde{\sigma}\neq\sigma} G_{\sigma\tilde{\sigma}} \sum_i w_i \psi^{(\tilde{\sigma})}(\boldsymbol{x} + \boldsymbol{c}_i \Delta t)\boldsymbol{c}_i \Delta t. \tag{9.121}$$

Here, $G_{\sigma\tilde{\sigma}} \equiv G_{\tilde{\sigma}\sigma}$ is the interaction strength of fluids σ and $\tilde{\sigma}$, and $\psi^{(\sigma)}$ is the pseudopotential of component σ, e.g. (9.102) or (9.103). Alternatively, we can allow the sum to run over all pairs, including $\tilde{\sigma} = \sigma$, and set $G_{\sigma\sigma} = 0$. In principle, there can be an arbitrary number of fluid components, but the SC model is mostly used for systems with two components ($S = 2$ as illustrated in Fig. 9.11). To achieve (partially) immiscible fluids, the interaction between components must be repulsive and hence the coupling strengths $G_{\sigma\tilde{\sigma}}$ positive. Different components can have different pseudopotentials.

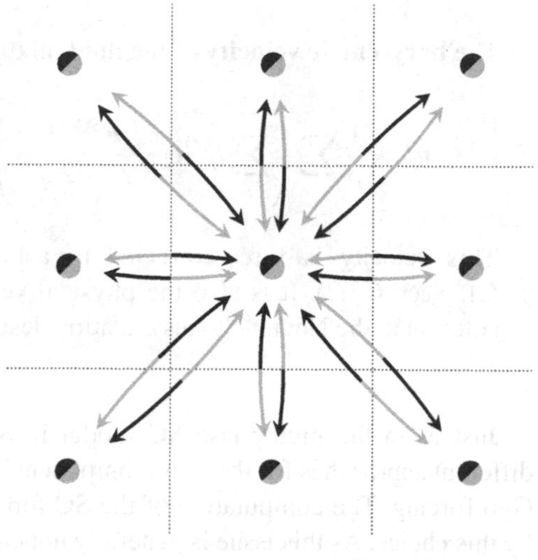

Fig. 9.11 Illustration of the interaction forces in a two-component system. At each lattice node, there exist two fluid components (*grey and black*). The black component interacts with the grey component at neighbouring cells and vice versa. These force pairs obey Newton's third law, illustrated by the *double-headed arrows*

In the multicomponent model, we need a set of populations $f_i^{(\sigma)}$ for each component σ. Each of these sets of populations obeys the standard LBGK equation

$$f_i^{(\sigma)}(\boldsymbol{x}+\boldsymbol{c}_i\Delta t, t+\Delta t) = f_i^{(\sigma)}(\boldsymbol{x}, t) - \frac{f_i^{(\sigma)}(\boldsymbol{x}, t) - f_i^{\mathrm{eq}(\sigma)}(\boldsymbol{x}, t)}{\tau^{(\sigma)}}\Delta t + \left(1 - \frac{\Delta t}{2\tau^{(\sigma)}}\right) F_i^{(\sigma)}(\boldsymbol{x}, t)\Delta t \qquad (9.122)$$

with its own relaxation time $\tau^{(\sigma)}$ (and therefore viscosity $\nu^{(\sigma)}$) and forcing terms $F_i^{(\sigma)}$ as defined below. We will briefly discuss the extension to the MRT collision operator in Sect. 9.4.

Now we have to discuss how to include the SC forces in the LBE and how to choose the equilibrium distributions when there is more than one LBGK equation.

9.3.3.2 Fluid Velocity in the Multicomponent Model

The most important change with respect to the single-component LBM involves the equilibrium distributions $f_i^{\mathrm{eq}(\sigma)}(\rho^{(\sigma)}, \boldsymbol{u}^{\mathrm{eq}(\sigma)})$. They are still given by (3.54) where the density ρ has to be replaced by the component density $\rho^{(\sigma)}$. However, it is not immediately clear which velocity $\boldsymbol{u}^{\mathrm{eq}(\sigma)}$ to use for the equilibrium distributions since we have more than one set of populations now.

We can define several velocities. The *bare component velocity* is given by

$$\boldsymbol{u}^{(\sigma)} = \frac{1}{\rho^{(\sigma)}}\sum_i f_i^{(\sigma)}\boldsymbol{c}_i, \quad \rho^{(\sigma)} = \sum_i f_i^{(\sigma)}. \qquad (9.123)$$

The **barycentric velocity** of the fluid mixture reads [83]

$$\boldsymbol{u}_{\mathrm{b}} = \frac{1}{\rho}\sum_{\sigma}\left(\sum_{i} f_i^{(\sigma)}\boldsymbol{c}_i + \frac{\boldsymbol{F}^{\mathrm{SC}(\sigma)}\Delta t}{2}\right), \quad \rho = \sum_{\sigma}\rho^{(\sigma)}. \tag{9.124}$$

This velocity is force-corrected to achieve second-order time accuracy (cf. Sect. 6.3.2). It is also the physical velocity that has to be taken as the solution to the Navier-Stokes equation describing the fluid mixture [69].

Just as in the multiphase SC model in Sect. 9.3.2, there are essentially two different approaches for the multicomponent LB algorithm: Shan-Chen forcing and Guo forcing. The computation of the SC force density $\boldsymbol{F}^{\mathrm{SC}(\sigma)}$ itself is not affected by this choice. As this issue is generally not carefully addressed in the literature, we discuss it in more detail here.

Shan-Chen forcing: In the original works [68, 69, 71], the equilibrium velocity of component σ was chosen as

$$\boldsymbol{u}^{\mathrm{eq}(\sigma)} = \boldsymbol{u}' + \frac{\tau^{(\sigma)}\boldsymbol{F}^{\mathrm{SC}(\sigma)}}{\rho^{(\sigma)}} \tag{9.125}$$

with a *common velocity* $\boldsymbol{u}'$ that is given by the weighted average

$$\boldsymbol{u}' = \frac{\sum_{\sigma}\frac{\rho^{(\sigma)}\boldsymbol{u}^{(\sigma)}}{\tau^{(\sigma)}}}{\sum_{\sigma}\frac{\rho^{(\sigma)}}{\tau^{(\sigma)}}}. \tag{9.126}$$

This expression becomes particularly simple if all relaxation times are identical: $\boldsymbol{u}' = \sum_{\sigma}\rho^{(\sigma)}\boldsymbol{u}^{(\sigma)}/\rho$. The fluid components interact (i) through the SC force in (9.121) and (ii) by sharing the same velocity $\boldsymbol{u}'$ in (9.126).

Exercise 9.14 The common velocity $\boldsymbol{u}'$ has to assume the form in (9.126) to ensure momentum conservation during collision in the absence of forces [69]. Start from (9.122) and show that this is indeed the case.

If this so-called SC forcing approach is used, the additional forcing terms $F_i^{(\sigma)}$ in (9.122) have to be set to zero (see also Sect. 6.4). This forcing is easy to implement, in particular when all components have the same viscosity. In fact, most published works about SC-based multicomponent systems follow this approach.

The problem with the SC forcing is that it has been shown to lead to τ-dependent surface tension [81], which is clearly an unphysical effect.

Guo forcing: A forcing approach that leads to viscosity-independent surface tension is the extension of Guo's forcing to multiple components. Sega et al. [83] carefully derived the correct algorithm. The first ingredient is to use the barycentric fluid velocity in (9.124) for *all* component equilibrium velocities:

$$\boldsymbol{u}^{\text{eq}(\sigma)} = \boldsymbol{u}_{\text{b}}. \tag{9.127}$$

Additionally, we have to specify the forcing terms $F_i^{(\sigma)}$ in (9.122). They are still given by (6.14), with the force replaced by the component SC force and the fluid velocity by the barycentric velocity [83]:

$$F_i^{(\sigma)} = w_i \left(\frac{c_{i\alpha}}{c_s^2} + \frac{\left(c_{i\alpha}c_{i\beta} - c_s^2\delta_{\alpha\beta}\right) u_{\text{b}\beta}}{c_s^4} \right) F_\alpha^{\text{SC}(\sigma)}. \tag{9.128}$$

Exercise 9.15 Show that (9.127) and (9.128) reduce to the standard Guo forcing presented in Sect. 6.3 if only a single fluid component exists ($S = 1$).

Shan-Chen and Guo forcing are equivalent to linear order in the force. Both methods differ by terms quadratic in $\boldsymbol{F}^{\text{SC}(\sigma)}$. These terms are responsible for the τ-dependence of the surface tension of the SC forcing method.

9.3.3.3 Component Forces

So far we have considered systems that are only subjected to SC interaction forces. Each component σ feels the SC force $\boldsymbol{F}^{\text{SC}(\sigma)}$ according to (9.121). Furthermore, there may be external forces $\boldsymbol{F}^{\text{ext}}$ that act on all components, e.g. gravity. In this case, these forces are distributed to the components according to their concentration:

$$\boldsymbol{F}^{\text{ext}(\sigma)} = \frac{\rho^{(\sigma)}}{\rho} \boldsymbol{F}^{\text{ext}}, \quad \rho = \sum_\sigma \rho^{(\sigma)}. \tag{9.129}$$

The total force felt by component σ is then given by

$$\boldsymbol{F}^{(\sigma)} = \boldsymbol{F}^{\text{SC}(\sigma)} + \boldsymbol{F}^{\text{ext}(\sigma)}. \tag{9.130}$$

9.3.3.4 Immiscible and (Partially) Miscible Fluids, Surface Tension

To simplify the discussion, let us consider a system with two fluid components and interaction strength G_{12}. For $G_{12} = 0$, both fluids interact only through their common velocity $\boldsymbol{u}'$ (in case of SC forcing) or $\boldsymbol{u}_{\text{b}}$ (in case of Guo forcing), but there is no interaction force and the system is an ideal fluid mixture and therefore completely miscible.

With increasing G_{12}, both components repel each other; above a critical value of G_{12}, both fluids finally separate and form an interface. The larger G_{12}, the thinner the interface region. This is shown in Fig. 9.12. If the repulsion was sufficiently

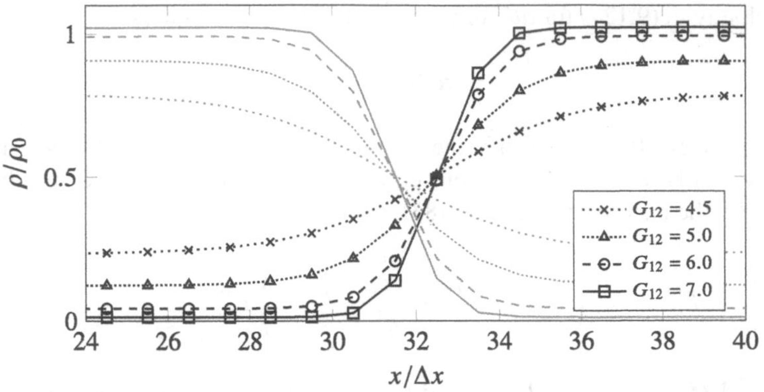

Fig. 9.12 Density profiles across the interface between two immiscible fluids. The interface is located at $x = 32\Delta x$. *Black curves* show the density of one component, *grey curves* the density of the other component. Symbols are only shown for the black lines to improve readability. Increasing G_{12} leads to a sharper interface and a more pronounced demixing of the components. No interface forms for values $G_{12} \approx 4$ or smaller. Although the total fluid density is not constant across the interface, the pressure according to (9.131) is constant up to finite difference approximation errors (both datasets not shown). Results have been obtained by using the pseudopotential in (9.102) and Guo's forcing

strong, the fluids would completely demix (completely immiscible). However, in reality there is always a small amount of fluid 1 in the domain of fluid 2 and the other way around (partially miscible). Particularly with diffuse interface methods, it is not possible to achieve completely immiscible fluids. In case of the SC model, the SC forces would become so large and numerically stiff at some point that component densities would become negative and the algorithm unstable [12].

In the miscible limit, where G_{12} is finite but sufficiently small, the system is characterised by mutual diffusion of the components. Shan and Doolen [69, 84] have thoroughly investigated the diffusion characteristics and the mixture viscosity in those systems. Miscible systems with a minority component, i.e. $\rho^{(\sigma)} \ll \rho^{(\tilde{\sigma})}$, can also be simulated with the methods presented in Chap. 8.

The SC model is particularly powerful when the components are immiscible. For sufficiently large G_{12}, when demixing occurs and interfaces between components appear, surface tension is an emergent feature of the SC model. Unfortunately, like in the multiphase model in Sect. 9.3.2, it is not clear *a priori* what the value of the surface tension γ is for a given choice of the interaction parameter G_{12}. We can pursue exactly the same strategy as in Sect. 9.3.2 and perform the Young-Laplace test to find the relation between G_{12} and γ for a given choice of pseudopotential. In order to undertake the Young-Laplace test, we need to know the pressure of the multicomponent fluid.

The **equation of state for the multicomponent SC model in the continuum limit** is [69]

$$p = c_s^2 \sum_\sigma \rho^{(\sigma)} + \frac{c_s^2 \Delta t^2}{2} \sum_{\sigma,\tilde{\sigma}} G_{\sigma\tilde{\sigma}} \psi^{(\sigma)} \psi^{(\tilde{\sigma})}. \tag{9.131}$$

The first term on the right-hand side reflects the ideal gas properties of the components while the second term denotes the interaction between them. It is this second term that can lead to phase separation of the components.

The Young-Laplace procedure is similar to that in Sect. 9.3.2. A droplet of one fluid is placed in the other fluid. Due to the surface tension and the curved droplet surface, the pressure from (9.131) in the droplet interior is larger then the exterior pressure. Measuring the pressure difference and the droplet radius allows us to obtain the surface tension in the usual way.

9.3.3.5 Boundary Conditions and Contact Angle

One of the most important applications of the multicomponent model is the simulation of flows in porous media where the flow behaviour is dominated by the interaction of the fluid components with the solid phase. These situations are commonly encountered in oil recovery and in the textile, pharmaceutical and food industries [13, 85, 86].

Like in the case of a multiphase fluid in Sect. 9.3.2, the interaction of a multicomponent fluid with a solid wall requires two main ingredients: (i) the no-slip condition and (ii) a wetting condition. The former is normally realised by the simple bounce-back method for all fluid components (cf. Sect. 5.3.3). For the latter we follow a similar approach as for a multiphase fluid in (9.116). In order to achieve the desired contact angle θ at the solid surface, different fluid components have to interact differently with the solid. The contact angle satisfies (9.9) where gas "g" and liquid "l" have to be replaced by fluid 1 and fluid 2.

Martys and Chen [71] proposed an interaction force between fluid and adjacent solid nodes:

$$\boldsymbol{F}^{s(\sigma)}(\boldsymbol{x}) = -G_{\sigma s} \rho^{(\sigma)}(\boldsymbol{x}) \sum_i w_i s(\boldsymbol{x} + \boldsymbol{c}_i \Delta t) \boldsymbol{c}_i \Delta t. \tag{9.132}$$

Here, $s(\boldsymbol{x})$ is an indicator function that assumes the values 0 and 1 for fluid and solid nodes, respectively, and $G_{\sigma s}$ is the interaction strength between fluid component σ and the solid boundary. For a wetting fluid, the interaction should be attractive and therefore $G_{\sigma s} < 0$ [87]. Accordingly, non-wetting fluids should have a positive value of $G_{\sigma s}$. Taking a binary fluid with a given value of G_{12} as an example, the contact angle θ can be tuned by varying the two free parameters G_{1s} and G_{2s}.

We can find different fluid-surface force models in the literature, e.g. [12, 88, 89]. Chen et al. [12] proposed a form that is a direct extension of (9.116):

$$\boldsymbol{F}^{s(\sigma)}(\boldsymbol{x}) = -G_{\sigma s}\psi^{(\sigma)}(\boldsymbol{x})\sum_{i}^{\text{solid}} w_i \psi(\rho_s)\boldsymbol{c}_i \Delta t. \tag{9.133}$$

This model includes a "solid density" ρ_s as another degree of freedom. The solid density can be used to tweak the contact angle, see [88, 89] for example. If ρ_s is constant everywhere on the solid, it can be absorbed in the definition of $G_{\sigma s}$. In that case, and choosing $\psi^{(\sigma)} = \rho^{(\sigma)}$, (9.132) and (9.133) are equivalent.

Depending on the exact details of the chosen fluid-solid interaction model, it is generally necessary to run a series of simulations to establish the relation between the desired contact angle θ and the simulation parameters, i.e. G_{12}, G_{1s}, G_{2s} and ρ_s. Huang et al. [87] carefully investigated the behaviour of the contact angle in a binary system subject to the force in (9.132). They suggested a simple equation to predict the contact angle *a priori*. Their results indicate that the solid interaction parameters G_{1s} and G_{2s} should be similar in magnitude, $G_{1s} \approx -G_{2s}$, although it is in principle possible to choose different values.

To extract the contact angle from a simulation, we proceed in the same way as in Sect. 9.3.2. The only difference is that the number of free parameters controlling the wetting properties is larger in the multicomponent model.

9.4 Limitations and Extensions

All multiphase and multicomponent models are challenging to develop and usually show a number of limitations. We discuss the most common and most important limitations of the free-energy and Shan-Chen models and some remedies that have been suggested. These include spurious currents (cf. Sect. 9.4.1), restricted density ratio (cf. Sect. 9.4.2), limited surface tension range (cf. Sect. 9.4.3) and viscosity ratio restrictions (cf. Sect. 9.4.4). The section is concluded with a non-exhaustive list of extensions in Sect. 9.4.5, showing the breadth of applications that can be tackled with LB-based multiphase and multicomponent models. We treat the free energy and the Shan-Chen models side by side as their limitations are of a similar nature.

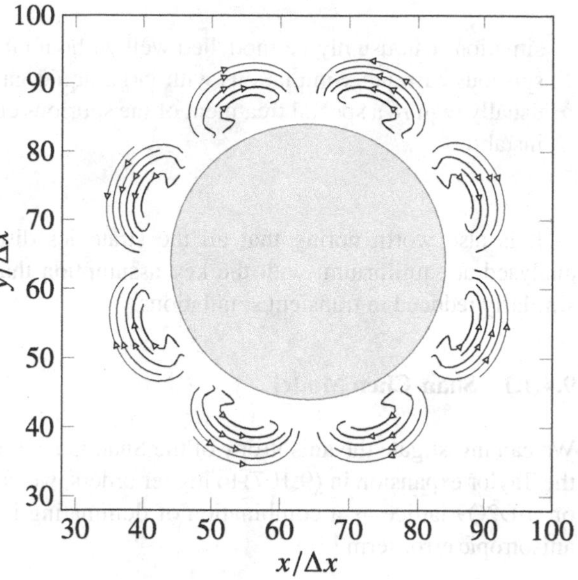

Fig. 9.13 Spurious currents appearing in the standard multiphase Shan-Chen model. A droplet with radius $20\Delta x$ is located at the centre of a 2D domain with size $128\Delta x \times 128\Delta x$. The interaction parameter is $G = -5.0$, and the relaxation time is $\tau = \Delta t$. The pseudopotential is $\psi = 1 - \exp(-\rho)$. Velocities with the magnitude less than $10^{-3}\Delta x/\Delta t$ are eliminated from the plot

9.4.1 Spurious Currents and Multirange Forces

When simulating a steady droplet, we expect zero fluid velocity everywhere. However, numerical simulations often reveal a different picture showing microcurrents (also *spurious* or *parasitic currents*) near the droplet interface. This is an unphysical and therefore undesirable effect. If the magnitude of the spurious currents is large, they can lead to numerical instability. Spurious currents are a well-known problem for LB and non-LB methods (see [11] and references therein). A recent review of spurious currents in LBM is available in [53].

Spurious currents appear both in the free-energy and the Shan-Chen models, both in multiphase and multicomponent applications [11]. Figure 9.13 shows the steady spurious currents near the surface of a droplet simulated with the Shan-Chen multiphase model. A similar profile for spurious currents is also observed for the free-energy model [51, 53].

Spurious currents are caused by numerical approximations of the surface tension force [12, 90]. For example, the surface tension force at the interface of a steady circular droplet should always point towards the centre of the droplet. If the numerical discretisation is not perfectly isotropic, we expect tangential force components that drive the spurious currents.

In many cases, the **characteristic flow** (e.g. due to droplet formation or phase separation) is significantly faster than the spurious currents. Those

(continued)

situations can usually be modelled well without taking additional care of the spurious currents. Simulations with large liquid-gas density ratios, however, usually require a special treatment of the spurious currents to avoid numerical instability.

It is also worth noting that all the strategies discussed in the following are analysed at equilibrium, with the key assumption that the spurious velocities are similarly reduced in transient simulations.

9.4.1.1 Shan-Chen Model

We can investigate the anisotropy of the Shan-Chen model analytically. Continuing the Taylor expansion in (9.107) to higher orders, we can write the Shan-Chen force on a D2Q9 lattice as a combination of dominating isotropic contributions and an anisotropic error term [74]:

$$\boldsymbol{F}^{\mathrm{SC}}(\boldsymbol{x}) = \underbrace{-G\psi(\boldsymbol{x})\left(\frac{1}{3}\nabla\psi(\boldsymbol{x}) + \frac{1}{18}\nabla\Delta\psi(\boldsymbol{x}) + \frac{1}{216}\nabla\Delta^2\psi(\boldsymbol{x})\right)}_{\text{isotropic}} + \boldsymbol{F}^{\mathrm{aniso}}. \tag{9.134}$$

The isotropic terms (up to fourth order) lead to radial forces, while the anisotropic term (fifth order),

$$\boldsymbol{F}^{\mathrm{aniso}} \propto G\psi(\boldsymbol{x})\left(\hat{\boldsymbol{e}}_x\partial_x^5 + \hat{\boldsymbol{e}}_y\partial_y^5\right)\psi(\boldsymbol{x}), \tag{9.135}$$

gives rise to a tangential force component and therefore the spurious currents in Fig. 9.13.

As the spurious currents are caused by the discretisation of the force, possible improvements could aim at the Shan-Chen force and its discretisation. One approach is to use a special mean-value approximation of the surface tension force [91]. Another solution is to improve the force isotropy. This approach is called *multirange* as it involves larger numerical stencils involving lattice nodes at greater distances (cf. Fig. 9.14).

The simplest multirange interaction force can be written with two interaction parameters G_1 and G_2 [74]:

$$\boldsymbol{F}^{\mathrm{SC}}(\boldsymbol{x}) = -\psi(\boldsymbol{x})\left[G_1\sum_{i\in\mathrm{b}1} w_i\psi(\boldsymbol{x}+\boldsymbol{c}_i\Delta t)\boldsymbol{c}_i\Delta t + G_2\sum_{i\in\mathrm{b}2} w_i\psi(\boldsymbol{x}+\boldsymbol{c}_i\Delta t)\boldsymbol{c}_i\Delta t\right] \tag{9.136}$$

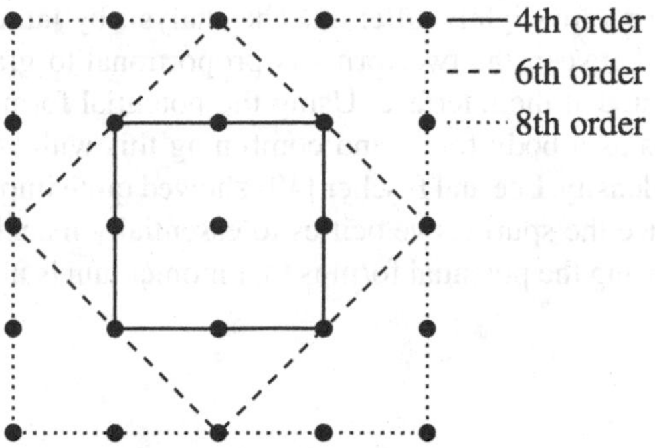

Fig. 9.14 Illustration of multirange stencils for a 2D Shan-Chen model. Increasing the range and therefore the number of interacting lattice nodes leads to an increasing isotropy order. The standard Shan-Chen model in (9.105) involves only the D2Q9 neighbours (4th order, *solid line*). The weights for the more isotropic numerical stencils are specified in [74]

where the first sum runs over belt 1 (b1, solid line in Fig. 9.14) and the second over belt 2 (b2, dotted line in Fig. 9.14). Including more and more belts of interacting lattice nodes decreases the magnitude of the spurious currents. We will come back to the multirange model in Sect. 9.4.3.

Despite its advantages, the multirange model is computationally more expensive. Also, boundary conditions need to be modified as there can now be several layers of solid nodes interacting with a fluid node. Apart from its primary function to increase isotropy, the multirange model has been employed to simulate emulsions with non-Newtonian rheology and non-coalescing droplets [92–94].

9.4.1.2 Free-Energy Model

Spurious currents appear in the free-energy model as well, and the root is the inexactness of discretised numerical stencils for approximating derivatives in the density or order parameter. Generally there are two common strategies which have been proposed to reduce spurious velocities in free-energy models. The first strategy, akin to the Shan-Chen model, is to improve the isotropy of the stencils for the derivatives [49–51], often at the expense of more computational time. Pooley and Furtado [51] showed that a good choice of stencil can lead to a reduction in spurious velocities by an order of magnitude.

As shown in Sect. 9.2, we have several choices to implement the non-ideal terms in the pressure tensor which account for the physics of surface tension. So the second strategy is to consider which form of the non-ideal terms can be discretised with least error. Numerical evidence shows that the forcing approach tends to perform better in comparison to the pressure tensor approach [51, 55]. For the forcing approach, we have seen that the non-ideal terms can be written either in the so-called *pressure* or *potential form*. While they are identical analytically, their discretised forms are slightly different. The analysis by Jamet et al. [54] shows that the difference between the two forms is proportional to gradients of density, which are exacerbated at the interface. Using the potential form for implementing the non-ideal terms as a body force, and combining this with isotropic numerical derivatives for the density, Lee and Fischer [49] showed quite impressively that such a strategy can reduce the spurious velocities to essentially machine precision. One valid criticism on using the potential form is that momentum is no longer conserved exactly [95, 96].

9.4.2 Equation of State and Liquid-Gas Density Ratio

In some multiphase applications it is desirable to achieve large liquid-gas density ratios. Realistic density ratios seen in nature can be of the order of 10^3; this poses significant challenges for numerical stability. For example, the Shan-Chen multiphase model with its equation of state from (9.111) becomes unstable at $G \lesssim -7.0$ with a density ratio around 70 [79]. For the standard free-energy model, the situation is even worse because the surface tension force $k\rho\nabla\Delta\rho$ becomes unstable in the presence of large density differences, and the density ratio is limited to around 10. This instability is caused by the numerical stiffness of the forces.

The gas and liquid coexistence densites are determined by the equation of state through the Maxwell area construction rule (cf. Sect. 9.1.1). Thus, a key strategy to increasing the liquid-gas density ratio is by changing the equation of state. For the Shan-Chen model, the inclusion of an arbitrary equation of state requires redefinition of the pseudopotential function [78]. Starting from (9.111), we find

$$p_{\mathrm{b}}(\rho) = c_{\mathrm{s}}^2\rho + \frac{c_{\mathrm{s}}^2\Delta t^2 G}{2}\psi^2(\rho) \quad \Longrightarrow \quad \psi(\rho) = \sqrt{\frac{2}{c_{\mathrm{s}}^2\Delta t^2|G|}\left(p_{\mathrm{b}}(\rho) - c_{\mathrm{s}}^2\rho\right)}. \tag{9.137}$$

The interaction strength G effectively cancels when substituting the pseudopotential in (9.137) into the Shan-Chen force. One should especially make sure that the expression under the root remains always positive [12]. This is possible by adopting a special form of equilibrium functions [70].

In principle, any equation of state can be incorporated into the Shan-Chen model. However, it is usually suggested to use the Peng-Robinson or Carnahan-Starling equations of state because they can reach a density ratio up to 10^3 and the obtained densities have smaller deviations from their predicted values from the Maxwell area construction rule. The van der Waals equation of state usually yields a larger violation of thermodynamic consistency and is not generally recommended [78].

It is possible to make the Shan-Chen model thermodynamically consistent [97]. Such models, however, suffer from instability due to the discretisation of the term $k\rho\nabla\Delta\rho$. This term is potentially unstable when the liquid-gas density ratio (and therefore the density gradients) are large.

Recently, Lycett-Brown and Luo [77] showed that the liquid-gas coexistence behaviour of the Shan-Chen model can be significantly improved when additional forcing terms are considered. They performed a Chapman-Enskog analysis up to third order in the force and derived correction terms that remove or improve a number of shortcomings of the Shan-Chen model (see also Sect. 9.4.3).

For the free-energy model, different equations of state are easily integrated by modifying the equilibrium functions (cf. (9.36)) or by including the non-ideal terms as forcing terms (cf. (9.43) and (9.44)). Only changing the equation of state, however, is inadequate to reach density ratios of the order of 10^3.

To improve the stability of multiphase free-energy models, several successful approaches have been developed which share a common thread: they exploit two distribution functions. Inamuro et al. [98] used one distribution for the density and another distribution for the predicted velocity without pressure gradient. Every LB step is then followed by a pressure correction step which is relatively expensive computationally. Lee and Lin [44] also used two distribution functions to track the density and pressure, respectively. Lee and Lin's method does not require a pressure correction step. However, its implementation is quite complex as it requires discretisations of many first- and second-order derivatives, as well as different representations of the surface tension force. Zheng et al. [99] developed a scheme which is effectively a "hybrid" between the multiphase and multicomponent model. In their scheme, one distribution function is assigned to the density field and for solving the hydrodynamic equations of motion. The other distribution belongs to an order parameter for tracking the liquid-gas interface. In contrast to Sect. 9.2.2 for the standard binary model where the density is required to be a constant, the fluid density is allowed to vary between the phases in Zheng et al.'s method.

More recently, Karlin and co-workers [45, 100] developed an entropic LB scheme for multiphase flow and achieved large density ratios without the need to implement a pressure correction or an additional set of distribution functions. In entropic LBM, entropy balance is approximated in the relaxation step at each node, which helps stabilise the liquid-gas interface.

Multiphase flows are often characterised by **low capillary and low Reynolds numbers**. One example is Bretherton flow of long bubbles or fingers in a microchannel. In this case the liquid-gas density ratio is not the governing non-dimensional number. As a consequence, the **exact density ratio does not have to be matched** and a simple multiphase or a multicomponent model can be employed. In many situations, it is actually possible to **replace a multiphase model by a multicomponent model with much lower liquid-gas density ratio** [101, 102], thus avoiding numerical instability and undesirable condensation/evaporation.

9.4.3 Restrictions on the Surface Tension

In practical applications, the numerically accessible range of the surface tension is limited. Large surface tensions can lead to instability, and this issue often goes hand-in-hand with spurious velocities discussed in Sect. 9.4.1. The magnitude of the spurious velocities typically increases with the surface tension.

For standard multiphase and multicomponent free-energy models, the highest achievable surface tension is no more than 0.1 (lattice units). The standard Shan-Chen model also has a similar limitation; a surface tension of ~ 0.1 is achievable for $G \approx -7$ beyond which simulations become unstable.

In interface-governed flows, the important dimensionless parameters include the capillary number $\mathrm{Ca} = \eta u/\gamma$ (u is a characteristic velocity) and the Bond number $\mathrm{Bo} = \rho g \ell_{\mathrm{S}}^2/\gamma$ (g is the gravitational acceleration and ℓ_{S} is a characteristic length scale, e.g. the system size). While the numerical range of the surface tension in LBM is limited, the relevant dimensionless numbers can, to some extent, be varied by changing the other parameters (cf. Sect. 7.3.5).

In the context of surface tension, an advantage of free-energy models is that the equation of state, the surface tension, and the interface width can all be varied independently. This is not the case for the *original* Shan-Chen model where the parameter G determines both the equation of state in (9.111) and the surface tension force $\propto G\psi(\rho)\nabla\Delta\psi(\rho)$. As such, low values of the surface tension are often linked to a large interface width and loss of immiscibility (in case of multicomponent fluids).

To decouple the equation of the state from the surface tension, the Shan-Chen multirange approach from Sect. 9.4.1 can be used. Starting from (9.136) and a Taylor expansion as in Sect. 9.3.2, one can show that the equation of state and the surface tension change to [103]

$$
\begin{aligned}
p_{\mathrm{b}}(\rho) &= c_{\mathrm{s}}^2\rho + \frac{c_{\mathrm{s}}^2\Delta t^2 A_1}{2}\psi^2(\rho), \\
\gamma &= -\frac{c_{\mathrm{s}}^4 A_2}{2}\int\left(\frac{\mathrm{d}\psi}{\mathrm{d}x}\right)^2\mathrm{d}x
\end{aligned}
\tag{9.138}
$$

with $A_1 = G_1 + 2G_2$ and $A_2 = G_1 + 8G_2$. By changing G_1 and G_2 accordingly, the equation of state and surface tension can be modified independently.

Exercise 9.16 Show the validity of (9.138), following the derivation outlined in the previous paragraph.

Recent progress in understanding the role of the Shan-Chen force makes it possible to change the equation of state (and therefore the liquid-gas density ratio), surface tension and interface width independently and over a wider range than previously possible [77].

9.4.4 *Viscosity Ratio and Collision Operator*

Multicomponent problems often involve fluids with different kinematic viscosities (or densities). Miscible multicomponent fluids are characterised by their Schmidt number (ratio of mass and momentum diffusivity). Modelling these systems can be challenging [104]. The original Shan-Chen model, for example, is limited to a viscosity ratio of about 5 [105]. For mixtures with density ratios other than unity, section 6 in the review [12] contains an overview of recent progress in the field.

Due to the well-known restrictions of the BGK collision operator, it is generally recommended to use MRT for fluid mixtures with large viscosity ratio. Porter et al. [104] achieved a kinematic viscosity ratio of up to 1000 in the Shan-Chen model by using MRT and enhanced force isotropy. MRT also helps to reach higher Reynolds numbers in bubble simulations [106]. While the BGK operator leads to a fixed Schmidt number, the MRT collision operator can be employed to change mass and momentum diffusivity independently [107]. Other works involving the MRT collision operator (both for multicomponent and multiphase, and free energy and Shan-Chen) include [81, 108–111].

It is worth mentioning that Zu and He [112] suggested a multi*component* model with density ratio; a feature that is usually neglected in other works.

9.4.5 *What Else Can Be Done with These Models?*

In this chapter we have deliberately focussed on two-phase and two-component flows, both using the free-energy and the Shan-Chen approaches. The ideas developed here can be extended in many different directions, and these are areas of current active research. We will now highlight some examples, inevitably selective, of interesting problems.

The simplest extension to models described here is to introduce *more fluid components*, and in recent years particularly ternary systems [6, 43, 113, 114] have attracted growing interest. In fact, we are not limited to "normal" fluids. It is also possible to extend the model to include *surfactants* [115, 116]. Surfactants are amphiphilic molecules; one end of the molecule is hydrophilic (likes water; dislikes oil), the other hydrophobic (dislikes water; likes oil). Thus, surfactants tend to sit at the interface between water and oil, and they tend to reduce the water-oil surface tension. Surfactants are foundational for many industries, from oil recovery to food and consumer products.

The complexity of fluid dynamical problems is tightly related to the *boundary conditions*. There is a large literature base covering systems with, e.g., free surfaces [117–119], droplet spreading on solid surfaces [120, 121] and Leidenfrost droplets [122]. The LBM is excellent for handling tortuous boundary conditions, such as for flow in porous materials [123, 124]. The wetting boundary conditions can also be extended to cases where the solid surfaces are mobile, thus allowing the simulation of colloidal particles and polymers at fluid-fluid interfaces [83, 89, 125–128].

Many investigations only focus on steady droplets for which conditions at the domain boundaries do not play an important role. Contrarily, despite some progress [129], *open boundary conditions* for multiphase flows have not yet been thoroughly investigated. Those conditions are important for problems such as droplet formation and manipulation in microfluidic channels [130, 131]. Pressure boundary conditions pose a particular challenge as they have to be combined with the modified pressure due to the non-ideal equation of state. In simulations this can manifest as unexpected condensation or evaporation.

Throughout this chapter we have mostly neglected *phase change*. LB models have been successfully applied to systems with evaporation [132, 133], solidification [134, 135] and even chemical reactions, e.g. at liquid interfaces [136].

The free-energy approach is particularly popular in the physics community. Different choices for the free energy can allow for *new physics*, and there is a wide range of problems in complex fluids where hydrodynamics is important. Including curvature energy into the gradient terms allow the study of lamellar phases [137] and vesicles [138, 139]. The bulk free energy can be modified as well to add more complex equations of state, e.g. for liquid crystals [140]. The descriptions of these physical phenomena enter the Navier-Stokes equation through the pressure tensor, which in turn can be implemented in the LBE through the equilibrium distribution functions or forcing terms.

9.5 Showcases

We discuss two common multiphase/multicomponent applications and explain how to simulate them using LB simulations: droplet collisions in Sect. 9.5.1 and wetting on structured surfaces in Sect. 9.5.2. Both applications are of great relevance for today's engineering challenges, such as inkjet printing and functional surfaces.

9.5.1 Droplet Collisions

One important application of multiphase LBM is the collision of droplets, in particular in an ambient gas phase. This phenomenon occurs in nature, e.g. cloud formation, and in many industrial areas, such as ink-jet printing and spray combustion in internal combustion engines. A better understanding of droplet collision helps improving these industrial processes. For example, the coalescence of droplets impinging on paper affects the quality of ink-jet printing. There are many experimental works

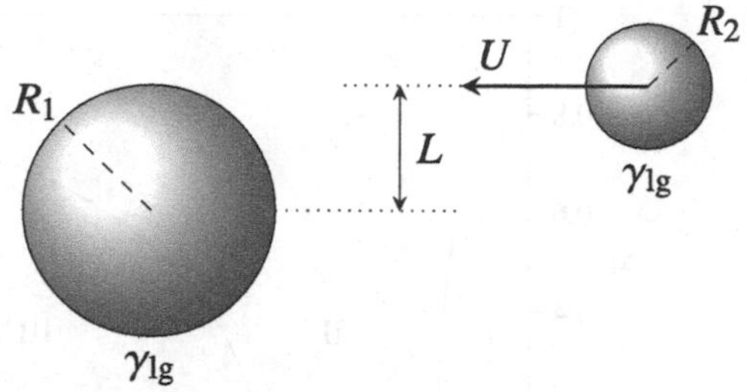

Fig. 9.15 Sketch of two colliding droplets with radii R_1 and R_2. This collision is shown in the rest frame of the larger droplet, having the smaller droplet approaching with velocity U. Collisions usually happen with a finite off-centre distance L

studying droplet collisions [141, 142]. It was found that this phenomenon exhibits a rich map of collision modes as detailed below.

Collisions of two droplets are characterised by a number of physical parameters, such as the droplet radii R_1 and R_2, impact velocity U, off-centre distance L, density ρ and viscosity η of both the liquid droplets (denoted by "l") and the surrounding gas (denoted by "g"), and surface tension γ (cf. Fig. 9.15). From those physical parameters we can construct relevant non-dimensional groups [141, 143]:

$$\begin{aligned}
\mathrm{We} &= \frac{U^2(m_1+m_2)}{4\pi\gamma\left(R_1^2+R_2^2\right)}, && \text{(symmetric Weber number)} \\
\mathrm{Re} &= \frac{\rho_l U(R_1+R_2)}{\eta_l}, && \text{(Reynolds number)} \\
\mathrm{B} &= \frac{L}{R_1+R_2}, && \text{(impact factor)} \\
&\frac{R_1}{R_2}, \frac{\eta_l}{\eta_g}, \frac{\rho_l}{\rho_g}. && \text{(radius, density and viscosity ratios)}
\end{aligned} \tag{9.139}$$

Many works concentrate on droplets with equal sizes ($R_1 = R_2 = R$). Thus, the Weber and Reynolds numbers simplify to $\mathrm{We} = \rho_l RU^2/(3\gamma)$ and $\mathrm{Re} = 2\rho_l RU/\eta_l$, respectively.

Depending on the value of the collision parameters, different regimes can be identified [141, 144]. Figure 9.16 shows a droplet collision map obtained from a large number of experiments. This map can be explained by considering the roles of surface tension and inertia, and by using the illustrations in Fig. 9.17.

1. **Coalescence after minor deformation.** If the kinetic energy is small, the gas between droplets is able to drain and the droplet deformation is small. Thus, the gas film does not lead to a strong repulsion between the droplets. When the droplets are close enough, coalescence happens through the van der Waals force.
2. **Bouncing.** Increasing the kinetic energy (higher Weber number), the gas between the droplets is not able to drain in time. A high pressure builds up between the droplets, and the kinetic energy is temporarily stored in the deformed surfaces. The droplets are pushed back and bounce before coalescence can occur.

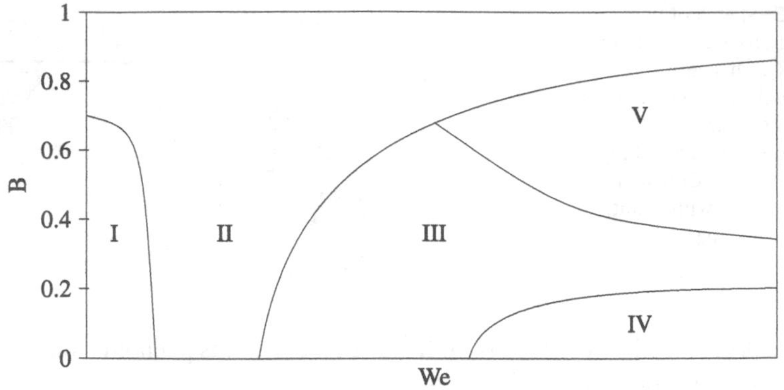

Fig. 9.16 Collision map for droplets of equal size. Depending on the Weber number and the impact parameter B, the following regimes can be found: (I) coalescence after minor deformation, (II) bouncing, (III) coalescence after substantial deformation, (IV) reflective separation and (V) stretching separation. See the text for more explanation and Fig. 9.17 for illustrations

3. **Coalescence after substantial deformation.** With further increase of the kinetic energy, the droplets are substantially deformed. The pressure at the centre of the film between the droplets is higher than at the outer film region, and it pushes gas out of the gap. After the gas in the film has sufficiently drained, the droplets coalesce. At the initial stage of coalescence, the newly formed droplet has a torus-like shape. Surface tension drives the droplet towards a spherical shape. This leads to droplet oscillations that are eventually damped by viscous dissipation.
4. **Reflexive separation.** For even higher Weber number, the droplet oscillation is not efficiently damped. Due to the large kinetic energy in the system, two liquid pockets form at opposite ends that drive the droplet apart along the original collision axis. Surface tension is too weak to balance the inertial force, and the droplet breaks up into two smaller droplets. Sometimes more droplets are formed as the filament between the droplets becomes unstable and breaks up into several satellite droplets.
5. **Stretching separation.** If the impact parameter B is large, droplet separation can happen without oscillations as significant parts of the droplets do not interact with each other. Instead, these droplet parts continue moving in their original directions. While moving, a filament forms between the droplets. Depending on the droplet parameters, the filament can either stabilise or disintegrate by forming small satellite droplets.

The collision map in Fig. 9.16 does not include the viscosity and density ratios; in fact, these parameters also have a strong effect on the collision outcome. For example, there is no bouncing regime for water droplets colliding in air, but bouncing exists for hydrocarbon droplets in air [141]. For more detailed parameter studies we refer to experimental works [141–143] and numerical studies [144–146].

To conclude this section, we now focus on the role of LBM in droplet collisions. Numerical studies allow us to obtain important information that is difficult to

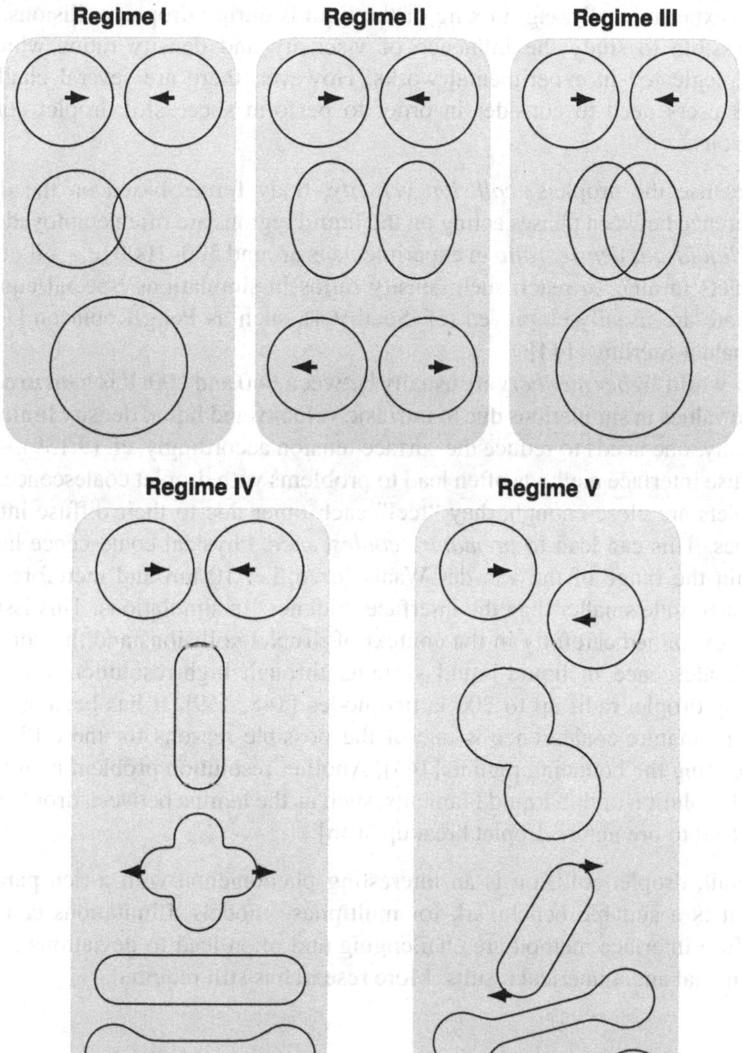

Fig. 9.17 Droplet collision regimes and mechanisms. See Fig. 9.16 for more details

observe experimentally, e.g. mixing of the liquids during droplet collisions. Also, it is possible to study the influence of viscosity and density ratios which are usually neglected in experimental works. However, there are several challenges that LB users need to consider in order to perform successful droplet collision simulations:

- To realise the droplets' *collision velocity*, body forces based on the density difference between phases acting on the liquid regions are often employed [147].
- The *liquid-gas density ratio* in experiments is around 500–1000 (e.g. oil or water droplets in air). To reach such density ratios in simulations, special equations of state are usually employed (cf. Sect. 9.4), such as Peng-Robinson [146] or Carnahan-Sterling [144].
- Real-world *Weber numbers* are usually between ~ 0 and 100. It is hard to achieve large values in simulations due to intrinsic velocity and liquid density limitations. Usually, one needs to reduce the surface tension accordingly, cf. (9.139).
- Diffuse interface methods often lead to problems with droplet coalescence. If the droplets are close enough, they "feel" each other due to their diffuse interface shapes. This can lead to *premature coalescence*. Physical coalescence happens within the range of the van der Waals force, i.e. 10 nm and therefore orders of magnitude smaller than the interface thickness in simulations. This issue has been examined carefully in the context of droplet collisions and the conditions for coalescence in liquid-liquid systems through high-resolution simulations having droplet radii up to 200 lattice nodes [148, 149]. It has been suggested that premature coalescence is one of the possible reasons for the difficulty of simulating the bouncing regime [144]. Another resolution problem manifests in the dissolution of thin liquid filaments, such as the lamina between droplets. This can lead to premature droplet breakup [146].

Overall, droplet collision is an interesting phenomenon with a rich parameter space. It is a suitable benchmark for multiphase models. Limitations caused by the diffuse interface method are challenging and often lead to deviations between experimental and numerical results. More research is still required.

9.5.2 Wetting on Structured Surfaces

One of the primary advantages of the LBM is its ability to handle complex boundary conditions. This advantage remains for multiphase and multicomponent flows. Variation in the surface wettability can be implemented easily by applying different values of the phenomenological parameter h in (9.64) and (9.83) for the free-energy multiphase and multicomponent models, or by varying the solid density ρ_s and the solid-fluid interaction parameters $G_{\sigma s}$ for the Shan-Chen models at different surface lattice sites.

Here we will highlight how LBM can be used to model drops spreading on chemically and topographically patterned surfaces. There are two complementary perspectives why such modelling is of great scientific interest. First, any real surface is never perfectly smooth and chemically homogeneous; indeed, surface heterogeneities are an important consideration in many areas, including oil recovery, fluid filtration and capillary action in plants [150, 151]. Secondly, it is becoming increasingly feasible to fabricate surfaces with roughness and heterogeneities in a controlled and reproducible manner. Thus, instead of being viewed as a problem, surface patterning has now become a versatile part of a designer toolbox to control the shapes and dynamics of liquid droplets and interfacial flows [152, 153].

9.5.2.1 Chemical Patterning

We first look at a drop spreading on a chemically patterned surface. For a homogeneous surface, the droplet's final state is a spherical cap with a contact angle equalling the Young angle, as illustrated in Exercise 9.6. This is not the case for heterogeneous surfaces, however.

Figure 9.18 shows simulation results of drops on a chemically patterned substrate. The surface is lined with hydrophilic and hydrophobic stripes with Young angles of 45° and 105° and widths of $8\Delta x$ and $24\Delta x$, respectively. The drop volumes have been chosen so that their final diameters were comparable to the stripe width.

Simulations and experiments show that the final drop shape is selected by the initial impact position and velocity. If the drop can touch two neighbouring hydrophilic stripes as it spreads, it will reach the "butterfly" configuration, Fig. 9.18a. Otherwise, it will retract back to the "diamond" pattern, spanning a single stripe, Fig. 9.18b. Both states are free energy minima but one of the two is a metastable minimum: which one is metastable depends on the exact choice of the physical parameters. For more detailed discussion we refer the readers to dedicated articles on this topic [32, 154, 155].

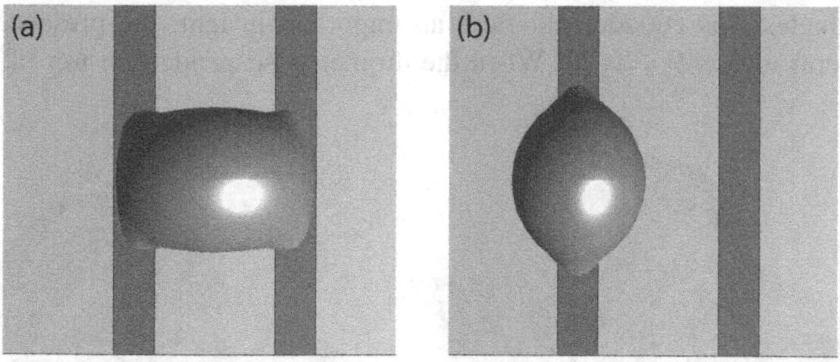

Fig. 9.18 Drops spreading on chemically striped surfaces. Hydrophilic (45°) and hydrophobic (105°) stripes are shown in *dark and light grey*, respectively. The drop shapes depend on the initial impact position and velocity, either spanning across (**a**) two or (**b**) one hydrophilic stripes

More complex chemical heterogeneities can be modelled in LB simulations. Such strategies have been proposed, for example, to control drop position in inkjet printing [156] or to control flow in open microfluidic platforms [157].

9.5.2.2 Topographical Patterning: Superhydrophobic Surfaces

In addition to chemical heterogeneities, surface roughness is important for determining the wetting properties of a solid surface. A prime example is the so-called *superhydrophobic* surface [158]. On a smooth hydrophobic surface, the highest contact angle that can be achieved is of the order of 120°, which is attainable for fluorinated solids (e.g. teflon). When the hydrophobic surface is made rough, however, higher contact angles are possible. The most famous example of a superhydrophobic surface is the lotus leaf (superhydrophobicity is often called the *lotus effect*) [159], but many other natural materials, such as butterfly wings, water strider legs, and duck feathers also exhibit this property. We are also now able to fabricate synthetic superhydrophobic surfaces [160, 161].

It is possible to distinguish two ways in which a drop can behave on a rough surface. One possibility is for the drop to be suspended on top of the surface roughness, as shown in Fig. 9.19a. The droplet effectively sees a composite of liquid-solid and liquid-gas areas. We use Φ to denote the area fraction of the liquid-solid contact (and hence $1 - \Phi$ is the area fraction of the liquid-gas contact). If the length scale of the patterning is much smaller than the drop size, the effective liquid-solid surface tension is the weighted average $\Phi\, \gamma_{sl} + (1 - \Phi)\, \gamma_{lg}$. The gas-solid surface tension is $\Phi\, \gamma_{sg}$. Substituting these into Young's equation, (9.9), gives us the *Cassie-Baxter formula* [162]

$$\cos\theta_{CB} = \Phi \cos\theta - (1 - \Phi) \tag{9.140}$$

where θ is the contact angle if the surface was smooth, and θ_{CB} is the effective contact angle. This equation provides an important insight: the presence of the second term means $\theta_{CB} > \theta$. When the droplet is suspended on top of what is

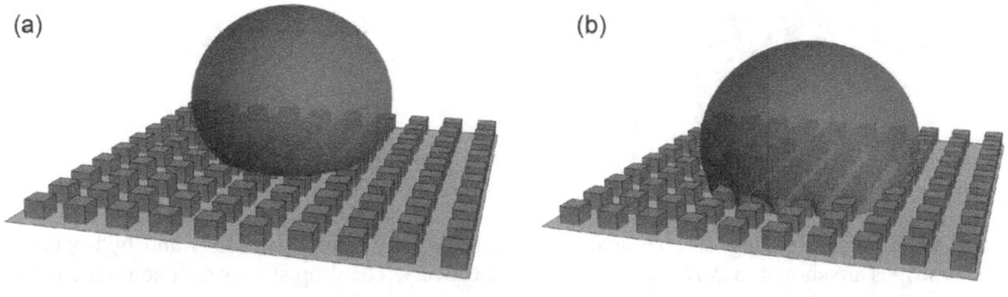

Fig. 9.19 Final states of drops spreading on topographically patterned surfaces. The material (if the surface was smooth) contact angle is 120° and 80° for panels (**a**) and (**b**) respectively. In (**a**) the drop is suspended on top of the corrugations, while it penetrates the posts in (**b**)

effectively an air mattress, it slides very easily across the surface. The drag reduction property of superhydrophobic surfaces is also highly superior. Slip lengths as high as several microns have been reported [163].

The suspended state is not always stable. For example, as we lower the material's contact angle, the liquid usually penetrates the corrugation and fills the space in between the posts, as shown in Fig. 9.19b. In such a case, both the liquid-solid and gas-solid contact areas are increased by a roughness factor r, which is the ratio of total area of the textured surface to its projected area. The effective contact angle θ_W is therefore given by the *Wenzel equation* [164]

$$\cos\theta_W = r\cos\theta. \tag{9.141}$$

This equation suggests that a hydrophilic surface will appear more hydrophilic in the presence of roughness, and similarly a hydrophobic surface will appear more hydrophobic. Compared to the case where the liquid is suspended, fluid drag is strongly increased when liquid penetrates the corrugation. In this collapsed state, the fluid interface is also strongly pinned by the surface corrugations.

Let us end this section by discussing typical considerations for simulating wetting on structured surfaces using LBM:

- It is important to get the hierarchy of length scales correctly. Both for chemically and topographically patterned surfaces, we have (i) the interface width, (ii) the pattern size, and (iii) the drop radius. Ideally we want to make sure that drop radius $\gg$ pattern size $\gg$ interface width, which can be demanding computationally. In practice, a separation of length scales by an order of magnitude is adequate.
- There is, in fact, a fourth length scale corresponding to mechanisms for which the contact line can move. In diffuse interface models, the contact line moves due to an evaporation-condensation mechanism for the multiphase model [41, 165], and due to a diffusive mechanism for the multicomponent model [42, 88]. We refer the readers to [66] for a detailed analysis on the effect of varying the contact line slip length against the interface width.
- Similar to the previous subsection, capturing realistic density ratios of 500–1000 can be critical for the accuracy of the simulations, and it is therefore desirable. However, depending on the applications, we can often get away with much smaller density ratios, for example when we are only interested in equilibrium/final configurations, or when the inertial terms in the Navier-Stokes equation are irrelevant for the problem at hand. For the latter, usually getting the viscosity ratio right is more important.

References

1. A.K. Gunstensen, D.H. Rothman, S. Zaleski, G. Zanetti, Phys. Rev. A **43**(8), 4320 (1991)
2. D. Grunau, S. Chen, K. Eggert, Phys. Fluids A **5**(10), 2557 (1993)

3. M.M. Dupin, I. Halliday, C.M. Care, J. Phys. A: Math. Gen. **36**(31), 8517 (2003)
4. M. Latva-Kokko, D.H. Rothman, Phys. Rev. E **71**(5), 056702 (2005)
5. H. Liu, A.J. Valocchi, Q. Kang, Phys. Rev. E **85**(4), 046309 (2012)
6. S. Leclaire, M. Reggio, J.Y. Trépanier, J. Comput. Phys. **246**, 318 (2013)
7. P. Asinari, Phys. Rev. E **73**(5), 056705 (2006)
8. S. Arcidiacono, I.V. Karlin, J. Mantzaras, C.E. Frouzakis, Phys. Rev. E **76**(4), 046703 (2007)
9. T.J. Spencer, I. Halliday, Phys. Rev. E **88**(6), 063305 (2013)
10. J. Tölke, G.D. Prisco, Y. Mu, Comput. Math. Appl. **65**(6), 864 (2013)
11. L. Scarbolo, D. Molin, P. Perlekar, M. Sbragaglia, A. Soldati, F. Toschi, J. Comput. Phys. **234**, 263 (2013)
12. L. Chen, Q. Kang, Y. Mu, Y.L. He, W.Q. Tao, Int. J. Heat Mass Transfer **76**, 210 (2014)
13. H. Liu, Q. Kang, C.R. Leonardi, S. Schmieschek, A. Narváez, B.D. Jones, J.R. Williams, A.J. Valocchi, J. Harting, Comput. Geosci. pp. 1–29 (2015)
14. H. Huang, M.C. Sukop, X.Y. Lu, *Multiphase Lattice Boltzmann Methods: Theory and Applications* (Wiley-Blackwell, Hoboken, 2015)
15. C.E. Brennen, *Fundamentals of Multiphase Flows* (Cambridge University Press, Cambridge, 2005)
16. J. Bibette, F.L. Calderon, P. Poulin, Reports Progress Phys. **62**(6), 969 (1999)
17. M.J. Blunt, Current Opinion Colloid Interface Sci. **6**(3), 197 (2001)
18. A. Faghri, Y. Zhang, *Transport Phenomena in Multiphase Systems* (Elsevier, Amsterdam, 2006)
19. C.D. Han, *Multiphase Flow in Polymer Processing* (Academic Press, New York, 1981)
20. A. Gunther, K.F. Jensen, Lab Chip **6**, 1487 (2006)
21. C. Wang, P. Cheng, Int. J. Heat Mass Transfer **39**(17), 3607 (1996)
22. S. Blundell, K.M. Blundell, *Concepts in Thermal Physics* (Oxford University Press, Oxford, 2006)
23. M. Doi, T. Ohta, J. Chem. Phys. **95**(2), 1242 (1991)
24. T.G. Mason, Current Opinion Colloid Interface Sci. **4**(3), 231 (1999)
25. R. Osserman, *A Survey of Minimal Surfaces* (Dover Publications, New York, 1986)
26. L.D. Landau, E.M. Lifshitz, *Fluid Mechanics* (Pergamon Press, Oxford, 1987)
27. P.G. de Gennes, F. Brochard-Wyart, D. Quéré, *Capillarity and Wetting Phenomena: Drops, Bubbles, Pearls, Waves* (Springer, New York, 2004)
28. J.S. Rowlinson, B. Widom, *Molecular Theory of Capillarity* (Oxford University Press, Oxford, 1989)
29. J.L. Barrat, L. Bocquet, Phys. Rev. Lett. **82**, 4671 (1999)
30. D.M. Huang, C. Sendner, D. Horinek, R.R. Netz, L. Bocquet, Phys. Rev. Lett. **101**, 226101 (2008)
31. A. Lafuma, Quéré, Nat. Mat. **2**, 457–460 (2003)
32. J. Léopoldés, A. Dupuis, D.G. Bucknall, J.M. Yeomans, Langmuir **19**(23), 9818 (2003)
33. Z. Wang, J. Yang, B. Koo, F. Stern, Int. J. Multiphase Flow **35**(3), 227 (2009)
34. G. Tryggvason, B. Bunner, A. Esmaeeli, D. Juric, N. Al-Rawahi, W. Tauber, J. Han, S. Nas, Y.J. Jan, J. Comput. Phys. **169**(2), 708 (2001)
35. C.S. Peskin, Acta Numerica **11**, 479–517 (2002)
36. J. van der Walls, J. Stat. Phys. **20**(2), 197 (1979)
37. D. Anderson, G. McFadden, A. Wheeler, Annu. Rev. Fluid Mech. **30**, 139 (1998)
38. J.G. Kirkwood, F.P. Buff, J. Chem. Phys. **17**(3), 338 (1949)
39. M.R. Swift, W.R. Osborn, J.M. Yeomans, Phys. Rev. Lett. **75**, 830 (1995)
40. M.R. Swift, E. Orlandini, W.R. Osborn, J.M. Yeomans, Phys. Rev. E **54**, 5041 (1996)
41. A.J. Briant, A.J. Wagner, J.M. Yeomans, Phys. Rev. E **69**, 031602 (2004)
42. A.J. Briant, J.M. Yeomans, Phys. Rev. E **69**, 031603 (2004)
43. C. Semprebon, T. Krüger, H. Kusumaatmaja, Phys. Rev. E **93**(3), 033305 (2016)
44. T. Lee, C.L. Lin, J. Comput. Phys. **206**(1), 16 (2005)
45. A. Mazloomi M., S.S. Chikatamarla, I.V. Karlin, Phys. Rev. E **92**(2), 023308 (2015)
46. G. Gompper, S. Zschocke, Europhys. Lett. **16**(8), 731 (1991)

47. V.M. Kendon, M.E. Cates, I. Pagonabarraga, J.C. Desplat, P. Bladon, J. Fluid Mech. **440**, 147 (2001)
48. A. Wagner, Q. Li, Physica A Stat. Mech. Appl. **362**(1), 105 (2006)
49. T. Lee, P.F. Fischer, Phys. Rev. E **74**, 046709 (2006)
50. T. Seta, K. Okui, JFST **2**(1), 139 (2007)
51. C.M. Pooley, K. Furtado, Phys. Rev. E **77**, 046702 (2008)
52. Z. Guo, C. Zheng, B. Shi, Phys. Rev. E **65**, 46308 (2002)
53. K. Connington, T. Lee, J. Mech. Sci. Technol. **26**(12), 3857 (2012)
54. D. Jamet, D. Torres, J. Brackbill, J. Comput. Phys. **182**(1), 262 (2002)
55. A.J. Wagner, Int. J. Modern Phys. B **17**(01n02), 193 (2003)
56. T. Inamuro, N. Konishi, F. Ogino, Comput. Phys. Commun. **129**(1), 32 (2000)
57. D.J. Holdych, D. Rovas, J.G. Georgiadis, R.O. Buckius, Int. J. Modern Phys. C **09**(08), 1393 (1998)
58. H. Kusumaatmaja, J. Léopoldés, A. Dupuis, J. M. Yeomans, Europhys. Lett. **73**(5), 740 (2006)
59. J.W. Cahn, J. Chem. Phys. **66**(8), 3667 (1977)
60. K.F. Riley, M.P. Hobson, S.J. Bence, *Mathematical Methods for Physics and Engineering (3rd edition): A Comprehensive Guide* (CUP, Cambridge, 2006)
61. H. Kusumaatmaja, J.M. Yeomans, in *Simulating Complex Systems by Cellular Automata*, ed. by J. Kroc, P.M. Sloot, A.G. Hoekstra, Understanding Complex Systems (Springer, New York, 2010), chap. 11, pp. 241–274
62. A. Lamura, G. Gonnella, J.M. Yeomans, Europhys. Lett. **45**(3), 314 (1999)
63. Q. Li, A.J. Wagner, Phys. Rev. E **76**(3), 036701 (2007)
64. J.W. Cahn, C.M. Elliott, A. Novick-Cohen, Eur. J. Appl. Math. **7**, 287 (1996)
65. J. Zhu, L.Q. Chen, J. Shen, V. Tikare, Phys. Rev. E **60**, 3564 (1999)
66. H. Kusumaatmaja, E.J. Hemingway, S.M. Fielding, J. Fluid Mech. **788**, 209 (2016)
67. J.J. Huang, C. Shu, Y.T. Chew, Phys. Fluids **21**(2) (2009)
68. X. Shan, H. Chen, Phys. Rev. E **47**(3), 1815 (1993)
69. X. Shan, G. Doolen, J. Stat. Phys. **81**(1), 379 (1995)
70. J. Zhang, F. Tian, Europhys. Lett. **81**(6), 66005 (2008)
71. N.S. Martys, H. Chen, Phys. Rev. E **53**(1), 743 (1996)
72. X. Shan, H. Chen, Phys. Rev. E **49**(4), 2941 (1994)
73. M.C. Sukop, D.T. Thorne, *Lattice Boltzmann Modeling: An Introduction for Geoscientists and Engineers* (Springer, New York, 2006)
74. M. Sbragaglia, R. Benzi, L. Biferale, S. Succi, K. Sugiyama, F. Toschi, Phys. Rev. E **75**(026702), 1 (2007)
75. M. Sbragaglia, H. Chen, X. Shan, S. Succi, Europhys. Lett. **86**(2), 24005 (2009)
76. J. Bao, L. Schaefer, Appl. Math. Model. **37**(4), 1860 (2013)
77. D. Lycett-Brown, K.H. Luo, Phys. Rev. E **91**, 023305 (2015)
78. P. Yuan, L. Schaefer, Phys. Fluids **18**(042101), 1 (2006)
79. A. Kuzmin, A. Mohamad, S. Succi, Int. J. Mod. Phys. C **19**(6), 875 (2008)
80. R. Benzi, L. Biferale, M. Sbragaglia, S. Succi, F. Toschi, Phys. Rev. E **74**(2), 021509 (2006)
81. Z. Yu, L.S. Fan, Phys. Rev. E **82**(4), 046708 (2010)
82. X. Shan, Phys. Rev. E **81**(4), 045701 (2010)
83. M. Sega, M. Sbragaglia, S.S. Kantorovich, A.O. Ivanov, Soft Matter **9**(42), 10092 (2013)
84. X. Shan, G. Doolen, Phys. Rev. E **54**(4), 3614 (1996)
85. J. Yang, E.S. Boek, Comput. Math. Appl. **65**(6), 882 (2013)
86. H. Liu, Y. Zhang, A.J. Valocchi, Phys. Fluids **27**(5), 052103 (2015)
87. H. Huang, D.T. Thorne, M.G. Schaap, M.C. Sukop, Phys. Rev. E **76**(6), 066701 (2007)
88. S. Chibbaro, Eur. Phys. J. E **27**(1), 99 (2008)
89. F. Jansen, J. Harting, Phys. Rev. E **83**(4), 046707 (2011)
90. I. Ginzburg, G. Wittum, J. Comp. Phys. **166**(2), 302 (2001)
91. A. Kupershtokh, D. Medvedev, D. Karpov, Comput. Math. Appl. **58**(5), 965 (2009)
92. R. Benzi, M. Sbragaglia, S. Succi, M. Bernaschi, S. Chibbaro, J. Chem. Phys. **131**(10), 104903 (2009)

93. R. Benzi, M. Bernaschi, M. Sbragaglia, S. Succi, Europhys. Lett. **91**(1), 14003 (2010)
94. S. Chibbaro, G. Falcucci, G. Chiatti, H. Chen, X. Shan, S. Succi, Phys. Rev. E **77**(036705), 1 (2008)
95. Z. Guo, C. Zheng, B. Shi, Phys. Rev. E **83**, 036707 (2011)
96. D. Chiappini, G. Bella, S. Succi, F. Toschi, S. Ubertini, Commun. Comput. Phys. **7**, 423 (2010)
97. A. Kuzmin, A. Mohamad, Comp. Math. Appl. **59**, 2260 (2010)
98. T. Inamuro, T. Ogato, S. Tajima, N. Konishi, J. Comp. Phys. **198**, 628 (2004)
99. H. Zheng, C. Shu, Y. Chew, J. Comput. Phys. **218**(1), 353 (2006)
100. A. Mazloomi M, S.S. Chikatamarla, I.V. Karlin, Phys. Rev. Lett. **114**, 174502 (2015)
101. A. Kuzmin, M. Januszewski, D. Eskin, F. Mostowfi, J. Derksen, Chem. Eng. J. **171**, 646 (2011)
102. A. Kuzmin, M. Januszewski, D. Eskin, F. Mostowfi, J. Derksen, Chem. Eng. J. **178**, 306 (2011)
103. A. Kuzmin, Multiphase simulations with lattice Boltzmann scheme. Ph.D. thesis, University of Calgary (2010)
104. M.L. Porter, E.T. Coon, Q. Kang, J.D. Moulton, J.W. Carey, Phys. Rev. E **86**(3), 036701 (2012)
105. Q. Kang, D. Zhang, S. Chen, Adv. Water Resour. **27**(1), 13 (2004)
106. Z. Yu, H. Yang, L.S. Fan, Chem. Eng. Sci. **66**(14), 3441 (2011)
107. M. Monteferrante, S. Melchionna, U.M.B. Marconi, J. Chem. Phys. **141**(1), 014102 (2014)
108. K. Premnath, J. Abraham, J. Comput. Phys. **224**, 539 (2007)
109. Z.H. Chai, T.S. Zhao, Acta. Mech. Sin. **28**(4), 983 (2012)
110. D. Zhang, K. Papadikis, S. Gu, Int. J. Multiphas. Flow **64**, 11 (2014)
111. K. Yang, Z. Guo, Sci. Bull. **60**(6), 634 (2015)
112. Y.Q. Zu, S. He, Phys. Rev. E **87**(4), 043301 (2013)
113. H. Liang, B.C. Shi, Z.H. Chai, Phys. Rev. E **93**(1), 013308 (2016)
114. Y. Fu, S. Zhao, L. Bai, Y. Jin, Y. Cheng, Chem. Eng. Sci. **146**, 126 (2016)
115. H. Chen, B.M. Boghosian, P.V. Coveney, M. Nekovee, Proc. R. Soc. Lond. A **456**, 2043 (2000)
116. M. Nekovee, P.V. Coveney, H. Chen, B.M. Boghosian, Phys. Rev. E **62**(6), 8282 (2000)
117. S. Bogner, U. Rüde, Comput. Math. Appl. **65**(6), 901 (2013)
118. D. Anderl, S. Bogner, C. Rauh, U. Rüde, A. Delgado, Comput. Math. Appl. **67**(2), 331 (2014)
119. S. Bogner, R. Ammer, U. Rüde, J. Comput. Phys. **297**, 1 (2015)
120. M. Gross, F. Varnik, Int. J. Mod. Phys. C **25**(01), 1340019 (2013)
121. X. Frank, P. Perré, H.Z. Li, Phys. Rev. E **91**(5), 052405 (2015)
122. Q. Li, Q.J. Kang, M.M. Francois, A.J. Hu, Soft Matter **12**(1), 302 (2015)
123. C. Pan, M. Hilpert, C.T. Miller, Water Resour. Res. **40**(1), W01501 (2004)
124. E.S. Boek, M. Venturoli, Comput. Math. Appl. **59**(7), 2305 (2010)
125. J. Onishi, A. Kawasaki, Y. Chen, H. Ohashi, Comput. Math. Appl. **55**(7), 1541 (2008)
126. A.S. Joshi, Y. Sun, Phys. Rev. E **79**(6), 066703 (2009)
127. T. Krüger, S. Frijters, F. Günther, B. Kaoui, J. Harting, Eur. Phys. J. Spec. Top. **222**(1), 177 (2013)
128. K.W. Connington, T. Lee, J.F. Morris, J. Comput. Phys. **283**, 453 (2015)
129. Q. Luo, Z. Guo, B. Shi, Phys. Rev. E **87**(063301), 1 (2013)
130. Z. Yu, O. Hemminger, L.S. Fan, Chem. Eng. Sci. **62**, 7172 (2007)
131. H. Liu, Y. Zhang, J. Appl. Phys. **106**(3), 1 (2009)
132. T. Munekata, T. Suzuki, S. Yamakawa, R. Asahi, Phys. Rev. E **88**(5), 052314 (2013)
133. R. Ledesma-Aguilar, D. Vella, J.M. Yeomans, Soft Matter **10**(41), 8267 (2014)
134. D. Sun, M. Zhu, S. Pan, D. Raabe, Acta Mater. **57**(6), 1755 (2009)
135. R. Rojas, T. Takaki, M. Ohno, J. Comput. Phys. **298**, 29 (2015)
136. P.R. Di Palma, C. Huber, P. Viotti, Adv. Water Resour. **82**, 139 (2015)
137. G. Gonnella, E. Orlandini, J.M. Yeomans, Phys. Rev. E **58**, 480 (1998)
138. Q. Du, C. Liu, X. Wang, J. Comput. Phys. **198**(2), 450 (2004)

139. J.S. Lowengrub, A. Rätz, A. Voigt, Phys. Rev. E **79**, 031926 (2009)
140. C. Denniston, E. Orlandini, J.M. Yeomans, Phys. Rev. E **63**, 056702 (2001)
141. J. Qian, C. Law, J. Fluid Mech. **331**, 59 (1997)
142. N. Ashgriz, J. Poo, J. Fluid Mech. **221**, 183 (1990)
143. C. Rabe, J. Malet, F. Feuillebois, Phys. Fluids **22**(047101), 1 (2010)
144. D. Lycett-Brown, K. Luo, R. Liu, P. Lv, Phys. Fluids **26**(023303), 1 (2014)
145. T. Inamuro, S. Tajima, F. Ogino, Int. J. Heat Mass Trans. **47**, 4649 (2004)
146. A. Moqaddam, S. Chikatamarla, I. Karlin, Phys. Fluids **28**(022106), 1 (2016)
147. A.E. Komrakova, D. Eskin, J.J. Derksen, Phys. Fluids **25**(4), 042102 (2013)
148. O. Shardt, J.J. Derksen, S.K. Mitra, Langmuir **29**, 6201 (2013)
149. O. Shardt, S.K. Mitra, J.J. Derksen, Langmuir **30**, 14416 (2014)
150. H. Kusumaatmaja, C.M. Pooley, S. Girardo, D. Pisignano, J.M. Yeomans, Phys. Rev. E **77**, 067301 (2008)
151. J. Murison, B. Semin, J.C. Baret, S. Herminghaus, M. Schröter, M. Brinkmann, Phys. Rev. Appl. **2**, 034002 (2014)
152. A.A. Darhuber, S.M. Troian, Annu. Rev. Fluid Mech. **37**(1), 425 (2005)
153. H. Gau, S. Herminghaus, P. Lenz, R. Lipowsky, Science **283**(5398), 46 (1999)
154. M. Brinkmann, R. Lipowsky, J. Appl. Phys. **92**(8), 4296 (2002)
155. H.P. Jansen, K. Sotthewes, J. van Swigchem, H.J.W. Zandvliet, E.S. Kooij, Phys. Rev. E **88**, 013008 (2013)
156. A. Dupuis, J. Léopoldés, D.G. Bucknall, J.M. Yeomans, Appl. Phys. Lett. **87**(2), 024103 (2005)
157. S. Wang, T. Wang, P. Ge, P. Xue, S. Ye, H. Chen, Z. Li, J. Zhang, B. Yang, Langmuir **31**(13), 4032 (2015)
158. D. Quéré, Annu. Rev. Mater. Res. **38**(1), 71 (2008)
159. W. Barthlott, C. Neinhuis, Planta **202**(1), 1 (1997)
160. J. Bico, C. Marzolin, D. Quéré, Europhys. Lett. **47**(2), 220 (1999)
161. A. Tuteja, W. Choi, M. Ma, J.M. Mabry, S.A. Mazzella, G.C. Rutledge, G.H. McKinley, R.E. Cohen, Science **318**(5856), 1618 (2007)
162. A.B.D. Cassie, S. Baxter, Trans. Faraday Soc. **40**, 546 (1944)
163. C.H. Choi, C.J. Kim, Phys. Rev. Lett. **96**, 066001 (2006)
164. R.N. Wenzel, J. Phys. Colloid Chem. **53**(9), 1466 (1949)
165. F. Diotallevi, L. Biferale, S. Chibbaro, G. Pontrelli, F. Toschi, S. Succi, Euro. Phys. J. Special Topics **171**(1), 237 (2009)

Chapter 10
MRT and TRT Collision Operators

Abstract After reading this chapter, you will have a solid understanding of the general principles of multiple-relaxation-time (MRT) and two-relaxation-time (TRT) collision operators. You will know how to implement these and how to choose the various relaxation times in order to increase the stability, the accuracy, and the possibilities of lattice Boltzmann simulations.

The BGK model is an elegant way to simplify the collision operator of the Boltzmann equation. However, the simplicity and efficiency of the lattice Boltzmann BGK collision operator comes at the cost of reduced accuracy (in particular for large viscosities) and stability (especially for small viscosities). Multiple-relaxation-time (MRT) collision operators offer a larger number of free parameters that can be tuned to overcome these problems.

After a short introduction to the main concept of these operators in Sect. 10.1, we will discuss the advantages of performing relaxation in moment rather than in population space (Sect. 10.2). These ideas lead to the general MRT algorithm (Sect. 10.3). We discuss two different approaches, namely the Hermite polynomial and the Gram-Schmidt procedures, demonstrated for the D2Q9 lattice (Sect. 10.4). Results for the common 3D lattices are given in Appendix A.6. We also address the inclusion of forces into the MRT collision operator (Sect. 10.5). The two-relaxation-time (TRT) collision operator (Sect. 10.6) can be viewed as a reduced MRT model which is nearly as simple as the BGK model but which still offers significant advantages. We conclude the chapter with an overview of practical guidelines as to how to choose the collision model and the relaxation rates (Sect. 10.7).

10.1 Introduction

Under certain conditions, the BGK collision operator approaches its limits of accuracy and/or stability. For example, stability issues can often occur for large Reynolds number simulations. In order to avoid increasing the grid resolution, a common approach to reach large Reynolds numbers is to reduce the viscosity or to increase the magnitude of the fluid velocity $\boldsymbol{u}$. As discussed in Sect. 4.4, small

T. Krüger et al., *The Lattice Boltzmann Method*, Graduate Texts in Physics,
DOI 10.1007/978-3-319-44649-3_10

viscosities (i.e. relaxation times $\tau/\Delta t$ close to $1/2$) and large velocity magnitudes ($u \not\ll c_s$) can lead to stability problems. In the end, the only viable solution for the BGK collision operator may be to increase the grid size, making the simulations more computationally expensive.

Furthermore, the accuracy of simulations based on the BGK collision operator varies with the relaxation time τ. This is the case for both the bulk LB algorithm (see Sect. 4.5) and some boundary conditions. For example, the wall location depends on the numerical value of τ when the standard bounce-back algorithm is used (see Sect. 5.3.3). This issue can be easily demonstrated by simulating a Poiseuille flow [1–3]. The error is found to increase with τ [4, 5]. This is a problem especially for porous media simulations where the apparent porosity strongly depends on the exact wall location [6]. Ideally, one wants to obtain so-called viscosity-independent numerical solutions [7] which only depend on the physical parameters (such as the Reynolds number), but not the relaxation time τ.

The LBM is also known to violate *Galilean invariance* [8, 9]. Galilean invariance means that physical phenomena occur in the same way in all inertial systems, i.e. in systems moving with constant velocities relatively to each other. The reason is that the second-order Hermite series expansion in (3.54) is not sufficient to guarantee Galilean invariance [10], causing an $O(u^3)$ error term in the macroscopic dynamics as shown in Sect. 4.1.

These examples indicate that something else beyond the hydrodynamic level influences the overall accuracy and stability of the LBM as a Navier-Stokes solver. We have to understand that the relaxation time τ is a kinetic parameter, and its relation to viscosity has to be analysed *via* the Chapman-Enskog analysis (Sect. 4.1). Contrarily, all other parameters (e.g. velocity, lattice size) are directly connected to macroscopic quantities. Essentially, this is the reason why the numerical BGK solutions (and their stability/accuracy) depend on τ [7, 11–13].

Since τ enters the LBE only through the collision operator, the underlying idea is to revise and improve the collision step to increase the accuracy and stability of LB simulations. A general requirement for the collision operator to recover the Navier-Stokes equation is the conservation of density and momentum. So far, we have considered the simplest form of such an operator: the BGK model. However, as we will see in the following, it is possible to construct collision operators that have more degrees of freedom than the BGK operator and can be used to improve accuracy and stability. The models discussed here are the multiple-relaxation-time (MRT) and the two-relaxation-time (TRT) collision operators.

Let us now present the main idea behind the MRT and TRT collision operators, starting from the BGK collision operator. The BGK collision operator uses only one relaxation rate $\omega = 1/\tau$ for all populations, i.e. $\Omega_i = -\omega(f_i - f_i^{eq})$. The first naive idea is to introduce different collision rates for each population: $\Omega_i = -\omega_i(f_i - f_i^{eq})$. However, it is easy to see that one needs certain constraints to satisfy the conservation laws for density and momentum (in the absence of forces):

$$
\begin{aligned}
0 = \sum_i \Omega_i = -\sum_i \omega_i \left(f_i - f_i^{\text{eq}}\right) \qquad &\Longrightarrow \qquad \sum_i \omega_i f_i = \sum_i \omega_i f_i^{\text{eq}}, \\
\mathbf{0} = \sum_i \Omega_i \mathbf{c}_i = -\sum_i \omega_i \left(f_i - f_i^{\text{eq}}\right) \mathbf{c}_i \qquad &\Longrightarrow \qquad \sum_i \omega_i f_i \mathbf{c}_i = \sum_i \omega_i f_i^{\text{eq}} \mathbf{c}_i.
\end{aligned}
\tag{10.1}
$$

Equation (10.1) is satisfied in the case of the BGK collision operator ($\omega_i = \omega$). For a general situation with distinct values of ω_i, however, the situation is more complicated and not obvious. This model is called MRT-L model [7, 13, 14], but we will skip it due to its complexity.

The next, slightly more sophisticated, idea is to use velocity *moments* (Sect. 1.3.5). Let us take another careful look at (10.1) for the BGK operator:

$$
\begin{aligned}
0 &= \sum_i \Omega_i = -\sum_i \omega \left(f_i - f_i^{\text{eq}}\right) = -\omega \left(\rho - \rho^{\text{eq}}\right), \\
\mathbf{0} &= \sum_i \Omega_i \mathbf{c}_i = -\sum_i \omega \left(f_i \mathbf{c}_i - f_i^{\text{eq}} \mathbf{c}_i\right) = -\omega \left(\mathbf{j} - \mathbf{j}^{\text{eq}}\right).
\end{aligned}
\tag{10.2}
$$

Here, the density ρ and momentum $\mathbf{j}$ are the zeroth and first velocity moments. In the BGK model, all moments are relaxed with a single relaxation rate ω. However, the moments can in principle be relaxed with individual rates. This is the basic idea behind the MRT collision operators, i.e. to individually control the different moments' relaxation to achieve better accuracy and stability.

The major change compared to the BGK model is that MRT collisions are performed in *moment space*. Thus, we first have to map the populations f_i to moment space, then collide (i.e. relax) the moments toward equilibrium, and finally map the relaxed moments back to population space. We will investigate the underlying mathematics in Sect. 10.2.

There are also other advanced LB collision operators which we will not cover in this chapter. One example is the regularised collision operator [15–17], which is based on using a reconstructed $f_i^{(1)}$ instead of f_i^{neq} in a BGK-like collision operator; it can also be equivalently expressed as a type of MRT collision operator [18]. Another example is entropic collision operators, which are based on defining a lattice version of the $\mathcal{H}$ function described in Sect. 1.3.6 and ensuring that collisions only ever *decrease* $\mathcal{H}$ [19, 20]. This can be seen as ensuring that entropy always increases in collisions, thus ensuring that many cases of instability cannot occur. Another MRT-like method is the central moment (also known as cascaded) method, where the moments are taken in a reference frame moving with the macroscopic velocity [21, 22]. Similar to the central moment method is the newer cumulant method, which individually relaxes cumulants, a set of fully independent properties of the distribution f_i, rather than its moments [23].

10.2 Moment Space and Transformations

We will now sketch the general mathematical procedure behind the MRT collision operators. In particular, transformations from population to moment space and back will be covered.

In Sect. 3.4 (cf. (3.57)) we introduced moments as certain summations over populations with Hermite polynomials:

$$\boldsymbol{a}^{(n)} = \sum_i \boldsymbol{H}_i^{(n)} f_i. \tag{10.3}$$

We can use a similar definition here and say that the moment m_k in a DdQq velocity set can be found through a $q \times q$ matrix as:

$$m_k = \sum_{i=0}^{q-1} M_{ki} f_i \qquad \text{for } k = 0, \dots, q-1. \tag{10.4}$$

This equation can be rewritten in vector-matrix form:

$$\boldsymbol{m} = \boldsymbol{M}\boldsymbol{f}, \quad \boldsymbol{m} = \begin{pmatrix} m_0 \\ \vdots \\ m_{q-1} \end{pmatrix}, \quad \boldsymbol{M} = \begin{pmatrix} M_{0,0} & \cdots & M_{0,q-1} \\ \vdots & \ddots & \vdots \\ M_{q-1,0} & \cdots & M_{q-1,q-1} \end{pmatrix}, \quad \boldsymbol{f} = \begin{pmatrix} f_0 \\ \vdots \\ f_{q-1} \end{pmatrix}. \tag{10.5}$$

The matrix $\boldsymbol{M}$ with elements M_{ki} can also be interpreted as a q-tuple of row vectors $\boldsymbol{M}_k(M_{k,0}, \dots, M_{k,q-1})$, with $m_k = \boldsymbol{M}_k \cdot \boldsymbol{f}$. Generally speaking, one obtains q moments m_k from the q populations f_i through the *linear* transformation from *population space* to *moment space* in (10.5).

The basic idea behind the MRT collision operator is to relax *moments* (rather than populations) with individual rates. By carefully choosing the transformation matrix $\boldsymbol{M}$, the obtained moments m_k can be made to directly correspond to hydrodynamic moments. Thus, it is possible to affect those hydrodynamic terms (density, momentum, momentum flux tensor, etc.) individually by choosing different distinct relaxation rates. In contrast, the BGK collision operator relaxes all moments with one relaxation rate $\omega = 1/\tau$.

So far, we have not yet specified how the MRT operator looks like in detail, but we do know that the BGK operator is the special case of the MRT collision operator with identical relaxation rates, all equal to ω. We can use this relation to derive the MRT operator from the BGK operator.

First we write the LBGK equation in vector form:

$$\boldsymbol{f}(\boldsymbol{x} + \boldsymbol{c}_i \Delta t, t + \Delta t) - \boldsymbol{f}(\boldsymbol{x}, t) = -\omega \left(\boldsymbol{f}(\boldsymbol{x}, t) - \boldsymbol{f}^{\text{eq}}(\boldsymbol{x}, t) \right) \Delta t. \tag{10.6}$$

The collision step is not changed when multiplied by the identity matrix $\boldsymbol{I} = \boldsymbol{M}^{-1}\boldsymbol{M}$ (assuming that $\boldsymbol{M}$ can be inverted):

$$\begin{aligned} \boldsymbol{f}(\boldsymbol{x}+\boldsymbol{c}_i\Delta t, t+\Delta t) - \boldsymbol{f}(\boldsymbol{x}, t) &= -\boldsymbol{M}^{-1}\boldsymbol{M}\omega\left[\boldsymbol{f}(\boldsymbol{x}, t) - \boldsymbol{f}^{\mathrm{eq}}(\boldsymbol{x}, t)\right]\Delta t \\ &= -\boldsymbol{M}^{-1}\omega\left[\boldsymbol{M}\boldsymbol{f}(\boldsymbol{x}, t) - \boldsymbol{M}\boldsymbol{f}^{\mathrm{eq}}(\boldsymbol{x}, t)\right]\Delta t \\ &= -\boldsymbol{M}^{-1}\omega\boldsymbol{I}\left[\boldsymbol{m}(\boldsymbol{x}, t) - \boldsymbol{m}^{\mathrm{eq}}(\boldsymbol{x}, t)\right]\Delta t \\ &= -\boldsymbol{M}^{-1}\boldsymbol{S}\left[\boldsymbol{m}(\boldsymbol{x}, t) - \boldsymbol{m}^{\mathrm{eq}}(\boldsymbol{x}, t)\right]\Delta t. \end{aligned} \tag{10.7}$$

Here, we have introduced a diagonal matrix $\boldsymbol{S} = \omega\boldsymbol{I} = \mathrm{diag}(\omega, \ldots, \omega)$ and the equilibrium moment vector $\boldsymbol{m}^{\mathrm{eq}} = \boldsymbol{M}\boldsymbol{f}^{\mathrm{eq}}$.

Let us take a detailed look at (10.7). The expression $\boldsymbol{S}(\boldsymbol{m} - \boldsymbol{m}^{\mathrm{eq}})$ is the relaxation of all *moments* with one rate ω. The result is then multiplied by the inverse matrix $\boldsymbol{M}^{-1}$. This step represents the transformation from moment space back to population space. The left-hand-side of (10.7) is the usual streaming step.

The fact that the matrix $\boldsymbol{S}$ contains only one parameter ω reflects that the BGK operator is a single-relaxation model. It is straightforward to introduce individual collision rates for every moment, which leads to a *relaxation matrix*

$$\boldsymbol{S} = \begin{pmatrix} \omega_0 & 0 & \ldots & 0 \\ 0 & \omega_1 & \ldots & 0 \\ \vdots & \vdots & \ddots & \vdots \\ 0 & 0 & \ldots & \omega_{q-1} \end{pmatrix}. \tag{10.8}$$

Using such a relaxation matrix $\boldsymbol{S}$ to relax the moments as $\boldsymbol{S}(\boldsymbol{m} - \boldsymbol{m}^{\mathrm{eq}})$ represents the core of an MRT collision operator.

The basic idea behind the MRT collision operator is a **mapping from population to moment space** *via* a transformation matrix $\boldsymbol{M}$. This allows moments rather than populations to be relaxed with individual rates (in the form of a relaxation matrix $\boldsymbol{S}$). The relaxed moments are then transformed back to population space where streaming is performed as usual. This procedure is shown in Fig. 10.1.

In Sect. 10.3 we will present the MRT algorithm more thoroughly, but first we must discuss the moments and their meaning. We know that the density and the momentum are conserved. Those conserved quantities could act as moments in the MRT framework.

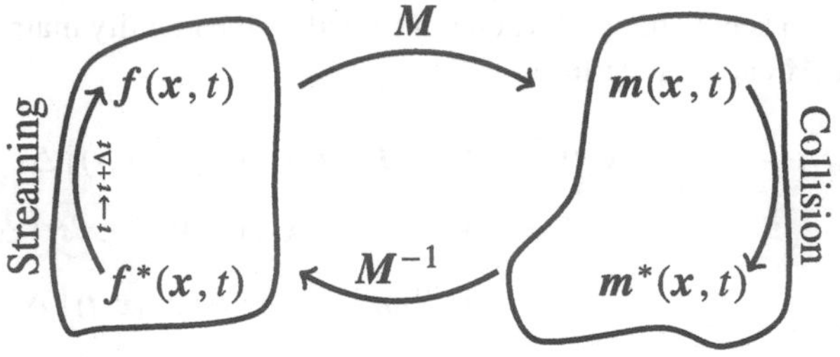

Fig. 10.1 Schematic representation of the MRT LBM. Populations are mapped to moment space where collision (relaxation) takes place. Relaxed moments are transformed back to population space before streaming. The superscript * denotes post-collision values

For example, the zeroth moment corresponds to the density:

$$m_0 = \rho = \sum_i f_i. \tag{10.9}$$

By expanding the zeroth order in terms of populations $m_0 = \sum_i M_{0,i} f_i$, one can find the entire first row of the matrix $\boldsymbol{M}$ as $M_{0,i} = 1$. Since the first row is related to the density ρ, we introduce the notation $M_{\rho,i} = M_{0,i}$ $(i = 0, \ldots, q-1)$.

Next we have momentum, which has two components in 2D and three in 3D. For the x-component we can write:

$$m_1 = j_x = \rho u_x = \sum_i f_i c_{ix}. \tag{10.10}$$

A similar equation holds for $m_2 = j_y$ and also for $m_3 = j_z$ in 3D. From this we can find the components of the matrix $\boldsymbol{M}$ corresponding to the momentum: $M_{j_\alpha,i} = c_{i\alpha}$ with $\alpha \in \{x, y\}$ in 2D and $\alpha \in \{x, y, z\}$ in 3D.

For the transformation matrix $\boldsymbol{M}$ we have specified so far only three row vectors in 2D and four row vectors in 3D. To specify the other moments (and therefore the other missing rows of the matrix $\boldsymbol{M}$), different approaches can be used. The two most common approaches are covered below. Another one is to reconstruct vectors of the matrix $\boldsymbol{M}$ from the tensor product of velocity vectors for one-dimensional lattices. This approach is used for entropic lattice Boltzmann models [24], and it is not covered here.

- **Hermite polynomials:** the most straightforward approach is to employ Hermite polynomials which are also used in the discretisation of the continuous Boltzmann equation as shown in Sect. 3.4. The moments can be taken as the Hermite polynomials $\boldsymbol{H}^{(n)}(\boldsymbol{c}_i)$. Density corresponds to the zeroth order polynomial, and the various components of momentum are in the first-order polynomial. Other Hermite moments are at least of second order in velocity. Despite its straightforwardness, this approach is not very common. It is mainly used for the D2Q9 lattice [25–27].

- **Gram-Schmidt procedure:** this is based on the idea of constructing a set of orthogonal vectors. We start the process with the vectors for the known moments (density ρ and momentum $\boldsymbol{j}$). The next step is to take a combination of the velocity vectors $c_{i\alpha}$ of appropriate order and find the coefficients in such a way that the resulting vector is orthogonal to all previously found ones. By repeating this process, one can construct the entire matrix M_{ki} consisting of orthogonal row vectors $\boldsymbol{M}_k$.

Each approach is thoroughly presented in Sect. 10.4 where we will construct $\boldsymbol{M}$ for the D2Q9 lattice. The corresponding Gram-Schmidt matrices for D3Q15 and D3Q19 can be found in Appendix A.6.

Both approaches provide consistent ways to obtain all moments m_k. The linear independence of vectors guarantees that $\boldsymbol{M}$ can be inverted, i.e. $\boldsymbol{M}^{-1}$ exists. Moreover, the corresponding vectors $\boldsymbol{M}_k$ are mutually orthogonal. Orthogonality guarantees that the hydrodynamic moments m_k are uniquely defined. As a consequence, moments can be independently relaxed, which in turn provides the necessary flexibility to tune accuracy and stability.

Though vectors are orthogonal in both bases, there is a difference. For the Gram-Schmidt vectors (denoted by G), orthogonality means

$$\sum_{i=0}^{q-1} {}^{\mathrm{G}}M_{ki}\, {}^{\mathrm{G}}M_{li} = 0 \quad \text{for } k \neq l. \tag{10.11}$$

In the case of the Hermite polynomials (denoted by H), the vectors are orthogonal when weighted with the lattice weights w_i:

$$\sum_{i=0}^{q-1} w_i\, {}^{\mathrm{H}}M_{ki}\, {}^{\mathrm{H}}M_{li} = 0 \quad \text{for } k \neq l. \tag{10.12}$$

Now that we have discussed the moment space, we will discuss the individual steps of the MRT algorithm in the next section.

10.3 General MRT Algorithm

Once we know the transformation matrix $\boldsymbol{M}$ and the relaxation matrix $\boldsymbol{S}$, which includes all the relaxation rates ω_k, we can perform the MRT algorithm as sketched in Fig. 10.1.[1] Explicitly, this algorithm can be written as

$$\boldsymbol{f}(\boldsymbol{x}+\boldsymbol{c}_i\Delta t, t+\Delta t) - \boldsymbol{f}(\boldsymbol{x}, t) = -\boldsymbol{M}^{-1}\boldsymbol{S}\boldsymbol{M}\left[\boldsymbol{f}(\boldsymbol{x}, t) - \boldsymbol{f}^{\mathrm{eq}}(\boldsymbol{x}, t)\right]\Delta t, \tag{10.13}$$

[1]In the following we will only explain the force-free algorithm; the inclusion of forces is discussed in Sect. 10.5.

and it can be logically split up into the following steps:

1. **Compute conserved moments:** The density ρ and momentum $\boldsymbol{j} = \rho\boldsymbol{u}$ can be obtained from the pre-collision populations as usual:

$$\rho = \sum_i f_i, \quad \rho\boldsymbol{u} = \sum_i f_i\boldsymbol{c}_i. \tag{10.14}$$

2. **Transform to moment space:** The moments m_k are obtained from the populations f_i *via*

$$\boldsymbol{m} = \boldsymbol{M}\boldsymbol{f}, \quad m_k = \sum_i M_{ki} f_i. \tag{10.15}$$

3. **Compute equilibrium moments:** The equilibrium moments can be straightforwardly computed as $\boldsymbol{m}^{\text{eq}} = \boldsymbol{M}\boldsymbol{f}^{\text{eq}}$. Alternatively, one can also construct equilibrium moments more precisely and efficiently from the known moments ρ and $\rho\boldsymbol{u}$ using a general polynomial representation:

$$m_k^{\text{eq}} = \rho \sum_{l,m,n} a_{k,lmn} u_x^l u_y^m u_z^n. \tag{10.16}$$

We discuss how to obtain the coefficients $a_{k,lmn}$ below. Exact forms of the equilibrium moments are given in Sect. 10.4.1 for the Hermite polynomial approach and in Sect. 10.4.2 for the Gram-Schmidt approach.
4. **Collide:** Once the moments $\boldsymbol{m}$ and $\boldsymbol{m}^{\text{eq}}$ have been calculated, collision is performed in moment space in a BGK relaxation manner:

$$m_k^\star = m_k - \omega_k \left(m_k - m_k^{\text{eq}}\right)\Delta t. \tag{10.17}$$

Note that collision does not have to be performed for the conserved moments since they are not changed during collision, i.e. $\rho = \rho^{\text{eq}}$ and $\boldsymbol{j} = \boldsymbol{j}^{\text{eq}}$. An exception is the presence of mass or momentum sources (e.g. forces which will be discussed in Sect. 10.5).
5. **Transform to population space:** After relaxation, the post-collision populations are obtained using the inverse transformation:

$$f_i^\star = \sum_k M_{ik}^{-1} m_k^\star. \tag{10.18}$$

Given a matrix $\boldsymbol{M}$, its inverse $\boldsymbol{M}^{-1}$ can be computed directly using any mathematical package.
6. **Stream:** Finally, streaming is performed in population space:

$$f_i(\boldsymbol{x} + \boldsymbol{c}_i\Delta t, t + \Delta t) = f_i^\star(\boldsymbol{x}, t). \tag{10.19}$$

Let us provide a few comments about the algorithm above. For the efficient computation of the equilibrium moments m_k^{eq} one needs to find the coefficients $a_{k,lmn}$. Recall that the equilibrium populations f_i^{eq} from (3.54) are given by conserved moments ρ and $\rho \boldsymbol{u}$ as

$$f_i^{\mathrm{eq}} = w_i \rho \left(1 + \frac{c_{i\alpha} u_\alpha}{c_s^2} + \frac{1}{2c_s^4}\left(c_{i\alpha} c_{i\beta} - c_s^2\right) u_\alpha u_\beta\right). \tag{10.20}$$

The "brute force" matrix-vector multiplication, i.e. $m_k^{\mathrm{eq}} = \sum_i M_{ki} f_i^{\mathrm{eq}}$ [28], provides the macroscopic equilibrium moments up to second order in velocity, i.e. $l+n+m \leq 2$ in (10.16). However, if we multiply each row vector $\boldsymbol{M}_k$ in the matrix $\boldsymbol{M}$ with the equilibrium function $\boldsymbol{f}^{\mathrm{eq}}$ analytically, we can find the coefficients $a_{k,lmn}$. Thus, we can find analytical expressions for the equilibrium moments as functions of the density ρ and velocity $\boldsymbol{u}$. This approach is much more efficient than using the brute force approach in every time step.

One limitation of the "brute force" method is that velocity terms in the equilibrium moments are only up to second order in $\boldsymbol{u}$ due to the lack of higher-order terms in f_i^{eq}. Sometimes it is desirable to include terms of third or even higher orders in velocity, for example to improve accuracy, stability or Galilean invariance [10]. In Chap. 3 we presented the expression for Hermite polynomial moments (cf. (3.57)). We also mentioned that only Hermite polynomials up to the second order in velocity are required to satisfy the conservation rules. However, one can calculate higher-order Hermite polynomial equilibrium moments. Those equilibrium moments and their corresponding Hermite polynomials can be included into the MRT approach. We underline here that those equilibrium moments are known exactly due to the Gauss quadrature rule.

There is one caveat. Only a few higher-order discretised Hermite polynomials are distinct. This is connected to the degeneracy of the velocity set, which we discussed in Sect. 4.2.1; i.e., $c_{i\alpha}^3 = c_{i\alpha}(\frac{\Delta x}{\Delta t})^2$ for the common velocity sets presented in this book. Thus, as we will see later in Sect. 10.4.1, only equilibrium moments up to the fourth order in velocity are constructed in this approach.

In general, any MRT vector $\boldsymbol{M}_k$ (not necessarily obtained through Hermite polynomials or Gram-Schmidt procedures) can be represented as a **linear combination of Hermite vectors**: $\boldsymbol{M}_l = \sum_k b_{lk}\, {}^{\mathrm{H}}\boldsymbol{M}_k$. For each of the Hermite vectors, we know exactly the associated equilibrium moment from Sect. 3.4.4 (see also (3.66)). This allows construction of the equilibrium moments for an arbitrary matrix $\boldsymbol{M}$.

Surprisingly, the MRT model can be implemented with only 15–20% computational overhead compared to the BGK implementation [29]. For example, the calculation of the equilibrium moments for MRT has the same overhead as the calculation of the equilibrium functions for BGK. This is because of the density and the momentum being calculated through populations, which are then similarly used to calculate the equilibrium moments. Also, the components of $\boldsymbol{M}$ and $\boldsymbol{M}^{-1}$ are known and constant, so the calculation of moments $\boldsymbol{m} = \boldsymbol{M}\boldsymbol{f}$ and reverting collided moments $\boldsymbol{M}^{-1}\boldsymbol{m}$ back to populations can be done efficiently through vector-matrix multiplication. It is also possible to compute the non-equilibrium populations first, followed by the application of the matrix $\boldsymbol{M}^{-1}\boldsymbol{S}\boldsymbol{M}$. Streaming is the same operation for the BGK and the MRT collision operators.

10.4 MRT for the D2Q9 Velocity Set

We will now present the MRT model for the D2Q9 velocity set in more detail. Note that the matrices that will be presented in this section depend on the ordering of the velocities. We will follow the ordering of Sect. 3.4.7, but different choices of velocity order can be found in the literature, which would lead to matrices with differently ordered columns. In this section we will also simplify the notation by adopting "lattice units", so that $\Delta x = \Delta t = 1$ and $c_s^2 = \frac{1}{3}$.

In Sect. 10.4.1 we will discuss the Hermite polynomial procedure, which also allows us to find the exact form of equilibrium moments beyond the second order in $\boldsymbol{u}$. In Sect. 10.4.2, we will investigate the Gram-Schmidt procedure. Finally we will discuss both methods in Sect. 10.4.3.

10.4.1 Hermite Polynomials

In the Hermite approach, denoted by H, the moment row vectors ${}^{\mathrm{H}}\boldsymbol{M}_k$ correspond to Hermite polynomials of the LBM velocities $\boldsymbol{c}_i$. For D2Q9, we can write

$$\begin{gathered}
{}^{\mathrm{H}}M_{\rho,i} = H^{(0)} = 1, \\
{}^{\mathrm{H}}M_{j_x,i} = H^{(1)}_x = c_{ix}, \qquad {}^{\mathrm{H}}M_{j_y,i} = H^{(1)}_y = c_{iy}, \\
{}^{\mathrm{H}}M_{p_{xx},i} = H^{(2)}_{xx} = c_{ix}c_{ix} - c_s^2, \qquad {}^{\mathrm{H}}M_{p_{yy},i} = H^{(2)}_{yy} = c_{iy}c_{iy} - c_s^2, \\
{}^{\mathrm{H}}M_{p_{xy},i} = H^{(2)}_{xy} = c_{ix}c_{iy}, \\
{}^{\mathrm{H}}M_{\gamma_x,i} = H^{(3)}_{xyy} = c_{ix}c_{iy}^2 - c_s^2 c_{ix}, \qquad {}^{\mathrm{H}}M_{\gamma_y,i} = H^{(3)}_{yxx} = c_{iy}c_{ix}^2 - c_s^2 c_{iy}, \\
{}^{\mathrm{H}}M_{\gamma,i} = H^{(4)}_{xxyy} = c_{ix}^2 c_{iy}^2 - c_s^2 c_{ix}^2 - c_s^2 c_{iy}^2 + c_s^4.
\end{gathered} \tag{10.21}$$

We have introduced several subscripts denoting the nine moments ρ, j_x, j_y, p_{xx}, p_{xy}, p_{yy}, γ_x, γ_y and γ. While the first six are the components of the macroscopic density, momentum vector and stress (or pressure) tensor, the additional moments γ_x, γ_y and γ are higher-order moments that do not affect the Navier-Stokes level hydrodynamics. They are sometimes called "non-hydrodynamic" or "ghost" moments [30, 31]. The nine vectors ${}^{\mathrm{H}}\boldsymbol{M}_k$ in (10.21) are orthogonal according to the condition specified in (10.12).

Substituting the numerical values of the D2Q9 velocity set, including $c_s^2 = \frac{1}{3}$, (10.21) can be written as the transformation matrix ${}^{\mathrm{H}}\boldsymbol{M}$:

$$
{}^{\mathrm{H}}\boldsymbol{M} = \begin{pmatrix} {}^{\mathrm{H}}\boldsymbol{M}_{\rho} \\ {}^{\mathrm{H}}\boldsymbol{M}_{j_x} \\ {}^{\mathrm{H}}\boldsymbol{M}_{j_y} \\ {}^{\mathrm{H}}\boldsymbol{M}_{P_{xx}} \\ {}^{\mathrm{H}}\boldsymbol{M}_{P_{yy}} \\ {}^{\mathrm{H}}\boldsymbol{M}_{P_{xy}} \\ {}^{\mathrm{H}}\boldsymbol{M}_{\gamma_x} \\ {}^{\mathrm{H}}\boldsymbol{M}_{\gamma_y} \\ {}^{\mathrm{H}}\boldsymbol{M}_{\gamma} \end{pmatrix} = \begin{pmatrix} 1 & 1 & 1 & 1 & 1 & 1 & 1 & 1 & 1 \\ 0 & 1 & 0 & -1 & 0 & 1 & -1 & -1 & 1 \\ 0 & 0 & 1 & 0 & -1 & 1 & 1 & -1 & -1 \\ -\frac{1}{3} & \frac{2}{3} & -\frac{1}{3} & \frac{2}{3} & -\frac{1}{3} & \frac{2}{3} & \frac{2}{3} & \frac{2}{3} & \frac{2}{3} \\ -\frac{1}{3} & -\frac{1}{3} & \frac{2}{3} & -\frac{1}{3} & \frac{2}{3} & \frac{2}{3} & \frac{2}{3} & \frac{2}{3} & \frac{2}{3} \\ 0 & 0 & 0 & 0 & 0 & 1 & -1 & 1 & -1 \\ 0 & -\frac{1}{3} & 0 & \frac{1}{3} & 0 & \frac{2}{3} & -\frac{2}{3} & -\frac{2}{3} & \frac{2}{3} \\ 0 & 0 & -\frac{1}{3} & 0 & \frac{1}{3} & \frac{2}{3} & \frac{2}{3} & -\frac{2}{3} & -\frac{2}{3} \\ \frac{1}{9} & -\frac{2}{9} & -\frac{2}{9} & -\frac{2}{9} & -\frac{2}{9} & \frac{4}{9} & \frac{4}{9} & \frac{4}{9} & \frac{4}{9} \end{pmatrix}. \tag{10.22}
$$

The equilibrium moments are simple functions of the density ρ and velocity $\boldsymbol{u}$. Using (3.43), the equilibrium moments are calculated exactly:

$$
\begin{aligned}
\rho^{\mathrm{eq}} &= \rho, & j_x^{\mathrm{eq}} &= \rho u_x, & j_y^{\mathrm{eq}} &= \rho u_y, \\
P_{xx}^{\mathrm{eq}} &= \rho u_x^2, & P_{yy}^{\mathrm{eq}} &= \rho u_y^2, & P_{xy}^{\mathrm{eq}} &= \rho u_x u_y, \\
\gamma_x^{\mathrm{eq}} &= \rho u_x u_y^2, & \gamma_y^{\mathrm{eq}} &= \rho u_x^2 u_y, & \gamma^{\mathrm{eq}} &= \rho u_x^2 u_y^2.
\end{aligned} \tag{10.23}
$$

In this way, we can avoid having to compute $\boldsymbol{m}^{\mathrm{eq}}$ from $\boldsymbol{M}\boldsymbol{f}^{\mathrm{eq}}$ in each time step.

The matrix ${}^{\mathrm{H}}\boldsymbol{M}^{-1}$ can be easily obtained by inverting ${}^{\mathrm{H}}\boldsymbol{M}$ analytically using any computer algebra system or by numerical matrix inversion.[2] We thus find:

$$
{}^{\mathrm{H}}\boldsymbol{M}^{-1} = \begin{pmatrix} \frac{4}{9} & 0 & 0 & -\frac{2}{3} & -\frac{2}{3} & 0 & 0 & 0 & 1 \\ \frac{1}{9} & \frac{1}{3} & 0 & \frac{1}{3} & -\frac{1}{6} & 0 & -\frac{1}{2} & 0 & -\frac{1}{2} \\ \frac{1}{9} & 0 & \frac{1}{3} & -\frac{1}{6} & \frac{1}{3} & 0 & 0 & -\frac{1}{2} & -\frac{1}{2} \\ \frac{1}{9} & -\frac{1}{3} & 0 & \frac{1}{3} & -\frac{1}{6} & 0 & \frac{1}{2} & 0 & -\frac{1}{2} \\ \frac{1}{9} & 0 & -\frac{1}{3} & -\frac{1}{6} & \frac{1}{3} & 0 & 0 & \frac{1}{2} & -\frac{1}{2} \\ \frac{1}{36} & \frac{1}{12} & \frac{1}{12} & \frac{1}{12} & \frac{1}{12} & \frac{1}{4} & \frac{1}{4} & \frac{1}{4} & \frac{1}{4} \\ \frac{1}{36} & -\frac{1}{12} & \frac{1}{12} & \frac{1}{12} & \frac{1}{12} & -\frac{1}{4} & -\frac{1}{4} & \frac{1}{4} & \frac{1}{4} \\ \frac{1}{36} & -\frac{1}{12} & -\frac{1}{12} & \frac{1}{12} & \frac{1}{12} & \frac{1}{4} & -\frac{1}{4} & -\frac{1}{4} & \frac{1}{4} \\ \frac{1}{36} & \frac{1}{12} & -\frac{1}{12} & \frac{1}{12} & \frac{1}{12} & -\frac{1}{4} & \frac{1}{4} & -\frac{1}{4} & \frac{1}{4} \end{pmatrix}. \tag{10.24}
$$

[2] Alternatively, because of the orthogonality condition (10.12) one can represent each column of $\boldsymbol{M}^{-1}$ through the corresponding row vector of $\boldsymbol{M}$. We leave this as an exercise for an interested reader.

The last missing piece of the puzzle is the relaxation matrix ${}^{\mathrm{H}}\boldsymbol{S}$. It is almost diagonal. One can show through the Chapman-Enskog procedure that in order to recover the Navier-Stokes equation the relaxation matrix ${}^{\mathrm{H}}\boldsymbol{S}$ should have the following form for D2Q9 [32]:

$$
{}^{\mathrm{H}}\boldsymbol{S} = \begin{pmatrix}
0 & 0 & 0 & 0 & 0 & 0 & 0 & 0 & 0 \\
0 & 0 & 0 & 0 & 0 & 0 & 0 & 0 & 0 \\
0 & 0 & 0 & 0 & 0 & 0 & 0 & 0 & 0 \\
0 & 0 & 0 & \frac{\omega_\zeta+\omega_\nu}{2} & \frac{\omega_\zeta-\omega_\nu}{2} & 0 & 0 & 0 & 0 \\
0 & 0 & 0 & \frac{\omega_\zeta-\omega_\nu}{2} & \frac{\omega_\zeta+\omega_\nu}{2} & 0 & 0 & 0 & 0 \\
0 & 0 & 0 & 0 & 0 & \omega_\nu & 0 & 0 & 0 \\
0 & 0 & 0 & 0 & 0 & 0 & \omega_{\gamma_\alpha} & 0 & 0 \\
0 & 0 & 0 & 0 & 0 & 0 & 0 & \omega_{\gamma_\alpha} & 0 \\
0 & 0 & 0 & 0 & 0 & 0 & 0 & 0 & \omega_\gamma
\end{pmatrix}. \tag{10.25}
$$

The relaxation rates for conserved moments (density and momentum) are arbitrary; they are chosen as zero here.

A Chapman-Enskog analysis of the resulting MRT collision operator $\boldsymbol{\Omega} = -{}^{\mathrm{H}}\boldsymbol{M}^{-1}\,{}^{\mathrm{H}}\boldsymbol{S}\,{}^{\mathrm{H}}\boldsymbol{M}$ is briefly covered in Appendix A.2.3. Through the analysis we recover a macroscopic Navier-Stokes equation

$$
\partial_t(\rho u_\alpha) + \partial_\beta(\rho u_\alpha u_\beta) = -\partial_\alpha p + \partial_\beta\left[\eta\left(\partial_\beta u_\alpha + \partial_\alpha u_\beta\right) + \left(\eta_{\mathrm{B}} - \frac{2\eta}{3}\right)\delta_{\alpha\beta}\partial_\gamma u_\gamma\right], \tag{10.26}
$$

with neglected $O(u^3)$ error terms and pressure p, shear viscosity η, and bulk viscosity η_{B} given by

$$
p = \rho c_s^2, \quad \eta = \rho c_s^2\left(\frac{1}{\omega_\nu} - \frac{1}{2}\right), \quad \eta_{\mathrm{B}} = \rho c_s^2\left(\frac{1}{\omega_\zeta} - \frac{1}{2}\right) - \frac{\eta}{3}. \tag{10.27}
$$

Unlike in the BGK model, shear and bulk viscosity can be chosen independently. This is a clear advantage of MRT over BGK. In particular, when the Reynolds number is large (and the shear viscosity small), increasing the bulk viscosity can stabilise simulations by attenuating any spurious pressure waves more quickly. (The effect of viscosity on pressure waves will be covered further in Sect. 12.1.)

For the higher order moments γ_x and γ_y, the same relaxation rate ω_{γ_α} is taken to satisfy isotropy. Therefore, there are four independent relaxation rates. Only two of them are related to physical quantities. The relaxation rates ω_{γ_α} and ω_γ do not appear in the leading-order terms, but they can be tuned to increase accuracy and stability [10, 27, 30].

In conclusion, the above expressions for the matrices ${}^{\mathrm{H}}\boldsymbol{M}$, ${}^{\mathrm{H}}\boldsymbol{M}^{-1}$ and ${}^{\mathrm{H}}\boldsymbol{S}$, together with the equilibrium moments in (10.23), can be used directly in the general MRT implementation detailed in Sect. 10.3.

10.4.2 Gram-Schmidt Procedure

The Gram-Schmidt approach is more widely used in the LB literature than the Hermite polynomial approach. The underlying idea of the Gram-Schmidt procedure is to construct a set of orthogonal vectors to form the transformation matrix $\boldsymbol{M}$. Note that we seek these vectors to be directly orthogonal without any corresponding weights unlike the Hermite approach in (10.12).

We start with a set of q linearly independent vectors $\{\boldsymbol{v}_n\}$ with $n = 0, \dots, q-1$. These vectors are generally not mutually orthogonal. The Gram-Schmidt procedure maps these vectors to a set of q other vectors $\{\boldsymbol{u}_n\}$ which are all mutually orthogonal, through the following process:

$$\begin{aligned} \boldsymbol{u}_0 &= \boldsymbol{v}_0, \\ \boldsymbol{u}_1 &= \boldsymbol{v}_1 - \boldsymbol{u}_0 \frac{\boldsymbol{u}_0 \cdot \boldsymbol{v}_1}{\boldsymbol{u}_0 \cdot \boldsymbol{u}_0}, \\ \boldsymbol{u}_2 &= \boldsymbol{v}_2 - \boldsymbol{u}_0 \frac{\boldsymbol{u}_0 \cdot \boldsymbol{v}_2}{\boldsymbol{u}_0 \cdot \boldsymbol{u}_0} - \boldsymbol{u}_1 \frac{\boldsymbol{u}_1 \cdot \boldsymbol{v}_2}{\boldsymbol{u}_1 \cdot \boldsymbol{u}_1}, \\ &\dots \\ \boldsymbol{u}_{q-1} &= \boldsymbol{v}_{q-1} - \sum_{i=0}^{q-2} \boldsymbol{u}_i \frac{\boldsymbol{u}_i \cdot \boldsymbol{v}_{q-1}}{\boldsymbol{u}_i \cdot \boldsymbol{u}_i}. \end{aligned} \tag{10.28}$$

This algorithm essentially projects each vector $\boldsymbol{v}_n$ onto a direction which is orthogonal to that of all previous vectors $\boldsymbol{u}_{m<n}$.

Exercise 10.1 Show that all vectors $\{\boldsymbol{u}_n\}$ are mutually orthogonal.

We seek the initial vector set $\{\boldsymbol{v}_n\}$ in the form $c_{ix}^n c_{iy}^m$. Notice that the pair of coefficients $(n, m) = (0, 0)$ is the vector $(1, 1, 1, 1, 1, 1, 1, 1, 1)$ related to density ρ. Pairs $(n, m) = (0, 1)$ and $(n, m) = (1, 0)$ give us vectors $(0, 1, 0, -1, 0, 1, -1, -1, 1)$ and $(0, 0, 1, 0, -1, 1, 1, -1, -1)$ related to x- and y-components of momentum $\boldsymbol{j}$. Those vectors are already orthogonal in the sense of (10.11).

The next step is to find the remaining six vectors for D2Q9. We have already shown in Sect. 4.2.1 that there are three distinct polynomials of second order corresponding to $(n, m) = (1, 1),\ (2, 0),\ (0, 2)$, two third-order polynomials corresponding to $(n, m) = (2, 1),\ (1, 2)$, and one fourth-order polynomial corresponding to $(n, m) = (2, 2)$. This gives nine linearly independent vectors in total.

Now we can perform the Gram-Schmidt procedure to reduce these initial vectors to a set of nine mutually orthogonal vectors. We will only show the result of the procedure. Taking the notation ${}^{\mathrm{G}}\boldsymbol{M}_k$ for the final Gram-Schmidt vectors, the set of orthogonal vectors is as follows [9]:

$$\begin{gathered}
{}^{\mathrm{G}}M_{\rho,i} = 1, \qquad {}^{\mathrm{G}}M_{e,i} = -4 + 3\left(c_{ix}^2 + c_{iy}^2\right), \\
{}^{\mathrm{G}}M_{\epsilon,i} = 4 - \tfrac{21}{2}\left(c_{ix}^2 + c_{iy}^2\right) + \tfrac{9}{2}\left(c_{ix}^2 + c_{iy}^2\right)^2, \\
{}^{\mathrm{G}}M_{j_x,i} = c_{ix}, \qquad {}^{\mathrm{G}}M_{q_x,i} = \left(-5 + 3\left(c_{ix}^2 + c_{iy}^2\right)\right) c_{ix}, \\
{}^{\mathrm{G}}M_{j_y,i} = c_{iy}, \qquad {}^{\mathrm{G}}M_{q_y,i} = \left(-5 + 3\left(c_{ix}^2 + c_{iy}^2\right)\right) c_{iy}, \\
{}^{\mathrm{G}}M_{p_{xx},i} = c_{ix}^2 - c_{iy}^2, \qquad {}^{\mathrm{G}}M_{p_{xy},i} = c_{ix}c_{iy}.
\end{gathered} \tag{10.29}$$

The moments ${}^{\mathrm{G}}\boldsymbol{M}_k\boldsymbol{f}$ correspond to the physical quantities; i.e., ρ is the density, j_x and j_y are components of momentum flux, q_x and q_y correspond to the energy flux components, e and ϵ correspond to the energy and the energy squared, p_{xx} and p_{xy} correspond to the diagonal and off-diagonal components of the stress tensor. Replacing the velocities $c_{i\alpha}$ by their numerical values, we can write the Gram-Schmidt transformation matrix:

$${}^{\mathrm{G}}\boldsymbol{M} = \begin{pmatrix} {}^{\mathrm{G}}\boldsymbol{M}_\rho \\ {}^{\mathrm{G}}\boldsymbol{M}_e \\ {}^{\mathrm{G}}\boldsymbol{M}_\epsilon \\ {}^{\mathrm{G}}\boldsymbol{M}_{j_x} \\ {}^{\mathrm{G}}\boldsymbol{M}_{q_x} \\ {}^{\mathrm{G}}\boldsymbol{M}_{j_y} \\ {}^{\mathrm{G}}\boldsymbol{M}_{q_y} \\ {}^{\mathrm{G}}\boldsymbol{M}_{p_{xx}} \\ {}^{\mathrm{G}}\boldsymbol{M}_{p_{xy}} \end{pmatrix} = \begin{pmatrix} 1 & 1 & 1 & 1 & 1 & 1 & 1 & 1 & 1 \\ -4 & -1 & -1 & -1 & -1 & 2 & 2 & 2 & 2 \\ 4 & -2 & -2 & -2 & -2 & 1 & 1 & 1 & 1 \\ 0 & 1 & 0 & -1 & 0 & 1 & -1 & -1 & 1 \\ 0 & -2 & 0 & 2 & 0 & 1 & -1 & -1 & 1 \\ 0 & 0 & 1 & 0 & -1 & 1 & 1 & -1 & -1 \\ 0 & 0 & -2 & 0 & 2 & 1 & 1 & -1 & -1 \\ 0 & 1 & -1 & 1 & -1 & 0 & 0 & 0 & 0 \\ 0 & 0 & 0 & 0 & 0 & 1 & -1 & 1 & -1 \end{pmatrix}. \tag{10.30}$$

The next step is to find the equilibrium moments $\boldsymbol{m}^{\mathrm{eq}}$ in the Gram-Schmidt basis. One way is the rather complicated linear wave number analysis as presented in [9]. Another route followed here is to represent the Gram-Schmidt vectors by Hermite vectors whose equilibrium moments are already known. In fact, the Gram-Schmidt vectors can be written as linear combinations of the Hermite vectors:

$$
\begin{aligned}
{}^{\mathrm{G}}M_i^{(\rho)} &= {}^{\mathrm{H}}M_i^{(\rho)}, \qquad {}^{\mathrm{G}}M_i^{(e)} = -2 + 3\left({}^{\mathrm{H}}M_i^{(p_{xx})} + {}^{\mathrm{H}}M_i^{(p_{yy})}\right), \\
{}^{\mathrm{G}}M_i^{(\epsilon)} &= 9\,{}^{\mathrm{H}}M_i^{(\gamma)} - 3\left({}^{\mathrm{H}}M_i^{(p_{xx})} + {}^{\mathrm{H}}M_i^{(p_{yy})}\right) + {}^{\mathrm{H}}M_i^{(\rho)}, \\
{}^{\mathrm{G}}M_i^{(j_x)} &= {}^{\mathrm{H}}M_i^{(j_x)}, \qquad {}^{\mathrm{G}}M_i^{(q_x)} = 3\,{}^{\mathrm{H}}M_i^{(\gamma_x)} - {}^{\mathrm{H}}M_i^{(j_x)}, \\
{}^{\mathrm{G}}M_i^{(j_y)} &= {}^{\mathrm{H}}M_i^{(j_y)}, \qquad {}^{\mathrm{G}}M_i^{(q_y)} = 3\,{}^{\mathrm{H}}M_i^{(\gamma_y)} - {}^{\mathrm{H}}M_i^{(j_y)}, \\
{}^{\mathrm{G}}M_i^{(p_{xx})} &= {}^{\mathrm{H}}M_i^{(p_{xx})} - {}^{\mathrm{H}}M_i^{(p_{yy})}, \qquad {}^{\mathrm{G}}M_i^{(p_{xy})} = {}^{\mathrm{H}}M_i^{(p_{xy})}.
\end{aligned}
\tag{10.31}
$$

Thus, one can work out the corresponding Gram-Schmidt equilibrium moments as function of density and velocity:

$$
\begin{aligned}
&\rho^{\mathrm{eq}} = \rho, \qquad e^{\mathrm{eq}} = \rho - 3\rho\left(u_x^2 + u_y^2\right), \qquad \epsilon^{\mathrm{eq}} = 9\rho u_x^2 u_y^2 - 3\rho\left(u_x^2 + u_y^2\right) + \rho, \\
&j_x^{\mathrm{eq}} = \rho u_x, \qquad q_x^{\mathrm{eq}} = 3\rho u_x^3 - \rho u_x, \qquad j_y^{\mathrm{eq}} = \rho u_y, \\
&q_y^{\mathrm{eq}} = 3\rho u_y^3 - \rho u_y, \quad p_{xx}^{\mathrm{eq}} = \rho\left(u_x^2 - u_y^2\right), \quad p_{xy}^{\mathrm{eq}} = \rho u_x u_y.
\end{aligned}
\tag{10.32}
$$

The inverse matrix ${}^{\mathrm{G}}\boldsymbol{M}^{-1}$ can be obtained from ${}^{\mathrm{G}}\boldsymbol{M}$ by the help of any mathematical package:

$$
{}^{\mathrm{G}}\boldsymbol{M}^{-1} = \begin{pmatrix}
\frac{1}{9} & -\frac{1}{9} & \frac{1}{9} & 0 & 0 & 0 & 0 & 0 & 0 \\
\frac{1}{9} & -\frac{1}{36} & -\frac{1}{18} & \frac{1}{6} & -\frac{1}{6} & 0 & 0 & \frac{1}{4} & 0 \\
\frac{1}{9} & -\frac{1}{36} & -\frac{1}{18} & 0 & 0 & \frac{1}{6} & -\frac{1}{6} & -\frac{1}{4} & 0 \\
\frac{1}{9} & -\frac{1}{36} & -\frac{1}{18} & -\frac{1}{6} & \frac{1}{6} & 0 & 0 & \frac{1}{4} & 0 \\
\frac{1}{9} & -\frac{1}{36} & -\frac{1}{18} & 0 & 0 & -\frac{1}{6} & \frac{1}{6} & -\frac{1}{4} & 0 \\
\frac{1}{9} & \frac{1}{18} & \frac{1}{36} & \frac{1}{6} & \frac{1}{12} & \frac{1}{6} & \frac{1}{12} & 0 & \frac{1}{4} \\
\frac{1}{9} & \frac{1}{18} & \frac{1}{36} & -\frac{1}{6} & -\frac{1}{12} & \frac{1}{6} & \frac{1}{12} & 0 & -\frac{1}{4} \\
\frac{1}{9} & \frac{1}{18} & \frac{1}{36} & -\frac{1}{6} & -\frac{1}{12} & -\frac{1}{6} & -\frac{1}{12} & 0 & \frac{1}{4} \\
\frac{1}{9} & \frac{1}{18} & \frac{1}{36} & \frac{1}{6} & \frac{1}{12} & -\frac{1}{6} & -\frac{1}{12} & 0 & -\frac{1}{4}
\end{pmatrix}.
\tag{10.33}
$$

The relaxation matrix in the Gram-Schmidt basis assumes the diagonal form

$$
{}^{\mathrm{G}}\boldsymbol{S} = \mathrm{diag}\left(0, \omega_e, \omega_\epsilon, 0, \omega_q, 0, \omega_q, \omega_\nu, \omega_\nu\right).
\tag{10.34}
$$

Zero relaxation rates above are for the conserved moments of density and momentum, ω_e and ω_ν are connected with bulk and shear viscosities, and ω_ϵ and ω_q are free parameters to tune. As shown in Sect. A.2.3, this recovers the Navier-Stokes equation

$$\partial_t(\rho u_\alpha) + \partial_\beta(\rho u_\alpha u_\beta) = -\partial_\alpha p + \partial_\beta \left[\eta \left(\partial_\beta u_\alpha + \partial_\alpha u_\beta \right) + \left(\eta_{\mathrm{B}} - \frac{2\eta}{3} \right) \delta_{\alpha\beta} \partial_\gamma u_\gamma \right], \tag{10.35}$$

with neglected $O(u^3)$ error terms. Pressure p, shear viscosity η, and bulk viscosity η_{B} are given by

$$p = \rho c_{\mathrm{s}}^2, \quad \eta = \rho c_{\mathrm{s}}^2 \left(\frac{1}{\omega_\nu} - \frac{1}{2} \right), \quad \eta_{\mathrm{B}} = \rho c_{\mathrm{s}}^2 \left(\frac{1}{\omega_e} - \frac{1}{2} \right) - \frac{\eta}{3}. \tag{10.36}$$

10.4.3 Discussion of MRT Approaches

MRT vectors: each velocity set has spurious invariants that are not found with a continuous velocity space. This was covered in Sect. 4.2.1. As a result of such invariants, only a limited number of independent Hermite polynomials can be obtained for discretised velocity sets. This number does not necessarily coincide with the number of velocity directions [33]. For D2Q9 and D3Q27 the number of Hermite polynomials equals the number of velocities, but for D3Q15 or D3Q19 this is not the case. Missing vectors have to be constructed, and this is usually done with the Gram-Schmidt procedure. This is one of the reasons why the Gram-Schmidt approach is more widely used than the Hermite approach.

Equilibrium moments: only the Hermite approach provides a relatively easy way to find analytical higher-order equilibrium moments than those found directly from the second-order equilibrium as $m_k^{\mathrm{eq}} = \boldsymbol{M}_k \boldsymbol{f}^{\mathrm{eq}}$. Thus, any arbitrary set of MRT vectors can be represented as linear combinations of Hermite vectors to obtain the equilibrium moments. This approach is significantly simpler than obtaining the moments from a linear wavenumber analysis [9].

In both approaches, some analytical equilibrium moments m_k^{eq} contain third- or fourth-order terms in $\boldsymbol{u}$. In contrast, the equilibrium moments found directly as $m_k^{\mathrm{eq}} = \boldsymbol{M}_k \boldsymbol{f}^{\mathrm{eq}}$ are of second order, yet they are still sufficient to recover the Navier-Stokes equation. Thus, one common approach is to neglect all equilibrium moment terms of order u^3 and u^4 [9]. Although there are some indications that keeping higher-order terms can improve stability [27, 34], these are only minor improvements.

One difference between the two approaches is that the equilibrium moments in the Hermite polynomial approach are in increasing order in the macroscopic velocity $\boldsymbol{u}$. In the Gram-Schmidt approach, however, polynomials of different order are mixed, as can be seen in (10.32).

Relaxation matrix: We have seen that the Gram-Schmidt relaxation matrix ${}^{\mathrm{G}}\boldsymbol{S}$ in (10.34) is diagonal, while the Hermite relaxation matrix ${}^{\mathrm{H}}\boldsymbol{S}$ in (10.25) is only nearly diagonal. This is an additional reason why the Gram-Schmidt approach is normally preferred.

We will discuss the choice of the relaxation rates for both relaxation matrices ${}^{\mathrm{H}}\boldsymbol{S}$ and ${}^{\mathrm{G}}\boldsymbol{S}$ and provide more practical advice on improving stability and accuracy in Sect. 10.7. However, we have already seen one clear advantage: the MRT collision operator allows tuning shear and bulk viscosities independently. Increasing the bulk viscosity can often improve stability by damping underresolved hydrodynamic artefacts [29, 35, 36].

10.5 Inclusion of Forces

So far we have assumed that the momentum is conserved during collision. This is no longer the case when external forces are present. We will now give a brief overview of how to include forces in the MRT collision model.

Let us start by recalling forcing in the BGK model:

$$f_i(\boldsymbol{x}+\boldsymbol{c}_i, t+1) = f_i(\boldsymbol{x}, t) - \frac{\Delta t}{\tau}\left(f_i(\boldsymbol{x}, t) - f_i^{\mathrm{eq}}(\boldsymbol{x}, t)\right) + S_i(\boldsymbol{x}, t)\Delta t. \tag{10.37}$$

As thoroughly discussed in Chap. 6, the effect of an external force density $\boldsymbol{F}^{\mathrm{ext}}$ can be included in the equilibrium distributions f_i^{eq} and/or the additional source term S_i. This depends on the chosen forcing model. For MRT, the following implementations are common.

Simple Forcing: The simplest approach is to approximate S_i by

$$S_i^{\mathrm{simple}} = 3w_i \boldsymbol{F}^{\mathrm{ext}} \cdot \boldsymbol{c}_i. \tag{10.38}$$

One can easily add this force contribution to the right-hand-side of the MRT equation:

$$\boldsymbol{f}^\star = \boldsymbol{f} - \boldsymbol{M}^{-1}\boldsymbol{S}\left(\boldsymbol{m} - \boldsymbol{m}^{\mathrm{eq}}\right)\Delta t + \boldsymbol{S}^{\mathrm{simple}}\Delta t. \tag{10.39}$$

Note that the equilibrium velocity used to calculate the equilibrium moments m_k^{eq} is not shifted, i.e. $\rho\boldsymbol{u}^{\mathrm{eq}} = \sum_i f_i\boldsymbol{c}_i$. However, the Navier-Stokes equation is reproduced *via* the Chapman-Enskog analysis with the shifted macroscopic physical velocity according to $\rho\boldsymbol{u}^{\mathrm{p}} = \sum_i f_i\boldsymbol{c}_i + \frac{\Delta t}{2}\boldsymbol{F}^{\mathrm{ext}}$ (cf. Chap. 6).

Guo Forcing: Guo's model [37] for the BGK collision operator leads to

$$S_i^{\text{Guo}} = \left(1 - \frac{\Delta t}{2\tau}\right) w_i \left(\frac{\boldsymbol{c}_i - \boldsymbol{u}^{\text{p}}}{c_{\text{s}}^2} + \frac{(\boldsymbol{c}_i \cdot \boldsymbol{u}^{\text{p}})\,\boldsymbol{c}_i}{c_{\text{s}}^4}\right) \cdot \boldsymbol{F}^{\text{ext}} = \left(1 - \frac{\Delta t}{2\tau}\right) F_i, \tag{10.40}$$

where we have introduced the abbreviation F_i. Again, the physical velocity is given by $\rho \boldsymbol{u}^{\text{p}} = \sum_i f_i \boldsymbol{c}_i + \frac{\Delta t}{2} \boldsymbol{F}^{\text{ext}}$. Furthermore, the equilibrium velocity is also shifted in the same way: $\rho \boldsymbol{u}^{\text{eq}} = \sum_i f_i \boldsymbol{c}_i + \frac{\Delta t}{2} \boldsymbol{F}^{\text{ext}}$. In the MRT model, one can use a similar form for the force contribution [2, 38]. The relaxed moments then become

$$m_k^\star = m_k - \omega_k \Delta t \left(m_k - m_k^{\text{eq}}\right) + \left(1 - \frac{\omega_k \Delta t}{2}\right) \boldsymbol{M}_k \boldsymbol{F}. \tag{10.41}$$

The forcing term F_i is first transformed to momentum space, and then its contribution to the corresponding moment is multiplied by $1 - \omega_k \Delta t/2$. After relaxation, the moments are transformed back to population space *via* $\boldsymbol{f}^\star = \boldsymbol{M}^{-1} \boldsymbol{m}^\star$, followed by the normal streaming. For D2Q9, the result of the matrix-vector multiplication $\boldsymbol{MF}$ in (10.41) is given by [2, 39]:

$$^{\text{H}}\boldsymbol{MF} = \begin{pmatrix} 0 \\ F_x^{\text{ext}} \\ F_y^{\text{ext}} \\ 2F_x^{\text{ext}} u_x^{\text{p}} \\ 2F_y^{\text{ext}} u_y^{\text{p}} \\ F_x^{\text{ext}} u_y^{\text{p}} + F_y^{\text{ext}} u_x^{\text{p}} \\ 0 \\ 0 \\ 0 \end{pmatrix}, \quad {}^{\text{G}}\boldsymbol{MF} = \begin{pmatrix} 0 \\ 6\left(F_x^{\text{ext}} u_x^{\text{p}} + F_y^{\text{ext}} u_y^{\text{p}}\right) \\ -6\left(F_x^{\text{ext}} u_x^{\text{p}} + F_y^{\text{ext}} u_y^{\text{p}}\right) \\ F_x^{\text{ext}} \\ -F_x^{\text{ext}} \\ F_y^{\text{ext}} \\ -F_y^{\text{ext}} \\ 2F_x^{\text{ext}} u_x^{\text{p}} - 2F_y^{\text{ext}} u_y^{\text{p}} \\ F_y^{\text{ext}} u_x^{\text{p}} + F_x^{\text{ext}} u_y^{\text{p}} \end{pmatrix}. \tag{10.42}$$

Other Forcing Schemes: An implicit formulation of the force inclusion is given in [40, 41]. This approach is similar to Guo's. Another MRT force implementation can be found in [42] where only the momentum is shifted without a need to introduce the source term S_i.

10.6 TRT Collision Operator

The two-relaxation-time (TRT) model (Sect. 10.6.1) combines the algorithmic simplicity of the BGK operator and improved accuracy and stability properties of the more general MRT model. We discuss the implementation of TRT in Sect. 10.6.2.

10.6.1 Introduction

The MRT model provides a great deal of flexibility in tuning the relaxation of individual moments. However, the number of free MRT parameters may be large and confusing from the perspective of an average LBM user. For example, for D2Q9 MRT there are two free relaxation rates to tune, and for D3Q15 and D3Q19 MRT, presented in Appendix A.6, there are even more free parameters. As these relaxation rates do not affect the macroscopic equations, it is not obvious how these rates should be tuned to achieve better accuracy or stability.

So far optimal MRT parameters have only been obtained numerically through parameter studies. One such example is the three-dimensional shear wave propagation study [29], which found the parameter values $\omega_e \Delta t = 1.19$, $\omega_\epsilon \Delta t = \omega_\pi \Delta t = 1.4$, $\omega_q \Delta t = 1.2$ and $\omega_m \Delta t = 1.98$ to give optimal stability for D3Q19 MRT (cf. Appendix A.6). However, in more general situations it is almost impossible or at least impractical to perform a parameter study or a linear stability analysis to optimise accuracy and/or stability.

Fortunately, there exists a collision model that combines the accuracy and stability advantages of the MRT model and the simplicity of the BGK model: the two-relaxation-time (TRT) model [1, 7, 43] requires only two relaxation rates. The first relaxation rate is related to the shear viscosity, the other is a free parameter. Having only two rates significantly simplifies the mathematical analysis of accuracy and stability.

The **TRT model** is a simplification of the MRT model. All moments related to even-order polynomials in velocity (i.e. $\rho, p_{xx}, p_{yy}, p_{xy}$, etc.) are relaxed with a rate ω^+, and odd-order moments (i.e. $j_x, j_y, \gamma_x, \gamma_y$, etc.) with another rate ω^-.

Two relaxation rates can cure the BGK collision operator's "disease" where the accuracy errors depend on the relaxation rate ω. Previously in Sect. 4.5 we presented the complicated dependency of the accuracy error on $\tau = \frac{1}{\omega}$. The spatial accuracy error is proportional to $(\tau - \Delta t/2)^2$; and the time accuracy error also depends on τ. Because the viscosity depends on τ (i.e. ω), the error is said to be dependent on viscosity. Especially, high viscosities ($\omega \Delta t \to 0$) drastically deteriorate accuracy. Performing simulations across different values of τ while keeping all other parameters fixed (i.e. grid number, macroscopic velocity, etc.), may give significantly different errors. Usually, to guarantee consistent errors across simulations one needs to simultaneously tune the relaxation rate ω, the grid number N and to a smaller extent the macroscopic velocity $\boldsymbol{u}$. In contrast, for the TRT collision operator it can be shown that the accuracy error depends on a certain combination of the two relaxation rates ω^+ and ω^-.

The so-called **magic parameter** Λ characterises the truncation error and stability properties of the TRT model. It is a function of the TRT relaxation rates ω^+ and ω^-:

$$\Lambda = \left(\frac{1}{\omega^+\Delta t} - \frac{1}{2}\right)\left(\frac{1}{\omega^-\Delta t} - \frac{1}{2}\right). \tag{10.43}$$

Thus, controlling the accuracy is drastically simplified with the TRT collision operator. If one needs to guarantee the same accuracy errors for simulations where the viscosity changes, then there is no need to tune a number of parameters like the grid number and the macroscopic velocity. The only thing required is to keep Λ fixed while changing the viscosity through ω^+ by correspondingly changing the free parameter ω^- [14].

One can show [7, 11, 43, 44] that the higher-order truncation errors of the steady-state TRT model scale with $\Lambda - \frac{1}{12}$ (third-order error) and $\Lambda - \frac{1}{6}$ (fourth-order error). Thus, there exist specific values of Λ that improve accuracy and stability. We will get back to specific numerical values of Λ in Sect. 10.7. An example of an optimal parameter choice leading to TRT's superiority is shown by a simulation of diffusion in Sect. 8.6.1.

10.6.2 Implementation

We will now provide the algorithmic details of TRT and show that it can be as computationally efficient as the BGK model, while offering more control over accuracy and stability.

Naive implementation: as we have already mentioned, one can use the MRT model with ω^+ for all even-order moments and ω^- for all odd-order moments. This implementation is as computationally "heavy" as the original MRT model, but it is a good starting point for optimising MRT relaxation rates.

Standard implementation: the TRT model is based on the decomposition of populations into symmetric and antisymmetric parts. Any LB velocity set is always symmetric, i.e. for any given velocity $\boldsymbol{c}_i$ there is always a velocity $\boldsymbol{c}_{\bar{i}} = -\boldsymbol{c}_i$. This is also true for $\boldsymbol{c}_0 = \boldsymbol{0}$. Using these notations one can decompose the populations and equilibrium populations into their symmetric and antisymmetric parts:

$$\begin{aligned} f_i^+ &= \frac{f_i + f_{\bar{i}}}{2}, & f_i^- &= \frac{f_i - f_{\bar{i}}}{2}, \\ f_i^{eq+} &= \frac{f_i^{eq} + f_{\bar{i}}^{eq}}{2}, & f_i^{eq-} &= \frac{f_i^{eq} - f_{\bar{i}}^{eq}}{2}, \end{aligned} \tag{10.44}$$

with equilibrium distributions from (3.54). For the rest population we have $f_0^+ = f_0$ and $f_0^- = 0$. The same correspondingly applies to the equilibrium value f_0^{eq}. From (10.44) we can now decompose the populations in terms of their symmetric and antisymmetric parts:

$$\begin{aligned} f_i &= f_i^+ + f_i^-, & f_{\bar{i}} &= f_i^+ - f_i^-, \\ f_i^{\mathrm{eq}} &= f_i^{\mathrm{eq}+} + f_i^{\mathrm{eq}-}, & f_{\bar{i}}^{\mathrm{eq}} &= f_i^{\mathrm{eq}+} - f_i^{\mathrm{eq}-}. \end{aligned} \tag{10.45}$$

The **standard TRT model** can be implemented similarly to the BGK model:

$$\begin{aligned} f_i^\star &= f_i - \omega^+ \Delta t \left(f_i^+ - f_i^{\mathrm{eq}+} \right) - \omega^- \Delta t \left(f_i^- - f_i^{\mathrm{eq}-} \right), \\ f_i(\boldsymbol{x} + \boldsymbol{c}_i \Delta t, t + \Delta t) &= f_i^\star(\boldsymbol{x}, t). \end{aligned} \tag{10.46}$$

This algorithm is independent of the chosen lattice (e.g. D2Q9, D3Q19).

The TRT model solves the Navier-Stokes equation with kinematic shear viscosity

$$\nu = c_s^2 \left(\frac{1}{\omega^+ \Delta t} - \frac{1}{2} \right). \tag{10.47}$$

Thus, ω_- is a free parameter. We have already seen that it can be used to choose certain values for Λ in order to improve accuracy or stability (cf. Chap. 4 and Sect. 10.7). Note, however, that unlike the MRT model, the TRT model does not allow setting the bulk viscosity independently of the shear viscosity.

The standard implementation is equivalent to the naive implementation but it is much more computationally efficient [45].

Efficient implementation: this is similar to the standard implementation, except that it takes advantage of the symmetric structure of TRT. The collision step is performed for just one half of the velocity set, and the other half follows automatically by symmetry considerations, with no need for computation. So compared to BGK, in TRT the added computational operations required to determine the even and odd parts of populations is compensated by the halved number of operations for the collision step of TRT. Thus, this TRT implementation is as efficient as BGK. More details can be found in the literature [6].

10.7 Overview: Choice of Collision Models and Relaxation Rates

There is a hierarchy of collision operators. MRT, as the most general relaxation model, can be reduced to the TRT model by taking one relaxation rate for all odd-order moments and another for all even-order moments. The TRT model can be further reduced to the BGK model if only one relaxation rate is used.

Which model should be employed in which situation, and how should the relaxation rates be chosen? We will now provide a simple guide on how to select the collision model and the relaxation rates.

10.7.1 BGK Model

There is only a single relaxation rate $\omega = 1/\tau$ in the BGK model. It is therefore the simplest choice, and the most commonly used model. The relaxation rate is determined by the chosen viscosity. This model has accuracy and stability properties determined by the assigned viscosity value, which leads to the deficiencies that are thoroughly discussed in Chap. 4.

We recommend that **LBM newcomers should start with the BGK model**. The first step is to master non-dimensionalisation and parameter selection strategies for the BGK model (Chap. 7). The next step is to understand its limitations. If the BGK deficiencies are severe for the given problem, the user should move on to TRT.

10.7.2 TRT Model

The TRT model provides more flexibility and control than the BGK model. At the same time, it is still relatively easy to mathematically analyse the TRT model.

There are some situations where TRT should definitely be chosen instead of BGK, for example in systems with large boundary areas (e.g. porous media). However, one has to understand that the TRT model itself is limited, and that it cannot cure all deficiencies of the BGK model. There are still truncation errors in TRT and they can be large for certain simulations. Sometimes it is not possible to improve stability by simply switching from the BGK to the TRT collision operator.

As the central part of the TRT model we have two relaxation rates ω^+ and ω^- and the magic parameter

$$\Lambda = \left(\frac{1}{\omega^+ \Delta t} - \frac{1}{2}\right)\left(\frac{1}{\omega^- \Delta t} - \frac{1}{2}\right). \tag{10.48}$$

While ω^+ controls the kinematic viscosity *via* (10.47), Λ can be used to control accuracy and stability. There are certain values of Λ that show distinctive properties:

- $\Lambda = \frac{1}{12}$ cancels the third-order spatial error, leading to optimal results for pure advection problems.
- $\Lambda = \frac{1}{6}$ cancels the fourth-order spatial error, providing the most accurate results for the pure diffusion equation.
- $\Lambda = \frac{3}{16}$ results in the boundary wall location implemented *via* bounce-back for the Poiseuille flow exactly in the middle between horizontal walls and fluid nodes [1].
- $\Lambda = \frac{1}{4}$ provides the most stable simulations [46, 47].

We recommend that the user should **experiment with different values** of Λ for a given problem. The list above provides a starting point. When changing the viscosity *via* ω^+, Λ should be kept fixed by adapting the value of ω^-. If the underlying problems can still not be solved, the next step is to move on to MRT.

10.7.3 MRT Model

MRT is the most general and advanced of the relaxation models, with the largest number of free parameters to tune accuracy and stability. It also allows choosing the bulk viscosity independently of the shear viscosity. However, it is also more difficult to code and understand, and it requires more computational resources if it is not coded efficiently.

The full power of the MRT model can only be exploited after careful testing. For example, the optimal relaxation rates to simulate lid-driven cavity flow with the D3Q15 model are $\omega_e \Delta t = 1.6$, $\omega_\epsilon \Delta t = 1.2$, $\omega_q \Delta t = 1.6$ and $\omega_m \Delta t = 1.2$ as numerically obtained through a linear von Neumann analysis [29]. (More details for the D3Q15 Gram-Schmidt model can be found in Appendix A.6). Such an analysis for more complex problems is impractical. As general advice, the initial set of MRT relaxation rates could be the equivalent rates obtained from a tested TRT model. Thus, once ω^+ and ω^- are determined, all even-order moments are relaxed with

ω^+ and all odd-order moments with ω^-. After this, individual rates can be changed to further improve the simulation accuracy and stability.

Unfortunately there is no ready-to-use recipe for choosing MRT relaxation rates different from the TRT rates. One general suggestion is to increase the bulk viscosity by decreasing the corresponding relaxation rate. This can suppress some underresolved numerical artefacts to improve stability and accuracy of simulations [35], and spurious sound waves are additionally suppressed (cf. Chap. 12). Some authors also choose $\omega_k \Delta t = 1$ for non-hydrodynamic modes so that they relax instantaneously to their equilibrium values [28].

As a final suggestion for the typical situation when one wants to keep numerical errors independent of viscosity in MRT simulations, a number of parameter products $\Gamma_i \Gamma_j$ should be kept constant, where a general parameter is $\Gamma_i = \frac{1}{\omega_i \Delta t} - \frac{1}{2}$ [6, 26]. For example, the following parameter products should be kept fixed for D2Q9:

- Hermite approach: $\Gamma_\nu \Gamma_{\gamma_\alpha}$, $\Gamma_\gamma \Gamma_{\gamma_\alpha}$, and $\Gamma_\xi \Gamma_{\gamma_\alpha}$
- Gram-Schmidt approach: $\Gamma_\nu \Gamma_q$, $\Gamma_e \Gamma_q$, and $\Gamma_\epsilon \Gamma_q$.

References

1. I. Ginzburg, F. Verhaeghe, D. d'Humières, Commun. Comput. Phys. **3**, 519 (2008)
2. A. Kuzmin, Multiphase simulations with lattice Boltzmann scheme. Ph.D. thesis, University of Calgary (2010)
3. X. He, Q. Zou, L.S. Luo, M. Dembo, J. Stat. Phys. **87**(1–2), 115 (1997)
4. I. Ginzbourg, P.M. Adler, J. Phys. II France **4**(2), 191 (1994)
5. I. Ginzburg, D. d'Humières, Phys. Rev. E **68**, 066614 (2003)
6. S. Khirevich, I. Ginzburg, U. Tallarek, J. Comp. Phys. **281**, 708 (2015)
7. D. d'Humières, I. Ginzburg, Comput. Math. Appl. **58**, 823 (2009)
8. Y.H. Qian, Y. Zhou, Europhys. Lett. **42**(4), 359 (1998)
9. P. Lallemand, L.S. Luo, Phys. Rev. E **61**(6), 6546 (2000)
10. P.J. Dellar, J. Comput. Phys. **259**, 270 (2014)
11. G. Silva, V. Semiao, J. Comput. Phys. **269**, 259 (2014)
12. B. Servan-Camas, F. Tsai, Adv. Water Resour. **31**, 1113 (2008)
13. I. Ginzburg, Phys. Rev. E **77**, 066704 (2008)
14. I. Ginzburg, Adv. Water Resour. **28**(11), 1171 (2005)
15. J. Latt, Hydrodynamic limit of lattice Boltzmann equations. Ph.D. thesis, University of Geneva (2007)
16. J. Latt, B. Chopard, Math. Comput. Simul. **72**(2–6), 165 (2006)
17. R. Zhang, X. Shan, H. Chen, Phys. Rev. E **74**, 046703 (2006)
18. A. Montessori, G. Facucci, P. Prestininzi, A. La Rocca, S. Succi, Phys. Rev. E **89**, 053317 (2014)
19. B.M. Boghosian, J. Yepez, P.V. Coveney, A. Wagner, Proc. R. Soc. A **457**(2007), 717 (2001)
20. S. Ansumali, I.V. Karlin, H.C. Öttinger, Europhys. Lett. **63**(6), 798 (2003)
21. M. Geier, A. Greiner, J. Korvink, Phys. Rev. E **73**(066705), 1 (2006)
22. Y. Ning, K.N. Premnath, D.V. Patil, Int. J. Num. Meth. Fluids **82**(2), 59 (2015)
23. M. Geier, M. Schönherr, A. Pasquali, M. Krafczyk, Comput. Math. Appl. **70**(4), 507 (2015)
24. I. Karlin, P. Asinari, Physica A **389**(8), 1530 (2010)
25. R. Adhikari, S. Succi, Phys. Rev. E **78**(066701), 1 (2008)
26. P.J. Dellar, J. Comp. Phys. **190**, 351 (2003)

27. A. Kuzmin, A. Mohamad, S. Succi, Int. J. Mod. Phys. C **19**(6), 875 (2008)
28. F. Higuera, S. Succi, R. Benzi, Europhys. Lett. **9**(4), 345 (1989)
29. D. d'Humières, I. Ginzburg, M. Krafczyk, P. Lallemand, L.S. Luo, Phil. Trans. R. Soc. Lond. A **360**, 437 (2002)
30. P.J. Dellar, Phys. Rev. E **65**(3) (2002)
31. R. Benzi, S. Succi, M. Vergassola, Phys. Rep. **222**(3), 145 (1992)
32. P. Asinari, Phys. Rev. E **77**(056706), 1 (2008)
33. R. Rubinstein, L.S. Luo, Phys. Rev. E **77**(036709), 1 (2008)
34. D.N. Siebert, L.A. Hegele Jr., P.C. Philippi, Phys. Rev. E **77**, 026707 (2008)
35. P. Asinari, I. Karlin, Phys. Rev. E **81**(016702), 1 (2010)
36. P. Dellar, Phys. Rev. E **64**(3) (2001)
37. Z. Guo, C. Zheng, B. Shi, Phys. Rev. E **65**, 46308 (2002)
38. A. Kuzmin, Z. Guo, A. Mohamad, Phil. Trans. Royal Soc. A **369**, 2219 (2011)
39. G. Silva, V. Semiao, J. Fluid Mech. **698**, 282 (2012)
40. S. Mukherjee, J. Abraham, Comput. Fluids **36**, 1149 (2007)
41. K. Premnath, J. Abraham, J. Comput. Phys. **224**, 539 (2007)
42. P. Lallemand, L.S. Luo, Phys. Rev. E **68**, 1 (2003)
43. I. Ginzburg, F. Verhaeghe, D. d'Humières, Commun. Comput. Phys. **3**, 427 (2008)
44. I. Ginzburg, Commun. Comput. Phys. **11**, 1439 (2012)
45. I. Ginzburg, Adv. Water Resour. **28**(11), 1196 (2005)
46. I. Ginzburg, D. d'Humières, A. Kuzmin, J. Stat. Phys. **139**, 1090 (2010)
47. A. Kuzmin, I. Ginzburg, A. Mohamad, Comp. Math. Appl. **61**, 1090 (2011)

Chapter 11
Boundary Conditions for Fluid-Structure Interaction

Abstract After reading this chapter, you will have insight into a large number of more complex lattice Boltzmann boundary conditions, including advanced bounce-back methods, ghost methods, and immersed boundary methods. These boundary conditions will allow you to simulate things like curved boundaries, flows in media with sub-grid porosity, rigid but moveable objects immersed in the fluid, and even flows with deformable objects such as red blood cells.

Boundary conditions play a paramount role in hydrodynamics. Chapter 5 concerns itself with the definition and conceptual introduction of boundary conditions, and it provides an overview of boundary conditions for relatively simple solid geometries, flow inlets and outlets and periodic systems. Here, we turn our attention to resting and moving boundaries with complex shapes (Sect. 11.1). It is nearly impossible to give an exhaustive overview of all available boundary conditions for fluid-structure interaction in the LBM. We will therefore focus on the most prominent examples: bounce-back methods in Sect. 11.2, extrapolation methods in Sect. 11.3 and immersed-boundary methods in Sect. 11.4. We provide a list of comparative benchmark studies and an overview of the strengths and weaknesses of the discussed boundary conditions in Sect. 11.5.

11.1 Motivation

Many works about boundary conditions in the LBM assume flat, resting and rigid boundaries. We have reviewed a selection of those methods in Chap. 5. But our experience tells us that only a small number of boundaries in fluid dynamics obey these assumptions. In reality, most boundaries are curved, some can move and others are deformable. Prominent examples are porous media, curved surfaces of cars and planes in aerodynamics, suspensions (e.g. clay, slurries) or deformable objects (e.g. cells, wings, compliant containers). Analytical solutions are often impossible to obtain, which makes computer simulations an indispensible tool. This challenge led to a remarkable variety of proposed methods to model complex boundaries in LB simulations.

T. Krüger et al., *The Lattice Boltzmann Method*, Graduate Texts in Physics,
DOI 10.1007/978-3-319-44649-3_11

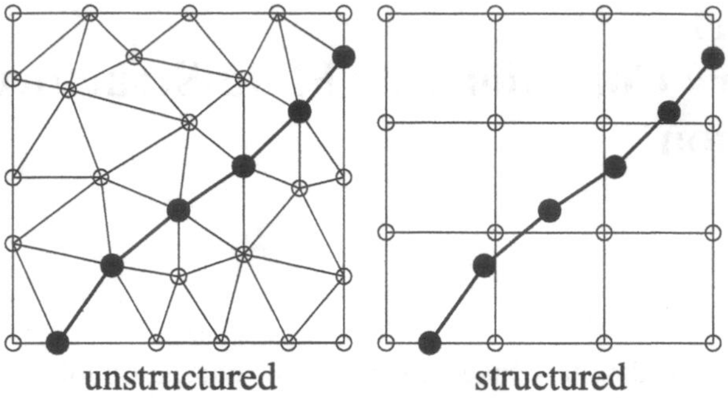

Fig. 11.1 Unstructured and structured meshes. The same boundary problem (*solid black circles connected with thick lines*) can be treated by, for example, an unstructured (*left*) or structured (*right*) approach. The former requires remeshing if the boundary moves, the latter leads to interpolations or extrapolations near the boundary

In order to accurately describe a complex domain, there are essentially two options (Fig. 11.1). The first approach is to formulate the problem in a coordinate system which fits the shape of the boundary. This leads to curvilinear or body-fitted meshes where the boundary treatment itself is trivial. However, this way we lose the advantages of the simple cartesian grid. For example, if the boundary shape changes in time, the curvilinear coordinate system also changes or remeshing becomes necessary. This can be a challenging and time-consuming task [1]. The alternative is to retain the cartesian structure of the bulk geometry, but then we have to introduce special procedures to account for the complex shape of the boundary which does generally not conform with the underlying lattice structure. In the end, this leads to interpolation and extrapolation boundary schemes.

Since most LB algorithms take advantage of the cartesian grid, the second route is usually preferred. First, it is easier to correct only the behaviour of the boundary nodes than touching all bulk nodes. Secondly, remeshing of the bulk involves interpolations in the entire numerical domain, which can lead to detrimental numerical viscosities (hyperviscosities) and a loss of exact mass/momentum conservation. More details about LB for non-cartesian geometries (i.e. curvilinear structured meshes or unstructured meshes) are provided in, e.g., [2–7]. It is therefore less harmful to use interpolations only in the vicinity of the boundaries. In this chapter we will exclusively address boundary treatments of the second kind, where the underlying lattice structure is not changed.

There are different types of problems which are typically encountered in connection with off-lattice boundaries. We can identify three main categories:

- stationary rigid obstacles (e.g. porous media, microfluidic devices, flow over stationary cylinders)
- moving rigid obstacles (e.g. suspensions of non-deformable particles, oscillating cylinder, rotating turbine blades)

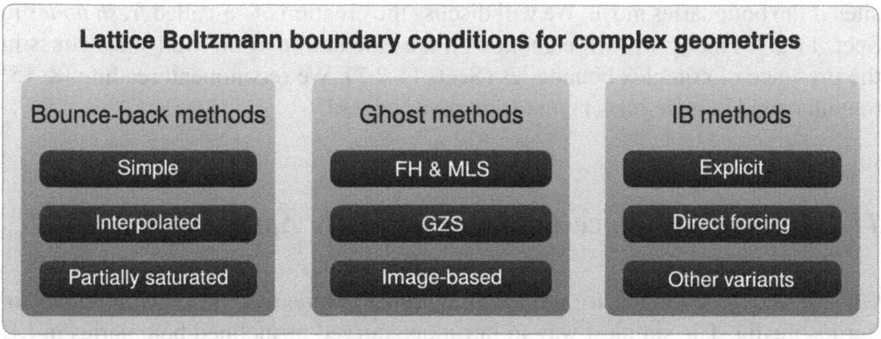

Fig. 11.2 Overview of boundary conditions for complex geometries in LB simulations as presented in this chapter. We can roughly distinguish between bounce-back, ghost and immersed-boundary (IB) methods. Each of them has several flavours. A large selection of those is covered in the following sections

- moving deformable obstacles (e.g. flexible wings, red blood cells, compliant channels)

No single numerical boundary treatment works best for all of them. It is therefore worth to properly categorise the problem first, identify the main challenge and then "shop around" and look for the most suitable boundary treatment for the problem at hand. This chapter helps the reader to understand what the differences of the available methods are, when they are applicable and what their advantages and disadvantages are.

There exists a zoo of curved boundary conditions for LB simulations. We can only cover the most popular ones in any depth, but we will provide references to a wider range of boundary conditions in passing. Figure 11.2 shows an overview of the boundary conditions discussed here. For the sake of compactness, we only consider single-phase fluids. Note that everything said in this chapter does equally apply to 2D and 3D systems.

11.2 Bounce-Back Methods

The most famous and certainly easiest boundary condition for LB simulations is bounce-back (Sect. 5.3.3). Many researchers believe that its locality, simplicity and efficiency should be retained even in the presence of complex boundary shapes. Therefore, the obvious way is to approximate a curved boundary by a staircase (Sect. 11.2.1). This can lead to some problems, in particular a reduction of the numerical accuracy. For that reason, improved and interpolated bounce-back schemes have been proposed (Sect. 11.2.2). Another variant to account for complex geometries is the partially saturated method (Sect. 11.2.3). A problem related to staircase and interpolated bounce-back BCs is the destruction and creation of fluid sites if the boundaries move. We will discuss the creation of so-called *fresh nodes* in Sect. 11.2.4. Finally we will elaborate on the calculation of the wall shear stress in the presence of complex boundaries (Sect. 11.2.5). We recommend reading [8–15] to understand bounce-back methods in greater detail.

11.2.1 Simple Bounce-Back and Staircase Approximation

One of the motivations to simulate complex geometries is to study flows in porous media. The simplest way to introduce curved or inclined boundaries in LB simulations is through a *staircase approximation* of the boundary and the bounce-back scheme, often called *simple bounce-back* (SBB, Sect. 5.3.3). This is illustrated in Fig. 11.3. The advantages are obvious: everything lives on the lattice, and SBB is fast and easy to implement. The problem becomes more complex when the boundaries can move, which requires the destruction and creation of fluid sites (Sect. 11.2.4).

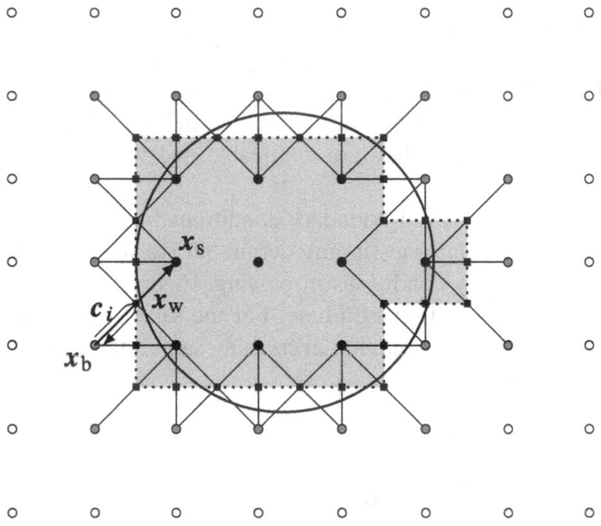

Fig. 11.3 Staircase approximation of a circle. A circle (here with an unrealistically small radius $r = 1.8\Delta x$) can be discretised on the lattice by identifying exterior fluid nodes (*white circles*), external boundary nodes (*grey circles*) and interior solid nodes (*black circles*) first. Any lattice link $\boldsymbol{c}_i$ connecting a boundary and a solid node is a cut link (*lines*) with a wall node (*solid squares*) in the middle. The resulting staircase shape is shown as a *grey-shaded area*. Populations moving along cut links $\boldsymbol{c}_i$ (defined as pointing inside the solid) from $\boldsymbol{x}_b$ to $\boldsymbol{x}_s$ are bounced back at $\boldsymbol{x}_w$

11.2.1.1 Revision of the Halfway Bounce-Back Method

In the following we will only consider the halfway bounce-back scheme: an incoming (post-collision) population $f_i^\star(\boldsymbol{x}_\mathrm{b}, t)$ which would propagate through a wall from a boundary node[1] $\boldsymbol{x}_\mathrm{b}$ to a solid node $\boldsymbol{x}_\mathrm{s} = \boldsymbol{x}_\mathrm{b} + \boldsymbol{c}_i \Delta t$ is instead reflected halfway to the solid node at the wall location $\boldsymbol{x}_\mathrm{w} = \boldsymbol{x}_\mathrm{b} + \frac{1}{2}\boldsymbol{c}_i \Delta t$ at time $t + \frac{1}{2}\Delta t$ and returns to $\boldsymbol{x}_\mathrm{b}$ as

$$f_{\bar{i}}(\boldsymbol{x}_\mathrm{b}, t + \Delta t) = f_i^\star(\boldsymbol{x}_\mathrm{b}, t) - 2w_i \rho \frac{\boldsymbol{u}_\mathrm{w} \cdot \boldsymbol{c}_i}{c_\mathrm{s}^2} \tag{11.1}$$

where $\boldsymbol{u}_\mathrm{w} = \boldsymbol{u}(\boldsymbol{x}_\mathrm{w}, t + \frac{1}{2}\Delta t)$ is the velocity of the wall, ρ is the fluid density at $\boldsymbol{x}_\mathrm{w}$ and $\bar{i}$ is defined by $\boldsymbol{c}_{\bar{i}} = -\boldsymbol{c}_i$. In practical implementations, ρ is often taken as the fluid density at $\boldsymbol{x}_\mathrm{b}$ or the average fluid density instead (cf. Sect. 5.3.3).

The halfway bounce-back scheme requires detection of all lattice links $\boldsymbol{c}_i$ intersecting the boundary. If the boundary is stationary, this has to be done only once.

We can compute the momentum exchange at the wall based on the incoming and bounced back populations alone by using the momentum exchange algorithm (MEA, Sect. 5.4.3). Here we will briefly revise the MEA for the simple bounce-back method. The first step is to evaluate the incoming and bounced back populations $f_i^\star$ and $f_{\bar{i}}$ at each boundary link. The total momentum exchange between the fluid and the solid during one time step is given by (5.79):

$$\begin{aligned} \Delta \boldsymbol{P} &= \Delta x^3 \sum_{\boldsymbol{x}_\mathrm{w}, i} \Big(f_i^\star(\boldsymbol{x}_\mathrm{w} - \tfrac{1}{2}\boldsymbol{c}_i \Delta t, t) + f_{\bar{i}}(\boldsymbol{x}_\mathrm{w} - \tfrac{1}{2}\boldsymbol{c}_i \Delta t, t + \Delta t) \Big) \boldsymbol{c}_i \\ &= \Delta x^3 \sum_{\boldsymbol{x}_\mathrm{w}, i} \left(2 f_i^\star(\boldsymbol{x}_\mathrm{w} - \tfrac{1}{2}\boldsymbol{c}_i \Delta t, t) - 2 w_i \rho \frac{\boldsymbol{u}_\mathrm{w} \cdot \boldsymbol{c}_i}{c_\mathrm{s}^2} \right) \boldsymbol{c}_i \end{aligned} \tag{11.2}$$

where the sum runs over all incoming links $\boldsymbol{c}_i$ (pointing from the fluid into the solid) intersecting the wall at $\boldsymbol{x}_\mathrm{w}$. Accordingly, the total angular momentum exchange during one time step is

$$\Delta \boldsymbol{L} = \Delta x^3 \sum_{\boldsymbol{x}_\mathrm{w}, i} \left(2 f_i^\star(\boldsymbol{x}_\mathrm{w} - \tfrac{1}{2}\boldsymbol{c}_i \Delta t, t) - 2 w_i \rho \frac{\boldsymbol{u}_\mathrm{w} \cdot \boldsymbol{c}_i}{c_\mathrm{s}^2} \right) (\boldsymbol{x}_\mathrm{w} - \boldsymbol{R}) \times \boldsymbol{c}_i. \tag{11.3}$$

$\boldsymbol{R}$ is a reference point. If the torque acting on a particle is computed, the reference point is the particle's centre of mass.

---

[1] We use the same notation as in Chap. 5: solid nodes are inside the obstacle, boundary nodes are outside the obstacle but have at least one solid neighbour, and fluid nodes are those without a solid neighbour (see Fig. 11.3).

The MEA works for any geometry approximated by the simple bounce-back scheme, including the staircase shown in Fig. 11.3. The tedious part is the identification of all cut links pointing from a boundary to a solid node and obtaining the wall velocity $\boldsymbol{u}_\mathrm{w}$ at each of the wall locations $\boldsymbol{x}_\mathrm{w}$.

11.2.1.2 Stationary Boundaries

There is a large number of applications where the flow in a complex stationary geometry has to be simulated. Examples are flows in porous media or blood flow in the vascular system (Fig. 11.4). Those geometries can by obtained by, for example, CT or MRI scans. Due to the large surface and complex shape of those geometries, it is preferable to use a simple and fast boundary condition algorithm, such as SBB.

Back in the 90s, Ginzburg and Adler [9] presented a very careful analysis of halfway SBB with several important conclusions. A more updated discussion of this work can be found in [15]. Apart from developing general theoretical tools to study boundary conditions, one contribution was the understanding of the numerical mechanism leading to the velocity slip at the wall. The exact location where the no-slip condition is satisfied is not a pre-determined feature; it rather depends on the specific choice of the relaxation rate(s). Furthermore, the above-mentioned defect is anisotropic with respect to the underlying lattice structure, i.e. the slip velocity depends on the way the boundary is inclined. This is confirmed for a Poiseuille flow in an horizontal channel where the wall can only be *exactly* located midway between lattice nodes if $\tau/\Delta t = 1/2 + \sqrt{3/16} \approx 0.933$ (cf. Sect. 5.3.3). Contrarily, for a diagonal channel, $\tau/\Delta t = 1/2 + \sqrt{3/8} \approx 1.11$ has to be chosen. With

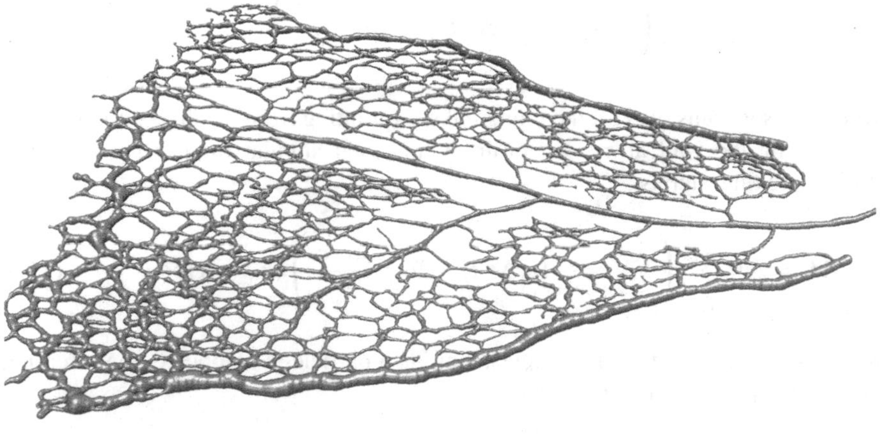

Fig. 11.4 Visualisation of a segment of the blood vessel network in a murine retina which has been used for LB simulations. This example shows the complexity of the involved boundaries. Original confocal microscope images courtesy of Claudio A. Franco and Holger Gerhardt. Luminal surface reconstruction courtesy of Miguel O. Bernabeu and Martin L. Jones. For more details see [16]

the LBGK method this τ-dependence leads to a viscosity-dependent slip velocity artefact, which is highly undesired from the physical viewpoint.

While in complex geometries the wall intersects the lattice at different positions, the SBB enforces the no-slip condition to be fixed at all lattice links. This leads to the staircase representation of the boundary, where the exact location of the no-slip surface will further depend on the choice of the relaxation time τ. In other words, the effective shape and location of the "numerical" wall will not agree with the expected boundary.

The aforementioned problems are particularly harmful in narrow domains where the distance between walls can be of the order of a few lattice units, for example in porous media flows or solid particles in suspensions. Hereby, using TRT/MRT collision operators with a properly tuned set of relaxation rates [12, 15, 17, 18], rather than BGK, helps controlling this situation. When τ is chosen close to 1 and the fluid domain is sufficiently resolved, SBB with BGK leads to acceptable results, though [19].

We also have to note that biological geometries obtained from CT or MRI scans are usually not very accurate in the first place. The resolution of those imaging techniques can be of the same order as the pore or channel size so that it may be nonsensical trying to increase the resolution of the numerical domain or choose more accurate boundary conditions. This means that SBB, although generally not the most recommendable solution, is still a good choice given the large geometrical modelling error in many applications. Furthermore, it is worth mentioning that SBB is exactly mass-conserving when used for stationary geometries of any shape; an advantage only a few higher-order accurate boundary conditions can claim (the reasons are explained in [20]). Therefore, before setting up a simulation, one should always ask whether the boundary condition is really the limiting factor in terms of accuracy.

Using **simple bounce-back (SBB)** for complex geometries generally leads to two sources of error:

1. geometrical discretisation error (modelling error) by approximating a complex shape by a staircase,
2. artificial and anisotropic slip caused by the choice of the relaxation rate(s), leading to a viscosity-dependent effect when is BGK used.

The advantages of SBB (mass conservation, ease of implementation, locality) explain why it is still a popular method.

11.2.1.3 Rigid Moving Particles

So far we have only addressed stationary boundaries with SBB. In the 90s researchers became interested in the simulation of suspensions *via* LBM. This requires the treatment of multiple rigid particles with translational and rotational degrees of freedom. One of the problems of earlier computational suspension models was the numerical cost which scaled with the square or cube of the particle number [10]. Ladd [10, 21, 22] introduced an LB-based model for suspensions of rigid particles with hydrodynamic interactions whose numerical cost scales linearly with the particle number.

Particle suspensions give rise to a plethora of physical effects and phenomena. In this section we will only focus on the algorithmic details. For physical results we refer to review articles about LB-based suspension simulations [23, 24] and the references therein.

For the sake of brevity we will not discuss lubrication forces which become necessary at high particle volume fractions. There exist several articles dealing with lubrication corrections in LB simulations [24–26]. The review by Ladd and Verberg [23], which we generally recommend to read, also descibes the use of thermal fluctuations for the simulation of Brownian motion in suspensions. Aidun and Clausen [24] have published a review about LBM for complex flows, which is an excellent starting point to learn about more recent developments.

In the following we will outline Ladd's [10, 22] idea of how to use SBB for suspensions. See also [25] for a compact and [23] for an extensive review of Ladd's method. Note that the particles in Ladd's algorithm are filled with fluid in order to avoid destruction and recreation of fluid nodes when the particles move on the lattice. The dynamics of the interior fluid is therefore fully captured. One can imagine this like a can filled with liquid concrete in an exterior fluid rather than the same can with set (and therefore solid) concrete. This is different compared to Aidun's model [27] which we will briefly describe at the end of this section.

The first step is to start with a distribution of suspended spherical particles on the lattice. For each particle, it is straightforward to work out which lattice nodes are located inside and outside of a particle (cf. Fig. 11.3). There is no conceptual difficulty in extending the model to non-spherical particles; but it will generally be more demanding to identify interior and exterior lattice nodes.

The next step is to identify all lattice links between boundary and solid nodes, i.e. those links cut by any particle surface. For moving boundaries, the list of those links has to be updated whenever the boundary configuration on the lattice changes. Generally one has to update the list every time step before propagation is performed. In the following, let $\boldsymbol{x}_\mathrm{b}$ be the location of a boundary node and $\boldsymbol{x}_\mathrm{s} = \boldsymbol{x}_\mathrm{b} + \boldsymbol{c}_i \Delta t$ the location of a solid node just inside the particle. The boundary link is then located at $\boldsymbol{x}_\mathrm{w} = \boldsymbol{x}_\mathrm{b} + \frac{1}{2}\boldsymbol{c}_i \Delta t$ (cf. Fig. 11.3).

Now we have to compute the wall velocity $\boldsymbol{u}_{\mathrm{w}}$ at each link $\boldsymbol{x}_{\mathrm{w}}$. From the known linear velocity $\boldsymbol{U}$ and angular velocity $\boldsymbol{\Omega}$ of the particle we can obtain

$$\boldsymbol{u}_{\mathrm{w}} = \boldsymbol{U} + \boldsymbol{\Omega} \times (\boldsymbol{x}_{\mathrm{w}} - \boldsymbol{R}) \tag{11.4}$$

where $\boldsymbol{R}$ is the particle's centre of mass.

With the known wall velocity at each link, we can compute the momentum exchange and therefore the value of all bounced-back populations. We have to take into account that a particle in Ladd's method is filled with fluid, as explained earlier, and all interior nodes participate in collision and propagation as well. This means that there are also populations streaming from the interior nodes at $\boldsymbol{x}_{\mathrm{s}}$ to exterior nodes at $\boldsymbol{x}_{\mathrm{b}}$. These populations have to be bounced back at $\boldsymbol{x}_{\mathrm{w}}$, too. While the populations streaming from the outside to the particle's interior are described by (11.1), we now also have to consider those populations streaming from the inside towards the exterior:

$$f_i(\boldsymbol{x}_{\mathrm{s}}, t + \Delta t) = f_{\bar{i}}^{\star}(\boldsymbol{x}_{\mathrm{s}}, t) - 2w_{\bar{i}}\rho\frac{\boldsymbol{u}_{\mathrm{w}} \cdot \boldsymbol{c}_{\bar{i}}}{c_{\mathrm{s}}^2} = f_{\bar{i}}^{\star}(\boldsymbol{x}_{\mathrm{s}}, t) + 2w_i\rho\frac{\boldsymbol{u}_{\mathrm{w}} \cdot \boldsymbol{c}_i}{c_{\mathrm{s}}^2}. \tag{11.5}$$

Equation (11.1) and (11.5) express that the two populations hitting a boundary link from both sides exchange a certain amount of momentum, $2w_i\rho\boldsymbol{u}_{\mathrm{w}} \cdot \boldsymbol{c}_i/c_{\mathrm{s}}^2$. This operation is obviously mass-conserving since f_i gains exactly the loss of $f_{\bar{i}}$ (or the other way around) so that the sum of both populations moving along the same link in different directions is not changed by the interaction with the boundary, at least as long the same density ρ is used in both equations. Ladd uses the average fluid density, and not the local density, for ρ.

Effectively, we can view the momentum transferred from the exterior to the interior fluid as the momentum transferred from the exterior fluid to the particle. In order to obey the global momentum and angular momentum conservation, we therefore have to update the particle momentum and angular momentum by summing up all transferred contributions. Equation (11.2) and (11.3) provide the total momentum $\Delta\boldsymbol{P}$ and angular momentum $\Delta\boldsymbol{L}$ transferred during one time step, but we have to take into account that each link has to be counted twice: once for all populations coming from the outside and once for all populations coming from the inside. This is necessary because the interior fluid participates in collision and propagation and therefore the momentum exchange.

The simplest way to update the particle properties is the forward Euler method, but more accurate and more stable methods are available, e.g. implicit time integration [23]. At each time step, the velocity and angular velocity are updated according to

$$\boldsymbol{U}(t + \Delta t) = \boldsymbol{U}(t) + \frac{\Delta\boldsymbol{P}}{M}, \quad \boldsymbol{\Omega}(t + \Delta t) = \boldsymbol{\Omega}(t) + \boldsymbol{I}^{-1} \cdot \Delta\boldsymbol{L} \tag{11.6}$$

where M and $\boldsymbol{I}$ are the particle's mass and tensor of inertia. The centre of mass is then moved according to the old or new velocity.[2] If the particles are spherical, their orientation does not have to be updated. For non-spherical particles, however, the situation is different, and several authors have suggested algorithms for this case. Aidun et al. [27], for example, use a fourth-order Runge-Kutta integration to update the particle orientation. Qi [28] employed quaternions to capture the particle orientation and a leap-frog time integration. It is noteworthy that Ladd [10] does not follow the simple scheme in (11.6). He instead averages the momentum and angular momentum transfer over two time steps before updating the particle properties. The reason for this is to reduce the undesired effect of so-called *staggered momenta* which are an artefact of lattice-based methods. We refer to [10] for a more thorough discussion of this issue (see also Sect. 5.3.3).

Ladd's algorithm [10] can be summarised in the following way:

1. Find the particle discretisation on the lattice (Fig. 11.3).
2. Identify all boundary links and compute $\boldsymbol{u}_{\mathrm{w}}$ by applying (11.4).
3. Perform collision on *all* nodes since particles are filled with fluid.
4. Propagate the populations. If a population moves along a boundary link, bounce-back this population *via* (11.1) or (11.5).
5. Compute the total momentum and angular momentum exchange according to (11.2) and (11.3).
6. Update the particle configuration, for example *via* (11.6).
7. Go back to step 1 for the next time step. There is no need to treat nodes crossing a boundary in a special way.

It is interesting to note why Ladd has chosen a link-based (halfway) rather than a node-based (fullway) bounce-back method. The simple explanation is that the link-based bounce-back leads to a "somewhat higher resolution" [10] for the same discretisation since there are more cut links than solid nodes near the particle surface. This can be easily seen in Fig. 11.3.

Ladd [10] pointed out that his method has a few disadvantages. First, the dynamics of the fluid inside the particles can affect the particle dynamics at higher Reynolds numbers where the interior fluid cannot any more be approximated by an effectively rigid medium. Furthermore, Ladd's method is limited to situations where the particle density is larger than the fluid density. Aidun et al. [27] proposed an alternative method with one major distinctive feature: the absence of fluid inside the particles. Therefore, in Aidun's approach, only the exterior fluid contributes to the momentum and angular momentum exchange in (11.2) and (11.3). Removing the fluid from the interior solves both disadvantages of Ladd's method, but it also introduces a new complexity: what happens when lattice nodes change their identity (fluid nodes become solid nodes and the other way around) when the particles move? We will come back to this point in Sect. 11.2.4. In contrast to Ladd's approach,

[2] The velocities are usually small so that the exact form of the position update is not very important.

Aidun's method does not obey global mass conservation [27]. Yin et al. [29] provide a detailed comparative study on the performance of both models.

We emphasise that several researchers have further improved the methods presented above. For example, Lorenz et al. [30], Clausen and Aidun [31] and Wen et al. [32] proposed modified versions of the momentum exchange to improve Galilean invariance.

We have only discussed *link-based* BB schemes in this section. It is possible to implement node-based BB schemes for complex geometries where the boundary velocity is enforced directly on lattice nodes, though. Behrend [33] and Gallivan et al. [34] provide discussions of the node-based BB approach. In Sect. 11.2.3 we will present partially saturated methods which are also built on the node-based BB scheme.

As pointed out by Han and Cundall [35], the **simple bounce-back (SBB) applied to moving boundaries** has its limitations compared to higher-order schemes, such as the partially saturated method (Sect. 11.2.3). This becomes most obvious when the particles are rather small (a few Δx in diameter) and move on the lattice. Eventually, the user has to decide whether the focus lies on the ease of implementation or level of accuracy. In the former case, SBB can be recommended. In the latter, a smoother boundary condition should be implemented.

11.2.2 Interpolated Bounce-Back

We will now cover interpolated bounce-back (IBB) methods which are suitable to describe curved and inclined boundaries with a higher accuracy than SBB. We emphasise the conceptual difference between IBB schemes and extrapolation-based methods (Sect. 11.3). While the idea of the former is to interpolate populations in the fluid region to perform bounce-back at a curved wall, the motivation for the latter is to create a virtual (ghost) fluid node inside the solid to compute the populations streaming out of the wall. We generally recommend reading [12, 17, 18, 36] for thorough reviews of IBB methods.

11.2.2.1 Basic Algorithm

In 2001, Bouzidi et al. [11] proposed the IBB approach for curved boundaries. The IBB is *second-order* accurate for arbitrary boundary shapes and therefore reduces the modelling error of the staircase bounce-back method which is only first-order accurate for non-planar boundaries.

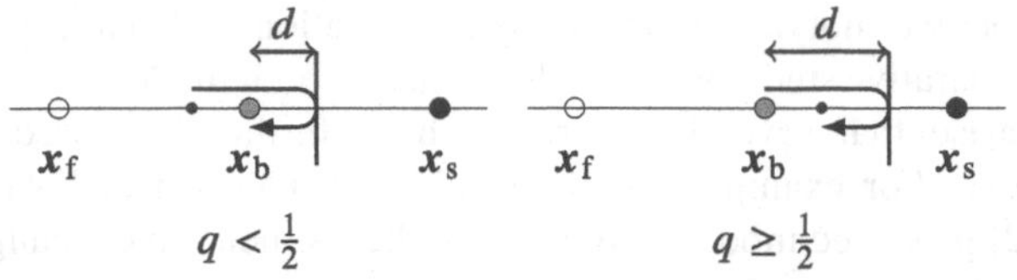

Fig. 11.5 Interpolated bounce-back cases. For $q < \frac{1}{2}$ (*left*), the distance d between the wall (*vertical line*) and the boundary node at x_b is smaller than half a lattice spacing. Interpolations are required to construct the post-collision population (*small black circle*). For $q \geq \frac{1}{2}$ (*right*), d is larger than half a lattice spacing and the endpoint of the streaming population (*small black circle*) lies between the wall and x_b. x_s denotes a solid node behind the wall, and x_f is a fluid node required for the interpolation

The basic idea of IBB is to include additional information about the wall location during the bounce-back process. A boundary link c_i generally intersects the wall at a distance d between 0 and $|c_i|\Delta t$ measured from the boundary node (Fig. 11.5). We define $q = d/(|c_i|\Delta t) \in [0, 1)$ as the reduced wall location and note that $q = \frac{1}{2}$ holds for simple bounce-back. In the following, we introduce three lattice nodes with locations x_b, $x_s = x_b + c_i\Delta t$ and $x_f = x_b - c_i\Delta t$ as shown in Fig. 11.5. x_b and x_s are neighbouring boundary and solid nodes which are located on either side of the wall, and x_f is the nearest fluid node beyond x_b.

The starting point of IBB is to assume that any population f_i moves a distance $|c_i|\Delta t$ during propagation. If the population hits a wall which is modelled by the halfway bounce-back, f_i first travels a distance $|c_i|\Delta t/2$ from the original boundary node to the wall and then another distance $|c_i|\Delta t/2$ back to the boundary node after bounce-back. If the wall is not located halfway between lattice nodes, $q \neq \frac{1}{2}$, f_i cannot reach another lattice node. Therefore, the origin of the population is chosen such that f_i exactly reaches a lattice node. This requires interpolation to find the post-collision value of f_i as illustrated in Fig. 11.5.

Bouzidi et al. [11] proposed a linear interpolation to construct the *a priori* unknown bounced back population $f_{\bar{i}}(x_b, t + \Delta t)$ from the post-collision values of the known populations at x_b and x_f. The algorithm for any cut link at a resting wall reads

$$f_{\bar{i}}(x_b, t + \Delta t) = \begin{cases} 2q f_i^\star(x_b, t) + (1 - 2q) f_i^\star(x_f, t) & q \leq \frac{1}{2} \\ \frac{1}{2q} f_i^\star(x_b, t) + \frac{2q-1}{2q} f_{\bar{i}}^\star(x_b, t) & q \geq \frac{1}{2} \end{cases}. \quad (11.7)$$

Exercise 11.1 Show that both cases in (11.7) reduce to simple bounce-back for $q = \frac{1}{2}$.

There are several remarks:

- The reason for having different expressions for $q < \frac{1}{2}$ and $q \geq \frac{1}{2}$ is to ensure that $f_{\bar{i}}(x_b, t + \Delta t)$ is always non-negative (given that the post-collision populations on the right-hand-side of (11.7) are positive), which improves the stability of the algorithm.

- Bouzidi et al. [11] have also proposed a quadratic interpolation which improves results, but the method remains *second-order* accurate. This extension requires a second fluid node $\boldsymbol{x}_{\mathrm{ff}} = \boldsymbol{x}_{\mathrm{b}} - 2\boldsymbol{c}_i \Delta t$.
- The boundary slip for linear and quadratic interpolations still depends on the collision relaxation rate(s). However, unlike with the SBB, the adoption of TRT/MRT collision operators does not allow absolute control of this error; the IBB thereby always presents some degree of viscosity dependence. For strategies to render IBB *exactly* viscosity-independent we refer to [15, 37, 38].
- The IBB algorithm, like simple bounce-back, is completely decoupled from the collision step and can therefore be combined with any collision operator.
- Due to its non-local implementation, the IBB may lead to problems in very narrow gaps (e.g. in porous media simulations) where there are not enough fluid nodes between neighbouring walls to apply (11.7) (two nodes required) or the quadratic interpolation (three nodes required). The number of required nodes may be reduced by one with a judicious choice of pre- and post-collision populations within the IBB algorithm, cf. [15, 17]. For example, the IBB with linear interpolations can be written in local form. Chun and Ladd [18] have addressed this issue and proposed an alternative local boundary scheme which works in general situations and corrects some defects of the IBB, e.g. its viscosity dependence.
- Due to its interpolations, IBB is generally not mass-conserving. There exist approaches to (partially) remedy this shortcoming, e.g. [36].

The interpolated bounce-back algorithm can be summarised as follows:

1. Identify all links penetrating a wall and compute their reduced distance q. If the boundary configuration does not change in time, this has to be done only once.
2. Collide on all fluid and boundary nodes. This will provide $f_i^\star(\boldsymbol{x}_{\mathrm{b}}, t)$, $f_{\bar{i}}^\star(\boldsymbol{x}_{\mathrm{b}}, t)$ and $f_i^\star(\boldsymbol{x}_{\mathrm{f}}, t)$.
3. Compute all $f_{\bar{i}}(\boldsymbol{x}_{\mathrm{b}}, t + \Delta t)$ from (11.7).
4. Propagate all remaining populations.
5. Go back to step 1.

The extension of the momentum exchange algorithm to the IBB is straightforward. According to Bouzidi et al. [11], the momentum exchange for a resting boundary link in Fig. 11.5 is

$$\Delta \boldsymbol{p}_i = \Delta x^3 \left(f_i^\star(\boldsymbol{x}_{\mathrm{b}}, t) - f_{\bar{i}}(\boldsymbol{x}_{\mathrm{b}}, t + \Delta t) \right) \boldsymbol{c}_i, \tag{11.8}$$

no matter the value q of the link. An alternative to (11.5), which improves its accuracy, can be found in [17].

11.2.2.2 Moving Boundaries

Most IBB applications in the literature deal with rigid boundaries (which can either be stationary or move on the lattice). There is only a small number of works featuring the IBB for deformable boundaries, e.g. [39]. The immersed boundary method (Sect. 11.4) is a more common approach in those situations. We will therefore not discuss deformable boundaries here.

Lallemand and Luo [40] extended the IBB to moving boundaries. The first ingredient is Ladd's algorithm: the term $-2w_i\rho\boldsymbol{u}_{\mathrm{w}}\cdot\boldsymbol{c}_i/c_s^2$ has to be added to the right-hand-side of (11.7), where $\boldsymbol{u}_{\mathrm{w}}$ is the wall velocity at the intersection point. Also, since there is generally no fluid inside the solid in the IBB framework, a refill mechanism has to be implemented when boundaries move and uncover new (fresh) fluid nodes. We will get back to this in Sect. 11.2.4. Lallemand and Luo's important finding is that the motion of a cylinder on the lattice and the subsequent destruction and creation of fresh nodes leads to some fluctuations of the drag coefficient. This is one of the largest disadvantages of the IBB for moving boundaries; a problem which can be reduced by using an advanced fresh node treatment (Sect. 11.2.4) or the immersed boundary method (Sect. 11.4).

11.2.2.3 Extended and Alternative Methods

Several other second-order accurate bounce-back-based boundary conditions for arbitrary geometries have been proposed. The following list shows a selection of those methods and their most notable properties.

- Yu et al. [41] presented a unified scheme of Bouzidi's algorithm which does not require separate treatment of the regions $q < \frac{1}{2}$ and $q \geq \frac{1}{2}$. Otherwise Yu's and Bouzidi's approaches lead to similar results, including viscosity-dependent slip (not easily solved by TRT/MRT collision models) and violation of mass conservation.
- Ginzburg and d'Humières [17] proposed the so-called *multireflection* boundary condition as an enhanced IBB. The key feature of this method is that it determines the coefficients of the interpolation, rather than heuristically, using the second-order Chapman-Enksog expansion on the interpolated populations. This way, it guarantees the closure condition reproduced by the mesoscopic populations are in exact agreement with the intended hydrodynamic condition. For the reasons explained in Chap. 5, the multireflection method is generally constructed to be formally third-order accurate, and that can be achieved in different ways: the original multireflection scheme [17] adopts a (non-local) interpolation process over five populations, while the more recent MLI scheme [15, 37, 38] only operates over three populations (belonging to the same node) and supplements this information with a (non-local) link-wise second-order finite difference approximation of the hydrodynamic quantity of interest. Both these variants can be implemented in two nodes only [15, 37, 38] and, also in both cases,

the algorithm shall consider a post-collision correction term, which guarantees the method's higher-order accuracy and consistency (i.e. viscosity independence with TRT/MRT collision models). A drawback of the multireflection technique, common to the generality of interpolation-based schemes, is the possible violation of mass conservation.

- Chun and Ladd [18] proposed a method based on the interpolation of the equilibrium distribution. It has the advantage that it requires only one node—in contrast to two or three nodes in the standard linear and quadratic IBB scheme. Similar to the SBB and multireflection schemes, this method guarantees the exact wall location is viscosity-independent with TRT/MRT collision models. This approach is particularly suitable for the time-dependent simulation of geometries with narrow gaps between solids, e.g. for porous media. Mass conservation is generally violated, just as in the majority of interpolation schemes.
- Kao and Yang [36] suggested an interpolation-free method based on the idea of local grid refinement in order to improve the mass conservation of the boundary condition.
- Yin and Zhang [42] presented another improved bounce-back scheme. Their idea was to use Ladd's momentum correction term and linearly interpolate the fluid velocity between a nearby boundary node and the wall location (which can be anywhere between two lattice nodes) to obtain the fluid velocity midway between boundary and solid nodes. This promising method shares common disadvantages with other interpolated bounce-back schemes: violation of mass conservation and viscosity-dependent wall location.

The **interpolated bounce-back (IBB) method** is a common extension of the simple bounce-back scheme for rigid resting or moving obstacles with complex shapes. IBB is second-order accurate but introduces an important weakness: the viscosity-dependent boundary slip is not easily corrected with TRT/MRT collision operators. Furthermore, due to the involved interpolations, IBB is not mass-conserving. Even so, due to its intuitive working principle and relatively simple implementation, the IBB is often the method of choice for improving the SBB accuracy in describing stationary complex geometries.

11.2.3 Partially Saturated Bounce-Back

Now we present the so-called *partially saturated* method (PSM), also known as *grey* LB model or *continuous bounce-back*, where a lattice node can be a pure fluid, a pure solid or a mixed (partially saturated) node as shown in Fig. 11.6. Interestingly,

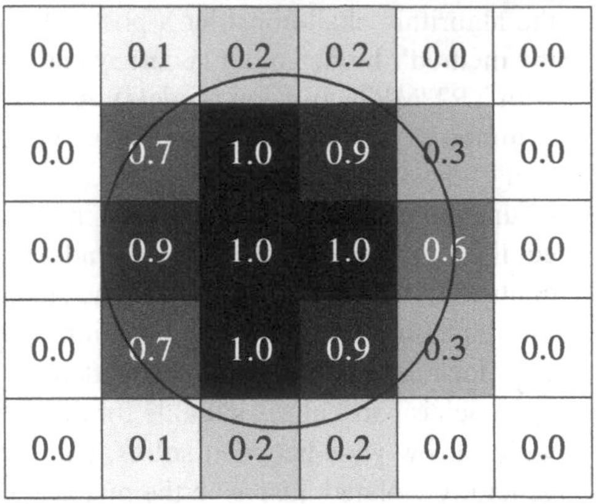

Fig. 11.6 Partially saturated bounce-back. A spherical particle (*circle*) covers a certain amount of each lattice cell. White corresponds to no coverage, black to full coverage. The solid fraction $0 \leq \epsilon \leq 1$ for each cell is shown up to the first digit. Lattice nodes (not shown here) are located at the centre of lattice cells

there exist two research communities which do not seem to interact strongly. The first applies the PSM to simulations of flows in porous media with heterogeneous permeability [43–45]. The other community is interested in suspension flows; they employ the PSM to map the sharp surface of an immersed structure onto the lattice [13, 35, 46, 47]. For the sake of brevity and since both approaches are technically similar, we only elaborate on the latter application. We emphasise that the PSM must not be confused with immersed boundary schemes (Sect. 11.4) which are, according to our definition, fundamentally different in nature. As demonstrated in a series of studies, e.g. [44, 45, 48, 49], the way PSM works can be considered equivalent to the standard LBE with an added friction force. The magnitude of this force varies locally, depending on the nodal fluid/solid fraction. This results in a continuous accommodation of the solution from open (fluid) to very impermeable (solid) regions. Hence, in PSM the nature of the wall can be understood as an interface condition, separating nodes of contrasting properties [48–50].

11.2.3.1 Basic Algorithm

In 1998, Noble and Torczynski [46] presented a bounce-back-based approach, later investigated more thoroughly by Strack and Cook [13], to approximate complex boundaries *on lattice nodes*. The central part of the PSM algorithm is a modified LBGK equation:

$$f_i(\boldsymbol{x} + \boldsymbol{c}_i\Delta t, t + \Delta t) = f_i(\boldsymbol{x}, t) + (1 - B)\Omega_i^{\mathrm{f}} + B\Omega_i^{\mathrm{s}} \tag{11.9}$$

where

$$\Omega_i^{\mathrm{f}} = -\frac{f_i(\boldsymbol{x}, t) - f_i^{\mathrm{eq}}(\boldsymbol{x}, t)}{\tau} \tag{11.10}$$

is the standard BGK collision operator for fluid (f) nodes and

$$\Omega_i^{\mathrm{s}} = \left(f_i(\boldsymbol{x},t) - f_i^{\mathrm{eq}}(\rho,\boldsymbol{u})\right) - \left(f_i(\boldsymbol{x},t) - f_i^{\mathrm{eq}}(\rho,\boldsymbol{u}_{\mathrm{s}})\right) \tag{11.11}$$

is the collision operator for solid (s) nodes. $\boldsymbol{u}$ is the local fluid velocity and $\boldsymbol{u}_{\mathrm{s}}$ is the velocity of the boundary at point $\boldsymbol{x}$. B is a weighting parameter defined by [46]

$$B(\epsilon,\tau) = \frac{\epsilon\left(\tau - \frac{1}{2}\right)}{(1-\epsilon) + \left(\tau - \frac{1}{2}\right)} \tag{11.12}$$

where $0 \leq \epsilon \leq 1$ is the solid fraction of the node. It can be shown that $B(\epsilon)$ increases monotonically between 0 for $\epsilon = 0$ (pure fluid node) and 1 for $\epsilon = 1$ (pure solid node) for any fixed value of $\tau > \frac{1}{2}$. The essential idea is to surrender any shape details of the off-lattice boundary and use an on-lattice volume fraction ϵ instead.

Exercise 11.2 Show that the collision operator in (11.9) is mass-conserving by computing $\sum_i \Omega_i^{\mathrm{f}}$ and $\sum_i \Omega_i^{\mathrm{s}}$.

Note that the PSM assumes the standard BGK form for $B = 0$ and describes a bounce-back of the non-equilibrium for $B = 1$. A mixed collision and bounce-back scheme is performed for partially saturated nodes ($0 < \epsilon < 1$), which are only found in direct boundary neighbourhood (Fig. 11.6).

Force and torque acting on the boundary can be computed from

$$\begin{aligned} \boldsymbol{f} &= \frac{\Delta x^3}{\Delta t} \sum_{\boldsymbol{x}_n} B(\boldsymbol{x}_n) \sum_i \Omega_i^{\mathrm{s}}(\boldsymbol{x}_n)\boldsymbol{c}_i, \\ \boldsymbol{T} &= \frac{\Delta x^3}{\Delta t} \sum_{\boldsymbol{x}_n} B(\boldsymbol{x}_n)(\boldsymbol{x}_n - \boldsymbol{R}) \times \sum_i \Omega_i^{\mathrm{s}}(\boldsymbol{x}_n)\boldsymbol{c}_i, \end{aligned} \tag{11.13}$$

respectively, where the $\boldsymbol{x}_n$ are all lattice nodes in contact with the solid (including all interior nodes), i.e. those nodes with $\epsilon > 0$, i runs over all lattice directions at a given position $\boldsymbol{x}_n$ and $\boldsymbol{R}$ is the location of the centre of mass of the solid. Updating the solid's momentum and angular momentum according to (11.13) guarantees overall momentum conservation.

It is worth mentioning that Zhou et al. [51] have combined the node-based method with Lees-Edwards BCs, which is relevant for the simulation of large bulk systems. Furthermore, Chen et al. [47] have recently proposed a combination of the PSM and a ghost method (Sect. 11.3) to improve the no-slip condition at the boundary surface. Yu et al. [52] proposed another variant taking into account a mass-conserving population migration process in the vicinity of moving walls.

11.2.3.2 Advantages and Limitations

The implementation of this algorithm is relatively straightforward. If the boundary is stationary, as for example encountered for a porous medium, ϵ can be computed once at each lattice site. For moving boundaries, ϵ has to be updated, which is the most challenging aspect of the PSM. Also, the correspondence between ϵ and the actual boundary shape is not trivial and requires some calibration [44, 45, 49]. For example, Han and Cundall [35] use a sub-cell method to estimate ϵ for a given lattice site while Chen et al. [47] employ a cut-cell approach. Apart from updating ϵ, no additional measures have to be taken when objects are moving on the lattice. In particular, fresh nodes appearing on the rear of a moving obstacle do not have to be treated in a special way. Neither are fluid nodes destroyed when they are covered by the advancing boundary. Since the interior fluid is never destroyed and still participates in the collisions described by Ω_i^{s}, mass and momentum are conserved.

In reality, curved boundaries in the PSM are nothing more than a sophisticated staircase (cf. Fig. 11.6). In the PSM, there is no information about the distance between lattice nodes and boundaries; instead, the local fluid filling fraction is considered. It is easy to imagine that many different boundary configurations can lead to the same filling fraction. Therefore, the PSM fails to capture the correct shape of the boundary.

Strack and Cook [13] performed careful 3D benchmark tests of the PSM. The authors report a significantly smoother motion when the weight $B(\epsilon, \tau)$ is used, rather than just falling back to a staircase approximation of the boundary. This is mostly due to the smooth uncovering of fluid nodes which have previously been solid nodes and *vice versa*. However, the smoothness of the observables (velocity, force and torque) depends on the accurate computation of the solid ratio ϵ.

Later, Han and Cundall [35] investigated the resolution sensitivity of the PSM and Ladd's BB scheme (Sect. 11.2.1) in 2D. They found that both methods are comparably accurate in terms of the drag coefficient of relatively large circles (diameter $\approx 10\Delta x$). However, for diameters as small as $4 - 5\Delta x$, the PSM is superior, in particular when the objects are moving on the lattice.

The PSM has a number of advantages. The first is that one does not face the fresh node problem (Sect. 11.2.4) which causes some trouble in most of the other BB variants. Moreover, the PSM, unlike IBB, is exactly mass conserving. Another advantage is the absence of interpolations to enforce the boundary condition. This makes the PSM a promising candidate for dense suspensions and porous media with small pore sizes.

However, when used for suspension flow, the PSM has so far mostly been applied to very simple geometries like circles in 2D or spheres in 3D. Although it is possible to construct more complex geometries by assembling several circles or spheres [35], additional work is necessary to make the PSM more attractive for moving, arbitrarily shaped boundaries. (Chen et al. [47] provide a short discussion of algorithms which can be used for more complex shapes.) Also, by sacrificing the treatment of the exact boundary shape, one cannot expect that the no-slip condition is accurately satisfied at the boundary [48, 49]. More investigations of the accuracy of the PSM for simple and complex boundary shapes would certainly be beneficial. Furthermore, the PSM in its present form is not suitable for the simulation of deformable boundaries or thin shells with fluid on both sides (unlike interpolated bounce-back, for example).

The **partially saturated method (PSM)** is a node-based method. PSM is exactly mass-conserving and does not require the treatment of fresh fluid nodes. The disadvantage is the difficulty of finding correct values for the solid fraction near solid boundaries, which is effectively limiting this method to stationary geometries (where the solid fraction has to be computed only once) or to spherical particles.

11.2.4 Destruction and Creation of Fluid Nodes

When boundaries move, it happens from time to time that lattice nodes cross the boundary, either from the fluid to the inside of the boundary or *vice versa*. If the interior of the boundary is not filled with a fluid, the former event requires the *destruction*, the latter the *creation* of a fluid site as shown in Fig. 11.7. Newly created nodes are also called *fresh nodes*. This applies to most methods described in Sect. 11.2 and also Sect. 11.3, but not to Ladd's method (Sect. 11.2.1) or the partially saturated method (Sect. 11.2.3) where nodes are neither created nor destroyed. Generally the number of fluid and solid nodes is not conserved when boundaries move on the lattice.

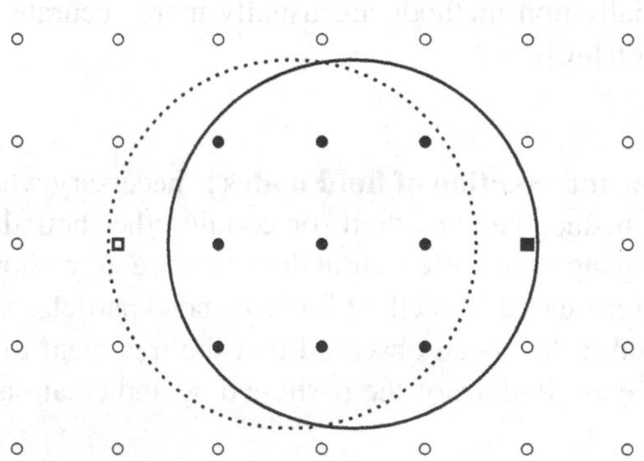

Fig. 11.7 Creation and destruction of fluid nodes. A particle is moving from its previous position (*dotted circle*) to its new position (*solid circle*). As a consequence, one fresh fluid node (*open square*) appears behind the particle and a fluid node is destroyed (*solid square*) at the front of the particle. Fluid and solid nodes are shown as *open and solid symbols*, respectively

Destruction is straightforward [27]: the state of the site is switched from "fluid" to "solid", and its momentum and angular momentum are transferred to the solid. For a destroyed fluid node at $\boldsymbol{x}_\mathrm{d}$ with density ρ and velocity $\boldsymbol{u}$, the particles receives a momentum contribution $\rho\boldsymbol{u}\Delta x^3$ and an angular momentum contribution $(\boldsymbol{x}_\mathrm{d}-\boldsymbol{R})\times(\rho\boldsymbol{u})\Delta x^3$ where $\boldsymbol{R}$ is the particle's centre of mass. Finally, the fluid information of the site at $\boldsymbol{x}_\mathrm{d}$ is omitted.

The inverse process, creation of a fluid site, is more difficult because the density, velocity and even all populations are unknown at first. The simplest approach to initialise a fresh node at point $\boldsymbol{x}_\mathrm{f}$ and time t is to estimate the density as the average of the neighbouring fluid sites [27],

$$\rho_\mathrm{f} = \rho(\boldsymbol{x}_\mathrm{f}, t) = \frac{1}{N_\mathrm{f}} \sum_i \rho(\boldsymbol{x}_\mathrm{f} + \boldsymbol{c}_i \Delta t, t), \tag{11.14}$$

where the sum runs only over those N_f neighbouring sites which are fluid. The velocity $\boldsymbol{u}_\mathrm{f}$ of the fresh node is computed from the known boundary velocity of the obstacle at the same point *via* (11.4). The populations f_i are then initialised with their equilibrium values $f_i^\mathrm{eq}(\rho_\mathrm{f}, \boldsymbol{u}_\mathrm{f})$. As additional step, the momentum and angular momentum of the fresh node have to be subtracted from the solid.

Although this approach is easy to implement, two disadvantages are obvious: the total mass in the system is generally not conserved, and the non-equilibrium contribution of the fresh node is missing, which can lead to distortions of the flow field.

Chen et al. [14] compared the above-mentioned approach and three other algorithms to initialise fresh nodes. One of those relies on extrapolation of populations from neighbouring fluid sites [40]. The other two approaches, both first described in [14], are based on the consistent initialisation [53] (see also Sect. 5.5). From benchmark tests involving moving cylinders, the authors come to the conclusion that the consistent initialisation methods are usually more accurate than interpolation [27] or extrapolation [40].

The **destruction and creation of fluid nodes** is necessary when the standard or interpolated bounce-back method (or certain other boundary conditions) are used for moving boundaries. Boundary treatments without the need for this consideration are Ladd's method for suspended particles and the partially saturated method. It has been observed that the treatment of fresh nodes is crucial to reduce oscillations of the particle drag and creation of detrimental sound waves.

11.2.5 Wall Shear Stress

We conclude this section with a more general discussion of the wall shear stress that is useful for most LB boundary conditions.

Several diseases of the circulatory system are assumed to be linked to pathological levels or changes of the shear stress at the arterial wall (see, e.g., [54] and references therein). Two prominent examples are atherosclerosis or aneursym formation. In recent years, a growing number of scientists became interested in the LB modelling of blood flow in realistic blood vessel geometries. Apart from finding and implementing reasonable boundary conditions, a key question is how the wall shear stress (WSS) can be computed and how accurate the obtained values are.

We will provide a brief review of the comparatively small number of publications in this field, but before that a few words about the WSS are necessary. The WSS is tightly connected to the momentum exchange at the boundary. Evaluating (11.2) is not sufficient to find the WSS, though. The reason is that WSS is a local quantity and not an integrated property of the entire surface of the boundary.

In order to find the WSS, the boundary location *and* orientation have to be known at each point of interest. Assuming that $\boldsymbol{x}_\mathrm{w}$ is a point on the wall, we denote $\hat{\boldsymbol{n}}$ the wall normal vector at $\boldsymbol{x}_\mathrm{w}$ pointing inside the fluid domain. For a given fluid stress tensor $\boldsymbol{\sigma}$ at $\boldsymbol{x}_\mathrm{w}$, we first define the *traction* vector as $\boldsymbol{T} = \boldsymbol{\sigma} \cdot \hat{\boldsymbol{n}}$. It is the force acting on an infinitesimal, oriented wall area element $\mathrm{d}\boldsymbol{A} = \mathrm{d}A\,\hat{\boldsymbol{n}}$. The WSS vector $\boldsymbol{\tau}$ is the tangential component of the traction vector, i.e. we have to subtract the normal component of the traction:

$$\boldsymbol{\tau} = \boldsymbol{T} - (\boldsymbol{T} \cdot \hat{\boldsymbol{n}})\hat{\boldsymbol{n}}, \quad \tau_\alpha = \sigma_{\alpha\beta}\hat{n}_\beta - (\sigma_{\beta\gamma}\hat{n}_\beta\hat{n}_\gamma)\hat{n}_\alpha. \tag{11.15}$$

The subtracted normal component $(\boldsymbol{T} \cdot \hat{\boldsymbol{n}})\hat{\boldsymbol{n}}$ contributes to the wall pressure. It is common to report only the magnitude of the WSS vector, simply called the WSS $\tau = |\boldsymbol{\tau}|$ (not to be confused with the BGK relaxation time).

We distinguish three typical situations:

1. The normal vector $\hat{\boldsymbol{n}}$ is known everywhere on the boundary, but the geometry is approximated by a staircase boundary (simple bounce-back).
2. The geometry is only known as a staircase, and no additional information about the normal vector $\hat{\boldsymbol{n}}$ is available.
3. A higher-order boundary conditions is used for the boundaries (e.g. IBB). We will not consider this case here.

An example for the first case is the discretisation of a known boundary geometry where a surface tessellation is converted to a staircase surface. Here we still have access to the original normal vectors, but the LB simulation is only aware of the staircase. The second case is important when voxel data is directly converted into a staircase geometry, without *a priori* knowledge of the boundary normals. This means that we first need to estimate $\hat{\boldsymbol{n}}$ from the known data before we can compute the WSS.

Stahl et al. [55] were the first authors to provide a careful investigation of the behaviour of the WSS in LB simulations with staircase geometries. They presented a scheme to obtain the unknown normal vectors $\hat{\boldsymbol{n}}$ from the flow field. This is straightforward in 2D where the no-penetration condition requires $\boldsymbol{u} \cdot \hat{\boldsymbol{n}} = 0$ for the fluid velocity $\boldsymbol{u}$ at the wall. From the known velocity $\boldsymbol{u}$ we can easily compute the unknown $\hat{\boldsymbol{n}}$ up to its sign. This is more complicated but possible in 3D, where additional information is required (see [55] for details). The authors then investigated the accuracy of the stress computation near the boundary in an inclined Poiseuille flow. They found strongly anisotropic behaviour: the error is minimum for walls aligned with one of the major lattice directions (i.e. when the wall is flat), but larger errors for arbitrarily inclined walls. This error, however, decreases when the stress is evaluated farther away from the wall. Therefore the authors suggested to measure the fluid stress a few lattice sites away from the wall. All in all, it requires relatively high resolutions (several $10\Delta x$) to estimate the WSS reasonably well in a staircase geometry, which makes this method unfeasible when the average channel diameter is small.

Later, Matyka et al. [56] proposed a different scheme to obtain a better estimate of the unknown normal vectors $\hat{\boldsymbol{n}}$. Their idea was to compute a weighted average of the staircase information of the neighbouring lattice nodes. The advantage of their approach is that it is based on the geometry alone; it is independent of the flow field. Furthermore, the authors showed that the WSS error is dominated by the flow field error and not by an inaccurate approximation of the normal vector. The flow field error in turn is caused by the staircase approximation (modelling error), which can only be decreased by using a more accurate boundary condition for the LBM.

Pontrelli et al. [54] used a finite-volume LBM to compute the WSS in a small artery with a realistic endothelial wall profile. Unfortunately the authors did not provide a benchmark test of the WSS accuracy in their setup. It would be interesting to investigate whether the finite-volume LBM is able to mitigate the shortcomings of the regular lattice with staircase approximation.

Very recently, Kang and Dun [57] studied the accuracy of the WSS in inclined 2D Poiseuille and Womersley flows for BGK and MRT collision operators and for SBB and IBB at the walls. One of the basic results is that, in channels *aligned* with a major lattice axis, the WSS converges with a first-order rate upon grid refinement when it is evaluated at the fluid layer closest to the wall and with a second-order rate when the result is extrapolated to the wall location. This is no surprise since the distance of the last fluid layer and the wall itself converges to zero with first-order rate. The authors report similar results for BGK and MRT for their chosen parameter range. When the flow in an *inclined* channel is simulated, the choice of the boundary condition plays a significant role. IBB leads to errors which are about one order of magnitude smaller than for SBB. Moreover, MRT leads to slightly better results than BGK.

The choice of wall boundary condition critically affects the quality of **wall shear stress** estimates. The best results are obtained when a curved boundary condition is used as this increases the accuracy of the flow field and the fluid stress tensor in direct vicinity of the wall. Further research is necessary to develop improved boundary conditions for more accurate WSS computations in complex geometries.

11.3 Ghost Methods

We will now present LB boundary methods which require extrapolations. A typical scenario is the extrapolation of fluid properties at virtual nodes within a solid body. These so-called *ghost* nodes then participate in collision and propagation like normal fluid nodes. This process produces those populations which stream out of the solid and would otherwise be unknown. After providing some definitions in Sect. 11.3.1, we discuss three distinct classes of extrapolation-based boundary conditions: the Filippova-Hänel and Mei-Luo-Shyy methods (Sect. 11.3.2), the Guo-Zheng-Shi method (Sect. 11.3.3) and comparably novel image-based ghost methods (Sect. 11.3.4). Some recommended articles are [58–62].

11.3.1 Definitions

In order to understand the motivation for the boundary conditions presented in this section, we first have to define certain terms and understand their implications.

- *Extrapolation* in the present context means that the known information of a quantity (e.g. velocity) within a geometrical region is used to approximate the quantity outside this region. The known region is typically the fluid region whereas the interior of a solid is unknown. For example, if we know the velocity u at points x and $x' = x + \Delta x$ (which may be inside the fluid) but not at $x'' = x' + \Delta x'$ (which may be inside the solid), we can still approximate it by *assuming* a linear behaviour and write

$$u(x'') = u(x') + \frac{u(x') - u(x)}{\Delta x} \Delta x'. \tag{11.16}$$

  Higher-order extrapolations require more known data points (usually $n+1$ points for extrapolation of order n). Extrapolations can often lead to instability (in particular when Δx in (11.16) is small) and loss of accuracy.

- *Fictitious domain* methods are based on the idea that the solution of a problem in a given (usually complex) domain Ω can be simpler when instead a substitute problem in a larger (and simpler) domain Ω' with $\Omega \subset \Omega'$ is solved. This obviously means that information in the complement region $\Omega' \setminus \Omega$ has to be constructed. This usually involves extrapolations.
- *Ghost* methods are a special case of fictitious domain methods where virtual fluid nodes (ghost nodes) are created inside the solid region close to the boundary. Extrapolations of fluid and boundary properties are used to reconstruct the ghost nodes. These nodes then participate in collision and propagation in the normal way in order to supply the boundary nodes with otherwise missing populations. Unfortunately these methods are sometimes denoted as *sharp-interface immersed-boundary methods*, although they hardly share any similarities with the immersed boundary method originally introduced by Peskin (Sect. 11.4).

Revisiting the bounce-back boundary conditions presented in Sect. 11.2, we can make the following comments:

- Standard and interpolated bounce-back do not involve any extrapolation or ghost nodes. Although one can implement both methods with the help of nodes which are on the solid side of the boundary, these nodes are only used for memory storage purposes and do not qualify these methods as ghost methods.
- The partially saturated method uses nodes which are inside the solid, but no extrapolations are required to create them. The reason is that the fluid is simply kept in the interior without the need to reconstruct it at every time step.

Therefore, neither of the bounce-back-based boundary conditions in Sect. 11.2 is an extrapolation or ghost method.

We present three extrapolation-based methods in more detail: the Filippova-Hänel and Mei-Luo-Shyy methods, the Guo-Zheng-Shi method and image-based ghost methods. Apart from this, the method by Verschaeve and Müller [63] as an extension of [64] to curved boundaries is yet another alternative which we will, however, not discuss in detail. In short, the underlying idea of [63] is to have boundary nodes in both the fluid and solid regions and to interpolate and extrapolate fluid properties, respectively. The equilibrium distributions are reconstructed from the density and velocity, the non-equilibrium distributions from the velocity gradient.

11.3.2 Filippova-Hänel (FH) and Mei-Luo-Shyy (MLS) Methods

In 1998, Filippova and Hänel [65] proposed the first LB boundary condition for curved geometries (FH method) using extrapolations. They assume the following situation: a population $f_i^\star(\boldsymbol{x}_\mathrm{b}, t)$ propagates towards a wall located between a boundary node at $\boldsymbol{x}_\mathrm{b}$ and a solid node at $\boldsymbol{x}_\mathrm{s} = \boldsymbol{x}_\mathrm{b} + \boldsymbol{c}_i \Delta t$. The wall is located at

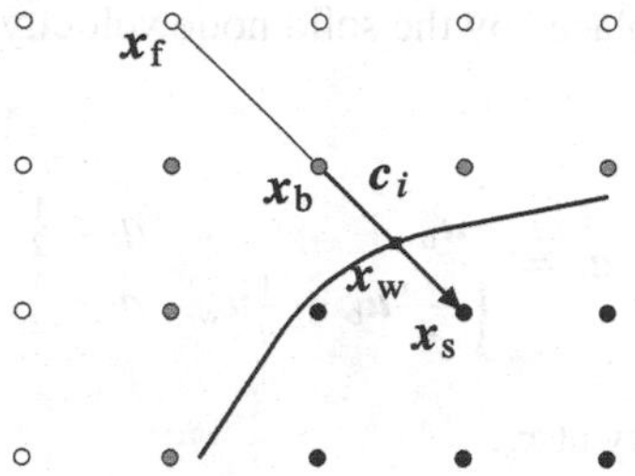

Fig. 11.8 Filippova and Hänel boundary condition. A link c_i is cut by a curved boundary at x_w (*solid square*). Fluid, boundary and solid nodes are shown as *white, grey and black circles*, respectively. Information about the velocity at the solid node x_s is required to find the post-streaming population $f_i(x_b)$. The original method [65] requires the velocity at x_b only, while the improved method [58] uses the velocity at x_f as well

$x_w = x_b + qc_i\Delta t$ as illustrated in Fig. 11.8. The question is how to find the missing population $f_i(x_b, t + \Delta t)$.

11.3.2.1 Original Method by Filippova and Hänel (FH)

Filippova and Hänel [65] suggested the equation

$$f_i(x_b, t + \Delta t) = (1 - \chi)f_i^\star(x_b, t) + \chi f_i^{eq}(x_s, t) - 2w_i\frac{u_w \cdot c_i}{c_s^2} \tag{11.17}$$

for each direction c_i crossing a wall. Here, u_w is the wall velocity (i.e. the velocity of the wall at the intersection point x_w, cf. Fig. 11.8) and χ a weighting factor (with the dimensionless BGK relaxation time τ):

$$\chi = \begin{cases} \frac{1}{\tau}\frac{2q-1}{1-\frac{1}{\tau}} & q < \frac{1}{2} \\ \frac{1}{\tau}(2q-1) & q \geq \frac{1}{2} \end{cases}. \tag{11.18}$$

Exercise 11.3 Show that (11.17) reduces to simple bounce-back for $q = \frac{1}{2}$.

Equation (11.17) is essentially an interpolation of populations at x_b and x_s, but we still have to investigate the shape of the required equilibrium term $f_i^{eq}(x_s, t)$. Filippova and Hänel [65] construct the "equilibrium distribution in the rigid nodes" from

$$f_i^{eq}(x_s, t) = w_i\left(\frac{p(x_b, t)}{c_s^2} + \frac{c_i \cdot u_s}{c_s^2} + \frac{(c_i \cdot u_b)^2}{2c_s^4} - \frac{u_b \cdot u_b}{2c_s^2}\right). \tag{11.19}$$

where we have used the abbreviations $u_b = u(x_b, t)$ and $u_s = u(x_s, t)$. This is nearly the standard incompressible equilibrium evaluated at x_b, with the only exception that the fluid velocity $\boldsymbol{u}_\mathrm{b}$ is replaced by the solid node velocity $\boldsymbol{u}_\mathrm{s}$ in the linear term. The authors suggested

$$\boldsymbol{u}_\mathrm{s} = \begin{cases} \boldsymbol{u}_\mathrm{b} & q < \frac{1}{2} \\ \frac{q-1}{q}\boldsymbol{u}_\mathrm{b} + \frac{1}{q}\boldsymbol{u}_\mathrm{w} & q \geq \frac{1}{2} \end{cases} \tag{11.20}$$

to find the missing velocity at $\boldsymbol{x}_\mathrm{s}$.

This deserves a few comments:

- For $q \geq \frac{1}{2}$, the solid node velocity is obtained by *extrapolating* the velocity at $\boldsymbol{x}_\mathrm{s}$ from $\boldsymbol{x}_\mathrm{b}$ and $\boldsymbol{x}_\mathrm{w}$.
- Since the extrapolation would lead to unstable results for $q \to 0$, the authors fall back to $\boldsymbol{u}_\mathrm{s} = \boldsymbol{u}_\mathrm{b}$ for $q < \frac{1}{2}$.
- The choice of the incompressible equilibrium also explains why the fluid density does not appear in the momentum exchange term on the right-hand-side of (11.17): in the incompressible method the density is constant and typically set to unity.

Although the FH method reduces to simple bounce-back for $q = \frac{1}{2}$, it is conceptually different from interpolated bounce-back (IBB, Sect. 11.2.2) which also reduces to simple bounce-back for $q = \frac{1}{2}$. For any q-value, only one fluid node is required in (11.17), which makes the FH method more local than IBB. The FH method requires an extrapolation for $q \geq \frac{1}{2}$, IBB does not.

11.3.2.2 Improvements by Mei, Luo and Shyy (MLS)

The FH method has the major disadvantage that the weight χ diverges for $\tau \to 1$ and $q < \frac{1}{2}$, which leads to instability. Mei et al. [58] therefore analysed the FH method and its stability properties in detail and proposed an improved version (MLS method). The starting point for the improvement is to realise that there are different ways to construct the term $f_i^\mathrm{eq}(\boldsymbol{x}_\mathrm{s}, t)$. The authors proposed new expressions for $q < \frac{1}{2}$:

$$\boldsymbol{u}_\mathrm{s} = \boldsymbol{u}_\mathrm{f}, \quad \chi = \frac{2q-1}{\tau - 2}, \tag{11.21}$$

where $\boldsymbol{u}_\mathrm{f} = \boldsymbol{u}(\boldsymbol{x}_\mathrm{f}, t)$ and $\boldsymbol{x}_\mathrm{f} = \boldsymbol{x}_\mathrm{b} - \boldsymbol{c}_i \Delta t$ is the location of the fluid node beyond the boundary node (cf. Fig. 11.8). The expressions for $q \geq \frac{1}{2}$ remain untouched.

Mei et al. [58] showed that this modification indeed improves the stability of the original FH method, but they are also sacrificing its locality as a boundary and a fluid node are required. The authors further mention that the above expressions are only strictly valid for stationary flows. They therefore suggested a higher-order extrapolation at $\boldsymbol{u}_\mathrm{s}$ for transient flows.

It is important to note that most follow-up works in the literature employ the improved [58] rather than the original [65] implementation. In the following we summarise some notable progress:

- Mei et al. [66] were the first to perform a thorough comparative evaluation of the momentum exchange algorithm (MEA) and stress integration in the context of the MLS method to obtain drag and lift coefficients at stationary curved boundaries. They found that the stress integration is much more demanding in terms of implementation effort and computing time while the MEA is still relatively accurate. Mei et al. therefore recommend to use the MEA.
- Like other interpolation- and extrapolation-based approaches, the FH and MLS methods suffer from a violation of mass conservation. Therefore, Bao et al. [67] analysed the mechanism responsible for the mass leakage in those boundary treatments and presented an improved mass-conserving method.
- Wen et al. [59] extended the MEA [66] to moving boundaries.

11.3.3 Guo-Zheng-Shi (GZS) Method

Guo et al. [60] proposed yet another extrapolation based LB boundary condition for curved boundaries (GZS). The problem is the same as in Sect. 11.3.2 and Fig. 11.8. In particular, the cut link $\boldsymbol{c}_i$ points into the solid.[3]

The question is how to find $f_{\bar{i}}(\boldsymbol{x}_\mathrm{b}, t + \Delta t)$. The GZS method uses a fictitious fluid node at $\boldsymbol{x}_\mathrm{s}$ which is assigned an equilibrium value

$$f_{\bar{i}}^{\mathrm{eq}}(\boldsymbol{x}_\mathrm{s}, t) = f_{\bar{i}}^{\mathrm{eq}}(\rho_\mathrm{s}, \boldsymbol{u}_\mathrm{s}) \tag{11.22}$$

where $f_i^{\mathrm{eq}}(\rho, \boldsymbol{u})$ is the standard incompressible equilibrium. Note that the only ficticious nodes are those solid nodes which are directly connected to a boundary node by a lattice vector $\boldsymbol{c}_i$.

The authors approximate the density at the solid site by its neighbour value: $\rho_\mathrm{s} = \rho(\boldsymbol{x}_\mathrm{s}, t) = \rho(\boldsymbol{x}_\mathrm{b}, t)$. For the velocity, they suggested

$$\boldsymbol{u}_\mathrm{s} = \begin{cases} q\boldsymbol{u}^{(1)} + (1-q)\boldsymbol{u}^{(2)} & q < \frac{3}{4} \\ \boldsymbol{u}^{(1)} & q \geq \frac{3}{4} \end{cases} \tag{11.23}$$

---

[3] Guo et al. [60] defined $\boldsymbol{c}_i$ exactly the other way around.

where $\boldsymbol{u}^{(1)}$ and $\boldsymbol{u}^{(2)}$ are extrapolations using the nodes at $\boldsymbol{x}_\mathrm{b}$ and $\boldsymbol{x}_\mathrm{f}$, respectively:

$$\begin{aligned} \boldsymbol{u}^{(1)} &= \frac{(q-1)\boldsymbol{u}_\mathrm{b} + \boldsymbol{u}_\mathrm{w}}{q}, \\ \boldsymbol{u}^{(2)} &= \frac{(q-1)\boldsymbol{u}_\mathrm{f} + 2\boldsymbol{u}_\mathrm{w}}{1+q}. \end{aligned} \tag{11.24}$$

This means that $\boldsymbol{u}_\mathrm{s}$ can be different for each considered link $\boldsymbol{c}_i$ crossing a boundary; it is therefore not a property of the position $\boldsymbol{x}_\mathrm{s}$ alone. When q is large enough, an extrapolation from the closest fluid node at $\boldsymbol{x}_\mathrm{b}$ is sufficiently stable, but for smaller q-values an extrapolation from the fluid node at $\boldsymbol{x}_\mathrm{f}$ becomes necessary.

Now, apart from the equilibrium, the GZS method also involves the non-equilibrium populations at the solid node. The authors proposed the extrapolation

$$f_i^\mathrm{neq}(\boldsymbol{x}_\mathrm{s}, t) = \begin{cases} q f_i^\mathrm{neq}(\boldsymbol{x}_\mathrm{b}, t) + (1-q) f_i^\mathrm{neq}(\boldsymbol{x}_\mathrm{f}, t) & q < \frac{3}{4} \\ f_i^\mathrm{neq}(\boldsymbol{x}_\mathrm{b}, t) & q \geq \frac{3}{4} \end{cases}. \tag{11.25}$$

The GZS algorithm includes the following steps:

1. Find q for a cut link connecting a boundary node $\boldsymbol{x}_\mathrm{b}$ and a solid node $\boldsymbol{x}_\mathrm{s}$.
2. Reconstruct the populations of the fictitious nodes by

$$f_i(\boldsymbol{x}_\mathrm{s}, t) = f_i^\mathrm{eq}(\boldsymbol{x}_\mathrm{s}, t) + f_i^\mathrm{neq}(\boldsymbol{x}_\mathrm{s}, t), \tag{11.26}$$

   where the equilibrium and non-equilibrium parts are computed from (11.22) and (11.25).
3. Collide on all fluid/boundary nodes *and* fictitious nodes.[4]
4. Stream populations from all fluid/boundary nodes *and* fictitious nodes to their fluid neighbours. In particular, $f_i^\star$ streams from the fictitious to the boundary node and provides the missing value for $f_{\bar{i}}(\boldsymbol{x}_\mathrm{b}, t + \Delta t)$.
5. Go back to step 1.

According to Guo et al. [60], the present method has advantages over the methods in Sect. 11.3.2. First, while the FH and MLS methods assume a slowly varying flow field, the GZS method only requires a low Mach number flow which can be unsteady. Secondly, the GZS scheme is more stable than the MLS approach.

We would also like to mention that Guo et al. [60] view FH and MLS as improved bounce-back methods. Although that statement is certainly not wrong (the functional form for the missing population $f_{\bar{i}}(\boldsymbol{x}_\mathrm{b}, t+\Delta t)$ is similar to the standard and interpolated bounce-back expressions), the FH, MLS and GZS methods all require

[4] In the original paper [60], the fictitious populations are already constructed in their post-collision state $f_i^\star(\boldsymbol{x}_\mathrm{s}, t) = f_i^\mathrm{eq}(\boldsymbol{x}_\mathrm{s}, t) + (1 - \frac{1}{\tau}) f_i^\mathrm{neq}(\boldsymbol{x}_\mathrm{s}, t)$. In this case, collision on fictitious nodes is of course not additionally performed.

extrapolations and are ghost-like methods. They are therefore conceptually different from the bounce-back methods presented in Sect. 11.2 which are all extrapolation-free.

11.3.4 Image-Based Ghost Methods

Only in 2012, Tiwari and Vanka [61] developed a ghost-fluid boundary condition for the LBM which is based on the so-called *image* method. The idea of their boundary condition is illustrated in Fig. 11.9.

The algorithm consists of the following steps:

1. Identify all required ghost nodes $\boldsymbol{x}_{\mathrm{s}}$. Those are all solid nodes which are connected to at least one boundary node along a lattice vector $\boldsymbol{c}_i$.
2. For each ghost node $\boldsymbol{x}_{\mathrm{s}}$ find the closest point $\boldsymbol{x}_{\mathrm{w}}$ on the wall. We define $\boldsymbol{n} = \boldsymbol{x}_{\mathrm{w}} - \boldsymbol{x}_{\mathrm{s}}$ as the outward-pointing normal vector at the wall. Note that $|\boldsymbol{n}|$ is the distance of the ghost node from the wall and $\boldsymbol{n}$ is generally not aligned with any of the lattice vectors $\boldsymbol{c}_i$.
3. The next step is to find the *image point* $\boldsymbol{x}_{\mathrm{i}}$ in the fluid:

$$\boldsymbol{x}_{\mathrm{i}} = \boldsymbol{x}_{\mathrm{s}} + 2\boldsymbol{n} = \boldsymbol{x}_{\mathrm{w}} + \boldsymbol{n}. \tag{11.27}$$

   For a stationary boundary, steps 1–3 have to be performed only once.
4. Interpolate the required fluid properties (velocity and density) at the image point $\boldsymbol{x}_{\mathrm{i}}$ to obtain $\boldsymbol{u}_{\mathrm{i}}$ and ρ_{i}. The interpolation process is somewhat tedious as it depends on whether interpolation support points are located in the fluid or on the wall. We refer to [61] for a detailed discussion.

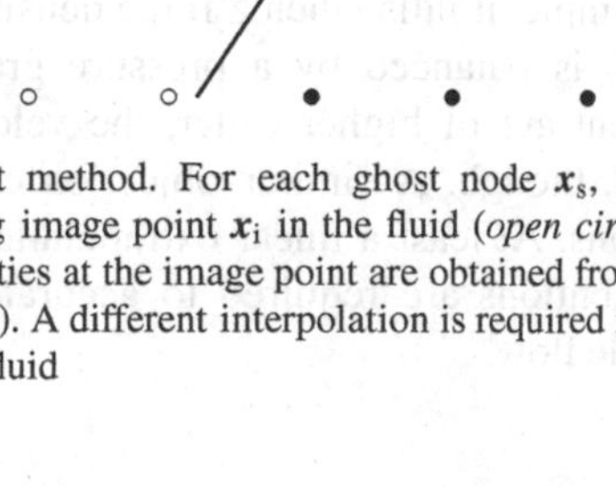

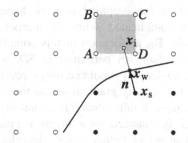

Fig. 11.9 Image-based ghost method. For each ghost node $\boldsymbol{x}_{\mathrm{s}}$, the closest wall point $\boldsymbol{x}_{\mathrm{w}}$ is computed. The corresponding image point $\boldsymbol{x}_{\mathrm{i}}$ in the fluid (*open circles*) is constructed along the normal vector $\boldsymbol{n}$. Fluid properties at the image point are obtained from an interpolation in the grey region (fluid nodes *A*, *B*, *C*, *D*). A different interpolation is required if not all interpolation support points are located within the fluid

5. Extrapolate velocity and density along the normal $\boldsymbol{n}$ to the ghost node. Tiwari and Vanka [61] used

$$\boldsymbol{u}_{\mathrm{s}} = 2\boldsymbol{u}_{\mathrm{w}} - \boldsymbol{u}_{\mathrm{i}}, \quad \rho_{\mathrm{s}} = \rho_{\mathrm{i}}. \tag{11.28}$$

The first equation refers to a linear extrapolation of the velocity, the second to a zero density gradient at the wall.
6. Compute the equilibrium distributions at the ghost nodes from $f_i^{\mathrm{eq}}(\boldsymbol{x}_{\mathrm{s}}, t) = f_i^{\mathrm{eq}}(\rho_{\mathrm{s}}, \boldsymbol{u}_{\mathrm{s}})$.
7. The non-equilibrium distributions $f_i^{\mathrm{neq}}(\boldsymbol{x}_{\mathrm{s}}, t)$ are obtained like the fluid density: interpolate them at $\boldsymbol{x}_{\mathrm{i}}$ first, then apply $f_i^{\mathrm{neq}}(\boldsymbol{x}_{\mathrm{s}}, t) = f_i^{\mathrm{neq}}(\boldsymbol{x}_{\mathrm{i}}, t)$.
8. Combine the equilibrium and non-equilibrium distributions at the ghost nodes and perform the propagation step, followed by the collision step.
9. Go back to step 1.

This algorithm deserves a few remarks:

- In contrast to the previous methods in this section, $\boldsymbol{u}_{\mathrm{s}}$ does not depend on the considered link $\boldsymbol{c}_i$; it is rather a unique property of each ghost node.
- According to [61], the extrapolated values of density, velocity and non-equilibrium distributions are post-collision rather than pre-collision. This is unusual since the moments (density, velocity, stress) are normally computed after the previous streaming and before the next collision step.
- The boundary condition is based on the hydrodynamic fields rather than the populations. This allows implementing Neumann boundary conditions. The authors for example demonstrated the applicability of their method for inlet and outlet boundary conditions [61].
- Extrapolation along normal vectors as in step 5 avoids typical stability issues encountered with other extrapolation methods.
- Although being trivial for circular or spherical boundary segments, finding the image point can be tedious for complex boundary shapes. Also the interpolation at the image points is complicated if not all interpolation support points are within the fluid domain. The application of this boundary condition to moving boundaries of complex shape, in particular in 3D, is therefore difficult and expensive. Tiwari and Vanka [61] simulated only circular boundaries in 2D.
- The assumption of a zero density gradient across the boundary is a gross over-simplification. For example, it fails when a force density along the extrapolation direction exists which is balanced by a pressure gradient [62]. Since errors in the pressure gradient are of higher order, the velocity profile may still be second-order accurate, though. A similar objection can be made for the non-equilibrium distributions. At least a linear extrapolation for the density and the non-equilibrium distributions are required to accurately capture second-order flows like the Poiseuille flow.

Several extensions and improvements of the algorithm have been proposed in the meantime:

- Khazaeli et al. [68] followed a similar route as Tiwari and Vanka [61] to impose higher-order boundary conditions for coupled fluid-heat problems in the two-population LBM.
- Mohammadipoor et al. [62] followed the same line as [61] and extended the approach of Zou and He [69] to curved boundaries.
- Pellerin et al. [70] proposed an image-based method that relies only on equilibrium distributions.

There exist several **extrapolation-based** boundary conditions for the LBM. These methods are conceptually more difficult than bounce-back-based methods. A common algorithmic complication all these methods share with the interpolated bounce-back method is the detection of boundary points (either on cut links or closest points to ghost nodes). In practice, these methods are quite unhandy for moving boundaries of complicated shape although they are promising candidates for highly accurate boundary conditions when properly applied. More research is required to make extrapolation-based method more attractive for moving objects with non-trivial shape.

11.4 Immersed Boundary Methods

The immersed boundary method (IBM) [71–73] is older than LBM, but the combination of both was not suggested before 2004 [74] (Sect. 11.4.1). The basic idea of the IBM is to approximate a boundary by a set of off-lattice marker points that affect the fluid only *via* a force field. An interpolation stencil is introduced to couple the lattice and the marker points (Sect. 11.4.2). This allows a relatively simple implementation of complex boundaries. There are several IBM variants, for example explicit (Sect. 11.4.3) or direct-forcing (Sect. 11.4.4) for rigid boundaries and explicit IBM for deformable boundaries (Sect. 11.4.5). We also mention a series of other related boundary conditions which are less commonly used (Sect. 11.4.6). We recommend reading [75, 76] for introductions and investigations of the IBM in conjunction with the LBM.

11.4.1 Introduction

Boundary conditions in the LBM are usually treated on the population level, i.e. the populations f_i are manipulated or constructed in such a way that the desired values for pressure and velocity (or their derivatives) are obtained at the boundary. This applies to all boundary conditions discussed in Chap. 5 and in the present chapter up to this point.

There is, however, a completely different way to enforce boundary conditions which was available long before anybody knew of the LBM. In 1972, Peskin proposed the *immersed boundary method* (IBM) in his dissertation [71], followed by an article in 1977 [72]. Peskin's idea was to use the force density $\boldsymbol{F}(\boldsymbol{x}, t)$ in the Navier-Stokes equation to mimic a boundary condition. To this end, $\boldsymbol{F}(\boldsymbol{x}, t)$ has to be computed such that the fluid behaves *as if* there was a boundary with desired properties (e.g. no-slip). When correctly applied, this approach can be used to recover immersed rigid or deformable objects with nearly arbitrary shape. Since the boundary condition exists only on the Navier-Stokes level (*via* the force density $\boldsymbol{F}(\boldsymbol{x}, t)$), IBM is not aware of the populations f_i.

The IBM provides a number of advantages. The main advantage is its front-tracking character, i.e. the shape of the boundary is directly known and does not have to be reconstructed (as in phase-field or level-set approaches). Neither do intersection points have to be computed (as required for nearly all boundary conditions presented in this chapter so far). The IBM can be combined with any Navier-Stokes solver which supports external forcing, such as the LBM. The IBM is relatively simple to implement and, if done so properly, its numerical overhead is small. Moving and deformable boundaries can be realised without remeshing. It has to be noted that fluid exists on both sides of an IBM surface. In particular, closed surfaces are filled with fluid.

The original IBM does not take any consideration of the kinetic origin of the LBM as it only operates on the Navier-Stokes level. Still, the combination of the IBM and the LBM, also called immersed-boundary-lattice-Boltzmann method (IB-LBM), first proposed by Feng and Michaelides [74], has become a popular application. It therefore deserves a somewhat thorough introduction in this book, together with some recent developments and related approaches. We cannot provide an exhaustive coverage of the IBM in general, though. Readers who are interested in the IBM independently of the LBM should read the seminal paper by Peskin [73] and the review by Mittal and Iaccarino [77].

There is some dissent in the literature what "immersed boundary method" actually means and how it is defined. Some people use it for nearly all methods where a boundary is immersed in a fluid, including, for example, fictitious domain methods (Sect. 11.3). Here, we define those methods as immersed boundary methods which involve, on the one hand, an Eulerian grid and Lagrangian markers and, on the other hand, some kind of velocity interpolation and force spreading as devised by Peskin [73], but there is no clear distinction between the IBM and related methods. Another way of putting this is the following.[5] In IBM we have marker points without mass that move exactly with the fluid. Through some mechanical model (e.g. a constitutive model for a deformable membrane or a penalty force for a rigid body), we can compute forces at these points which we apply to the fluid directly, rather than to the mechanical model itself.

In the remainder of this section we will focus on the IBM combined with LBM using the BGK collision operator. However, several authors recently pointed out that the MRT or TRT collision operators can bring additional advantages by reducing undesired velocity slip at the immersed boundaries [78, 79]. We will not discuss those extensions further.

11.4.2 Mathematical Basis

We will now review the original IBM, discuss its mathematical properties and show its basic numerical algorithm.

11.4.2.1 Eulerian and Lagrangian Systems

Mathematically, the basis of the IBM is an *Eulerian* and a *Lagrangian* system. The former is represented by a fixed regular grid on which the fluid lives and the Navier-Stokes equations are solved. The latter is an ensemble of marker points $\{\boldsymbol{r}_j\}$. They can be (nearly) arbitrarily distributed in space, as long as they are sufficiently dense (see below). These markers represent discrete surface points of the boundary and are generally allowed to move: $\boldsymbol{r}_j = \boldsymbol{r}_j(t)$. We therefore have to distinguish between two node systems (Fig. 11.10) with the following properties:

1. The Eulerian grid defined by the LBM lattice nodes (coordinates designated by $\boldsymbol{x}$) is regular and stationary.
2. The immersed boundary marker points $\boldsymbol{r}_j(t)$ are Lagrangian nodes. They are not bound to the Eulerian grid and can move in space.

If the boundary is rigid, one would ideally fix relative distances such that $|\boldsymbol{r}_{jk}(t)| = |\boldsymbol{r}_j(t) - \boldsymbol{r}_k(t)| = \text{const}$. This is often not achievable, and a somewhat softened condition $|\boldsymbol{r}_{jk}(t)| \approx \text{const}$ is used instead. For deformable boundaries, a relative marker motion is actually desired.

It may or may not be necessary to connect neighbouring markers. Most implementations of rigid boundary conditions do not require connected markers, while all deformable algorithms require some kind of surface tessellation which involves defining the markers *and* their connectivity (surface mesh). This mesh is called *nonconforming* as it does not have to be aligned with the lattice of the LBM. The

[5] Thanks to Eric Lorenz for suggesting this description.

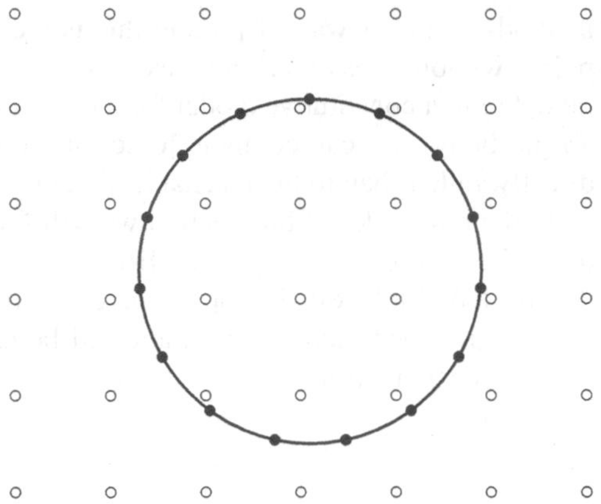

Fig. 11.10 Cylinder with boundary markers arbitrarily positioned in the regular fluid domain. The Eulerian mesh (fluid nodes, *open circles*) and the Lagrangian mesh (boundary, *solid nodes*) are independent. No intersections of lattice links with the boundary have to be computed

main advantage of the IBM is that the complex shape of the boundary is not related to the lattice structure and no intersection points have to be computed. This makes the IBM particularly useful for deformable boundaries (Sect. 11.4.5).

The decomposition of the geometry into two coordinate systems brings up the important question how to couple the dynamics of the boundary and the fluid. We need a bi-directional coupling where the fluid has to know about the presence of the boundary and *vice versa*. It is therefore required to communicate some information between both node systems through *velocity interpolation* and *force spreading*.

11.4.2.2 Continuous Governing Equations

We start with a fully continuous description and later turn our attention to the discretised version. In the following we assume the validity of the no-slip boundary condition, which is the first key idea of the IBM. It implies that each point of the surface $\boldsymbol{r}(t)$ and the ambient fluid at position $\boldsymbol{r}$ have to move with the same velocity:

$$\dot{\boldsymbol{r}}(t) = \boldsymbol{u}(\boldsymbol{r}(t), t). \tag{11.29}$$

The time dependence on the right-hand side of (11.29) is, on the one hand, caused by the variation of $\boldsymbol{u}$ itself and, on the other hand, by the boundary moving and therefore seeing different parts of the flow field. We can rewrite (11.29) as

$$\dot{\boldsymbol{r}}(t) = \int \mathrm{d}^3x\, \boldsymbol{u}(\boldsymbol{x}, t)\delta(\boldsymbol{x} - \boldsymbol{r}(t)) \tag{11.30}$$

where $\delta(\boldsymbol{x} - \boldsymbol{r}(t))$ is Dirac's delta distribution. Equation (11.30) is the first of two governing equations of the as yet continuous IBM. We will later see that the discretised version of (11.30) requires *velocity interpolation*. Note that we write all equations for 3D applications, but everything said (unless otherwise stated) can be directly applied in 2D as well.

The second governing equation describes the momentum exchange between the boundary and the fluid. In the IBM picture, we are interested in the force the boundary surface exerts on the nearby fluid, rather than the other way around. Let us assume we know the force density (per area) $\boldsymbol{F}_A(\boldsymbol{r}(t), t)$ everywhere on the boundary surface. Therefore, $\boldsymbol{F}_A\, \mathrm{d}^2 r$ is the force acting on a small area element $\mathrm{d}^2 r$.[6] The force density (per volume) that the fluid feels due to the presence of the immersed boundary can then be written as

$$\boldsymbol{F}(\boldsymbol{x}, t) = \int \mathrm{d}^2 r\, \boldsymbol{F}_A(\boldsymbol{r}(t), t)\delta(\boldsymbol{x} - \boldsymbol{r}(t)). \tag{11.31}$$

The delta distribution is the same as in (11.30). Equation (11.31) essentially means that the Lagrangian boundary force is spread to the Eulerian fluid. Therefore this equation is called *force spreading*.

Note that $\boldsymbol{F}(\boldsymbol{x}, t)$ is singular. Since the delta distribution $\delta(\boldsymbol{x} - \boldsymbol{r}(t))$ is 3D, but the integration is only 2D along the boundary, $\boldsymbol{F}(\boldsymbol{x}, t)$ is singular when crossing the boundary in normal direction. This marks the defining difference between velocity interpolation (which is non-singular) and force spreading.

Equation (11.30) and (11.31) are the basic IBM equations in their continuous form. We will now show their discretised versions which can be used in computer simulations.

11.4.2.3 Discretised Governing Equations

Since the velocity field is only known at discrete lattice sites, the integral in equation (11.30) cannot be exactly computed in a lattice-based simulation. The same holds for (11.31). Instead, both integrals have to be replaced by sums with a suitably chosen discretisation of the delta distribution.

Peskin [73] provided a full derivation of a general set of equations for the IBM. We will restrict ourselves, for the sake of brevity and clarity, to the final set of equations based on the assumption that the fluid in the entire volume is homogeneous, in particular its density and viscosity. We further assume that the markers are massless, which means that the boundary has the same density as the surrounding fluid.

[6] Remember that the boundary in 3D is a 2D surface.

The **discretised IBM equations** read

$$\dot{\boldsymbol{r}}_j(t) = \sum_{\boldsymbol{x}} \Delta x^3 \boldsymbol{u}(\boldsymbol{x}, t) \Delta(\boldsymbol{r}_j(t), \boldsymbol{x}) \tag{11.32}$$

and

$$\boldsymbol{F}(\boldsymbol{x}, t) = \sum_j \boldsymbol{f}_j(t) \Delta(\boldsymbol{r}_j(t), \boldsymbol{x}). \tag{11.33}$$

The fluid is discretised as an Eulerian lattice with coordinates $\boldsymbol{x}$, the boundary is approximated by an ensemble of markers at $\boldsymbol{r}_j(t)$. Here, $\boldsymbol{u}$ and $\boldsymbol{F}$ are velocity and force density (per volume) on the lattice, $\dot{\boldsymbol{r}}_j$ and $\boldsymbol{f}_j$ are the velocity of and the total force acting on the markers. Velocity interpolation in (11.32) and force spreading in (11.33) are the central IBM equations. Both require an appropriate kernel function (or stencil) Δ, as discussed below.

It is important to realise that $\boldsymbol{f}_j(t)$ is the total force (not force density) acting on node j at position $\boldsymbol{x}_j(t)$. Apart from the no-slip condition discussed above, it is one of the key ideas of the IBM that the force $\boldsymbol{f}_j(t)$ is first computed in the Lagrangian system and then spread to the lattice. This brings up the central question how $\boldsymbol{f}_j(t)$ can be found in the first place. In fact, this depends strongly on the chosen kind of the IBM. We will discuss this problem in the upcoming sections. Let us for now simply assume that all forces $\boldsymbol{f}_j(t)$ are known at each time step.

We further emphasise that $\boldsymbol{F}(\boldsymbol{x}, t)$ is the only mechanism through which the fluid is aware of the presence of the boundary; there is otherwise no direct boundary condition for the fluid. Once we know $\boldsymbol{F}(\boldsymbol{x}, t)$, we can use one of the forcing schemes described in Chap. 6 to update the fluid. To the LBM, the IBM force density is not at all different from gravity (although gravity is usually homogeneous and constant).

11.4.2.4 Kernel Functions

The function $\Delta(\boldsymbol{r}_j, \boldsymbol{x})$ is a suitably discretised version of the Dirac delta distribution and another key ingredient of any IBM. It is in most cases simplified by assuming $\Delta(\boldsymbol{r}_j, \boldsymbol{x}) = \Delta(\boldsymbol{r}_j - \boldsymbol{x})$, i.e. it is only a function of the distance vector $\boldsymbol{r}_j - \boldsymbol{x}$ rather than a more general function of $\boldsymbol{r}_j$ and $\boldsymbol{x}$ (see [80] for a kernel without this simplification). It is not directly obvious which functions $\Delta(\boldsymbol{r}_j - \boldsymbol{x})$ qualify as valid interpolation and spreading kernels. While Peskin [73] explains the procedure to find suitable discretisations in detail and an overview of interpolation function can be found in [76, 81], we will only provide the basic ideas and final results.

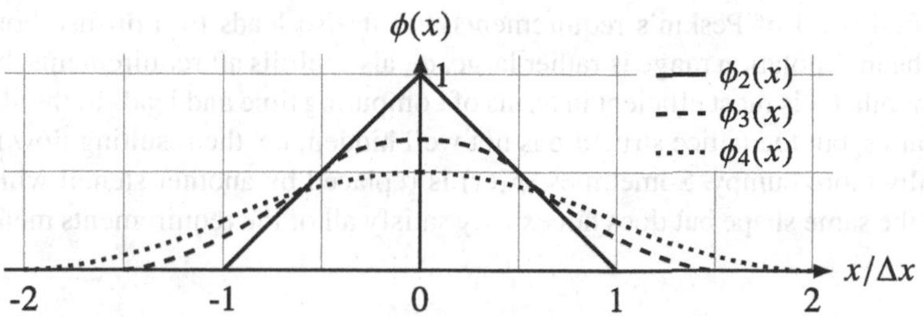

Fig. 11.11 IBM interpolation stencils ϕ_2, ϕ_3, and ϕ_4. The total kernel range is two, three and four lattice sites, respectively

The fundamental claims and restrictions are:

- Interpolation and spreading should be short-ranged. This is required to reduce computational overhead by making the number of summands in (11.32) and (11.33) as small as possible.
- Momentum and angular momentum have to be identical when evaluated either in the Eulerian or the Lagrangian system (same speed and rotation in both systems).
- Lattice artefacts ("bumpiness" of the interpolation when boundaries move) should be suppressed as much as possible.
- The kernel has to be normalised: $\sum_{\boldsymbol{x}} \Delta x^3 \Delta(\boldsymbol{x}) = 1$.

It is convenient to factorise the kernel function as $\Delta(\boldsymbol{x}) = \phi(x)\phi(y)/\Delta x^2$ in 2D and $\Delta(\boldsymbol{x}) = \phi(x)\phi(y)\phi(z)/\Delta x^3$ in 3D, i.e. each major coordinate axis contributes independently. This is not essential but simplifies the procedure. Peskin [73] derived a series of stencils which are also shown in Fig. 11.11. Those kernels read

$$\phi_2(x) = \begin{cases} 1 - |x| & (0 \leq |x| \leq \Delta x) \\ 0 & (\Delta x \leq |x|) \end{cases}, \tag{11.34}$$

$$\phi_3(x) = \begin{cases} \frac{1}{3}\left(1 + \sqrt{1 - 3x^2}\right) & 0 \leq |x| \leq \frac{1}{2}\Delta x \\ \frac{1}{6}\left(5 - 3|x| - \sqrt{-2 + 6|x| - 3x^2}\right) & \frac{1}{2}\Delta x \leq |x| \leq \frac{3}{2}\Delta x \\ 0 & \frac{3}{2}\Delta x \leq |x| \end{cases}, \tag{11.35}$$

$$\phi_4(x) = \begin{cases} \frac{1}{8}\left(3 - 2|x| + \sqrt{1 + 4|x| - 4x^2}\right) & 0 \leq |x| \leq \Delta x \\ \frac{1}{8}\left(5 - 2|x| - \sqrt{-7 + 12|x| - 4x^2}\right) & \Delta x \leq |x| \leq 2\Delta x \\ 0 & 2\Delta x \leq |x| \end{cases}. \tag{11.36}$$

The integer index denotes the number of lattice nodes required for interpolation and spreading along each coordinate axis (Fig. 11.11). Therefore, the stencils require 2^d, 3^d and 4^d lattice sites in d dimensions, respectively.

ϕ_4 fulfills all of Peskin's requirements, but it also leads to a diffuse boundary since the interpolation range is rather large. ϕ_3 also fulfills all requirements, but it is less smooth. ϕ_2 is most efficient in terms of computing time and leads to the sharpest boundaries, but the lattice structure is not well hidden, i.e. the resulting flow field is generally more bumpy. Sometimes, $\phi_4(x)$ is replaced by another stencil which has nearly the same shape but does not exactly satisfy all of the requirements mentioned above:

$$\phi_4'(x) = \begin{cases} \frac{1}{4}\left(1+\cos(\frac{\pi x}{2})\right) & 0 \leq |x| \leq 2\Delta x \\ 0 & 2\Delta x \leq |x| \end{cases}. \tag{11.37}$$

We also note that additional, smoothed representations of the delta distribution have been proposed [82].

11.4.2.5 General IB-LBM Algorithm

Without specifying yet how the forces $\boldsymbol{f}_j$ acting on the boundary nodes are obtained, we can still jot down a simple IB-LBM algorithm. It consists of the following sub-steps:

1. Compute the Lagrangian forces $\boldsymbol{f}_j(t)$ from the current boundary configuration $\{\boldsymbol{r}_j(t)\}$. This is a model-dependent step which still remains to be discussed.
2. Spread the Lagrangian forces $\boldsymbol{f}_j(t)$ to the lattice *via* (11.33) to obtain the Eulerian force density $\boldsymbol{F}(\boldsymbol{x}, t)$. See also Fig. 11.12.
3. Compute the uncorrected (pre-collision) velocity $\boldsymbol{u}(\boldsymbol{x}, t)$ from $\boldsymbol{u} = \sum_i f_i \boldsymbol{c}_i / \rho$.
4. Perform the LB algorithm (computing equilibrium distributions, collision and propagation) with forcing (Chap. 6), using $\boldsymbol{u}(\boldsymbol{x}, t)$ and $\boldsymbol{F}(\boldsymbol{x}, t)$ as input. If other forces, such as gravity, are present, the total force is the sum of all these contributions. Note that the choice of an accurate forcing scheme (e.g. Guo et al. [83]) is important. This is often ignored in the literature.

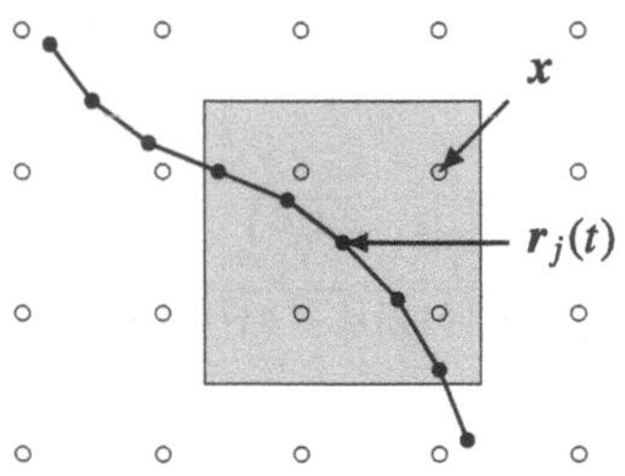

Fig. 11.12 Interpolation and spreading. The lattice velocity is interpolated at $\boldsymbol{r}_j(t)$. For this operation, all lattice nodes within the grey region are required (here ϕ_2 is used). The force density at a given lattice node $\boldsymbol{x}$ is the sum of all contributions from those nodes $\boldsymbol{r}_j(t)$ whose interpolation box covers $\boldsymbol{x}$

5. As we know from Chap. 6, the physical fluid velocity during the time step is given by the first moment of the populations and a force correction:

$$\boldsymbol{u}_{\mathrm{f}}(\boldsymbol{x},t) = \boldsymbol{u}(\boldsymbol{x},t) + \frac{\boldsymbol{F}(\boldsymbol{x},t)\Delta t}{2\rho(\boldsymbol{x},t)}. \tag{11.38}$$

Leaving the force correction out can lead to significant stability (and accuracy) problems.
6. Interpolate the fluid velocity $\boldsymbol{u}_{\mathrm{f}}(\boldsymbol{x},t)$ at the Lagrangian node positions *via* (11.32) to obtain $\dot{\boldsymbol{r}}_j(t)$. See also Fig. 11.12.
7. Advect the boundary nodes (usually by the explicit forward Euler method) to find the new boundary configuration:

$$\boldsymbol{r}_j(t+\Delta t) = \boldsymbol{r}_j(t) + \dot{\boldsymbol{r}}_j(t)\Delta t. \tag{11.39}$$

There exist different explicit time integration schemes [84, 85], though.
8. Go back to step 1 for the next time step.

Note that the time steps for the LBM and the marker position update are identical, i.e. in the standard IB-LBM there can only be one marker update per LB time step.

Not all IBM flavours follow this algorithm. There are several approaches (and algorithms) to deal with rigid boundaries. We will get back to those later.

Once the discretised kernel functions $\Delta(\boldsymbol{x})$ have been implemented, the rules for computing the forces $\boldsymbol{f}_j$ have been defined and the initial boundary node locations $\boldsymbol{r}_j(t=0)$ are known, the simulation can be executed. The real challenge is normally hidden behind the models providing the required forces $\boldsymbol{f}_j$. We will get back to this point in the following sections.

For the sake of **efficiency**, note that is it very easy to implement a naive IBM which is, despite being mathematically correct, horribly inefficient. Equation (11.32) clearly shows that the sum should run over the lattice neighbours of a given boundary node. For a boundary node $\boldsymbol{r}_j$, it is easy to identify the neighbouring lattice sites. However, (11.33) suggests to go the other way around and to identify all boundary markers in the vicinity of a given lattice site. This can be extremely expensive, in particular when the Eulerian lattice is large. A small trick can make the computational effort for interpolation and spreading identical though. In order to do so, we run over all known boundary markers $\boldsymbol{r}_j$ and compute the fraction of the force density a neighbouring lattice site $\boldsymbol{x}$ would receive:

$$\delta_j \boldsymbol{F}(\boldsymbol{x}) = \boldsymbol{f}_j(t)\Delta(\boldsymbol{r}_j(t),\boldsymbol{x}). \tag{11.40}$$

(continued)

Here, $\delta_j \boldsymbol{F}(\boldsymbol{x})$ is the contribution to $\boldsymbol{F}(\boldsymbol{x})$ due to the presence of $\boldsymbol{r}_j$ alone. All these contributions are simply summed and the correct total force density $\boldsymbol{F}(\boldsymbol{x}) = \sum_j \delta_j \boldsymbol{F}(\boldsymbol{x})$ is automatically obtained in the end. This also highlights the conceptual difference between interpolation and spreading. In fact, Tryggvason et al. [86] give the helpful advice (where "front" means "boundary" in our case):

> When information is transferred between the front and the fixed grid, it is always easier to go from the front to the grid and not the other way around. Since the fixed grid is structured and regular, it is very simple to determine the point on the fixed grid that is closest to a given front position.

11.4.2.6 Implications of the Combination of IBM and LBM

Although the IBM is just another way to impose boundary conditions on the Navier-Stokes level, the populations f_i are completely unimpressed by the presence of the Lagrangian marker points. In particular, the f_i simply penetrate any closed IB surface. This is not problematic as long as one is only interested in the no-slip condition of the velocity $\boldsymbol{u}$ and one does not care what the populations are doing. But we can already see that the IBM is not an ideal approach when one wants to keep, for example, two fluids separate on two sides of a membrane.

Another observation is that the fluid usually fills the entire space, including the regions inside any boundary. This significantly simplifies things but can also lead to additional difficulties. For example, it has been shown that the dynamics of the interior fluid can have an effect on the dynamics of the exterior fluid if the immersed boundaries are rotating [87]. There are ways around this, for example by adding *interior* marker points. In the following, we will not discuss methods with interior markers, such as direct-forcing/fictitious-domain methods [88].

11.4.2.7 Distribution of Markers in Space

One open question is how to distribute the markers $\boldsymbol{r}_j$ in space initially. For 2D problems this answer is easy to answer: define a 1D chain of markers with a given mutual distance d on the boundary. Each marker knows which one is its left and right neighbour.

The choice of d is a more delicate issue. On the one hand, intuitively, d cannot be too large because otherwise there are "holes" in the boundary and fluid can flow between markers. On the other hand, too small a value for d can lead to problems as well [89]. This is due to the peculiarities of the IBM algorithm: the marker position update relies on the interpolated fluid velocity. If two markers are very close, $d \ll \Delta x$, they essentially see the same fluid environment and move with the same velocity. Markers which are too close can therefore not be separated again (or only with a lot of effort) and they can stick together. It is usually recommended to choose d somewhere between 0.5 and one lattice constant, but some authors even choose $d \approx 2\Delta x$. We will get back to this point in Sect. 11.4.3.

The situation is much more complicated in 3D where boundaries are generally curved 2D surfaces. One has to distribute the markers such that the mutual distance of any pair of neighbours is approximately the same. This can be a tedious task for general surface shapes and is one of the biggest challenges when applying the IBM in 3D. Furthermore, the node connectivity (i.e. an unstructured mesh) is required for deformable boundaries and additional constraints may apply in those situations (e.g. the resulting triangular face elements should be as equilateral as possible). Here, we can only give some starting points for further literature studies:

- For simple geometries of high symmetry (spheres, red blood cells), one can start from an *icosahedron* and subdivide each triangular surface element into n^2 ($n > 1$ being an integer) triangular elements [85, 90]. The markers are radially or tangentially shifted to approximate the desired boundary shape.
- In the *minimum potential approach* [91] a fixed number of markers is initially randomly distributed on the surface. Markers interact via repulsive forces and move along the surface until the system has found an energetic minimum. The resulting marker configuration can then be used in simulations as initial boundary discretisation.
- Feng and Michaelides [91] presented another approach to distribute markers on a sphere by defining parallel segments containing equidistant nodes.
- There exist free meshing tools which can cope with more complicated boundary shapes, for example [92, 93].

11.4.2.8 Accuracy and Convergence

One shortcoming of the IBM is that the velocity interpolation does not generally maintain the solenoidal properties of the fluid. Even if the fluid solver is perfectly divergence-free (which LBM usually cannot claim), the interpolated velocity may not be divergence-free. The consequence is that the volume of an enclosed region can change in time.

Furthermore, the IBM is formally a first-order accurate boundary condition [73]. There seems to be some dispute in the literature about the actual convergence rate, though. While Peng and Luo [94] report second-order convergence, other authors observed only first-order convergence for the velocity field [95, 96].

Related to the question of accuracy and convergence is the apparent size of particles and/or apparent location of walls modelled with the IBM. Several authors, e.g. [85, 91, 97], have reported that particles appear to be larger than they actually are. Instead of the input radius r, a larger radius $r + \delta r$ is observed where δr is somewhere between $0.2\Delta x$ and $0.5\Delta x$, depending on the chosen stencil (a kernel with wider support usually leads to a larger δr). In order to model a sphere of actual radius r, Feng and Michaelides [91] suggested to distribute markers on a sphere with radius

$$r_b = \sqrt[3]{\frac{r^3 + (r - \Delta x)^3}{2}} \tag{11.41}$$

instead. This finding is important for the modelling of particle suspensions or porous media where the rheology strongly depends on the volume fraction and porosity. We will get back to the convergence and apparent wall location in Sect. 11.4.3.

Once again, we see that there is no free lunch. The advantages of the standard IBM (ease of implementation, no need to find boundary intersections, no treatment of fresh fluid nodes required), which explain the IBM's popularity, are challenged by inferior accuracy and convergence compared to other boundary conditions. Note, however, that there have been efforts to make the IB-LBM more accurate [78, 79, 97].

11.4.3 Explicit Feedback IBM for Rigid Boundaries

We show a simple way to compute the nodal forces $\boldsymbol{f}_j$ for (nearly) rigid boundaries. This *explicit* IBM is easy to implement but shows weak stability properties. After discussing the algorithm we use the explicit IBM to model Poiseuille flow and demonstrate the convergence and boundary location issues within the IBM. Note that the explicit IBM does not work very well for unsteady flows as it takes some time for the marker points to respond to the flow.

11.4.3.1 Algorithm

A rigid body is defined by $|\boldsymbol{r}_j(t) - \boldsymbol{r}_k(t)| = \text{const}$ for any two points $\boldsymbol{r}_j$ and $\boldsymbol{r}_k$ of the body. The simplest way to approximate rigid objects with the IBM is to model the boundary as a collection of marker points $\boldsymbol{r}_j(t)$ which are individually connected by an elastic spring to their reference locations $\boldsymbol{r}_j^{(0)}(t)$. Feng and Michaelides [74] first proposed this idea within the framework of IB-LBM in 2004. While the virtual reference locations obey the rigidity condition exactly, $|\boldsymbol{r}_j^{(0)}(t) - \boldsymbol{r}_k^{(0)}(t)| = \text{const}$, the real markers are allowed to deviate slightly from this condition.

The magnitude of the undesired body deformation can be controlled by springs with strength κ. We can then explicitly compute the marker "penalty" force $\boldsymbol{f}_j$ from a function like

$$\boldsymbol{f}_j(t) = -\kappa \delta \boldsymbol{r}_j(t), \quad \delta \boldsymbol{r}_j(t) = \boldsymbol{r}_j(t) - \boldsymbol{r}_j^{(0)}(t) \tag{11.42}$$

at each time step so that the required nodal forces $\boldsymbol{f}_j$ are known. Contrarily, Feng and Michaelides [74] proposed a form similar to

$$\boldsymbol{f}_j(t) = \begin{cases} 0 & |\delta \boldsymbol{r}_j(t)| = 0 \\ -\kappa \frac{\delta \boldsymbol{r}_j(t)}{|\delta \boldsymbol{r}_j(t)|} & |\delta \boldsymbol{r}_j(t)| > 0 \end{cases}. \tag{11.43}$$

In the example shown below, we use another penalty force:

$$\boldsymbol{f}_j(t) = -\kappa \frac{d}{\Delta x} \delta \boldsymbol{r}_j(t). \tag{11.44}$$

The difference to (11.42) is that the *force per node* is weighted by the average distance d between the nodes. This guarantees that increasing the number of markers (and therefore decreasing d) does not increase the *total force* at the boundary. In any case, the IBM algorithm in Sect. 11.4.1 is employed: in step 1, the forces $\boldsymbol{f}_j(t)$ are obtained *via* one of the approaches shown above.

Using the explicit penalty IBM, each marker point is allowed to be slightly carried away from its reference position. Each point applies a penalty force as discussed above. This force then tends to pull the marker back towards its reference position. After a few time steps (given a steady flow), a marker point will reach an "equilibrium position" where the force it exerts on the fluid is just enough to keep the fluid, and therefore itself, in place. It has then achieved a no-slip condition locally.

Ideally, the exact form of the penalty force should not be important, but it depends on the chosen parameter values whether this is actually the case. For example, if κ is too small, the undesired deformation becomes too large, and if κ is too large, the simulation can become unstable. A clear disadvantage of this method is that the optimum range for κ has to be obtained and that a small time step may be necessary. It is not possible to achieve perfectly rigid boundaries with an explicit IBM algorithm.

Finally, we distinguish between three fundamental cases:

1. The rigid body is fixed in space. All reference points $\boldsymbol{r}_j^{(0)}$ are stationary, and their positions do not have to be updated. The Poiseuille flow in the example below belongs to this category.
2. The body is rigid, and its motion is externally prescribed. This is similar to the first case, but the marker point positions $\boldsymbol{r}_j^{(0)}$ are updated according to the a priori known velocity.
3. The body is rigid but can move *freely* in space. This means that the reference points $\boldsymbol{r}_j^{(0)}$ have to be updated according to the equations of motion of a rigid body. In contrast to the second case, this requires the momentum and angular momentum exchange to be integrated on the surface of the body to find the total force and torque acting on the body. Updating the marker positions of rigid bodies can be complicated. We will not discuss details here and instead refer to the literature [74, 87, 91, 98].

11.4.3.2 Stationary Boundary: Poiseuille Flow

We simulate a force-driven Poiseuille flow along the x-axis in 2D with (nearly) rigid boundaries as shown in Fig. 11.13. The gravitational force density driving the flow is $F = 10^{-5}$, and the fluid domain consists of $N_x \times N_y = 19 \times 20$ nodes on a D2Q9 lattice. The BGK collision operator with $\tau = \Delta t$ is used. We approximate both walls by lines of markers with mutual distance d as free parameter and employ the penalty force in (11.44). The distance between the IBM walls is $D = 15.3\Delta x$. The spring constant κ is the second free parameter. Simulations are run until the velocity profile is stationary.

We have chosen a prime number for N_x and a non-integer for D to reduce the symmetry of the problem and therefore avoid situations which may accidentally have small numerical errors. Note, however, that the chosen benchmark problem is still highly idealised. Typical IBM applications involve moving curved boundaries with complex shapes. The purpose of this exercise is to get an initial feeling for the IBM simulation parameters.

The first task is to investigate the effect of the Lagrangian mesh spacing $d/\Delta x$. We keep $\kappa = \Delta t$ fixed and vary d for two interpolations stencils, ϕ_2 and ϕ_4. As error measure we take the largest value of u_y in the simulation, normalised by the Poiseuille peak velocity $\hat{u}_x$. Note that ideally we expect $u_y = 0$ everywhere. The results are shown in Fig. 11.14a. The ϕ_2-errors are larger than the ϕ_4-errors for $d > \Delta x$, but they are smaller for $d < \Delta x$. A resonance effect with vanishing u_y can be seen for $d = \Delta x$ and $d = 0.5\Delta x$. In those situations the problem is highly symmetric as the system is x-periodic after a single lattice unit. Generally we conclude that d should not be larger than $1.5\Delta x$. Cheng et al. [76] reported a similar

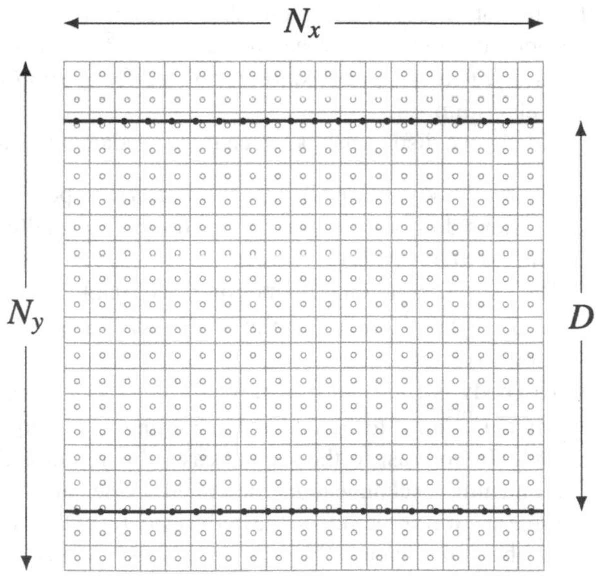

Fig. 11.13 Setup of the Poiseuille flow problem. The lattice size is $N_x \times N_y = 19\Delta x \times 20\Delta x$, and the distance between the IBM walls is $D = 15.3\Delta x$. In this particular example, $d = 0.95\Delta x$ is chosen

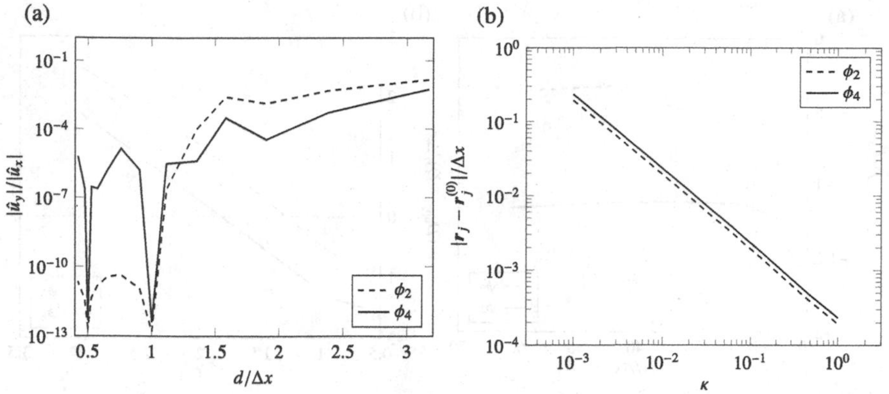

Fig. 11.14 Benchmark results showing the effect of mesh spacing d and penalty parameter κ on the accuracy of the explicit IBM for Poiseuille flow. (**a**) Sensitivity to mesh spacing. (**b**) Sensitivity to penalty parameter

observation. For this simple example, ϕ_2 provides significantly better results than ϕ_4, but this observation should certainly not be generalised to arbitrary situations. The resonance effect is expected to disappear for more complex geometries with curved boundaries.

In the second test, we set $d = \Delta x$ and vary the penalty parameter κ. Due to the explicitness of the algorithm, the Lagrangian nodes are slightly dragged by the fluid along the x-axis until the penalty force balances the drag force. We show the displacement of the Lagrangian nodes as function of penalty parameter κ in Fig. 11.14b. As expected, the displacement is inversely proportional to the penalty parameter. For $\kappa = 1$, the displacement is less than 0.1% of a lattice spacing Δx, which should be sufficient for most applications. We found that $\kappa > 3$ leads to instability. Concluding, $d = \Delta x$ and $\kappa = 1$ are reasonable choices for the current problem; we will keep these values for the final tests. Note, however, that different flow configurations may require different parameter values for optimum results.

We now investigate the apparent boundary location and the convergence rate of the IBM. For that purpose, we perform a grid refinement study. We only vary the system size, but keep $d = \Delta x$, $\kappa = 1$ and $\tau = \Delta t$ fixed (diffusive scaling). As a consequence, the gravitational force density F scales with $(D/\Delta x)^{-3}$ and the expected peak velocity $\hat{u}_x$ with $(D/\Delta x)^{-1}$ (cf. Chap. 7). For each simulation, we fit a parabola to the flow field in the central region between $\pm D/2$ and compute the appararent channel diameter D_{app}. Figure 11.15a shows the mismatch of the channel diameter, $D_{\text{app}} - D$, as function of resolution. Obviously the channel appears to be smaller than expected, which also leads to a reduced peak velocity compared to its expected values (not shown here). The mismatch is larger for ϕ_4 than for ϕ_2. Futhermore, the diameter mismatch does not significantly depend on the resolution. This means that the mismatch cannot be removed by increasing the resolution, which leads only to a first-order convergence rate of the velocity error.

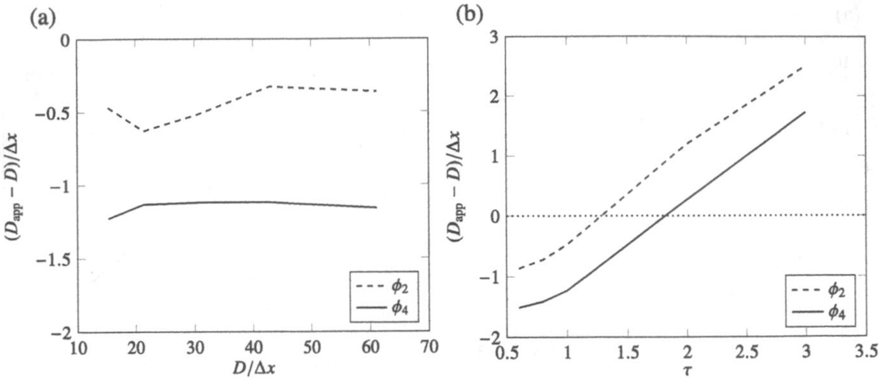

Fig. 11.15 Channel diameter mismatch as function of spatial resolution and LBM relaxation time τ. (**a**) Sensitivity to resolution. (**b**) Sensitivity to relaxation time

In the final test we investigate how the wall location mismatch depends on the relaxation time τ. We now vary τ at fixed resolution. Figure 11.15b reveals that D_{app} is a function of τ. Depending on the value of τ, the channel can appear smaller or larger than expected. While the exact values depend on the choice of the interpolation stencil, we can conclude that the appararent channel diameter increases roughly linearly with τ for $\tau > \Delta t$. This is a highly undesirable effect that has been discussed by several authors [76, 79, 99]. It has recently been suggested to use the MRT [78] or TRT [79] collision operators to resolve this problem. We will not discuss these approaches here.

As already concluded at the end of Sect. 11.4.2, the IBM accuracy is typically inferior to other available boundary conditions. Care has to be taken when the exact channel diameter or particle size (for suspension simulations) is important. In the end, it can take a significant amount of work to make sure that an IBM code is working reliably.

11.4.4 Direct-Forcing IB-LBM for Rigid Boundaries

The explicit penalty IBM for rigid boundaries has a major disadvantage: it involves a free parameter whose choice affects the stability and accuracy. We seek an alternative implementation without a free parameter. This means that the IB force has to be computed directly from the flow field. Therefore, we call this class of methods *direct-forcing IB-LBM*.

Feng and Michaelides [91, 100] originally combined the direct forcing IBM with the LBM. A number of alternative direct-forcing IB-LBMs have been proposed since then. We can distinguish between three different approaches, each with different levels of accuracy and numerical cost:

1. implicit IBM
2. multi-direct-forcing IBM (iterative)
3. direct-forcing IBM (explicit)

We cannot present and compare all available approaches in depth. Instead, we will start from the underlying hydrodynamic problem and show which steps are necessary to construct a reliable parameter-free IB-LBM.

11.4.4.1 Background

We assume a rigid boundary described by a number of marker points at positions $\boldsymbol{r}_j$. The markers have known velocities $\dot{\boldsymbol{r}}_j = \boldsymbol{u}_\text{b}(\boldsymbol{r}_j)$, the desired boundary velocity. In most situations, the boundary is resting, but there is no fundamental difficulty with moving (translating and rotating) boundaries.

Before collision, the fluid velocity on a lattice node $\boldsymbol{x}$ is

$$\boldsymbol{u}(\boldsymbol{x}) = \frac{1}{\rho(\boldsymbol{x})}\sum_i f_i(\boldsymbol{x})\boldsymbol{c}_i. \tag{11.45}$$

In the absence of a force, collision leaves the momentum and velocity invariant. The only mechanism that can change the fluid velocity during collision is a body force $\boldsymbol{F}$:

$$\boldsymbol{u}^\star(\boldsymbol{x}) = \frac{1}{\rho(\boldsymbol{x})}\sum_i f_i^\star(\boldsymbol{x})\boldsymbol{c}_i = \frac{1}{\rho(\boldsymbol{x})}\sum_i f_i(\boldsymbol{x})\boldsymbol{c}_i + \frac{\boldsymbol{F}(\boldsymbol{x})\Delta t}{\rho(\boldsymbol{x})}. \tag{11.46}$$

As usual, a star denotes post-collision quantities. Furthermore, we drop the time t because everything which follows is happening within a single time step.

We know that the physical fluid velocity during a time step is the average of the pre- and post-collision velocities [83]:

$$\boldsymbol{u}_\text{f}(\boldsymbol{x}) = \frac{\boldsymbol{u}(\boldsymbol{x}) + \boldsymbol{u}^\star(\boldsymbol{x})}{2} = \boldsymbol{u}(\boldsymbol{x}) + \frac{\boldsymbol{F}(\boldsymbol{x})\Delta t}{2\rho(\boldsymbol{x})}. \tag{11.47}$$

The central idea of any direct-forcing IB-LBM is to construct the force $\boldsymbol{F}(\boldsymbol{x})$ in such a way that $\boldsymbol{u}_\text{f}(\boldsymbol{x})$ matches the known boundary velocity $\boldsymbol{u}_\text{b}(\boldsymbol{r}_j)$ at the marker positions to satisfy the no-slip condition. As we will now see, this is a non-trivial task and explains why there is a number of direct-forcing variants in the literature.

Since we are working in the framework of the IBM, the boundary velocity is known at the positions of the boundary markers: $\boldsymbol{u}_\text{b}(\boldsymbol{r}_j) = \dot{\boldsymbol{r}}_j$. This means that we have to interpolate $\boldsymbol{u}_\text{f}(\boldsymbol{x})$ at the boundary markers $\boldsymbol{r}_j$ to obtain $\boldsymbol{u}_\text{f}(\boldsymbol{r}_j)$ first, then find the required boundary force $\boldsymbol{f}_j$ and finally spread the force back to the lattice to obtain $\boldsymbol{F}(\boldsymbol{x})$.

The difficulty is caused be the non-local velocity interpolation and force spreading. In order to compute $\boldsymbol{f}_j$, we require the velocity on all lattice nodes $\boldsymbol{x}$ close to $\boldsymbol{r}_j$. In return, we have to spread $\boldsymbol{f}_j$ back to those lattice nodes. However, many lattice nodes $\boldsymbol{x}$ participate in interpolation and spreading of more than one marker $\boldsymbol{r}_j$ *at the same time*. This means that (11.47) has to be solved *simultaneously* on all lattice nodes to guarantee a consistent solution.

11.4.4.2 Implicit IB-LBM

In 2009, Wu and Shu [101] proposed the *implicit velocity correction-based IB-LBM* which is probably the most accurate and consistent way to enforce the no-slip condition at a rigid boundary with the IB-LBM. We will only provide the derivation of the algorithm. Benchmark tests can be found in [101–103].

The basic idea of the implicit IB-LBM is to consider the required force density $\boldsymbol{F}(\boldsymbol{x})$ as the unknowns for which the problem has to be solved in such a way that the no-slip condition is satisfied. Since the unknowns $\boldsymbol{F}(\boldsymbol{x})$ depend on the current and the desired flow field, the problem is implicit.

The first step is to write the physical fluid velocity in (11.47) as

$$\boldsymbol{u}_{\mathrm{f}}(\boldsymbol{x}) = \boldsymbol{u}(\boldsymbol{x}) + \delta\boldsymbol{u}(\boldsymbol{x}) \tag{11.48}$$

where $\boldsymbol{u}(\boldsymbol{x})$, given by (11.45), is the known uncorrected velocity and $\delta\boldsymbol{u}(\boldsymbol{x}) = (\boldsymbol{F}(\boldsymbol{x})\Delta t)/(2\rho(\boldsymbol{x}))$ is the unknown velocity correction required to achieve the desired no-slip condition.

Now, we use the IBM relations in (11.32) and (11.33) to link Eulerian and Lagrangian quantities. We can express the Eulerian correction terms $\delta\boldsymbol{u}(\boldsymbol{x})$ by their Lagrangian counterparts $\delta\boldsymbol{u}(\boldsymbol{r}_j)$:

$$\delta\boldsymbol{u}(\boldsymbol{x}) = \sum_j \delta\boldsymbol{u}(\boldsymbol{r}_j)\Delta(\boldsymbol{r}_j, \boldsymbol{x}). \tag{11.49}$$

Note that $\delta\boldsymbol{u}(\boldsymbol{r}_j)$ is proportional to the unknown Lagrangian force $\boldsymbol{f}_j$.

Let $\boldsymbol{u}_{\mathrm{b}}(\boldsymbol{r}_j)$ be the desired boundary velocity imposed on the Lagrangian nodes. The aim is construct an Eulerian flow field $\boldsymbol{u}_{\mathrm{f}}(\boldsymbol{x})$ which, interpolated at the boundary nodes, equals the boundary velocity $\boldsymbol{u}_{\mathrm{b}}(\boldsymbol{r}_j)$:

$$\boldsymbol{u}_{\mathrm{b}}(\boldsymbol{r}_j) = \sum_{\boldsymbol{x}} \boldsymbol{u}_{\mathrm{f}}(\boldsymbol{x})\Delta(\boldsymbol{r}_j, \boldsymbol{x}). \tag{11.50}$$

To achieve this, we combine (11.48), (11.49) and (11.50):

$$u_b(r_j) = \sum_x \left[u(x) + \sum_k \delta u(r_k) \Delta(r_k, x) \right] \Delta(r_j, x). \tag{11.51}$$

The only unknowns in this equation are the desired correction terms $\delta u(r_k)$. We can rewrite this equation in the following way:

$$\underbrace{\sum_k \left[\sum_x \Delta(r_k, x) \Delta(r_j, x) \right]}_{A_{jk}} \underbrace{\delta u(r_k)}_{X_k} = \underbrace{u_b(r_j) - \sum_x u(x) \Delta(r_j, x)}_{B_j} \tag{11.52}$$

or, in simple matrix-vector notation, as $AX = B$.

The vectors X and B have N elements, and A is an $N \times N$-matrix where N is the number of marker points, i.e. j and k run from 1 to N. The elements of A are functions of the node positions r_j only, depending on the choice of the IBM stencil Δ. Finding the unknowns X *via* $X = A^{-1}B$ requires inversion of A. Obviously the matrix A can be large, with N typically ranging from several 10^2 to several 10^4 or 10^5.

Concluding, the implicit IB-LBM algorithm works as follows:

1. Compute the matrix A and its inverse A^{-1} from the known node positions r_j. See [101] for details.
2. Stream the populations to obtain $f_i(x)$ and compute the density and uncorrected velocity from $\rho u = \sum_i f_i c_i$.
3. Using the known boundary velocity $u_b(r_j)$ and the uncorrected fluid velocity $u(x)$, solve the matrix equation, (11.52), for the unknown corrections $\delta u(r_j)$.
4. Spread $\delta u(r_j)$ to the Eulerian grid *via* (11.49).
5. Compute the desired force density from $\delta u(x) = (F(x)\Delta t)/(2\rho(x))$.
6. Perform collision with forcing.
7. If the boundary is stationary, i.e. all boundary velocities obey $u_b(r_j) = 0$, go back to step 2 for the next time step.
8. If the boundary is not stationary, update the positions r_j and velocities $\dot{r}_j = u_b(r_j)$. The position update may be enforced (e.g. oscillating cylinder) or a consequence of fluid stresses (e.g. freely moving cylinder). In the latter case, a suitable time integrator has to be chosen [74, 87, 91, 98].
9. Go back to step 1 for the next time step.

Note that the re-computation of the matrix A and its inverse at each time step for non-stationary boundaries can be expensive when N is large. Therfore, alternative approaches, such as multi-direct forcing, which are computationally more efficient and conceptually simpler, have been suggested.

11.4.4.3 Multi Direct-Forcing IB-LBM

The aim of *multi direct-forcing IB-LBM* is to avoid the construction and inversion of the matrix $\boldsymbol{A}$ of the implicit IB-LBM, while keeping its consistency. Kang and Hassan [75] provided an exhaustive overview of the multi direct-forcing IB-LBM. Since the underlying idea is similar to that of the implicit IB-LBM, we only provide the algorithm and a few comments.

Instead of constructing and inverting a large matrix $\boldsymbol{A}$, the multi direct-forcing method relies on an iterative approach to satisfy the no-slip condition at all markers $\boldsymbol{r}_j$ simultaneously. Again, the underlying idea is to take advantage of the velocity correction in (11.47). The algorithm of the multi direct-forcing approach can be summarised as follows:

1. Set iteration counter m to 0.
2. Stream the populations to obtain $f_i(\boldsymbol{x})$ and compute the density and uncorrected velocity from $\rho\boldsymbol{u}^{(m)} = \sum_i f_i\boldsymbol{c}_i$.
3. Interpolate $\boldsymbol{u}^{(m)}(\boldsymbol{x})$ at the boundary marker locations $\boldsymbol{r}_j$ to obtain $\boldsymbol{u}^{(m)}(\boldsymbol{r}_j)$.
4. Increment iteration counter m by 1.
5. Compute the Lagrangian correction force from [75]

$$\boldsymbol{f}_j^{(m)} = 2\rho\frac{\boldsymbol{u}_\text{b}(\boldsymbol{r}_j) - \boldsymbol{u}^{(m-1)}(\boldsymbol{r}_j)}{\Delta t}. \tag{11.53}$$

6. Spread $\boldsymbol{f}_j^{(m)}$ to the Eulerian lattice to obtain $\boldsymbol{F}^{(m)}(\boldsymbol{x})$.
7. Correct previous Eulerian velocity according to

$$\boldsymbol{u}^{(m)}(\boldsymbol{x}) = \boldsymbol{u}^{(m-1)}(\boldsymbol{x}) + \frac{\boldsymbol{F}^{(m)}(\boldsymbol{x})\Delta t}{2\rho(\boldsymbol{x})}. \tag{11.54}$$

8. Repeat steps 3–7 until m reaches a pre-defined limit m_max or until $\boldsymbol{u}^{(m)}(\boldsymbol{r}_j)$ converges to $\boldsymbol{u}_\text{b}(\boldsymbol{r}_j)$.
9. Use the total correction force

$$\boldsymbol{F}(\boldsymbol{x}) = \sum_{m=1}^{m_\text{max}} \boldsymbol{F}^{(m)}(\boldsymbol{x}) \tag{11.55}$$

   in the collision step.
10. Go back to step 1 for the next time step.

Kang and Hassan [75] compared results of benchmark tests for different iteration numbers up to $m_{\max} = 20$. They found that $m_{\max} = 5$ is a reasonable compromise of accuracy and efficiency. Since the iteration involves only those lattice nodes close to the boundary, the additional computational cost is relatively low.

11.4.4.4 Explicit, Non-iterative Direct-Forcing IB-LBM

As pointed out by Kang and Hassan [75], a non-iterative direct-forcing scheme can be obtained as special case of the multi direct-forcing method in the previous section. Setting $m_{\max} = 1$ leads to a simple explicit scheme that does not require expensive matrix inversions or iterations. This special case is commonly denoted "direct-forcing" IB-LBM, although the implicit and iterative methods are, strictly speaking, also direct-forcing methods.[7]

There exist different flavours of non-iterative direct-forcing IB-LBM (see for example [79]). However, it is obvious that this method will generally not give results of a comparable accuracy and consistency compared to implicit or iterative schemes.

11.4.5 Explicit IBM for Deformable Boundaries

The first works utilising the deformable IB-LBM for flowing deformable red blood cells were published in 2007 by several groups [104–106]. The overall algorithm follows the layout described in Sect. 11.4.2. The step from the IBM algorithm for rigid boundaries as presented in Sect. 11.4.3 to deformable boundaries is straightforward. Instead of finding suitable penalty forces to keep the boundary deformation as small as possible, one has to use forces which arise from elastic surface stresses due to the (desired) deformation of the boundary. This requires two additional ingredients: (i) a constitutive model for the boundary deformation and (ii) a surface mesh (i.e. markers *and* their connectivity) to evaluate the boundary deformation.

Sui et al. [107] were the first to present a 3D model for elastic particles (capsules, red blood cells) in an LB simulation with well-defined constitutive behaviour and a finite-element method to find the elastic membrane forces. Krüger et al. [85] later investigated the effect of the choice of interpolation stencil and distance between neighbouring Lagrangian nodes on the deformation of a capsule in shear flow. The IB-LBM for elastic problems has been applied to, for example, viscous flow over a flexible sheet [96] and dense suspensions of red blood cells [108] (see also Fig. 11.16).

[7] Remember that "direct forcing" means that there are no free parameters, such as the elasticity κ of the explicit method in Sect. 11.4.3.

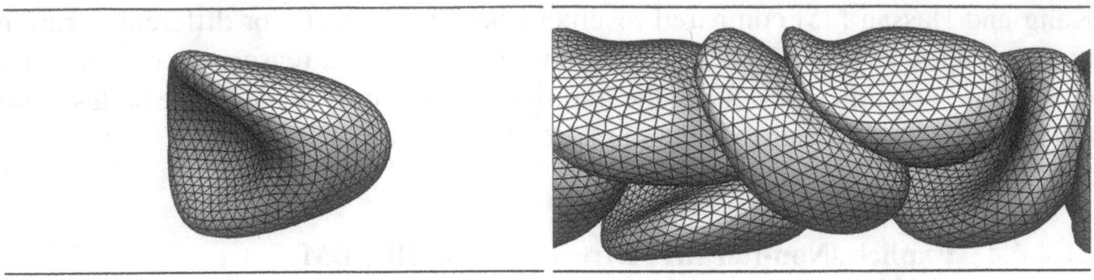

Fig. 11.16 Flow of a single (*left*) and multiple (*right*) red blood cells in a straight tube (indicated by *solid horizontal lines*) with circular cross-section. The tube diameter is slightly larger than that of an undeformed red blood cell. The Lagrangian mesh consists of 998 nodes and 2000 triangular elements. The simulations are based on the model presented in [108]

The IBM provides a major advantage over other boundary conditions for the LBM when it comes to deformable objects. Since IBM boundaries in the original implementation are intrinsically deformable, it is relatively simple to turn this presumed disadvantage into an advantage for problems where the deformability is actually desired. Allowing Lagrangian markers to move with the fluid and distributing forces to the fluid is the natural algorithm of the IBM and lends itself to problems where the fluid causes structure deformation and the structure "reacts" elastically. Applying any of the other boundary conditions presented in this chapter to deformable boundaries is significantly more difficult.

11.4.5.1 Constitutive Models

The *constitutive model* contains all the physics of the boundary deformation. Its choice is independent of the IBM algorithm itself and has to be defined by the user. In the end, the IBM expects the marker forces $\boldsymbol{f}_j$, but the IBM itself is unable to provide them. Boundaries are mostly considered hyperelastic (i.e. the dynamics can be fully described in terms of an energy density) or viscoelastic. In the former case, the marker forces depend only on the current deformation state, in the latter case the forces depend both on the deformation state and its rate of change.

There exists a large variety of hyperelastic and viscoelastic models for deformable boundaries. The problem of finding and implementing an appropriate constitutive model is highly problem-specific. We cannot delve into details here; this could easily fill a book on its own. The most commonly used hyperelastic models for red blood cells are briefly discussed in [107].

For simplicity, we will assume that hyperelastic models can be written in the form

$$f_j(t) = f_j(\{r_k(t)\}) \tag{11.56}$$

and viscoelastic models as

$$f_j(t) = f_j(\{r_k(t)\}, \{\dot{r}_k(t)\}). \tag{11.57}$$

This means that the instantaneous marker forces are (arbitrarily complicated) functions of all current boundary marker positions and, if viscoelastic, of all current boundary marker velocities. Once these laws have been specified, they can be coded and used to find the forces $\boldsymbol{f}_j$ for a given deformation state at every time step.

Example 11.1 A simple hyperelastic constitutive model which can be used in 2D and 3D is (dropping the time dependence for simplicity)

$$\boldsymbol{f}_j(\{\boldsymbol{r}_k\}) = -\kappa \sum_{k \neq j} \frac{d_{jk} - d_{jk}^{(0)}}{d_{jk}^{(0)}} \frac{\boldsymbol{d}_{jk}}{d_{jk}}, \quad \boldsymbol{d}_{jk} = \boldsymbol{r}_k - \boldsymbol{r}_j, \quad d_{jk} = |\boldsymbol{d}_{jk}| \tag{11.58}$$

where κ is an elastic modulus, the sum runs over all next neighbours of marker j and $d_{jk}^{(0)}$ is the equilibrium distance between markers j and k. This example shows that not only the markers, but also their connectivity is an important part of the problem description. In many situations, additional constraints are necessary, for instance conservation of the total volume or surface of a boundary.

In most cases of hyperelastic boundaries one first defines an elastic energy density $\epsilon(\{\boldsymbol{r}_j(t)\})$. The force acting on node j can then be recovered by applying the principle of virtual work,

$$\boldsymbol{f}_j = \frac{\partial \epsilon(\{\boldsymbol{r}_k\})}{\partial \boldsymbol{r}_j} A_j, \tag{11.59}$$

where A_j is the area related to marker j, e.g. its Voronoi area. More details are provided in [85, 107].

11.4.6 Additional Variants and Similar Boundary Treatments

The previous sections cover the most prominent flavours of the IB-LBM. This is, however, not the end of the rope. There are more variations on the market, some of which we want to mention in the following.

There are a number of fluid-structure interaction approaches which share some features with the IBM (in particular the existence of off-lattice markers or a Lagrangian mesh and kernel functions for velocity interpolation and/or force spreading), but their algorithms reveal distinct differences. Schiller [109] recently revisited those algorithms and pointed out their mathematical similarity.

1. Ahlrichs and Dünweg [110] introduced a dissipative coupling method for LBM and molecular dynamics (MD), which has been further analysed by Caiazzo and Maddu [95] and recently reviewed by Dünweg and Ladd [111] and is used in the open-source package ESPResSo [112], mostly for polymer simulations. Lagrangian markers are allowed to move with a different velocity than the velocity of the fluid at the location of the marker (obtained by velocity interpolation). A finite slip velocity results in a drag force acting on the marker whose magnitude is controlled *via* a numerical drag coefficient. An equal but opposite force is exerted on the fluid by spreading it to the Eulerian lattice. Additionally, the markers may experience external or interaction forces. The marker update is treated by higher-order MD, which requires the introduction of another model parameter, a finite marker mass. An advantage of this approach is that the time step for the update of the markers is decoupled from the LB time step, which can be exploited to implement more stable time-integration schemes; a freedom which is not available for the conventional IB-LBM algorithm. Disadvantages are that the no-slip condition is not strictly satisfied and that two model parameters are required (drag coefficient and marker mass).
2. The momentum-exchange-based IB-LBM, as proposed for rigid boundaries [113, 114] and recently extended to flexible boundaries [115], uses a different approach to obtain the force density acting on the fluid. The basic idea is to interpolate the LB populations (rather than the velocity) to find their value at the location of the Lagrangian markers. The bounce-back scheme is then applied on the Lagrangian mesh to find the momentum exchange and therefore the marker force $\boldsymbol{f}_j$ which is then distributed to the lattice *via* the standard force spreading operation. As a consequence, there is no need for user-defined penalty parameters (for rigid boundaries) and the markers can move independently of the fluid motion. However, this may lead to a violation of the no-slip condition.[8]
3. Wu and Aidun [116] proposed the so-called *external-boundary force* (EBF) which can be used both as alternative to the direct-forcing IBM for rigid boundaries and for deformable objects. Similar to the other two examples above, the most notable difference to the IBM is that the markers are not directly advected by the fluid. Instead, a relative slip velocity is permitted which is counteracted by a fluid-solid interaction force, which is essentially a penalty force. Note that this force does not require a free parameter like the dissipative

[8]While the authors of [113] use bounce-back to obtain the momentum exchange at the boundary markers, the populations on the lattice are not directly affected by this bounce-back procedure. Although the momentum exchange is correctly obtained, there is no strong mechanism enforcing the local no-slip condition. Therefore, streamlines may penetrate the boundary.

coupling does. Also here, the allowance of a relative slip velocity may lead to problems with the no-slip condition.

Concluding, we can say that, although it is desirable to decouple the motion of the marker points from the fluid motion (as this allows higher-order and therefore potentially more stable time integration schemes), the no-slip condition at the boundary cannot be strictly enforced at the same time. The reason is that all the methods discussed above [110, 113, 115, 116] are explicit with respect to the fluid velocity computation. Still, if the precise realisation of the no-slip condition is not the primary goal, the above methods are attractive alternatives and overcome some of the disadvantages of the more conventional IB schemes.

It is also worth mentioning that the IBM can be used to model thermal boundary conditions. In 2010, Jeong et al. [117] combined the IBM with a thermal LBM to simulate flows around bluff bodies with heat transfer. Seta [118] later improved the thermal IB-LBM by analysing the governing equations through a Chapman-Enskog analysis. Another IBM variant that has apparently not yet been combined with the LBM is the so-called penalty IBM (p-IBM) [119–121] for flexible boundaries. It involves two set of Lagrangian markers: one interacting with the fluid, the other used for the calculation of the Lagrangian forces. Both marker sets are coupled by springs generating penalty forces.

11.5 Concluding Remarks

The number of available boundary conditions for the LBM is overwhelming, and it can be a daunting task to grasp the implications of those schemes. In the following we list a series of publications which provide comparative studies of boundary conditions for curved geometries. This should help to understand the relative performance of certain boundary methods for a given flow geometry.

- Ginzburg and d'Humières [17] compared simple (Sect. 11.2.1) and interpolated bounce-back (Sect. 11.2.2) with the multireflection method [17] in a number of stationary situations (inclined Couette and Poiseuille flows, flow over single cylinder and array of cylinders, impulsively started cylinder and moving sphere in a cylinder). They conclude that the multireflection method is more accurate than the linear interpolated bounce-back method.
- Pan et al. [12] compared the performance of simple bounce-back, interpolated bounce-back (both linear and quadratic interpolations) and multireflection boundary conditions for porous media simulations. They also included an analysis of the effect of collision operator (BGK *vs.* MRT) on the permeability of idealised porous media. Their main finding is that the permeability is generally viscosity-dependent through an unphysical dependence on the relaxation rate(s). Especially the combination of BGK and simple bounce-back leads to a strongly increasing permeability with viscosity, an effect caused by the increasing slip velocity at the boundary. They conclude that the combination of simple bounce-

back with MRT is consistently better than with BGK. The reason is that the no-slip condition can be much better controlled in the MRT framework when the viscosity is large. Using interpolated rather than simple bounce-back can also improve the accuracy of the simulations.

- Peng and Luo [94] investigated the relative performance of interpolated bounce-back and a direct-forcing immersed boundary method (IBM, Sect. 11.4.4). They considered steady and unsteady flows about a stationary rigid cylinder in 2D. Their major findings are that both methods require roughly the same computing time, the interpolated bounce-back is more accurate, but that the IBM is easier to implement.
- Chen et al. [47] have combined the ghost method (Sect. 11.3.4) with the partially saturated method (PSM, Sect. 11.2.3) in order to remove spurious pressure oscillations. The immediate consequence is that the algorithm becomes significantly more complicated than Noble's and Torczynski's [46] original one: interpolations become necessary to find the ghost node properties and a treatment of fresh nodes is required. This is a clear disadvantage compared to the original method [46] where the fresh node problem is naturally avoided. The conclusion is that the combined method yields better results than the original PSM or the interpolated bounce-back method, especially for moving obstacles at smaller resolutions. According to Chen et al., PSM is recommended when code simplicity and efficiency are desired, while the combined method should be favoured for high-accuracy applications.
- Chen et al. [14] recently conducted a thorough comparison of three bounce-back schemes (standard, interpolated and unified interpolation), two IBM variants (explicit and implicit direct forcing) and three additional methods. The authors were primarily interested in acoustic problems involving sound wave generation from moving bodies due to the fresh node treatment. The authors found that the IBM is more suitable for moving boundaries than the interpolated bounce-back when fresh nodes are involved.
- Nash et al. [122] compared the accuracy of simple and interpolated bounce-back, the Guo-Zheng-Shi extrapolation method (GZS, Sect. 11.3.3) and the Junk-Yang method in non-grid-aligned Poiseuille, Womersley and Dean flows at moderate Reynolds numbers (up to 300). The authors found that the Junk-Yang method shows poor stability in the selected parameter range. The linear interpolated bounce-back and the GZS methods have comparable accuracy (with a second-order convergence) although the latter becomes unstable for the highest Reynolds numbers tested. For the situation of interest (flow in inclined channels with moderate Reynolds number), the authors conclude that interpolated bounce-back is the best all-around option, although simple bounce-back (despite its first-order convergence) may be the method of choice when code development time is at a premium.

Everything said up to this point applies to *rigid* boundaries. It seems that the IBM and its related methods is still the most convenient approach for *deformable* boundaries (cf. Sects. 11.4.5 and 11.4.6). The reason is that all other boundary conditions presented in this chapter require an accurate *local* momentum exchange algorithm to compute the local stresses in the deformable material.[9] This is normally a very challenging and expensive problem that is elegantly circumvented by the IBM.

We can generally conclude that all existing boundary conditions claim their own compromise of accuracy, stability and efficiency/ease of implementation. Furthermore, some boundary conditions perform better in stationary situations, others when the boundaries are moving. It is up to the user to identify the requirements before choosing one of the many available boundary conditions. There is no best boundary treatment for all possible scenarios. We hope that this chapter sheds some light on the plethora of boundary conditions and helps the reader to find a suitable scheme for a given problem.

References

1. S. Haeri, J.S. Shrimpton, Int. J. Multiphas. Flow **40**, 38 (2012)
2. X. He, G. Doolen, J. Comput. Phys. **134**, 306 (1997)
3. P. Lallemand, L.S. Luo, Phys. Rev. E **61**(6), 6546 (2000)
4. T. Lee, C.L. Lin, J. Comput. Phys. **171**(1), 336 (2001)
5. Z. Guo, T.S. Zhao, Phys. Rev. E **67**(6), 066709 (2003)
6. N. Rossi, S. Ubertini, G. Bella, S. Succi, Int. J. Numer. Meth. Fluids **49**(6), 619 (2005)
7. H. Yoshida, M. Nagaoka, J. Comput. Phys. **257, Part A**, 884 (2014)
8. R. Cornubert, D. d'Humières, D. Levermore, Physica D **47**, 241 (1991)
9. I. Ginzburg, P.M. Adler, J. Phys. II France **4**(2), 191 (1994)
10. A.J.C. Ladd, J. Fluid Mech. **271**, 285 (1994)
11. M. Bouzidi, M. Firdaouss, P. Lallemand, Phys. Fluids **13**, 3452 (2001)
12. C. Pan, L.S. Luo, C.T. Miller, Comput. Fluids **35**(8-9), 898 (2006)
13. O.E. Strack, B.K. Cook, Int. J. Numer. Meth. Fluids **55**(2), 103 (2007)
14. L. Chen, Y. Yu, J. Lu, G. Hou, Int. J. Numer. Meth. Fluids **74**(6), 439 (2014)
15. S. Khirevich, I. Ginzburg, U. Tallarek, J. Comp. Phys. **281**, 708 (2015)
16. M.O. Bernabeu, M.L. Jones, J.H. Nielsen, T. Krüger, R.W. Nash, D. Groen, S. Schmieschek, J. Hetherington, H. Gerhardt, C.A. Franco, P.V. Coveney, J. R. Soc. Interface **11**(99), 20140543 (2014)
17. I. Ginzburg, D. d'Humières, Phys. Rev. E **68**, 066614 (2003)
18. B. Chun, A.J.C. Ladd, Phys. Rev. E **75**, 066705 (2007)
19. X. He, Q. Zou, L.S. Luo, M. Dembo, J. Stat. Phys. **87**(1–2), 115 (1997)
20. I. Ginzburg, D. d'Humières, J. Stat. Phys. **84**, 927 (1996)
21. A.J.C. Ladd, Phys. Rev. Lett. **70**(9), 1339 (1993)
22. A.J.C. Ladd, J. Fluid Mech. **271**, 311 (1994)
23. A.J.C. Ladd, R. Verberg, J. Stat. Phys. **104**(5–6), 1191 (2001)
24. C.K. Aidun, J.R. Clausen, Annu. Rev. Fluid Mech. **42**, 439 (2010)
25. N.Q. Nguyen, A.J.C. Ladd, Phys. Rev. E **66**(4), 046708 (2002)
26. E.J. Ding, C.K. Aidun, J. Stat. Phys. **112**(3–4), 685 (2003)
27. C.K. Aidun, Y. Lu, E.J. Ding, J. Fluid Mech. **373**, 287 (1998)

[9] For rigid objects it is sufficient to compute the total momentum and angular momentum transfer. Soft objects, however, deform locally. This local deformation depends on the local stresses.

28. D. Qi, J. Fluid Mech. **385**, 41 (1999)
29. X. Yin, G. Le, J. Zhang, Phys. Rev. E **86**(2), 026701 (2012)
30. E. Lorenz, A. Caiazzo, A.G. Hoekstra, Phys. Rev. E **79**(3), 036705 (2009)
31. J.R. Clausen, C.K. Aidun, Int. J. Multiphas. Flow **35**(4), 307 (2009)
32. B. Wen, C. Zhang, Y. Tu, C. Wang, H. Fang, J. Comput. Phys. **266**, 161 (2014)
33. O. Behrend, Phys. Rev. E **52**(1), 1164 (1995)
34. M.A. Gallivan, D.R. Noble, J.G. Georgiadis, R.O. Buckius, Int. J. Numer. Meth. Fluids **25**(3), 249–263 (1997)
35. Y. Han, P.A. Cundall, Int. J. Numer. Meth. Fluids **67**(3), 314–327 (2011)
36. P.H. Kao, R.J. Yang, J. Comput. Phys. **227**(11), 5671 (2008)
37. I. Ginzburg, F. Verhaeghe, D. d'Humières, Commun. Comput. Phys. **3**, 427 (2008)
38. I. Ginzburg, F. Verhaeghe, D. d'Humières, Commun. Comput. Phys. **3**, 519 (2008)
39. X. Descovich, G. Pontrelli, S. Melchionna, S. Succi, S. Wassertheurer, Int. J. Mod. Phys. C **24**(05), 1350030 (2013)
40. P. Lallemand, L.S. Luo, J. Comput. Phys. **184**(2), 406 (2003)
41. D. Yu, R. Mei, W. Shyy, in *41st Aerospace Sciences Meeting and Exhibit*, 2003-953 (AIAA, New York, 2003)
42. X. Yin, J. Zhang, J. Comput. Phys. **231**(11), 4295 (2012)
43. O. Dardis, J. McCloskey, Phys. Rev. E **57**(4), 4834 (1998)
44. S.D.C. Walsh, H. Burwinkle, M.O. Saar, Comput. Geosci. **35**(6), 1186 (2009)
45. J. Zhu, J. Ma, Adv. Water Resour. **56**, 61 (2013)
46. D.R. Noble, J.R. Torczynski, Int. J. Mod. Phys. C **09**(08), 1189 (1998)
47. L. Chen, Y. Yu, G. Hou, Phys. Rev. E **87**(5), 053306 (2013)
48. I. Ginzburg, Adv. Water Resour. **88**, 241 (2016)
49. I. Ginzburg, G. Silva, L. Talon, Phys. Rev. E **91**, 023307 (2015)
50. H. Yoshida, H. Hayashi, J. Stat. Phys. **155**, 277 (2014)
51. G. Zhou, L. Wang, X. Wang, W. Ge, Phys. Rev. E **84**(6), 066701 (2011)
52. H. Yu, X. Chen, Z. Wang, D. Deep, E. Lima, Y. Zhao, S.D. Teague, Phys. Rev. E **89**(6), 063304 (2014)
53. R. Mei, L.S. Luo, P. Lallemand, D. d'Humières, Comput. Fluids **35**(8-9), 855 (2006)
54. G. Pontrelli, C.S. König, I. Halliday, T.J. Spencer, M.W. Collins, Q. Long, S. Succi, Med. Eng. Phys. **33**(7), 832 (2011)
55. B. Stahl, B. Chopard, J. Latt, Comput. Fluids **39**(9), 1625–1633 (2010)
56. M. Matyka, Z. Koza, Ł. Mirosław, Comput. Fluids **73**, 115 (2013)
57. X. Kang, Z. Dun, Int. J. Mod. Phys. C p. 1450057 (2014)
58. R. Mei, L.S. Luo, W. Shyy, J. Comput. Phys. **155**(2), 307 (1999)
59. B. Wen, H. Li, C. Zhang, H. Fang, Phys. Rev. E **85**(1), 016704 (2012)
60. Z.L. Guo, C.G. Zheng, B.C. Shi, Phys. Fluids **14**, 2007 (2002)
61. A. Tiwari, S.P. Vanka, Int. J. Numer. Meth. Fluids **69**(2), 481 (2012)
62. O.R. Mohammadipoor, H. Niazmand, S.A. Mirbozorgi, Phys. Rev. E **89**(1), 013309 (2014)
63. J.C.G. Verschaeve, B. Müller, J. Comput. Phys. **229**, 6781 (2010)
64. J. Latt, B. Chopard, O. Malaspinas, M. Deville, A. Michler, Phys. Rev. E **77**(5), 056703 (2008)
65. O. Filippova, D. Hänel, J. Comput. Phys. **147**, 219 (1998)
66. R. Mei, D. Yu, W. Shyy, L.S. Luo, Phys. Rev. E **65**(4), 041203 (2002)
67. J. Bao, P. Yuan, L. Schaefer, J. Comput. Phys. **227**(18), 8472 (2008)
68. R. Khazaeli, S. Mortazavi, M. Ashrafizaadeh, J. Comput. Phys. **250**, 126 (2013)
69. Q. Zou, X. He, Phys. Fluids **9**, 1591 (1997)
70. N. Pellerin, S. Leclaire, M. Reggio, Comput. Fluids **101**, 126 (2014)
71. C.S. Peskin, Flow patterns around heart valves: A digital computer method for solving the equations of motion. Ph.D. thesis, Sue Golding Graduate Division of Medical Sciences, Albert Einstein College of Medicine, Yeshiva University (1972)
72. C.S. Peskin, J. Comput. Phys. **25**(3), 220 (1977)
73. C.S. Peskin, Acta Numerica **11**, 479–517 (2002)
74. Z.G. Feng, E.E. Michaelides, J. Comput. Phys. **195**(2), 602 (2004)

75. S.K. Kang, Y.A. Hassan, Int. J. Numer. Meth. Fluids **66**(9), 1132 (2011)
76. Y. Cheng, L. Zhu, C. Zhang, Commun. Comput. Phys. **16**(1), 136 (2014)
77. R. Mittal, G. Iaccarino, Annu. Rev. Fluid Mech. **37**, 239 (2005)
78. J. Lu, H. Han, B. Shi, Z. Guo, Phys. Rev. E **85**(1), 016711 (2012)
79. T. Seta, R. Rojas, K. Hayashi, A. Tomiyama, Phys. Rev. E **89**(2), 023307 (2014)
80. B.E. Griffith, X. Luo, D.M. McQueen, C.S. Peskin, Int. J. Appl. Mech. **01**, 137 (2009)
81. X. Wang, L.T. Zhang, Comput. Mech. **45**(4), 321 (2010)
82. X. Yang, X. Zhang, Z. Li, G.W. He, J. Comput. Phys. **228**(20), 7821 (2009)
83. Z. Guo, C. Zheng, B. Shi, Phys. Rev. E **65**, 46308 (2002)
84. S.K. Doddi, P. Bagchi, Int. J. Multiphas. Flow **34**(10), 966 (2008)
85. T. Krüger, F. Varnik, D. Raabe, Comput. Method. Appl. **61**(12), 3485 (2011)
86. G. Tryggvason, B. Bunner, A. Esmaeeli, D. Juric, N. Al-Rawahi, W. Tauber, J. Han, S. Nas, Y.J. Jan, J. Comput. Phys. **169**(2), 708 (2001)
87. K. Suzuki, T. Inamuro, Comput. Fluids **49**(1), 173 (2011)
88. D. Nie, J. Lin, Commun. Comput. Phys. **7**(3), 544 (2010)
89. S.K. Doddi, P. Bagchi, Phys. Rev. E **79**(4), 046318 (2009)
90. S. Ramanujan, C. Pozrikidis, J. Fluid Mech. **361**, 117 (1998)
91. Z.G. Feng, E.E. Michaelides, Comput. Fluids **38**(2), 370 (2009)
92. CGAL, Computational Geometry Algorithms Library. http://www.cgal.org
93. Gmsh: a three-dimensional finite element mesh generator with built-in pre- and post-processing facilities. http://www.geuz.org/gmsh
94. Y. Peng, L.S. Luo, Prog. Comput. Fluid Dyn. **8**(1), 156 (2008)
95. A. Caiazzo, S. Maddu, Comput. Math. Appl. **58**(5), 930 (2009)
96. L. Zhu, G. He, S. Wang, L. Miller, X. Zhang, Q. You, S. Fang, Comput. Math. Appl. **61**(12), 3506 (2011)
97. Q. Zhou, L.S. Fan, J. Comput. Phys. **268**, 269 (2014)
98. O. Shardt, J.J. Derksen, Int. J. Multiphase Flow **47**, 25 (2012)
99. G. Le, J. Zhang, Phys. Rev. E **79**(2), 026701 (2009)
100. Z.G. Feng, E.E. Michaelides, J. Comput. Phys. **202**(1), 20 (2005)
101. J. Wu, C. Shu, J. Comput. Phys. **228**(6), 1963 (2009)
102. J. Wu, C. Shu, Y.H. Zhang, Int. J. Numer. Meth. Fluids **62**(3), 327 (2010)
103. J. Wu, C. Shu, Int. J. Numer. Meth. Fluids **68**(8), 977 (2012)
104. P. Bagchi, Biophys. J. **92**(6), 1858 (2007)
105. M.M. Dupin, I. Halliday, C.M. Care, L. Alboul, L.L. Munn, Phys. Rev. E **75**(6), 066707 (2007)
106. J. Zhang, P.C. Johnson, A.S. Popel, Phys. Biol. **4**(4), 285 (2007)
107. Y. Sui, Y. Chew, P. Roy, H. Low, J. Comput. Phys. **227**(12), 6351 (2008)
108. T. Krüger, M. Gross, D. Raabe, F. Varnik, Soft Matter **9**(37), 9008 (2013)
109. U.D. Schiller, Comput. Phys. Commun. **185**(10), 2586 (2014)
110. P. Ahlrichs, B. Dünweg, Int. J. Mod. Phys. C **09**(08), 1429 (1998)
111. B. Dünweg, A.J.C. Ladd, in *Advances in Polymer Science* (Springer, Berlin, Heidelberg, 2008), pp. 1–78
112. I. Cimrák, M. Gusenbauer, I. Jančigová, Comput. Phys. Commun. **185**(3), 900 (2014)
113. X.D. Niu, C. Shu, Y.T. Chew, Y. Peng, Phys. Lett. A **354**(3), 173 (2006)
114. Y. Hu, H. Yuan, S. Shu, X. Niu, M. Li, Comput. Math. Appl. **68**(3), 140 (2014)
115. H.Z. Yuan, X.D. Niu, S. Shu, M. Li, H. Yamaguchi, Comput. Math. Appl. **67**(5), 1039 (2014)
116. J. Wu, C.K. Aidun, Int. J. Numer. Meth. Fl. **62**(7), 765–783 (2009)
117. H.K. Jeong, H.S. Yoon, M.Y. Ha, M. Tsutahara, J. Comput. Phys. **229**(7), 2526 (2010)
118. T. Seta, Phys. Rev. E **87**(6), 063304 (2013)
119. Y. Kim, M.C. Lai, J. Comput. Phys. **229**(12), 4840 (2010)
120. W.X. Huang, C.B. Chang, H.J. Sung, J. Comput. Phys. **230**(12), 5061 (2011)
121. W.X. Huang, C.B. Chang, H.J. Sung, J. Comput. Phys. **231**(8), 3340 (2012)
122. R.W. Nash, H.B. Carver, M.O. Bernabeu, J. Hetherington, D. Groen, T. Krüger, P.V. Coveney, Phys. Rev. E **89**(2), 023303 (2014)

Chapter 12
Sound Waves

Abstract After reading this chapter, you will understand the fundamentals of sound propagation in a viscous fluid as they apply to lattice Boltzmann simulations, and you will know why sound waves in these simulations do not necessarily propagate according to the "speed of sound" lattice constant. You will have insight into why sound waves can appear spontaneously in lattice Boltzmann simulations and know how to create sound waves artificially in your simulations. Additionally, you will know about special boundary conditions that minimise the reflection of sound waves back into the system, allowing you to avoid reflected sound waves polluting the simulation results.

We previously saw in Chap. 4 that the lattice Boltzmann method behaves on the macroscopic scale according to the continuity equation and the *compressible* Navier-Stokes equation. As a natural consequence of this, any fluid phenomenon that is in keeping with these equations can also be captured in a lattice Boltzmann simulation.

One such phenomenon is that of sound waves, which follows from the continuity equation and the Euler equation, the latter of which is an approximated version of the compressible Navier-Stokes equation. Indeed, if we plot the pressure field of an LB simulation right after its startup we can often reveal sound waves caused by the initial conditions of the simulation. Most simulations contain sound waves to some degree or other, even though their presence may be unintended. For this reason, everyone who performs LB simulations can benefit from knowing a little about sound waves.

Even so, sound in LB simulations has been a neglected subfield compared to e.g. incompressible and multiphase/multicomponent flows. The first detailed treatment of LB sound waves came as late as 1998 [1]. However, since 2007 several theses have been published on this topic [2–9], especially on the subtopics of aeroacoustics[1] and boundary conditions that do not reflect sound waves.

This chapter starts with a brief general introduction to acoustics in Sect. 12.1, as some knowledge on acoustics is required to read the rest of the chapter. Section 12.2 deals with how sound propagates in LB simulations, Sect. 12.3 covers how sound

[1]Typically in the wider sense of the word *aeroacoustics*, i.e. as the interaction of sound and flow.

T. Krüger et al., *The Lattice Boltzmann Method*, Graduate Texts in Physics,
DOI 10.1007/978-3-319-44649-3_12

may appear or be deliberately imposed in simulations, and Sect. 12.4 introduces boundary conditions that do not reflect sound waves.[2]

12.1 Background: Sound in Viscous Fluids

This section will give a quick introduction to where sound waves come from and how they propagate, with the aim of giving readers without a special background in acoustics some required insight for the rest of this chapter.[3] For those readers desiring a deeper insight, there are many books available that all provide a thorough grounding in acoustics [10–14].

Sound waves follow as a consequence of the conservation equations of a compressible fluid. The scientific field of acoustics is largely based on the linearised Euler equations, which result in a linear, ideal and lossless wave equation. Here we will instead base the discussion on the linearised Navier-Stokes equations, which gives a wave equation with viscous sound attenuation.[4] The reason for this slightly more complex choice is that the sound waves that we observe in LB simulations are typically heavily affected by fluid viscosity. Indeed, we will soon see that we will typically need to choose a relaxation time $\tau/\Delta t$ very close to $1/2$ when simulating sound below the high ultrasonic range. Otherwise, the simulations require an enormous numerical grid, or the sound waves will be attenuated far too quickly.

To linearise the conservation equations, we assume that the field variables of density ρ, pressure p, and velocity $\boldsymbol{u}$ are of the form of an infinitesimal perturbation around a constant rest state,[5]

$$\begin{aligned} \rho(\boldsymbol{x},t) &= \rho_0 + \rho'(\boldsymbol{x},t), \\ p(\boldsymbol{x},t) &= p_0 + p'(\boldsymbol{x},t), \\ \boldsymbol{u}(\boldsymbol{x},t) &= \boldsymbol{0} + \boldsymbol{u}'(\boldsymbol{x},t). \end{aligned} \tag{12.1}$$

Here, the subscripted zeros indicate the rest state constants while the perturbations are indicated as primed quantities.

[2] These non-reflecting BCs are in contrast to pressure and velocity BCs, which reflect sound waves back into the system.

[3] Deeper insight in acoustics is often missing in the LB literature, as most LB researchers are more interested in incompressible fluid mechanics than in acoustics.

[4] The basic LB model which gives isothermal Navier-Stokes behaviour is only affected by this viscous sound attenuation. It does not allow for attenuation due to heat conduction or molecular relaxation, the latter of which dominates at audible frequencies in e.g. air and seawater. We will touch on these attenuation mechanisms in Sect. 12.1.5.

[5] An LB model for the linearised conservation equations was discussed briefly in Sect. 4.3.1.

Inserting these linearised variables into the conservation equation and neglecting terms that are nonlinear in the infinitesimal fluctuations, we end up with

$$\partial_t \rho + \rho_0 \partial_\alpha u_\alpha = 0, \tag{12.2a}$$

$$\rho_0 \partial_t u_\alpha + \partial_\alpha p = \partial_\beta \sigma'_{\alpha\beta}. \tag{12.2b}$$

The viscous stress tensor is unchanged by the linearisation:

$$\sigma'_{\alpha\beta} = \eta \left(\partial_\beta u_\alpha + \partial_\alpha u_\beta - \tfrac{2}{3}\delta_{\alpha\beta}\partial_\gamma u_\gamma \right) + \eta_{\mathrm{B}} \delta_{\alpha\beta}\partial_\gamma u_\gamma. \tag{12.2c}$$

The equation of state is also very important for sound waves. We use here the isothermal equation of state,

$$p = \rho c_{\mathrm{s}}^2; \tag{12.2d}$$

not only is it exact for the basic LBE, but as we discussed previously in Sect. 1.1.3, it also works as a linear approximation to other equations of state given that entropy changes are small.

12.1.1 The Viscous Wave Equation

The wave equation is quite simple to derive from these equations in the ideal Euler equation case (i.e. $\sigma'_{\alpha\beta} = 0$), and the derivation is only slightly more difficult if we carry along the viscous stress tensor. To begin, we find the sum ∂_t(12.2a)–∂_α(12.2b), which results in

$$\partial_t^2 \rho - \nabla^2 p = -\left(\tfrac{4}{3}\eta + \eta_{\mathrm{B}}\right) \nabla^2 \left(\partial_\alpha u_\alpha\right) = \left(\tfrac{4}{3}\nu + \nu_{\mathrm{B}}\right) \nabla^2 \left(\partial_t \rho\right). \tag{12.3}$$

The simplification of the viscous stress in the last equality uses (12.2a). In addition, we can assume that the viscosities are nearly constant, with only an infinitesimal variation.

At this point we nearly have the wave equation already. The final piece of the puzzle is the equation of state, (12.2d), with which we can replace the density ρ with p/c_{s}^2.[6]

[6]It is also possible to use the equation of state the other way around, which results in a wave equation with the density as the only remaining variable instead of the pressure.

Applying the equation of state in this way, we find the **viscous wave equation**

$$\frac{1}{c_s^2}\partial_t^2 p - (1 + \tau_{\text{vi}}\partial_t)\,\nabla^2 p = 0, \tag{12.4a}$$

where the effect of viscosity on sound propagation is expressed through the **viscous relaxation time**

$$\tau_{\text{vi}} = \frac{1}{c_s^2}\left(\frac{4}{3}\nu + \nu_{\text{B}}\right). \tag{12.4b}$$

The **ideal wave equation**, which is used more widely in the literature and is typically derived from the Euler equation, can be found from this simply by assuming $\tau_{\text{vi}} \simeq 0$,

$$\frac{1}{c_s^2}\partial_t^2 p - \nabla^2 p = 0. \tag{12.4c}$$

Exercise 12.1

(a) Show that the generic one-dimensional function $p(c_s t \mp x)$, known as d'Alembert's solution, is a solution of the ideal wave equation, (12.4c). *Hint: Insert the solution into the wave equation and apply the chain rule.*
(b) Show that this solution propagates at the speed of sound, i.e. $\pm c_s$.

Exercise 12.2

(a) Show that the incompressible fluid model, its linearised equations being $\partial_\gamma u_\gamma = 0$ and $\rho_0 \partial_t u_\alpha + \partial_\alpha p = \eta \partial_\beta \partial_\beta u_\alpha$, does *not* support sound waves. *Hint: Show that a wave equation cannot be derived from its equations.*
(b) Show that the "incompressible" LBE model actually *does* support sound waves due to its artificial compressibility, by deriving a wave equation from (4.45).

The viscous relaxation time can be seen as a characteristic time for the stabilisation of viscous processes. For physical fluids, it is typically on the order of 10^{-10} s for gases and 10^{-12} s for liquids [10]. Inserting the LBE viscosities into (12.4b) using the isothermal bulk viscosity $\eta_{\text{B}} = 2\eta/3$ found in Sect. 4.1.4, we can relate the viscous relaxation time with the BGK relaxation time as

$$\tau_{\text{vi}} = 2\left(\tau - \Delta t/2\right). \tag{12.5}$$

12.1.2 The Complex-Valued Representation of Waves

The linearity of the wave equation lets us employ some neat tricks. Steady-state solutions of a single angular frequency ω, related to the wave period T as $\omega = 2\pi/T$, are commonly expressed using complex-valued notation. This complex-valued approach has been widely used in acoustics since Lord Rayleigh introduced it in 1877 [15, 16]. If we know the solution of a particular case for any frequency ω, a corresponding time-domain solution can be found through an inverse Fourier transform [10, 11].

The simplest solution to the wave equation is an infinite plane wave, which is a one-dimensional solution that is not localised in space or time. In **complex form**, this solution of the ideal (i.e. $\tau_{\mathrm{vi}} = 0$) wave equation is

$$\hat{p}(x,t) - p_0 = \hat{p}'(x,t) = \hat{p}^\circ \mathrm{e}^{\mathrm{i}(\omega_0 t \mp k_0 x)} . \tag{12.6}$$

In this notation, we indicate complex-valued variables by carets and plane wave amplitudes by circles. As can be found by comparison with Exercise 12.1, the angular frequency ω_0 and the wavenumber k_0 are related as $\omega_0/k_0 = c_s$. (The real-valued frequency and wavenumber for the ideal wave equation are indicated by subscripted zeroes.) The complex amplitude $\hat{p}^\circ$ has both a magnitude $|\hat{p}^\circ|$ and a phase φ_p, i.e. $\hat{p}^\circ = |\hat{p}^\circ|\mathrm{e}^{\mathrm{i}\varphi_p}$.

Of course, the **physical pressure** is not complex. It can be found directly from the **real part of the complex-valued pressure**,

$$p'(x,t) = \Re\left(\hat{p}'(x,t)\right) = \left|\hat{p}^\circ\right| \cos\left(\omega_0 t \mp k_0 x + \varphi_p\right) . \tag{12.7}$$

The motivation for using complex notation is that these complex quantities are much easier to handle than the corresponding real expressions. We can augment an originally real-valued field with an imaginary part, perform calculations on the result, and retrieve the real part of the result, because the wave equation is *linear*, and taking the real part is also a linear operation. These operations commute. On the other hand, we must be careful when using the complex notation in nonlinear equations. As an example, $\Re(\hat{p}^2) \neq \left(\Re(\hat{p})\right)^2$.

We can also simulate complex quantities directly in LB simulations by letting artificial sound sources (which will be described later in Sect. 12.3) emit complex-valued distribution functions. However, this should only be used with the linearised equilibrium described earlier in Sect. 4.3.1 as linearity is a prerequisite for the complex wave notation. This approach has been used to simplify the analysis of sound wave simulations [17].

Exercise 12.3 Show by insertion that (12.6) and (12.7) are both valid solutions of the ideal wave equation, (12.4c).

Exercise 12.4 Show that the plane wave expressed in (12.6) has a *period* (i.e. the time before the wave repeats itself) of $T = 2\pi/\omega_0$ and a *wavelength* (i.e. the distance between successive wave peaks) of $\lambda = 2\pi/k_0$, and that it propagates at the speed of sound $c_s = \omega_0/k_0$.

The corresponding plane wave solution to the **viscous wave equation** (i.e. for $\tau_{vi} \neq 0$) is similar, but has a **complex wavenumber** $\hat{k}$ **and/or frequency** $\hat{\omega}$. For a wave propagating in the $+x$ direction in a viscous fluid, a plane wave looks like

$$\hat{p}'(x,t) = \hat{p}^{\circ} e^{i\left(\hat{\omega}t - \hat{k}x\right)} = \hat{p}^{\circ} e^{-\alpha_t t} e^{-\alpha_x x} e^{i(\omega t - kx)}. \tag{12.8}$$

Here, the complex wavenumber and frequency have been split into their real and imaginary parts,

$$\hat{\omega} = \omega + i\alpha_t, \quad \hat{k} = k - i\alpha_x. \tag{12.9}$$

The αs represent **attenuation coefficients**, while the **phase speed** of the sound wave is given by $c_p = \omega/k$, which may differ from the ideal sound speed $c_s = \omega_0/k_0$. This means that sound propagation in viscous fluids is dispersive: the discrepancy between c_p and c_s increases with frequency and τ_{vi}, as found in Exercise 12.5.

Wave dispersion is often seen in LB simulations; unintended sound waves usually move at a speed noticably different from c_s. However, the dispersion in LB simulations is also partly due to numerical error, and the rest of the dispersion differs from what is predicted by the Navier-Stokes equation, as we will see in Sect. 12.2.

Exercise 12.5

(a) Show by insertion that (12.8) being a solution of (12.4a) implies a *dispersion relation* that relates the frequency and wavenumber as

$$\frac{\hat{\omega}^2}{\hat{k}^2} = c_s^2 \left(1 + i\hat{\omega}\tau_{vi}\right). \tag{12.10}$$

(b) In which limits can this dispersion relation become the ideal dispersion relation $c_s = \omega_0/k_0$?

12.1.3 Simple One-Dimensional Solutions: Free and Forced Waves

For an attenuated sound wave, typically either the frequency or the wavenumber is complex; not both. For instance, let us take the semi-infinite system $0 \leq x$. If the boundary condition at $x = 0$ is $\hat{p}'(0, t) = \hat{p}^{\circ} e^{i\omega_0 t}$, i.e. the pressure oscillates with a frequency ω_0 and constant amplitude $|\hat{p}^{\circ}|$, the resulting wave in the rest of the system must have the same frequency $\hat{\omega} = \omega_0$,

$$\hat{p}'(x, t) = \hat{p}^{\circ} e^{-\alpha_x x} e^{i(\omega_0 t - kx)}. \tag{12.11a}$$

Following the nomenclature of Truesdell [18], this is a *forced wave*. In general, forced waves are radiated (or *forced*) by a source oscillating at a constant amplitude.

Forced waves are an idealisation of physically attainable waves. For instance, the wave above could be a high-frequency sound wave in a duct (neglecting the marginal effect of no-slip boundaries).

Using the same nomenclature, a *free wave* is a wave which is not emitted by any source, but which is attenuated as it propagates. Let us say that at $t = 0$ we have a plane wave of constant amplitude and infinite extent, i.e. $\hat{p}'(x, 0) = \hat{p}^{\circ} e^{-ik_0 x}$. For such a case, the resulting plane wave is

$$\hat{p}'(x, t) = \hat{p}^{\circ} e^{-\alpha_t t} e^{i(\omega t - k_0 x)}. \tag{12.11b}$$

Free waves are not as relevant or even physically realisable in the same way as forced waves. Even so, they may serve as useful simulation benchmarks since their periodicity in space can be captured perfectly using a periodic boundary condition.[7]

Exercise 12.6

(a) Show from (12.10) that the complex wavenumber for the forced plane wave and the complex frequency for the free plane wave are respectively

$$\frac{\hat{k}}{k_0} = \frac{1}{\sqrt{1 + i\omega_0 \tau_{vi}}}, \tag{12.12a}$$

$$\frac{\hat{\omega}}{\omega_0} = i\frac{\omega_0 \tau_{vi}}{2} + \sqrt{1 - \left(\frac{\omega_0 \tau_{vi}}{2}\right)^2}, \tag{12.12b}$$

with k_0 and ω_0 related through $c_s = \omega_0 / k_0$.

[7] Free waves are also key in the von Neumann stability analyses described in Sect. 4.4.

(b) Show from Taylor expansion of these equations that the nondimensionalised attenuation and dispersion for forced and free waves are, to lowest order,

$$\text{Forced:} \qquad \frac{\alpha_x}{k_0} = \frac{\omega_0 \tau_{\text{vi}}}{2}, \qquad \frac{c_{\text{p}}}{c_{\text{s}}} = \frac{k_0}{k} = 1 + \frac{3}{8}(\omega_0 \tau_{\text{vi}})^2, \tag{12.13a}$$

$$\text{Free:} \qquad \frac{\alpha_t}{\omega_0} = \frac{\omega_0 \tau_{\text{vi}}}{2}, \qquad \frac{c_{\text{p}}}{c_{\text{s}}} = \frac{\omega}{\omega_0} = 1 - \frac{1}{8}(\omega_0 \tau_{\text{vi}})^2. \tag{12.13b}$$

Generally, the dimensionless number $\omega_0 \tau_{\text{vi}}$, which we can call the *acoustic viscosity number*, indicates the effect of viscosity on sound propagation at a given frequency. The higher the value of $\omega_0 \tau_{\text{vi}}$, the higher the degree of viscous attenuation and dispersion.

Exercise 12.7 Let us take a look at how the acoustic viscosity number scales. We know that $\omega_0 \sim c_{\text{s}}/\lambda$ and that $\nu \sim \nu_{\text{B}} \sim c_{\text{s}} \ell_{\text{mfp}}$, ℓ_{mfp} being the mean free path. From this, show that

$$\omega_0 \tau_{\text{vi}} \sim \frac{\ell_{\text{mfp}}}{\lambda}, \tag{12.14}$$

i.e. that the acoustic viscosity number represents an acoustic Knudsen number.

From (12.13) we see that for both forced and free waves, the attenuation coefficient (α_x and α_t, respectively) to lowest-order increases with the square of the frequency ω_0. (Due to the typically low values of $\omega_0 \tau_{\text{vi}}$, the lowest order is typically the only one felt.) However, free and forced waves experience different dispersion according to the isothermal Navier-Stokes model; the wave speed *increases* with frequency for *forced* waves, and *decreases* with frequency for *free* waves.

The attenuation and dispersion given above are derived from the Navier-Stokes equation, which can be seen as a continuum approximation to the Boltzmann equation. Higher-order fluid models such as the Burnett equation generally predict the same viscous attenuation to lowest order, but make different predictions for terms of higher order in $\omega_0 \tau_{\text{vi}}$ [7, 19]. One thing that generally holds for all these fluid models is that in a series expansion of $\hat{k}/k_0$ or $\hat{\omega}/\omega_0$ odd powers in $\omega_0 \tau_{\text{vi}}$ are related to attenuation while even powers are related to dispersion.

Let us take a quick look at **typical values** of $\omega_0 \tau_{\text{vi}}$ for **audible sound**. At everyday conditions, the ideal speed of sound in air is around $c_{\text{s}} \simeq 340\,\text{m/s}$, the kinematic shear viscosity is $\nu \simeq 1.5 \times 10^{-5}\,\text{m}^2/\text{s}$ and the bulk viscosity is $\nu_{\text{B}} \simeq 0.61\nu$ [11, 13, 14]. For a high but audible frequency of 10 kHz, we get an acoustic viscosity number of $\omega_0 \tau_{\text{vi}} = 1.6 \times 10^{-5}$. From Chap. 7, we know that this dimensionless number will be the same in any system of units.

Now, what would it require to achieve such a number in lattice units? Just to have a number, let us assume a reasonable lattice wavelength of $\lambda = 10$,

(continued)

which corresponds to an angular frequency of $\omega_0 = 2\pi c_s/\lambda \approx 0.36$. Relating τ_{vi} to τ using (12.5), we find that this acoustic viscosity number, given $\lambda = 10$, requires $\tau/\Delta t = 0.500022$. This incredibly low value of $\tau/\Delta t$ is hard to achieve. In principle it can be increased by increasing the lattice wavelength, but that is equivalent with increasing the resolution which may lead to a simulation domain with unfeasibly many nodes. On the other hand, it has been shown possible to simulate at least *some* acoustic phenomena for the inviscid case of $\tau/\Delta t = 0.5$ with both accuracy and stability using MRT or MRT-like collisions [7, 17]. Generally, **low values of τ are very important for LB acoustics.**

In some cases, it may be possible to simulate larger acoustic systems at higher values of τ, by using a much higher acoustic viscosity number $\omega_0\tau_{vi}$ than what would be realistic and compensating for the additional viscous attenuation in the received pressure wave [20].

12.1.4 Time-Harmonic Waves: The Helmholtz Equation

The wave equation can be simplified considerably in the steady-state case where the entire field is harmonically varying in time, which means that the field varies with time as $e^{i\omega_0 t}$. Such a time variation allows the simplifying substitution $\partial_t \to i\omega_0$. Consequently, the viscous wave equation, (12.4a), can be reduced to the viscous Helmholtz equation

$$\nabla^2\hat{p}' + \hat{k}^2\hat{p}' = 0, \tag{12.15}$$

with the wavenumber $\hat{k} = k_0/\sqrt{1 + i\omega_0\tau_{vi}}$. (In the more commonly seen inviscid case, $\tau_{vi} \to 0$ and the wavenumber is real, i.e. $\hat{k}^2 = k_0^2$.)

The Helmholtz equation holds universally for a harmonic time variation no matter the spatial shape of the wave; it may be plane, cylindrical, spherical, a sound beam, or any other shape, and the complex wavenumber $\hat{k}$ remains the same. This shows that the wavenumber of a forced plane wave, described in Sect. 12.1.3, equals that of any other viscously affected wave.

The Helmholtz equation is a steady-state equation and time does not enter into it except through the underlying assumption that the fields vary as $e^{i\omega_0 t}$. On the other hand, LB simulations are inherently time-dependent. This means that LB simulations can only approximate Helmholtz equation solutions, and then only in cases with time-harmonic sound sources that are left to radiate sound until a near-steady state is reached.

12.1.5 Other Attenuation and Absorption Mechanisms

In isothermal lattice Boltzmann simulations, shear and bulk viscosity are typically the only causes of sound wave attenuation. However, in real fluids, there are other mechanisms that cause attenuation and absorption of sound waves in free fields and for wave-surface interaction, respectively. We will not go into these mechanisms in too much depth here; they are described more thoroughly in the acoustics literature [7, 10, 11, 13].

One free-field mechanism is heat conduction, which does not appear in isothermal LB fluids. In short, sound wave peaks experience an increase not only in pressure and density, but also in temperature. Sound wave troughs experience a corresponding *decrease* in temperature. This temperature inhomogenity results in heat conduction. This reduces the temperature differences, and consequently the pressure differences, between peaks and troughs; the wave amplitude decreases.

Heat conduction causes a wave attenuation which, similarly to the viscous attenuation, is proportional to the square of the frequency. In gases, this heat conduction typically causes an attenuation of similar strength to that from viscosity. In liquids, attenuation from heat conduction is typically negligible. For free-field attenuation, the importance of viscosity compared to heat conduction scales with the Prandtl number as $\mathcal{O}(\mathrm{Pr})$ [10].

Additionally, when a sound wave interacts with a solid boundary, the viscous friction and heat transfer between the two will also cause a partial absorption of the sound wave, in particular at high frequencies [10, 13].

Another free-field mechanism is molecular relaxation processes, which is linked to transfer of energy between various forms [7, 10, 11]. In short, a passing sound wave will temporarily change the temperature of the fluid. As we saw in Sect. 1.3, the temperature is directly proportional to the translational energy of the fluid molecules. Increasing the translational energy in this way puts it out of equilibrium with inner forms of energy in the fluid. In a polyatomic gas, these are the vibrational and rotational energies of the molecules. In seawater, the increase in temperature changes the equilibrium point of a chemical reaction between solutes in the water. In both cases, translational energy is equilibrated with these inner energies through a relaxation process. This reduces the temperature differences (and consequently the pressure differences) between peaks and troughs, weakening the pressure wave as it propagates.

At audible frequencies, this tends to completely dominate viscous and thermal attenuation in typically encountered fluids such as air or seawater. However, this mechanism is dependent on the rate at which translational energy is converted to inner energies and back.[8] Therefore, it is less effective at very high frequencies where the inner energies cannot keep up and remain nearly constant, or *frozen*.

[8] These rates can depend on the presence of other molecules. For instance, water molecules act as a catalyst for the transfer between translational and vibrational energy in air [21]. Humidity therefore has a surprisingly large effect on sound attenuation in air.

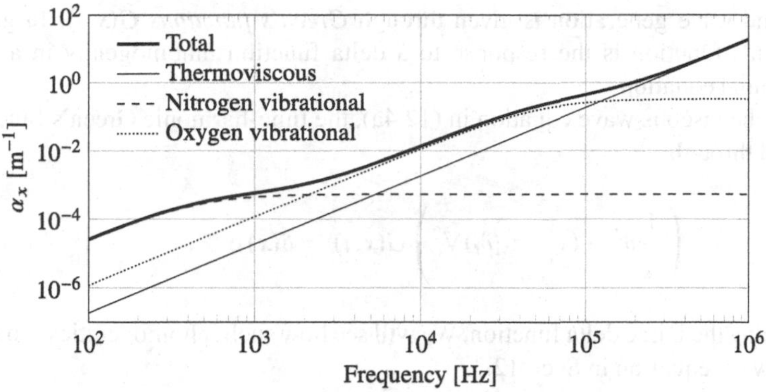

Fig. 12.1 Contributions to the total sound attenuation for air at conditions of 293.15 K, atmospheric pressure, and 70 % relative humidity [11], according to ISO 9613-1:1993

The relative importance of vibrational energy to sound attenuation in air, compared to a combined attenuation due to viscosity and heat conduction, is shown in Fig. 12.1. In the audible range, the thermoviscous attenuation is typically *negligible*.

The most important source of sound absorption in day-to-day situations, however, is sound waves' interaction with surfaces that are not perfectly rigid. This can be represented through the surface property of *normal impedance* $\hat{Z}_n = \hat{p}/\hat{u}_n$, where $\hat{u}_n$ is the normal velocity of the surface for an applied pressure $\hat{p}$ [10]. Higher impedance, corresponding to a harder and less moveable surface, typically results in the surface absorbing less sound. For instance, a room with hard concrete walls will be more reverberant (i.e., sound will linger for a longer time) than a similar room with softer wooden walls.

Now, how can these mechanisms be captured in LB simulations? Thermal attenuation is something we get for free in thermal LB models, and the attenuation in such LB simulations matches theory [22]. Vibrational relaxation, on the other hand, is a more complex effect that cannot be captured so easily. The only published attempt at this time of writing was very simple and limited itself to a single relaxation process and monofrequency sound. It achieved good accuracy but poor stability [7]. Similarly, only a little has been published on impedance BCs for the LBM [23].

12.1.6 Simple Multidimensional Waves: The Green's Function

The wave and Helmholtz equations discussed above only describe how *existing* waves propagate, not how they are *generated*. One simple mathematical description of sound wave generation is given through *Green's functions* $\hat{G}(\boldsymbol{x}, t)$. In general, a Green's function is the response to a delta function inhomogenity in a partial differential equation.

For the viscous wave equation in (12.4a), the time-harmonic Green's function is defined through

$$\left(\frac{1}{c_\mathrm{s}^2}\partial_t^2 - (1 + \tau_\mathrm{vi}\partial_t)\nabla^2\right)\hat{G}(\boldsymbol{x}, t) = \delta(\boldsymbol{x})\,\mathrm{e}^{\mathrm{i}\omega_0 t}, \tag{12.16}$$

$\delta(\boldsymbol{x})$ being the Dirac delta function. We will see how such inhomogenities can appear in the wave equation in Sect. 12.3.

The solution to this inhomogeneous equation depends on the number of spatial dimensions and represents the **simplest type of wave** for that number of dimensions. For a **1D** space, the Green's function is a **plane wave**. For **2D** space, it is a **cylindrical wave**. For **3D** space, it is a **spherical wave**. Explicitly, these solutions are [12]

$$\text{1D:} \qquad \hat{G}(x, t) = \frac{1}{2\mathrm{i}\hat{k}}\mathrm{e}^{\mathrm{i}(\omega_0 t - \hat{k}|x|)}, \tag{12.17a}$$

$$\text{2D:} \qquad \hat{G}(\boldsymbol{x}, t) = \frac{1}{4\mathrm{i}}H_0^{(2)}(\hat{k}|\boldsymbol{x}|)\,\mathrm{e}^{\mathrm{i}\omega_0 t}, \tag{12.17b}$$

$$\text{3D:} \qquad \hat{G}(\boldsymbol{x}, t) = \frac{1}{4\pi|\boldsymbol{x}|}\mathrm{e}^{\mathrm{i}(\omega_0 t - \hat{k}|\boldsymbol{x}|)}. \tag{12.17c}$$

In these equations $|\boldsymbol{x}|$ is the distance to the origin (often denoted as r in the literature), and $H_n^{(2)}$ is an n-th order Hankel function of the second kind, which is a particular superposition of Bessel functions.

It is also possible to define time-impulsive Green's functions through the alternate inhomogenity term $\delta(\boldsymbol{x})\delta(t)$. However, these have no similarly simple analytical solutions for two dimensions or for sound waves affected by viscosity. In any case, such time-impulsive Green's functions can be indirectly found through an inverse Fourier transformation of time-harmonic Green's functions.

Knowing the response of a point inhomogenity also lets us find the pressure field from a spatially distributed inhomogenity, meaning that the equation

$$\left(\frac{1}{c_\mathrm{s}^2}\partial_t^2 - (1 + \tau_\mathrm{vi}\partial_t)\nabla^2\right)\hat{p}(\boldsymbol{x}, t) = \hat{\mathcal{T}}(\boldsymbol{x})\,\mathrm{e}^{\mathrm{i}\omega_0 t}, \tag{12.18a}$$

is solved by

$$\hat{p}'(\boldsymbol{x},t) = \int \hat{\mathcal{T}}(\boldsymbol{y})\,\hat{G}(\boldsymbol{x}-\boldsymbol{y},t)\,\mathrm{d}^3y, \tag{12.18b}$$

where $\boldsymbol{x}$ is the "receiver" point and we integrate over all possible "source" points $\boldsymbol{y}$.

These delta function inhomogenities radiate waves equally in every direction, which means that they are *monopole* sources. However, inhomogenities that radiate directively, for instance as *dipoles* and *quadrupoles*, are also possible. Such inhomogenities are represented as tensor terms with spatial derivatives applied to them [7, 24, 25], i.e. as

$$\left(\frac{1}{c_s^2}\partial_t^2 - (1+\tau_{\mathrm{vi}}\partial_t)\nabla^2\right)\hat{p}(\boldsymbol{x},t) = \mathrm{e}^{\mathrm{i}\omega_0 t}\big[\hat{\mathcal{T}}(\boldsymbol{x}) + \partial_\alpha\hat{\mathcal{T}}_\alpha(\boldsymbol{x}) + \partial_\alpha\partial_\beta\hat{\mathcal{T}}_{\alpha\beta}(\boldsymbol{x})\big]. \tag{12.19a}$$

It can be shown [7, 24, 25] that this equation is solved by

$$\hat{p}'(\boldsymbol{x},t) = \int \big[\hat{\mathcal{T}}(\boldsymbol{y})\hat{G}(\boldsymbol{x}-\boldsymbol{y},t) + \hat{\mathcal{T}}_\alpha(\boldsymbol{y})\partial_\alpha\hat{G}(\boldsymbol{x}-\boldsymbol{y},t) + \hat{\mathcal{T}}_{\alpha\beta}(\boldsymbol{y})\partial_\alpha\partial_\beta\hat{G}(\boldsymbol{x}-\boldsymbol{y},t)\big]\,\mathrm{d}^3y, \tag{12.19b}$$

where the spatial derivatives in the integrand are always operating on $\boldsymbol{x}$. The first spatial derivative of the Green's function corresponds to dipole radiation and the second spatial derivative of the Green's function to quadrupole radiation.

Exercise 12.8

(a) Using the 3D Green's function in (12.17c), show for the far field (i.e. for large $|\boldsymbol{x}|$), that $\partial_\alpha G(\boldsymbol{x},t)$ has a dipole variation in space, i.e. that it varies as $x_\alpha/|\boldsymbol{x}|$.
(b) Similarly, show that $\partial_\alpha\partial_\beta G(\boldsymbol{x},t)$ has a quadrupole variation in space, i.e. that it varies as $x_\alpha x_\beta/|\boldsymbol{x}|^2$.

12.2 Sound Propagation in LB Simulations

If we would look at the sound waves that typically appear by accident in LB simulations, one of the first two things we would notice is that they are attenuated fairly quickly and that they do not propagate at a speed c_s. For the latter point, we already established in Sect. 12.1.3 that waves in a viscous medium are dispersive at high frequencies.[9]

Additionally, **sound waves do not propagate in numerical solvers exactly as predicted by the NSEs**. There are several reasons for this:

- Due to **discretisation error**, waves (or Fourier components of waves) that are more heavily discretised, meaning that they have fewer points per wavelength, will propagate differently to well-resolved waves.
- Numerical solvers typically discretise space into elements or a grid, and it is not possible to discretise space in a fully isotropic way. Therefore, sound propagation is typically slightly **anisotropic**, again especially at short wavelengths.

These points hold fully for the lattice Boltzmann method, though there are some LB-specific issues as well:

- The LBM **discretises velocity space**, and it is not possible to do this fully isotropically. Even without the discretisation in space and time, this makes sound propagation anisotropic, though this effect is only significant for $\omega_0\tau_{vi} > 0.1$ [28].
- At second order and higher in series expansions in $\omega_0\tau_{vi}$, the **Boltzmann equation predicts different sound propagation** to the NSEs [19]. It is similar for the LBE: To $O(\omega_0\tau_{vi})$ the attenuation is the same as for Navier-Stokes, but the dispersion (which to lowest order is $O(\omega_0\tau_{vi})^2$) is different.[10]

Consequently, the sound attenuation and dispersion in LB simulations depends both on the physical model (i.e. the discrete-velocity Boltzmann equation) and on the purely numerical discretisation. As we will see in the following sections, it is possible to tell these effects apart. Additionally, the anisotropy of each velocity set may also uniquely affect LB sound propagation.

One classic method for analysing the sound propagation predicted by fluid models goes back to Stokes' first article on his fluid momentum equation that we now know as the Navier-Stokes equation [29]. In this method, sometimes known as *linearisation analysis*, we assume that the unknowns (i.e. density, pressure, and fluid velocity) are varying on the form of an infinitesimal plane wave around a rest state,

[9]For strong sound waves non-linear effects also affect the local wave speed. These non-linear effects can be reproduced in LB simulations [26, 27], though we will not go into non-linear sound in this book.

[10]This is not to say that the sound propagation is the same in the LBE as in the Boltzmann equation. Remember that the LBE is a minimal discretisation of the Boltzmann equation, only detailed enough such that correct Navier-Stokes behaviour is reproduced.

e.g. $\hat{p} = p_0 + \hat{p}^{\circ} e^{i(\hat{\omega}t - \hat{k}x)}$. In the fluid model's governing partial differential equations, space and time derivatives can be applied directly to these unknown quantities since their behaviour in space and time is known. From the resulting derivative-free equations, dispersion relations such as (12.10) which relate the frequency $\hat{\omega}$ and wavenumber $\hat{k}$ can be found.

Such linearisation analyses have also been applied to the Boltzmann equation [30–32]. However, they were not able to determine a closed-form dispersion relation, instead ending up with a non-convergent power series in quantities of $O(\omega_0 \tau_{vi})$. Thankfully, the finite velocity space of the discrete-velocity Boltzmann equation and the lattice Boltzmann method make them more amenable to this kind of analysis.

12.2.1 Linearisation Method

To analyse LB sound propagation, a linearisation analysis similar to the von Neumann analysis mentioned in Sect. 4.4 can be used. Similarly to what we saw in Sect. 12.1.2, the unknowns f_i are assumed to be varying on plane wave form. However, exact space and time derivatives do not exist in any numerical solver with discretised space and time, which complicates the analysis somewhat. Instead, we must relate the unknowns f_i over several nodes and time steps.

A number of such analyses have been performed in the literature [7, 27, 28, 33–39]. However, the most thorough and detailed analyses have unfortunately only been performed for free waves. As discussed in Sect. 12.1.3, free waves are not usually relevant in practice beyond benchmarks and stability analyses, and knowing how free waves behave is not sufficient to know how forced waves behave [18].

In this section we will go through a simple linearisation analysis of sound propagation for the D1Q3 velocity set given in Table 3.2. This will result in expressions for the attenuation and dispersion of sound in the LBM. This analysis does not take into account the anisotropy of each higher-dimensional velocity set. Even so, the end results hold for propagation along the x-, y-, or z-axis in higher-dimensional velocity sets of which D1Q3 is a one-dimensional projection as discussed in Sect. 3.4.7.

The process behind this analysis, which will now be presented, is not really necessary to understand the results, which are presented in Sect. 12.2.2. For notational simplicity, this analysis uses lattice units, i.e. $\Delta t = 1$ and $\Delta x = 1$.

In this linearisation analysis we assume that the solution is an infinitesimally weak plane sound wave on top of a rest state. Consequently, we can use a linearised equilibrium like in Sect. 4.3.1,

$$f_i^{\text{eq}} = w_i \left(\rho_0 + \hat{\rho}' + \frac{\rho_0 \hat{u}'}{c_s^2} c_i \right). \tag{12.20}$$

Since we are basing this analysis on the one-dimensional D1Q3 velocity set, this linearised equilibrium distribution is also one-dimensional.

We define the distribution function $\{f_i\}$ to be split into two parts:

$$\hat{f}_i(x,t) = F_i^{\mathrm{eq}} + \hat{f}_i'(x,t) = F_i^{\mathrm{eq}} + \hat{f}_i^{\circ} \mathrm{e}^{\mathrm{i}(\hat{\omega}t - \hat{k}x)}. \tag{12.21}$$

Here, we define F_i^{eq} to be the quiescent (i.e. rest state) component of the distribution function. $\hat{f}_i'$ is defined to be the infinitesimal plane-wave disturbance on top, and $\hat{f}_i^{\circ}$ is the plane-wave amplitude. As the F_i^{eq} component is constant and unaffected by the fluctuations $\hat{f}_i'$ due to linearity, it remains in a permanent state of both macroscopic and mesoscopic equilibrium. Mathematically, these definitions can be expressed as

$$\begin{aligned} \begin{bmatrix} \sum_i \hat{f}_i(x,t) \\ \sum_i c_i \hat{f}_i(x,t) \end{bmatrix} &= \begin{bmatrix} \sum_i F_i^{\mathrm{eq}} \\ \sum_i c_i F_i^{\mathrm{eq}} \end{bmatrix} + \begin{bmatrix} \sum_i \hat{f}_i^{\circ}(x,t) \\ \sum_i c_i \hat{f}_i^{\circ}(x,t) \end{bmatrix} \mathrm{e}^{\mathrm{i}(\hat{\omega}t - \hat{k}x)} \\ = \begin{bmatrix} \hat{\rho}(x,t) \\ \rho_0 \hat{u}(x,t) \end{bmatrix} &= \begin{bmatrix} \rho_0 \\ 0 \end{bmatrix} + \begin{bmatrix} \hat{\rho}^{\circ} \\ \rho_0 \hat{u}^{\circ} \end{bmatrix} \mathrm{e}^{\mathrm{i}(\hat{\omega}t - \hat{k}x)}. \end{aligned} \tag{12.22}$$

The distribution function $\{f_i\}$ has the same zeroth and first moments as its corresponding equilibrium $\{f_i^{\mathrm{eq}}\}$. Therefore, the linearised equilibrium in (12.20) can similarly be split into a quiescent component and a plane wave fluctuation component,

$$F_i^{\mathrm{eq}} = \rho_0 w_i, \qquad \hat{f}_i^{\circ\mathrm{eq}} = w_i \left(\hat{\rho}^{\circ} + \frac{\rho_0 \hat{u}^{\circ}}{c_s^2} c_i \right). \tag{12.23}$$

Inserting this split equilibrium distribution into the LBE, the quiescent component is cancelled out everywhere and we are left with an LBE only for the fluctuation itself,

$$\hat{f}_i'(x + c_i, t + 1) = \left(1 - \tfrac{1}{\tau}\right) \hat{f}_i'(x,t) + \tfrac{1}{\tau} \hat{f}_i'^{\mathrm{eq}}(x,t). \tag{12.24}$$

From the above definitions we can perform the substitution $\hat{f}_i'(x,t) = \hat{f}_i^{\circ} \mathrm{e}^{\mathrm{i}(\hat{\omega}t - \hat{k}x)}$. Subsequently explicitly expanding the moments in the equilibrium distribution and dividing the resulting LBE by $\mathrm{e}^{\mathrm{i}(\hat{\omega}t - \hat{k}x)}$, we find

$$\hat{f}_i^{\circ} \mathrm{e}^{\mathrm{i}(\hat{\omega} - \hat{k}c_i)} = \left(1 - \tfrac{1}{\tau}\right) \hat{f}_i^{\circ} + \frac{w_i}{\tau} \left[\hat{f}_1^{\circ} + \hat{f}_0^{\circ} + \hat{f}_2^{\circ} + 3 \left(\hat{f}_1^{\circ} - \hat{f}_2^{\circ} \right) c_i \right]. \tag{12.25}$$

This corresponds to three explicit equations, one for each i. These relate the three unknown f_is with each other and with $\hat{\omega}$ and $\hat{k}$. After some algebra, these can be put in the form of an eigenvalue problem $\hat{A}f^\circ = e^{i\hat{\omega}}f^\circ$, which in explicit form is

$$\hat{A}\hat{f}^\circ = \frac{1}{3}\begin{bmatrix} e^{-i\hat{k}}(3-1/\tau) & e^{-i\hat{k}}/2\tau & -e^{-i\hat{k}}/\tau \\ 2/\tau & 3-1/\tau & 2/\tau \\ -e^{i\hat{k}}/\tau & e^{i\hat{k}}/2\tau & e^{i\hat{k}}(3-1/\tau) \end{bmatrix}\begin{bmatrix} \hat{f}_2^\circ \\ \hat{f}_0^\circ \\ \hat{f}_1^\circ \end{bmatrix} = e^{i\hat{\omega}}\begin{bmatrix} \hat{f}_2^\circ \\ \hat{f}_0^\circ \\ \hat{f}_1^\circ \end{bmatrix}. \tag{12.26}$$

Now, let us remember what we are really looking for here: a **dispersion relation** that relates the unknowns $\hat{\omega}$ and $\hat{k}$. This can be found from the characteristic polynomial of the above eigenvalue problem,

$$\det\left(\hat{A} - Ie^{i\hat{\omega}}\right) = g(\hat{\omega}, \hat{k}, \tau) = 0. \tag{12.27}$$

Here, $g(\hat{\omega}, \hat{k}, \tau) = 0$ is the analytical dispersion relation that is our goal. However, this relation is **very cumbersome**; it should not be dealt with by hand, nor will it be reproduced explicitly here.

12.2.2 Linearisation Results

With the help of a computer algebra system, the dispersion relation in (12.27) can be solved for either $\hat{k}$ or $\hat{\omega}$, given some assumption on the relation between the two. In particular, we can determine exactly how D1Q3 waves propagate for the aforementioned forced and free wave cases, where the frequency or the wavenumber are real-valued, respectively.

12.2.2.1 Forced Waves

For forced waves, the frequency is real-valued, i.e. $\hat{\omega} = \omega_0$, and we can solve the dispersion relation to find an exact, though extremely unwieldy, analytical wavenumber $\hat{k}$. There are two solutions for $\hat{k}$, corresponding to propagation in the $+x$ and the $-x$ direction. The $+x$-propagating solution is:

$$\hat{k} = i\ln\left[\frac{3\tau(\zeta^2 - \zeta + 1 - \zeta^{-1}) + \zeta - 2 + \zeta^{-1}(3 + \sqrt{3\Xi})}{4 + 6\tau(\zeta - 1) - 2\zeta}\right], \tag{12.28a}$$

where the shorthands $\zeta = e^{i\omega_0}$ and

$$\Xi = (\zeta + 1)(\zeta - 1)^2(\tau\zeta + 1 - \tau)(3\tau\zeta^2 - \zeta + 3 - 3\tau) \tag{12.28b}$$

have been used.

While this solution is exact for D1Q3 and a good approximation[11] for those velocity sets that can be projected to D1Q3, it is a mess to look at. It is difficult to get any feeling from this solution for how the wavenumber behaves.

Instead, we can find something simpler and clearer by series expanding this and cleverly arranging the result as

$$\begin{aligned}\frac{\hat{k}}{k_0} = &\left[1 + \tfrac{1}{12}\omega_0^2 + \tfrac{13}{480}\omega_0^4 + O(\omega_0^6)\right] - i\tfrac{1}{2}(\omega_0\tau_{vi})\left[1 + \tfrac{5}{12}\omega_0^2 + O(\omega_0^4)\right] \\ &- \tfrac{5}{8}(\omega_0\tau_{vi})^2\left[1 + \tfrac{13}{20}\omega_0^2 + O(\omega_0^4)\right] + i\tfrac{13}{16}(\omega_0\tau_{vi})^3\left[1 + O(\omega_0^2)\right] \\ &+ O([\omega_0\tau_{vi}]^4).\end{aligned} \tag{12.29}$$

This arrangement nests a series expansion in ω_0, which here represents the numerical resolution, in a series expansion in the dimensionless acoustic viscosity number $\omega_0\tau_{vi}$, which represents the physical effect of viscosity on sound propagation in the model. From (12.9) we know that the imaginary terms represent sound wave attenuation while the real terms represent dispersion.

To make the expansion more clear, consider letting $\omega_0 \to 0$ while keeping $\omega_0\tau_{vi}$ constant. This is equivalent to having an infinitely refined resolution for sound waves. This limit removes the discretisation error in space and time, resulting in the same wave propagation as for the discrete-velocity Boltzmann equation [28]. Conversely, letting $\omega_0\tau_{vi} \to 0$ while keeping ω_0 finite is equivalent to ideal sound propagation unaffected by dissipative effects. Only the discretisation error in space and time remains.

12.2.2.2 Free Waves

For free waves, the wavenumber is real-valued, i.e. $\hat{k} = k_0$, and we can solve the dispersion relation for the unknown $\hat{\omega}$. This actually admits three solutions [7]: two for plane wave propagation in each direction and one non-propagating dissipative mode. All three solutions are significantly more cumbersome than even (12.28a),

[11] Again, it is exact if we disregard anisotropy.

and we will jump straight to the approximate series expansion form,

$$\begin{aligned}\frac{\hat{\omega}}{\omega_0} &= \left[1 - \tfrac{1}{36}k_0^2 - \tfrac{1}{1440}k_0^4 + O(k_0^6)\right] + \mathrm{i}\tfrac{1}{2}(\omega_0\tau_{\mathrm{vi}})\left[1 + O(k_0^4)\right] \\ &+ \tfrac{1}{8}(\omega_0\tau_{\mathrm{vi}})^2\left[1 - \tfrac{1}{4}k_0^2 + O(k_0^4)\right] + \mathrm{i}\tfrac{1}{8}(\omega_0\tau_{\mathrm{vi}})^3\left[1 + O(k_0^2)\right] \\ &+ O([\omega_0\tau_{\mathrm{vi}}]^4).\end{aligned} \tag{12.30}$$

In this case, $\omega_0\tau_{\mathrm{vi}}$ still represents the physical effect of viscosity in the model while numerical resolution is represented by k_0.

12.2.2.3 Discussion

For both forced and free waves, we have seen that the time and space discretisation error first appears at second order in the parameter describing the numerical resolution. This shows that LB sound wave propagation is second-order accurate.

The results presented from this analysis are highly dependent on the specifics of the LBE. Take for example the model described in Sect. 4.3.3, where the equilibrium f_i^{eq} can be altered in order to change the equation of state. A linearisation analysis of this model found that changing the equilibrium affects every order of the time and space discretisation error [27]. Additionally, this change affects the continuous model, represented by $\omega_0\tau_{\mathrm{vi}}$ for $\omega_0 \to 0$, at $O(\omega_0\tau_{\mathrm{vi}})^2$ and presumably also higher orders. Similar results have also been reported elsewhere in the literature [34, 37].

On the other hand, the results from this derivation accurately describe the wave propagation for the classic isothermal equilibrium, even for non-plane forced waves, as explained in Sect. 12.1.4. This will be demonstrated in Sect. 12.3 for a case with 2D cylindrical waves and a simple MRT operator.

In the same continuous-model limit, we can compare (12.29) and (12.30) with the corresponding results in (12.13) for the isothermal Navier-Stokes model. We see that to lowest order, the attenuation is consistent while the dispersion is not. In fact, it can be found that all terms above $O(\omega_0\tau_{\mathrm{vi}})$ deviate from the isothermal Navier-Stokes model [7]. This deviation becomes significant at $\omega_0\tau_{\mathrm{vi}} \approx 0.1$ [7, 28].

Example 12.1 We have from (12.5) that $\tau_{\mathrm{vi}} = 2(\tau - \Delta t/2)$ for the BGK collision operator, so that for $\tau/\Delta t = 1$ we get $\tau_{\mathrm{vi}} = \Delta t$. Having $\omega_0\tau_{\mathrm{vi}} < 0.1$, so that waves still propagate more or less according to the Navier-Stokes model, thus requires a sound wave period of $T = 2\pi/\omega_0 > 62.8\,\Delta t$. This high resolution requirement for $\tau/\Delta t = 1$ underlines the importance of choosing low numerical viscosities for LB simulations where sound is physically relevant.

The analysis performed here has some clear limits: it assumes that there is no background flow, it is performed for one specific equilibrium, it is limited to D1Q3 (i.e. the anisotropy of other velocity sets[12] is lost), and it does not take into account various possible choices of collision operator. There exist articles that take these effects into account in D2Q9-based linearisation analyses [34, 37], though their analyses are unfortunately limited to free waves and are by necessity *far* more complex.

12.3 Sources of Sound

There are many reasons why sound waves may appear in LB simulations:

- **Intentional setup:** The initial conditions of the simulation can be set up to contain sound waves. Free waves can be initialised with one wavelength in a periodic system [1, 7, 26, 27, 35, 40, 41]. Forced waves can be initialised with an inhomogeneous (e.g. Gaussian) density distribution [5, 42–47].
- **Unintentional setup:** When setting up a flow field as an initial condition for a simulation, this field is seldom perfect for the case at hand. Typically, this initial condition will be a mix of an incompressible flow field and a sound wave field. However, generally separating a low-Ma field into incompressible flow and sound is an unsolved, and perhaps *unsolvable*, problem,[13] which makes it hard to avoid including sound waves in initial flow conditions of simulations.[14]
- **Aerodynamic noise:** Sound can come from unsteady flow fields and their interaction with surfaces. Indeed, this is the main source of e.g. aircraft noise. As this behaviour is a consequence of the fluid conservation equations, it can also be captured in LB simulations.
- **Artificial sound sources:** It is also possible to make sound sources in simulations that generate sound artificially during the simulation. These can either be inside the domain or be implemented as boundary conditions.

In this section we will focus on the last two cases. They are tied together through the theory of sound sources which we will now look into.

[12]For forced waves in the continuous model using the D2Q9 velocity set, anisotropy first occurs at $O(\omega_0\tau_{\mathrm{vi}})^3$. Thus, attenuation and dispersion are both isotropic to lowest order, though anisotropy becomes significant at around $\omega_0\tau_{\mathrm{vi}} \approx 0.2$ [7, 28].

[13]Even so, some filtering techniques have been developed to approximately separate the sound field from the total flow field [48–51].

[14]Initial flow conditions for LB simulations are described in Sect. 5.5.

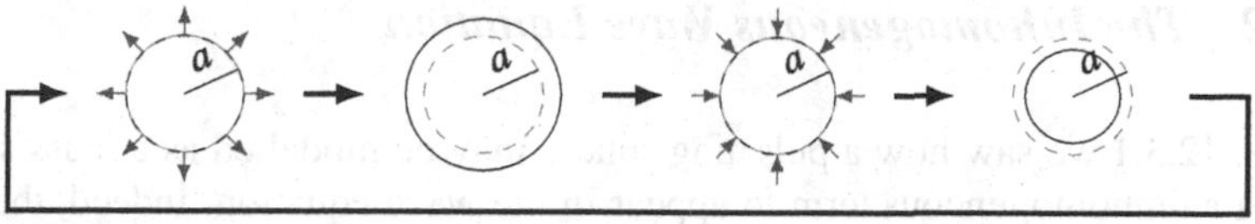

Fig. 12.2 Exaggerated sketch of four stages of a pulsating sphere of rest radius *a*. *From left to right:* Expanding, fully expanded, contracting, and fully contracted. The surface velocity is indicated by *arrows*

12.3.1 Example: The Pulsating Sphere

One of the most basic examples of a sound source within a fluid is a pulsating sphere which continuously expands and contracts at a single frequency ω_0. For a sphere with a rest radius of a centred at $\boldsymbol{x} = 0$, such a pulsation is typically modelled by a velocity boundary condition $u_r(|\boldsymbol{x}| = a, t) = u^\circ e^{i\omega_0 t}$, u_r being the fluid velocity in the radial direction and u° being the sphere's velocity amplitude. The sphere's pulsation is sketched in Fig. 12.2.

This velocity boundary condition leads to a mass displacement per time unit of $Qe^{i\omega t} = 4\pi a^2 \rho_0 u_0 e^{i\omega t}$. In three dimensions it also leads to the generation of the pressure wave [10]

$$\hat{p}'(\boldsymbol{x}, t) = \frac{i\omega Q}{4\pi |\boldsymbol{x}|} e^{i(\omega_0 t - \hat{k}|\boldsymbol{x}|)}. \tag{12.31}$$

Alternatively, we can model this mass displacement as a *mass source* at $\boldsymbol{x} = \boldsymbol{0}$ through an inhomogeneous linearised continuity equation

$$\partial_t \rho + \rho_0 \partial_\alpha u_\alpha = Q\delta(\boldsymbol{x}) e^{i\omega t}, \tag{12.32}$$

which leads to a wave equation

$$\left(\frac{1}{c_s^2}\partial_t^2 - (1 + \tau_{vi}\partial_t)\nabla^2\right)\hat{p}'(\boldsymbol{x}, t) = i\omega Q\delta(\boldsymbol{x}) e^{i\omega t}, \tag{12.33}$$

which, from the Green's functions in Sect. 12.1.6, has a solution identical to (12.31).

From this we can tell that pulsating objects scattered throughout the fluid can be modelled as mass sources. Similarly, oscillating objects (i.e. objects moving side-to-side) can be modelled as forces throughout the fluid.

Now, how is this relevant to LB simulations? In Sect. 12.3.3 we will come back to how such a simple source can be implemented in LB simulations.

12.3.2 The Inhomogeneous Wave Equation

In Sect. 12.3.1 we saw how a pulsating sphere may be modelled as a mass source, causing an inhomogeneous term to appear in the wave equation. Indeed, there are several possible sources of such inhomogeneous terms which we will now look at.

Instead of using approximate, linearised, forceless conservation equations, the wave equation may also be derived from the Navier-Stokes-level conservation equations

$$\partial_t \rho + \partial_\alpha \rho u_\alpha = Q(\boldsymbol{x}, t), \tag{12.34a}$$

$$\partial_t \left(\rho u_\alpha\right) + \partial_\beta \left(\rho u_\alpha u_\beta\right) = -\partial_\alpha p + \partial_\beta \sigma'_{\alpha\beta} + f_\alpha(\boldsymbol{x}, t), \tag{12.34b}$$

where $Q(\boldsymbol{x}, t)$ is a time-varying distribution of mass sources throughout the physical domain.

From these, an *exact* inhomogeneous wave equation for the *density* can be derived

$$\left(\frac{1}{c_s^2}\partial_t^2 - \nabla^2\right) c_s^2 \rho'(\boldsymbol{x}, t) = \partial_t Q(\boldsymbol{x}, t) - \partial_\alpha f_\alpha(\boldsymbol{x}, t) + \partial_\alpha \partial_\beta T_{\alpha\beta}(\boldsymbol{x}, t). \tag{12.35}$$

Each of the terms on the right-hand side represents a type of sound source. The last term contains the *Lighthill stress tensor*

$$T_{\alpha\beta} = \rho u_\alpha u_\beta + (p' - c_s^2 \rho')\delta_{\alpha\beta} - \sigma'_{\alpha\beta}. \tag{12.36}$$

The latter two terms in the Lighthill stress tensor do not typically contribute as sources of sound [25]. However, the second is linked to the deviation from the linearised isentropic equation of state,[15] while the third is linked to viscous attenuation as discussed earlier in this chapter.

The first term, which comes from the nonlinear term on the left-hand side of the momentum equation, *does* contribute. The contribution from the inhomogeneous term $\partial_\alpha \partial_\beta (\rho u_\alpha u_\beta)$ is typically strongest when there is a strongly fluctuating underlying flow field, such as a turbulent field.

From an impulsive Green's function approach, similar to the time-harmonic Green's function approach described in Sect. 12.1.6, an exact solution for the inhomogeneous density wave equation (12.35) can be found as an integral of the

[15]This term is zero for the isothermal equation of state.

inhomogeneous terms over the entire domain [7, 24]:

$$\begin{aligned} c_s^2 \rho'(\boldsymbol{x}, t) = \int \Big[\big(\partial_t Q(\boldsymbol{y}, t)\big) G(\boldsymbol{x} - \boldsymbol{y}, t) - f_\alpha(\boldsymbol{y}, t) \partial_\alpha G(\boldsymbol{x} - \boldsymbol{y}, t) \\ + T_{\alpha\beta}(\boldsymbol{y}, t) \partial_\alpha \partial_\beta G(\boldsymbol{x} - \boldsymbol{y}, t) \Big] \mathrm{d}^3 y. \end{aligned} \tag{12.37}$$

From Sect. 12.1.6 we know that these three terms can be interpreted as monopole, dipole, and quadrupole terms, respectively. For the case where all the inhomogeneous terms in (12.35) are time-harmonic, the corresponding Green's function is shown in (12.17).[16]

The sound generated by the interaction of the flow field, in particular the pressure, with extended surfaces is a field of study in of itself. For more on this this we refer to the literature [24, 52–54].

12.3.3 Point Source Monopoles in LB Simulations

At this point we have introduced three sources of sound: the mass displacement Q from small pulsating structures makes monopole sound sources, time-varying inhomogeneous forces are dipole sound sources, and a strongly fluctuating fluid velocity represents a distributed quadrupole sound source.

Now, how do these inhomogeneous sound waves appear in LB simulations? The forcing can be imposed using any of the forcing schemes described in Chap. 6. The quadrupole source is a direct consequence of the Navier-Stokes equation, which means that even the most basic LBE is therefore able to simulate such sources without modification if the flow is resolved finely enough [55–57].

On the other hand, the mass fluctuation must be imposed. In this section we will describe how this can be done through an approach that can also be generalised further to dipoles and quadrupoles.

An early approach to monopole point sources in LB simulations was to replace the entire particle distribution in a node with an equilibrium distribution at an imposed density that fluctuates around the rest density ρ_0 [5, 58]. However, this method is not advisable for several reasons [17]: it will disturb the flow unphysically, and the relationship between the source and the radiated wave's amplitude and phase is unknown.

[16]Note again that there *is no* simple impulsive Green's function available for 2D cases [11].

A newer approach to LB **sound sources** is based on adding a **particle source term** j_i to the lattice Boltzmann equation,

$$f_i(\boldsymbol{x} + \boldsymbol{c}_i\Delta t, t + \Delta t) - f_i(\boldsymbol{x}, t) = j_i(\boldsymbol{x}, t) + \Omega_i(\boldsymbol{x}, t)\Delta t. \tag{12.38}$$

This term represents particles being added to or removed from the distribution function with a certain distribution in velocity space, physical space, and time. This can be contrasted with the mass source in (12.34a), which is distributed in space and time but not in *velocity* space.

The effect of the particle source term j_i is best seen through its moments, which we define similarly to the moments of f_i,

$$J = \sum_i j_i, \quad J_\alpha = \sum_i c_{i\alpha} j_i, \quad J_{\alpha\beta} = \sum_i c_{i\alpha} c_{i\beta} j_i. \tag{12.39}$$

Naturally, adding a source term to the LBE will affect its macroscopic behaviour. This behaviour can be determined by a Chapman-Enskog analysis as detailed in Sect. 4.1, resulting in mass and momentum conservation equations

$$\partial_t \rho + \partial_\beta(\rho u_\beta) = J - \tfrac{\Delta t}{2}\left(\partial_t J + \partial_\beta J_\beta\right), \tag{12.40a}$$

$$\begin{aligned}\partial_t(\rho u_\alpha) + \partial_\beta(\rho u_\alpha u_\beta) = {} & -\partial_\alpha p + \partial_\beta \sigma'_{\alpha\beta} + (1 - \tfrac{\Delta t}{2}\partial_t) J_\alpha - \partial_\beta\left(\tau J_{\alpha\beta}\right) \\ & + \partial_\beta\left(\tau - \frac{\Delta t}{2}\right)\left(\delta_{\alpha\beta} c_s^2 J + u_\alpha J_\beta + u_\beta J_\alpha - u_\alpha u_\beta J\right),\end{aligned} \tag{12.40b}$$

where the viscous stress tensor is, as usual, $\sigma'_{\alpha\beta} = \rho c_s^2(\tau - \Delta t/2)(\partial_\beta u_\alpha + \partial_\alpha u_\beta)$. These conservation equations have gained source terms given by the moments of the particle source term j_i. The derivative terms with the $\Delta t/2$ prefactor are a result of the velocity space discretisation error of the LBE [7]. Disregarding these terms, the moment J directly takes the place of Q in (12.34a).

Many of these additional terms in the conservation equations are small, being of order $(\tau - \Delta t/2)uj$. Neglecting these, we can derive a viscous wave equation as done in Sect. 12.1.1, resulting in

$$\begin{aligned}\left[\frac{1}{c_s^2}\partial_t^2 - (1 + \tau_{\mathrm{vi}}\partial_t)\nabla^2\right] p = {} & \left(1 - \tfrac{\Delta t}{2}\partial_t\right)\partial_t J - \partial_\alpha J_\alpha \\ & + \tau \partial_\alpha \partial_\beta J_{\alpha\beta} - \left(\tau - \tfrac{\Delta t}{2}\right)\partial_\alpha\partial_\beta\left(3\delta_{\alpha\beta} c_s^2 J\right).\end{aligned} \tag{12.41}$$

Example 12.2 Let us look at the radiated wave from a point source at $\boldsymbol{x} = \mathbf{0}$, which from (12.35) and (12.37) is

$$\hat{p}'(\boldsymbol{x},t) = \left[\left(\mathrm{i}\omega_0 + \frac{\omega_0^2}{2}\right)J\right]\hat{G}(\boldsymbol{x},t) - J_\alpha\partial_\alpha\hat{G}(\boldsymbol{x},t)$$
$$+\left[\tau J_{\alpha\beta} - \left(\tau - \frac{\Delta t}{2}\right)3\delta_{\alpha\beta}c_s^2 J\right]\partial_\alpha\partial_\beta\hat{G}(\boldsymbol{x},t). \tag{12.42}$$

The J moment affects monopoles and quadrupoles, the J_α moment only affects dipoles, and the $J_{\alpha\beta}$ moment only affects quadrupoles.

In general, we can choose j_i cleverly to tailor which moments are non-zero and which are zero. For monopoles, a simple assumption is that mass appears at equilibrium, i.e.

$$j_i(\boldsymbol{x},t) = w_i Q(\boldsymbol{x},t), \tag{12.43}$$

which from (3.60) results in the moments

$$J = \sum_i j_i = Q, \quad J_\alpha = \sum_i c_{i\alpha} j_i = 0, \quad J_{\alpha\beta} = \sum_i c_{i\alpha}c_{i\beta} j_i = c_s^2\delta_{\alpha\beta}Q. \tag{12.44}$$

It might seem contradictory to choose a monopole source that also affects the quadrupole moment. However, this choice is ideal as it causes the quadrupole moment to cancel against the discretisation error term for low values of $\tau/\Delta t$. (As explained in Sect. 12.1.3, such low values are generally necessary to use in LB acoustics.)

Let us show where this cancellation comes from: using the harmonic 2D and 3D Green's functions in (12.17), it can be found that (12.42) becomes

$$\partial_\alpha\partial_\alpha\hat{G}(\boldsymbol{x},t) = -\hat{k}^2\hat{G}(\boldsymbol{x},t) \tag{12.45}$$

in both cases. Consequently, since (12.44) shows that $J_{\alpha\beta} = c_s^2\delta_{\alpha\beta}J$ and $J_\alpha = 0$, and since $(\hat{k}c_s) \approx \omega_0$, for $j_i = w_i Q(t)\delta(\boldsymbol{x})$ and $\tau/\Delta t \to \frac{1}{2}$ (12.42) becomes simply

$$\hat{p}'(\boldsymbol{x},t) = \mathrm{i}\omega_0 Q\hat{G}(\boldsymbol{x},t), \tag{12.46}$$

just as it should be for a monopole point source.

To summarise, we can add sound sources to LB simulations through a particle source term j_i as in (12.38). By setting $j_i(\boldsymbol{x},t) = w_i Q(\boldsymbol{x},t)$, we can create

(continued)

monopole sound sources. These sources can either be single point sources that radiate as (12.46) or distributions of such point sources around the domain.

The same approach can be expanded to create dipole and quadrupole sources. Indeed, j_i can be decomposed into a basis of a monopole and various dipole and quadrupole sources. We will not go into this further here, but rather refer to the literature [7, 17].

As an example of the monopole point source in action, Fig. 12.3 compares the field of a D2Q9 monopole sound source simulated at $\tau/\Delta t = 1/2$ using both the BGK operator and a simple MRT operator with immediate relaxation (i.e. relaxation times of 1) of the non-hydrodynamic moments. Taking the analytical wavenumber from (12.28a), we find excellent agreement for MRT collisions while the BGK case shows spurious oscillations around the analytical solution. It should also be pointed out that there is a singularity in the analytical solution at $\boldsymbol{x} = \boldsymbol{0}$, which means that the simulated solution can never match the analytical solution in the immediate area around the source point.

Finally, we should briefly mention *precursors*, a wave phenomenon that appears in dispersive media. If sources are immediately switched on at reasonably high viscosities, precursors will appear [59], manifesting as bumps at the first wavefront. Additionally, if sources are immediately turned on in simulations at very low viscosities, errors may be visible along the first wavefront. In such cases, sound sources should be turned on smoothly. More details on this can be found elsewhere [7].

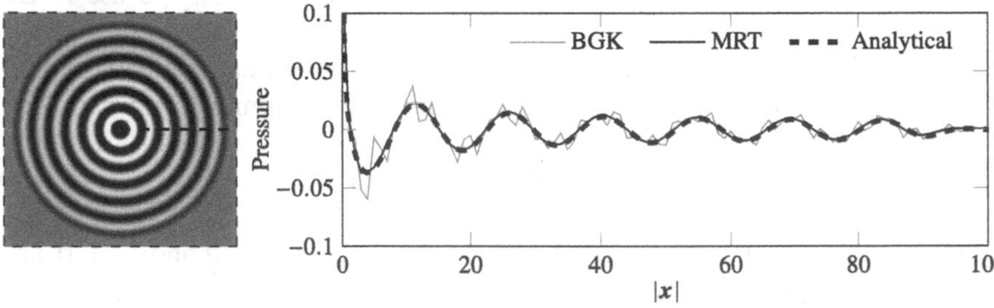

Fig. 12.3 *Left*: Snapshot of the pressure wave radiated by a smoothly switched-on monopole source, using a simple MRT operator at $\tau/\Delta t = 1/2$. *Right*: Comparison of the radial pressure along the dashed line on the left with the pressure predicted by theory, for MRT and BGK operators at $\tau/\Delta t = 1/2$. Taken from [7]

12.4 Non-reflecting Boundary Conditions

The most commonly used boundary conditions for inlets and outlets in lattice Boltzmann simulations specify the flow field's velocity or density. Such boundary conditions are far from ideal for flows containing sound waves, as they will reflect sound waves back into the fluid [60]. Consequently, any sound waves generated in the simulated system will not *exit* the system, and this may pollute the simulation results. In particular, the density field may be strongly polluted [6] as its variations are typically quite weak for steady flows, where they scale as $\rho'/\rho_0 = O(\mathrm{Ma}^2)$ [61]. (And of course, when the density field is polluted, the pressure field is polluted accordingly, as $p' = c_s^2\rho'$.) It is also desirable that the boundary conditions allow other wavelike phenomena, such as *vorticity waves* and *entropy waves* [13] to exit the system smoothly [45].

This is not an LB specific problem, but a general problem for compressible flow solvers. To solve it, a variety of different BC approaches that reflect much less sound have been proposed [62]. Some work has been done on adapting such non-reflecting BCs (or *NRBCs*) to lattice Boltzmann simulations [6, 45, 47, 63–65], but at this point there has been little work done on comparing different approaches against each other [9].

There are **two main approaches** to NRBCs. The first is **characteristic boundary conditions** (CBCs), where the fluid equations are decomposed in boundary nodes in a way that allows supressing waves being reflected back from the boundary. The second approach is **absorbing layers**, where the simulated domain is bounded by a layer several nodes thick. In an absorbing layer, the LBM is modified in such a way that incoming waves are absorbed as they pass through the layer without being reflected back.

Absorbing layers are generally more demanding than CBCs, as they add more nodes to the system that need to be computed similarly to the existing ones. However, better results may be achieved by absorbing layers [9].

12.4.1 *Reflecting Boundary Conditions*

To understand the point of NRBCs, we must first comprehend how sound waves are reflected by simpler boundary conditions. These BCs typically impose a constant value for the fluid pressure or the fluid velocity. In this section we will look at what happens when an incoming sound wave, which represents a fluctuation around this constant pressure or velocity, meets such a BC. For simplicity, we will restrict ourselves to the basic one-dimensional case of a plane sound wave hitting the boundary at normal incidence. The reflection of non-plane waves is a more difficult topic which we will not go into here.

First, however, we must know how the pressure and the velocity in a sound wave are coupled. From Exercise 12.1 we already know that the pressure of a one-dimensional wave propagating in the $\pm x$ direction can be represented as $p'(c_s t \mp x)$. From Euler's momentum equation, the relation to the fluid velocity $u'(c_s t \mp x)$ can be determined:

Exercise 12.9 From the linearised one-dimensional Euler's equation $\rho_0 \partial_t u' = -\partial_x p$, show that the pressure and velocity are coupled as

$$p'(c_s t \mp x) = \pm \rho_0 c_s u'(c_s t \mp x). \tag{12.47}$$

Hint: Use the chain rule with the argument $c_s t \mp x$.

In the following examples, we will consider boundary conditions imposed at $x = 0$, with a known incoming sound wave $p'_i(c_s t - x)$ and an unknown reflected sound wave $p'_r(c_s t + x)$, the physical domain being $x \leq 0$.

Let us first consider a *pressure* boundary condition, which imposes $p = p_0$ at $x = 0$. Imposing this constant pressure means imposing a zero pressure fluctuation, $p' = 0$, at $x = 0$, so that

$$p'(0,t) = p'_i(c_s t - 0) + p'_r(c_s t + 0) = 0. \tag{12.48a}$$

The unknown reflected wave p_r must therefore be the inverse of the incoming wave, i.e.

$$p'_r(c_s t + x) = -p'_i(c_s t - x). \tag{12.48b}$$

Second, let us consider a *velocity* boundary condition, which imposes $u = u_0$, i.e. $u' = 0$, at $x = 0$. The incoming and reflected fluid velocities are linked as

$$u'(0,t) = u'_i(c_s t - 0) + u'_r(c_s t + 0) = 0, \tag{12.49a}$$

and from (12.47) this links the incoming and reflected pressures as

$$p'_r(c_s t + x) = p'_i(c_s t - x). \tag{12.49b}$$

Both for a pressure and a velocity BC, **the incoming wave is transformed into an identically shaped reflected wave of the same amplitude**. For the pressure BC, the reflected wave has the opposite sign, while it has the same sign for the velocity BC. These two cases are shown in Fig. 12.4, which also

(continued)

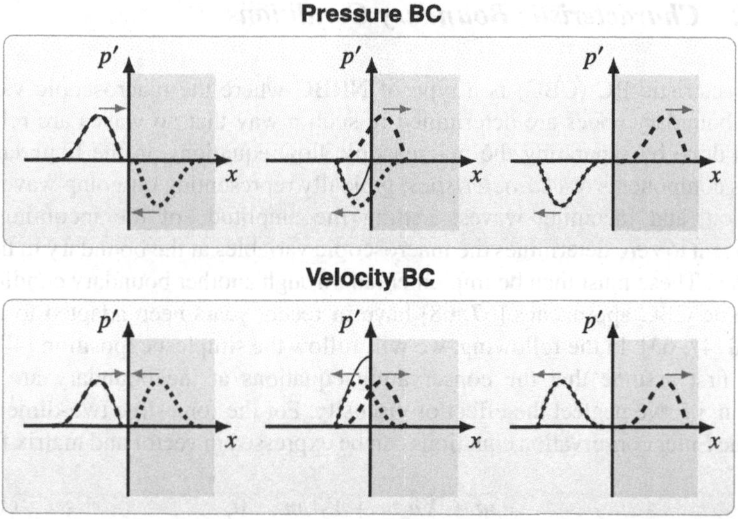

Fig. 12.4 For two different BCs, the incoming pressure pulse $p'_{\rm i}(c_{\rm s}t-x)$ (*dashed*) hits a boundary at $x=0$ and a reflected pulse $p'_{\rm r}(c_{\rm s}t+x)$ (*dotted*) is sent back. The total pressure $p'_{\rm i}+p'_{\rm r}$ (*solid*) in the physical domain ($x\leq 0$) is also shown. (The non-physical "mirror" domain ($x>0$) is indicated by a *darker colour*)

shows the non-physical "mirror" domain $x>0$ from which the reflected waves come.

Consequently, any LB BCs that enforce constant pressure or constant velocity along a boundary will reflect sound waves back into the system, regardless of the specific implementation of these BCs.

In a more realistic case, such as a sound wave hitting a building wall, there will not just be an incoming and a reflected sound wave: as most people will have noticed, sound can also be *transmitted* into the wall, and transmitted again from the wall to the air on the other side. Considering only normal incidence, the efficiency of this transmission between two media depends on their *characteristic impedance* $Z=\rho_0 c_{\rm s}=\pm p'/u'$ [10]. If the impedances of both media are equal, the transmission is perfect, meaning that the incoming and transmitted waves are equal and that nothing is reflected at the boundary. The same reflections as for the pressure and velocity BCs are recovered if the impedance of the second medium is $Z=0$ and $Z\to\infty$, respectively.[17]

[17]Such transmission between two media of different impedances has also been simulated correctly using the LBM, including weak interfacial effects not predicted by the hydrodynamic equations [66].

12.4.2 Characteristic Boundary Conditions

A characteristic BC (CBC) is a type of NRBC where the macroscopic variables on the boundary nodes are determined in such a way that no waves are reflected. This is done by separating the macroscopic flow equations on the boundary into various components or *characteristics*, typically representing outgoing waves, pure advection, and incoming waves. Setting the amplitude of the incoming wave component to zero determines the macroscopic variables at the boundary in the next time step. These must then be implemented through another boundary condition.

Classic CBC approaches [67, 68] have in recent years been adapted to the LB method [47, 63]. In the following, we will follow the simplest exposition [47].

We first assume that the conservation equations at the boundary are nearly Eulerian, i.e. we neglect the effect of viscosity. For the force-free two-dimensional case, the Euler conservation equations can be expressed in vector and matrix form as

$$\partial_t \boldsymbol{m} + \boldsymbol{X}\partial_x \boldsymbol{m} + \boldsymbol{Y}\partial_y \boldsymbol{m} = 0, \tag{12.50a}$$

using the fluid variable vector $\boldsymbol{m} = (\rho, u_x, u_y)^{\mathrm{T}}$ and the matrices

$$\boldsymbol{X} = \begin{bmatrix} u_x & \rho & 0 \\ c_s^2/\rho & u_x & 0 \\ 0 & 0 & u_x \end{bmatrix}, \qquad \boldsymbol{Y} = \begin{bmatrix} u_y & 0 & \rho \\ 0 & u_y & 0 \\ c_s^2/\rho & 0 & u_y \end{bmatrix}. \tag{12.50b}$$

As this system of equations is hyperbolic, these matrices are diagonalisable as $\boldsymbol{X} = \boldsymbol{P}_x^{-1}\boldsymbol{\Lambda}_x\boldsymbol{P}_x$ and $\boldsymbol{Y} = \boldsymbol{P}_y^{-1}\boldsymbol{\Lambda}_y\boldsymbol{P}_y$, where $\boldsymbol{\Lambda}_x$ and $\boldsymbol{\Lambda}_y$ are diagonal matrices containing the eigenvalues of $\boldsymbol{X}$ and $\boldsymbol{Y}$,

$$\begin{aligned} \boldsymbol{\Lambda}_x &= \mathrm{diag}(\lambda_{x,1}, \lambda_{x,2}, \lambda_{x,3}) = \mathrm{diag}(u_x - c_s, u_x, u_x + c_s), \\ \boldsymbol{\Lambda}_y &= \mathrm{diag}(\lambda_{y,1}, \lambda_{y,2}, \lambda_{y,3}) = \mathrm{diag}(u_y - c_s, u_y, u_y + c_s). \end{aligned} \tag{12.51a}$$

The diagonalisation matrices are given by

$$\begin{aligned} \boldsymbol{P}_x &= \begin{bmatrix} c_s^2 & -c_s\rho & 0 \\ 0 & 0 & 1 \\ c_s^2 & c_s\rho & 0 \end{bmatrix}, \quad \boldsymbol{P}_x^{-1} = \begin{bmatrix} \frac{1}{2c_s^2} & 0 & \frac{1}{2c_s^2} \\ -\frac{1}{2\rho c_s} & 0 & \frac{1}{2\rho c_s} \\ 0 & 1 & 0 \end{bmatrix}, \\ \boldsymbol{P}_y &= \begin{bmatrix} c_s^2 & 0 & -c_s\rho \\ 0 & 1 & 0 \\ c_s^2 & 0 & c_s\rho \end{bmatrix}, \quad \boldsymbol{P}_y^{-1} = \begin{bmatrix} \frac{1}{2c_s^2} & 0 & \frac{1}{2c_s^2} \\ 0 & 1 & 0 \\ -\frac{1}{2\rho c_s} & 0 & \frac{1}{2\rho c_s} \end{bmatrix}. \end{aligned} \tag{12.51b}$$

The physical meaning of this diagonalisation can be seen in a y-invariant case where $\partial_y \boldsymbol{m} = \boldsymbol{0}$. Left-multiplying (12.50a) with $\boldsymbol{P}_x$ results in

$$\boldsymbol{P}_x \partial_t \boldsymbol{m} + \boldsymbol{\Lambda}_x \boldsymbol{P}_x \partial_x \boldsymbol{m} = 0. \tag{12.52}$$

A subsequent definition $\mathrm{d}\boldsymbol{n} = \boldsymbol{P}_x \mathrm{d}\boldsymbol{m}$ leads to the three equations

$$\partial_t n_i + \lambda_{x,i} \partial_x n_i = 0. \tag{12.53}$$

These equations are mathematically identical to the advection equation, describing propagation of the quantities n_i, the propagation speeds being given by the eigenvalues $\lambda_{x,i}$. From (12.51), these propagation speeds in turn describe the combined effect of advection and sound propagation in the $-x$ direction, pure advection, and advection and sound propagation in the $+x$ direction. The quantities n_i can consequently be interpreted as amplitudes of the corresponding components of the flow field.

Why this decomposition into characteristics is relevant for non-reflecting BCs can now be seen. If we can **enforce** $n_i = 0$ **at the boundary** for the characteristics that represent **sound entering the system**, we can in principle ensure that sound waves smoothly **exit the system without reflection**. Let us now look at how to implement this.

The x-derivative term in (12.50a) can be expressed as

$$\boldsymbol{X} \partial_x \boldsymbol{m} = \boldsymbol{P}_x^{-1} \boldsymbol{\Lambda}_x \boldsymbol{P}_x \partial_x \boldsymbol{m} = \boldsymbol{P}_x^{-1} \mathcal{L}_x, \tag{12.54}$$

defining a characteristic vector as $\mathcal{L}_{x,i} = \lambda_{x,i} \sum_j P_{x,ij} \partial_x m_j$. Each component $\mathcal{L}_{x,i}$ is proportional to the amplitude of one of the three characteristics. Explicitly, the components of $\mathcal{L}_x$ are

$$\begin{aligned} \mathcal{L}_{x,1} &= (u_x - c_\mathrm{s}) \left[c_\mathrm{s}^2 \partial_x \rho - c_\mathrm{s} \rho \partial_x u_x \right], \\ \mathcal{L}_{x,2} &= (u_x) \left[\partial_x u_y \right], \\ \mathcal{L}_{x,3} &= (u_x + c_\mathrm{s}) \left[c_\mathrm{s}^2 \partial_x \rho + c_\mathrm{s} \rho \partial_x u_x \right]. \end{aligned} \tag{12.55}$$

At x-boundaries we want to enforce a *modified* characteristic vector with elements

$$\mathcal{L}'_{x,i} = \begin{cases} \mathcal{L}_{x,i} & \text{for outgoing characteristics,} \\ 0 & \text{for incoming characteristics.} \end{cases} \tag{12.56}$$

Outgoing characteristics are those where the eigenvalue $\lambda_{x,i}$ corresponds to propagation out of the system (e.g. $\lambda_{x,3} = u_x + c_s$ at the rightmost boundary). For incoming characteristics, the eigenvalue corresponds to propagation into the system (i.e. $\lambda_{x,1} = u_x - c_s$ at the same boundary). In this way, we ensure that the outgoing characteristics are undisturbed while the incoming characteristics carry nothing back into the system.

A similar characteristic vector can be defined for the y-derivatives as $\mathcal{L}_{y,i} = \lambda_{y,i} \sum_j P_{y,ij} \partial_y m_j$. The only difference to $\mathcal{L}_{x,i}$ given in (12.55) is that all the x and y indices are switched. We similarly want to enforce a modified characteristic vector $\mathcal{L}'_y$ at the y boundaries.

There are several different approaches to evolving the fluid variables at the x-boundary that enforce $\mathcal{L}_x$. They can be generalised using a modified version of (12.50a) as

$$x\text{-boundary:} \qquad \partial_t \boldsymbol{m} = -\boldsymbol{P}_x^{-1} \mathcal{L}'_x - \gamma \boldsymbol{Y} \partial_y \boldsymbol{m}. \tag{12.57a}$$

Here, $\gamma = 1$ corresponds to the original approach of Thompson [67], while the LB CBC of Izquierdo and Fueyo used an one-dimensional approach with $\gamma = 0$ that does not include any y-contribution [63]. Heubes et al. found the choice of $\gamma = 3/4$ to be superior [47]. In the same generalised approach, y-boundaries and the x- and y-boundary conditions at corners may be treated as

$$y\text{-boundary:} \qquad \partial_t \boldsymbol{m} = -\gamma \boldsymbol{X} \partial_x \boldsymbol{m} - \boldsymbol{P}_y^{-1} \mathcal{L}'_y, \tag{12.57b}$$

$$\text{corner:} \qquad \partial_t \boldsymbol{m} = -\boldsymbol{P}_x^{-1} \mathcal{L}'_x - \boldsymbol{P}_y^{-1} \mathcal{L}'_y. \tag{12.57c}$$

To determine the time derivatives $\partial_t \boldsymbol{m}$, the characteristic vectors must be estimated according to (12.56) and (12.55). This requires estimating the spatial derivatives. For derivatives *across* a boundary (e.g. x-derivatives at an x-boundary), this can be done through the one-sided second-order finite difference approximations

$$(\partial_x m_i)(x) \approx \frac{\mp 3m_i(x) \pm 4m_i(x \pm \Delta x) \mp m_i(x \pm 2\Delta x)}{2\Delta x}, \tag{12.58a}$$

where the upper and lower signs correspond to forward and backward difference approximations, respectively. For derivatives *along* a boundary (i.e. y-derivatives at an x-boundary), the second-order central difference approximation

$$(\partial_x m_i)(x) \approx \frac{m_i(x + \Delta x) - m_i(x - \Delta x)}{2\Delta x}, \tag{12.58b}$$

is appropriate.

With the spatial derivatives in place, we have a known approximation of the macroscopic variables' time derivative $\partial_t \boldsymbol{m}$ on every edge of the system. To determine these macroscopic variables for the next time step, a simple approach is a forward Euler one where

$$m_i(\boldsymbol{x}, t + \Delta t) \approx m_i(\boldsymbol{x}, t) + \Delta t \partial_t m_i(\boldsymbol{x}, t). \tag{12.59}$$

In one benchmark, this simple first-order approach performed near-identically to a higher-order Runge-Kutta approach [47], which suggests that the error of this approximation is typically dominated by other sources of error in the CBC method.

With a known approximation of each macroscopic variable m_i on the non-reflecting boundary at the next time step, these macroscopic variables may be implemented through any boundary condition that allows specifying ρ, u_x, and u_y. The simplest choice, which has also been made in the literature [9, 47], is to simply replace the distribution function f_i at the boundary with an equilibrium distribution f_i^{eq} determined by the macroscopic variables ρ, u_x, and u_y, like in the equilibrium scheme discussed in Sect. 5.3.4.

When implementing this boundary condition in code, no collisions are required in the CBC boundary nodes. After the macroscopic variables have been determined, but before streaming, the f_is in the CBC boundary nodes are replaced using predetermined macroscopic variables $m_i(t)$, and the macroscopic variables $m_i(t + \Delta t)$ for the next time step are determined.

As an example of how CBCs are significantly less reflective than velocity boundaries, CBC boundaries as described above are compared against no-slip bounce-back in Fig. 12.5. In the simulation, a pulse was initialised as a Gaussian-distributed density $\rho'(\boldsymbol{x}, t = 0) = 10^{-4} \mathrm{e}^{-(\boldsymbol{x}-\boldsymbol{x}_0)^2/10}$, $\boldsymbol{x}_0$ being the centre of the system. The simulation was run using the BGK collision operator with $\tau/\Delta t = 0.51$, and the CBC used $\gamma = 0.75$. Note however that the CBCs are not perfect; weak reflections can be seen from these boundaries.

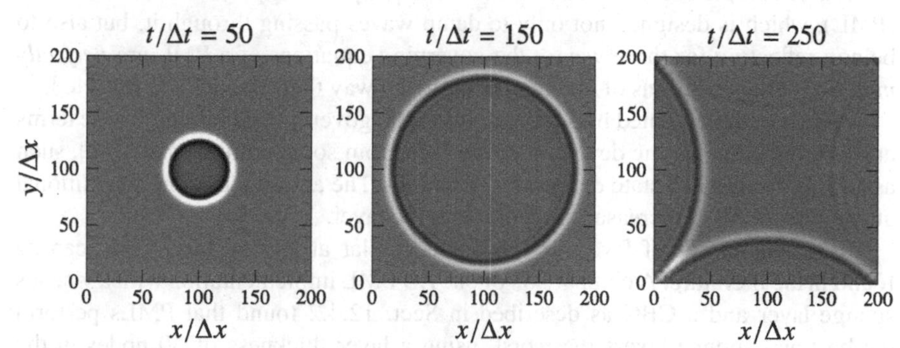

Fig. 12.5 Reflection of a sound pulse from bounce-back boundaries (*bottom and left*) and characteristic BC boundaries (*top and right*)

12.4.3 Absorbing Layers

A characteristic boundary condition (CBC), as described in Sect. 12.4.2, is a type of a non-reflecting BC (NRBC) that only affects the nodes adjacent to a boundary. A different approach is taken by *absorbing layer*-type NRBCs, where the simulation domain is bounded by a layer of many nodes of thickness. In this layer, the LBM is modified in a way that attenuates sound waves as they pass through the layer, with as much attenuation and as little reflection as possible. The performance of an absorbing layer typically increases with its thickness.

At the outer edge of the layer, another BC must be chosen to close the system. Ideally this should be a CBC or similar in order to reduce the reflection at the edge of the system (in addition to the wave absorption in the layer) [62]. In this case, waves are not only attenuated as they pass through the layer, but only a small part of the wave that hits the edge of the system is reflected back in, and that part is further attenuated as it passes *back* through the layer into the simulation domain.

We will not treat absorbing layers in any depth here, as the worthwhile methods are somewhat complex and do not offer as much additional physical insight as CBCs.

The simplest absorbing layer is the *viscous sponge layer*, where the simulated system is bounded by a layer with a higher viscosity than the simulation domain [46, 69]. This viscosity is typically smoothly increasing from the viscosity of the interior domain to a very high viscosity near the edge of the system. Viscous sponge layers are very simple to implement; the value of τ must be made a local function of space inside the layer, but otherwise the LBM proceeds as normal. On the other hand, viscous sponge layers are not as effective at being non-reflective as other types of NRBCs [9]; sound waves may be reflected as they pass through the layer itself [62].

A more complex but more effective absorbing layer is the *perfectly matched layer* (PML), which is designed not only to damp waves passing through it, but also to be non-reflecting for these waves; the governing equations of a PML are *perfectly matched* to the equations of the interior in such a way that sound is not reflected.

PMLs are implemented by adding terms to the governing equations. These terms are determined using the deviation of the field from some nominal state [62], such as an equilibrium rest state defined by ρ_0 and $\boldsymbol{u}_0$. The amplitude of these additional attenuating terms is increased smoothly into the layer.

LBM adaptations of PMLs [64, 65] and similar absorbing layers [45] can be found in the literature. A comparison of one LB PML implementation with a viscous sponge layer and a CBC as described in Sect. 12.4.2 found that PMLs perform the best and sponge layers the worst, using a layer thickness of 30 nodes in the comparison [9].

12.5 Summary

Fluid models represented by a set of conservation equations, such as the Euler model or Navier-Stokes model, support sound waves if they fulfill certain criteria: they must have a mass conservation equation of the form $\partial\rho/\partial t + \nabla \cdot (\rho \boldsymbol{u})$, a momentum equation with a pressure gradient, and an equation of state relating the changes in pressure and density. If those criteria are fulfilled, a wave equation can be derived as in Sect. 12.1.1, which indicates that the fluid model in question supports sound waves.

The Euler and Navier-Stokes models both fulfill these criteria and therefore support sound waves. On the other hand, the *incompressible* Navier-Stokes model presented in Sect. 1.1.2 does *not*: its assumption of constant density modifies the continuity equation to $\nabla \cdot \boldsymbol{u}$ and breaks the link between changes in pressure and changes in density. (Interestingly, the incompressible LB model described in Sect. 4.3.2 *does* fulfill the criteria and therefore supports sound waves.)

As discussed in Sect. 12.3, there are several ways that sound waves can appear in simulations. Sound waves can be present in the initial conditions of a simulation, either intentionally or unintentionally. Additionally, sound can be generated by unsteady flow fields, either by their interaction with surfaces or in areas of rapid fluid velocity variation such as turbulent areas. As these sound generation mechanisms follow from the fluid model's equations, they are not in any way specific to LB simulations; sound waves can generally appear spontaneously in compressible fluid simulations. (Artificial sound sources can also be put in LB simulations, as discussed in Sect. 12.3.)

These sound waves are typically reflected from velocity or density boundary conditions back into the simulated system as discussed in Sect. 12.4. This pollutes the flow field of the solution; in particular, the density and pressure fields. In order to ensure that the sound waves exit the system smoothly, special *non-reflecting* boundary conditions must be applied such as characteristic boundary conditions, viscous layers, or perfectly matched layers.

Sound waves in LB simulations do not propagate in quite the same way as in theory, as discussed in Sect. 12.2. There are two reasons for this. The first is the second-order discretisation error in space and time. The second is the fact that the Boltzmann equation, and its discrete-velocity version that the LBM is based on, predicts a different attenuation and dispersion in some cases. This deviation can be quantified through the acoustic viscosity number $\omega_0 \tau_{\mathrm{vi}} \sim \mathrm{Kn}$, where ω_0 is the angular frequency and $\tau_{\mathrm{vi}} = (4\nu/3 + \nu_{\mathrm{B}})/c_{\mathrm{s}}^2$ is the viscous relaxation time.

Disregarding the discretisation error, the sound wave propagation of the LBM starts to deviate significantly from that predicted by the Navier-Stokes model at $\omega_0 \tau_{\mathrm{vi}} \approx 0.1$ [28]. From $\omega_0 \tau_{\mathrm{vi}} \approx 0.2$, anisotropy can also start becoming significant.

Generally, choosing $\tau/\Delta t$ values as close to $1/2$ as possible is important for LB sound simulations; this was also discussed in Sect. 12.1.3. However, as shown in Fig. 12.3, this can push the limits of the BGK collision operator and can thus require choosing another collision operator such as the TRT or MRT operators described in Chap. 10.

As a concluding remark, sound waves are to some degree inevitable in unsteady LB simulations, and it therefore benefits researchers who use the LBM to know something about sound waves, regardless of whether they are interested in sound-related applications or not.

References

1. J.M. Buick, C.A. Greated, D.M. Campbell, Europhys. Lett. **43**(3), 235 (1998)
2. A. Wilde, Anwendung des Lattice-Boltzmann-Verfahrens zur Berechnung strömungsakustischer Probleme. Ph.D. thesis, TU Dresden (2007)
3. S. Marié, Etude de la méthode Boltzmann sur réseau pour les simulations en aéroacoustique. Ph.D. thesis, Institut Jean le Rond d'Alembert, Paris (2008)
4. A.R. da Silva, Numerical studies of aeroacoustic aspects of wind instruments. Ph.D. thesis, McGill University, Montreal (2008)
5. E.M. Viggen, The lattice Boltzmann method with applications in acoustics. Master's thesis, Norwegian University of Science and Technology (NTNU), Trondheim (2009)
6. M. Schlaffer, Non-reflecting boundary conditions for the lattice Boltzmann method. Ph.D. thesis, Technical University of Munich (2013)
7. E.M. Viggen, The lattice Boltzmann method: Fundamentals and acoustics. Ph.D. thesis, Norwegian University of Science and Technology (NTNU), Trondheim (2014)
8. M. Hasert, Multi-scale lattice Boltzmann simulations on distributed octrees. Ph.D. thesis, RWTH Aachen University (2014)
9. S.J.B. Stoll, Lattice Boltzmann simulation of acoustic fields, with special attention to non-reflecting boundary conditions. Master's thesis, Norwegian University of Science and Technology (NTNU), Trondheim (2014)
10. L.E. Kinsler, A.R. Frey, A.B. Coppens, J.V. Sanders, *Fundamentals of Acoustics*, 4th edn. (Wiley, New York, 2000)
11. D.T. Blackstock, *Fundamentals of Physical Acoustics* (Wiley, New York, 2000)
12. P.M. Morse, K.U. Ingard, *Theoretical Acoustics* (McGraw-Hill Book Company, New York, 1968)
13. A.D. Pierce, *Acoustics* (The Acoustical Society of America, New York, 1989)
14. S. Temkin, *Elements of Acoustics* (Wiley, New York, 1981)
15. Lord Rayleigh, *The Theory of Sound*, vol. 1, 1st edn. (Macmillan and Co., London, 1877)
16. Lord Rayleigh, Proc. Lond. Math. Soc. **s1-9**(1), 21 (1877)
17. E.M. Viggen, Phys. Rev. E **87**(2) (2013)
18. C. Truesdell, J. Rat. Mech. Anal. **2**(4), 643 (1953)
19. M. Greenspan, in *Physical Acoustics*, vol. IIA, ed. by W.P. Mason (Academic Press, New York, 1965), pp. 1–45
20. E.M. Salomons, W.J.A. Lohman, H. Zhou, PLOS ONE **11**(1), e0147206 (2016)
21. N.H. Johannesen, J.P. Hodgson, Reports Prog. Phys. **42**(4), 629 (1979)
22. Y. Li, X. Shan, Phil. Trans. R. Soc. A **369**(1944), 2371 (2011)
23. C. Sun, F. Paot, R. Zhang, D.M. Freed, H. Chen, Commun. Comput. Phys. **13**(1), 757 (2013)
24. M.S. Howe, *Theory of Vortex Sound* (Cambridge University Press, Cambridge, 2003)
25. M.J. Lighthill, Proc. R. Soc. Lond. A **211**(1107), 564 (1952)
26. J.M. Buick, C.L. Buckley, C.A. Greated, J. Gilbert, J. Phys. A **33**, 3917 (2000)
27. E.M. Viggen, Phys. Rev. E **90**, 013310 (2014)
28. E.M. Viggen, Commun. Comput. Phys. **13**(3), 671 (2013)
29. G.G. Stokes, Trans. Cambridge Phil. Soc. **8**, 287 (1845)

30. C.S. Wang Chang, G.E. Uhlenbeck, in *Studies in Statistical Mechanics*, vol. V (North-Holland Publishing, Amsterdam, 1970)
31. J. Foch, G. Uhlenbeck, Phys. Rev. Lett. **19**(18), 1025 (1967)
32. J. Foch, M. Losa, Phys. Rev. Lett. **28**(20), 1315 (1972)
33. J.D. Sterling, S. Chen, J. Comput. Phys. **123**(1), 196 (1996)
34. P. Lallemand, L.S. Luo, Phys. Rev. E **61**(6), 6546 (2000)
35. P. Dellar, Phys. Rev. E **64**(3) (2001)
36. A. Wilde, Comput. Fluids **35**, 986 (2006)
37. T. Reis, T.N. Phillips, Phys. Rev. E **77**(2), 026702 (2008)
38. S. Marié, D. Ricot, P. Sagaut, J. Comput. Phys. **228**(4), 1056 (2009)
39. E.M. Viggen, Phil. Trans. R. Soc. A **369**(1944), 2246 (2011)
40. D. Haydock, J.M. Yeomans, J. of Phys. A **34**, 5201 (2001)
41. J.M. Buick, J.A. Cosgrove, J. Phys. A **39**(44), 13807 (2006)
42. X.M. Li, R.C.K. Leung, R.M.C. So, AIAA J. **44**(1), 78 (2006)
43. X.M. Li, R.M.C. So, R.C.K. Leung, AIAA J. **44**(12), 2896 (2006)
44. A.R. da Silva, G.P. Scavone, J. Phys. A **40**(3), 397 (2007)
45. E.W.S. Kam, R.M.C. So, R.C.K. Leung, AIAA J. **45**(7), 1703 (2007)
46. E. Vergnault, O. Malaspinas, P. Sagaut, J. Comput. Phys. **231**(24), 8070 (2012)
47. D. Heubes, A. Bartel, M. Ehrhardt, J. Comput. Appl. Math. **262**, 51 (2014)
48. W. de Roeck, M. Baelmans, W. Desmet, AIAA J. **46**(2), 463 (2008)
49. C. Silva, F. Nicoud, S. Moreau, in *16th AIAA/CEAS Aeroacoustics Conference* (2010)
50. S. Sinayoko, A. Agarwal, Z. Hu, J. Fluid Mech. **668**, 335 (2011)
51. S. Sinayoko, A. Agarwal, J. Acoust. Soc. Am. **131**(3), 1959 (2012)
52. N. Curle, Proc. R. Soc. A **231**, 505 (1955)
53. J.E. Ffowcs Williams, D.L. Hawkings, Phil. Trans. R. Soc. A **264**(1151), 321 (1969)
54. K.S. Brentner, F. Farassat, AIAA J. **36**(8), 1379 (1998)
55. T. Colonius, S.K. Lele, Progress Aerospace Sci. **40**(6), 345 (2004)
56. M. Wang, J.B. Freund, S.K. Lele, Annu. Rev. Fluid Mech. **38**(1), 483 (2006)
57. E. Garnier, N. Adams, P. Sagaut, *Large Eddy Simulation for Compressible Flows*. Scientific Computation (Springer, New York, 2009)
58. H. Yu, K. Zhao, Phys. Rev. E **61**(4), 3867 (2000)
59. D.T. Blackstock, J. Acoust. Soc. Am. **41**(5), 1312 (1967)
60. S. Izquierdo, P. Martínez-Lera, N. Fueyo, Comp. Math. Appl. **58**(5), 914 (2009)
61. P.A. Thompson, *Compressible-Fluid Dynamics* (McGraw-Hill, New York, 1972)
62. T. Colonius, Annu. Rev. Fluid Mech. **36**, 315 (2004)
63. S. Izquierdo, N. Fueyo, Phys. Rev. E **78**(4) (2008)
64. M. Tekitek, M. Bouzidi, F. Dubois, P. Lallemand, Comp. Math. Appl. **58**(5), 903 (2009)
65. A. Najafi-Yazdi, L. Mongeau, Comput. Fluids **68**, 203 (2012)
66. M.M. Tekitek, M. Bouzidi, F. Dubois, P. Lallemand, Prog. Comput. Fluid Dyn. **8**(1–4), 49 (2008)
67. K.W. Thompson, J. Comput. Phys. **68**(1), 1 (1987)
68. T. Poinsot, S. Lele, J. Comput. Phys. **101**, 104 (1992)
69. E. Vergnault, O. Malaspinas, P. Sagaut, J. Acoust. Soc. Am. **133**(3), 1293 (2013)

Part IV
Numerical Implementation of the Lattice Boltzmann Method

Chapter 13
Implementation of LB Simulations

Abstract After reading this chapter, you will understand the fundamentals of high-performance computing and how to write efficient code for lattice Boltzmann method simulations. You will know how to optimise sequential codes and develop parallel codes for multi-core CPUs, computing clusters, and graphics processing units. The code listings in this chapter allow you to quickly get started with an efficient code and show you how to optimise your existing code.

This chapter presents a tutorial-style guide to high performance computing and the efficient implementation of LBM algorithms. In Sect. 13.1 we discuss the selection of appropriate programming languages and review floating point arithmetic. We then cover essential optimisation strategies in Sect. 13.2, including the simplification of arithmetic expressions and loop optimisations (Sect. 13.2.1), the use of automatic optimisation features during compilation (Sect. 13.2.2), and memory caches (Sect. 13.2.3). This section also covers how to assess the performance of code (Sect. 13.2.4). Section 13.3 then introduces a simple implementation of a Taylor-Green vortex decay simulation (Sect. 13.3.1), and the performance of this code is subsequently analysed and optimised (Sect. 13.3.2). Several optimisations related to the LBM algorithm itself are presented in Sect. 13.3.4. Next, we turn our attention to parallel computing (Sect. 13.4). Section 13.4.1 presents the use of OpenMP for writing code for multiple cores and CPUs that share access to a common memory system. Section 13.4.2 describes programming for clusters of interconnected computers with MPI. Finally, Sect. 13.4.3 presents programming for graphics processing units. Complete code examples accompany this book.[1]

13.1 Introduction

When developing simulation software, researchers have a broad range of options for the tools they can use to program and the hardware that carries out the computations. Often the choice of hardware and software is pragmatic: the most readily available

[1] https://github.com/lbm-principles-practice

T. Krüger et al., *The Lattice Boltzmann Method*, Graduate Texts in Physics,
DOI 10.1007/978-3-319-44649-3_13

laptop or desktop hardware and whichever programming language the researcher knows best. For many problems, this simple choice is adequate, but when simulation times reach several hours to perhaps months, researchers will undoubtedly consider whether they could write a code that finishes faster and what would be the most efficient way to develop such a code. The purpose of this chapter is to provide an introductory to intermediate level guide to implementing LBM simulations on modern high performance computing platforms, with a special emphasis on parallel programming. It is intended to be accessible to readers without significant prior training in scientific computing, and therefore introduces readers to many aspects of development tools and system architectures.

This chapter is not intended to show readers how to develop a code with *the* shortest possible execution time. Many optimisations that are possible for a particular problem are not relevant to other problems, and it is unlikely that many readers will benefit from a highly tuned code for the specific problem we use as an example throughout this chapter. Similarly, many optimisation decisions that are needed to eke out the last possible improvements depend on the details of the processors and memory that run the software, and it is difficult to adequately handle the wide range of platforms that readers may be using. Finally, it is often impractical to dedicate significant time to optimising software for a small improvement in performance.

A **4% improvement** to the speed of a simulation that already takes **one day to finish** reduces the execution time by **only one hour**. In contrast, a **programming mistake** that causes **2–10 times worse performance** can make an otherwise feasible project appear impractical. We therefore focus on issues that have the potential to make such a significant impact on performance, and explain why certain subtle differences in code can cause major differences in the time a program takes to finish.

In this chapter, we use a model problem, the decay of a Taylor-Green vortex, to demonstrate the key ideas that are important for writing codes that run efficiently on a single core of a contemporary conventional central processing unit (CPU), multiple CPU cores on one computer, clusters of many interconnected computers, and graphics processing units (GPUs). Named for their initial use in accelerating algorithms for generating computer graphics, GPUs were later extended to facilitate more general computational tasks, and they were quickly adopted in the LBM community due to their numerous computational cores and fast memories.

13.1.1 Programming Languages and Development Tools

Programming languages are often classified as either interpreted or compiled based on the way that the operations described in source code are eventually carried out by a processor. To run code written in an interpreted language, a program called the interpreter reads the source code file and carries out the operations it describes. In contrast, the source code for a program written in a compiled language is first "translated" (i.e. compiled using a program called a compiler) to produce a binary executable file that contains the sequence of elementary operations (instructions) that a processor will later execute when the program is started within an operating system. In principle, the fastest programs can be written by avoiding compilation and directly specifying the sequence of instructions that the processor will carry out, but few projects justify the time and effort required by this approach.

For many tasks, interpreted languages provide convenient features that outweigh their performance drawbacks and allow programmers to write programs faster. Such features include automatic memory management, conversion between data types, easy manipulation of complex data structures, and flexibility with function definitions. These conveniences increase memory use and reduce the speed of execution because the exact operation that a statement describes, or even whether the statement is meaningful, is not known until all the statements preceding it have been completed. For example, in a hypothetical interpreted language the statement `b = a+2` could involve integer addition (if it follows `a = 1`), floating point addition (after `a = 3.14`), addition of a constant to each element of an array (after `a = [1,2,3]`), text concatenation (after `a = "Price: $"`), or produce an error (if the variable `a` has not been defined).

Where needed, performance-sensitive portions of programs can be written in a compiled language and compiled to generate a library file that is then used within an interpreted language. The developers of interpreted languages strive to provide the conveniences of interpreted languages with minimal impact on performance, making these languages competitive for many computational tasks. As a result, the line between the two classes of programming languages is blurred because it is based on how a language is implemented rather than the characteristics of the language itself (its syntax and semantics). One could write a compiler for what is usually considered an interpreted language, and this is how some interpreters are designed.

Since they reduce programming effort and development time, interpreted languages are preferable for computational tasks that do not take a significant amount of time to complete. For example, a program that finishes in less than a few minutes, perhaps a comparison of boundary condition implementations in a small domain, could be conveniently programmed with an interpreted language. Interpreted languages, such as Python and MATLAB (or the GNU alternative Octave), are also often used to post-process the results of simulations. In fact, these languages were used to generate many of the figures in this book. In the code samples accompanying this book, we have included codes in Python and MATLAB for the same problem that the code in this chapter solves.

This chapter does not examine performance optimisation for interpreted languages, but we offer here one key guideline for writing efficient computational code for **interpreted languages: avoid explicit loops where possible**. Instead, programmers should use efficient built-in features for matrix multiplication and vector/array manipulation rather than writing loops to perform these tasks.

For example, in MATLAB, one should write

```
c = a+b;
```

to add the components of two arrays together rather than

```
for n = 1:length(a)
    c(n) = a(n) + b(n);
end
```

Similarly, functions that operate directly on arrays should be used instead of one function call for each element of the array. Clever array indexing can also be used to avoid a loop. To compute a central finite difference first derivative in MATLAB, the single statement

```
diffy = 0.5/dx*(y(3:end)-y(1:end-2));
```

runs faster than

```
for n = 2:length(y)-1
    diffy(n-1) = 0.5/dx*(y(n+1)-y(n-1));
end
```

Exercise 13.1 Which of the codes for a central difference uses more memory? Why?

The guidelines we describe later in this chapter for compiled code also apply to interpreted code, specifically those regarding memory use, but the ability to control the organisation of memory is rarely available in these languages.

Exercise 13.2 Run the provided sample Python or MATLAB code for a Taylor-Green vortex decay computation or code written in an interpreted language of your choice. How does the speed compare with the performance results given later in this chapter for the compiled code?

The differences in performance of compiled programs written in different languages are more subtle and sometimes controversial. Two languages, C and Fortran, remain common in high performance computing, and the choice of the language for a project depends on the project's history and programmers' knowledge. Reasons for variations in performance in specific tests include programmers' familiarity with the languages' features, the languages' default rules for organising variables in memory, the quality of the compiler (for the particular processor architecture used to run the program), and the configuration options specified during compilation. In general, well-written codes in compiled languages will outperform interpreted codes significantly, likely by factors from two to two orders of magnitude.

The code in this chapter is written in C++, but it could be easily converted to a purely C code because it uses only a few features of C++ for convenience. The optimisation strategies discussed in this chapter, in particular those concerning memory access, can be applied to other compiled languages. We present these optimisation methods and cover some of the differences between C and Fortran in Sect. 13.2.

In this chapter, we focus on the software development tools and hardware architectures that are commonly encountered in contemporary academic high performance computing (HPC) systems. We use the open source GNU compiler collection (GCC), which includes compilers for C, C++, and Fortran. For parallel programming, we use OpenMP and MPI, of which open source implementations are available. We run the codes on processors with 64 bit x86 architecture, which are readily available from Intel and AMD and are found in most current laptops, desktops, and servers. For GPU programming, we use the devices and proprietary development tools from NVIDIA. Computing clusters are usually shared by many researchers, may be located far from the researchers' work places, and are therefore used by remote access over the Internet. The operating systems on these clusters are often Unix or Unix-like, such as Linux, and we make use of some utilities available on these systems in the code examples. These utilities are also available in a terminal in Mac OS X (which is a Unix system) and can be installed on Windows.

All the code in this chapter is presented as plain text that can be written and edited with any text editor, and we show the commands that would be typed to compile and run the code. Naturally, readers may work with their preferred editors and integrated development environments (IDEs). In these environments, the options for the compilers can be specified in various configuration windows.

To assist readers who are unfamiliar with programming in C or C++, Appendix A.9 reviews the main features of these languages that are used in the code presented in this chapter and accompanying the book. We do not make use of the object-oriented programming features of C++ and do not aim to provide an object-oriented framework for LBM simulations. We focus on writing a simple and efficient code that readers could adapt to their own problems or incorporate into a framework.

To avoid ambiguity, especially in memory bandwidth calculations, we express large memory quantities with units based on powers of two. While in some contexts a kilobyte might refer to 1000 or $1024 = 2^{10}$ bytes, a kibibyte denotes the "binary" version of a kilobyte and is abbreviated as KiB, i.e. $1\ \text{KiB} = 2^{10}$ bytes $=$ 1024 bytes. Similarly, we use mebibyte (MiB) and gibibyte (GiB) instead of megabyte and gigabyte, with 1 MiB $= 2^{20}$ bytes $\approx 1.05 \times 10^6$ bytes and 1 GiB $= 2^{30}$ bytes $\approx 1.07 \times 10^9$ bytes.

13.1.2 Floating Point Arithmetic

Modern computing hardware efficiently performs computations with rational approximations of real numbers called floating point numbers. Such numbers can be written as $s \times v \times 2^p$, where $s = \pm 1$ is the sign, v is a non-negative integer, and p is an integer, with both v and p being bounded (represented with a finite number of binary digits, i.e. bits). We omit here the details of how these numbers are represented in binary, though this usually follows the IEEE 754 standard [1]. In this chapter, we use exclusively double precision floating point values in which a total of 64 bits (8 bytes) are used to store one such value. IEEE 754 double precision numbers have 53 bits of precision, and the exponent p can range from -1022 to 1023, which provides about 16 decimal digits of precision for numbers from 10^{-308} to 10^{308}. In some situations, single precision values (32 bits) are sufficient and appealing because they occupy half the memory and processors can perform computations with them faster.

Floating point numbers and arithmetic operations with them have several key properties with important consequences. Clearly, neither irrational numbers nor rational numbers with prime factors other than two in their denominator can be represented exactly. Therefore, an exact representation of $1/2$ is possible, but $1/3$ must be approximated as $\frac{1}{4} + \frac{1}{16} + \frac{1}{64} + \frac{1}{256} + \frac{1}{1024} + \ldots$, truncated to the available precision. The results of arithmetic operations are generally inexact and can be interpreted as being the outcome of rounding the exact result to the closest available floating point number.

The accuracy of arithmetic operations and the representation of real numbers is characterised by a number called the **machine epsilon**. For double precision, it is $2^{-52} \approx 2.22 \times 10^{-16}$, and this is the smallest number that can be added to 1 to yield a value that differs from 1.

Operations with floating point numbers are in general not associative or distributive due to the rounding of intermediate results. Due to these rounding errors, the results of computations should not be tested for exact equality (using the equality operator `==`) with an exact expected value or a value computed in a different way. Instead, the difference between two floating point values should be compared with a threshold that is reasonable for that situation, i.e. by using `abs(a-b) < tol` instead of `a==b`.

Exercise 13.3 Compute $1 - \sum_{i=1}^{n} \frac{1}{n}$ for several values of n. For which values of n is the result exactly 0? Why? How much do the results differ from 0 and how does the error increase with n? Compare results for double and single precision calculations.

Another caveat regarding programming with floating point numbers concerns the use of numerical constants in code. In C and C++, numbers written in code are interpreted as integers if they have no decimal part and as double precision values if one is present. For example, `double d = 3/2;` is interpreted as assignment of the result of *integer* division of the integers 3 and 2 to a double. The result is a value of 1 in `d`. To obtain 1.5 in `d`, one must write `3.0/2.0`, `3/2.0`, or `3.0/2` because the presence of at least one floating point quantity in the division triggers the use of floating point division (and automatic conversion of the integer to a `double`). Readers should therefore familiarise themselves with the rules for conversion between different numerical data types from a standard language reference.

13.1.3 *Taylor-Green Vortex Decay*

All the codes in this chapter solve the same problem, the decay of a Taylor-Green vortex. This choice allows us to compare the performance of the codes and better understand how the code needs to be changed to run efficiently on single cores (Sect. 13.3), multiple cores (Sect. 13.4.1), clusters (Sect. 13.4.2), and GPUs (Sect. 13.4.3). The details of this flow problem are provided in Appendix A.3. It involves the decay over time of a particular initial velocity and pressure distribution, shown in Fig. A.1, in a fully periodic two-dimensional domain. We use a D2Q9 velocity set (Sect. 3.4.7) and the BGK collision operator (Sect. 3.5.3).

Since an analytical solution is available, we can compare it with the numerical solutions, and we will use the parallel codes to perform a convergence study with domain sizes from 32×32 to 4096×4096 lattice nodes. The results are presented in Sect. 13.5, and they show second-order convergence. In this chapter, we focus on the programming aspects of simulating large domains efficiently, and readers may find further information about the convergence of LBM simulations in Sect. 4.5.

13.2 Optimisation

When writing code for numerical methods, it is generally sufficient for us to work with a highly simplified conceptual model of how a computer carries out the sequence of operations we specify. We might imagine that a program is stored within the memory of the computer as a sequence of instructions. Other regions of memory store the data that the program manipulates. This memory, which contains the instructions and data, is a large ordered list of numbers, and instructions can refer to these numbers by their index (or address). Each instruction takes a particular amount of time, during which the computer's processor may need to load numbers from addresses in memory specified within the instruction and save the result to another address. When all the tasks required by an instruction are completed, the processor moves on to the next instruction, and the process repeats indefinitely. A clock signal, which switches its state several billion times a second, synchronises the operation of a processor. The tasks that a processor carries out include arithmetic operations, movement of data between locations in memory, and selection of which instruction will be performed next by moving forward or backward in the instruction sequence based on the result of a previous operation.

To ensure we accomplish a computational task in as little time as possible, we have two main options within this simple model of computing. The first option is to devise a **better algorithm**, one that achieves the same result in fewer steps. The second is to perform optimisation at the level of the instructions by finding ways to use one or more **faster instructions** in the place of one that takes more time to finish.

Though in many cases the simple model is adequate to guide the minimisation of execution time, the details of computer architecture are much more complex and can have a significant impact on performance. Many scientific computing applications, such as CFD, involve the processing of large data sets. For such applications, the **details of how memory is used** are critical.

Modern processors have a hierarchy of memory types ranging from small and fast memories located on the same chip as the processor to large but slow memories residing on other chips or devices. Circuits on the processor chips, such as registers and caches, provide small amounts of memory with minimal access delays. Off-chip hardware devices, such as random access memory (RAM) and hard drives, have much larger capacities but access is significantly slower. For example, an instruction that operates on data held in registers can be completed within several clock cycles, while access to RAM can incur a delay of several hundred clock cycles.

Algorithms need to be implemented in ways that maximise the use of fast memory. One common strategy is to load a portion of memory from a slow location to a faster one, process this segment of data, and then move on to the next portion. Caches, the topic of Sect. 13.2.3, automatically assist in improving memory access by loading data before it is likely to be used. Care must be taken, however, to implement algorithms in ways that take advantage of such features rather than interfering with them and reducing performance.

Internally, processors employ a variety of optimisations to speed up the execution of instruction sequences. For example, when one instruction does not depend on the outcome of an instruction that precedes it, this instruction might be executed without waiting for the previous one to finish. This optimisation is possible when the instructions use different computational circuits, such as those for integer and floating point arithmetic or different types of arithmetic such as multiplication and division. Due to such optimisations and the details of memory architecture, the time needed to execute an instruction depends not only on the instruction being executed, but also on the nature of the instructions that preceded it and the memory locations they accessed.

Most modern computers, from mobile phones to supercomputers, have multiple processors. When these processors are combined on one chip or consist of several connected chips in one package, they are called cores and share some resources, such as memory access channels and caches. Many contemporary workstations and servers have several multi-core processors, and systems with 20 or more cores in total are readily available. In this section, we consider only optimisation of the tasks that occur on one core. Parallel programming that takes advantage of multiple cores and processors operating simultaneously is discussed in Sect. 13.4.1 and Sect. 13.4.2.

The optimisation of computer codes is a highly complex topic due to the wide range of techniques that can be employed and cases that need to be considered. The sections that follow provide an introduction and overview of the main topics, starting with the basic methods for simplifying algorithms and the evaluation of mathematical expressions in Sect. 13.2.1. This is followed in Sect. 13.2.2 by a discussion of the automatic optimisation features of modern compilers that analyse code and transform it to generate better sequences of machine instructions. The behaviour and use of memory caches is then presented in Sect. 13.2.3.

13.2.1 Basic Optimisation

In this section, we review some of the most common strategies used to implement algorithms in ways that efficiently utilise computational resources and minimise execution time. Since code can be adjusted in numerous ways while attempting to optimise it, programmers should carefully consider how they spend their time optimising a program. Completely eliminating a sequence of operations that accounts for 1% of total execution time is clearly less useful than a 10% reduction in the execution time of a task that takes 50% of the total time. It is therefore beneficial to first understand what tasks in an algorithm contribute the most to its execution time. Software such as `gprof` is available to assist in this type of assessment, called profiling, of a program.

The first, and perhaps simplest, optimisation strategy is to **avoid unnecessary repetition of sequences of computations** by storing their results in memory the first time they are computed. This simplification is quite intuitive and involves replacing, for example, `v1 = f1(x,g(x));`

(continued)

`v2 = f2(x,g(x));` with `gx = g(x); v1 = f1(x,gx); v2 = f2(x,gx);`. The expression or function `g(x)` must always yield the same result for identical values of `x` and therefore cannot involve variables that are modified during previous calls to the function or depend on external events or variables (such as the system time). The identification of suitable expressions and their replacement with a new variable is called **common subexpression elimination**.

When evaluation of `g(x)` is time-consuming and the variable `g` can be stored in memory that is quick to access, the benefit of this simplification is clear. However, when the common subexpression is trivial, such as perhaps `m+n` for integers `m` and `n`, and the availability of fast memory for temporary variables is scarce, programmers must compare the time required to recompute the expression each time it is needed or load a pre-computed value from memory. When the decision is not obvious but has the potential to make a noticeable improvement, one may proceed empirically and compare execution times with and without the potential optimisation; otherwise programmers should focus on optimising other parts of the code. In addition to improving execution time, the elimination of common subexpressions can also make code easier to understand and maintain because changes to the subexpression need to be made in only one place.

A second common type of optimisation is to **exploit the mathematical properties of expressions** to evaluate them more efficiently. The possibility of such optimisations depends on the details of the expressions involved, and identifying simplifications may require specific insights into the nature of the expressions.

We consider here the evaluation of polynomials as an example that is often encountered in scientific computing. Evaluating the polynomial $p(x) = \sum_{i=0}^{n} q_i x^i$ in the form it has been written here requires n addition operations and $n(n+1)/2$ multiplication operations if each power of x is evaluated separately by repeated multiplication. Recognising that $x^i = x \times x^{i-1}$ decreases the number of multiplications required to $2n - 1$, but this is still not the optimal approach. Instead, one should rewrite the polynomial as $p(x) = q_0 + x(q_1 + x(q_2 + x(\ldots)))$. Evaluation of a polynomial in this way is called Horner's method, and it requires n additions and n multiplications. On architectures that provide a combined (fused) multiplication and addition operation, n such operations are required.

Even though the mathematical identities used to transform expressions are exact when the numbers involved are real, transformed expressions will in general yield slightly different results when floating point arithmetic is used due to rounding errors. Optimisation of numerical methods must therefore be performed with caution and awareness of the effects of runtime optimisations on the accuracy of results.

Other optimisations deal with how features of higher-level programming languages are implemented as low-level machine instructions. Functions (or subroutines) are available in many languages and they allow sections of code to be separated to avoid repetition, enhance readability, and facilitate implementation of recursive algorithms. Depending on the language and compiler, several tasks need to be performed when a function starts and finishes. For example, memory may need to be allocated for use by the variables in the function and freed for subsequent reuse when the function completes its task.

When the task performed by a function takes sufficiently little time that the time needed to set up and complete the function call is significant, it is useful to avoid this overhead, especially when the function is used frequently. In C and C++, the function qualifier `inline` indicates to the compiler that a function should not be implemented as a regular function, but instead any calls to this function should be replaced with the definition of the function. Thus, an inline function assists in organising code without hindering performance.

In many algorithms, computational tasks are repeated for every element of a data set, and this is often implemented with a `for` loop. We consider here three common **optimisations of `for` loops**: loop **unrolling**, loop **peeling**, and loop **combining**.

When the inner block of a `for` loop finishes quickly, the overhead of incrementing a counter and checking if it has exceeded a bound constitutes a noticeable portion of the execution time of the loop. Loop unrolling involves replacing the iterations of a `for` loop with explicit repetition of the inner block. For example,

```
for(int i = 0; i < N; ++i)
    short_task(i);
```

would be replaced with

```
short_task(0);
short_task(1);
short_task(2);
//
short_task(N-1);
```

When `N` is large, the binary file produced by compilation becomes unreasonably large due to repetition of the required instructions. However, a loop does not need to be unrolled completely to improve performance. One may repeat the inner block several times to reduce the overhead of the loop, for example as follows:

```
for(int i = 0; i < N; i = i+3)
{
    short_task(i);
    short_task(i+1);
    short_task(i+2);
}
```

Though one would typically pick a number of repetitions that divides `N` evenly, in some cases the number of explicit repetitions is set by other considerations. When `N` is not divisible by the number of repetitions, the additional iterations that are necessary can be executed after the `for` loop. One should not add `if` statements within the loop to check if each repetition inside the `for` loop should be performed!

The second optimisation of `for` loops, loop peeling, deals with the handling of special cases in loops. In general, the use of `if` statements inside a `for` loop should be evaluated carefully to ensure that checking the conditional expression frequently does not reduce performance unnecessarily. For example, the inner block of this `for` loop is clearly inefficient, especially when `check_special` is expensive to evaluate and is rarely `true`:

```
for(int i = 0; i < N; ++i)
{
    if(check_special(i))
        handle_special(i);
    else
        short_task(i);
}
```

When it is easy to determine which `i` are special, we can avoid expensive calls to `check_special()` by writing `if(i == special)` instead of `if(check_special(i))`. However, it would be even better to write

```
for(int i = 0; i < special; ++i)
    short_task(i);

handle_special(special);

for(int i = special+1; i < N; ++i)
    short_task(i);
```

assuming that only one particular index needs to be treated differently. When the special cases occur at the beginning or end of a `for` loop, the relevant iterations can be "peeled" off of the main loop. For example, a loop that estimates the first derivative of uniformly spaced data with spacing `h` might be written as

```
diff[0] = (data[1]-data[0])/h;
for(int i = 1; i < N-1; ++i)
```

```
    diff[i] = (data[i+1]-data[i-1])/(2.0*h);
diff[N-1] = (data[N-1]-data[N-2])/h;
```

instead of the less efficient alternative

```
for(int i = 0; i < N; ++i)
{
    if(i < 1) // handle first
        diff[i] = (data[i+1]-data[i])/h;
    else if(i == N-1) // handle last
        diff[i] = (data[i]-data[i-1])/h;
    else
        diff[i] = (data[i+1]-data[i-1])/(2.0*h);
}
```

The third optimisation considers the relationship between several `for` loops. It seems intuitive that

```
for(int i = 0; i < N; ++i)
    task_a(i);
for(int i = 0; i < N; ++i)
    task_b(i);
```

could be replaced with

```
for(int i = 0; i < N; ++i)
{
    task_a(i);
    task_b(i);
}
```

to avoid the overhead of one of the loops, assuming that `task_b(i)` does not require data derived from later iterations of `task_a`. This optimisation is called loop combining, and it can save more time than only the loop overhead when `task_a` generates intermediate results that can be re-used in `task_b`. However, as will be discussed in Sect. 13.2.3, keeping tasks separated is sometimes beneficial, for example when `task_b` interferes with the caching of data for `task_a`. As is the case with many other potential optimisations, the net benefit of this optimisation depends on a variety of factors, making optimisation an iterative process that involves repeated profiling and modification of code.

13.2.2 Automatic Optimisation During Compilation

Modern compilers incorporate a wide range of algorithms that analyse the code being compiled to produce sequences of machine instructions that execute as quickly as possible. Detailed optimisation of arithmetic expressions is one task that may often be left for the compiler to perform automatically, and in this section we examine how and when to use compilers' automatic optimisation features. The extent to which compilers attempt to optimise code is specified on the command line during compilation. For example, GCC has three pre-configured levels of optimisation. The lowest level performs only optimisations that do not increase compilation time significantly. Higher levels of optimisation use additional strategies, but require more time and memory to analyse and optimise the code. The option for selecting an optimisation level is `-On` where `n` is an integer from 0, which requests no optimisation and fastest compilation, to 3, the highest and most computationally expensive level. To compile the source code in the file `source_code.cpp` to generate the program `sim`, one can use the command `g++ source_code.cpp -o sim` to compile without optimisation (the default) or use `g++ -O3 source_code.cpp -o sim` to compile with level 3 optimisation. Specific optimisations can be enabled or defaults can be disabled with special options that start with `-f`.

Programmers sometimes attempt optimisations that are better left to compilers, which have been developed by experts in finding quick ways to accomplish common tasks. Consider, for example, a programmer who is considering replacing division of an integer by 2^p with a shift right by p places[2] to avoid a time consuming division instruction. Such optimisations that take advantage of the binary representation of numbers and many others are part of the automatic optimisations that compiler developers have implemented.

While many programmers are likely aware that shifting can be used instead of division by powers of two, many more optimisations of simple arithmetic operations are possible. We show here one example that illustrates the extent of the "tricks" that compilers perform automatically: division of an integer by an integer constant is implemented as a combination of multiplication and bit shifting [2]. With this algorithm, division of an unsigned 32 bit integer by 5 is performed as a multiplication by the 32 bit constant 3435973837 followed by a right shift by 2 of the upper 32 bits of the 64 bit product. The use of the upper 32 bits of the product is effectively a right shift of the 64 bit product by 32 places. The algorithm therefore performs multiplication by 3435973837 followed by division by 2^{34}. The selection of the magic number and the required number of shifts follows from noting that $3435973837/2^{34} = 1/5 + 1/85899345920 = 0.2 + O(10^{-11})$. Though we do not go into the proof here, the precision of this approximation of $1/5$ is sufficient to obtain the correct result for all 32 bit unsigned integers.

The main message of this example is that most programmers should spend time thinking about the algorithms they are implementing and how they

(continued)

[2]This is analogous to dividing decimal numbers by powers of 10 by "shifting" right.

use memory rather than focusing on the details of the instructions used to implement operations and the binary representation of the data they are processing. **Instruction-level optimisation can be left to the compiler**, allowing programmers to focus on writing code that is easier to understand.

Though compilers can, in general, convert mathematical expressions to efficient sequences of machine instructions, two special topics that restrict the extent of automatic optimisation are important for numerical algorithms. The first issue occurs due to the availability of pointers[3] in languages such as C and C++. Though useful for managing memory, pointers restrict the optimisations that a compiler can perform because of the possibility that two different pointer variables refer to the same location in memory (i.e. the pointer variables contain the same memory address). This is called aliasing because the same value in memory can be read and modified in different ways. The reduction in possible optimisation occurs because the code must be compiled in such a way that the result is correct whether or not some pointers refer to the same locations in memory.

Listing 13.1 Example function for demonstrating the effects of pointer aliasing.

```
void aliasfunc(double *a, double *b)
{
    *b = *a + 1.0;
    *a = *a + 2.0;
}
```

Consider, for example, the function `void aliasfunc(double *a, double *b)` in Listing 13.1. When optimising, the compiler must ensure that the result is correct whether or not `a` and `b` hold the same memory address. In other words, the result must be correct whether the programmer uses `aliasfunc(p,q)` or `aliasfunc(p,p)`. The quantity `*a` must therefore be loaded twice from memory because the first statement, which changes `*b`, might change the value of `*a` that should be used in the second statement. The compiler cannot load `*a` into a temporary variable once and then add the two constants and store the results to their destinations. In many cases, however, this optimisation is desirable because the functions are never used with identical arguments.

[3]Pointers are variables that hold the address of another variable. See Appendix A.9.6 for more details.

One way to circumvent the reduction in automatic optimisation is to use a temporary variable to unambigously indicate to the compiler how the code is supposed to behave in all cases. This modification is shown in Listing 13.2

Listing 13.2 An alternative to the function in Listing 13.1 that avoids the aliasing problem.

```
void aliasfunc(double *a, double *b)
{
    double temp = *a;
    *b = temp + 1.0;
    *a = temp + 2.0;
}
```

This is a reasonable strategy when the pointer refers to a single value or a small array, but it is unreasonable for a programmer to define a new variable for every element of a large array and impossible to do so when the size of the array is not known. For such cases, compilers have options that allow programmers to indicate when pointers are guaranteed to refer to different memory locations. In the C99 standard, the `restrict` keyword serves this purpose. Standard C++ does not have such a keyword, but GCC provides `__restrict__` as a custom extension of the language that serves the same purpose as the C99 keyword. An example of the use of this keyword is shown in Listing 13.3.

Listing 13.3 Use of the `__restrict__` keyword to indicate absence of aliasing.

```
void aliasfunc(double * __restrict__ a,
               double * __restrict__ b)
{
    *b = *a + 1.0;
    *a = *a + 2.0;
}
```

The problem of aliasing does not exist in languages without pointers, such as versions of Fortran prior to Fortran 90. In such languages, more optimisations are possible by default, and this is one of the reasons why Fortran has been considered a fast language for scientific computing.

The second main issue that restricts the extent of optimisation relates to the specifications for floating point arithmetic (most often the IEEE 754 standard [1]). Consider the function in Listing 13.4 that computes `a*c + b*c` for floating point values `a`, `b`, and `c`. By default, compilers do not simplify expressions in ways that change the propagation of error. As a result, the expression `a*c + b*c` is not simplified to `(a+b)*c` to save a multiplication operation unless we ask the compiler to do so.

Listing 13.4 Example function used to illustrate the effects of various optimisation options.

```
double multiply_add(double a, double b, double c)
{
    return a*c + b*c;
}
```

When compiled without optimisation (the default for GCC), the resulting machine instructions [4] are those shown in Listing 13.5.

Listing 13.5 Disassembly of the result of compiling Listing 13.4 without optimisation (default) or specifying the compiler option `-O0` to explicitly disable optimisation

```
push   rbp
mov    rbp,rsp
movsd  QWORD PTR [rbp-0x8],xmm0
movsd  QWORD PTR [rbp-0x10],xmm1
movsd  QWORD PTR [rbp-0x18],xmm2
movsd  xmm0,QWORD PTR [rbp-0x8]
movapd xmm1,xmm0
mulsd  xmm1,QWORD PTR [rbp-0x18]
movsd  xmm0,QWORD PTR [rbp-0x10]
mulsd  xmm0,QWORD PTR [rbp-0x18]
addsd  xmm0,xmm1
movsd  QWORD PTR [rbp-0x20],xmm0
mov    rax,QWORD PTR [rbp-0x20]
mov    QWORD PTR [rbp-0x20],rax
movsd  xmm0,QWORD PTR [rbp-0x20]
pop    rbp
ret
```

The inefficiency of the unoptimised compilation is particularly noteworthy: many unnecessary transfers between memory and registers are performed. After the addition instruction **addsd**, **xmm0** is saved to memory, the value in memory is transferred to the register **rax**, back to memory, and finally from memory back to **xmm0**.

In comparison, only four instructions are generated when optimisation is enabled, as shown in Listing 13.6. As expected, there are two multiplication instructions and one addition.

[4] For readers unfamiliar with assembly language or the instructions shown here for a typical modern 64 bit Intel processor, **push** and **pop** are instructions that save and retrieve their parameter from the "stack," a special memory region where data can be stored temporarily. The instruction **mov** `dst,src` copies the contents of `src` to `dst` where `src` and `dst` may be locations in memory or registers. **QWORD PTR** `[addr]` refers to the contents of the quadword (four words, which is eight bytes) at the location `addr` in memory. Numbers written as `0xhh` represent the value `hh` in base 16 (hexadecimal). The symbols **rax**, **rbp**, and **rsp** denote 64 bit general purpose registers, and **xmm0**, **xmm1**, and **xmm2** are registers for floating point values. Note that these are 128 bit floating point registers that can store two double precision values or four single precision values, but in this code only the lower 64 bits are used. The instruction **movsd** `dst,src` means "move scalar double" and copies `src` to `dst` using only the lowest 64 bits if a register is specified. **movapd** `dst,src` moves the full 128 bit value from `src` to `dst`. The instructions **addsd** `dst,src` and **mulsd** `dst,src` are scalar addition and multiplication instructions, repectively, that store the result of adding/multiplying `dst` and `src` to `dst`. The function's parameters are provided in the registers **xmm0-2** and its result is returned in **xmm0**. Execution continues in the calling function after the instruction **ret**.

Listing 13.6 Disassembly of the result of compiling Listing 13.4 with the options `-O1`, `-O2`, or `-O3`

```
mulsd  xmm0,xmm2
mulsd  xmm1,xmm2
addsd  xmm0,xmm1
ret
```

Two compiler options are relevant to the simplification of floating point arithmetic expressions. Quoting from the documentation of GCC (4.9.2), these options are:

- `-fassociative-math`
  Allow re-association of operands in series of floating-point operations. This violates the ISO C and C++ language standard by possibly changing computation result. [...]
  The default is `-fno-associative-math`.
- `-freciprocal-math`
  Allow the reciprocal of a value to be used instead of dividing by the value if this enables optimisations. For example `x / y` can be replaced with `x * (1/y)`, which is useful if `(1/y)` is subject to common subexpression elimination. Note that this loses precision and increases the number of flops operating on the value.
  The default is `-fno-reciprocal-math`.

Both of these options are enabled by the option `-funsafe-math-optimizations`, which in turn is enabled by `-ffast-math`. The option `-Ofast` permits the compiler to disregard strict standards compliance and includes the optimisations enabled by `-O3` and `-ffast-math`. These optimisation options cannot be used when an algorithm relies on having error be propagated in a particular way. The machine instructions generated after compilation with `-Ofast` or `-O3` with `-ffast-math` are shown in Listing 13.7, and there is only one multiplication instruction, as expected.

Listing 13.7 Disassembly of the result of compiling Listing 13.4 with the options `-Ofast` or `-O3` together with `-ffast-math`

```
addsd  xmm0,xmm1
mulsd  xmm0,xmm2
ret
```

Due to differences in error propagation, the results of simulations obtained with `-Ofast` or `-ffast-math` will not in general be identical with the results computed by code compiled with different optimisation options. When comparing the results of floating point operations, it is important to take into account the presence of rounding error.

Recent processors from both AMD and Intel as well as GPUs provide a fused multiply add instruction that computes `a+b*c` in one step, rounding the result only at the end. The generation of executables with these instructions can be enabled through compiler options that specify the processor architecture on which the code will run. Due to variations in compilation settings and processor architecture, the same code when compiled with different (possibly non-standards-compliant) optimisations enabled may yield numerically different results due to differences in rounding and the order of evaluation of expressions. On one system the expression `a+b*c` might result in one multiplication then one addition, while another computes the result in one step. Programmers must therefore be cautious when comparing the results of codes compiled for different architectures.

The difference between the unoptimised (Listing 13.5) and optimised (Listing 13.6) compilation outputs is striking. The difference between Listings 13.6 and 13.7 is important for LBM implementations because the computation of the equilibrium distributions involves similar expressions.

Though convenient for debugging because it prevents the compiler from eliminating unnecessary functions and variables, **compilation without optimisation should not be used** for actual simulations. The resulting executables run too slowly, wasting resources that are often shared on clusters. Considering the excessive memory access in the version without optimisation, optimisation should also be enabled when checking the benefits of changes to algorithms and memory layout. In some cases, however, the compiler may make an optimisation that adversely affects the memory access pattern. Therefore, some tuning of compiler options, such as using a lower optimisation level, may be needed to obtain the fastest outcome.

13.2.3 Memory Caches

Consider a typical CPU with a 3 GHz clock of which one core can perform 4 double precision arithmetic operations per clock cycle. The theoretical peak computational speed of this core is therefore 12 GFlops (12×10^9 floating point operations per second) in double precision. Suppose that the operation being performed is an addition or multiplication of two values whose sum is stored to a third variable, all of which must eventually be loaded from and stored to RAM (rather than CPU registers). Given that each double precision value is 8 bytes, computation at the peak speed would require reading 179 GiB/s and writing 89 GiB/s for a total required memory bandwidth of 268 GiB/s. In comparison, memory transfer rates are typically 10–25 GiB/s for CPU systems. For example, one channel of DDR4 2400 memory provides a peak theoretical rate just under 18 GiB/s, so four channels would provide 72 GiB/s.

Multiple cores increase demand for memory bandwidth, while multiple memory channels help satisfy this demand. Nonetheless, memory access rates are significantly slower than possible processing speeds. The speed at which memory can be read and written therefore determines the performance of algorithms that perform relatively few operations per byte they read or write. As we will see later, LBM simulations typically fall in this category, and therefore implementations must use memory carefully. Since they can have more recent memory systems, newer models of inexpensive laptops may run memory-intensive programs faster than older high-end workstations and servers that have an earlier generation of memory architecture.

The mismatch between memory access and computation speeds led to the development of **caches**, small but fast memories that temporarily store the data the processor is using. In contemporary CPUs, several levels of caches are present between the processor and RAM: each level of cache therefore mediates either the memory transfers between the processor and another cache, between two caches, or between a cache and the RAM chips. The smallest and fastest of these caches are located on the same chips as the CPUs and can be accessed **within several clock cycles** compared with **several hundred cycles** for RAM.

In general, caches operate under two assumptions:

- Recently accessed memory is likely to be re-used. Code that satisfies this assumption is said to exhibit temporal locality in its memory access pattern, meaning that little time elapses between consecutive uses of the data and it is therefore read/written at nearby points in time.
- Memory residing in addresses adjacent to those of a recent access are likely to be used soon. Code that satisfies this assumption has a memory access pattern that exhibits spatial locality.

Based on these assumptions about how programs use memory, caches work in the background to keep processors supplied with the data they need. While they are beneficial "on average," caches cannot help in some cases and will interfere in others. Programmers therefore need to understand how caches behave to be able to write codes that access memory as quickly as possible.

Caches improve performance in the following way. When the CPU initiates a read or write of a memory address that is available in cache, which is called a cache hit, the cache handles the request and supplies or stores the relevant data. This process occurs significantly faster than reading or writing data directly from/to RAM chips. A cache miss occurs when the CPU requests a memory address that is not available in a cache. In this case the cache requests from RAM not only the desired number of bytes at the requested address but also the adjacent bytes that form what is called a *cache line*. If needed, previously-cached data is written back to RAM to make space available for the new data.

A cache line is the smallest unit of memory that is transferred between caches and the main RAM. A typical size is 64 bytes (512 bits), which can hold 16 32-bit integers, 16 single-precision floating point values, 8 "long" 64-bit integers, or

8 double precision floating point values. Access to any byte in memory results in the entire corresponding cache line being loaded into the cache from RAM. An algorithm with good spatial locality of its memory accesses will then use all the memory that was loaded. In contrast, consecutive access to, for example, 32 bit integers (4 bytes) at addresses that are 64 bytes apart causes very poor performance: 64 bytes are loaded for each 4 bytes that are used. To achieve good spatial locality, programmers must ensure that data is organised appropriately in memory and accessed in a sensible order.

Temporal locality of memory accesses improves peformance because a cache line remains unchanged in the cache until this cache line needs to hold a different line from RAM. The conditions that cause the eviction of a cache line back to main RAM depend on how the cache is organised. Consider, for the purpose of illustration, a 1024 byte cache with 16 lines that store 64 bytes. In a direct-mapped cache, which is one possible cache design, every address in memory can reside in only one location in the cache. For this hypothetical 1 KiB cache, one can imagine that all memory is split into 1024 byte blocks. Whenever an address in the first 64 bytes of a 1024 byte block is accessed, these 64 bytes are loaded into the first cache line. In general, an access of *any* byte in the nth group of 64 bytes of a 1024 byte block results in the use of the nth cache line.

An advantage of these direct-mapped caches is that the task of checking whether a particular address is in the cache is quick to perform because any address can only reside in one line. The cache only needs to check whether that line holds the right data. This is also a disadvantage because consecutive accesses to memory addresses that map to the same cache line lead to poor performance. In other words, reading data at the address $1024n + 8$ after reading $1024m + 56$, where n and m are integers, requires use of the same cache line. (The offsets 8 and 56, as long as they are less than 64, do not change which cache line is used.) This problem is called cache thrashing and occurs when one memory transaction uses the same cache line as a recent transaction, forcing the previous data to be evicted from the cache and returned to main RAM.

Another type of cache organisation, called a fully-associative cache, avoids this problem at the expense of requiring more resources to check whether a particular address is available. In a fully-associative cache, each cache line can store any block of 64 bytes (whose first byte is at an address that is a multiple of 64). Cache lines are only evicted when no more space is available in the cache.

An n-way set associative cache organisation lies in between the two extremes of direct mapped and fully-associative caches. In such caches, each memory address may be stored in any of n potential lines in the cache. Such caches offer a trade-off between the benefits of avoiding cache thrashing and the logic circuits required to check which addresses are in the cache.

As an example, suppose the previous 1024 byte cache is 2-way set associative. We then imagine memory to be split into 512 byte blocks instead of 1024 bytes for the direct-mapped cache. A memory access to the nth set of 64 bytes within a 512 byte block can be loaded to either of two possible lines in the cache. With this cache design, reading memory at the address $512n + 8$ after $512m + 56$ does not

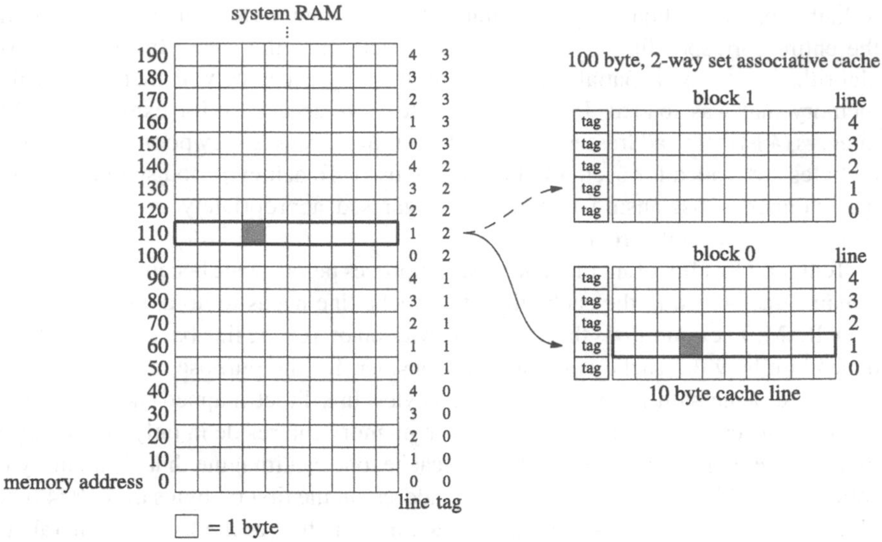

Fig. 13.1 Diagram of a hypothetical cache designed around the decimal representation of memory addresses. This cache holds 100 bytes, uses 10-byte lines, and is 2-way set associative. When read, the shaded byte at the address 113 would be loaded into line 1 in one of the 2 blocks, whichever is available or holds the oldest data (which must first be evicted back to RAM if it was modified while it resided in the cache). The tags identify which data in RAM is held in each cache line

cause the cache to evict the line that was used for $512m + 56$. However, a third access to the same set of 64 bytes within a 512 byte block, for example $512p + 24$, would cause a previous line to be evicted. Thus, the likelihood of cache thrashing is reduced, but for each memory access the cache controller has more work to do: it must check what is stored in two potential places. Both direct mapped and fully associative caches are rare in current CPU architectures, and varying levels of set associativity are used instead.

To help explain the organisation and operation of n-way set associative caches, Fig. 13.1 shows a hypothetical cache designed with lines that hold 10 bytes. The left portion illustrates the bytes stored in RAM grouped into rows of 10 bytes. The starting addresses of each row are therefore multiples of 10. Whenever the CPU needs to read a byte in RAM, the whole row containing that byte will be loaded into a cache line. The destination of that row depends on the starting address of that row. The last digit (or the required number of bits for binary systems) of the address is dropped, and the upper digits modulo 5 (the number of lines in each block, which is the total number of lines in the cache divided by n) gives the line number that will be used.

Of the two blocks, the block that holds the oldest data in the required cache line will be used to store new data. First, however, the data in this line would be written back to memory if needed. When it fills a cache line, the cache also stores a tag that identifies the the starting address of the memory that was loaded into the line.

This identifies the data in the cache line, and it is used to test whether data requested by the CPU is available in the cache. The tag is the starting address divided by the number of bytes in a block, and dropping the fractional part.

Modern CPUs have a **hierarchy of caches**. The first cache level, the **L1 cache**, is the fastest, smallest, has the lowest level of associativity, and is physically closest to the CPU core it assists. The L1 cache mediates access to an **L2 cache** that, though slower, is larger and designed to satisfy a higher proportion of memory requests. The L2 cache, which may be dedicated to one core or shared by several, in turn provides access to an **L3 cache**, which is the largest and has the highest associativity. The L3 cache is usally shared by all cores in a processor, and it interfaces with the **main system memory**.

Writing to cached memory deserves special attention because several strategies are used to update cache and main memory and maintain their consistency. With the first option, called write-through, data written to cache is also always written to main memory (or the next highest cache level). Appealing because it guarantees consistency, this strategy is disadvantageous when the same memory location is written frequently. With the second option, called write-back, data is only written back to main memory when a cache line is evicted and was modified while residing in the cache. This option offers better performance because it avoids unnecessarily frequent writing to memory, but maintaining a consistent memory state for multiple CPUs is challenging. Consider what happens when one CPU needs to read data that was modified while it was in a cache belonging to another CPU: the processors need to communicate to ensure that they are aware of changes to cached data.

These two writing policies assume that the address being written is already in the cache. When it is not, two strategies are available. The first is to simply write the value directly to memory and only load it into cache if it is subsequently read. The second strategy is to first read the destination address, load the cache line, and then write to cache (and also main memory if the cache uses the write-through strategy). Caches in the most common current CPUs follow the second option due to the assumption of temporal locality and that therefore a memory address what was recently written is likely to be read soon as well.

Programmers can conceptually assume that writing to an uncached location causes **(1) eviction** and writing to memory (if required) of the cache line that will be occupied, **(2) reading** the cache line that will be written to, **(3) writing to cache**, and **(4) writing to main memory** once the cache line is evicted by a future memory access. Though somewhat simplified, this model of cache

(continued)

is sufficient for the purposes of this chapter. Interested readers are invited to look at the technical documentation from the manufacturers of the CPUs they use or other reference materials. Special instructions can be used to bypass the caching mechanism to improve writing performance (see Sect. 13.3.4).

Multi-core CPUs and multi-CPU systems present many challenges for cache design. The different cache levels are often shared by the cores of a CPU. For example, each core may have its own L1 cache, pairs of cores may share L2 caches, and all cores on one chip may share an L3 cache. The organisation, size, and communication features of caches are important differences between the CPUs from different manufacturers and differentiate the performance of laptop, personal desktop, and high performance server-class CPUs. For example, the author of this chapter has a laptop with a dual core CPU with 128 KiB of 8-way L1 cache split among the cores, 512 KiB of 8-way L2 cache also split, and 3 MiB of 12-way L3 cache that is shared by the cores. In comparison, a recent Intel server-class CPU with 18 cores has 1152 KiB L1, 4.5 MiB L2, and 45 MiB of L3 cache. We will see later when optimising the LBM code how performance depends on the sizes of the caches.

The optimal use of available caches is a complex topic, and we do not examine caches in greater detail in this chapter. Interested readers are invited to pursue strategies to further optimise the use of caches, both in code for a single core (see also Sect. 13.3.4) and multiple cores and processors (Sect. 13.4.1 and Sect. 13.4.2). Helpful tutorials that describe more about cache architecture can be found in references [3] and [4].

13.2.4 Measuring Performance

To assess the impact of optimisations and determine where bottlenecks exist before optimising code, programmers need to know what functions to use to accurately measure time intervals. This is achieved by calling a function that provides the time elapsed since an arbitrary event, such as when the operating system was started, and finding the difference between the values before and after running the code being tested. Common functions that provide the local time and date are not suitable for this task. They are not sufficently precise (micro- or nanosecond accuracy is needed), and do not always increase monotonically because the system might synchronise its time with an external server, for example. Instead, high-precision timers need to be used. The interfaces that provide access to these timers and the nature of the timers that are available are unfortunately not standard and vary between platforms. A sample function `double seconds()` that returns the number of seconds since an arbitrary reference time is provided in the code accompanying this book, and it uses different functions and libraries when compiled for Unix, Windows, and Mac OS X systems.

The timing function is used in the example code to calculate the speed at which lattice nodes are updated and the apparent memory transfer rate. The net rate at which the code uses memory is the total number of nodes updated in the simulation multiplied by twice (read and write once per update) the number of bytes that each node's populations occupy ($9 * 8 = 72$ bytes for a D2Q9 lattice and double precision) and divided by the time taken. The following code shows these calculations:

```
double start = seconds();

// take NSTEPS simulation time steps
for(int n = 0; n < NSTEPS; ++n)
{
  // main simulation loop
}

double end = seconds();
double runtime = end-start;

size_t nodes_updated = NSTEPS*size_t(NX*NY);

// calculate speed in million lattice updates per second
double speed = nodes_updated/(1e6*runtime);

// calculate memory access rate in GiB/s
double bytesPerGiB = 1024.0*1024.0*1024.0;
double bandwidth = nodes_updated*(2*ndir)
                   *sizeof(double)/(runtime*bytesPerGiB);
```

Note that a large integer type, `size_t`, is used in this code because the quantities involved can be rather large: 2000 updates of a 2048×2048 domain is about 8 billion node updates, twice the maximum number that can be stored in an unsigned 32 bit integer.

13.3 Sequential Code

13.3.1 Introductory Code

We start with a relatively simple code with few optimisations to illustrate the capabilities of automatic optimisation and point out which areas of the code are most important to optimise manually.

To introduce how the code is structured, we begin with the `main()` function, which can be found in the file `main.cpp` and is shown in Listing 13.8. In this listing and those that follow, some details, such as output statements and the code for measuring performance, have been omitted for clarity. The full code is included with this book. Note that the program, as presented, does nothing useful because the computed values are never used and an optimising compiler would likely generate an empty program that truly does nothing. In the actual code, the computed values are used when the density and velocity fields are saved at the end and the error between the computed and analytical flow and pressure fields is calculated *and displayed*.[5]

Listing 13.8 `main()` function for the Taylor-Green vortex decay simulation

```
int main(int argc, char* argv[])
{
  // allocate memory
  double *f1  = (double*) malloc(mem_size_ndir);
  double *f2  = (double*) malloc(mem_size_ndir);
  double *rho = (double*) malloc(mem_size_scalar);
  double *ux  = (double*) malloc(mem_size_scalar);
  double *uy  = (double*) malloc(mem_size_scalar);

  // compute Taylor-Green flow at t=0
  // to initialise rho, ux, uy fields.
  taylor_green(0,rho,ux,uy);

  // initialise f1 as equilibrium for rho, ux, uy
  init_equilibrium(f1,rho,ux,uy);

  // main simulation loop; take NSTEPS time steps
  for(unsigned int n = 0; n < NSTEPS; ++n)
  {
    // stream from f1 storing to f2
    stream(f1,f2);

    // calculate post-streaming density and velocity
    compute_rho_u(f2,rho,ux,uy);

    // perform collision on f2
    collide(f2,rho,ux,uy);

    // swap pointers
    double *temp = f1;
    f1 = f2;
    f2 = temp;
  }

  // deallocate memory
```

[5]Only calculating these values is not enough; they must be used somehow or the compiler will discard the unnecessary calculations.

```
    free(f1);  free(f2);
    free(rho); free(ux); free(uy);

    return 0;
}
```

Several constants are used in the `main()` function and other functions in the code. These variables are declared as constants (with the qualifier `const`) to enable compiler optimisations such as the evaluation of expressions involving constants at compilation time. They are declared as global variables, which means they can be used in any function. The parameters are:

- The scaling factor for convergence studies and the domain size. The factor `scale` is used to perform a convergence study in which the Reynolds number of the flow is kept constant while decreasing the grid size (Sect. 13.1.3 and Sect. 13.5).
```
const unsigned int scale = 1;
const unsigned int NX = 32*scale;
const unsigned int NY = NX;
```
- The number of directions in the lattice
```
const unsigned int ndir = 9;
```
- The memory size (in bytes) for the populations and scalar values
```
const size_t mem_size_ndir    = sizeof(double)
        *NX*NY*ndir;
const size_t mem_size_scalar = sizeof(double)*NX*NY;
```
- The lattice weights
```
const double w0 = 4.0/9.0;  // zero weight
const double ws = 1.0/9.0;  // adjacent weight
const double wd = 1.0/36.0; // diagonal weight
```
- Arrays of the lattice weights and direction components
```
const double wi[] = {w0,ws,ws,ws,ws,wd,wd,wd,wd};
const int dirx[] = {0,1,0,-1, 0,1,-1,-1, 1};
const int diry[] = {0,0,1, 0,-1,1, 1,-1,-1};
```
- The kinematic viscosity ν and the corresponding relaxation parameter τ
```
const double nu = 1.0/6.0;
const double tau = 3.0*nu+0.5;
```
- The maximum flow speed
```
const double u_max = 0.04/scale;
```
- The fluid density
```
const double rho0 = 1.0;
```
- The number of time steps in the simulation
```
const unsigned int NSTEPS = 200*scale*scale;
```

The first task in the `main()` function is to allocate memory for the two particle populations (`f1` and `f2`) and the scalar density (`rho`) and velocity components (`ux` and `uy`). Memory is allocated as contiguous blocks for each quantity and the address of the first entry in each block is stored in the corresponding pointer. The use of multidimensional arrays or array objects provided by libraries is generally not recommended because their use may introduce inefficiencies such as increased memory requirements (for padding or due to automatic resizing) or increased overhead for every memory access (bounds checking or multiple pointer dereferencing). Two dimensional array coordinates are converted to linear indices using the two functions in Listing 13.9. These functions are declared as `inline` to hint to the compiler that they should be expanded where they are used to avoid the overhead of an actual function call.

For scalar variables, the nodes are numbered consecutively along the x direction and increase in multiples of the domain width in the y direction. The populations are laid out as consecutive blocks of $NX \times NY$ doubles for each direction, and each block is indexed in the same way as the scalar variables. This can be seen by re-writing the expression $NX \times (NY \times d + y) + x$ as $(NX \times NY \times d) + (NX \times y + x)$. As will be discussed in detail in the section that follows, this choice is not optimal, and improving it is the main opportunity for optimising this code.

Listing 13.9 Functions for computing linear array indexes from two-dimensional coordinates

```
inline size_t scalar_index(unsigned int x, unsigned int y)
{
  return NX*y+x;
}

inline size_t field_index(unsigned int x, unsigned int y,
    unsigned int d)
{
  return NX*(NY*d+y)+x;
}
```

After the memory has been allocated, it is initialised. We first compute the density and velocity components for the Taylor-Green vortex flow using the functions in Listing 13.10. The first function computes the solution at a particular position and time while the second fills the density and velocity variables with the values for the whole domain. The task was split into two functions because the first function is re-used in the code that computes the error between the numerical and exact solutions.

Listing 13.10 Functions used to compute the exact solution for Taylor-Green vortex decay

```
void taylor_green(unsigned int t,
                  unsigned int x, unsigned int y,
                  double *r, double *u, double *v)
{
  double kx = 2.0*M_PI/NX;
  double ky = 2.0*M_PI/NY;
  double td = 1.0/(nu*(kx*kx+ky*ky));

  double X = x+0.5;
  double Y = y+0.5;
  double ux = -u_max*sqrt(ky/kx)*cos(kx*X)*sin(ky*Y)
              *exp(-1.0*t/td);
  double uy =  u_max*sqrt(kx/ky)*sin(kx*X)*cos(ky*Y)
              *exp(-1.0*t/td);
```

```
    double P = -0.25*rho0*u_max*u_max
                *( (ky/kx)*cos(2.0*kx*X)
                  +(kx/ky)*cos(2.0*ky*Y) )
                *exp(-2.0*t/td);
    double rho = rho0+3.0*P;

    *r = rho;
    *u = ux;
    *v = uy;
  }

  void taylor_green(unsigned int t, double *r,
                    double *u, double *v)
  {
    for(unsigned int y = 0; y < NY; ++y)
    for(unsigned int x = 0; x < NX; ++x)
    {
      size_t sidx = scalar_index(x,y);
      taylor_green(t,x,y,&r[sidx],&u[sidx],&v[sidx]);
    }
  }
```

Next, the particle population `f1` is initialised with the equilibrium populations for the initial values of the density and velocity. The function that performs this task, `init_equilibrium`, is presented in Listing 13.11.

Listing 13.11 Function for initialising a particle population with the equilibrium values for the specified density and velocity

```
  void init_equilibrium(double *f, double *r,
                        double *u, double *v)
  {
    for(unsigned int y = 0; y < NY; ++y)
    {
      for(unsigned int x = 0; x < NX; ++x)
      {
        double rho = r[scalar_index(x,y)];
        double ux  = u[scalar_index(x,y)];
        double uy  = v[scalar_index(x,y)];

        for(unsigned int i = 0; i < ndir; ++i)
        {
          double cidotu = dirx[i]*ux + diry[i]*uy;
          f[field_index(x,y,i)] =
              wi[i]*rho*(1.0 + 3.0*cidotu
                        +4.5*cidotu*cidotu
                        -1.5*(ux*ux+uy*uy));
        }
      }
    }
  }
```

With all the required variables initialised, the next part of the `main()` function is the `for` loop that performs each time step of the simulation. This loop performs four tasks: (1) streaming the populations in `f1` along the lattice's directions and storing the result in the temporary populations in `f2` (Listing 13.12), (2) computing the post-streaming values of the density and velocity (Listing 13.13), (3) performing the collision operation (relaxation to equilibrium) on `f2` (Listing 13.14), and (4) exchanging the pointers `f1` and `f2` so that in the next iteration the roles of the memory designated by `f1` and `f2` are reversed.

Listing 13.12 Function that performs streaming of the populations in a fully periodic domain, reading from `f_src` and storing to `f_dst`

```
void stream(double *f_src, double* f_dst)
{
  for(unsigned int y = 0; y < NY; ++y)
  {
    for(unsigned int x = 0; x < NX; ++x)
    {
      for(unsigned int i = 0; i < ndir; ++i)
      {
        // enforce periodicity
        // add NX to ensure that value is positive
        unsigned int xmd = (NX+x-dirx[i]) % NX;
        unsigned int ymd = (NY+y-diry[i]) % NY;

        f_dst[field_index(x,y,i)] =
                        f_src[field_index(xmd,ymd,i)];
      }
    }
  }
}
```

Listing 13.13 Function that computes the density and velocity of the provided populations `f`

```
void compute_rho_u(double *f, double *r,
                   double *u, double *v)
{
  for(unsigned int y = 0; y < NY; ++y)
  {
    for(unsigned int x = 0; x < NX; ++x)
    {
      double rho  = 0.0;
      double ux   = 0.0;
      double uy   = 0.0;

      for(unsigned int i = 0; i < ndir; ++i)
      {
        rho  += f[field_index(x,y,i)];
        ux   += dirx[i]*f[field_index(x,y,i)];
        uy   += diry[i]*f[field_index(x,y,i)];
      }
```

```
        r[scalar_index(x,y)] = rho;
        u[scalar_index(x,y)] = ux/rho;
        v[scalar_index(x,y)] = uy/rho;
      }
    }
  }
```

Listing 13.14 Function that performs the collision operation on the particle populations using pre-computed density and velocity values

```
void collide(double *f, double *r, double *u, double *v)
{
  // useful constants
  const double tauinv = 2.0/(6.0*nu+1.0); // 1/tau
  const double omtauinv = 1.0-tauinv;     // 1 - 1/tau

  for(unsigned int y = 0; y < NY; ++y)
  {
    for(unsigned int x = 0; x < NX; ++x)
    {
      double rho = r[scalar_index(x,y)];
      double ux  = u[scalar_index(x,y)];
      double uy  = v[scalar_index(x,y)];

      for(unsigned int i = 0; i < ndir; ++i)
      {
        // calculate dot product
        double cidotu = dirx[i]*ux + diry[i]*uy;

        // calculate equilibrium
        double feq = wi[i]*rho*(1.0 + 3.0*cidotu
                                 +4.5*cidotu*cidotu
                                 -1.5*(ux*ux+uy*uy));

        //relax to equilibrium
        f[field_index(x,y,i)] =
                      omtauinv*f[field_index(x,y,i)]
                      +tauinv*feq;
      }
    }
  }
}
```

Finally, at the end of the `main` function, the memory allocated for all the arrays is released.

The code was compiled with GCC version 4.9.2 with several levels of automatic optimisation and run on a workstation with an Intel Xeon W3550 CPU and 12 GiB of DDR3 RAM running at 1066 MHz in a triple channel configuration for a total theoretical maximum transfer rate of 24 GiB/s (8 GiB/s per memory channel). The performance results for several domain sizes are presented in Table 13.1. The speed of the code is reported in millions of lattice updates per second (Mlups). This is the

Table 13.1 Performance of the introductory code compiled with several optimisation levels and domains of varying sizes

Domain size	Time steps	Memory (MiB)	Optimisation	Speed (Mlups)	Relative speed
32 × 32	100000	0.14	`-O0`	2.3	0.2
			`-O1`	11.0	1
			`-O3`	12.2	1.1
			`-Ofast`	14.8	1.3
128 × 128	5000	2.25	`-O0`	2.2	0.2
			`-O1`	9.7	1
			`-O3`	10.9	1.1
			`-Ofast`	13.9	1.4
256 × 256	2000	9	`-O0`	2.1	0.3
			`-O1`	6.6	1
			`-O3`	8.2	1.2
			`-Ofast`	10.8	1.6
512 × 512	500	36	`-O0`	2.1	0.4
			`-O1`	5.3	1
			`-O3`	6.2	1.2
			`-Ofast`	8.4	1.6
2048 × 2048	25	576	`-O0`	2.0	0.4
			`-O1`	5.1	1
			`-O3`	5.6	1.1
			`-Ofast`	8.0	1.6

number of nodes in the domain multiplied by the number of time steps performed and divided by the runtime. For each domain size, the number of time steps was chosen so that the runtime was at most one minute and the average of three runs is reported.

The most obvious result is the exceedingly poor performance without optimisation: five times slower than with the lowest optimisation level. As noted in Sect. 13.2 (Listing 13.5), this is not surprising since the compiler makes no effort to remove unnecessary instruction sequences. With optimisation, the improvement in performance between levels 1 and 3 is about 10–20% and another 30% improvement is achieved by using `-Ofast`.

In general, the speed decreases as the domain size increases. The performance drops noticeably when the memory allocated exceeds the size of the 8 MiB L3 cache on this CPU. For the largest domain, the speed is approximately half the speed of small domains. The reasons for the dependence of the performance on the cache size and ways to improve this, are discussed in the next section (Sect. 13.3.2).

13.3.2 Optimising the Introductory Code

The code introduced in the previous section presents several opportunities for optimisation. We start by noticing one relatively minor optimisation: the rest populations f_0 should not need to be copied during streaming. To avoid this unnecessary memory access, we use a separate variable `f0` for the rest populations and keep two variables, `f1` and `f2`, for f_{1-8}. The indices for the `f0` array are computed in the same was as for the scalars.

The second optimisation follows from noticing that the values of the populations stored for every node during streaming are those that are then read during the collision step. Consequently, the three functions `stream`, `compute_rho_u`, and `collide` may be combined into one function, `stream_collide_save`, that accesses memory significantly less frequently. The code for this function is shown in Listing 13.15.

This combined function includes several less important optimisations. The function includes a boolean parameter `save` that is used to indicate whether the moments should be written to memory. For simulations that save the intermediate density and velocity fields at regular intervals, the required memory writing is avoided for the majority of the time steps taken. The second optimisation is the unrolling of all `for` loops that iterate over the nine populations (cf. (3.12) and (3.65)). In addition to avoiding the overhead of the loop, this allows many terms to be dropped in the evaluation of the equilibrium populations. Finally, several common factors are stored in temporary variables to assist the evaluation of the equilibrium values. The function `field0_index` for indexing into the array of rest populations is the same as the `scalar_index` function.

Note that after combining the streaming and collision operations into one step it is no longer beneficial to keep the rest populations in a separate variable because now all populations are read and written by the combined function. However, we keep the rest populations separate because this is useful for parallel versions of the code (Sect. 13.4.2) in which only the non-rest populations need to be shared between the processors working on different portions of the simulation domain.

Listing 13.15 Function that performs streaming, computation of moments, and collision in one step

```
void stream_collide_save(double *f0, double *f1, double *f2,
                         double *r, double *u, double *v,
                         bool save)
{
  // useful constants
  const double tauinv = 2.0/(6.0*nu+1.0); // 1/tau
  const double omtauinv = 1.0-tauinv;     // 1 - 1/tau

  for(unsigned int y = 0; y < NY; ++y)
  {
    for(unsigned int x = 0; x < NX; ++x)
    {
```

```
unsigned int xp1 = (x+1)%NX;
unsigned int yp1 = (y+1)%NY;
unsigned int xm1 = (NX+x-1)%NX;
unsigned int ym1 = (NY+y-1)%NY;

// direction numbering scheme
// 6 2 5
// 3 0 1
// 7 4 8

double ft0 = f0[field0_index(x,y)];

// load populations from adjacent nodes
double ft1 = f1[fieldn_index(xm1,y,  1)];
double ft2 = f1[fieldn_index(x,  ym1,2)];
double ft3 = f1[fieldn_index(xp1,y,  3)];
double ft4 = f1[fieldn_index(x,  yp1,4)];
double ft5 = f1[fieldn_index(xm1,ym1,5)];
double ft6 = f1[fieldn_index(xp1,ym1,6)];
double ft7 = f1[fieldn_index(xp1,yp1,7)];
double ft8 = f1[fieldn_index(xm1,yp1,8)];

// compute moments
double rho = ft0+ft1+ft2+ft3+ft4+ft5+ft6+ft7+ft8;
double rhoinv = 1.0/rho;

double ux = rhoinv*(ft1+ft5+ft8-(ft3+ft6+ft7));
double uy = rhoinv*(ft2+ft5+ft6-(ft4+ft7+ft8));

// only write to memory when needed
if(save)
{
  r[scalar_index(x,y)] = rho;
  u[scalar_index(x,y)] = ux;
  v[scalar_index(x,y)] = uy;
}

// now compute and relax to equilibrium
// note that
// feq_i = w_i rho [1 + 3(ci . u)
//                  +(9/2) (ci . u)^2 - (3/2) (u.u)]
//       = w_i rho [1 - 3/2 (u.u)
//                  +(ci . 3u) + (1/2) (ci . 3u)^2]
//       = w_i rho [1 - 3/2 (u.u)
//                  +(ci . 3u)(1 + (1/2) (ci . 3u))]

// temporary variables
double tw0r = tauinv*w0*rho; //    w[0]*rho/tau
double twsr = tauinv*ws*rho; // w[1-4]*rho/tau
double twdr = tauinv*wd*rho; // w[5-8]*rho/tau

double omusq = 1.0-1.5*(ux*ux+uy*uy); // 1-(3/2)u.u

double tux = 3.0*ux;
```

```
            double tuy = 3.0*uy;

            f0[field0_index(x,y)] = omtauinv*ft0 + tw0r*(omusq);

            double cidot3u = tux;
            f2[fieldn_index(x,y,1)] = omtauinv*ft1
                     + twsr*(omusq + cidot3u*(1.0+0.5*cidot3u));
            // ... similar expressions for directions 2-4

            cidot3u = tux+tuy;
            f2[fieldn_index(x,y,5)] = omtauinv*ft5
                     + twdr*(omusq + cidot3u*(1.0+0.5*cidot3u));
            // ... similar expressions for directions 6-8
        }
    }
}
```

Optimising the memory access pattern to improve cache utilisation has a **greater impact** on performance than the memory, loop, and arithmetic optimisations presented so far.

To understand why the use of cache can be improved, let us consider two consecutive iterations of the innermost `for` loop that updates the simulation domain (Listing 13.15), which is the loop that iterates along the x direction. Figure 13.2 illustrates how the memory addresses that are read and written for the non-rest populations are spread across the memory address space when the linear array index of f_{d} at the node with 2D coordinates `x` and `y` is `NX*(NY*d+y)+x`. In this figure and those that follow for different memory layouts, each box represents the memory occupied by one double precision value (8 bytes) and the symbols show which direction's population is stored in that location. The expression in the bottom left corner gives the address of the first box that is shown in the diagram. Memory addresses increase first from left to right then bottom to top. The offsets relative to the location of the bottom left box are shown along the bottom and left sides of the diagrams. Boxes shaded in light grey show the memory locations read/written for the first node and those shown in dark grey are for the next node. The box with a dashed outline shows the position of the data for the first node (at the coordinates $(x-1, y)$) and the box with a thicker outline shows the next node (located at (x, y)).

With the memory layout illustrated in Fig. 13.2, every double precision value read from or written to memory is located immediately after the value that was read/written for each population in the previous iteration of the loop. This memory layout therefore allows good spatial and temporal locality of memory accesses, but the locality is limited to the values for each direction. When each node is updated, there is a good chance that the required memory locations were already loaded into a cache line during the updating of a previous node. However, the value of f_1, for

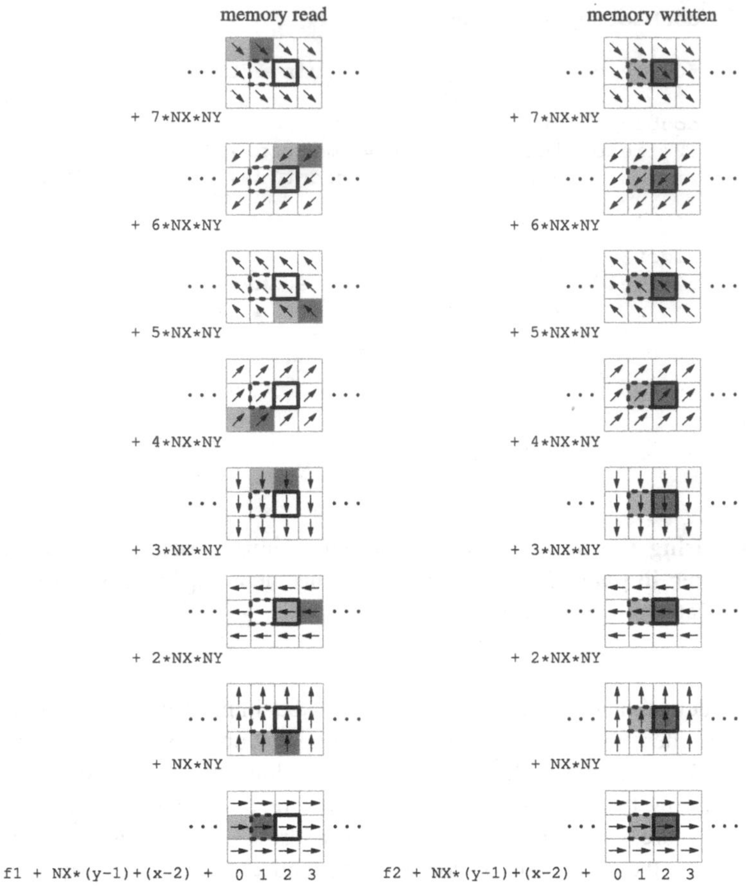

Fig. 13.2 Diagram showing the memory locations read (*left*) and written (*right*) during the updating of two consecutive nodes when the populations $f_d(x, y)$ are stored at the linear indices `NX*(NY*d+y)+x`. The data for the node at $(x - 1, y)$ is outlined with a *dashed line*, and a *thick outline* shows the data for the node at (x, y)

example, only remains in the cache until it is needed in the next iteration if the reads of $f_{2\text{–}9}$ and the writes of all populations did not cause too many accesses to addresses with the same lowest bits. This is particularly problematic for the values that are written because all the f_i for one node are separated by `8*NX*NY` bytes. When this is a multiple of the cache block size (see Sect. 13.2.3), as happens when the number of nodes in the domain is a sufficiently high power of two, all these writes use the same line within a block and only a cache with high set associativity can retain enough values to be useful. A thorough analysis of the conditions when the values loaded into cache in one step survive until the next step is complex and depends on the alignment of the allocated memory and the sizes, levels of set associativity, and write modes of the caches that are available.

For this memory layout, non-power-of-two domain sizes are beneficial, and a 131×128 domain, for example, can be updated faster than a 128×128 domain. Padding, i.e. adding unused memory to prevent alignment of memory addresses, can be used to alleviate this problem. To do this, one would allocate enough memory and compute indexes as if the domain size were 131×128 but only use the values for a 128×128 domain.

Exercise 13.4 Edit the code so that it uses different variables, perhaps `NX_MEM` and `NY_MEM`, for the domain dimensions in the allocation and indexing functions. For a 128×128 domain, how does adding padding at the end of each row, i.e. `NX_MEM = NX + PAD_X` where `PAD_X` is the number of additional doubles, affect the simulation speed? What is the optimal amount of padding?

Before moving on to the best memory layout, let us consider briefly the worst, which is illustrated in Fig. 13.3. In this layout, compared with the previous (Fig. 13.2), the roles of x and y have been exchanged. Therefore the use of this layout is equivalent to using the previous and exchanging the order of the two `for` loops so that the inner loop iterates along the y direction. With this worst memory access pattern, spatial and temporal locality of memory accesses is lost. Values loaded into the cache can only be used if they survive the completion of the entire inner `for` loop. The main message here is that **the memory layout must match the structure of the `for` loops.**

Programmers who choose multidimensional arrays instead of linear arrays should be aware of how they are organised. In C, C++, and Python, multidimensional arrays are in row-major order, meaning that values in the same row (values with the same first index) occupy consecutive locations in memory. Fortran and MATLAB use column-major order, in which the values of a column (second index) occupy consecutive memory locations. An easy way to remember the correct pairing of memory layout and `for` loop order is that consecutive iterations of the innermost loop should access consecutive locations in memory. Another alternative is to use one `for` loop over all the nodes in the domain and compute the two-dimensional coordinates from the linear index where needed.

Exercise 13.5 Re-write the main `for` loop that updates the simulation domain so that it takes the form `for(int k = 0; k < NX*NY; ++k)`. Compute the linear indices of the adjacent lattice nodes, needed for streaming, from `k`. Does this change improve performance? Do you think it enhances or worsens the readability of the code?

Of the six possible ways to map a three-dimensional array with indices `x`, `y`, and `d` to a linear index, the best method for the code in this section is shown in Fig. 13.4. The different geometry of the figure compared to the previous two quickly emphasises the difference in layout. In this layout, the populations for

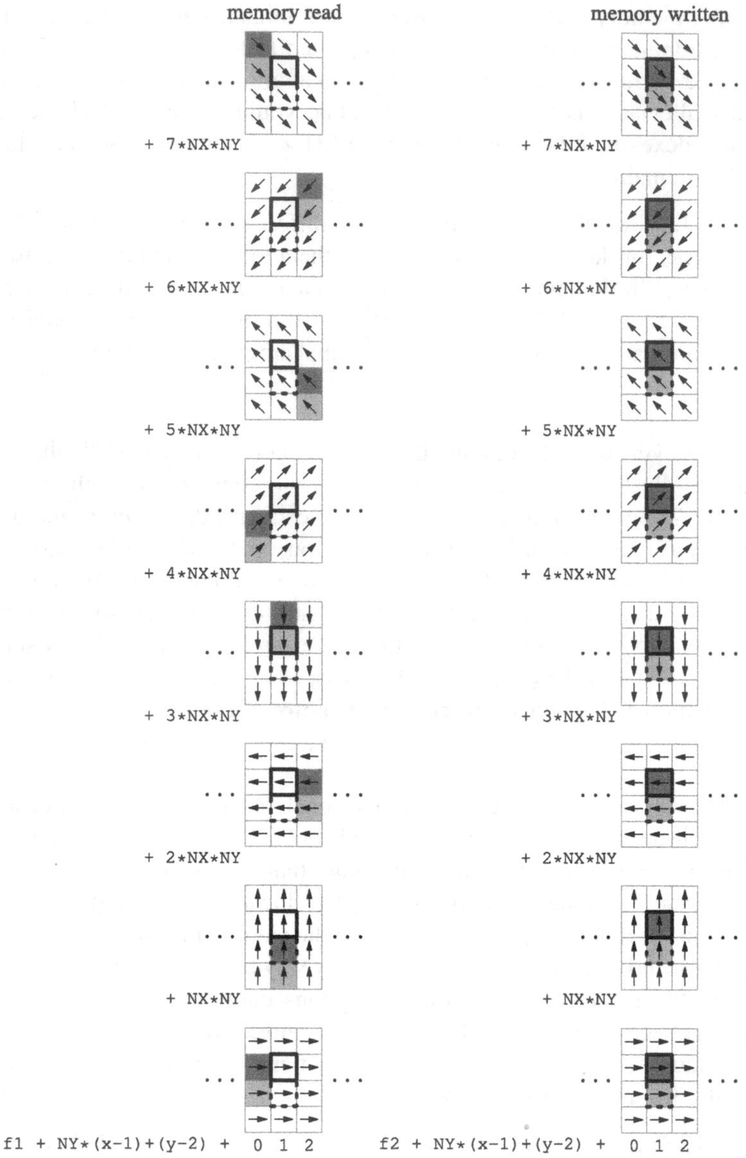

Fig. 13.3 Diagram showing the memory locations read (*left*) and written (*right*) during the updating of two consecutive nodes when the populations $f_d(x, y)$ are stored at the linear indices `NY*(NX*d+x)+y`. Notice how none of the data read or written for the second node is located beside the data for the first node, which prevents use of cached data

each node are stored in consecutive locations in memory, and the populations for the node with coordinates (x, y) follow the populations for $(x - 1, y)$. This is achieved by storing f_d for the node with coordinates `x` and `y` at the linear

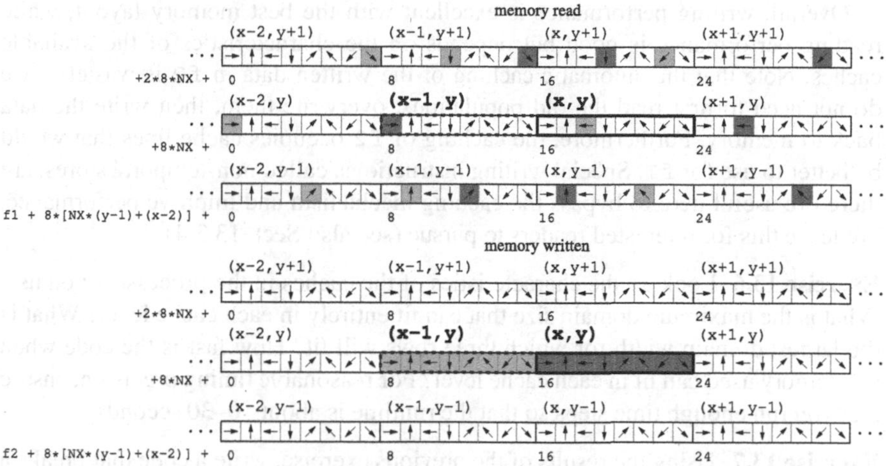

Fig. 13.4 Diagram showing the memory locations read (*top*) and written (*bottom*) during the updating of two consecutive nodes when the populations $f_d(x, y)$ are stored at the linear indices `(ndir-1)*(NX*y+x)+(d-1)`

array index `(ndir-1)*(NX*y+x) + (d-1)` where `ndir` is 9, the number of directions in the D2Q9 velocity set. The benefit of the new layout on writing performance is clear: the values are written in perfectly consecutive order from the first direction's population in the first node to the last direction's population in the last node.

The benefits for reading are more complex. As the domain is updated first along *x* and then with increasing *y*, the reading of the three directions "↙", "↓", and "↘" causes loading of all the populations at the corresponding nodes into cache. Note that for the 9 direction model with 8 non-zero directions, all 8 values fit in one 64 byte cache line (a typical size in current CPUs). Though alignment of the allocated memory so that the address of the first ("→") direction is a multiple of 64 bytes is not guaranteed, in practice this occurs in the tests we present. It is therefore the read of the "↙" direction at (`x+1`,`y+1`) when updating the node at (`x`,`y`) that loads the whole population (excluding the zero direction) at (`x+1`,`y+1`). If the cache can hold the populations for three rows of the domain (the previous, the current, and the next), this whole population will not need to be read again: the directions "↙", "↓", and "↘" of the nodes in row `y` are used to update row `y-1`, "→" and "←" are used for row `y`, and "↗", "↑", and "↖" are used for row `y+1`. When the cache is insufficient to hold a complete row of the domain, Fig. 13.4 shows that three new reads of 8 doubles (64 bytes) are needed for every node (compared with only one new read in the best case). Of the populations loaded, three ("↙", "←", and "↖" in column `x+1`) are used to update the node at (`x`,`y`), two ("↑" and "↓") are used for the node (`x+1`,`y`), and three ("↗", "→", and "↘") are used for the node (`x+2`,`y`). Thus, in the worst case, $8/24 = 1/3$ of the doubles loaded in each step of the inner `for` loop are used within a span of three iterations of this loop.

Overall, writing performance is excellent with the best memory layout, while reading performance is good but depends on the characteristics of the available caches. Note that the automatic caching of the written data in `f2` is wasteful: we do not need to first read the old populations, overwrite them, then write the data back to memory. Furthermore, the caching of `f2` occupies cache lines that would be better to use for `f1`. Special writing instructions, called non-temporal stores, are therefore useful here to bypass the caching mechanism and improve performance. We leave this for interested readers to pursue (see also Sect. 13.3.4).

Exercise 13.6 Look up the characteristics of the caches in the processors you use. What is the maximum domain size that can fit entirely in each cache level? What is the largest domain width for which three rows will fit? How fast is the code when all memory used can fit in each cache level? For reasonable timing precision, ensure that you run enough time steps so that the runtime is about 10–30 seconds.

Exercise 13.7 Using the results of the previous exercise, write a code that breaks a large simulation domain into subdomains that fit into one of the cache levels. Does this division of the domain improve the speed of the simulation? Which cache level is best to target? You may find it useful to reorganise the memory layout so that the memory for each subdomain is in a contiguous block.

Exercise 13.8 The analysis of the use of cache memory in this section considered only interior nodes away from the periodic boundaries. What happens at the boundaries? Is any of the required data still located in adjacent memory addresses?

To assess the benefits of improving the memory layout, we first consider the initial implementation and change only the memory layout. The same workstation with an Intel Xeon W3550 processor was used as for the unoptimised version (Sect. 13.3.1). This quad-core processor has one 32 KiB 4-way set associative L1 instruction cache per core, one 32 KiB 8-way set associative L1 data cache per core, one 256 KiB 8-way set associative L2 cache per core, and one 8 MiB 16-way set associative L3 cache shared by all four cores.

The simulation speeds using the unoptimised initial code for the three memory layouts considered in this section and several domain sizes are shown in Fig. 13.5a. Using the best layout approximately doubles the simulation speed throughout the range of domain sizes considered. Note the significant difference in speed between the worst and best layouts when the domain is large: **picking the right memory layout yields 15 times faster simulations** than choosing the worst layout!

The results for the optimised code, shown in Fig. 13.5b, reveal an interesting difference in performance relative to the unoptimised code. For small domain sizes, the speed with the best memory layout is roughly the same as that for the unoptimised code. The major difference in these codes is their

(continued)

memory use: the optimised code does not perform a separate streaming step to avoid a pair of reads and writes of the populations. The similar performance of the codes (in fact the optimised code runs slightly slower for the smallest domain) suggests that when the simulation data can fit in the L2 cache, the penalty of excess memory accesses is small. **As the domain size is increased**, however, **the optimised version maintains roughly the same speed**, while the unoptimised version and the worse memory layouts show rapid degradation in performance.

As for the initial code, compiling without automatic optimisation is a bad choice. As shown in Table 13.2, the optimised code performs about 4 times faster than the previous code also compiled without optimisation (Table 13.1). This is not, however, a practically useful comparison because enabling automatic optimisation increases the speed about three times. Unlike the results with the initial code, the different

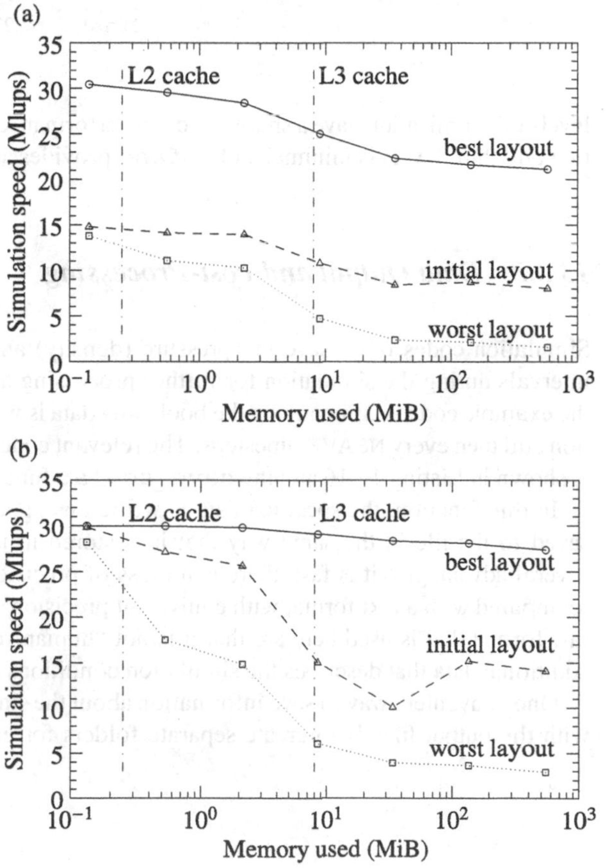

Fig. 13.5 Simulation speeds with the three memory layouts considered in this section for several domain sizes using (**a**) the initial code and (**b**) the optimised code. For all cases the code was compiled with the optimisation flag `-Ofast`. The *dashed vertical lines* show the sizes of the L2 and L3 caches on the processor that executed the code

Table 13.2 Performance of the optimised code for several domain sizes and compiled with different optimisation options. Compare with the performance of the initial code in Table 13.1

Domain size	Time steps	Memory (MiB)	Optimisation	Speed (Mlups)	Relative speed
32 × 32	100000	0.13	`-O0`	8.7	0.3
			`-O1`	26.4	1
			`-O3`	27.5	1.0
			`-Ofast`	30.0	1.1
128 × 128	5000	2.1	`-O0`	8.7	0.3
			`-O1`	26.5	1
			`-O3`	27.5	1.0
			`-Ofast`	29.8	1.1
512 × 512	500	34	`-O0`	8.5	0.3
			`-O1`	25.5	1
			`-O3`	25.8	1.0
			`-Ofast`	28.4	1.1
2048 × 2048	25	544	`-O0`	8.5	0.4
			`-O1`	24.6	1
			`-O3`	24.9	1.0
			`-Ofast`	27.4	1.1

levels of optimisation have a small effect on performance: the difference between the first and third levels is minimal and `-Ofast` provides at most a 10% improvement.

13.3.3 Data Output and Post-Processing

Simulation codes often save the pressure (density) and velocity fields at regular intervals during the simulation for further processing after the simulation ends. In the example code accompanying the book, this data is written to files after initialisation and then every `NSAVE` timesteps. The relevant code from the `main()` function is shown in Listing 13.16 and the `save_scalar` function is in Listing 13.17.

In this function, the specified data is written using a raw format, i.e. the data is saved to the file in the same way that it is stored in memory. This choice offers several advantages: it is fast, there is no loss of precision, and the file size is small (compared with a text format with equivalent precision). Some disadvantages of the raw format that is used here are that it is not "human-readable," and it includes no additional data that describes the simulation conditions or geometry.

One convenient way to save information about the simulation conditions together with the output files is to create separate folders for each simulation run, display the simulation conditions in the code, and save the code's output to a file.[6] For the previous benchmarks, the saving of density and velocity components both to memory and to disk was disabled to avoid interference with the timing.

Listing 13.16 Code in the main **for** loop for saving simulation data at regular time intervals

```
for(unsigned int n = 0; n < NSTEPS; ++n)
{
  bool save = (n+1)%NSAVE == 0;

  stream_collide_save(f0,f1,f2,rho,ux,uy,save);

  if(save)
  {
    save_scalar("rho",rho,n+1);
    save_scalar("ux", ux, n+1);
    save_scalar("uy", uy, n+1);
  }

  // ...
}
```

Listing 13.17 The save_scalar function that saves the contents of the supplied scalar field to a file

```
void save_scalar(const char *name, double *scalar,
                 unsigned int n)
{
  // assume reasonably-sized file names
  char filename[128];
  char format[16];

  // compute maximum number of digits
  int ndigits = floor(log10((double)NSTEPS)+1.0);

  // generate format string
  // file name format is name0000nnn.bin
  sprintf(format,"%%s%%0%dd.bin",ndigits);
  sprintf(filename,format,name,n);

  // open file for writing
  FILE *fout = fopen(filename,"wb+");
```

6On a command line, we can do this with output redirection. For example, `./sim > sim.out` on a Unix command line (or `sim.exe > sim.out` in Windows) runs the program `sim` and saves its output to the text file `sim.out`. The output is not shown on the screen. To both display and save the output we can use (on Unix systems) the `tee` command: `./sim | tee sim.out` where we have used a pipe, |, to send the output of one program to the input of another, in this case `tee`.

```
    // write data
    fwrite(scalar,1,mem_size_scalar,fout);

    // close file
    fclose(fout);
}
```

The data produced by CFD simulations can be analysed in many ways to reveal and present insights about the problems being studied. Below are listed some of the tools commonly used to analyse and visualise simulation data. Some of the listed software can directly use the output of the example code, while for others readers will need to use libraries or write their own codes to save the data in the required formats. Some popular software packages for data processing are:

- ParaView, an application for interactive visualisation that uses the Visualisation Toolkit (VTK) libraries. Users can write code that exports their data in a VTK format and then load it in ParaView.
- MATLAB, a platform for computation and data processing. The `fread` command can be used to load the output of the example code in this chapter.
- Spreadsheets. Text formats, such as comma separated value (csv) files are convenient for small data sets, but are inefficient for large data sets, for which binary files are preferable.
- Python and its many libraries that are available for performing computations (e.g. SciPy) and generating graphics (e.g. matplotlib).
- Packages for LaTeX such as TikZ, PGFPlots, and Asymptote. These packages are often used to produce figures for publications, including many of the figures in this book. Asymptote can also be used by itself outside of LaTeX documents as a general purpose language, like Python or MATLAB, and to generate vector and 3D graphics.

Simulations can easily produce very large datasets that become impractical to analyse. Saving the whole simulation domain every time step is usually unnecessary, and could quickly fill many hard drives. Some quantities of interest, however, can be efficiently computed during the simulation, eliminating the need to save and process data later.

In the example code, we are performing a convergence study and could compute the error in the results by loading the output data. Instead, we compute the errors in the solution as the simulation runs and the required data is still in memory. Listing 13.18 shows the function that computes the kinetic energy in the flow and the L_2 errors in the density and velocity as compared with the analytical solution (we re-use here the same function that computes the initial conditions). This function simply iterates over the simulation domain, accumulating several sums. The four computed values are then saved in an array. Though simple in this version, this function will be used to demonstrate several concepts in parallel programming in Sect. 13.4.

Listing 13.18 The `compute_flow_properties` function that computes the kinetic energy of the flow and the L_2 errors between the computed and analytical quantities of density and velocity components

```
void compute_flow_properties(unsigned int t, double *r,
    double *u, double *v, double *prop)
{
    // prop must point to space for 4 doubles:
    // 0: energy
    // 1: L2 error in rho
    // 2: L2 error in ux
    // 3: L2 error in uy

    double E = 0.0;

    double sumrhoe2 = 0.0;
    double sumuxe2 = 0.0;
    double sumuye2 = 0.0;

    double sumrhoa2 = 0.0;
    double sumuxa2 = 0.0;
    double sumuya2 = 0.0;

    for(unsigned int y = 0; y < NY; ++y)
    {
      for(unsigned int x = 0; x < NX; ++x)
      {
        double rho = r[scalar_index(x,y)];
        double ux  = u[scalar_index(x,y)];
        double uy  = v[scalar_index(x,y)];
        E += rho*(ux*ux + uy*uy);

        double rhoa, uxa, uya;
        taylor_green(t,x,y,&rhoa,&uxa,&uya);

        sumrhoe2 += (rho-rhoa)*(rho-rhoa);
        sumuxe2  += (ux-uxa)*(ux-uxa);
        sumuye2  += (uy-uya)*(uy-uya);

        sumrhoa2 += (rhoa-rho0)*(rhoa-rho0);
        sumuxa2  += uxa*uxa;
        sumuya2  += uya*uya;
      }
    }

    prop[0] = E;
    prop[1] = sqrt(sumrhoe2/sumrhoa2);
    prop[2] = sqrt(sumuxe2/sumuxa2);
    prop[3] = sqrt(sumuye2/sumuya2);
}
```

13.3.4 LBM Algorithm Optimisations

We present here as exercises two directions that readers can pursue to further improve the efficiency of their LBM implementations. The first option is to enhance the use of memory: by carefully updating the populations we can avoid keeping two sets in memory. This is particularly useful for architectures with restricted memory capacity, such as GPUs (see Sect. 13.4.3).

Exercise 13.9 Implement one or more of the single-lattice streaming methods as reviewed in [5] and compare the speed with that of the code presented in this chapter. Can you achieve comparable or better speed while using half the memory? The article also considers the use of non-temporal stores, which are special instructions that bypass the caches when writing data to memory. Does the use of non-temporal stores improve the code in this chapter?

The second opimisation involves the special case of a relaxation time equal to one. In this case, the post-collision populations are the equilibrium populations and therefore do not depend on the populations in the previous time step. We therefore only need to store the macroscopic variables and compute the populations only where they are needed from the macroscopic quantities. This approach is discussed in [6].

Exercise 13.10 Write an optimised code for the case of $\tau = 1$. Store only the density and velocity data at each node in memory. To perform streaming and collision at each lattice node, read the densities and velocities at adjacent nodes, compute the populations that propagate from these adjacent nodes to the current node, and compute and store the new density and velocity at that node. How much does performance improve over the un-optimised case? Can you reduce the number of times that the density and velocity data for each node are read?

13.4 Parallel Computing

Parallel computing is the use of computational hardware that can perform multiple operations simultaneously, and there are two main motives for using parallel computing for simulations.

The strongest and perhaps most obvious motive is to finish simulations faster. When the speed of simulations is increased, simulations of larger physical domains (or higher resolutions for the same physical size) and more simulations at different conditions become feasible to perform in a given amount of time. Furthermore, additional physical phenomena can be modelled and functions can be evaluated more accurately (for example by using more terms of a series).

The second motive for parallelisation is to overcome resource contraints in a cost-effective manner. When the memory required by a simulation far exceeds the capacity of typical computers, one could purchase or create a specialised system with more memory. For some simulation algorithms, however, higher performance can be achieved at a lower cost by instead connecting several low-capacity systems with each other via a high speed network. This is the case for LBM simulations because their performance is largely determined by the available memory bandwidth (the rate at which memory can be read or written), which increases as more processors with independent memory systems are added to the network.

The central challenge in parallel computing is determining how to allocate tasks to available resources that can perform them simultaneouly. It is here that we can see the differences between algorithms that are amenable to parallelisation and those that are not. Ideally, an algorithm would finish n times faster or process n times more data in a fixed time when n times more processors are used. However, if every step in an algorithm uses the outcome of the previous step, this algorithm cannot make use of additional resources and must run sequentially. Fortunately, this is the worst case scenario and it is not encountered often.

In the LBM example we consider in this chapter, the task of updating each node of the domain requires only one component of the distribution at each adjacent node in the previous time step. Parallelisation of such LBM simulations is therefore straightforward: the simulation domain is divided into several subdomains, and each subdomain is updated independently. If needed, data for the directions that cross subdomain boundaries are shared between the devices performing the computations for each subdomain.

Parallel processing can be achieved in many different ways. Basic parallel operations are incorporated into the individual CPU cores, such as vector arithmetic instructions that perform the same operation on 4 quantities that form a vector. CPUs can also simultaneously execute commands that require different circuits, for example integer and floating point arithmetic. Use of such features is usually automatic, handled during compilation of code or during the execution of instructions by the CPU. Provided that they do not compete for access to external resources, the cores within CPUs and the separate (multi-core) CPUs in computers can perform arbitrary operations simultaneously. Some compilers and languages provide features that automatically identify tasks that can execute in parallel and handle the allocation of these tasks to available resources, but this automation is not yet complete and programmers may need to provide information to guide and optimise the automatic parallelisation.

The programming techniques and tools used to program parallel system depend on how the processors and memories that make up a parallel computer are organised. In shared memory systems, numerous cores share access to a single large memory. The CPUs operate independently and communicate through variables in the shared memory. A benefit of this arrangement is that communication is fast and all processors can access the whole data set that is being processed. However, the shared memory is organised in a hierarchy and cores can read and write some portions of memory faster than others. For example, each CPU may have a direct connection to a portion of the installed RAM chips. Access to data stored on these chips is fast, while access to data on other chips is slower because it involves memory controllers belonging to other CPUs. Since the number of cores usually exceeds the number of memory access channels, limited total memory bandwidth remains an obstable to high parallel efficiency (per core) of memory-intensive algorithms. In general, shared memory systems are better for tasks that benefit from the ability of each core to access the whole data set, eliminating the need for cores to share large quantities of data between each other over a network.

In contrast, distributed memory systems consist of multiple computers with independent memory systems that communicate via a network. The advantage of such systems is their independent memories, making them beneficial for parallel computing tasks that do not require significant communication of data between processors. These systems are highly appropriate and useful for LBM simulations, in which the updating of each subdomain requires only a small fraction of the data from adjacent subdomains.

In addition to shared and distributed memory systems that combine conventional processors, systems with specialised co-processors for parallel computing are also available. These co-processors are controlled by a conventional CPU host system and are optimised for performing a particular task. In this chapter, we will consider GPUs, which have hundreds to thousands of arithmetic units that are limited to performing the same operation simultaneously on different data. These devices also have fast access to dedicated memories that are separate from their host system's memory.

Modern clusters combine often all three of these types of parallel processing: They consist of thousands of networked servers each having several multi-core CPUs and one or more GPUs. The sections that follow in this chapter describe programming for the three situations, and readers can combine the techniques into programs suitable for hybrid clusters. Section 13.4.1 introduces the use of OpenMP to facilitate programming for a multi-core multi-CPU environment with shared memory. Use of OpenMP will allow readers to take advantage of the multiple cores that are most likely already available on their laptops, desktop workstations, or servers. Programming for a cluster of servers with a high-speed network is introduced in Sect. 13.4.2. The concepts and code presented in this section can be used on high performance clusters with hundreds to hundreds of thousands of CPUs. Finally, Sect. 13.4.3 introduces the use of GPUs.

13.4.1 Multithreading and OpenMP

On computers with one single-core CPU, multitasking operating systems provide the illusion that several programs are executing simultaneously. Such operating systems achieve this illusion by rapidly switching between programs many times per second: they interrupt the running program, save the state of the CPU, load the state of the CPU for another program, and then resume execution of that program.

On a system with one or more multi-core CPUs, the operating system schedules programs to run truly simultaneously on the available processors. To utilise multiple cores for computational tasks, one may run several programs that communicate with each other. This approach, however, could be inefficient and introduces several programming complexities because operating systems isolate programs to maintain security and allocate resources. Many tasks that benefit from parallel execution, including the computational tasks we are interested in, do not require this isolation and are hampered by it.

Operating systems provide mechanisms for performing multitasking with less overhead than full programs. **Threads** are units of code that, like a program, are scheduled by the operating system to run simultaneously or share time on the available processors. Unlike a full program, they inherit some of the resources of their parent program, such as its memory address space. The common address space is particularly important because it allows threads to share pointers and easily access the same regions of memory.

To facilitate programming and hide the operating system-specific details of setting up and managing multiple threads, the OpenMP Architecture Review Board (ARB) publishes and maintains the specifications of a collection of tools called OpenMP (Open MultiProcessing) for C, C++, and Fortran [7]. The specifications describe a collection of special statements called directives, a library of functions, and a set of environment variables that together describe how code execution is split among threads and how data is shared between them. The OpenMP ARB does not provide implementations of these tools. Instead, compiler authors choose whether to support OpenMP directives and provide the required libraries. Support for OpenMP is included in many compilers, both commercial and free, for a wide range of operating systems and computer architectures. In this chapter, we consider features that are available at least since version 3.1 of OpenMP [8]. At the time of writing, a more recent version with additional features is also available [9].

13.4.1.1 OpenMP Directives

The OpenMP directives that are used to describe when and how threads are created to execute blocks of code are structured as special pre-processor directives in C and C++ and as comments in Fortran. In this section we consider only OpenMP for C and C++. Pre-processor directives are special statements in code that are not a part of the language but rather provide additional information to the compiler. OpenMP directives begin with `#pragma omp`, which is followed by the name of the directive and then a list of options that are called clauses. A long directive, such as one with many clauses, can be split across several lines by using the line continuation character \ at the end of each line that continues on the next line.

This and the following sections introduce the directives and clauses that are essential to using OpenMP and those that are used in a sample parallel LBM code (Sect. 13.4.1.4). Readers may consult the OpenMP specifications [8] or publications about them for a full reference of all features. Many online tutorials, such as [10] are also available.

The most important directive in OpenMP is the **`parallel` directive** that indicates that the subsequent block of code is to be run by several threads. Regions of sequential code, i.e. code that is outside a `parallel` block, are executed by what is called the master thread.

Upon starting the `parallel` block, OpenMP creates additional threads to form a team of threads, all of which execute that block of code. The threads are assigned an identification number starting with zero for the master thread. Logic within the `parallel` block can use the thread's identification number, provided by the function `int omp_get_thread_num()`, to perform a different task in each thread. The number of threads that are executing the block can be obtained with the `int omp_get_num_threads()` function. Listing 13.19 provides a minimal example of the use of these functions and the `parallel` directive. Upon reaching the end of the `parallel` section, the master thread by default waits until all threads finish and then continues with the sequential code after the `parallel` block.

Listing 13.19 Example code showing the use of the **`parallel`** directive

```
// ...
// sequential code

#pragma omp parallel
{
  // block of code to be executed by several threads

  // get identification number of thread
  int id = omp_get_thread_num();

  // get number of threads executing block
  int threadcount = omp_get_num_threads();

  // determine task to perform
  // based on id and threadcount
}

// sequential code
// ...
```

The number of threads used to execute a `parallel` section may be specified in several ways. In order of precedence from highest to lowest, the methods are:

1. the `num_threads` clause of the `parallel` section,
2. the number specified by the most recent call to `void omp_set_num_threads(int)`,
3. the value of the `OMP_NUM_THREADS` environment variable, and finally
4. an implementation-specific default that is often the number of available processors.

Since it is unlikely that a specific number of threads is best for all systems that a program may run on, the use of the `num_threads` clause is generally not advised. It is more convenient to be able to specify the number of threads at the time the code is executed either by using the environment variable `OMP_NUM_THREADS` or through a configuration option of the program (perhaps a command line option or an option in a configuration file) and a call to `omp_set_num_threads(int)`.

A multithreaded version of the sequential LBM code (Sect. 13.3.2) could now be implemented using only the `parallel` directive. For example, the `for` loops that iterate over the nodes in the domain could be modified so that each thread simultaneously updates a portion of the domain as determined by its identification number. However, it is more convenient and elegant to use specialised directives to achieve this task.

> The **`parallel for` directive**, shown in Listing 13.20, automates the allocation of iterations of `for` loops to different threads. The `omp parallel for` directive is a shortcut for an `omp for` directive inside a `parallel` block.

Listing 13.20 Example of OpenMP **`parallel for`** directive

```
#pragma omp parallel for
for(int n = 0; n < N_MAX; ++n)
{
  // iterations executed by different threads
}
```

Like specifying the number of threads for a `parallel` block, clauses or environment variables can be used to indicate the way that the iterations of the `for` loop are split among the threads. With these options, programmers can ensure that the threads are allocated an equal share of the work. When work is shared evenly, processor time is not wasted waiting for some threads to finish. The clauses for the three types of scheduling are:

- `schedule(static, chunk)` where `static` is the name of the scheduling method (in this case the literal text `static`) while `chunk` must be an integer.

This must be an explicitly-written integer, not a variable. With `static` scheduling, the threads are allocated `chunk` iterations in order by their identification number. This allocation can be imagined to be the same as dealing cards: `chunk` iterations are "dealt" to each thread until none remain. Naturally, the last set of iterations dealt to a thread will be smaller than `chunk` if the total number of iterations is not a multiple of `chunk`. If `chunk` is not specified, the iterations are split as evenly as possible among the threads. This type of scheduling is useful when the workload in each iteration is known to be equal, in which case the threads can be expected to finish their work in the same amount of time when given the same number of iterations to perform.

- `schedule(dynamic,chunk)` In `dynamic` scheduling, each thread first performs the number of iterations specified by `chunk`. When a thread completes its initial allocation, it is assigned the next set of `chunk` iterations until all have been completed. The default value for `chunk` in this case is one. This type of scheduling is useful when workload varies between iterations because it allows threads that happen to be assigned quick iterations to perform more of them.
- `schedule(guided,chunk)` This type of scheduling is similar to `dynamic` scheduling, but the number of iterations allocated to a thread that completes its initial allocation decreases over time, staying proportional to the number of iterations that remain to be performed. For this scheduling option, the parameter `chunk` specifies the minimum number of iterations that can be allocated to a thread and defaults to one.

Another possible clause, `schedule(auto)` defers selection of the iteration allocation scheme to the compiler or OpenMP libraries. For our purposes, `static` scheduling is sufficient because every iteration of the outermost `for` loops that we will parallelise performs an equal amount of computations. If the `schedule(runtime)` clause is used, the `OMP_SCHEDULE` environment variable determines how the threads execute the iterations of the loop. The format of the `OMP_SCHEDULE` variable is a string with the name of the scheduling method optionally followed by a comma and the chunk size. The third alternative for setting the scheduling method is the `void omp_set_schedule(omp_sched_t,int)` function. As for setting the number of threads, the clause has the highest precedence, followed by the function call, and the environment variable has lowest precedence.

Now that we have seen how to designate blocks of code that will be executed in parallel by several threads, we need to consider how the threads use the variables that they share access to and how data is shared between parallel and sequential sections of code. This is the topic of the next section, Sect. 13.4.1.2.

13.4.1.2 Data Sharing

The `parallel` and `parallel for` directives presented in Sect. 13.4.1.1 accept clauses that specify how threads handle variables that all threads can access because they are declared outside a `parallel` block. Wrong specifications are a common reason for incorrect behaviour of a multithreaded OpenMP program, and therefore this topic deserves special attention.

Each thread has its own copy of every variable that is declared inside a `parallel` block. These variables cannot be accessed after the block ends, and the data in them is lost. Two options are available for variables that are declared before a `parallel` region and are used inside it: the threads may all share access to the one copy of that variable that is used in the sequential regions of the code, or they may each have their own copy. Variables are designated as being shared among threads with the `shared(variable_list)` clause, where `variable_list` is a comma separated list of variable names. When variables are shared, threads may read or write to these variables at any time. Therefore care must be taken to ensure that the memory accesses occur in the correct order.

Consider, for example, the statement `x = x+1` as it is being executed by two threads with `x` being a shared variable. The outcome is ill-defined. The two threads might initially read the same value then both write the same result, or one thread might read `x` after the other has already written it, causing the variable to be incremented twice. To specify the intended outcome of such situations, OpenMP has a `#pragma omp critical` directive that defines sections of code than can only be executed by one thread at a time. Threads that reach a `critical` section while another is executing it must wait until the executing thread finishes. An example that shows the use of this directive is given in Listing 13.21.

Listing 13.21 Example showing the use of the `critical` directive. This code computes the sum of the threads' identification numbers.

```
int x = 0;
#pragma omp parallel shared(x)
{
  int id = omp_get_thread_num();
  #pragma omp critical
  {
    x = x+id;
  }
}
printf("x = %d\n",x);
```

The `private(variable_list)` clause is the second option for variables declared outside a `parallel` block. Threads have their own versions of all variables in the `variable_list`, and all these variables must be initialised in every thread. Alternatively, variables may be listed in a `firstprivate(variable_list)` clause to request that the value in each thread be automatically initialised with the value of the variable prior to entering the `parallel` section. These `private` variables can be thought of as entirely different variables that only happen to share the same name as a variable outside the `parallel` block. Any data stored in `private` variables will not affect the variable with the same name that exists outside the `parallel` block—this variable is unchanged by the execution of the `parallel` block.

For convenience, one may also specify that variables not listed in other clauses are by default `shared` using the clause `default(shared)`. However, this is not recommended because it may lead to an incorrect classification for an overlooked variable. Instead, beginners are advised to use `default(none)` which requires each variable used inside the `parallel` block to be explicitly listed in a data sharing clause. If a variable is accidentally omitted, compilation will end with an error, forcing the programmer to evaluate how that variable should be handled.

The data sharing clauses of `parallel for` directives are the same as those for `parallel` directives together with two additional clauses. The first is the `lastprivate(variable_list)` clause that indicates that variables will be treated as private in all threads but the value computed in the final iteration of the loop will be saved to the corresponding variable with the same name outside the `parallel for` block. In other words, after execution of the `for` loop in parallel, the variable has the same value as it would have if the loop had been executed sequentially. Note that the final iteration is not necessarily the iteration that is executed chronologically last by the threads: the thread that is allocated the final iteration might finish earlier than other threads.

The second additional clause that is available in `parallel for` directives simplifies the implementation of many algorithms. In a simulation, one might need to compute perhaps the sum, product, or minimum/maximum of a quantity at every node in the domain, thereby summarising or "reducing" the values at every node to one value for the whole domain. Though one could write the required code to perform this task with the previously-presented OpenMP concepts, use of the `reduction` memory sharing method simplifies the task and helps avoid common errors.

The syntax for this second clause is `reduction(op:variable_list)` where `op` is one of the operators: `+`, `*`, `-`, `&`, `|`, `^`, `&&`, `||`, `min`, or `max`. The meanings of these operators are the same as the corresponding C/C++ arithmetic, bitwise, and logical operators. The latter two operators return the minimum or maximum value, respectively. For each variable listed in `variable_list`, every thread receives a private variable that is initialised appropriately for the chosen reduction operator (0 for `+` and 1 for `*`; the required initialiser for each operator is intuitive and readers may consult the OpenMP standard [8] for the details for other operators). At the end of the execution of the `parallel for` section, the value of the variable that can be used in the subsequent sequential code (like a shared variable) is computed by applying the specified operator to the initial value of the variable (before the start of the `parallel for` section) and all the private variables of each thread. An example showing the computation of $1 + \sum_{k=1}^{10} k$ and $2\prod_{k=1}^{10} k$ is shown in Listing 13.22. Note that reductions involving floating point arithmetic that are performed in parallel will not necessarily provide exactly the same result as a sequential reduction due to rounding error and the different order of evaluation.

Listing 13.22 Example showing the use of reduction operations with OpenMP

```
int sum = 1;
int product = 2;

#pragma omp parallel for default(none) \
   reduction(+:sum) reduction(*:product)
for(int y = 1; y <= 10; ++y)
{
  sum = sum + y;
  product = product*y;
}

printf("sum = %d\n",sum);
printf("product = %d\n",product);
```

13.4.1.3 Compiling and Running OpenMP Code

In GCC, support for OpenMP directives and linking with OpenMP libraries is enabled with the `-fopenmp` compiler flag. To compile a program contained in only one C++ source file, a possible command is

```
g++ -Ofast -fopenmp simulation.cpp -o sim
```

When the source code for a program is split among several files, all files that use OpenMP directives must be compiled with the `-fopenmp` option and the command that links the object files must also have `-fopenmp` to link with the OpenMP libraries. For example, the commands

```
g++ -Ofast -fopenmp source_openmp.cpp -o source_openmp.o
g++ -Ofast source_no_openmp.cpp -o source_no_openmp.o
g++ -fopenmp source_openmp.o source_no_openmp.o -o sim
```

compile one C++ file with OpenMP directives (`source_openmp.cpp`) and one without (`source_no_openmp.cpp`) then link the resulting object files and OpenMP libraries to generate the executable `sim`.

As discussed in the preceding sections (Sect. 13.4.1.1 and Sect. 13.4.1.2), several environment variables can be used to affect the behaviour of programs that use OpenMP. For reference, these may be set on the command line or in a shell script by using the command

```
export OMP_ENV_VAR=value
```

in the Bourne family of shells, which includes `sh` and `bash`, or the command

```
setenv OMP_ENV_VAR value
```

for C shells (`csh` and `tcsh`). In both these examples, `OMP_ENV_VAR` is the name of an OpenMP environment variable such as `OMP_NUM_THREADS` or `OMP_SCHEDULE`, and `value` is its new value. With the environment variables set as desired, the program may then be run as usual:

```
./sim
```

OpenMP codes can also be run on clusters with many more cores than typical workstations. For more information on using the scheduling software that manages resources on such clusters and find out how to request resources and schedule simulations for execution, see Sect. 13.4.2.6.

13.4.1.4 Multithreaded LBM Implementation

Modifying the previous simulation code (Sect. 13.3.2) to support multithreading with OpenMP is reasonably straightforward. First, we `#include` the `omp.h` header file that declares the OpenMP functions and variables. We then add some informative output about the OpenMP runtime environment to the `main` function:

```
printf("OpenMP information\n");
printf("  maximum threads: %d\n", omp_get_max_threads());
printf("       processors: %d\n", omp_get_num_procs());
printf("\n");
```

Next, we indicate which `for` loops will be executed by several threads. The most important loop to parallelise in this way is the loop that updates the populations f_i every time step in the `stream_collide_save` function:

```
void stream_collide_save(double *f0, double *f1, double *f2,
                         double *r, double *u, double *v,
                         bool save)
{
  const double tauinv = 2.0/(6.0*nu+1.0); // 1/tau
  const double omtauinv = 1.0-tauinv;     // 1 - 1/tau

  #pragma omp parallel for default(none) \
      shared(f0,f1,f2,r,u,v,save) schedule(static)
  for(unsigned int y = 0; y < NY; ++y)
  {
    for(unsigned int x = 0; x < NX; ++x)
    {
      // same code as in serial version
    }
  }
}
```

All the pointers and the boolean variable have been specified as `shared` because the same values need to be visible to all threads. Alternatively, these variables could also have been declared as `firstprivate` because their values are never changed. The performance implications of this choice are left to interested readers to investigate. Static scheduling is selected because the time to complete of each iteration of the loop is expected to be nearly the same though caching issues may cause some variation between iterations. Note that the `const` variables `tauinv` and `omtauinv` do not require a data sharing specification. Since they cannot be written to, they are automatically treated as shared. Note also that because the pointer variables themselves are not changed (only the memory at the location they point to is altered) they could be declared `const` in the function declaration. With this qualifier, the type of data sharing required for the pointer variables would not need to be indicated.

With only a single OpenMP directive, you can **parallelise the main domain update loop** of your simulation and make use of the multiple cores available on most computers.

Exercise 13.11 Of the two nested `for` loops, which is better to parallelise and why?

Exercise 13.12 Instead of allocating multiples of complete rows of the simulation domain to each thread, write an OpenMP program in which each thread updates a fraction of the domain in both dimensions (i.e. two dimensional domain decomposition instead of one dimensional decomposition). Does this strategy increase the speed of the simulation? Does your conclusion depend on the size of the simulation domain and the number of cores and CPUs on your system?

Exercise 13.13 Try different scheduling methods for the threads and vary their `chunk` sizes. Which method offers the best performance? How does performance relate to cache use? Does synchronisation, i.e. `#pragma omp barrier`, at the end of each row help prevent one core from getting too far ahead of others?

Though parallelisation of additional loops does not have a significant impact on the performance of the whole program, we do so to illustrate several features. The function that initialises the particle populations, `init_equilibrium`, runs only once, and therefore decreasing its runtime provides a negligible improvement in the total runtime of the program. Similarly, the functions that compute the exact solution and the error in the numerical solution (`taylor_green` and `compute_flow_properties`, respectively) run infrequently compared to the main domain update function. The outer `for` loops in `taylor_green` and `init_equilibrium` are parallelised identically to the `for` loop of `stream_collide_save`: all variables are shared and the scheduling is static.

The parallelisation of the `compute_flow_properties` function illustrates the use of `reduction` variables. For reference, the complete multithreaded version of this function is shown in Listing 13.23. Seven variables in this function are used to compute cumulative sums of quantities at every position in the domain. These variables are therefore listed in a `reduction` clause with the + operator. Each thread has a set of private variables, all automatically initialised to zero, that are used to compute the required sums for the nodes handled by each thread. The remainder of the function is the same as in the sequential case (c.f. Listing 13.18).

Listing 13.23 Multithreaded version of the `compute_flow_properties` function that illustrates the use of `reduction` variables

```
void compute_flow_properties(unsigned int t,
                             double *r, double *u, double *v,
                             double *prop)
{
  // prop must point to space for 4 doubles:
  // 0: energy
  // 1: L2 error in rho
  // 2: L2 error in ux
  // 3: L2 error in uy

  double E = 0.0; // kinetic energy

  double sumrhoe2 = 0.0; // sum of error squared in rho
  double sumuxe2  = 0.0; //                         ux
  double sumuye2  = 0.0; //                         uy

  double sumrhoa2 = 0.0; // sum of analytical rho squared
  double sumuxa2  = 0.0; //                   ux
  double sumuya2  = 0.0; //                   uy

  #pragma omp parallel for default(none) shared(t,r,u,v) \
      reduction(+:E,sumrhoe2,sumuxe2,sumuye2, \
                  sumrhoa2,sumuxa2,sumuya2) \
      schedule(static)
  for(unsigned int y = 0; y < NY; ++y)
  {
    for(unsigned int x = 0; x < NX; ++x)
    {
      double rho = r[scalar_index(x,y)];
      double ux  = u[scalar_index(x,y)];
      double uy  = v[scalar_index(x,y)];

      E += rho*(ux*ux + uy*uy);

      double rhoa, uxa, uya;
      taylor_green(t,x,y,&rhoa,&uxa,&uya);

      sumrhoe2 += (rho-rhoa)*(rho-rhoa);
      sumuxe2  += (ux-uxa)*(ux-uxa);
      sumuye2  += (uy-uya)*(uy-uya);

      sumrhoa2 += (rhoa-rho0)*(rhoa-rho0);
      sumuxa2  += uxa*uxa;
      sumuya2  += uya*uya;
    }
```

```
    }

    prop[0] = E;
    prop[1] = sqrt(sumrhoe2/sumrhoa2);
    prop[2] = sqrt(sumuxe2/sumuxa2);
    prop[3] = sqrt(sumuye2/sumuya2);
}
```

13.4.1.5 Performance Results

Figure 13.6 presents the speed of the OpenMP version of the simulation code. These simulations were run on the same quad-core Intel Xeon W3550 CPU that was used for the single core benchmarks (Sect. 13.3.2). To avoid competition for resources with other running programs, the system was first restarted and then only the simulations were run. Representative results are shown for four domain sizes, and each point is an average of three runs. For reference, a dashed line shows ideal parallel performance for which the speed would be proportional to the number of threads.

The actual improvement achieved by increasing the number of threads is lower than in the ideal case. Understanding the performance of multithreaded code on shared memory systems is complex because it depends on many system-specific details such as:

- the number of CPUs present,
- the number of cores per CPU,
- cache sizes,
- the sharing of caches between cores,

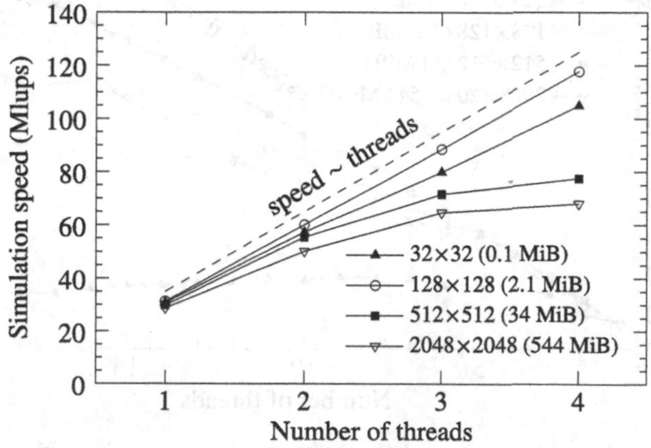

Fig. 13.6 Speed of multithreaded simulations with several domain sizes using 1 to 4 threads running on a quad-core Intel Xeon W3550 CPU

- the sharing of data between caches to avoid access to RAM,
- the number of memory channels between the CPUs and RAM, and
- how the operating system schedules the threads on the available cores and CPUs.

While the details of how performance depends on the domain size and number of threads are architecture-dependent, a generalisation is possible for memory-intense algorithms like LBM simulations: on typical contemporary systems, the available memory bandwidth will be exhausted when simulating large domains before the number of threads equals the number of cores.

In Fig. 13.6, we see that the speed increases nearly ideally for the two smaller domain sizes, but for the two larger domain sizes, the use of four threads provides only a minimal improvement over three threads. The loss of performance appears to coincide with the domain size exceeding the capacity of the L3 cache, which is shared by all cores on the CPU that was used. The lower speed of the smallest domain (32×32) compared to the next smallest (128×128) suggests an inefficiency in the sharing of data between the cores' caches. For comparison, Fig. 13.7 presents the speed of simulations run with 1 to 20 cores on a system with two 10-core processors.

Exercise 13.14 On multi-core and multi-processor systems, operating systems may allocate threads to run on different cores over time. Investigate how to force your operating system to always schedule each thread on a particular core. Does this improve performance? Why?

Exercise 13.15 Look up and compare the characteristics of the caches in the processors used for Figs. 13.6 and 13.7. Determine the sizes of the caches and how they are shared by the cores of each CPU. What roles do caches play in determining

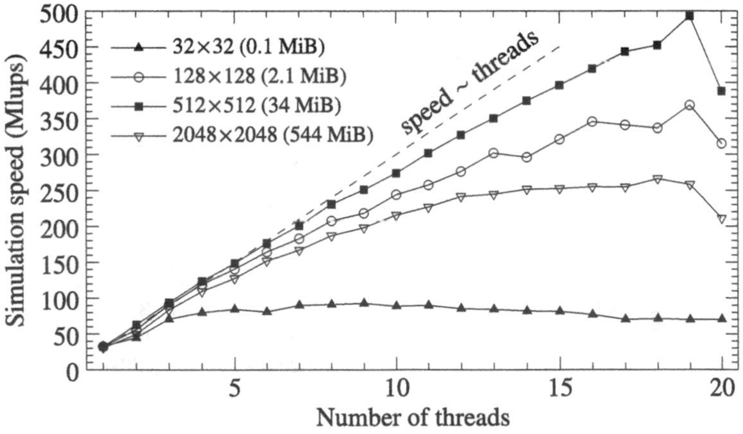

Fig. 13.7 Speed of multithreaded simulations with several domain sizes using 1 to 20 threads running on a system with two 10-core Intel Xeon E5-2660 v2 CPUs

how speed depends on domain size (memory transferred per domain update) in multithreaded programs?

A detailed discussion of the reasons for the observed performance of the multithreaded simulations is beyond the scope of this section. To reach higher speeds, we require multiple CPUs that do not compete for memory bandwidth. This is commonly achieved by connecting many otherwise independent computers, such as small form factor server-type units, through a high-speed and low-latency network. Programming such systems is the subject of Sect. 13.4.2.

13.4.2 Computing Clusters and MPI

Computing clusters can be conceptually understood as collections of many computers, often called nodes,[7] that communicate with each other but are otherwise independent. The need for high data transfer rates and low latencies, which are the delays in starting a data transfer, has led to the development of specialised interconnect hardware that transfers data between the nodes of a cluster. The most common choice of interconnect on contemporary supercomputers, InfiniBand, provides data transfer rates exceeding 10 Gbps (10^{10} bits per second) and latencies on the order of one microsecond. Since a variety of vendors provides InfiniBand and other high-performance interconnects for clusters, programming such systems and adapting code to run on different systems would be rather complex in the absence of a standard programming interface.

The Message Passing Interface (MPI) standard describes a collection of protocols, data structures, and routines for the processes in a parallel program to exchange data, called messages, with each other and synchronise their operations. Developed by the MPI Forum [11], this standard describes methods for performing these tasks that hide the details of the underlying hardware devices and operating systems that actually carry them out. Code written using MPI is therefore highly portable, and when written carefully, it can efficiently use the resources of an individual shared memory machine, a small cluster of computers, or the world's largest supercomputing clusters [12].

Since the first version of MPI (MPI-1.0) was released in 1994, the standard has undergone many revisions and updates to add new features that facilitate the efficient use of continually advancing hardware capabilities. The most recent version is MPI-3.1 (June 2015). In this section, however, we consider only features available since MPI-1, specifically version MPI-1.3 [13] from 2008. Implementations of this version are widely available, and it is sufficient for demonstrating the core capabilities of MPI and the use of clusters for parallel computing.

[7]Where the term "node" is potentially ambiguous, we use the more specific terms "computing node" and "lattice node" for clarity.

Many implementations of MPI are available from hardware and software vendors as well as open source versions by several groups. This chapter uses only standard MPI features, and the code that is presented has been tested with the open source Open MPI implementation [14]. This chapter does not cover the installation and configuration of clusters or MPI, and we assume readers have access to a working cluster. Interested readers without access to a company or academic cluster can try the code examples by purchasing resources from commercial cloud computing providers, such as Amazon[8].

In addition to the original reference documents from the MPI Forum [15], documentation is available from Open MPI [16], many online tutorials (such as [17]), and books (such as [18]).

13.4.2.1 MPI Concepts

When working with MPI, we write a single program that we then run numerous times simultaneously on one or more computers. Each of the processes is assigned an identification number called a rank, and it performs its own subset of the required calculations and coordinates its work and shares data with other processes by communicating with them over the interconnect. The MPI standard defines a set of functions, constants, and data structures [13] that assist the development of such a program. All names defined by MPI begin with `MPI_`. The functions perform a wide variety of tasks and can be divided into two main categories: functions for communication and functions for querying and modifying the state of MPI systems. The communication functions are further divided into two subcategories, point-to-point and collective communications, depending on how the processes that communicate are connected with each other. Of these two types of communications functions, point-to-point functions involve the sending and receiving of messages between two processes. Collective communication operations involve data transfers between a group of processes, such as sending data from one process to all others or receiving data in one process from the others. A sketch of the two types of communication methods is shown in Fig. 13.8, and concrete examples of the communications functions will be presented in Sect. 13.4.2.2.

The MPI functions for communication between processes include a parameter that selects the processes that will be involved in the data transfers. When an MPI program is launched on a cluster, the MPI-capable executable is started by many processes scattered across the cluster's nodes. In MPI-1, the number of running processes is constant and specified at launch time, while in later versions the number of processes can change during execution. The processes running an MPI program may be grouped to help manage work sharing and communication. For example, a simulation domain could be split into a multi-level hierarchy of subdomains, and some data might only need to be shared among the processes that handle a particular

[8]http://aws.amazon.com/hpc/

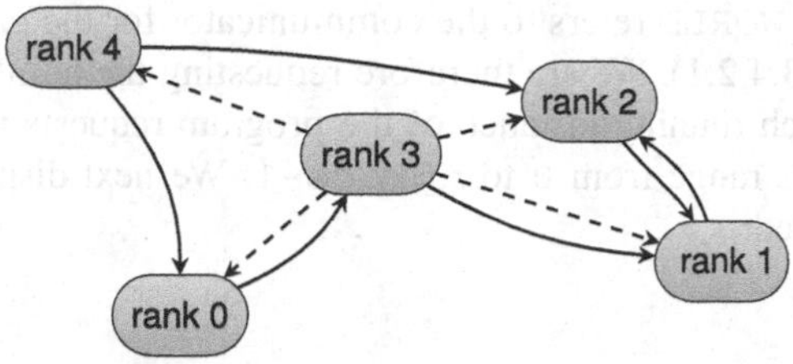

Fig. 13.8 Five processes, each identified by a unique rank, exchange information between each other using point-to-point communications (*solid arrows*) and a collective communication (*dashed arrows*). The *arrows* show the directions of the data transfers

subdomain. In this chapter we use a fixed number of processes and do not perform any grouping of them. In the MPI communication functions, we must therefore specify that we want communication to occur between all processes. The necessary parameter is a reference to an MPI object called a communicator that manages communications between the processes in a group. The required communicator for our purposes is the built-in communicator named `MPI_COMM_WORLD` that requests communication among all the processes in the program's "world."

The sections that follow present the MPI functions that are relevant to parallelising the LBM code and related MPI functions that are also available. Most MPI functions return error codes. Though useful for debugging, these values are discarded in the example code.

13.4.2.2 MPI LBM Implementation

We now proceed to modify the sequential version of the Taylor-Green vortex decay LBM code (Sect. 13.3.2) to create a parallel version using MPI.

The first step for implementing the MPI version of the LBM code is to add `#include<mpi.h>` at the beginning of any source files that use MPI. This loads the declarations for MPI functions and variables. Next, at the start of the `main` function, we initialise the MPI library with `MPI_Init` (this must be matched by a call to `MPI_Finalize(void)` in every process at the end of the program) and then save both the rank of the process and the number of processes that are running as follows:

```
int rank, nprocs;

// initialise MPI
MPI_Init(&argc,&argv);

// save rank of this process
MPI_Comm_rank(MPI_COMM_WORLD,&rank);

// save number of processes
MPI_Comm_size(MPI_COMM_WORLD,&nprocs);
```

Here `MPI_COMM_WORLD` refers to the communicator for the group of all running processes (Sect. 13.4.2.1). We are therefore requesting the number of processes in this group, and each running instance of the program requests its rank within this group. These ranks range from `0` to `nprocs-1`. We next display the number of processes being used:

```
if(rank == 0)
{
  printf("Simulating Taylor-Green vortex decay\n");
  // ...
  printf("\n");
  printf("MPI information\n");
  printf("          processes: %d\n",nprocs);
  printf("\n");
}
```

Without the initial conditional statement `if(rank == 0)` the output would be unnecessarily repeated by each process. The MPI tools used to launch the processes also manage their output, combining it so that users can see (or save to a file) the output from all processes. We next synchronise all the processes with the function call

```
MPI_Barrier(MPI_COMM_WORLD);
```

In each process, this function does not return until all processes have called it. This synchronisation can be thought of as a barrier because no process may proceed until all processes reach this point in the code. The reason for the synchronisation here is cosmetic. Without it, all processes with nonzero rank would skip the initial display statements and continue, causing their output to appear in between the initial output from rank 0.[9]

Next, we use each process's rank and the total number of processes to calculate which portion of the domain each process will handle. For simplicity, we consider only one dimensional domain decomposition into subdomains consisting of several complete rows of the domain along the x direction. Two variables keep track of the allocated portion: `rank_ny` stores the number of rows and `rank_ystart` is the y coordinate of the first row. The simulation domain is split as evenly as possible. If the size of the domain in the y direction is not divisible by the number of processes, each process with rank less than the remainder is assigned one additional row. An

[9]Depending on how the MPI implementation combines the output from the different processes, this synchronisation might not have the desired effect. Later output from rank 1, for example, might appear before any output from rank 0. If it is essential for the order of output to be synchronised, the data to be output from all ranks can be sent to one rank that displays it all in the correct order.

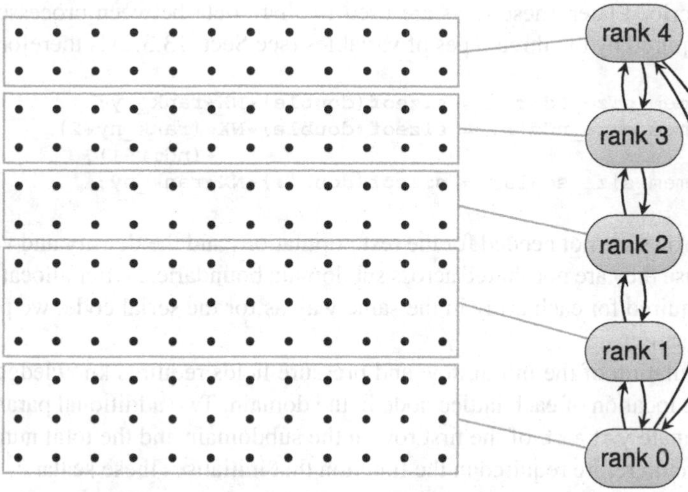

Fig. 13.9 An example showing how a periodic domain of 12 × 12 nodes would be split among five processes. Also shown is the point-to-point communication between "adjacent" processes (see Sect. 13.4.2.3)

example of such a split is shown in Fig. 13.9. The code for computing `rank_ny` and `rank_ystart` is:

```
if(rank < NY % nprocs) // ranks that are less than remainder
{
  rank_ny = NY/nprocs+1;
  rank_ystart = rank*rank_ny;
}
else
{
  rank_ny = NY/nprocs;
  rank_ystart = NY-(nprocs-rank)*rank_ny;
}
printf("Rank %d: %d nodes from y = %d to y = %d\n",
       rank,rank_ny,rank_ystart,rank_ystart+rank_ny-1);
```

In this domain decomposition strategy, the processes are assigned consecutive blocks of rows in order by their ranks. The final line of the code shows which rows of the simulation domain were allocated to each rank. It is important to note that in general the output will not appear in order by rank. Even on a cluster with identical computational nodes, the processes may run at different speeds for various reasons including the effects of other users' tasks running simultaneously on the same node or even the temperature of the system (CPUs may decrease their clock rates when they heat up to avoid overheating).

With the number of rows for each rank determined, the amount of memory required can now be calculated and allocated. Each process allocates two additional rows, one below the assigned portion of the simulation domain, and one above. As will be described later, these rows are used to share data between processes. The memory required by the three types of variables (see Sect. 13.3.2) is therefore:

```
size_t mem_size_0dir   = sizeof(double)*NX*rank_ny;
size_t mem_size_n0dir  = sizeof(double)*NX*(rank_ny+2)
                                         *(ndir-1);
size_t mem_size_scalar = sizeof(double)*NX*rank_ny;
```

Additional rows are not needed for the rest populations and the density and velocity fields because they are not shared across subdomain boundaries. After allocating the memory required for each array in the same way as for the serial code, we proceed with initialisation.

The calculation of the initial flow and pressure fields requires knowledge about the absolute location of each lattice node in the domain. Two additional parameters, the y coordinate `ystart` of the first row in the subdomain and the total number of rows `ny`, are therefore required in the function that initialises these scalars:

```
void taylor_green(unsigned int t,
                  double *r, double *u, double *v,
                  unsigned int ystart, unsigned int ny)
{
  for(unsigned int y = 0; y < ny; ++y)
  {
    for(unsigned int x = 0; x < NX; ++x)
    {
      taylor_green(t,x,ystart+y,
                   &r[scalar_index(x,y)],
                   &u[scalar_index(x,y)],
                   &v[scalar_index(x,y)]);
    }
  }
}
```

Comparing with Listing 13.10, the only differences are that the loop for `y` counts up to the number of rows handled by each process and `ystart` is added to the value of `y` that is passed to the function that calculates the Taylor-Green flow.

The populations f_i for each lattice node can be computed from the corresponding velocity and density values without knowledge of the absolute positions of the nodes. Therefore only the number of rows handled by the process is added as a parameter to the initialisation function:

```
void init_equilibrium(double *f0, double *f1,
                      double *r, double *u, double *v,
                      unsigned int ny)
{
  for(unsigned int y = 0; y < ny; ++y)
  {
    // ... same code as in serial version
  }
}
```

The omitted code is identical to that in Listing 13.11. However, a detail has been hidden in this implementation. Due to the additional row below each process's subdomain, the memory indexes are computed with `y+1` instead of `y`:

```
inline size_t fieldn_index(unsigned int x, int y,
                           unsigned int d)
{
  return (ndir-1)*(NX*(y+1)+x)+(d-1);
}
```

With this function, the rows handled by each process run from `0` to `rank_ny-1` and the ghost rows, used for communication, that are below and above the internal rows are accessed with the indices `-1` and `rank_ny`, respectively, using `fieldn_index(x,-1,d)` and `fieldn_index(x,rank_ny,d)`. One could alternatively rewrite the `for` loop in the `init_equilibrium` function as

```
for(unsigned int y = 1; y < ny+1; ++y)
{
  // ...
}
```

and leave `fieldn_index` unchanged. However, with the method we employ, the same indices `x` and `y` can be used to refer to the same locations in the particle population arrays as the density and velocity arrays. We leave optimisation of the memory address calculations to the compiler, specifically the avoidance of an extra addition operation to compute the address for each memory access.

With the arrays initialised, we turn our attention now to the main loop of the simulation. Though the saving of the scalar fields and computation of the error are also performed in this loop, we defer discussion of the implementation of these features to Sects. 13.4.2.4 and 13.4.2.5 because they involve special MPI features.

13.4.2.3 Blocking and Non-blocking Communications

Each process can update the populations f_i in rows `y=1` to `y=rank_ny-2` using data that resides in its own memory. However, to update the populations in the bottom (`y=0`) and top (`y=rank_ny-1`) rows, each process requires the incoming populations from rows that are computed by different processes. More specifically, to update its bottom row, each process requires data from the top row of the process that handles the lower subdomain. In the domain decomposition strategy we use, the lower subdomain is handled by the process with rank `rank-1`. Similarly, updating the top row requires data from the bottom row of the process with rank `rank+1`. Due to the periodic domain, the ranks of these adjacent processes must be computed as

```
int rankp1 = (rank+1) % nprocs;
int rankm1 = (nprocs+rank-1) % nprocs;
```

The communication performed between adjacent processes (processes with ranks that differ by one) is shown in Fig. 13.9.

The MPI specification describes two main classes of communication functions. The first class, called **blocking functions**, consists of functions that **wait until after the transfer has been completed** before allowing further execution (i.e. further execution is temporarily *blocked*). Blocking functions for sending data return only once it is safe to modify the memory containing the data to be sent without affecting the transfer. Note that the transfer to the recipient need not have been truly completed: the data may have been only copied to a temporary buffer. On the receiving side, a blocking function that receives data only returns once the transfer is truly finished and the memory containing the received data may be read. Functions in the second class of communication functions, called **nonblocking functions**, only initiate a transfer and return, **allowing execution to proceed** before the transfer is completed (i.e. further execution is *not* blocked).

Though slightly more complicated to use, nonblocking functions offer two advantages. First, they allow calculations that do not depend on the transferred data to be performed during the transfer. Second, programmers do not need to ensure that one process receives while the other sends to avoid deadlock situations in which the programs cannot proceed. These two advantages will be clarified once code that uses the two types of communication functions has been presented.

The two standard blocking communication functions are `MPI_Send` and `MPI_Recv`, which send and receive data:

```
int MPI_Send(const void *buf,
             int count, MPI_Datatype datatype,
             int dest, int tag, MPI_Comm comm)

int MPI_Recv(void *buf,
             int count, MPI_Datatype datatype,
             int source, int tag, MPI_Comm comm,
             MPI_Status *status)
```

In these functions, the first three parameters describe the nature of the transferred data. The source and destination memory locations of the data are specified by `buf`, the type of data transferred is given by `datatype`, and the number of elements of this type is `count`. In `MPI_Send`, `buf` is a pointer to the first variable that will be sent, while in `MPI_Recv` it is the address where the first received variable will be stored. Among other options given in the MPI standard, common choices for the data type are `MPI_INT`, `MPI_UNSIGNED`, `MPI_FLOAT`, and `MPI_DOUBLE`, which respectively match the C data types `int`, `unsigned int`, `float`, and `double`. In the code that follows, we transfer only `double`s and therefore use `MPI_DOUBLE`.

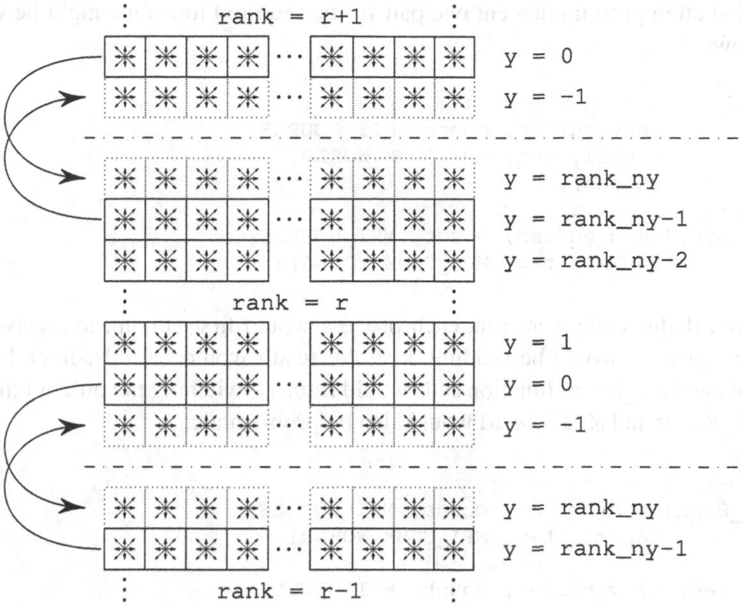

Fig. 13.10 Diagram showing the rows of data that need to be transferred across subdomain boundaries between processes. *Boxes with solid outlines* denote the nodes of the subdomain updated by each rank, while *boxes with dotted outlines* denote the extra rows used to store data from adjacent subdomains that are handled by different ranks

In `MPI_Send`, the parameter `count` specifies the number of variables of the specified type that are sent; in `MPI_Recv` it indicates the maximum number that can be received. In the code we present, the two `counts` are equal.

The next three parameters in the blocking communications functions provide additional information about the transfer. The parameters `dest` and `source` indicate the ranks of the processes within the communicator `comm` to which the data will be sent (`dest` for `MPI_Send`) and from which it will be received (`source` for `MPI_Recv`). On the receiving side, the special source `MPI_ANY_SOURCE` allows receiving data from any rank that is sending data to the receiving rank. In both functions, `tag` may be used by programmers to identify the nature of the transfer: a receive with a particular `tag` can only obtain data that was sent by a call to the sending function with the same `tag`. Alternatively, the tag `MPI_ANY_TAG` allows a function to receive data irrespective of the tag specified on the sending side. Finally, in `MPI_Recv`, `status` is a pointer to an `MPI_Status` structure that the function modifies to provide information about the transfer that was performed.

How can we use these two functions to perform the transfers required by the LBM implementation? In each rank we need to: receive the bottom row from `rankp1`, send the top row to `rankp1`, receive the top row from `rankm1`, and send the bottom row to `rankm1`. Figure 13.10 illustrates the transfers that need to be performed from the point of view of one process.

A first attempt to implement one pair of the required transfers might be written as follows:

```
// receive from above, i.e. rank+1
MPI_Recv(recv_buffer, count, MPI_DOUBLE,
         rankp1, tag, MPI_COMM_WORLD,
         status);
// send below, i.e. rank-1
MPI_Send(send_buffer, count, MPI_DOUBLE,
         rankm1, tag, MPI_COMM_WORLD);
```

However, if this code were run, each process would first attempt to receive data. Since no process would be sending data, the result would be a deadlock because none of the `MPI_Recv` function calls would return. Switching the order of the calls to `MPI_Recv` and `MPI_Send` would also be problematic:

```
// send below, i.e. rank-1
MPI_Send(send_buffer, count, MPI_DOUBLE,
         rankm1, tag, MPI_COMM_WORLD);
// receive from above, i.e. rank+1
MPI_Recv(recv_buffer, count, MPI_DOUBLE,
         rankp1, tag, MPI_COMM_WORLD,
         status);
```

Depending on details of the MPI implementation used, this version might succeed provided that the size of each transfer is smaller than an internal MPI buffer. In this case, each `MPI_Send` returns after copying its data to this temporary buffer. The `MPI_Recv`s may then run and complete the transfers. If, however, the size of the data to be sent exceeds the buffer size, the result is a deadlock: the `MPI_Send`s cannot place all the data into a buffer and they therefore stall, waiting for data to be removed from the buffers. The data is never removed from the buffers because the receiving functions only start after the sending functions finish.

To avoid such deadlocks, we must ensure that some processes receive while others send data. One way to achieve this is to have processes with even ranks first call `MPI_Send` while processes with odd ranks call `MPI_Recv`. Then the roles are reversed, and the even ranks receive data from the odd ranks. Neglecting the details of specifying all parameters, sample code for achieving this is:

```
if(rank % 2 == 0) // even ranks send then receive
{
  // send below, i.e. rank-1
  MPI_Send(send_buffer, count, MPI_DOUBLE,
           rankm1, tag, MPI_COMM_WORLD);
  // receive from above, i.e. rank+1
  MPI_Recv(recv_buffer, count, MPI_DOUBLE,
           rankp1, tag, MPI_COMM_WORLD,
           status);
}
else // odd ranks receive then send
```

```
{
  // receive from above, i.e. rank+1
  MPI_Recv(recv_buffer, count, MPI_DOUBLE,
         rankp1, tag, MPI_COMM_WORLD,
         status);
  // send below, i.e. rank-1
  MPI_Send(send_buffer, count, MPI_DOUBLE,
         rankm1, tag, MPI_COMM_WORLD);
}
```

MPI provides a simpler way to carry out the required transfers. The `MPI_Sendrecv` function combines a send and receive operation and is implemented within the MPI library in a way that prevents deadlocks. This function's parameters are those of `MPI_Send` (excluding the communicator) followed by those of `MPI_Recv`:

```
int MPI_Sendrecv(const void *sendbuf,
                 int sendcount, MPI_Datatype sendtype,
                 int dest, int sendtag,
                 void *recvbuf,
                 int recvcount, MPI_Datatype recvtype,
                 int source, int recvtag,
                 MPI_Comm comm, MPI_Status *status)
```

Using this function, the body of the main simulation loop starts with

```
MPI_Sendrecv(&f1[fieldn_index(0,rank_ny-1,1)],
             transfer_doubles,MPI_DOUBLE,
             rankp1,rank,
             &f1[fieldn_index(0,-1,1)],
             transfer_doubles,MPI_DOUBLE,
             rankm1,rankm1,
             MPI_COMM_WORLD,MPI_STATUS_IGNORE);

MPI_Sendrecv(&f1[fieldn_index(0, 0,1)],
             transfer_doubles,MPI_DOUBLE,
             rankm1,rank,
             &f1[fieldn_index(0,rank_ny,1)],
             transfer_doubles,MPI_DOUBLE,
             rankp1,rankp1,
             MPI_COMM_WORLD,MPI_STATUS_IGNORE);

stream_collide_save(f0,f1,f2,rho,ux,uy,save,rank_ny);
```

In this code, `transfer_doubles` is the number of `doubles` that are sent and received, which is initialised prior to the `for` loop as

```
size_t transfer_doubles = (ndir-1)*NX;
```

First each rank sends the top row of its subdomain "upward" to the process with rank `rankp1` and receives this row from rank `rankm1`, storing it in the row below the subdomain (`y=-1`). Next the bottom row (`y=0`) is sent downward to `rankm1` and the bottom row from `rankp1` is received and stored in the row above the subdomain (`y=rank_ny`). The tags for these transfers are the ranks of the sending processes.

Once these transfers have completed, the whole subdomain is updated. The only changes required in `stream_collide_save()` are an the additional parameter and a modified `for` loop, both changed exactly as for `init_equilibrium` (Listing 13.4.2.2). Since the subdomains, unlike the whole domain, are not periodic, the expressions for the row indices `yp1` and `ym1` are simplified to `y+1` and `y-1`, respectively.

Note that for simplicity we transfer more data than is necessary: only the populations whose directions cross subdomain boundaries actually need to be transferred. One way to avoid the unnecessary transfers is to copy the required populations into a temporary buffer, send this buffer, and then load the transferred populations into their corresponding locations on the receiving side. We leave implementing this and assessing the performance benefits as an exercise for interested readers.

The code that accompanies this book also includes an MPI version of the simulation code that uses nonblocking communications functions. In this version, we use the nonblocking alternatives to `MPI_Send()` and `MPI_Recv()`, which are `MPI_Isend()` and `MPI_Irecv()`:

```
// Start a nonblocking send
int MPI_Isend(const void *buf,
              int count, MPI_Datatype datatype,
              int dest, int tag,
              MPI_Comm comm, MPI_Request *request)

// Start a nonblocking receive
int MPI_Irecv(void *buf,
              int count, MPI_Datatype datatype,
              int source, int tag,
              MPI_Comm comm, MPI_Request *request)
```

The parameters are the same as those for the corresponding blocking versions, but exclude a pointer to an `MPI_Status` structure and include a pointer `request` to an `MPI_Request` structure that is used to identify initiated transfers, query their status, and wait for their completion.

These nonblocking functions return as soon as they have completed whatever tasks are needed to initiate the transfers and fill in the fields of the `MPI_Request` structure. To use the nonblocking communication functions, we first create an array of four `MPI_Requests` that are used to identify the four transfers that are subsequently performed, and also an array of four `MPI_Status` objects for the status of each transfer once it is completed:

```
MPI_Request reqs[4];
MPI_Status  stats[4];
```

The nonblocking version of the simulation's main `for` loop starts by initiating the required transfers:

```
MPI_Isend(&f1[fieldn_index(0,rank_ny-1,1)],
          transfer_doubles,MPI_DOUBLE,
          rankp1,rank,  MPI_COMM_WORLD,&reqs[0]);
MPI_Irecv(&f1[fieldn_index(0,-1,1)],
          transfer_doubles,MPI_DOUBLE,
          rankm1,rankm1,
          MPI_COMM_WORLD,&reqs[1]);

MPI_Isend(&f1[fieldn_index(0,0,1)],
          transfer_doubles,MPI_DOUBLE,
          rankm1,rank,
          MPI_COMM_WORLD,&reqs[2]);
MPI_Irecv(&f1[fieldn_index(0,rank_ny,1)]
          transfer_doubles,MPI_DOUBLE,
          rankp1,rankp1,
          MPI_COMM_WORLD,&reqs[3]);
```

Each call uses its own `MPI_Request` structure in the array `reqs`. With non-blocking functions unlike blocking functions, the order of the calls is irrelevant and deadlocks do not occur if the order is changed.

We may now overlap computations with the these communication operations, and update of the interior rows (`y` from `1` to `rank_ny-2` inclusive):

```
stream_collide_save(f0,f1,f2,rho,ux,uy,save,1,rank_ny-1);
```

The `stream_collide_save` function in the nonblocking version differs slightly from the one in the blocking version because the entire domain cannot be updated at once. Therefore the nonblocking version has two extra parameters, `ystart` and `yend`, that specify the range of rows to update as `ystart` to `yend-1` inclusive:

```
void stream_collide_save(double *f0, double *f1, double *f2,
                         double *r, double *u, double *v,
                         bool save,
                         unsigned int ystart,
                         unsigned int yend)
{
  for(unsigned int y = ystart; y < yend; ++y)
  {
    for(unsigned int x = 0; x < NX; ++x)
    {
      // ...
    }
  }
}
```

Before updating rows 0 and rank_ny-1, we must ensure that the transfers have finished. This can be achieved in several ways. The first is to use one call to MPI_Wait for every transfer that was initiated. The declaration of this function is

```
int MPI_Wait(MPI_Request *request, MPI_Status *status)
```

This function returns when the transfer identified by request has been completed. If the completed operation involved receiving data, the supplied MPI_Status structure is updated accordingly.

Several variants of MPI_Wait are useful when working with multiple pending transfers. These functions are MPI_Waitany, MPI_Waitsome, and MPI_Waitall. Both MPI_Waitany and MPI_Waitsome wait until at least one transfer has finished; the difference between them is that MPI_Waitany provides status information for only one completed transfer, while MPI_Waitsome provides information for all the completed transfers. The third variant, MPI_Waitall, is the one we use:

```
int MPI_Waitall(int count,
                MPI_Request *array_of_requests,
                MPI_Status *array_of_statuses)
```

The parameter count indicates how many MPI_Requests have been supplied by the pointer array_of_requests, and array_of_statuses is an array of MPI_Statuses that must be able to hold at least count of them. This variant waits until all the transfers complete and updates the MPI_Status array accordingly. We use this function as follows:

```
MPI_Waitall(4,reqs,stats);
```

With the transfers completed, we finally update the bottom and top rows:

```
stream_collide_save(f0,f1,f2,rho,ux,uy,save,0,1);
stream_collide_save(f0,f1,f2,rho,ux,uy,save,rank_ny-1,rank_ny);
```

One subtle but important performance issue deserves particular attention here. Although MPI_Isend and MPI_Irecv initiate the data transfers, MPI implementations do not necessarily continue the transfers concurrently with subsequent computations. In implementations that do not perform the communications automatically in the background (perhaps by the network hardware devices or a separate thread), the communications delay is incurred when MPI_Waitall is called. In this case, communications and computations do not truly overlap, eliminating any possible speed up compared to the blocking version.

To **ensure that nonblocking communications proceed in the background** while computations are performed, we must ensure that some MPI functions are called in between computations to allow the communications to progress. For this purpose we **use the MPI_Testall function**.

`MPI_Testall` performs a similar task as `MPI_Waitall`[10]: it checks the status of the specified transfers. Unlike `MPI_Waitall`, it only indicates if all transfers have been completed and does not wait for them to finish. This function is declared as:

```
int MPI_Testall(int count,
                MPI_Request *array of requests,
                int *flag,
                MPI_Status *array of statuses)
```

The parameters are the same as for `MPI_Waitall` except for `flag`, which the function sets to 0 unless all transfers have been completed. If all the transfers are finished, `flag` is set to 1 and the `MPI_status` variables are set accordingly. We use this function after every row is updated by modifying `stream_collide_save` as follows:

```
void
stream_collide_save_test(double *f0, double *f1, double *f2,
                         double *r, double *u, double *v,
                         bool save,
                         unsigned int ystart,
                         unsigned int yend,
                         int nr,
                         MPI_Request *reqs,
                         MPI_Status *stats)
{
  int com_finished = 0;
  for(unsigned int y = ystart; y < yend; ++y)
  {
    for(unsigned int x = 0; x < NX; ++x)
    {
      //...
    }
    if(com_finished == 0)
    {
      MPI_Testall(nr,reqs,&com_finished,stats);
    }
  }
}
```

Here `nr` is the number of requests, `reqs` points to an array of `MPI_Request` structures, and `stats` points to an array of `MPI_Status` structures. We use this modified function `stream_collide_save_test` instead of `stream_collide_save` after the calls to `MPI_ISend` and `MPI_IRecv` and before `MPI_Waitall`. We still include `MPI_Waitall` in case the transfers are not completed during the computations. The `com_finished` variable is used to

[10] `MPI_Testall` is a variant of `MPI_Test`. The variants of `MPI_Test` are analogous to those of `MPI_Wait`: `MPI_Testany`, `MPI_Testsome`, and `MPI_Testall`.

avoid wasting time by calling `MPI_Testall()` after the communications have already been completed.

In this code, the calls to `MPI_Testall()` appear to be superfluous. Their purpose however, is not to check the status of the communications. Rather, it is to ensure that the communications proceed as a side-effect of checking their status. In some MPI implementations, this is not required, so programmers should consult the documentation for the implementation they use and test their code.

13.4.2.4 Collective Communications

In addition to point-to-point communications functions (Sect. 13.4.2.3), MPI defines a set of functions that facilitate efficient sharing of data between all processes in a group. Some of these functions only transfer data between processes, while others also compute a function of the data received from other processes. These latter operations are called reductions (see also Sect. 13.4.1.2) and can be used, for example, to compute the sum of a value at every point in a simulation domain that has been split among the processes. In such an application of a reduction, each process would compute the sum over the nodes it handles, and then perform a reduction to compute the global sum of all the local sums.

Since the example MPI implementation of an LBM simulation does not use collective operations that only perform communication, in this section we examine what functions MPI defines for these features but do not consider the details of how to use them.

Several types of transfers between the member processes of a group are possible. A transfer of a dataset from one process to all others is called a broadcast, and it is performed by the `MPI_Bcast` function. A scatter operation (`MPI_Scatter`) is similar to a broadcast in that one process sends data to all others, but the data sent to each process may be different. The inverse of a scatter, in which one process receives data from each other process, is called a gather (`MPI_Gather`). When every process requires the same set of data from each other process, the `MPI_Allgather` function can be used to perform the same operation as gather, but with every process receiving the combined data set. Finally, every process can send different data to each other process through the use of the `MPI_Alltoall` function. Several variants of these functions allow each process to send a different amount of data to the recipients.

The example code uses reductions in two places. The first place is in the computation of the total energy of the flow field and the error between the numerical and exact solutions for the flow. For this purpose, we use the `MPI_Allreduce` function that performs a reduction and provides the result to every process. Its definition is:

```
int MPI_Allreduce(void* sendbuf, void* recvbuf, int count,
                  MPI_Datatype datatype, MPI_Op op,
                  MPI_Comm comm)
```

Except for the first two parameters, all processes must call the function with the same parameter values. The parameter `sendbuf` points to the first of `count` variables of type `datatype`. The communicator `comm` specifies which processes participate in the reduction. In every process, the *n*th value of `recvbuf` receives the result of the reduction applied to the *n*th values of `sendbuf` from every process. Users may specify a custom reduction operation or use one of the predefined options for `MPI_Op op`: `MPI_SUM` for a sum, `MPI_PROD` for a product, `MPI_MIN` for the minimum, `MPI_MAX` for the maximum, and several others that perform logical operations, bitwise operations, and minimum/maximum computations that also provide the location of the extrema.

The MPI version of the `compute_flow_properties` function is shown in Listing 13.24 (see Listing 13.18 for the sequential version). It requires several additional parameters (the rank of the process as `rank`, the y coordinate of the first row in the subdomain as `ystart`, and the number of rows in the subdomain as `ny`), computes local sums, and then performs a reduction to obtain the global sums.

Listing 13.24 The MPI version of the `compute_flow_properties` function uses a reduction to compute the kinetic energy in the flow domain and the L_2 errors of the density and velocity fields

```
void compute_flow_properties(unsigned int t,
                             double *r, double *u, double *v,
                             double *prop,
                             int rank,
                             unsigned int ystart,
                             unsigned int ny)
{
  // prop must point to space for 4 doubles:
  // 0: energy
  // 1: L2 error in rho
  // 2: L2 error in ux
  // 3: L2 error in uy

  // sums over nodes belonging to this process
  double local_sumdata[7];
  // global sums
  double global_sumdata[7];

  // initialise local sum values
  for(int i = 0; i < 7; ++i)
    local_sumdata[i] = 0.0;

  for(unsigned int y = 0; y < ny; ++y)
  {
    for(unsigned int x = 0; x < NX; ++x)
    {
      double rho = r[scalar_index(x,y)];
      double ux  = u[scalar_index(x,y)];
      double uy  = v[scalar_index(x,y)];
```

```
            // add to local sum of energy
            local_sumdata[0] += rho*(ux*ux + uy*uy);

            // compute exact solution at this location
            double rhoa, uxa, uya;
            taylor_green(t,x,ystart+y,&rhoa,&uxa,&uya);

            // add to local sums of errors
            local_sumdata[1] += (rho-rhoa)*(rho-rhoa);
            local_sumdata[2] += (ux-uxa)*(ux-uxa);
            local_sumdata[3] += (uy-uya)*(uy-uya);

            // add to local sums of exact solution
            local_sumdata[4] += (rhoa-rho0)*(rhoa-rho0);
            local_sumdata[5] += uxa*uxa;
            local_sumdata[6] += uya*uya;
        }
    }

    // compute global sums
    MPI_Allreduce(local_sumdata,global_sumdata,
                  7,MPI_DOUBLE,MPI_SUM,MPI_COMM_WORLD);

    // compute and store final values
    prop[0] = global_sumdata[0];
    prop[1] = sqrt(global_sumdata[1]/global_sumdata[4]);
    prop[2] = sqrt(global_sumdata[2]/global_sumdata[5]);
    prop[3] = sqrt(global_sumdata[3]/global_sumdata[6]);
}
```

The second use of a reduction is in the `main` function, where one is used to compute the total memory allocated by all processes. In this case we use `MPI_Reduce`:

```
int MPI_Reduce(void* sendbuf, void* recvbuf,
               int count, MPI_Datatype datatype,
               MPI_Op op, int root, MPI_Comm comm)
```

This function provides the result of the reduction in `recvbuf` only in the process whose rank is specified by the parameter `root`. The other parameters are the same as those for `MPI_Allreduce`. Prior to starting the main simulation loop, each process computes the amount of memory it allocated:

```
double bytesPerGiB = 1024.0*1024.0*1024.0;
size_t total_mem_bytes = mem_size_0dir
                       + 2*mem_size_n0dir
                       + 3*mem_size_scalar;
```

After the main simulation loop ends, we use MPI_Reduce:

```
size_t global_total_mem_bytes = 0;
MPI_Reduce(&total_mem_bytes,&global_total_mem_bytes,
           1,MPI_LONG_LONG_INT,MPI_SUM,0,MPI_COMM_WORLD);

if(rank == 0)
{
  printf("memory allocated: %.1f (MiB)\n",
         global_total_mem_bytes/bytesPerMiB);
}
```

Every process sends one MPI_LONG_LONG_INT[11] to rank 0, which receives the sum in the variable global_total_mem_bytes. Finally, rank 0 displays the total memory allocated by all processes.

For completeness, we mention the two other reduction-type operations that are available in MPI. These are MPI_Reduce_scatter that performs a reduction and scatters the result to all processes and MPI_Scan that computes cumulative reductions in which the value returned to rank *n* is the result of the reduction on the values from ranks 0 to *n* inclusive.

13.4.2.5 I/O

When an MPI program runs, the output data it generates will be scattered across many nodes. The processing of this data can be handled in several ways. Sometimes, combining the data from all the nodes into one file is particularly useful and the necessary data transfers are sufficiently small or infrequent the their impact on performance is negligible. In these cases, the full data sets from every rank can be sent to one rank that writes the data it receives to a single file.

One fairly obvious pitfall must be avoided when implementing this approach. The writing process most often cannot allocate enough memory at once to store the whole data set that is spread across all processes. Instead, the process should allocate as much data as is needed for each process individually (to avoid multiple allocations/deallocations, one may use a "maximum" reduction and allocate the maximum memory required by any process), and write the data immediately once it is received.

As an alternative to this approach, each process could sequentially write its data to the same file. In this case, however, the data structures that represent an open file on one process cannot be shared with other processes. Furthermore, special care must be taken to synchronise the processes and ensure that each correctly adds its data after the data from the previous process.

[11] This matches the size of size_t on the systems used for testing, but it is not portable and may need to be changed for other systems.

For **efficient performance**, it is preferable to **avoid unnecessary communication and synchronisation**. In the code for this section, we take the simplest approach to saving the data generated in each process: each process opens its own files to which it saves its portion of the density and velocity fields separately.

Though the density and velocity values could be stored together in one file per process, keeping the data separate makes combining the results from all the processes easier. To analyse the data after the simulation has run, one may write a program that loads the results for each process and then computes desired values, generates visualisations, or creates the input files required to perform these tasks with other software. Instead of using a custom program to combine the data, the "raw" data that the program saves can be combined with the common Unix command `cat`, which stands for "concatenate," as follows:

```
cat ux_010_r0.bin ux_010_r1.bin ux_010_r2.bin > ux_010.bin
```

This command combines the data from a hypothetical run of a program with three processes and stores the result in `ux_010.bin`. The `ux` in the file names is used to indicate that the files store the x component of the velocity field, the digits `010` indicate that the data is for the result of the 10th time step (or the 10th multiple of the number of time steps taken until data is saved), and the digit after `r` indicates the rank of the process that generated the data.

The MPI version of the function `save_scalar` that saves data differs minimally from the serial version (see Listing 13.17). The differences are: the addition of the process rank as a parameter, the inclusion of the process rank in the output file's name, the addition of a parameter for the number of bytes of memory to be saved (since it is not necessarily the same in each process), and the cosmetic use of `MPI_Barrier` at the end of the function to synchronise the text output from the processes.

This approach for saving data from parallel programs, in which each process saves its own file, has a special efficiency advantage. Each node of a cluster typically has fast local filesystems and shared network filesystems. When one combined data file is created for all processes, it must be saved in a slower shared filesystem that every process can access. In comparison, writes to local filesystems are significantly faster, and the local files could be examined during a simulation or after it finishes to determine whether a transfer to a shared network filesystem is justified (for example when the simulation was performed at the correct conditions for a phenomenon to occur). Transfers to a network filesystem could also be entirely avoided if postprocessing software can be run in a distributed fashion like the simulations are.

Though it is not the case for the code in this chapter, many simulations need to read data from files during initialisation or execution. The processes can all read from a common input file, or as for writing, the data for each process can be split into separate files. When this is inconvenient, one rank can read the required data and share it with other processes, using calls to `MPI_Scatter` or `MPI_Send`/`MPI_Isend`.

As an alternative to the file access methods presented so far, MPI-2 includes features for parallel reading and writing of shared files. The interface for using these features, which we do not cover in further detail, is similar to the interface for the communications functions described in this section. The interface provides, for example, functions that allow each process to read from and write to different locations in a file.

13.4.2.6 Compilation and Execution

MPI implementations and cluster administrators usually provide tools to assist the compilation and parallel execution of MPI programs. The `mpic++` program is useful for the compilation of C++ code. For example, the MPI progam `sim` can be generated from the source file `sim_mpi.cpp` using the command

```
mpic++ -O3 sim_mpi.cpp -o sim
```

This MPI compilation program is called a wrapper because it does not perform the compilation itself but rather runs a compiler and passes it the required options for finding header files and linking with the MPI libraries. To see which compiler is invoked and all the options that would be used to compile the code, we can use the option `-showme`. For example the command

```
mpic++ -showme -O3 sim_mpi.cpp -o sim
```

shows output that is similar to

```
g++ -O3 sim_mpi.cpp -o sim -I/path/to/include -pthread \
    -L/path/to/lib -lmpi_cxx -lmpi
```

This shows which additional libraries are linked and the paths that are searched for include and library files. Wrappers are also available for other languages, including C (`mpicc`) and Fortran (`mpif90`).

The tasks that users must perform to run an MPI program on a cluster depend on the configuration of the cluster and the software that is used to share the computing resources with many users. Readers are therefore advised to consult the documentation for their cluster or ask a system administrator. In general, cluster users submit requests for computing resources to resource management and job scheduling software that then run their program on a group of the cluster's nodes once the requested processing and memory resources are available. For example, on systems that use TORQUE (Terascale Open-source Resource and QUEue Manager) [19], the command

```
qsub -l nodes=32:ppn=2,pmem=2gb,walltime=12:00:00 \
     ./simulation_script.sh
```

requests two (`ppn` or processors per node) processors each on 32 nodes (`nodes`), for a total of 64 processes, and 2 GiB of RAM per process (`pmem`) for a maximum of 12 hours (`walltime`). The `qsub` command does not wait for the script to run and only places it in a queue where it waits until the requested resources become available. The scheduler runs the job script `simulation_script.sh` at a time when these resources are available for the specified duration. If the script does not finish within the chosen time limit, the scheduler terminates it.

A minimal job submission script is:

```
#!/bin/bash

# go to directory from which the script was submitted
cd $PBS_O_WORKDIR

# run MPI program
mpirun ./sim
```

The first command in this script switches the current directory to the one from which the script was submitted. The path to this directory is available in the environment variable `PBS_O_WORKDIR`, one of many variables set up by the job scheduling system to provide information about the system's configuration and the runtime environment. The next command, `mpirun`[12] sets up the execution environment for the MPI program `sim` and starts it as many times as is needed. On many systems the number of times is automatically determined from the options specified to `qsub`, in our case the product of the `nodes` and `ppn` settings. On other systems, it is necessary to explicitly specify the number of processes to start by using `mpirun -np N ./sim` where `N` is the number of processes.

The requested resources can also be specified through the use of special comments in the script file that start with `PBS`. For example, this job script requests the same resources as the previous `qsub` command:

```
#!/bin/bash
#PBS -l nodes=32:ppn=2
#PBS -l pmem=2gb
#PBS -l walltime=12:00:00

# go to directory from which the script was submitted
cd $PBS_O_WORKDIR

# run MPI program
mpirun ./sim
```

[12] `mpirun`, `mpiexec`, and `orterun` are synonyms in Open MPI.

This script would be submitted for scheduling using the command `qsub simulation_script.sh`. Options specified in the `qsub` command that submits a script override those specified inside the script.

In both cases, the job script `simulation_script.sh` must be executable, i.e. the user must have permission to execute the script file. This script file may be created as any plain text file, and then given execution permission with the command

```
chmod u+x simulation_script.sh
```

Commands for setting up the simulation environment and postprocessing the results can also be included in the job script.

In addition to requesting resources, job configuration settings allow users to receive status updates about the job by email, such as when it starts and ends. The output of the program (that would appear on screen when the program is run interactively) is saved to default files or those specified in the `qsub` command or job script.

13.4.2.7 Performance Results

We examine now the performance of the parallel code presented in this section. The code was run on up to 32 nodes of a cluster with two 8-core Xeon E5-2670 or E5-2680 processors per node, for a total of 16 cores on each node.

Two different methods were used for assigning processing cores to the simulation to show the effects of this choice on performance. In the first method, each process participating in the simulation ran on a different node. This choice can improve performance by giving each process independent access to memory, but it slows communication by requiring data transfers over the interconnect rather than within the memory of a node. The other process distribution that was used had all processes run on one node. While this choice reduces communication time, it can introduce competition for memory bandwidth.

In practical cluster usage for simulations involving many more processes than the number of cores that are available on any node, one will typically run simulations to occupy all processing cores on each of as many nodes as are needed. Alternatively, one can request the scheduler to pick any available processors no matter how they are distributed across the cluster. The choice between these two options is not always determined by performance. On a busy cluster, the simulation will likely start sooner if the scheduler can allocate any unused cores instead of waiting for whole nodes to become available. Users can therefore optimise in different ways to achieve the shortest time to completion of a simulation.

Exercise 13.16 Test the MPI and OpenMP (Sect. 13.4.1) versions of the code on the same system and varying numbers of processes. Which performs better? Why?

Exercise 13.17 Compare the performance of the MPI code when running it with the same total number of processes and varying numbers of processes on each node.

What is the optimal number of active processes per node? To prevent competition for memory bandwidth with other users' programs running on the same nodes, reserve whole nodes for the simulations and use only some of the available cores.

The benefits of parallelisation are typically assessed in two ways. When additional computing resources are used to reduce the computing time of a problem with a fixed size, we are interested in what is called the strong scaling performance of the code. In this case, we want to find out whether using n processes (or cores) reduces the execution time by a factor of n. The second type of performance assessment is called weak scaling and examines how execution time varies as the size of the problem increases in proportion to the number of processes used. In other words, we test whether n times as many processes can finish a problem that is n times larger in the same amount of time.

Both types of parallel performance analyses can be relevant to CFD simulations depending on the details of the problem being considered. For example, one might be interested in how many processes could be used to reduce the simulation time for a particular domain size to a practical value (strong scaling) or how much larger a domain could be computed with more processes (weak scaling).

Figures 13.11 and 13.12 present the weak and strong scaling analysis results for this section's MPI code. For the weak scaling results, the domains had a width of 2048 and a height of 8 times the number of processes, i.e. each process handles a 2048×8 subdomain. For the strong scaling analysis, the domain size was 2048×256. All simulations ran for 1000 time steps. Both figures present results for the two processor allocation methods (all processes on one node and one process per node) and MPI communication techniques (blocking and nonblocking).

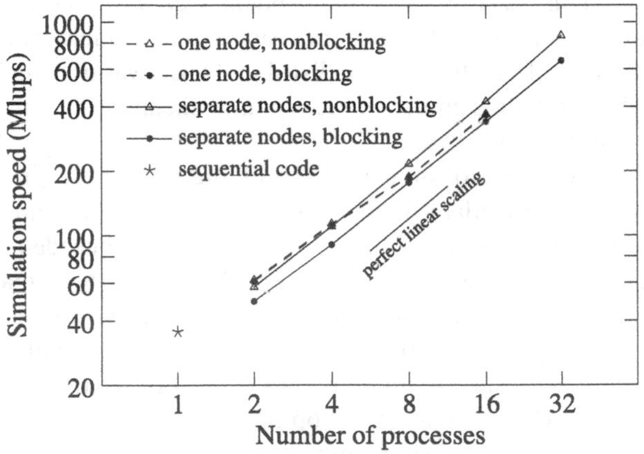

Fig. 13.11 Weak scaling performance of the simulation code run with 2 to 32 processes. *Dashed lines* show results for up to 16 processes on a single node, and *solid lines* show results for up to 32 processes all on separate nodes. The domains had a constant width of 2048 and a height that was 8 times the number of processes used. All values shown are averages of three runs. The speed for one process was obtained with the sequential code from Sect. 13.3.2

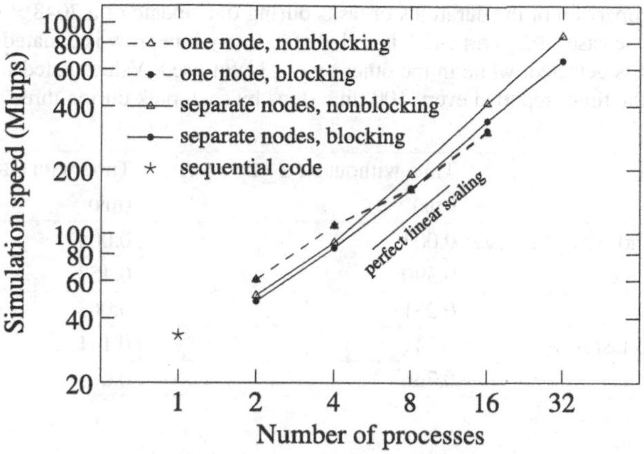

Fig. 13.12 Strong scaling performance of the simulation code run with 2 to 32 processes. *Dashed lines* show results for up to 16 processes on a single node, and *solid lines* show results for up to 32 processes all on separate nodes. The domain size was 2048 × 256 for all runs. The speed for one process was obtained with the sequential code from Sect. 13.3.2

When all the processes are all on one node, the difference between blocking and nonblocking communications is negligible because the time spent communicating, in this case by copying data between different locations in memory, cannot be hidden. The time lost to this copying is the same whether it is spent before updating the simulation domain, in between computations, or after the internal rows are updated. For up to four processes, running all processes on one node is fastest. For larger numbers of processors, it is beneficial to split the processes among separate nodes.

Overall, both the strong and weak scalings are nearly linear for up to 32 processes and the simulation speed is slightly higher with nonblocking rather than blocking communications. The parallel performance can be expected to remain good as the number of processes increases provided that the time for communications is reasonably less than the computing time. With 256 processes, each process would handle only one row, and the simulation speed would be determined by the communication speed rather than the processing speed.

Exercise 13.18 Optimise the MPI code presented in this section. One potential improvement is to transfer only those populations that cross subdomain boundaries. When implementing this improvement, avoid unnecessary copying of the useful populations into or from a temporary array by using derived datatypes (Hint: the functions `MPI_Type_vector` and `MPI_Type_contiguous` may be useful). Another optimisation is to allow the first and last rows to be updated as soon as the corresponding data transfers have completed instead of waiting for all to finish.

Exercise 13.19 Adapt the code to decompose the simulation domain in both dimensions. Try to write the new code so that it can handle as general a tiling of the simulation domain as possible. For example, add command line arguments that

Table 13.3 Comparison of the durations of tasks during one update of a 2048 × 64 domain with 8 processes. In one case `MPI_Testall` is called after every inner row is updated until the end of communications is detected, while in the other case it is not used. Values listed for each case are the averages of the times reported every 100 time steps by each rank during three runs of the code on a cluster

Task	Time without `MPI_Testall` (ms)	Time with `MPI_Testall` (ms)
`MPI_Isend` and `MPI_Irecv`	0.004	0.004
Update inner rows	0.390	0.453
`MPI_Waitall`	0.231	0.063
Update first and last rows	0.141	0.141
Total	0.766	0.661

specify the number of subdomains in each direction, or load parameters from a configuration file. Note that in one dimension the data that needs to be transferred is located in contiguous blocks of memory, while for the other dimension the data is scattered. Use derived datatypes, for example the functions `MPI_Type_vector` and `MPI_Type_contiguous`, to avoid copying this scattered data to/from a contiguous temporary array before/after transferring it. Compare the performance of the code with different subdomain geometries. What is the optimal decomposition for a given domain size and number of processes?

Exercise 13.20 For several large domain sizes, calculate how many computing nodes need to be used so that the memory required on each node fits in the processor's L3 cache. Compare performance with slightly fewer and slightly more nodes than this number.

To examine the effectiveness of using calls to `MPI_Testall` to ensure overlapping of communications and computations in the nonblocking version of the MPI code, the code was modified to measure and report the durations of the communication and computation steps during one update of the simulation domain every 100 time steps. Table 13.3 shows the averages of the reported times for three runs of a 2048 × 64 domain computed with 8 processes on separate nodes. For one of the two sets of three simulations, the call to `MPI_Testall` was commented out.

Overall, the removal of the `MPI_Testall` calls decreased the simulation speed from 198 to 171 Mlups. Comparing the two cases, the times to initiate nonblocking communications (total time spent in calls to `MPI_Isend` and `MPI_Irecv`) and update the first and last rows are the same, averaging 0.071 ms per row. Inner rows are those rows that can be updated without information from adjacent subdomains handled by different ranks, i.e. these are the second to the second to last rows of each subdomain. The additional 0.063 ms taken to update the inner rows due to the calls to `MPI_Testall` is offset by a reduction of 0.17 ms in the time taken by `MPI_Waitall`, which waits for the completion of all communications.

The averages, however, do not tell the complete story. Figure 13.13 shows the timing results reported by every rank during one simulation run with and without

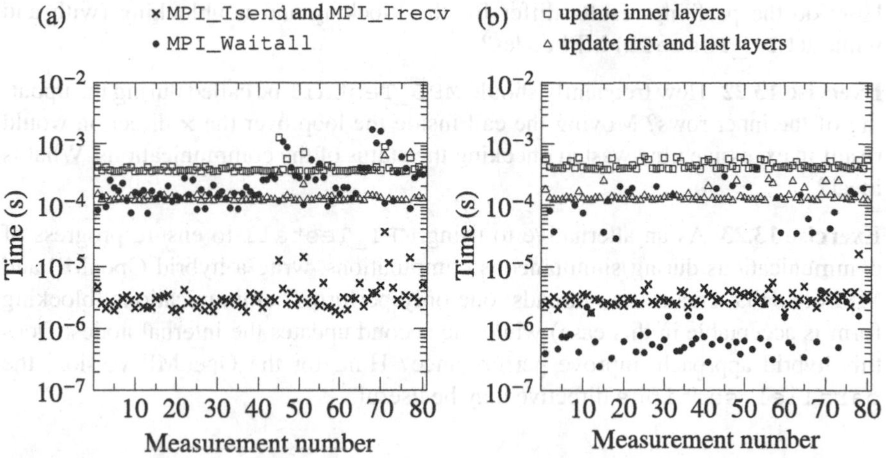

Fig. 13.13 Timing of the communication and computation operations during one domain update as measured on the 8 ranks participating in the simulations (**a**) without the use of `MPI_Testall` during the updating of internal rows and (**b**) with the use of `MPI_Testall`

calls to `MPI_Testall`. Each of the eight ranks provided measurements ten times during the simulations for a total of 80 measurements of each task. As expected, the time taken to update the first and last rows is nearly constant over time and the same for both cases. The time to initiate nonblocking communications is also equal and effectively negligible at 4 μs.

The biggest difference between the two cases is the time taken by the calls to `MPI_Waitall`. When `MPI_Testall` is used, nearly all these calls return in less than 1 μs, indicating the the communications were completed before the call to `MPI_Waitall` was reached; without `MPI_Testall` they take 0.1 ms or more. In some cases, however, the calls to `MPI_Waitall` take as long in the case with `MPI_Testall` as in the case without it. This suggests that occasionally a fast node completes its update of inner rows and reaches `MPI_Waitall` before a slower node is able to provide the information the faster node requires to proceed. Overall, the whole simulation domain cannot be updated faster than the slowest process can complete its subdomain. Over the 1000 time steps in the two simulations analysed in Fig. 13.13, the use of `MPI_Testall` increased the average simulation speed from 177 to 227 Mlups.

Exercise 13.21 Find out what tools for profiling[13] MPI programs are available on a cluster you can use. These tools show the time taken by the most time-consuming tasks in a program and are therefore useful for identifying performance bottlenecks.

[13]The analysis of how a program uses memory and computing resources is called profiling. Automatic profiling software typically reports the time taken by the most time-consuming functions in a program and is useful for optimisation.

How do the profiling results differ for the blocking and nonblocking (with and without `MPI_Testall`) MPI codes?

Exercise 13.22 How frequently should `MPI_Testall` be called during the updating of the inner rows? Moving the call inside the loop over the `x` direction would result in excessive time wasted checking the status of the communications. What is ideal?

Exercise 13.23 As an alternative to using `MPI_Testall` to ensure progress of communications during simultaneous computations, write a hybrid OpenMP and MPI code that creates two threads: one only performs communications (blocking form is acceptable in this case) while the second updates the internal nodes. Does this hybrid approach improve performance? Hint: for the OpenMP version, the `parallel sections` directive may be useful.

13.4.3 General Purpose Graphics Processing Units

Originally developed to accelerate the algorithms used to generate the graphics shown on computer screens, graphics processing units (GPUs) are now found in a wide range of devices from smart phones and tablets to workstations and supercomputers where they are used for algorithms far beyond those related to rendering graphics. In this section, we examine the use of GPUs for flow simulations with the lattice Boltzmann method.

Early GPUs accelerated two-dimensional drawing algorithms, such as those for drawing lines and filling polygons, performing these tasks faster than a CPU and freeing the CPU to perform other tasks. GPUs were later developed to implement algorithms for generating three-dimensional graphics: projecting the 3D coordinates of mesh vertices to the 2D coordinates of an image, determining which portions of a mesh are visible, and computing the colour of each pixel that is drawn from the colour data stored with the mesh and the locations and characteristics of light sources in the digital scene. In these two- and three-dimensional graphics applications, the same computations must be performed on different input data, such as the on-screen coordinates of the pixel being drawn or the three-dimensional coordinates of the vertices of all the triangles in a mesh. The circuits in GPUs are designed to efficiently perform such calculations in parallel to allow the content of a screen to be updated several tens to hundreds of times per second, making animations appear smooth.

Taking advantage of the fact that the same operation must be performed many times, a larger fraction of the transistors in GPUs than CPUs is dedicated to computational tasks. In contrast, CPUs use more transistors for control logic and caches to ensure that any sequence of instructions can be executed quickly whether or not it involves repetition of the same operation. To take advantage of the efficiency of GPU architectures for executing computations in parallel, programmers devised ways to map non-graphics algorithms to the capabilities of GPUs by assigning different meanings to the graphics data they process and the images they generate.

In comparison to early computing on GPUs, the capabilities of modern GPUs and software development tools greatly facilitate their use for general computational tasks. Double precision floating point arithmetic is now supported, and libraries with standard math functions, including the fast Fourier transform and common linear algebra operations, are available.

Code for GPU platforms may be written with common languages, such as C, C++, and Fortran, and compiled using vendor-specific interfaces from both major vendors of GPU devices (NVIDIA and AMD) and non-vendor-specific platforms, such as OpenCL [20]. While code written with the vendor-specific interfaces can only run on the corresponding vendor's devices, code written with OpenCL is not tied to a particular architecture. Access to GPU-based computing is also available in many high-level languages, such as Python, R, MATLAB, and Java. In these languages, objects that reside in GPU memory can be created and operations on these objects are performed on the GPU. Language extensions, such as OpenACC [21], are also being developed to automate the sharing of computational tasks between CPUs and GPUs in a hybrid computing environment.

In this section, we will study GPU programming with NVIDIA's Compute Unified Device Architecture (CUDA). There are two reasons for this choice. The first reason is that NVIDIA devices are most common on current supercomputers and research clusters. In the November 2015 list of the 500 highest-performance supercomputers [12], 68 systems used NVIDIA chips and only three used AMD devices. The number two ranked system, Titan in the United States, and the seventh, Piz Daint in Switzerland, have 18688 and 5272 NVIDIA K20X GPUs, respectively. These GPUs each have 6 GiB of memory, 2688 processing cores, a peak memory bandwidth of 232 GiB/s, and can perform a maximum of 1.31 trillion floating point operations per second (teraflops) in double precision. The second reason for using NVIDIA CUDA is that learning a lower-level interface, instead of the GPU features provided in a high-level language, requires readers to become more familiar with GPU architecture and therefore its benefits and limitations.

At the time of writing, the most recent version of the CUDA toolkit was 7.5, released in September 2015. CUDA programs can be run on a wide range of NVIDIA devices marketed for computer gaming, professional workstations, and high performance clusters. Though devices for gaming often have faster memory and more cores, those used in scientific computing have additional features such as better performance for double precision arithmetic and memory with automatic detection and correction of errors (error-correcting code, or ECC, memory). The features of GPUs have evolved rapidly in recent years, and the capability of NVIDIA devices is specified by a version number (or "compute capability"). Early devices are no longer supported in recent versions of the development tools, and we therefore consider only devices with capability 2.0 and higher in this section.

This section provides an overview of the NVIDIA architecture and programming interface, focussing on those features that are relevant to the LBM implementation presented in Sect. 13.4.3.4. Readers may find more detailed information in the documentation [22] and code samples [23, 24] from NVIDIA as well as a wide range of reference materials online and in books [25].

13.4.3.1 NVIDIA GPU Architecture

Conventional CPUs are generally classified as Single Instruction, Single Data (SISD) devices because the commands they execute each perform one operation on one set of input data. Modern CPUs also have Single Instruction, Multiple Data (SIMD) features: special commands that operate on multiple data elements, for example performing elementwise addition of the four components of two vectors in one step. NVIDIA describes their GPU-based parallel computing architecture as Single Instruction, Multiple Thread (SIMT), which differs significantly from SISD architecture and more subtly from SIMD.

In NVIDIA GPUs, several processing units called **multiprocessors** store the state and manage the execution of a large number of **threads**, up to several thousand. The host system uses the GPU device by supplying it with the code that every thread will run and specifying how many threads will run this code. The multiprocessor groups the threads it manages into **warps**,[14] which on current devices consist of **32 threads**. The multiprocessor's scheduling logic iterates through the warps it manages, progressing them through the stages of executing each instruction. Once the required data (in registers or off-chip memory) and computational resources (e.g. arithmetic units) are available to complete a warp's next command, **all threads in that warp simultaneously execute** the command.

This parallel execution of commands is the first main performance benefit of GPUs. The second is that the latencies of executing instructions and accessing memory, i.e. the time delays incurred in performing these tasks, are effectively "hidden" as long as the multiprocessor is handling a sufficient number of threads that one warp is always ready to execute its next command. Performance is therefore best when an algorithm can be split among many threads.

The preceding presentation of thread execution has glossed over one important topic, however: the handling of conditional statements. When the logic in several threads diverges, which occurs when the value of the conditional expression differs in these threads, the threads that do not need to follow the branch are marked as inactive. When a warp is ready to execute an instruction in a branch, only those threads that are active actually execute it. For efficient execution on GPUs, algorithms need to be structured so that threads usually follow the same execution paths. Otherwise, the benefits of parallel execution are lost.

[14] In textile weaving, a warp is a collection of parallel *threads* through which other thread, called the weft, is interlaced.

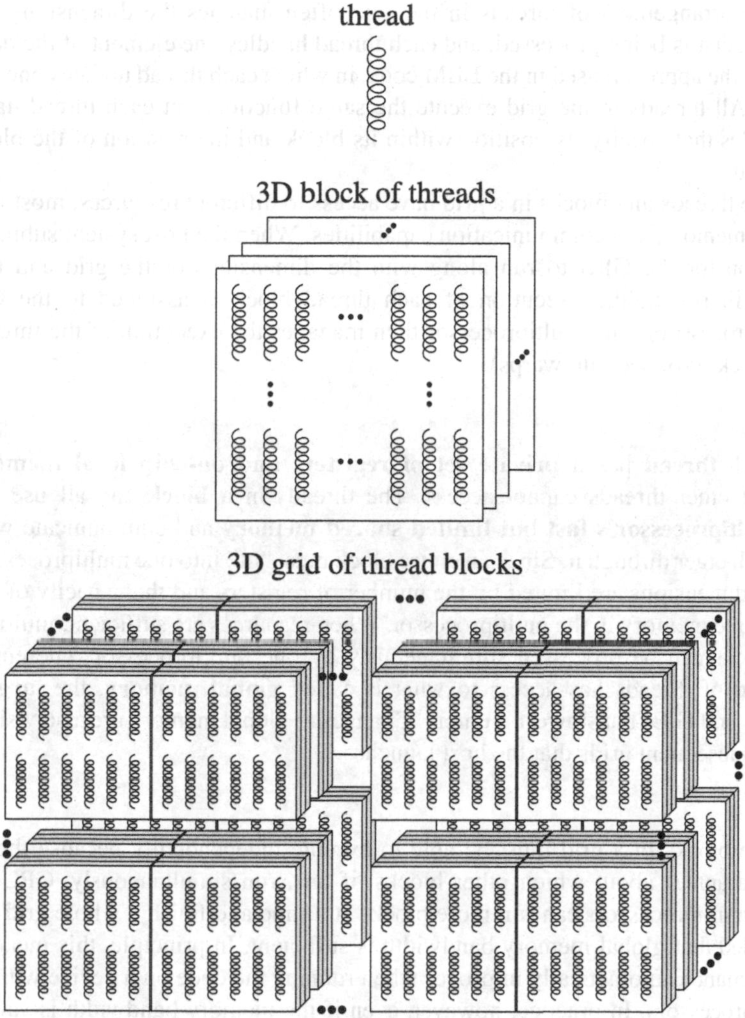

Fig. 13.14 Diagram of the logical organisation of threads run on a GPU into a three-level hierarchy. The threads in each level have access to different memory resources and communication capabilities

To allocate GPU resources to threads, the threads used to perform a task on a GPU are grouped into the **three-level hierarchy** illustrated in Fig. 13.14. This hierarchy consists of **threads** arranged into **blocks** that are arranged in a **grid**.

The arrangement of threads in the grid often matches the dimensions of the dataset that is being processed, and each thread handles one element of the dataset. This is the approach used in the LBM code, in which each thread updates one lattice node. All threads in the grid execute the same function, but each thread has two variables that specify its position within its block and the position of the block in the grid.

The threads and blocks in a grid have access to different resources, most importantly memory, and communication capabilities. When the host system submits the function for the GPU to run along with the dimensions of the grid and blocks that will run it, the execution of each thread block is assigned to the GPU's multiprocessors. The multiprocessor then manages the execution of the threads in the block (grouped into warps).

Each **thread** has a **private set of registers and on-chip local memory** that other threads cannot access. The threads in a **block** can all use the **multiprocessor's fast but limited shared memory** and communicate with each other through it. Since an entire block must "fit" into one multiprocessor, its dimensions are limited by the number of registers and the capacity of the shared memory in the multiprocessor. When resources are sufficient, multiple blocks can be processed simultaneously by one multiprocessor. The entire **grid** of threads has access to what is called **global memory**, the **largest** (several GiB) but **slowest** memory. Changes to global memory are also visible to subsequent grids that the host launches.

The blocks in a grid must be able to execute independently because there are no guarantees about which other blocks, if any, run simultaneously. GPUs with more multiprocessors can run more blocks at a time and finish a whole grid faster, provided that global memory bandwidth is sufficient. In principle, this means that performance automatically improves when running the code on a device with more multiprocessors. In practice, however, even if the memory bandwidth is sufficent, the dimensions of the blocks may need to be adjusted to take advantage of available resources and achieve optimal performance, especially when switching to a newer GPU with more registers. The Tesla M2070, one of the GPUs on which we run the LBM code, has 14 multiprocessors each with 32 cores for computations, for a total of 448 cores. It has 5375 MiB of global memory (with ECC enabled), 48 KiB of shared memory, and 32768 registers available per block. The Tesla K20, another GPU on which we test the code, has 13 multiprocessors with 192 cores per multiprocessor (2496 total cores), 4800 MiB global memory, 48 KiB shared memory, and 65536 registers per block.

13.4.3.2 Programming Model

In addition to providing many functions for performing special tasks on the host system and GPU devices, CUDA uses several extensions to the C language to specify whether functions will run on the host CPU or GPU and which variables will reside in the CPU's or GPU's memory. CUDA C also introduces a special syntax for launching the execution of a grid of threads.

The task that all the threads in a grid perform is described in a **function called a kernel**. These functions do not return a value and have the qualifier `__global__`. Within a kernel, the code may use several variables, similar to ranks in MPI and thread identification numbers in OpenMP, to determine the subset of data it will process. These variables are

- **`threadIdx`:** three-component integer vector (of type `uint3`, which has three `uint` components `x`, `y`, and `z`) with the position of the thread within its block
- **`blockIdx`:** three-component `uint3` vector with the position of the block within its grid
- **`blockDim`:** three-component vector (of type `dim3`, which like `uint3` has the `uint` components `x`, `y`, and `z`) with the dimensions of the block
- **`gridDim`:** three-component `dim3` vector with the dimensions of the grid

A kernel can therefore be thought of as the inner block of code inside nested `for` loops that iterate over the coordinates of each block in the grid and each thread in every block. This conceptual understanding of kernels is presented in Listing 13.25. The key feature of GPU programming, though, is that the **inner block is executed in parallel** by many threads and not serially as the pseudocode suggests.

Listing 13.25 Pseudocode presenting the organisation of threads into blocks that form a grid

```
// iterate over blocks in grid
for(int bz = 0; bz < gridDim.z; ++bz)
for(int by = 0; by < gridDim.y; ++by)
for(int bx = 0; bx < gridDim.x; ++bx)
  // iterate over threads in block
  for(int tz = 0; tz < blockDim.z; ++tz)
  for(int ty = 0; ty < blockDim.y; ++ty)
  for(int tx = 0; tx < blockDim.x; ++tx)
  { // executed in separate threads
    dim3 blockIdx(bx,by,bz)
    dim3 threadIdx(tx,ty,tz);

    // run body of kernel
  }
```

Simulation domains and the sizes of datasets are usually larger than the blocks used to process them. To convert the thread and block coordinates into unique three-dimensional indices `i`, `j`, and `k` for each thread, the expressions are:

```
int i = blockDim.x * blockIdx.x + threadIdx.x;
int j = blockDim.y * blockIdx.y + threadIdx.y;
int k = blockDim.z * blockIdx.z + threadIdx.z;
```

For data that is one or two dimensional, only the expressions for `i` or `i` and `j` are needed.

When launching a kernel, the code running on the host CPU system must specify the kernel to be launched, its arguments, the dimensions of the grid and its blocks, and the amount of shared memory required. For this purpose, CUDA C defines a triple angle bracket syntax:

```
kernel_name<<< gridDim, blockDim, sharedMem >>>(arg1, arg2);
```

where `gridDim` is a `dim3` vector with the dimensions of the grid, `blockDim` is a `dim3` vector with the dimensions of the blocks, and `size_t sharedMem` specifies the number of bytes of shared memory required by each block (default 0 when omitted). The dimensions of the grid and block may also be integers instead of `dim3` vectors when they are one-dimensional. A fourth optional argument is used for features that are not required for this section.

Kernels may also be launched from within a kernel, but we do not cover this topic here. This functionality is called dynamic parallelism and is available on devices with compute capability 3.5 or higher. The launching of a kernel by the host system is asynchronous: the launch command returns before execution is completed, freeing the CPU to perform other tasks and launch subsequent kernels. Special functions are available to wait for the completion of tasks on the GPU.

In addition to the `__global__` function type qualifier used for kernels, CUDA C defines two other qualifiers that specify whether functions will be used on the host CPU or an attached GPU. The qualifier `__host__` indicates that a function will be used by, and must therefore be compiled for, the host CPU, while `__device__` indicates that a function will be used on the GPU. Functions with the `__device__` qualifer can be called either from a kernel or a `__device__` function. Functions without a qualifier are assumed to be `__host__` functions. The `__device__` and `__host__` qualifiers can be used together to indicate that a function will be used on both the host and the GPU.

Arrays stored in shared memory are declared with the `__shared__` qualifier and can be used in `__device__` functions and kernels (`__global__` functions). The amount of shared memory available to the threads in a block is specified when the kernel is launched, and programmers must manage the use of shared memory within the kernels themselves. In this section, we only use one shared array of `doubles` whose location in memory is automatically initialised to be the start of the available shared memory. For codes with multiple arrays potentially of different types, we refer readers to NVIDIA's documentation for the details of specifying the starting address of each array and ensuring that alignment rules for these addresses are satisfied.

Since the threads in a block can all read and modify the same shared and global memory, it is necessary to ensure that threads reading from a location that was written by a different thread "see" the correct value. This is achieved with the `void __syncthreads()` function, which serves two purposes. When executed, this function does not return until all threads in the block reach it. Furthermore, once `__syncthreads()` returns, all modifications to global and shared memory started before the synchronisation are visible to all other threads.

13.4.3.3 Code Compilation and Execution

We assume readers have access to a computer with an NVIDIA GPU and the required drivers and development toolkit have been installed. The necessary downloads [26] and a quick start guide [27] are available online.

Code that uses CUDA-specific language features is stored in files with the `.cu` extension and compiled with the `nvcc` program, which is provided in the CUDA toolkit. This program automates the steps that generate the code that runs on the host system and GPU device. These steps include separating the portions of code that need to be compiled for the different architectures, combining code files from different sources (as specified by `#include` directives), invoking compilers for the separated code, and linking with libraries. The final binary executable file contains both the instructions that run on the host and the binaries that will be loaded to run on the GPU.

Usage of `nvcc` is fairly simple for the code presented in this section. The command used in the compilation script provided with the code that accompanies the book is:

```
nvcc -arch sm_20 -v --ptxas-options=-v -O3 main.cu -o sim
```

This command requests compilation of the source file `main.cu` with level 3 optimisation (`-O3`) to generate the executable named `sim`. Code that can run on devices with compute capability 2.0 (such as the Tesla M2070) or higher is generated by specifying the architecture option `-arch sm_20`. The two digits following `sm_` specify the major and minor numbers of the compute capability. For the Tesla K20 with compute capability 3.5, it should be changed to `sm_35`. We request verbose output for two steps in the compilation process with the options `-v --ptxas-options=-v`. As will be discussed later, this provides detailed information about how the kernels were compiled, including the number of registers required by each kernel. After successful compilation, the generated program can be run like any other program, in this case with the command `./sim`. When run on systems without a GPU device, the program will end with an error message indicating the absence of an available device.

To run programs that use GPUs on a cluster, readers should consult the documentation for the clusters they use to find out how to request GPU resources for the jobs they submit. The job scripts may also need to load CUDA libraries before starting the GPU-accelerated program. Clusters often have several dedicated nodes on which users may interactively compile and run code outside of the job scheduling and resource allocation system. Such nodes are useful for quick testing and debugging of code, but one should be aware when interpreting performance results that the resources may be used simultaneously by other cluster users.

13.4.3.4 GPU LBM Implementation

The organisation of the GPU version of the Taylor-Green vortex decay code is quite similar to that of the CPU version. The main differences are the need to allocate and free GPU memory, transfer data between the GPU device's memory and its host's memory, and invoke kernels to perform the steps in the LBM algorithm.

The code is divided into three files: `LBM.h` contains primarily variable and function declarations, `LBM.cu` contains the functions and kernels that implement the LBM algorithm, and `main.cu` manages memory and has the main simulation loop. `LBM.cu` is not compiled separately. Instead it is simply `#included` in `main.cu`.

The presentation of the code that follows omits some details about error checking. Interested readers may examine the full code that accompanies the book and check the documentation provided by NVIDIA for details about the functions used.

We start in `main.cu` by selecting a GPU device:

```
checkCudaErrors(cudaSetDevice(0));
int deviceId = 0;
checkCudaErrors(cudaGetDevice(&deviceId));
```

In this and the code that follows, `checkCudaErrors` is a macro[15] defined in `LBM.cu` that checks for errors and displays diagnostic information if one occurs. If the code executes on a system without a GPU, the program will exit with an error message. For the purpose of this chapter, this code does not perform a useful task and is only meant to illustrate the use of these functions. On systems with multiple GPU devices, the code could instead select a device with particular characteristics, perhaps one with the most available memory. On such multi-GPU systems, `cudaSetDevice` can be called throughout the code to specify which device will execute subsequent kernel invocations and allocate/free memory.

[15] A macro is a compiler shortcut that allows programmers to conveniently use a fragment of code in many places. When preparing code for compilation, the compiler system replaces the name of the macro with the corresponding code fragment.

Next we obtain and display information about the selected device:

```
cudaDeviceProp deviceProp;
checkCudaErrors(cudaGetDeviceProperties(&deviceProp,
                                        deviceId));

size_t gpu_free_mem, gpu_total_mem;
checkCudaErrors(cudaMemGetInfo(&gpu_free_mem,
                               &gpu_total_mem));

printf("CUDA information\n");
printf("       using device: %d\n", deviceId);
printf("               name: %s\n",deviceProp.name);
printf("    multiprocessors: %d\n",
                             deviceProp.multiProcessorCount);
printf(" compute capability: %d.%d\n",
                             deviceProp.major,deviceProp.minor);
printf("      global memory: %.1f MiB\n",
                             deviceProp.totalGlobalMem/bytesPerMiB);
printf("        free memory: %.1f MiB\n",
                             gpu_free_mem/bytesPerMiB);
printf("\n");
```

The first function, `cudaGetDeviceProperties`, fills in the `cudaDeviceProp` structure `deviceProp` with information about the device's capabilities, such as clock rates, maximum block and grid dimensions, and the available capacities of different types of memory. Some of this information is then displayed. Detailed information about available GPU devices on a system can also be obtained with the `deviceQuery` utility program that is provided with the NVIDIA CUDA development tools.

The next step is to allocate memory, both on the GPU and its host:

```
double *f0_gpu,*f1_gpu,*f2_gpu;
double *rho_gpu,*ux_gpu,*uy_gpu;
double *prop_gpu;
checkCudaErrors(cudaMalloc((void**)&f0_gpu,mem_size_0dir));
checkCudaErrors(cudaMalloc((void**)&f1_gpu,mem_size_n0dir));
checkCudaErrors(cudaMalloc((void**)&f2_gpu,mem_size_n0dir));
checkCudaErrors(cudaMalloc((void**)&rho_gpu,mem_size_scalar));
checkCudaErrors(cudaMalloc((void**)&ux_gpu,mem_size_scalar));
checkCudaErrors(cudaMalloc((void**)&uy_gpu,mem_size_scalar));
const size_t mem_size_props =
                    7*NX/nThreads*NY*sizeof(double);
checkCudaErrors(cudaMalloc((void**) &prop_gpu,
                           mem_size_props));

double *scalar_host  = (double*) malloc(mem_size_scalar);

size_t total_mem_bytes = mem_size_0dir
                       + 2*mem_size_n0dir
                       + 3*mem_size_scalar
                       + mem_size_props;
```

We follow the naming convention that pointers to memory on the GPU have the suffix _gpu while pointers to host memory have the suffix _host. The cudaError_t cudaMalloc(void**,size_t) function allocates memory on the GPU. The first parameter is a pointer to the pointer variable that receives the address of the start of the allocated memory, the second parameter is the requested number of bytes, and like other CUDA functions, the function returns a status code. Memory is organised in the same way as in the CPU code (Sect. 13.3.2): we have one array for the rest populations (f0_gpu), two arrays for the eight other particle populations (f1_gpu and f2_gpu), and one array each for the density (rho_gpu) and velocity components (ux_gpu and uy_gpu). The sizes of these arrays are defined in LBM.h as

```
const unsigned int ndir = 9;
const size_t mem_size_0dir   = sizeof(double)*NX*NY;
const size_t mem_size_n0dir  = sizeof(double)*NX*NY*(ndir-1);
const size_t mem_size_scalar = sizeof(double)*NX*NY;
```

One additional array, prop_gpu, is used for computing the energy and error of the flow solution. The expression for its size, mem_size_props, will be explained when we look at the gpu_compute_flow_properties kernel, which performs the computations of these scalars. We only need to allocate one scalar array on the host, for which we use the C function void* malloc(size_t). Finally, the total allocated GPU memory (in bytes) is stored in total_mem_bytes for future use.

Before initialising the GPU memory, we first create two event objects that will be used to measure the execution time of the code:

```
cudaEvent_t start, stop;
checkCudaErrors(cudaEventCreate(&start));
checkCudaErrors(cudaEventCreate(&stop));
```

The functions that initialise the simulation variables have the same names as their CPU counterparts, but instead of directly performing their tasks, they invoke kernels that carry out the required operations on the GPU. The purpose of organising the code in this way is to hide the details of launching kernels from the main function. Defined in LBM.cu, these functions are taylor_green(unsigned int t, double *r, double *u, double *v), which computes the density (stored in r) and velocity components (u and v) for a decaying Taylor-Green vortex at any time t, and init_equilibrium(double *f0, double *f1, double *r, double *u, double *v), which initialises the populations (in f0 and f1) with the equilibrium populations for the supplied density (r) and velocity components (u and v). The code for these functions, the GPU kernels they use, and a supporting GPU function are shown in Listing 13.26. The two CPU functions are effectively identical: they define the dimensions of the block and grid used to run the kernel, launch the kernel, and then check for any errors in starting the kernel.

It should be noted that kernel launches are asynchronous, meaning that the CPU code is free to continue with other tasks while the GPU kernel executes. A consequence of this, however, is that errors from previous kernel launches could be caught in unexpected places, which should be kept in mind when debugging. Error checking is performed with the `getLastCudaError` macro, which is defined in `LBM.cu`. This macro allows the error message to include the line number at which the error was found, which is useful for debugging. Errors in launching kernels occur when the dimensions of the block or grid, the number of registers, or the static memory required to satisfy the request exceed the device's limits.

Exercise 13.24 Try launching a kernel while requesting many more threads in each block than the maximum possible. First, compile a sample kernel and examine the compilation output to determine how many registers the kernel requires. What happens when this kernel is launched by a block with too many threads?

A simple block and grid geometry is used for all kernel launches in the GPU code: blocks are one-dimensional with `nThreads` threads. The threads of a block therefore update nodes with consecutive x coordinates; this has a significant impact on the choice of memory layout, which will be described later. For simplicity we assume the domain size in the x direction is a multiple of the number of threads in a block. The dimensions of the grid are a width equal to the domain width divided by the number of threads in a block and a height equal to the height of the domain. An example of this is shown in Fig. 13.15. It follows that each thread updates the node with the coordinates `x` and `y` determined as:

```
unsigned int x = blockIdx.x * blockDim.x + threadIdx.x;
unsigned int y = blockIdx.y;
```

Comparing with the CPU version (Sect. 13.3.1), we see that the majority of the code has been moved into the kernels. `for` loops that iterate over the nodes' coordinates are now absent: they have been replaced by the logical organisation of threads into a grid of blocks. In the kernel that computes the Taylor-Green flow, `gpu_taylor_green`, we simply call the `__device__` function `taylor_green_eval`, passing it the time, the coordinates of the node being updated, and the addresses to which it will store the density and velocity components. These values are subsequently used when the `gpu_init_equilibrium` kernel runs to compute the equilibrium populations. The code is the same as in the CPU version and includes the optimisations presented in Listing 13.15.

Listing 13.26 CPU functions, GPU kernels, and GPU function used to calculate a Taylor-Green vortex flow and initialise the populations

```
// forward declarations of kernels
__global__ void gpu_taylor_green(unsigned int,
                                 double*,double*,double*);
__global__ void gpu_init_equilibrium(double*,double*,
                                     double*,double*,double*);

__device__ void taylor_green_eval(unsigned int t,
                                  unsigned int x, unsigned int y,
                                  double *r, double *u, double *v)
{
  double kx = 2.0*M_PI/NX;
  double ky = 2.0*M_PI/NY;
  double td = 1.0/(nu*(kx*kx+ky*ky));

  double X = x+0.5;
  double Y = y+0.5;
  double ux = -u_max*sqrt(ky/kx)*cos(kx*X)*sin(ky*Y)
              *exp(-1.0*t/td);
  double uy =  u_max*sqrt(kx/ky)*sin(kx*X)*cos(ky*Y)
              *exp(-1.0*t/td);
  double P = -0.25*rho0*u_max*u_max
             *((ky/kx)*cos(2.0*kx*X)+(kx/ky)*cos(2.0*ky*Y))
             *exp(-2.0*t/td);
  double rho = rho0+3.0*P;

  *r = rho;
  *u = ux;
  *v = uy;
}

__host__ void taylor_green(unsigned int t,
                           double *r, double *u, double *v)
{
  // blocks in grid
  dim3  grid(NX/nThreads, NY, 1);
  // threads in block
  dim3  threads(nThreads, 1, 1);

  gpu_taylor_green<<< grid, threads >>>(t,r,u,v);
  getLastCudaError("gpu_taylor_green kernel error");
}

__global__ void gpu_taylor_green(unsigned int t,
                                 double *r, double *u, double *v)
{
  unsigned int y = blockIdx.y;
  unsigned int x = blockIdx.x*blockDim.x+threadIdx.x;

  taylor_green_eval(t,x,y,
                    &r[gpu_scalar_index(x,y)],
                    &u[gpu_scalar_index(x,y)],
```

```
                         &v[gpu_scalar_index(x,y)]);
}

__host__ void init_equilibrium(double *f0, double *f1,
                               double *r, double *u, double *v)
{
    // blocks in grid
    dim3  grid(NX/nThreads, NY, 1);
    // threads in block
    dim3  threads(nThreads, 1, 1);

    gpu_init_equilibrium<<< grid, threads >>>(f0,f1,r,u,v);
    getLastCudaError("gpu_init_equilibrium kernel error");
}

__global__ void gpu_init_equilibrium(double *f0, double *f1,
                               double *r, double *u, double *v)
{
    unsigned int y = blockIdx.y;
    unsigned int x = blockIdx.x*blockDim.x+threadIdx.x;

    double rho = r[gpu_scalar_index(x,y)];
    double ux  = u[gpu_scalar_index(x,y)];
    double uy  = v[gpu_scalar_index(x,y)];
    // temporary variables
    double w0r = w0*rho;
    double wsr = ws*rho;
    double wdr = wd*rho;
    double omusq = 1.0 - 1.5*(ux*ux+uy*uy);

    double tux = 3.0*ux;
    double tuy = 3.0*uy;

    f0[gpu_field0_index(x,y)]    = w0r*(omusq);

    double cidot3u = tux;
    f1[gpu_fieldn_index(x,y,1)]  =
                   wsr*(omusq + cidot3u*(1.0+0.5*cidot3u));
    cidot3u = tuy;
    f1[gpu_fieldn_index(x,y,2)]  =
                   wsr*(omusq + cidot3u*(1.0+0.5*cidot3u));
    // ... similar expressions for directions 2-4

    cidot3u = tux+tuy;
    f1[gpu_fieldn_index(x,y,5)]  =
                   wdr*(omusq + cidot3u*(1.0+0.5*cidot3u));
    // ... similar expressions for directions 6-8
}
```

In Listing 13.26, three functions are used to convert the two-dimensional coordinates of nodes in the simulation domain to one-dimensional array indices.

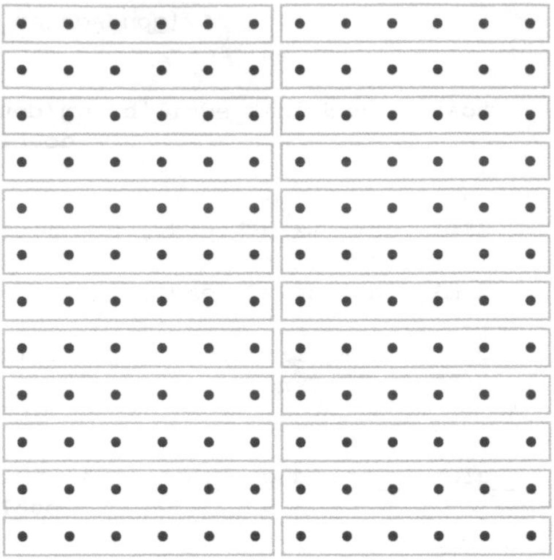

Fig. 13.15 An example showing a domain of 12 × 12 nodes split up into a 2 × 12 grid of blocks. Each block contains 6 threads that each update one node

These functions determine how memory is accessed as the domain is updated, and therefore have a significant impact on performance.

For the sequential CPU code (Sect. 13.3.2), we ensured that consecutive memory accesses were to consecutive locations in memory. For example, when updating any given node, the population f_3 is read after f_2, and should therefore be stored adjacent to it in memory. This pattern ensures that most memory accesses can be performed with data already loaded into a cache by previous accesses.

On GPUs, the multiprocessors that execute threads combine the memory accesses of each warp and only load the cache lines (Sect. 13.2.3) needed to satisfy these requests. For example, if the 32 threads in a warp all access `doubles` stored in consecutive locations, only two 128 byte transfers are needed (provided that the address of the first `double` is a multiple of 128). To take advantage of the enhanced performance due to this combining of memory accesses (also called coalesced memory access), we ensure that consecutively numbered threads access consecutive memory locations simultaneously.[16] The populations f_i for each direction are therefore arranged in memory first by their `x` index, then their `y` index, and then the direction number `d`. This arrangement matches the way that lattice nodes in the simulation domain are allocated to threads: nodes with consecutive positions in the x direction are handled by consecutively numbered threads.

The function for determining the linear index in the array of non-rest populations, `gpu_fieldn_index`, is defined as:

[16]This strict ordering of memory accesses is not necessary in general. The memory accesses within a warp are combined as long as they involve a contiguous block of memory regardless of the details of which threads access which locations in memory.

```
__device__ __forceinline__
size_t gpu_fieldn_index(unsigned int x, unsigned int y,
                        unsigned int d)
{
  return (NX*(NY*(d-1)+y)+x);
}
```

The functions for the rest populations (gpu_field0_index) and for scalar variables (gpu_scalar_index) are the same, with the data first arranged along x then y:

```
__device__ __forceinline__
size_t gpu_field0_index(unsigned int x, unsigned int y)
{
  return NX*y+x;
}

__device__ __forceinline__
size_t gpu_scalar_index(unsigned int x, unsigned int y)
{
  return NX*y+x;
}
```

These three functions have the __forceinline__ qualifier that specifies that the compiler should expand their definitions wherever they are used rather than generating an actual function. This allows the code to avoid the overhead of invoking a separate function each time an array index needs to be calculated.

With initialisation completed, the initial data is saved using functions that will be presented later, and we then proceed with the main loop of the simulation. The main loop and the initialisation code that precedes it are shown in Listing 13.27. Immediately before starting the main loop, we save the starting time of the loop in two ways for the purpose of illustrating these methods: using the system clock time and by recording an event on the GPU. The same tasks will be performed after the loop ends to determine the time spent performing all the time steps of the simulation. Inside the for loop, the update of the domain is performed in the stream_collide_save function that combines the tasks of streaming, collision, and moment computation. The remaining code provides output and swaps the pointers for the populations with non-zero directions.

Listing 13.27 Initialisation and main loop of the simulation

```
taylor_green(0,rho_gpu,ux_gpu,uy_gpu);
init_equilibrium(f0_gpu,f1_gpu,rho_gpu,ux_gpu,uy_gpu);
save_scalar("rho",rho_gpu,scalar_host,0);
save_scalar("ux", ux_gpu, scalar_host,0);
save_scalar("uy", uy_gpu, scalar_host,0);
report_flow_properties(0,rho_gpu,ux_gpu,uy_gpu,
                       prop_gpu,scalar_host);

double begin = seconds();
checkCudaErrors(cudaEventRecord(start,0));
```

```
for(unsigned int n = 0; n < NSTEPS; ++n)
{
  bool save = (n+1)%NSAVE == 0;

  stream_collide_save(f0_gpu,f1_gpu,f2_gpu,
                      rho_gpu,ux_gpu,uy_gpu,save);

  if(save)
  {
    save_scalar("rho",rho_gpu,scalar_host,n+1);
    save_scalar("ux", ux_gpu, scalar_host,n+1);
    save_scalar("uy", uy_gpu, scalar_host,n+1);

    // note: scalar_host is big enough
    //       by a factor of nThreads/7
    report_flow_properties(n+1,rho_gpu,ux_gpu,uy_gpu,
                           prop_gpu,scalar_host);
  }
}

  // swap pointers
  double *temp = f1_gpu;
  f1_gpu = f2_gpu;
  f2_gpu = temp;
}
```

Like the other "wrapper" functions, `stream_collide_save` simply launches a kernel, in this case `gpu_stream_collide_save`. This kernel reads the populations from adjacent nodes, computes the local moments, performs the collision operation, and saves the new populations. It only spends the memory bandwidth to save the moments if the argument `save` is `true`. This approach, in which streaming and collision are combined into one step, aims to minimise memory bandwidth and was also used for the CPU code (Sect. 13.3.2).

Listing 13.28 CPU "wrapper" function and the GPU kernel that it calls to perform one update of the simulation domain

```
void stream_collide_save(double *f0, double *f1, double *f2,
                         double *r, double *u, double *v,
                         bool save)
{
  // blocks in grid
  dim3  grid(NX/nThreads, NY, 1);
  // threads in block
  dim3  threads(nThreads, 1, 1);

  gpu_stream_collide_save<<< grid, threads >>>(f0,f1,f2,
                                              r,u,v,save);
  getLastCudaError("gpu_stream_collide_save kernel error");
}
```

```
__global__ void gpu_stream_collide_save(double *f0,
                                        double *f1, double *f2,
                                        double *r, double *u, double *v,
                                        bool save)
{
    unsigned int y = blockIdx.y;
    unsigned int x = blockIdx.x*blockDim.x+threadIdx.x;

    unsigned int xp1 = (x+1)%NX;
    unsigned int yp1 = (y+1)%NY;
    unsigned int xm1 = (NX+x-1)%NX;
    unsigned int ym1 = (NY+y-1)%NY;

    // 6 2 5
    // 3 0 1
    // 7 4 8

    double ft0 = f0[gpu_field0_index(x,y)];

    double ft1 = f1[gpu_fieldn_index(xm1,y,   1)];
    double ft2 = f1[gpu_fieldn_index(x,   ym1,2)];
    double ft3 = f1[gpu_fieldn_index(xp1,y,   3)];
    double ft4 = f1[gpu_fieldn_index(x,   yp1,4)];
    double ft5 = f1[gpu_fieldn_index(xm1,ym1,5)];
    double ft6 = f1[gpu_fieldn_index(xp1,ym1,6)];
    double ft7 = f1[gpu_fieldn_index(xp1,yp1,7)];
    double ft8 = f1[gpu_fieldn_index(xm1,yp1,8)];

    double rho = ft0+ft1+ft2+ft3+ft4+ft5+ft6+ft7+ft8;
    double rhoinv = 1.0/rho;

    double ux = rhoinv*(ft1+ft5+ft8-(ft3+ft6+ft7));
    double uy = rhoinv*(ft2+ft5+ft6-(ft4+ft7+ft8));

    if(save)
    {
      r[gpu_scalar_index(x,y)] = rho;
      u[gpu_scalar_index(x,y)] = ux;
      v[gpu_scalar_index(x,y)] = uy;
    }

    // temporary variables
    double tw0r = tauinv*w0*rho;
    double twsr = tauinv*ws*rho;
    double twdr = tauinv*wd*rho;
    double omusq = 1.0 - 1.5*(ux*ux+uy*uy);

    double tux = 3.0*ux;
    double tuy = 3.0*uy;

    f0[gpu_field0_index(x,y)]     = omtauinv*ft0
                        + tw0r*(omusq);

    double cidot3u = tux;
```

```
    f2[gpu_fieldn_index(x,y,1)]   = omtauinv*ft1
                + twsr*(omusq + cidot3u*(1.0+0.5*cidot3u));
    // ... similar expressions for directions 2-4

    cidot3u = tux+tuy;
    f2[gpu_fieldn_index(x,y,5)]   = omtauinv*ft5
                + twdr*(omusq + cidot3u*(1.0+0.5*cidot3u));
    // ... similar expressions for directions 6-8
}
```

Immediately after the end of the main `for` loop, we request the recording of an event. This request is asynchronous, so we then wait for completion of all pending tasks until the event is recorded. We may then compute the time elapsed between the `start` and `stop` events:

```
checkCudaErrors(cudaEventRecord(stop,0));
checkCudaErrors(cudaEventSynchronize(stop));
float milliseconds = 0.0f;
checkCudaErrors(cudaEventElapsedTime(&milliseconds,
                                     start,stop));
double end = seconds();
double runtime = end-begin;
double gpu_runtime = 0.001*milliseconds;

size_t doubles_read = ndir; // per node every time step
size_t doubles_written = ndir;
size_t doubles_saved = 3; // per node every NSAVE time steps

// note NX*NY overflows when NX=NY=65536
size_t nodes_updated = NSTEPS*size_t(NX*NY);
size_t nodes_saved   = (NSTEPS/NSAVE)*size_t(NX*NY);
double speed = nodes_updated/(1e6*runtime);

double bandwidth = (nodes_updated*(doubles_read
                                   + doubles_written)
                      + nodes_saved*(doubles_saved))
                   *sizeof(double)/(runtime*bytesPerGiB);

printf(" ----- performance information -----\n");
printf("  memory allocated (GPU): %.1f
    (MiB)\n",total_mem_bytes/bytesPerMiB);
printf(" memory allocated (host): %.1f
    (MiB)\n",mem_size_scalar/bytesPerMiB);
printf("               timesteps: %u\n",NSTEPS);
printf("           clock runtime: %.3f (s)\n",runtime);
printf("             gpu runtime: %.3f (s)\n",gpu_runtime);
printf("                   speed: %.2f (Mlups)\n",speed);
printf("               bandwidth: %.1f (GiB/s)\n",bandwidth);
```

To compute the rate at which memory is accessed when the code runs, we count the number of doubles read and written by the `gpu_stream_collide_save` kernel: 9 populations are read and written every time step, and 3 additional values (density and two velocity components) are written every `NSAVE` time steps. The total amount of memory read and written is then divided by the runtime to determine the value of `bandwidth`. In this accounting of the LBM implementation's memory transfers, we have neglected the transfers involved in computing the error in the solution for simplicity because this task is performed infrequently, and it is disabled for the performance assessment runs.

When saving simulation data, the data must first be transferred from the GPU's memory to the host's memory before it can be written to a file. Listing 13.29 presents the `save_scalar` function that is used to store the density and velocity fields at regular intervals during the simulations. Other than the need for a GPU-to-host memory transfer, the function is the same as the CPU version (Listing 13.17). In addition to the name of the scalar (`name`) and the time step (`n`), this function takes two pointer parameters: a pointer to the memory on the GPU and a pointer to host memory where the data can be temporarily stored. The transfer between GPU and host memory is performed with a call to the function

```
cudaError_t cudaMemcpy(void *dst, const void *src,
                       size_t count,
                       enum cudaMemcpyKind kind)
```

This function transfers `count` bytes from `src` to `dst`. The last parameter specifies whether `src` and `dst` reside in GPU or host memory. Here, we use `cudaMemcpyDeviceToHost`, which as its name suggests, specifies that the transfer will occur from the GPU to the host. Though not used in the code for this chapter, the other three options are: `cudaMemcpyDeviceToDevice`, `cudaMemcpyHostToDevice`, and `cudaMemcpyHostToHost`.

Listing 13.29 Function for transferring memory from the GPU to its host and saving it to a file

```
void save_scalar(const char* name, double *scalar_gpu,
                 double *scalar_host, unsigned int n)
{
  // assume reasonably-sized file names
  char filename[128];
  char format[16];

  int ndigits = floor(log10((double)NSTEPS)+1.0);

  // generate format string
  sprintf(format,"%%s%%0%dd.bin",ndigits);
  // generate file name
  sprintf(filename,format,name,n);

  // transfer memory from GPU to host
  checkCudaErrors(cudaMemcpy(scalar_host,scalar_gpu,
                             mem_size_scalar,
                             cudaMemcpyDeviceToHost));

  // open file, write to it, then close it
  FILE *fout = fopen(filename,"wb+");
```

```
  fwrite(scalar_host,1,mem_size_scalar,fout);

  fclose(fout);
}
```

The computations of the kinetic energy of the flow and the error between the numerical and exact solutions demonstrate the use of shared memory. For each of these values, a sum over all nodes in the simulation domain needs to be calculated, and shared memory is used to store temporary values. These tasks are performed in the `compute_flow_properties` CPU function that uses the kernel `gpu_compute_flow_properties`. The code for these two functions is shown in Listing 13.30. In `main.cu`, these functions are invoked through the function `report_flow_properties`, which first calls `compute_flow_properties` and then displays the results.

The `gpu_compute_flow_properties` kernel computes seven values at each node: the kinetic energy, the squares of the differences between the analytical and numerical density and two velocity components, and the squares of the three analytical values. The kernel starts by logically splitting its shared memory into seven contiguous arrays, one for each quantity. The threads first call the function `taylor_green_eval`, which was also used for initialisation, to compute the analytical density and velocity, and then each thread stores its seven computed values to the block's shared memory space. The threads are then synchronised, and the first thread of every block computes the sums of the values stored in shared memory, saving these partial sums to an array in global memory (`prop_gpu`). Within this global array, the seven values are stored in consecutive entries, with one set for each block in the grid.

Listing 13.30 CPU function and GPU kernel used to compute the kinetic energy of the flow and errors between the numerical and exact solutions

```
void compute_flow_properties(unsigned int t,
                             double *r, double *u, double *v,
                             double *prop,
                             double *prop_gpu,
                             double *prop_host)
{
  // prop must point to space for 4 doubles:
  // 0: energy
  // 1: L2 error in rho
  // 2: L2 error in ux
  // 3: L2 error in uy

  // blocks in grid
  dim3  grid(NX/nThreads, NY, 1);
  // threads in block
  dim3  threads(nThreads, 1, 1);

  gpu_compute_flow_properties<<< grid, threads,
          7*threads.x*sizeof(double) >>>(t,r,u,v,prop_gpu);
```

```
    getLastCudaError(
                    "gpu_compute_flow_properties kernel error");

    // transfer block sums to host memory
    size_t prop_size_bytes = 7*grid.x*grid.y*sizeof(double);
    checkCudaErrors(cudaMemcpy(prop_host,prop_gpu,
                               prop_size_bytes,
                               cudaMemcpyDeviceToHost));

    // initialise sums
    double E = 0.0;

    double sumrhoe2 = 0.0;
    double sumuxe2 = 0.0;
    double sumuye2 = 0.0;

    double sumrhoa2 = 0.0;
    double sumuxa2 = 0.0;
    double sumuya2 = 0.0;

    // finish summation with CPU
    for(unsigned int i = 0; i < grid.x*grid.y; ++i)
    {
      E += prop_host[7*i];
      sumrhoe2 += prop_host[7*i+1];
      sumuxe2  += prop_host[7*i+2];
      sumuye2  += prop_host[7*i+3];

      sumrhoa2 += prop_host[7*i+4];
      sumuxa2  += prop_host[7*i+5];
      sumuya2  += prop_host[7*i+6];
    }

    // compute and return final values
    prop[0] = E;
    prop[1] = sqrt(sumrhoe2/sumrhoa2);
    prop[2] = sqrt(sumuxe2/sumuxa2);
    prop[3] = sqrt(sumuye2/sumuya2);
}

__global__ void gpu_compute_flow_properties(unsigned int t,
                               double *r, double *u, double *v,
                               double *prop_gpu)
{
    unsigned int y = blockIdx.y;
    unsigned int x = blockIdx.x*blockDim.x+threadIdx.x;

    extern __shared__ double data[];

    // set up arrays for each variable
    // each array begins after the previous ends
    double *E     = data;
    double *rhoe2 = data +   blockDim.x;
    double *uxe2  = data + 2*blockDim.x;
```

```
    double *uye2  = data + 3*blockDim.x;
    double *rhoa2 = data + 4*blockDim.x;
    double *uxa2  = data + 5*blockDim.x;
    double *uya2  = data + 6*blockDim.x;

    // load density and velocity
    double rho = r[gpu_scalar_index(x,y)];
    double ux  = u[gpu_scalar_index(x,y)];
    double uy  = v[gpu_scalar_index(x,y)];

    // compute kinetic energy density
    E[threadIdx.x] = rho*(ux*ux + uy*uy);

    // compute analytical results
    double rhoa, uxa, uya;
    taylor_green_eval(t,x,y,&rhoa,&uxa,&uya);

    // compute terms for L2 error
    rhoe2[threadIdx.x] = (rho-rhoa)*(rho-rhoa);
    uxe2[threadIdx.x]  = (ux-uxa)*(ux-uxa);
    uye2[threadIdx.x]  = (uy-uya)*(uy-uya);

    rhoa2[threadIdx.x] = (rhoa-rho0)*(rhoa-rho0);
    uxa2[threadIdx.x]  = uxa*uxa;
    uya2[threadIdx.x]  = uya*uya;

    // synchronise data in shared memory
    __syncthreads();

    // only one thread proceeds
    if(threadIdx.x == 0)
    {
      // compute linear index for this block within grid
      size_t idx = 7*(gridDim.x*blockIdx.y+blockIdx.x);

      // initialise partial sums in global memory
      for(int n = 0; n < 7; ++n)
        prop_gpu[idx+n] = 0.0;

      // sum values for this block from shared memory
      for(int i = 0; i < blockDim.x; ++i)
      {
        prop_gpu[idx  ] += E[i];
        prop_gpu[idx+1] += rhoe2[i];
        prop_gpu[idx+2] += uxe2[i];
        prop_gpu[idx+3] += uye2[i];

        prop_gpu[idx+4] += rhoa2[i];
        prop_gpu[idx+5] += uxa2[i];
        prop_gpu[idx+6] += uya2[i];
      }
    }
  }
```

On the CPU side, the function `compute_flow_properties` first invokes the `gpu_compute_flow_properties` kernel. For this kernel launch, we use the third parameter in the triple angle bracket syntax that specifies the number of bytes of shared memory required by each block. In this case, we require seven times the number of thread in a block times the number of bytes in a double, i.e. `7*threads.x*sizeof(double)`. After the kernel is launched, we request a transfer of the array `prop_gpu`, which was previously allocated to have space for seven doubles for each block in the thread grid. This is why during initialisation the size of this array was computed as `7*NX/nThreads*NY*sizeof(double)`. The destination of the transfer is the CPU-side array `scalar_host`, which is larger than required to hold `prop_gpu` by a factor of `nThreads/7`. The memory transfer from GPU to CPU memory runs only after execution of the previous kernel is completed. The CPU function then proceeds to finish the summation, and it stores the final four values in the array `prop`.

Exercise 13.25 Improve the implementation of the reduction operations in `compute_flow_properties` that compute the kinetic energy of the flow and the error in the numerical solution. As a first step, implement the summation currently performed on the CPU with a second kernel. Next, modify the first kernel so that more threads participate in the summation within each block. For example, half the threads could first compute one set of partial sums, then one quarter of the threads compute the next partial sums, then one eighth of the threads compute another partial sum, and so on until one thread computes the final value of the sum. Are these summations limited by memory bandwidth or computation speed? How close to the maximum theoretical performance do your kernels reach? Can you eliminate the need for two kernels and efficiently complete the reduction with only one block? For devices with Kepler or later architecture, try using warp shuffle operations instead of shared memory. See the reduction example in the sample code provided with CUDA [24] for a discussion of optimisation strategies.

Finally, we clean up and release resources at the end of the `main` function in `main.cu` by

- destroying the event objects:

```
checkCudaErrors(cudaEventDestroy(start));
checkCudaErrors(cudaEventDestroy(stop));
```

- freeing all the memory allocated on the GPU and host:

```
checkCudaErrors(cudaFree(f0_gpu));
checkCudaErrors(cudaFree(f1_gpu));
checkCudaErrors(cudaFree(f2_gpu));
checkCudaErrors(cudaFree(rho_gpu));
checkCudaErrors(cudaFree(ux_gpu));
checkCudaErrors(cudaFree(uy_gpu));
checkCudaErrors(cudaFree(prop_gpu));
free(scalar_host);
```

- and releasing any resources associated with the GPU device:

```
cudaDeviceReset();
```

13.4.3.5 GPU Performance Optimisation and Results

To design and optimise code that takes advantage of the unique capabilities of GPUs, the performance of programs and their kernels needs to be assessed carefully. We summarise here the most important optimisations that are described in the CUDA C Best Practices Guide [28].

Many of the possible optimisations that are unique to GPU architecture are fairly intuitive to understand. Ideally, all the threads in a block, or at least within each warp, would perform the same sequence of instructions. Otherwise, the benefits of parallel execution are lost when threads must wait for other threads to finish their divergent execution paths.

The other important optimisations deal with memory use. In general, one should prefer to use faster memory wherever possible, i.e. shared memory instead of global memory, and avoid unnecessary transfers between GPU and host memory. Since the memory required by typically-sized LBM simulation domains exceeds the available shared memory, use of global memory is unavoidable. As discussed earlier, access to global memory should follow a pattern that allows the transfers requested by warps to be coalesced, or merged, into a small number of large transactions. Though not suitable for the data of the whole simulation domain, shared memory can be used as fast temporary storage space and for pre-loading data that would otherwise have to be accessed with a pattern that precludes coalescence. Furthermore, for some algorithms, it may be beneficial to use a kernel for a particular task to avoid a slow memory transfer even though a CPU might handle this task more efficiently.

In summary, avoid **divergent execution paths** in kernels, **uncoalesced access** to global memory, and **unnecessary transfers of memory** between the GPU and its host.

One way to assess the performance of a kernel is to consider the fraction of available computational resources that is used while it executes. This is called the occupancy, and it is the ratio of the maximum number of thread warps of a particular kernel than can run simultaneously to the maximum number of warps that a multiprocessor can handle. Calculation of the occupancy is complex due to the specifics of how registers are allocated and the available resources on each generation of devices. However, NVIDIA provides a spreadsheet and several API functions to perform these calculations.

Table 13.4 Performance of the GPU code on the Tesla M2070 (compute capability 2.0) and Tesla K20 (compute capability 3.5)

Domain size	Time steps	GPU memory (MiB)	Tesla M2070 (2.0) Speed (Mlups)	Tesla K20 (3.5) Speed (Mlups)
32×32	1638400	0.2	232	159
64×64	409600	0.6	507	558
128×128	102400	2.5	587	731
256×256	25600	10	612	859
512×512	6400	40	620	889
1024×1024	1600	162	635	897
2048×2048	400	647	669	895
4096×4096	100	2588	649	881

Using the verbose compilation output (Sect. 13.4.3.3), we find that the `gpu_stream_collide_save` is compiled to use 37 registers. On devices with compute capability 2.x, we have 32768 registers available for use by the threads in each block. Clearly, we could not run this kernel using blocks with 1024 threads, as this would require $37 \times 1024 = 37888$ registers in total.

When using 32 threads per block, we find that 8 blocks of this kernel, the maximum possible for this compute capability, can execute simultaneously on a multiprocessor. Therefore 8 warps (32 threads each) would be able to run on each multiprocessor out of a maximum of 48 for compute capability 2.x, an occupancy of only 16.6%. Though low, we find that the overall performance of the code, as measured by the utilisation of available memory bandwidth, is good. We leave studying the effects of changing the dimensions of the blocks used for launching the kernels in this section to interested readers.

In some cases it is beneficial to force the compiler to avoid using too many registers to improve occupancy. For such cases, one may use the `--maxrregcount` option of `nvcc` to force the compiler to ensure that kernels use fewer than the specified maximum number of registers and instead use other types of (slower) memory for temporary storage when needed.

The speed of the simulations in millions of lattice updates per second (Mlups) running on Tesla M2070 (compute capability 2.0) and K20 (compute capability 3.5) GPUs is listed in Table 13.4 for domain sizes from 32×32 to 4096×4096. For these simulations, the saving of data was disabled to assess only the time needed for computations. The error in the solution was calculated after the last time step to prevent automatic optimisation from simplifying away the main simulation loop. Comparing with the CPU performance data in Table 13.2, the simulations run more than 20 times faster on the M2070 and 30 times faster on the K20 for domains larger than 128×128. The number of timesteps was chosen so that the runtime of each simulation was about 3 seconds. In all cases, the grids used to run the kernels were one dimensional with 32 threads allocated along the x direction.

As for CPUs, an efficient memory access pattern is essential for achieving high performance. The Tesla M2070 has a 1.5 GHz memory clock, a 384 bit memory bus, and double data rate (DDR) memory, which means that two transfers can be performed per clock cycle. The theoretical maximum memory transfer rate (bandwidth) is therefore

$$\frac{1.5 \times 10^9 \text{ clock cycles}}{\text{s}} \times \frac{2 \text{ transfers}}{\text{clock cycle}} \times \frac{384 \text{ bits}}{\text{transfer}} \times \frac{1 \text{ byte}}{8 \text{ bits}} \times \frac{1 \text{ GiB}}{1024^3 \text{ bytes}} = 134 \text{ GiB/s}$$

The error detection and correction (ECC) system reduces memory bandwidth by 20% to 107 GiB/s. Since the updating of one node in a D2Q9 lattice requires reading and writing 9 doubles, a total of 144 bytes must be transferred for each node update. At a simulation speed of 669 Mlups, achieved for the 2048 × 2048 domain, memory is therefore transferred at a rate of 89.7 GiB/s, which is 84% of the theoretical maximum. The code is therefore quite efficient at using the available bandwidth.

If the **layout of the memory is flipped** (i.e. the two indices are exchanged in the expression for the linear index so that consecutively numbered threads no longer access consecutive memory locations), the **speed** of the 512 × 512 simulation, for example, **falls by a factor just over 10** to 64.2 Mlups.

The Tesla K20 has a 2.6 GHz memory clock and 320 bit bus, for which the maximum theoretical bandwidth is 194 GiB/s, about 45% more than that of the M2070. The maximum simulation speed achieved with this device, listed in Table 13.4, is 34% higher, slightly less than the increase in memory bandwidth. Better performance on the K20 can be achieved by increasing the number of threads in the blocks used to run the kernels. With 64 threads per block, the maximum simulation speed increases to 1021 Mlups on the K20.

Exercise 13.26 Determine the occupancy of the `gpu_stream_collide_save` kernel on the Tesla K20 or a GPU you can use. What is the highest possible occupancy and how much does performance improve when using the best number of threads per block?

More recent devices, for example the K40, have twice the memory capacity and bandwidth of the M2070. Simulations that require more memory or speed can be run on GPU clusters by using MPI as for CPU clusters (Sect. 13.4.2). The code in this section can also run on unconventional GPU platforms. For example, it runs on the NVIDIA Shield tablet, reaching up to 58 Mlups (57% of the available 13.8 GiB/s memory bandwidth without any tuning for this platform). This is almost double the speed of the optimised single-thread CPU version (Sect. 13.3.2). With upcoming generations of GPU devices expected to have memory bandwidths exceeding 1 TB/s, the maximum theoretical performance of a single device will reach 7 billion lattice updates per second for a D2Q9 lattice and 3 billion for D3Q19.

Due to the **high performance** they provide at **relatively low cost**, GPUs are likely to remain competitive for large-scale LBM simulations.

13.4.3.6 Further Reading

Over the past decade, the use of GPUs for LBM simulations has grown together with the performance and availability of these devices, and we provide here a brief survey of the literature describing the advances.

In 2003, Li et al. [29] presented a pre-CUDA implementation of a D3Q19 LBM simulation in which the LBM variables were stored as textures and rendering operations were used to implement the algorithm. The use of GPU clusters to overcome memory limitations and achieve higher simulation speeds was quickly recognised [30]. Soon after the release of CUDA in 2007, Tölke [31] reported a CUDA implementation of 2D LBM simulations running on an NVIDIA 8800 Ultra GPU (compute capability 1.0), which supported only single precision floating point and was limited to 768 MB of memory. This implementation used shared memory to assist propagation, and it reached 59% utilisation of the theoretical maximum memory bandwidth. Tölke and Krafczyk [32] described a similar implementation of a D3Q13 model and examined the drag on a sphere as a benchmark.

Mawson and Revell [33] provide a recent review of the developments in the use of GPUs for LBM simulations. They also examined the possibility of accelerating the LBM streaming operation by using new shuffle commands that allow quick sharing of data between the threads in a warp. These commands, which were first provided in the NVIDIA Kepler architecture (compute capability 3 and higher), did not improve performance, and the simpler "pull" algorithm for streaming was better. This is the approach shown in the code in this section, and it performs streaming by reading the density values from adjacent nodes that point towards the node being updated. In contrast, "pushing" performs streaming by saving the densities computed at a node to their destinations in adjacent nodes.

On early GPU devices, the conditions for achieving coalesced memory access were significantly more restrictive than they are now. Much of the early literature on the use of GPUs for LBM simulations examined how to implement the algorithm in a way that minimises performance losses due to non-coalesced memory access. These early devices (compute capability 1.X) are no longer supported by NVIDIA's compilation tools, and we do not consider their limitations further. When comparing performance results of their own codes and those reported in different papers, readers should be aware that many authors use single precision computations and devices without ECC memory, both of which increase processing speed.

GPUs are useful for multiphase flow simulations with lattice Boltzmann methods (Chap. 9). For example, the free-energy method for binary liquid systems was implemented on individual GPUs and multiple GPUs in a cluster to determine the critical conditions for coalescence of droplets in shear [34, 35] as well as the conditions for and characteristics of droplet breakup [36, 37]. GPUs on some of the world's largest supercomputers have been used to simulate a variety of other multiphase flow problems such as flow around a sphere using 96 GPUs on TSUBAME [38], multiphase flows in porous media on Mole-8.5 [39], and 4096 GPUs and 65536 CPUs together achieving over 200000 Mlups on Titan [40]. Several thousand of Titan's GPUs were also used to simulate flows of complex fluids with particles [41].

GPUs can also accelerate simulations of flows coupled with other physical phenomena. Obrecht et al. [42] describe an MRT LBM flow solver coupled with a finite-difference heat transfer solver running together on a GPU, and later extended to multiple GPUs in one host system [43, 44]. These authors earlier showed [45] that propagation can be efficiently performed directly in global memory without using shared memory (as in an earlier implementation [46]). They also found that the additional computations required for MRT collisions (in comparison to BGK collisions) require little time compared to the time consumed by memory transfers. This group has also worked on implementing thermal flow simulations with LBM on GPU clusters using CUDA and MPI [47].

The efficient implementation of boundary conditions is important for high overall performance especially for systems with large interfacial areas. In addition to the previously mentioned simulations of porous media, authors have examined optimisations for flows with free surfaces [48], curved boundaries [49], nonuniform grids [50], and the implementation of linearly interpolated bounceback boundary conditions [51]. As is generally the case for LB simulations, programmers need to pay attention to memory access patterns. Lattice nodes can be classified as either bulk (or fluid) nodes, at which a regular update is performed, and interfacial nodes, some of whose incoming densities are specified by boundary conditions. When the fraction of interfacial nodes is small, the speed of updating the bulk nodes should not be adversely affected by conditional expressions that test whether the nodes are bulk or interfacial nodes. For example, the kernel that updates the bulk nodes could skip all interfacial nodes or update them as if they were bulk nodes. This avoids divergent execution paths within thread blocks, and the interfacial nodes can then be updated correctly with a separate kernel. When the fraction of interfacial nodes is significant, as in porous media, the data describing the boundaries must be carefully organised and accessed to minimise the impact on the speed at which the domain is updated. Two strategies are often used [52]: an array that identifies the nature of each node or a pre-computed list of the fluid nodes.

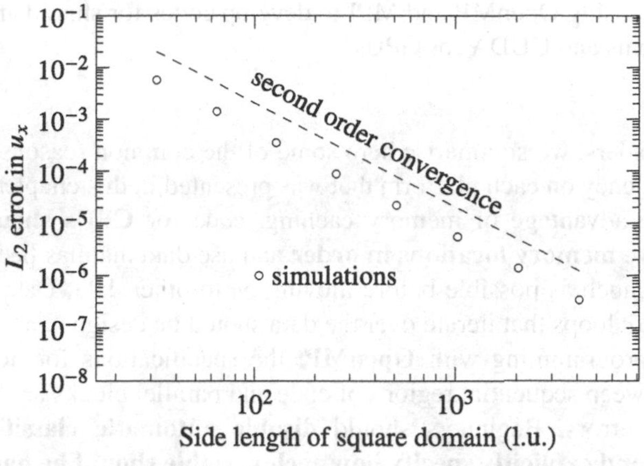

Fig. 13.16 Second-order convergence of the L_2 error in u_x for simulations of a decaying Taylor-Green vortex flow in a square domain. The domain sizes range from 32×32 to 4096×4096 lattice nodes. The L_2 errors are reported at $t/t_d = 1.93$

13.5 Convergence Study

At the start of this chapter, we set out to perform a convergence study for the decay of a Taylor-Green vortex flow (see Appendix A.3 for the definitions of the variables used in this section). Figure 13.16 shows the result of this study for domain sizes from $\ell_x \times \ell_y = 32 \times 32$ to 4096×4096 lattice nodes. As the grid is refined successively by a factor of two, we decrease u_0 by a factor of two (starting with $u_0 = 0.04$ for $\ell_x = 32$) to keep $\mathrm{Re}_\ell = u_0 \ell_x / \nu$ constant with $\nu = 1/6$. The number of time steps in each simulation was increased by a factor of 4 to simulate the same physical time in each case. The computational expense of the simulations (the total number of node updates) therefore increases by a factor of 16 for each doubling of the resolution.

The simulations were performed on a Tesla M2070 GPU, and the simulation of the largest domain completed 3 276 800 time steps in 23.5 hours, which corresponds to 649 Mlups. We clearly see that the L_2 error in the x velocity component exhibits the expected second-order convergence.

13.6 Summary

In this chapter, we have looked at writing efficient implementations of LB simulations. We examined the optimisation of arithmetic expressions and loops, automatic optimisation during compilation, and careful selection of memory access patterns to utilise memory caches correctly. We then turned our attention to parallel programming, using OpenMP and MPI to develop codes for shared and distributed memory systems and CUDA for GPUs.

To help readers, we summarise here some of the common reasons for errors and inefficiency on each platform that was presented in this chapter.

To take advantage of memory caching, code for **CPUs should access consecutive memory locations in order** and use data that has been recently loaded as much as possible before moving on to other data. Data structures and the `for` loops that iterate over the data should be designed accordingly.

When programming with **OpenMP**, the specifications for how data is shared between sequential regions of code and parallel blocks are a common source of errors. Beginners should **disable automatic classification of variables and explicitly specify how each variable should be handled**.

Dependencies in **MPI** communications between processes can cause lock-ups if they are structured incorrectly. Where possible, **prefer non-blocking communication methods over blocking methods** and use the MPI library functions to implement common communication patterns instead of writing custom codes with lower-level functions.

The threads that run kernels on **GPUs should access consecutive memory locations simultaneously**. The GPUs can then combine the threads' requests for memory so that fewer transfers need to be performed. Use blocks with as many threads as possible; the maximum is determined by the number of registers and the amount of shared memory that the kernel uses.

A recurring theme in this chapter was the significance of data transfer speeds. This topic was encountered frequently because LB simulations are generally memory-intensive rather than computing-intensive: more time is spent loading input data and storing results than carrying out the computations once data has been loaded. The dependence of LB simulation performance on memory transfer rates has broad implications. For example, researchers need not necessarily hesitate to use more computationally demanding algorithms such as TRT or MRT (Chap. 10) due to concerns about increasing simulation times. When algorithms are limited by the speed of memory access, additional computations can be "hidden" in the time that would otherwise be spent only waiting for data to be transferred. When implementing three dimensional simulations, the consequences of poor memory designs become more significant due to the increased number of populations stored at each node and the larger numbers of nodes required for high-resolution simulations.

This chapter used a single phase transient flow to demonstrate the key ideas behind writing efficient serial and parallel codes, and the concepts also apply to more complex simulations. In multiphase and multicomponent flow simulations (Chap. 9) as well as more general coupled flow and scalar transport simulations

(Chap. 8), two populations are often used. The data dependency arising from the coupling between them must be handled carefully. For example, when computing macroscopic variables and their gradients, care must be taken to reuse data as much as possible and minimise the amount of memory transferred. In all simulations, boundary conditions must be implemented efficiently and should not interfere with quick updating of bulk nodes. Simulations of porous media, in which boundary conditions must be applied at many nodes, require special treatment to ensure that the memory describing the nature of each node (bulk or interface) is used effectively. These situations can all be handled by applying the guidelines given in this chapter.

Advances in computing technology continue at a rapid pace, and researchers will undoubtedly take advantage of them for LB simulations. With development proceeding towards more parallel processing instead of faster individual cores, LBM is attractive for CFD due to its amenenability to parallisation. Upcoming generations of GPUs are expected to have terabyte (10^{12} bytes) per second memory bandwidths, and the use of processors with many cores (50 or more in one package together with inter-core communication systems) is increasing. With these and likely many other new technologies, supercomputers with exascale performance (10^{18} floating point operations per second) are expected to be available in the early 2020s. This chapter has presented the key tools and concepts needed to prepare readers to tackle the challenges of developing efficient codes for current and new platforms.

References

1. Institute of Electrical and Electronics Engineers. 754-2008 — IEEE standard for floating-point arithmetic (2008). http://dx.doi.org/10.1109/IEEESTD.2008.4610935
2. H.S. Warren Jr., *Hacker's Delight*, 2nd edn. (Addison-Wesley, Boston, 2013)
3. U. Drepper. What every programmer should know about memory (2007). https://www.akkadia.org/drepper/cpumemory.pdf
4. S. Chellappa, F. Franchetti, M. Püschel, in *Generative and Transformational Techniques in Software Engineering II: International Summer School, GTTSE 2007, Braga, Portugal, July 2–7, 2007. Revised Papers*, ed. by R. Lämmel, J. Visser, J. Saraiva (Springer, Berlin, Heidelberg, 2008), pp. 196–259
5. M. Wittmann, T. Zeiser, G. Hager, G. Wellein, Comput. Math. Appl. **65**, 924 (2013)
6. D.A. Bikulov, D.S. Senin, Vychisl. Metody Programm. **3**, 370 (2013). This article is in Russian.
7. OpenMP Architecture Review Board. About the OpenMP ARB and OpenMP.org. http://openmp.org/wp/about-openmp/
8. OpenMP Architecture Review Board. OpenMP application program interface (2011). http://www.openmp.org/mp-documents/OpenMP3.1.pdf. Version 3.1
9. OpenMP Architecture Review Board. OpenMP application programming interface (2015). http://www.openmp.org/mp-documents/openmp-4.5.pdf. Version 4.5
10. B. Barney. OpenMP. https://computing.llnl.gov/tutorials/openMP/
11. Message Passing Interface Forum. Message Passing Interface (MPI) Forum Home Page. http://www.mpi-forum.org/
12. TOP500. November 2015 TOP500 supercomputer sites. http://www.top500.org/lists/2015/11/
13. Message Passing Interface Forum. MPI: A Message-Passing Interface standard (2008). http://www.mpi-forum.org/docs/mpi-1.3/mpi-report-1.3-2008-05-30.pdf. Version 1.3

14. The Open MPI Project. Open MPI: Open Source High Performance Computing. https://www.open-mpi.org/
15. Message Passing Interface Forum. MPI documents. http://www.mpi-forum.org/docs/docs.html
16. The Open MPI Project. Open MPI documentation. https://www.open-mpi.org/doc/
17. B. Barney. Message Passing Interface (MPI). https://computing.llnl.gov/tutorials/mpi/
18. W. Gropp, E. Lusk, A. Skjellum, *Using MPI: Portable parallel programming with the Message-Passing Interface*, 3rd edn. (MIT Press, Cambridge, 2014)
19. Adaptive Computing, Inc. TORQUE resource manager. http://www.adaptivecomputing.com/products/open-source/torque/
20. Khronos Group. OpenCL. https://www.khronos.org/opencl/
21. OpenACC. Directives for accelerators. http://www.openacc.org/
22. NVIDIA. CUDA toolkit documentation. http://docs.nvidia.com/cuda/
23. NVIDIA. CUDA code samples. https://developer.nvidia.com/cuda-code-samples
24. NVIDIA. CUDA toolkit documentation. http://docs.nvidia.com/cuda/cuda-samples/
25. J. Sanders, E. Kandrot, *CUDA by Example: An Introduction to General Purpose GPU Programming* (Addison-Wesley, Boston, 2010)
26. NVIDIA. CUDA downloads. https://developer.nvidia.com/cuda-downloads
27. NVIDIA. CUDA quick start guide. http://docs.nvidia.com/cuda/pdf/CUDA_Quick_Start_Guide.pdf
28. NVIDIA. CUDA C best practices guide (2015). http://docs.nvidia.com/cuda/pdf/CUDA_C_Best_Practices_Guide.pdf
29. W. Li, X. Wei, A. Kaufman, Visual Comput. **19**, 444 (2003)
30. A. Kaufman, Z. Fan, K. Petkov, J. Stat. Mech. **2009**, P06016 (2009)
31. J. Tölke, Comput. Visual. Sci. **13**, 29 (2010)
32. J. Tölke, M. Krafczyk, Int. J. Comput. Fluid. D. **22**, 443 (2008)
33. M.J. Mawson, A.J. Revell, Comput. Phys. Commun. **185**, 2566 (2014)
34. O. Shardt, J.J. Derksen, S.K. Mitra, Langmuir **29**, 6201 (2013)
35. O. Shardt, S.K. Mitra, J.J. Derksen, Langmuir **30**, 14416 (2014)
36. A.E. Komrakova, O. Shardt, D. Eskin, J.J. Derksen, Int. J. Multiphase Flow **59**, 24 (2014)
37. A.E. Komrakova, O. Shardt, D. Eskin, J.J. Derksen, Chem. Eng. Sci. **126**, 150 (2015)
38. W. Xian, A. Takayuki, Parallel Comput. **37**, 521 (2011)
39. X. Li, Y. Zhang, X. Wang, W. Ge, Chem. Eng. Sci. **102**, 209 (2013)
40. J. McClure, H. Wang, J.F. Prins, C.T. Miller, W.C. Feng, in *Parallel and Distributed Processing Symposium, 2014 IEEE 28th International* (2014), pp. 583–592
41. A. Gray, A. Hart, O. Henrich, K. Stratford, Int. J. High Perform. C. **29**, 274 (2015)
42. C. Obrecht, F. Kuznik, B. Tourancheau, J.J. Roux, Comput. Fluids **54**, 118 (2012)
43. C. Obrecht, F. Kuznik, B. Tourancheau, J.J. Roux, Comput. Math. Appl. **65**, 252 (2013)
44. C. Obrecht, F. Kuznik, B. Tourancheau, J.J. Roux, Comput. Fluids **80**, 269 (2013)
45. C. Obrecht, F. Kuznik, B. Tourancheau, J.J. Roux, Comput. Math. Appl. **61**, 3628 (2011)
46. F. Kuznik, C. Obrecht, G. Rusaouen, J.J. Roux, Comput. Math. Appl. **59**, 2380 (2010)
47. C. Obrecht, F. Kuznik, B. Tourancheau, J.J. Roux, Parallel Comput. **39**, 259 (2013)
48. M. Schreiber, P. Neumann, S. Zimmer, H.J. Bungartz, Procedia Comput. Sci. **4**, 984 (2011)
49. H. Zhou, G. Mo, F. Wu, J. Zhao, M. Rui, K. Cen, Comput. Methods Appl. Mech. Eng. **225–228**, 984 (2011)
50. M. Schönherr, K. Kucher, M. Geier, M. Stiebler, S. Freudiger, M. Krafczyk, Comput. Math. Appl. **61**, 3730 (2011)
51. C. Obrecht, F. Kuznik, B. Tourancheau, J.J. Roux, Comput. Math. Appl. **65**, 936 (2013)
52. H. Liu, Q. Kang, C.R. Leonardi, S. Schmieschek, A. Narváez, B.D. Jones, J.R. Williams, A.J. Valocchi, J. Harting, Comput. Geosci. **20**, 777 (2016)

Appendix

A.1 Index Notation

Many equations in physics deal with vector quantities which have both a magnitude and an orientation in physical space. For instance, the simplified form of Newton's second law,

$$\boldsymbol{f} = m\boldsymbol{a}, \tag{A.1}$$

connects the vector quantity of force $\boldsymbol{f}$ and the vector quantity of acceleration $\boldsymbol{a}$, both having the same orientation.

Equations such as this can also be expressed more explicitly as three scalar equations, one for each spatial direction:

$$f_x = ma_x, \quad f_y = ma_y, \quad f_z = ma_z. \tag{A.2}$$

However, it is cumbersome to write all three equations in this way, especially when their only difference is that their index changes between x, y, and z. Instead, we can represent the same equation using only one generic index $\alpha \in \{x, y, z\}$ as

$$f_\alpha = ma_\alpha. \tag{A.3}$$

This style of notation, called *index notation*, retains the explicitness of the notation in (A.2) while remaining as brief as (A.1).

With simple vector equations like this, the advantage of index notation might not seem all that great. However, vectors are only first-order *tensors* (scalars being zeroth-order tensors). We can also apply index notation to a second-order tensor (or matrix) $\boldsymbol{A}$ by pointing to a generic element as $A_{\alpha\beta}$, α and β being two generic indices that may or may not be different. Higher-order tensors are equally explicit:

T. Krüger et al., *The Lattice Boltzmann Method*, Graduate Texts in Physics,
DOI 10.1007/978-3-319-44649-3

a generic element of the third-order tensor $\boldsymbol{R}$ is $R_{\alpha\beta\gamma}$. Indeed, this style of notation lets us immediately see the order of the tensor from the number of unique indices.

Another strength of index notation is that it allows the use of the *Einstein summation convention* where repeating the same index twice in a single term implies summation over all possible values of that index. Thus, the dot product of the vectors $\boldsymbol{a}$ and $\boldsymbol{b}$ can be expressed as

$$a_\alpha b_\alpha = \sum_\alpha a_\alpha b_\alpha = a_x b_x + a_y b_y + a_z b_z = \boldsymbol{a} \cdot \boldsymbol{b}. \tag{A.4}$$

The dot product can be expressed equally briefly in index and vector notation.

The dot product is expressed in index notation using only the Einstein summation convention, while the vector notation uses a specific, dedicated symbol "·" to express it. For the dyadic product,

$$\boldsymbol{A} = \boldsymbol{a} \otimes \boldsymbol{b} \quad \Leftrightarrow \quad A_{\alpha\beta} = a_\alpha b_\beta, \tag{A.5}$$

the vector notation requires yet another specific, dedicated symbol "⊗" while the index notation is explicit and clear: the $\alpha\beta$-component of the second-order tensor $\boldsymbol{A}$ equals the product of the α-component of the vector $\boldsymbol{a}$ and the β-component of the vector $\boldsymbol{b}$.

We may also use index notation to generalise coordinate notation: a general component of the spatial coordinate vector $\boldsymbol{x} = (x, y, z) = (x_1, x_2, x_3)$ can be written as x_α. In this way, we can also express, e.g., gradients in index notation:

$$\nabla \lambda(\boldsymbol{x}) \quad \Leftrightarrow \quad \frac{\partial \lambda(\boldsymbol{x})}{\partial x_\alpha} \quad \Leftrightarrow \quad \partial_\alpha \lambda(\boldsymbol{x}). \tag{A.6}$$

The third option is a common shorthand for derivatives, used throughout the literature and this book. Similarly, the time derivative can be expressed using the shorthand $\partial\lambda(t)/\partial t = \partial_t \lambda(t)$.

Most common vector and tensor operations can be conveniently expressed in index notation, as shown in Table A.1. One exception to this convenience is the always inconvenient cross product, which must be expressed using the *Levi-Civita symbol*

$$\varepsilon_{\alpha\beta\gamma} = \begin{cases} +1 & \text{if } (\alpha, \beta, \gamma) \text{ is } (1, 2, 3), (3, 1, 2) \text{ or } (2, 3, 1), \\ -1 & \text{if } (\alpha, \beta, \gamma) \text{ is } (3, 2, 1), (1, 3, 2) \text{ or } (2, 1, 3), \\ 0 & \text{if } \alpha = \beta, \beta = \gamma, \text{ or } \gamma = \alpha. \end{cases} \tag{A.7}$$

However, while the cross product is widely used in fields like electromagnetics, it is far less used for the topics covered in this book.

Table A.1 Examples of common operations in vector and index notation, including an index notation shorthand for derivatives

Operation	Vector notation	Index notation	Shorthand
Vector dot product	$\lambda = \boldsymbol{a} \cdot \boldsymbol{b}$	$\lambda = a_\alpha b_\alpha$	
Vector outer product	$\boldsymbol{A} = \boldsymbol{a} \otimes \boldsymbol{b}$	$A_{\alpha\beta} = a_\alpha b_\beta$	
Vector cross product	$\boldsymbol{c} = \boldsymbol{a} \times \boldsymbol{b}$	$c_\alpha = \varepsilon_{\alpha\beta\gamma} a_\beta b_\gamma$	
Tensor contraction	$\lambda = \boldsymbol{A} : \boldsymbol{B}$	$\lambda = A_{\alpha\beta} B_{\alpha\beta}$	
Gradient	$\boldsymbol{a} = \boldsymbol{\nabla} \lambda$	$a_\alpha = \partial \lambda / \partial x_\alpha$	$a_\alpha = \partial_\alpha \lambda$
Laplacian	$\Lambda = \nabla^2 \lambda$	$\Lambda = \partial^2 \lambda / (\partial x_\alpha \partial x_\alpha)$	$\Lambda = \partial_\alpha \partial_\alpha \lambda$
1st order tensor divergence	$\lambda = \boldsymbol{\nabla} \cdot \boldsymbol{a}$	$\lambda = \partial a_\alpha / \partial x_\alpha$	$\lambda = \partial_\alpha a_\alpha$
2nd order tensor divergence	$\boldsymbol{a} = \boldsymbol{\nabla} \cdot \boldsymbol{A}$	$a_\alpha = \partial A_{\alpha\beta} / \partial x_\beta$	$a_\alpha = \partial_\beta A_{\alpha\beta}$
3rd order tensor divergence	$\boldsymbol{A} = \boldsymbol{\nabla} \cdot \boldsymbol{R}$	$A_{\alpha\beta} = \partial R_{\alpha\beta\gamma} / \partial x_\gamma$	$A_{\alpha\beta} = \partial_\gamma R_{\alpha\beta\gamma}$

In this book, we use Greek indices for the Cartesian indices x, y, and z. We also use Roman indices such as i, j, and k for non-Cartesian indices; typically to index discrete velocities as e.g. $\boldsymbol{\xi}_i$. Einstein's summation convention is used *only* for the Cartesian indices.

A.2 Details in the Chapman-Enskog Analysis

A.2.1 Higher-Order Terms in the Taylor-Expanded LBE

In (4.5) we found the Taylor expansion of the $f_i(\boldsymbol{x} + \boldsymbol{c}_i \Delta t, t + \Delta t) - f_i(\boldsymbol{x}, t)$ terms in the LBE to be

$$\sum_{n=1}^{\infty} \frac{\Delta t^n}{n!} \left(\partial_t + c_{i\alpha} \partial_\alpha \right)^n . \tag{A.8}$$

We neglected terms at third order and higher in the subsequent analysis. If we can show that these terms are at least two orders higher in Kn than the lowest-order terms, this neglection is justified. This is because the ansatz was that it is only necessary to keep the two lowest orders in Kn. Let us take a closer look.

Recall that the Knudsen number is $\mathrm{Kn} = \ell_{\mathrm{mfp}}/\ell$, where ℓ_{mfp} is the mean free path and ℓ is a macroscopic length scale. Kn can be related to a similar ratio in times instead of lengths using the speed of sound c_s. As c_s is on the order of the mean particle speed in the gas [1], we find that the mean time between collisions is $\mathcal{T}_{\mathrm{mfp}} = O(\ell_{\mathrm{mfp}}/c_s)$. Additionally, we can define an acoustic time scale as $\mathcal{T}_{c_s} = \ell/c_s$, this being the time it takes for an acoustic disturbance to be felt across the length scale ℓ. Together, these two relations show that $\mathrm{Kn} = \ell_{\mathrm{mfp}}/\ell = O(\mathcal{T}_{\mathrm{mfp}}/\mathcal{T}_{c_s})$.

Collisions are the mechanism by which the distribution function relaxes to equilibrium, and relatively few collisions are required for this[1], so $\tau = O(\mathcal{T}_{\text{mfp}})$. From the space and time discretisation of Sect. 3.5, we know that $\Delta t = O(\tau)$: $\tau/\Delta t$ must be larger than 0.5 for reasons of linear stability, while we should avoid choosing $\tau/\Delta t \gg 1$ for reasons of accuracy, explained in Sect. 4.5.

The *acoustic* time scale $\mathcal{T}_{c_s} = \ell/c_s$ is typically shorter than the *advective* time scale $\mathcal{T}_u = \ell/u$ where u is a characteristic fluid velocity. Indeed, it can readily be found that $\mathcal{T}_{c_s}/\mathcal{T}_u = u/c_s = \text{Ma}$. Therefore, $\mathcal{T}_{\text{mfp}}/\mathcal{T}_{c_s} = O(\text{Kn})$ and $\mathcal{T}_{\text{mfp}}/\mathcal{T}_u = O(\text{Kn} \times \text{Ma})$; these two ratios scale at the same order in the Knudsen number.

With this said, we can now take another look at the terms in (4.5) and examine their order in the Knudsen number, allowing the characteristic time scale to be either advective or acoustic:

$$\begin{aligned}
\text{Advective:} \quad & O(\Delta t \partial_t f_i) \sim O(\tau/\mathcal{T}_u) \sim O(\text{Ma}\,\mathcal{T}_{\text{mfp}}/\mathcal{T}_{c_s}) \sim O(\text{Ma} \times \text{Kn}) \\
\text{Acoustic:} \quad & O(\Delta t \partial_t f_i) \sim O(\tau/\mathcal{T}_{c_s}) \sim O(\mathcal{T}_{\text{mfp}}/\mathcal{T}_{c_s}) \sim O(\text{Kn}) \\
& O(\Delta t c_{i\alpha} \partial_\alpha f_i) \sim O(\tau c_s/\ell) \sim O(\mathcal{T}_{\text{mfp}}/\mathcal{T}_{c_s}) \sim O(\text{Kn})
\end{aligned} \tag{A.9}$$

Consequently, we have that $\Delta t^n (\partial_t + c_{i\alpha}\partial_\alpha)^n f_i$ scales with Kn^n. Neglecting third- and higher-order terms in (4.5) is therefore consistent with our ansatz of keeping only terms of the two lowest orders in Kn.

A.2.2 The Moment Perturbation

To be able to find the macroscopic momentum equation through the Chapman-Enskog analysis in Sect. 4.1, we must determine the moment $\Pi^{(1)}_{\alpha\beta}$. This can be found from (4.10c) as

$$\Pi^{(1)}_{\alpha\beta} = -\tau \left(\partial_t^{(1)} \Pi^{\text{eq}}_{\alpha\beta} + \partial_\gamma^{(1)} \Pi^{\text{eq}}_{\alpha\beta\gamma} \right). \tag{A.10}$$

From this we would like to find an explicit expression for $\Pi^{(1)}_{\alpha\beta}$ through the macroscopic quantities ρ and $\boldsymbol{u}$ and their derivatives.

We already know the two equilibrium moments in (A.10) explicitly:

$$\Pi^{\text{eq}}_{\alpha\beta} = \rho u_\alpha u_\beta + \rho c_s^2 \delta_{\alpha\beta}, \quad \Pi^{\text{eq}}_{\alpha\beta\gamma} = \rho c_s^2 \left(u_\alpha \delta_{\beta\gamma} + u_\beta \delta_{\alpha\gamma} + u_\gamma \delta_{\alpha\beta} \right). \tag{A.11}$$

Since we would like the resulting momentum equation to be similar to the Euler and Navier-Stokes equations, all time derivatives in $\Pi^{(1)}_{\alpha\beta}$ should be eliminated. We can

[1]We show in Sect. 12.1.1 that the quantity of *viscous relaxation time* τ_{vi} is $O(\tau)$, and it is shown elsewhere [2] that $\tau_{\text{vi}} = O(\mathcal{T}_{\text{mfp}})$.

do this by rewriting (4.10) more explicitly as

$$\partial_t^{(1)}\rho = -\partial_\alpha^{(1)}(\rho u_\alpha), \quad \partial_t^{(1)}(\rho u_\alpha) = -\partial_\beta^{(1)}\left(\rho u_\alpha u_\beta + \rho c_s^2\delta_{\alpha\beta}\right). \tag{A.12}$$

We also need to make use of a corollary of the product rule; if ∂_* is a generic derivative and a, b, and c are generic variables, then

$$\partial_*(abc) = a\partial_*(bc) + b\partial_*(ac) - ab\partial_* c. \tag{A.13}$$

The following derivation is simplified by our use of the isothermal equation of state $p = c_s^2\rho$ where the pressure and density are linearly related through the constant c_s^2. In other cases, we would have to treat the pressure in a more complicated fashion. For monatomic gases, for example, we would need to express pressure changes using the conservation equation for translational energy [3].

We will now resolve the two equilibrium moment derivatives in (A.10) separately, starting with the one which is the simplest to resolve:

$$\begin{aligned}\partial_\gamma^{(1)}\Pi_{\alpha\beta\gamma}^{\text{eq}} &= \partial_\gamma^{(1)}\left(\rho c_s^2\left[u_\alpha\delta_{\beta\gamma} + u_\beta\delta_{\alpha\gamma} + u_\gamma\delta_{\alpha\beta}\right]\right)\\ &= c_s^2\left(\partial_\beta^{(1)}\rho u_\alpha + \partial_\alpha^{(1)}\rho u_\beta\right) + c_s^2\delta_{\alpha\beta}\partial_\gamma^{(1)}(\rho u_\gamma).\end{aligned} \tag{A.14}$$

The other equilibrium moment derivative is more complicated, and we will resolve it in steps. First of all, we apply (A.13) and find

$$\begin{aligned}\partial_t^{(1)}\Pi_{\alpha\beta}^{\text{eq}} &= \partial_t^{(1)}(\rho u_\alpha u_\beta + \rho c_s^2\delta_{\alpha\beta})\\ &= u_\alpha\partial_t^{(1)}(\rho u_\beta) + u_\beta\partial_t^{(1)}(\rho u_\alpha) - u_\alpha u_\beta\partial_t^{(1)}\rho + c_s^2\delta_{\alpha\beta}\partial_t^{(1)}\rho.\end{aligned} \tag{A.15a}$$

Then we apply (A.12) to replace the time derivatives and subsequently rearrange:

$$\begin{aligned}\partial_t^{(1)}\Pi_{\alpha\beta}^{\text{eq}} &= -u_\alpha\partial_\gamma^{(1)}\left(\rho u_\beta u_\gamma + \rho c_s^2\delta_{\beta\gamma}\right) - u_\beta\partial_\gamma^{(1)}\left(\rho u_\alpha u_\gamma + \rho c_s^2\delta_{\alpha\gamma}\right)\\ &\quad + u_\alpha u_\beta\partial_\gamma^{(1)}(\rho u_\gamma) - c_s^2\delta_{\alpha\beta}\partial_\gamma^{(1)}(\rho u_\gamma)\\ &= -\left[u_\alpha\partial_\gamma^{(1)}(\rho u_\beta u_\gamma) + u_\beta\partial_\gamma^{(1)}(\rho u_\alpha u_\gamma) - u_\alpha u_\beta\partial_\gamma^{(1)}(\rho u_\gamma)\right]\\ &\quad - c_s^2\left(u_\alpha\partial_\beta^{(1)}\rho + u_\beta\partial_\alpha^{(1)}\rho\right) - c_s^2\delta_{\alpha\beta}\partial_\gamma^{(1)}(\rho u_\gamma).\end{aligned} \tag{A.15b}$$

Finally, the bracketed terms can be simplified by using (A.12) in reverse, giving

$$\partial_t^{(1)}\Pi_{\alpha\beta}^{\text{eq}} = -\partial_\gamma^{(1)}(\rho u_\alpha u_\beta u_\gamma) - c_s^2\left(u_\alpha\partial_\beta^{(1)}\rho + u_\beta\partial_\alpha^{(1)}\rho\right) - c_s^2\delta_{\alpha\beta}\partial_\gamma^{(1)}(\rho u_\gamma). \tag{A.15c}$$

Now that we have explicit forms of the two equilibrium moment derivative terms in (A.14) and (A.15c), we insert them into (A.10). After using the product rule and having some terms cancel, we end up with the explicit expression

$$\Pi_{\alpha\beta}^{(1)} = -\tau\left[\rho c_s^2\left(\partial_\beta^{(1)} u_\alpha + \partial_\alpha^{(1)} u_\beta\right) - \partial_\gamma^{(1)}\left(\rho u_\alpha u_\beta u_\gamma\right)\right]. \tag{A.16}$$

The last term is an error term: it would have been entirely cancelled if $\Pi_{\alpha\beta\gamma}^{\mathrm{eq}}$ in (A.11) had contained the $\rho u_\alpha u_\beta u_\gamma$ term which it includes in the exact kinetic theory. The reason why this term is missing is that we have truncated the equilibrium distribution f_i^{eq} to $O(u^2)$. This truncation allows using smaller velocity sets like D2Q9, D3Q15, D3Q19, and D3Q27 without any undesirable anisotropy (cf. Sect. 4.2.1).

In other words, removing the $O(u^3)$ error term in (A.16) would require an extended lattice. This would slow down computations and make boundary conditions more difficult to deal with. However, recent work suggests that this error term can also be nearly cancelled by using a modified collision operator where τ depends on $\boldsymbol{u}$ [4].

A.2.3 Chapman-Enskog Analysis for the MRT Collision Operator

The Chapman-Enskog analysis in Sect. 4.1 assumes the use of the BGK collision operator. However, it is not that much more difficult to perform the analysis for the general multiple-relaxation-time (MRT) collision operator described in Chap. 10. We will here show how its Chapman-Enskog analysis differs from that in Sect. 4.1. (We will not give a full analysis here, only highlight the differences to the one given previously.) While this description will be mainly tied to the D2Q9 MRT collision operators presented in Sect. 10.4, we can use this approach for any MRT-based collision operator. At the end of this section we will point out the minor differences arising for the D3Q15 and D3Q19 results described in Sect. A.6.

The fundamental difference to the BGK analysis is that MRT collision operators can have different moments. The analysis relies on the three collision operator moments

$$\sum_i \Omega_i = 0, \quad \sum_i c_{i\alpha}\Omega_i = 0, \quad \sum_i c_{i\alpha}c_{i\beta}\Omega_i. \tag{A.17}$$

Of these moments, the first two are zero due to mass and momentum conservation in collisions. For the BGK collision operator, the third moment becomes $-(\Delta t/\tau)\Pi_{\alpha\beta}^{\mathrm{neq}}$, with $\Pi_{\alpha\beta} = \sum_i c_{i\alpha}c_{i\beta}f_i$. As we shall see, the results of the corresponding MRT analysis hinge on the differences in this moment.

We can find the second moment by left-multiplying an MRT collision operator $\boldsymbol{\Omega} = -\boldsymbol{M}^{-1}\boldsymbol{S}\boldsymbol{M}(\boldsymbol{f} - \boldsymbol{f}^{\mathrm{eq}})$ individually with the row vectors $\boldsymbol{M}_{\Pi_{xx}}$, $\boldsymbol{M}_{\Pi_{yy}}$ and $\boldsymbol{M}_{\Pi_{xy}}$

where $M_{\Pi_{\alpha\beta},i} = c_{i\alpha}c_{i\beta}$. (While it is feasible to find $\boldsymbol{SM}(\boldsymbol{f}-\boldsymbol{f}^{\text{eq}})$ analytically, $\boldsymbol{M}_{\Pi_{xx}}\boldsymbol{M}^{-1}$ *etc.* are best computed numerically.)

Thus, using the Hermite polynomial-based MRT approach from Sect. 10.4.1, we find after some algebra that

$$\sum_i c_{i\alpha}c_{i\beta}\Omega_i = -\omega_\nu \Pi^{\text{neq}}_{\alpha\beta} - \frac{\omega_\zeta - \omega_\nu}{2}\delta_{\alpha\beta}\Pi^{\text{neq}}_{\gamma\gamma}. \tag{A.18}$$

Thus, only the relaxation rates ω_ν and ω_ζ affect the macroscopic momentum equation at the Navier-Stokes level. If $\omega_\nu = \omega_\zeta$, this equation is equivalent with that of the BGK collision operator with $\omega_\nu = 1/\tau$.

Using the Gram-Schmidt approach in Sect. 10.4.2 instead, we find the same result as in (A.18) with $\omega_\zeta \to \omega_e$. Except for this tiny change in notation, the Chapman-Enskog procedure for the Hermite and Gram-Schmidt approaches are therefore identical.

Now, let us look at how the analysis itself differs from the previous BGK analysis. Using a generic collision operator Ω_i, the LBE after Taylor expansion and some algebra becomes

$$\Delta t\,(\partial_t + c_{i\alpha}\partial_\alpha) f_i = \Omega_i - \Delta t\,(\partial_t + c_{i\alpha}\partial_\alpha)\frac{\Delta t}{2}\Omega_i \tag{A.19}$$

instead of (4.7). Expanding f_i and the derivatives, the different moments at different orders in ϵ of this equation become as in (4.10) and (4.12), except that two equations have a few extra terms stemming from the $\Pi_{\alpha\beta}$ moment in (A.18):

$$\partial_t^{(1)}\Pi^{\text{eq}}_{\alpha\beta} + \partial_\gamma^{(1)}\Pi^{\text{eq}}_{\alpha\beta\gamma} = -\omega_\nu \Pi^{(1)}_{\alpha\beta} - \frac{\omega_\zeta - \omega_\nu}{2}\delta_{\alpha\beta}\Pi^{(1)}_{\gamma\gamma}, \tag{A.20a}$$

$$\partial_t^{(2)}(\rho u_\alpha) = -\partial_\beta\left[\left(1 - \frac{\omega_\nu\Delta t}{2}\right)\Pi^{(1)}_{\alpha\beta} - \frac{(\omega_\zeta - \omega_\nu)\Delta t}{4}\delta_{\alpha\beta}\Pi^{(1)}_{\gamma\gamma}\right]. \tag{A.20b}$$

Using the same procedure as in Sect. A.2.2, we can find from (A.20a) that

$$\Pi^{(1)}_{\alpha\beta} = -\frac{\rho c_s^2}{\omega_\nu}\left(\partial_\alpha u_\beta + \partial_\beta u_\alpha\right) - \frac{1}{2}\left(\frac{\omega_\zeta}{\omega_\nu} - 1\right)\delta_{\alpha\beta}\Pi^{(1)}_{\gamma\gamma}, \tag{A.21}$$

having neglected the $O(u^3)$ errors. We can make this more explicit by multiplying with $\delta_{\alpha\beta}$. As we are using the two-dimensional D2Q9 velocity set, $\delta_{\alpha\beta}\delta_{\alpha\beta} = \delta_{\gamma\gamma} = 2$, and after some rearranging we find

$$\Pi^{(1)}_{\gamma\gamma} = -\frac{2\rho c_s^2}{\omega_\zeta}\partial_\gamma u_\gamma. \tag{A.22}$$

When re-assembling the different orders in ϵ of the momentum equation, we find that the resulting viscous stress tensor $\sigma'_{\alpha\beta}$ is given by the right-hand side of (A.20b). After some algebra we find

$$\begin{aligned} \sigma'_{\alpha\beta} &= -\left(1 - \frac{\omega_\nu \Delta t}{2}\right) \Pi^{(1)}_{\alpha\beta} + \frac{(\omega_\zeta - \omega_\nu)\Delta t}{4} \delta_{\alpha\beta} \Pi^{(1)}_{\gamma\gamma} \\ &= \eta \left(\partial_\alpha u_\beta + \partial_\beta u_\alpha - \frac{2}{3} \delta_{\alpha\beta} \partial_\gamma u_\gamma \right) + \eta_{\mathrm{B}} \delta_{\alpha\beta} \partial_\gamma u_\gamma \end{aligned} \tag{A.23}$$

with the dynamic shear and bulk viscosities

$$\eta = \rho c_{\mathrm{s}}^2 \left(\frac{1}{\omega_\nu} - \frac{\Delta t}{2} \right), \quad \eta_{\mathrm{B}} = \rho c_{\mathrm{s}}^2 \left(\frac{1}{\omega_\zeta} - \frac{\Delta t}{2} \right) - \frac{\eta}{3}. \tag{A.24}$$

This result is valid for the Hermite and Gram-Schmidt D2Q9 MRT of Sect. 10.4, with $\omega_\zeta \to \omega_e$ in the Gram-Schmidt approach. We can easily confirm this by simulating a free sound wave as described in Sect. 12.1.3 and verifying that its amplitude decay varies with η and η_{B} as predicted. (Note that the amplitude may ripple around the expected exponential decay if the wave is not initialised perfectly [5].)

For the D3Q15 and D3Q19 MRT described in Sect. A.6, the analysis is the same apart from two minor differences. First, the D3Q15 and D3Q19 moments are both like in (A.18), except with a different coefficient $(\omega_e - \omega_\nu)/3$ in the last term. Additionally, $\delta_{\gamma\gamma} = 3$ instead of 2. These differences end up changing the bulk viscosity slightly, so that

$$\eta = \rho c_{\mathrm{s}}^2 \left(\frac{1}{\omega_\nu} - \frac{\Delta t}{2} \right), \quad \eta_{\mathrm{B}} = \frac{2}{3} \rho c_{\mathrm{s}}^2 \left(\frac{1}{\omega_e} - \frac{\Delta t}{2} \right). \tag{A.25}$$

This result can be verified in the same way as the D2Q9 result.

A.3 Taylor-Green Vortex Flow

The decaying *Taylor-Green vortex flow*, shown in Fig. A.1, solves the incompressible Navier-Stokes equations, (1.18). As this flow is known analytically, it is often used as a benchmark test for Navier-Stokes solvers.

The Taylor-Green flow is unsteady and fully periodic in a domain of size $\ell_x \times \ell_y$. Formulated in two spatial dimensions its velocity and pressure fields read

$$\boldsymbol{u}(\boldsymbol{x}, t) = u_0 \begin{pmatrix} -\sqrt{k_y/k_x} \cos(k_x x) \sin(k_y y) \\ \sqrt{k_x/k_y} \sin(k_x x) \cos(k_y y) \end{pmatrix} \mathrm{e}^{-t/t_{\mathrm{d}}} \tag{A.26}$$

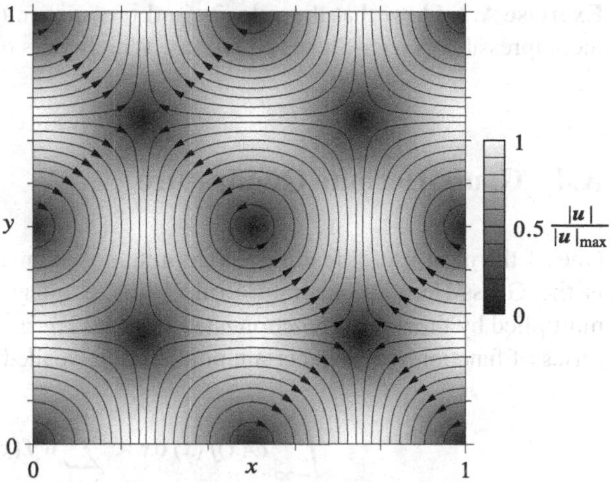

Fig. A.1 Structure of a Taylor-Green vortex flow with $k_x = k_y = 2\pi$ in $[0, 1] \times [0, 1]$. The flow maintains the same structure while decaying exponentially

and

$$p(\boldsymbol{x}, t) = p_0 - \rho\frac{u_0^2}{4}\left[\frac{k_y}{k_x}\cos(2k_x x) + \frac{k_x}{k_y}\cos(2k_y y)\right]\mathrm{e}^{-2t/t_\mathrm{d}}. \tag{A.27}$$

Here, u_0 is the initial velocity scale, $k_{x,y} = 2\pi/\ell_{x,y}$ are the components of the wave vector $\boldsymbol{k}$ and

$$t_\mathrm{d} = \frac{1}{\nu(k_x^2 + k_y^2)} \tag{A.28}$$

is the vortex decay time. The pressure average p_0 is arbitrary and does not enter the Navier-Stokes equations. The initial state is defined by $\boldsymbol{u}(\boldsymbol{x}, 0)$ and $p(\boldsymbol{x}, 0)$.

Exercise A.1 Show that the velocity field in (A.26) leads to a deviatoric stress tensor with components

$$\begin{aligned}
\sigma_{xx} &= 2\rho\nu u_0\sqrt{k_x k_y}\sin(k_x x)\sin(k_y y)\,\mathrm{e}^{-t/t_\mathrm{d}},\\
\sigma_{xy} &= \rho\nu u_0\left(\sqrt{k_x^3/k_y} - \sqrt{k_y^3/k_x}\right)\cos(k_x x)\cos(k_y y)\,\mathrm{e}^{-t/t_\mathrm{d}},\\
\sigma_{yx} &= \sigma_{xy},\\
\sigma_{yy} &= -\sigma_{xx}.
\end{aligned} \tag{A.29}$$

In particular this means that the stress tensor is symmetric and traceless. Furthermore, σ_{xy} and σ_{yx} vanish if $k_x = k_y$, i.e. if $\ell_x = \ell_y$. This means that $\ell_x \neq \ell_y$ should be chosen if one wants to investigate the accuracy of the off-diagonal stress tensor components.

Exercise A.2 Show that the velocity and pressure in (A.26) and (A.27) solve the incompressible Navier-Stokes equations. Which pairs of terms cancel each other?

A.4 Gauss-Hermite Quadrature

One of the most useful features of Hermite polynomials for numerical integration is the Gauss-Hermite quadrature rule [6]: the integral of any 1D function $f(x)$ multiplied by the weight function $\omega(x)$ (cf. (3.22)) can be approximated by a finite series of function values in certain points x_i, also called *abscissae*:

$$\int_{-\infty}^{\infty} \omega(x) f(x)\,\mathrm{d}x \approx \sum_{i=1}^{q} w_i f(x_i). \tag{A.30}$$

The accuracy of the integration depends on the values and number q of point values x_i. If one chooses x_i as the n roots of the Hermite polynomials of order n, i.e. $H^{(n)}(x_i) = 0$ and $q = n$, then it is guaranteed that any polynomial $P^{(N)}(x)$ of order $N = 2n - 1$ can be integrated exactly:

$$\int_{-\infty}^{\infty} \omega(x) P^{(N)}(x)\,\mathrm{d}x = \sum_{i=1}^{n} w_i P^{(N)}(x_i). \tag{A.31}$$

The weights can be found as [7]

$$w_i = \frac{n!}{\left(nH^{(n-1)}(x_i)\right)^2}. \tag{A.32}$$

The abscissae and weigths required to integrate polynomials up to fifth order ($N = 5$) are shown in Table A.2.

Example A.1 To integrate a third-order polynomial $P^{(3)}(x)$, one needs $n = 2$, and therefore the polynomial $H^{(2)}(x)$ with two abscissae points at ± 1 (cf. Table A.2).

Table A.2 Abscissae x_i and weights w_i for exact integration of polynomials up to fifth order

Number of abscissae	Polynomial degree	Abscissae	Weights
n	$N = 2n - 1$	x_i	w_i
1	1	0	1
2	3	± 1	$1/2$
3	5	0	$2/3$
		$\pm\sqrt{3}$	$1/6$

Together with the weights $w_{1,2} = 1/2$ for $H^{(2)}$, one obtains

$$\int_{-\infty}^{\infty} \omega(x)P^{(3)}(x)\,\mathrm{d}x = \frac{1}{2}P^{(3)}(+1) + \frac{1}{2}P^{(3)}(-1). \tag{A.33}$$

Those abscissae allow us to obtain the most common lattices for the LBM as we will describe below. More elaborate examples and advanced lattices can be found in the seminal work of Shan et al. [8] or in [9].

The extension to multiple dimensions is straightforward. Any real polynomial of order N in d-dimensional space can be written in the form

$$P^{(N)}(\boldsymbol{x}) = \sum_{N_1+\ldots+N_d\le N} a_{N_1\ldots N_d} x_1^{N_1}\cdots x_d^{N_d} \tag{A.34}$$

where the $a_{N_1\ldots N_d}$ are real coefficients and $\{N_1,\ldots,N_d\}$ are integers. A well-known example of a second-order polynomial in two dimensions is $P^{(2)}(\boldsymbol{x}) = \boldsymbol{x}\cdot\boldsymbol{x} = x_1^2 + x_2^2 + x_3^2$. Here, the mixed coefficients a_{12}, a_{23} etc. all vanish and $a_{11} = a_{22} = a_{33} = 1$. The integral of such a polynomial multiplied by the multidimensional weight function $\omega(\boldsymbol{x})$ can be written as sum of integrals with 1D weight functions $\omega(x)$:

$$\begin{aligned}\int \omega(\boldsymbol{x})P^{(N)}(\boldsymbol{x})\,\mathrm{d}^d x &= \int \omega(\boldsymbol{x})\sum a_{N_1\ldots N_d} x_1^{N_1}\cdots x_d^{N_d}\,\mathrm{d}^d x \\ &= \sum a_{N_1\ldots N_d}\prod_{j=1}^{d}\int \omega(x_j)x_j^{N_j}\,\mathrm{d}x_j,\end{aligned} \tag{A.35}$$

where we have used $\omega(\boldsymbol{x}) = \prod_{j=1}^{d}\omega(x_j)$. Each of the 1D integrals can now be decomposed using the Gauss-Hermite quadrature rule from (A.31):

$$\sum a_{N_1\ldots N_d}\prod_{j=1}^{d}\int \omega(x_j)x_j^{N_j}\,\mathrm{d}x_j = \sum a_{N_1\ldots N_d}\prod_{j=1}^{d}\sum_{i=1}^{n_j} w_{i,j}x_{i,j}^{N_j}. \tag{A.36}$$

Here, $x_{i,j}$ is the j-component of the i-th abscissa.

Let us assume that all 1D integrals are discretised using the same Hermite polynomial, i.e. $n_1 = \ldots = n_d = n$, $x_{i,1} = \ldots = x_{i,d} = x_i$ and $w_{i,1} = \ldots = w_{i,d} = w_i$. In this case, we can rewrite the product of sums:

$$\prod_{j=1}^{d}\sum_{i=1}^{n_j} w_{i,j}x_{i,j}^{N_j} = \sum_{i_1=1}^{n}\cdots\sum_{i_d=1}^{n} w_{i_1}\cdots w_{i_d}x_{i_1}^{N_1}\cdots x_{i_d}^{N_d}. \tag{A.37}$$

Introducing new multi-dimensional abscissae $\boldsymbol{x}_i = (x_{i_1}, \ldots, x_{i_D})$ and weights $w_i = w_{i_1} \cdots w_{i_D}$ allows us to obtain the multi-dimensional Gauss-Hermite quadrature rule:

$$\int \omega(\boldsymbol{x}) P^{(N)}(\boldsymbol{x})\, \mathrm{d}^d x = \sum_{i=1}^{n^d} w_i P^{(N)}(\boldsymbol{x}_i). \tag{A.38}$$

Example A.2 We demonstrate the multi-dimensional Gauss-Hermite quadrature rule by integrating the polynomial $P^{(3)}(\boldsymbol{x}) = x^2 y$ in 2D (we write x and y instead of x_1 and x_2). Since $N = 3$, we need two abscissae ($n = 2$) for each dimension, i.e. $n^d = 2^2 = 4$ in total. First we note that, for a 1D polynomial of third order, we get

$$\int w(x) x^N\, \mathrm{d}x = w_1 x_1^N + w_2 x_2^N = \frac{1}{2}(-1)^N + \frac{1}{2} 1^N \tag{A.39}$$

for $N \leq 3$. In the last step we have used the known weights and abscissae from Table A.2. It follows that

$$\begin{aligned} \int \frac{1}{2\pi} \mathrm{e}^{-(x^2+y^2)/2} x^2 y\, \mathrm{d}x\, \mathrm{d}y &= \int \frac{1}{\sqrt{2\pi}} \mathrm{e}^{-x^2/2} x^2\, \mathrm{d}x \int \frac{1}{\sqrt{2\pi}} \mathrm{e}^{-y^2/2} y\, \mathrm{d}y \\ &= \left(w_1 x_1^2 + w_2 x_2^2\right)(w_1 y_1 + w_2 y_2). \end{aligned} \tag{A.40}$$

Using $w_1 = w_2 = \frac{1}{2}$, $x_1 = y_1 = -1$ and $x_2 = y_2 = 1$, it is straightforward to show that the result is zero.

We have seen in Sect. 3.4 that one needs to integrate fifth-order polynomials in order to obtain Navier-Stokes behaviour. This implies $N = 5$ and $n = 3$. From Table A.2 we find that $n = 3$ leads to the abscissae 0 and $\pm\sqrt{3}$. After rescaling the velocities to get rid of the factor $\sqrt{3}$ (cf. Sect. 3.4.5) one obtains the D1Q3 lattice in Table 3.2.

It is now straightforward to build the corresponding 2D and 3D lattices *via* (A.37). In 2D and 3D we require $q = n^d = 3^2 = 9$ and $q = n^d = 3^3 = 27$ abscissae, respectively. As a result, we obtain the D2Q9 (cf. Table 3.3) and the D3Q27 (cf. Table 3.6) lattices [10].

Based on symmetry considerations, one can construct other lattices than D2Q9 and D3Q27. First of all, any integral of the form in (A.31) vanishes if the polynomial $P^{(N)}(x)$ contains only odd orders, e.g. $P^{(5)}(x) = 2x^5 - x^3 + x$. This can be generalised: if all monomials of a multi-dimensional polynomial contain at least one odd-order term, e.g. xy^2z^2 (which is odd in x) or $x^5y^3z^2$ (which is odd in x and y), the integral vanishes. This means that we can directly abandon all of those monomials since they do not contribute to the integral anyway. As a result, we keep only monomials of the form $x^{2a}y^{2b}z^{2c}$ where a, b and c are non-negative integers. Examples are x^2y^2 ($a = 1, b = 1, c = 0$) or $x^2y^4z^2$ ($a = 1, b = 2, c = 1$). Additionally, if our scope is limited to Navier-Stokes simulations we have to care about polynomials up to the fifth order only. The lowest-order even monomial containing x, y and z, however, is

Table A.3 Abscissae and weights for exact integration of 2D polynomials of up to fifth order

Number of abscissae	Abscissae	Weights	
q	x_i	w_i	
7	$(0, 0)$	1/2	
	$2\left(\cos\frac{m\pi}{3}, \sin\frac{m\pi}{3}\right)$	1/12	$m = 1, \ldots, 6$
9	$(0, 0)$	4/9	
	$\left(0, \pm\sqrt{3}\right), \left(\pm\sqrt{3}, 0\right)$	1/9	
	$\left(\pm\sqrt{3}, \pm\sqrt{3}\right)$	1/36	

$x^2y^2z^2$ and therefore already of sixth order. It is therefore possible to devise a lattice without the $(\pm1, \pm1, \pm1)$-velocities: D3Q19 in Table 3.5.

We are able to obtain different lattices with even fewer abscissae with other methods than symmetry arguments. For example, by using the abscissae of integrals with another weight function, one can construct D2Q7 (which is not on a square but on a hexagonal lattice) and D3Q13. These are the smallest possible sets in 2D and 3D that can still be used to solve the NSE. We skip the mathematical details and refer to [8, 9] instead.

Tables A.3 and A.4 contain the **multi-dimensional abscissae and weights** for the most common discretisations in 2D and 3D. All of these are **suitable for LB simulations of the NSE** after an appropriate renormalisation to obtain integer lattice velocity components (cf. Sect. 3.4.7).

A.5 Integration Along Characteristics for the BGK Operator

In Sect. 3.5.1, we presented a rather general scheme for the integration along characteristics where the collision operator was not yet specified. Here we will take a closer look ath the integration along characteristics with the BGK operator.

With the BGK collision operator it is possible to partially solve the continuous Boltzmann equation of the form

$$\frac{\partial f_i}{\partial t} + c_{i\alpha}\frac{\partial f_i}{\partial x_\alpha} = -\frac{f_i - f_i^{\text{eq}}}{\tau}. \tag{A.41}$$

Table A.4 Abscissae and weights for exact integration of 3D polynomials of up to fifth order

Number of abscissae	Abscissae	Weights	
q	x_i	w_i	
13	$(0, 0, 0)$	$2/5$	
	$(\pm r, \pm s, 0)$	$1/20$	$r^2 = (5 + \sqrt{5})/2$
	$(0, \pm r, \pm s)$	$1/20$	$s^2 = (5 - \sqrt{5})/2$
	$(\pm s, 0, \pm r)$	$1/20$	
15	$(0, 0, 0)$	$2/9$	
	$(\pm\sqrt{3}, 0, 0)$, $(0, \pm\sqrt{3}, 0)$, $(0, 0, \pm\sqrt{3})$	$1/9$	
	$(\pm\sqrt{3}, \pm\sqrt{3}, \pm\sqrt{3})$	$1/72$	
19	$(0, 0, 0)$	$1/3$	
	$(\pm\sqrt{3}, 0, 0)$, $(0, \pm\sqrt{3}, 0)$, $(0, 0, \pm\sqrt{3})$	$1/18$	
	$(\pm\sqrt{3}, \pm\sqrt{3}, 0)$, $(\pm\sqrt{3}, 0, \pm\sqrt{3})$, $(0, \pm\sqrt{3}, \pm\sqrt{3})$	$1/36$	
27	$(0, 0, 0)$	$8/27$	
	$(\pm\sqrt{3}, 0, 0)$, $(0, \pm\sqrt{3}, 0)$, $(0, 0, \pm\sqrt{3})$	$2/27$	
	$(\pm\sqrt{3}, \pm\sqrt{3}, 0)$, $(\pm\sqrt{3}, 0, \pm\sqrt{3})$, $(0, \pm\sqrt{3}, \pm\sqrt{3})$	$1/54$	
	$(\pm\sqrt{3}, \pm\sqrt{3}, \pm\sqrt{3})$	$1/216$	

To find the solution for f_i, we need to solve the following system in terms of the newly introduced variable ζ:

$$\frac{\mathrm{d}f_i}{\mathrm{d}\zeta} = \frac{\partial f_i}{\partial t}\frac{\mathrm{d}t}{\mathrm{d}\zeta} + \frac{\partial f_i}{\partial x_\alpha}\frac{\mathrm{d}x_\alpha}{\mathrm{d}\zeta} = -\frac{f_i(\zeta) - f_i^{\mathrm{eq}}(\zeta)}{\tau} \tag{A.42}$$

with

$$\frac{\mathrm{d}t}{\mathrm{d}\zeta} = 1, \quad \frac{\mathrm{d}x_\alpha}{\mathrm{d}\zeta} = c_{i\alpha}. \tag{A.43}$$

The dependence of f_i^{eq} on ζ enters through $f_i^{\mathrm{eq}}(\rho(\zeta), \boldsymbol{u}(\zeta))$ with the parametrisations $\rho(\zeta) = \rho(\boldsymbol{x}(\zeta), t(\zeta))$ and $\boldsymbol{u}(\zeta) = \boldsymbol{u}(\boldsymbol{x}(\zeta), t(\zeta))$.

We can easily integrate equation (A.43) to obtain the characteristics equation:

$$t = \zeta + t_0, \quad \boldsymbol{x} = \boldsymbol{c}_i\zeta + \boldsymbol{x}_0 \tag{A.44}$$

with $t_0 = t(\zeta = 0)$ and $\boldsymbol{x}_0 = \boldsymbol{x}(\zeta = 0)$.

To integrate equation (A.42), we consider the ODE

$$\frac{\mathrm{d}y(\zeta)}{\mathrm{d}\zeta} = g(\zeta)y(\zeta) + h(\zeta) \tag{A.45}$$

for $y(\zeta)$ with given coefficient functions $g(\zeta)$ and $h(\zeta)$. The solution can be obtained following the well-known *variation of constants*:

$$y(\zeta) = \mathrm{e}^{G(\zeta)} \left[C + \int_{\zeta_0}^{\zeta} \mathrm{e}^{-G(\zeta')} h(\zeta') \, \mathrm{d}\zeta' \right] \tag{A.46}$$

with

$$G(\zeta) = \int_{\zeta_0}^{\zeta} g(\zeta') \, \mathrm{d}\zeta' \tag{A.47}$$

and integration constants C and ζ_0.

Exercise A.3 Show that (A.46) solves (A.45).

We can recast (A.42) into the form of (A.45):

$$\frac{\mathrm{d}f_i(\zeta)}{\mathrm{d}\zeta} = -\frac{1}{\tau} f_i(\zeta) + \frac{f_i^{\mathrm{eq}}}{\tau}. \tag{A.48}$$

Now we identify $f_i(\zeta)$ with $y(\zeta)$, $-1/\tau$ with $g(\zeta)$ and $f_i^{\mathrm{eq}}(\zeta)/\tau$ with $h(\zeta)$. This leads to $G(\zeta) = -(\zeta - \zeta_0)/\tau$.

Using the integration limits ζ_0 and $\zeta = \zeta_0 + \Delta t$, i.e. integrating over one time step, we first obtain

$$f_i(\zeta_0 + \Delta t) = \mathrm{e}^{-\Delta t/\tau} \left[C + \frac{1}{\tau} \int_{\zeta_0}^{\zeta_0 + \Delta t} \mathrm{e}^{\zeta'/\tau} f_i^{\mathrm{eq}}(\zeta') \, \mathrm{d}\zeta' \right]. \tag{A.49}$$

We write $C = f_i(\zeta_0)$ and replace the ζ-dependence by introducing $\boldsymbol{x}$ and t again. Furthermore, we drop the index 0 from $\boldsymbol{x}_0$ and t_0 as these integration constants can be chosen arbitrarily:

$$f_i(\boldsymbol{x} + \boldsymbol{c}_i \Delta t, t + \Delta t) = \mathrm{e}^{-\Delta t/\tau} \left[f_i(\boldsymbol{x}, t) + \frac{1}{\tau} \int_t^{t+\Delta t} \mathrm{e}^{(t'-t)/\tau} f_i^{\mathrm{eq}}(\boldsymbol{x} + \boldsymbol{c}_i(t' - t), t') \, \mathrm{d}t' \right]. \tag{A.50}$$

This is the integral-form solution of the LBGK equation.

Now we want to discretise the integral in equation (A.50) to make it applicable in computer simulations. We have several options to achieve this discretisation, for instance forward Euler (first-order accurate) or the trapezoidal rule (second-order accurate).

If we choose a first-order approximation, we replace an integral of the form $\int_t^{t+\Delta t} g(t')\,\mathrm{d}t'$ by $g(t)\Delta t$. This results in

$$f_i(\boldsymbol{x}+\boldsymbol{c}_i\Delta t, t+\Delta t) = \mathrm{e}^{-\Delta t/\tau} f_i(\boldsymbol{x}, t) + \frac{\mathrm{e}^{-\Delta t/\tau}}{\tau} f_i^{\mathrm{eq}}(\boldsymbol{x}, t)\Delta t. \tag{A.51}$$

Expanding the exponentials and keeping only terms up to first order in Δt gives

$$f_i(\boldsymbol{x}+\boldsymbol{c}_i\Delta t, t+\Delta t) = \left(1-\frac{\Delta t}{\tau}\right) f_i(\boldsymbol{x}, t) + \frac{\Delta t}{\tau} f_i^{\mathrm{eq}}(\boldsymbol{x}, t) + O(\Delta t^2). \tag{A.52}$$

This is exactly the standard discretised LBGK equation, but it is only first-order accurate in time.

To achieve a second-order discretisation, we can approximate the integral with the trapezoidal rule: $\int_t^{t+\Delta t} g(t')\,\mathrm{d}t' \approx [g(t) + g(t+\Delta t)]\Delta t/2$. Therefore we have

$$\begin{aligned} f_i(\boldsymbol{x}+\boldsymbol{c}_i\Delta t, t+\Delta t) &= \mathrm{e}^{-\Delta t/\tau} f_i(\boldsymbol{x}, t) \\ &\quad + \frac{\mathrm{e}^{-\Delta t/\tau}\Delta t}{2\tau}\left(\mathrm{e}^{\Delta t/\tau} f_i^{\mathrm{eq}}(\boldsymbol{x}+\boldsymbol{c}_i\Delta t, t+\Delta t) + f_i^{\mathrm{eq}}(\boldsymbol{x}, t)\right). \end{aligned} \tag{A.53}$$

Again, we expand the exponentials, but this time we keep all terms up to second order in Δt:

$$\begin{aligned} f_i(\boldsymbol{x}+\boldsymbol{c}_i\Delta t, t+\Delta t) &= \left(1-\frac{\Delta t}{\tau}+\frac{\Delta t^2}{2\tau^2}\right) f_i(\boldsymbol{x}, t) \\ &\quad + \frac{\Delta t}{2\tau}\left[f_i^{\mathrm{eq}}(\boldsymbol{x}+\boldsymbol{c}_i\Delta t, t+\Delta t) + \left(1-\frac{\Delta t}{\tau}\right) f_i^{\mathrm{eq}}(\boldsymbol{x}, t)\right] + O(\Delta t^3). \end{aligned} \tag{A.54}$$

Our aim is to find a suitable transformation $f_i \to \bar{f}_i$ to bring (A.54) into the form of (3.76) with the BGK collision operator $\Omega_i = -(f_i - f_i^{\mathrm{eq}})/\tau$. In fact, this is possible by introducing the new population

$$\bar{f}_i = f_i - \frac{\Omega_i \Delta t}{2} = f_i + \frac{\left(f_i - f_i^{\mathrm{eq}}\right)\Delta t}{2\tau}. \tag{A.55}$$

As Ω_i conserves mass and momentum, this new population $\bar{f}_i$ has the same mass and momentum moments as f_i,

$$\begin{aligned} \sum_i \bar{f}_i &= \sum_i f_i - \sum_i \frac{\Omega_i \Delta t}{2} = \sum_i f_i = \rho, \\ \sum_i \bar{f}_i \boldsymbol{c}_i &= \sum_i f_i \boldsymbol{c}_i - \sum_i \frac{\Omega_i \boldsymbol{c}_i \Delta t}{2} = \sum_i f_i \boldsymbol{c}_i = \rho \boldsymbol{u}. \end{aligned} \tag{A.56}$$

After a series of algebraic manipulations [11] we arrive at

$$\bar{f}_i(\boldsymbol{x} + \boldsymbol{c}\Delta t, t + \Delta t) = \bar{f}_i(\boldsymbol{x}, t) - \frac{\bar{f}_i(\boldsymbol{x}, t) - f_i^{\text{eq}}(\boldsymbol{x}, t)}{\bar{\tau}} + O(\Delta t^3) \tag{A.57}$$

with a modified relaxation time

$$\bar{\tau} = \tau + \frac{\Delta t}{2}. \tag{A.58}$$

A.6 MRT for D3Q15, D3Q19, and D3Q27 Velocity Sets

We provide the matrices $\boldsymbol{M}$, the equilibrium moments m_k^{eq} and the collision rates ω_k for the most widespread 3D MRT models. The D3Q15 and D3Q19 models are based on the Gram-Schmidt procedure [12]. We only provide references for D3Q27 as it is not often used.

Note that the matrices presented here assume that the velocity sets are ordered as in Sect. 3.4.7. Different choices of velocity order would lead to matrices with differently ordered columns.

A.6.1 D3Q15

The D3Q15 Gram-Schmidt moments are

$$^{\text{G}}\boldsymbol{m} = (\rho, e, \epsilon, j_x, q_x, j_y, q_y, j_z, q_z, p_{xx}, p_{ww}, p_{xy}, p_{yz}, p_{zx}, m_{xyz})^{\top}. \tag{A.59}$$

These correspond to density, energy, energy squared, momentum, heat flux and momentum flux. The relaxation rates are

$$^{\text{G}}\boldsymbol{S} = \text{diag}(0, \omega_e, \omega_\epsilon, 0, \omega_q, 0, \omega_q, 0, \omega_q, \omega_\nu, \omega_\nu, \omega_\nu, \omega_\nu, \omega_\nu, \omega_m) \tag{A.60}$$

where collision rates of 0 are specified for the four conserved moments, i.e. density and momentum. (Note that collision rates of 0 for momentum are not suitable for simulating a force density $\boldsymbol{F}$ in the Navier-Stokes equation [13].)

The moment vectors are

$$^{\text{G}}M_{\rho,i} = 1, \quad ^{\text{G}}M_{e,i} = c_{ix}^2 + c_{iy}^2 + c_{iz}^2 - 2,$$

$$^{\text{G}}M_{\epsilon,i} = \frac{1}{2}(15(c_{ix}^2 + c_{iy}^2 + c_{iz}^2)^2 - 55(c_{ix}^2 + c_{iy}^2 + c_{iz}^2) + 32),$$

$$
\begin{aligned}
{}^{\mathrm{G}}M_{j_x,i} &= c_{ix}, \quad {}^{\mathrm{G}}M_{q_x,i} = \frac{1}{2}(5(c_{ix}^2 + c_{iy}^2 + c_{iz}^2) - 13)c_{ix}, \\
{}^{\mathrm{G}}M_{j_y,i} &= c_{iy}, \quad {}^{\mathrm{G}}M_{q_y,i} = \frac{1}{2}(5(c_{ix}^2 + c_{iy}^2 + c_{iz}^2) - 13)c_{iy}, \\
{}^{\mathrm{G}}M_{j_z,i} &= c_{iz}, \quad {}^{\mathrm{G}}M_{q_z,i} = \frac{1}{2}(5(c_{ix}^2 + c_{iy}^2 + c_{iz}^2) - 13)c_{iz}, \\
{}^{\mathrm{G}}M_{p_{xx},i} &= 3c_{ix}^2 - (c_{ix}^2 + c_{iy}^2 + c_{iz}^2), \quad {}^{\mathrm{G}}M_{p_{ww},i} = c_{iy}^2 - c_{iz}^2, \\
{}^{\mathrm{G}}M_{p_{xy},i} &= c_{ix}c_{iy}, \quad {}^{\mathrm{G}}M_{p_{yz},i} = c_{iy}c_{iz}, \quad {}^{\mathrm{G}}M_{p_{xz},i} = c_{ix}c_{iz}, \\
{}^{\mathrm{G}}M_{m_{xyz},i} &= c_{ix}c_{iy}c_{iz}.
\end{aligned}
\tag{A.61}
$$

They form the transformation matrix

$$
{}^{\mathrm{G}}\boldsymbol{M} = \begin{pmatrix}
1 & 1 & 1 & 1 & 1 & 1 & 1 & 1 & 1 & 1 & 1 & 1 & 1 & 1 & 1 \\
-2 & -1 & -1 & -1 & -1 & -1 & -1 & 1 & 1 & 1 & 1 & 1 & 1 & 1 & 1 \\
16 & -4 & -4 & -4 & -4 & -4 & -4 & 1 & 1 & 1 & 1 & 1 & 1 & 1 & 1 \\
0 & 1 & -1 & 0 & 0 & 0 & 0 & 1 & -1 & 1 & -1 & 1 & -1 & -1 & 1 \\
0 & -4 & 4 & 0 & 0 & 0 & 0 & 1 & -1 & 1 & -1 & 1 & -1 & -1 & 1 \\
0 & 0 & 0 & 1 & -1 & 0 & 0 & 1 & -1 & 1 & -1 & -1 & 1 & 1 & -1 \\
0 & 0 & 0 & -4 & 4 & 0 & 0 & 1 & -1 & 1 & -1 & -1 & 1 & 1 & -1 \\
0 & 0 & 0 & 0 & 0 & 1 & -1 & 1 & -1 & -1 & 1 & 1 & -1 & 1 & -1 \\
0 & 0 & 0 & 0 & 0 & -4 & 4 & 1 & -1 & -1 & 1 & 1 & -1 & 1 & -1 \\
0 & 2 & 2 & -1 & -1 & -1 & -1 & 0 & 0 & 0 & 0 & 0 & 0 & 0 & 0 \\
0 & 0 & 0 & 1 & 1 & -1 & -1 & 0 & 0 & 0 & 0 & 0 & 0 & 0 & 0 \\
0 & 0 & 0 & 0 & 0 & 0 & 0 & 1 & 1 & 1 & 1 & -1 & -1 & -1 & -1 \\
0 & 0 & 0 & 0 & 0 & 0 & 0 & 1 & 1 & -1 & -1 & -1 & -1 & 1 & 1 \\
0 & 0 & 0 & 0 & 0 & 0 & 0 & 1 & 1 & -1 & -1 & 1 & 1 & -1 & -1 \\
0 & 0 & 0 & 0 & 0 & 0 & 0 & 1 & -1 & -1 & 1 & -1 & 1 & -1 & 1
\end{pmatrix}.
\tag{A.62}
$$

The inverse matrix ${}^{\mathrm{G}}\boldsymbol{M}^{-1}$ can be found using a computer program.
The corresponding equilibrium moments ${}^{\mathrm{G}}m_k^{\mathrm{eq}}$ are

$$
\begin{aligned}
e^{\mathrm{eq}} &= -\rho + \rho(u_x^2 + u_y^2 + u_z^2), \quad \epsilon^{\mathrm{eq}} = \rho - 5\rho(u_x^2 + u_y^2 + u_z^2), \\
q_x^{\mathrm{eq}} &= -\frac{7}{3}\rho u_x, \quad q_y^{\mathrm{eq}} = -\frac{7}{3}\rho u_y, \quad q_z^{\mathrm{eq}} = -\frac{7}{3}\rho u_z, \\
p_{xx}^{\mathrm{eq}} &= 2\rho u_x^2 - \rho(u_y^2 + u_z^2), \quad p_{ww}^{\mathrm{eq}} = \rho(u_y^2 - u_z^2), \\
p_{xy}^{\mathrm{eq}} &= \rho u_x u_y, \quad p_{yz}^{\mathrm{eq}} = \rho u_y u_z, \quad p_{xz}^{\mathrm{eq}} = \rho u_x u_z, \\
m_{xyz}^{\mathrm{eq}} &= 0.
\end{aligned}
\tag{A.63}
$$

Note that the equilibrium moment ϵ^{eq} is not uniquely defined and can be tuned [12]. The resulting macroscopic stress tensor and viscosity are given in Sect. A.2.3.

Exercise A.4 Show that the equilibrium moments can be calculated as ${}^{\text{G}}m_k^{\text{eq}} = {}^{\text{G}}M_{ki}f_i^{\text{eq}}$ with f_i^{eq} from (3.54).

A.6.2 D3Q19

The D3Q19 Gram-Schmidt moments are

$$ {}^{\text{G}}\boldsymbol{m} = (\rho, e, \epsilon, j_x, q_x, j_y, q_y, j_z, q_z, p_{xx}, \pi_{xx}, p_{ww}, \pi_{ww}, p_{xy}, p_{yz}, p_{xz}, m_x, m_y, m_z)^\top . \tag{A.64} $$

There are additional moments compared to the D3Q15 model in Sect. A.6.1. They correspond to third-order (m_x, m_y, m_z) and fourth-order polynomials (π_{xx}, π_{ww}). The relaxation matrix reads

$$ {}^{\text{G}}\boldsymbol{S} = \text{diag}(0, \omega_e, \omega_\epsilon, 0, \omega_q, 0, \omega_q, 0, \omega_q, \omega_\nu, \omega_\pi, \omega_\nu, \omega_\pi, \omega_\nu, \omega_\nu, \omega_\nu, \omega_m, \omega_m, \omega_m). \tag{A.65} $$

The moment vectors are

$$ \begin{gathered}
{}^{\text{G}}M_{\rho,i} = 1, \quad {}^{\text{G}}M_{e,i} = 19(c_{ix}^2 + c_{iy}^2 + c_{iz}^2) - 30, \\
{}^{\text{G}}M_{\epsilon,i} = \big(21(c_{ix}^2 + c_{iy}^2 + c_{iz}^2)^2 - 53(c_{ix}^2 + c_{iy}^2 + c_{iz}^2) + 24\big)/2, \\
{}^{\text{G}}M_{j_x,i} = c_{ix}, \quad {}^{\text{G}}M_{q_x,i} = \big(5(c_{ix}^2 + c_{iy}^2 + c_{iz}^2) - 9\big)c_{ix}, \\
{}^{\text{G}}M_{j_y,i} = c_{iy}, \quad {}^{\text{G}}M_{q_y,i} = \big(5(c_{ix}^2 + c_{iy}^2 + c_{iz}^2) - 9\big)c_{iy}, \\
{}^{\text{G}}M_{j_z,i} = c_{iz}, \quad {}^{\text{G}}M_{q_z,i} = \big(5(c_{ix}^2 + c_{iy}^2 + c_{iz}^2) - 9\big)c_{iz}, \\
{}^{\text{G}}M_{p_{xx},i} = 3c_{ix}^2 - (c_{ix}^2 + c_{iy}^2 + c_{iz}^2), \\
{}^{\text{G}}M_{\pi_{xx},i} = \big(3(c_{ix}^2 + c_{iy}^2 + c_{iz}^2) - 5\big)\big(3c_{ix}^2 - (c_{ix}^2 + c_{iy}^2 + c_{iz}^2)\big), \\
{}^{\text{G}}M_{p_{ww},i} = c_{iy}^2 - c_{iz}^2, \quad {}^{\text{G}}M_{\pi_{ww},i} = \big(3(c_{ix}^2 + c_{iy}^2 + c_{iz}^2) - 5\big)\big(c_{iy}^2 - c_{iz}^2\big), \\
{}^{\text{G}}M_{p_{xy},i} = c_{ix}c_{iy}, \quad {}^{\text{G}}M_{p_{yz},i} = c_{iy}c_{iz}, \quad {}^{\text{G}}M_{p_{xz},i} = c_{ix}c_{iz} \\
{}^{\text{G}}M_{m_x,i} = (c_{iy}^2 - c_{iz}^2)c_{ix}, \quad {}^{\text{G}}M_{m_y,i} = (c_{iz}^2 - c_{ix}^2)c_{iy}, \quad {}^{\text{G}}M_{m_z,i} = (c_{ix}^2 - c_{iy}^2)c_{iz}.
\end{gathered} \tag{A.66} $$

They define the transformation matrix

$$
{}^{\mathrm{G}}\boldsymbol{M} = \begin{pmatrix}
1 & 1 & 1 & 1 & 1 & 1 & 1 & 1 & 1 & 1 & 1 & 1 & 1 & 1 & 1 & 1 & 1 & 1 & 1 \\
-30 & -11 & -11 & -11 & -11 & -11 & -11 & 8 & 8 & 8 & 8 & 8 & 8 & 8 & 8 & 8 & 8 & 8 & 8 \\
12 & -4 & -4 & -4 & -4 & -4 & -4 & 1 & 1 & 1 & 1 & 1 & 1 & 1 & 1 & 1 & 1 & 1 & 1 \\
0 & 1 & -1 & 0 & 0 & 0 & 0 & 1 & -1 & 1 & -1 & 0 & 0 & 1 & -1 & 1 & -1 & 0 & 0 \\
0 & -4 & 4 & 0 & 0 & 0 & 0 & 1 & -1 & 1 & -1 & 0 & 0 & 1 & -1 & 1 & -1 & 0 & 0 \\
0 & 0 & 0 & 1 & -1 & 0 & 0 & 1 & -1 & 0 & 0 & 1 & -1 & -1 & 1 & 0 & 0 & 1 & -1 \\
0 & 0 & 0 & -4 & 4 & 0 & 0 & 1 & -1 & 0 & 0 & 1 & -1 & -1 & 1 & 0 & 0 & 1 & -1 \\
0 & 0 & 0 & 0 & 0 & 1 & -1 & 0 & 0 & 1 & -1 & 1 & -1 & 0 & 0 & -1 & 1 & -1 & 1 \\
0 & 0 & 0 & 0 & 0 & -4 & 4 & 0 & 0 & 1 & -1 & 1 & -1 & 0 & 0 & -1 & 1 & -1 & 1 \\
0 & 2 & 2 & -1 & -1 & -1 & -1 & 1 & 1 & 1 & 1 & -2 & -2 & 1 & 1 & 1 & 1 & -2 & -2 \\
0 & -4 & -4 & 2 & 2 & 2 & 2 & 1 & 1 & 1 & 1 & -2 & -2 & 1 & 1 & 1 & 1 & -2 & -2 \\
0 & 0 & 0 & 1 & 1 & -1 & -1 & 1 & 1 & -1 & -1 & 0 & 0 & 1 & 1 & -1 & -1 & 0 & 0 \\
0 & 0 & 0 & -2 & -2 & 2 & 2 & 1 & 1 & -1 & -1 & 0 & 0 & 1 & 1 & -1 & -1 & 0 & 0 \\
0 & 0 & 0 & 0 & 0 & 0 & 0 & 1 & 1 & 0 & 0 & 0 & 0 & -1 & -1 & 0 & 0 & 0 & 0 \\
0 & 0 & 0 & 0 & 0 & 0 & 0 & 0 & 0 & 0 & 0 & 1 & 1 & 0 & 0 & 0 & 0 & -1 & -1 \\
0 & 0 & 0 & 0 & 0 & 0 & 0 & 0 & 0 & 1 & 1 & 0 & 0 & 0 & 0 & -1 & -1 & 0 & 0 \\
0 & 0 & 0 & 0 & 0 & 0 & 0 & 1 & -1 & -1 & 1 & 0 & 0 & 1 & -1 & -1 & 1 & 0 & 0 \\
0 & 0 & 0 & 0 & 0 & 0 & 0 & -1 & 1 & 0 & 0 & 1 & -1 & 1 & -1 & 0 & 0 & 1 & -1 \\
0 & 0 & 0 & 0 & 0 & 0 & 0 & 0 & 0 & 1 & -1 & -1 & 1 & 0 & 0 & -1 & 1 & 1 & -1
\end{pmatrix} . \tag{A.67}
$$

Also in this case, the inverse ${}^{\mathrm{G}}\boldsymbol{M}^{-1}$ can be found using a computer program.

The equilibrium moments are

$$
\begin{gathered}
e^{\mathrm{eq}} = -11\rho + 19\rho(u_x^2 + u_y^2 + u_z^2), \quad \epsilon^{\mathrm{eq}} = 3\rho - \frac{11}{2}\rho(u_x^2 + u_y^2 + u_z^2), \\
q_x^{\mathrm{eq}} = -\frac{2}{3}\rho u_x, \quad q_y^{\mathrm{eq}} = -\frac{2}{3}\rho u_y, \quad q_z^{\mathrm{eq}} = -\frac{2}{3}\rho u_z, \\
p_{xx}^{\mathrm{eq}} = 2\rho u_x^2 - \rho(u_y^2 + u_z^2), \quad \pi_{xx}^{\mathrm{eq}} = -\frac{1}{2}\left(2\rho u_x^2 - \rho(u_y^2 + u_z^2)\right), \\
p_{ww}^{\mathrm{eq}} = \rho(u_y^2 - u_z^2), \quad \pi_{ww}^{\mathrm{eq}} = -\frac{1}{2}\rho(u_y^2 - u_z^2), \\
p_{xy}^{\mathrm{eq}} = \rho u_x u_y, \quad p_{yz}^{\mathrm{eq}} = \rho u_y u_z, \quad p_{xz}^{\mathrm{eq}} = \rho u_x u_z, \\
m_x^{\mathrm{eq}} = 0, \quad m_y^{\mathrm{eq}} = 0, \quad m_z^{\mathrm{eq}} = 0.
\end{gathered} \tag{A.68}
$$

The resulting macroscopic stress tensor and viscosity are provided in Sect. A.2.3.

A.6.3 D3Q27

The D3Q27 model has the largest memory footprint among the common 3D models, and it is the most computationally demanding. The MRT formulation of the collision operator leads to additional computational overhead. Thus, although D3Q27 has the best isotropy properties, its MRT counterpart is not commonly used. One can construct the D3Q27 MRT model from velocity polynomials [14] and the Gram-Schmidt approach [15] that we have discussed in Sect. 10.2. Another alternative is to use a variation of the D3Q27 MRT model for the so-called cascaded lattice Boltzmann model [16, 17].

A.7 Planar Interface for the Free Energy Gas-Liquid Model

We show that the free energy multiphase model satisfies the Maxwell construction rule. To do this, let us examine the planar density profile. The stationary interface between gas and liquid phases is assumed at $x = 0$, and the density changes along the x-axis. Thus, we search the density profile in the form $\rho = \rho(x)$. Far away from the interface, we expect the fluid to assume the gas density $\rho_g = \rho(-\infty)$ and the liquid density $\rho_l = \rho(+\infty)$.

Since the interface is stationary, the momentum flux P_{xx} (cf. (9.30)) must be constant along the x-axis, i.e. $\mathrm{d}P_{xx}/\mathrm{d}x = 0$. Its value equals the bulk pressure p_0 far away from the interface where all density gradients are zero. These constraints result in

$$p_0 = p_b(\rho) + \frac{k}{2}\rho'^2 - k\rho\rho'', \tag{A.69a}$$

$$p_0 = p_b(\rho_g) = p_b(\rho_l), \tag{A.69b}$$

where the prime denotes the derivative with respect to x. Our aim is to solve this system of equations to find the values of the liquid and gas densities, ρ_l and ρ_g.

Exercise A.5 To find the density profile as function of the spatial coordinate x from (A.69a), we can introduce a substitution $z = \rho'^2$. This substitution is widely used when in a second-order ODE there is no explicit involvement of the independent variable, i.e. x. Show that the second derivative of the density obeys $\rho'' = \dot{z}/2$, where the dot denotes the derivative with respect to density ρ.

After the introduction of $z = \rho'^2$, changing the independent variable x to ρ, and performing some calculations, the ODE for the density profile becomes

$$\dot{z} - \frac{z}{\rho} = \frac{2}{k\rho}(p_b(\rho) - p_0). \tag{A.70}$$

This equation is of the form $\dot{z} + f(\rho)z = g(\rho)$ that has the solution

$$z(\rho) = \mathrm{e}^{-\int^{\rho} f(\tilde{\rho})\,\mathrm{d}\tilde{\rho}} \left(\int^{\rho} g(\tilde{\rho}) \mathrm{e}^{\int^{\tilde{\rho}} f(\tilde{\tilde{\rho}})\,\mathrm{d}\tilde{\tilde{\rho}}}\,\mathrm{d}\tilde{\rho} + C \right) \tag{A.71}$$

where C has to be found from the boundary conditions.

In our case, we identify $f(\rho) = -1/\rho$ and $g(\rho) = \frac{2}{k\rho}(p_{\mathrm{b}}(\rho) - p_0)$ and therefore

$$z(\rho) = \rho \left(\int^{\rho} \frac{2}{k\tilde{\rho}^2} \left(p_{\mathrm{b}}(\tilde{\rho}) - p_0 \right) \mathrm{d}\tilde{\rho} + C \right). \tag{A.72}$$

The boundary conditions are $z = \rho'^2 = 0$ far away from the interface where there are no density gradients, i.e. $z(\rho_{\mathrm{g}}) = 0$ and $z(\rho_{\mathrm{l}}) = 0$. This is only possible if z has the solution

$$z(\rho) = \frac{2\rho}{k} \int_{\rho_{\mathrm{g}}}^{\rho} (p_{\mathrm{b}}(\tilde{\rho}) - p_0) \frac{\mathrm{d}\tilde{\rho}}{\tilde{\rho}^2}. \tag{A.73}$$

The boundary condition $z(\rho_{\mathrm{l}}) = z(\rho_{\mathrm{g}}) = 0$ results is nothing else than the **Maxwell area construction rule for gas-liquid systems**:

$$\int_{\rho_{\mathrm{g}}}^{\rho_{\mathrm{l}}} (p_{\mathrm{b}}(\tilde{\rho}) - p_0) \frac{\mathrm{d}\tilde{\rho}}{\tilde{\rho}^2} = 0. \tag{A.74}$$

This is consistent as the gas-liquid model with the pressure tensor from (9.30) is obtained from the free-energy functional based on principles of thermodynamics.

If we want to find an expression for the density profile $\rho(x)$, we can use $z = (\mathrm{d}\rho/\mathrm{d}x)^2$ and solve the implicit integral equation assuming that the interface lies right in the middle between phases, i.e. $\bar{\rho} = \rho(x = 0) = (\rho_{\mathrm{g}} + \rho_{\mathrm{l}})/2$:

$$x = \int_{\bar{\rho}}^{\rho} \frac{\mathrm{d}\tilde{\rho}}{\sqrt{z(\tilde{\rho})}}. \tag{A.75}$$

If the equation of state $p_{\mathrm{b}}(\rho)$ includes a double-well potential as in (9.31), then the density profile will be of tanh-form as in (9.70).

A.8 Planar Interface for the Shan-Chen Liquid-Vapour Model

Appendix A.7 contains the free-energy calculations for the planar interface in a liquid-vapour system. Here we repeat these calculations in the context of the Shan-Chen (SC) model.

The SC pressure tensor from (9.112) reads (again setting $\Delta t = 1$ for simplicity)

$$P_{\alpha\beta}^{\text{SC}} = \left(c_s^2\rho + \frac{c_s^2 G}{2}\psi^2 + \frac{c_s^4 G}{4}(\nabla\psi)^2 + \frac{c_s^4 G}{2}\psi\Delta\psi \right)\delta_{\alpha\beta} - \frac{c_s^4 G}{2}(\partial_\alpha\psi)(\partial_\beta\psi). \tag{A.76}$$

We can distinguish between the equation of state $p_b(\rho) = c_s^2\rho + (c_s^2 G/2)\psi^2(\rho)$ from (9.111) that dictates the bulk behaviour and the other terms that contain derivatives of ψ and therefore are important near the interface between phases.

As in Appendix A.7, we consider a planar interface at $x = 0$. In mechanical equilibrium, the component P_{xx}^{SC} must be constant across the interface. This allows us to compute the density profile $\rho(x)$.

We can rewrite P_{xx}^{SC} as

$$P_{xx}^{\text{SC}} = c_s^2\rho + \frac{c_s^2 G}{2}\psi^2(\rho) - \frac{c_s^4 G}{4}\left(\dot{\psi}(\rho)\rho'\right)^2 + \frac{c_s^4 G}{2}\psi(\rho)\left(\ddot{\psi}(\rho)\rho' + \dot{\psi}(\rho)\rho''\right). \tag{A.77}$$

The prime denotes the derivative with respect to x, the dot the derivative with respect to the density ρ. Far away from the interface, both in the gas (g) and liquid (l) phases, gradients vanish and we find the bulk pressure

$$p_0 = p_b(\rho_g) = c_s^2\rho_g + \frac{c_s^2 G}{2}\psi^2(\rho_g) = p_b(\rho_l) = c_s^2\rho_l + \frac{c_s^2 G}{2}\psi^2(\rho_l). \tag{A.78}$$

Now we introduce $z = \rho'^2$ and perform steps similar to those detailed in Appendix A.7. This leads to

$$z(\rho) = \frac{4}{c_s^4 G}\frac{\psi(\rho)}{\dot{\psi}^2(\rho)}\int_{\rho_g}^{\rho}\left(p_0 - c_s^2\tilde{\rho} - \frac{c_s^2 G}{2}\psi^2(\tilde{\rho})\right)\frac{\dot{\psi}(\tilde{\rho})}{\psi^2(\tilde{\rho})}\mathrm{d}\tilde{\rho}. \tag{A.79}$$

To satisfy the boundary conditions $z(\rho_g) = 0$ and $z(\rho_l) = 0$ we need

$$\int_{\rho_g}^{\rho_l}\left(p_0 - c_s^2\tilde{\rho} - \frac{c_s^2 G}{2}\psi^2(\tilde{\rho})\right)\frac{\dot{\psi}(\tilde{\rho})}{\psi^2(\tilde{\rho})}\mathrm{d}\tilde{\rho} = 0. \tag{A.80}$$

We compare this expression with the Maxwell area construction rule:

$$\int_{\rho_g}^{\rho_l} \left(p_0 - c_s^2 \tilde{\rho} - \frac{c_s^2 G}{2} \psi^2(\tilde{\rho}) \right) \frac{d\tilde{\rho}}{\tilde{\rho}^2} = 0. \tag{A.81}$$

Obviously, the SC model can only reproduce thermodynamic consistency for

$$\frac{\dot{\psi}(\tilde{\rho})}{\psi^2(\tilde{\rho})} = \frac{1}{\tilde{\rho}^2} \tag{A.82}$$

which is solved by $\psi(\rho) = \rho$.

The final step is to find the density profile across the planar interface by solving the ordinary differential equation $\rho' = \sqrt{z}$ for $\rho = \rho(x)$:

$$x(\rho) = \int_{\bar{\rho}}^{\rho} \frac{d\tilde{\rho}}{\sqrt{z(\tilde{\rho})}}. \tag{A.83}$$

Here we have defined the x-axis in such a way that the interface location at $x = 0$ coincides with the average $\bar{\rho} = (\rho_l + \rho_g)/2$. In practise, however, obtaining a closed form for (A.83) is usually not possible.

A.9 Programming Reference

To assist readers who are unfamiliar with programming in C or C++, this appendix reviews the main features of these languages that are used in the code presented in Chap. 13 and accompanying the book. The code is written in standard C++ (1998). It uses only several features of C++ for convenience, such as function overloading (which allows functions to have the same name provided they have different parameters), and could be converted easily to a C code.

The text that constitutes a program, or a portion of one, is called source code. Source code for C and C++ does not need to follow strict formatting rules, and whitespace (spaces, tabs, and new lines) can be freely used to assist interpretation of the code.

Source code consists of a set of declarations that describe the units of the program and the tasks they perform. The source code for programs is often split among different files to separate it into portions that perform related tasks and to allow these portions of code to be used in different programs. The separate source files are combined into one complete program during compilation and linking (Sect. A.9.15).

One of the most important declarations in the source code of a program is for the `main()` function: this is where the program starts when the operating system runs it. In general, functions are named sequences of statements. They are used to group together the statements that accomplish a specific task and should have a descriptive name that describes what they do. Functions also allow programmers to repeat a task in different parts of a program without having to write out the same statements.

Simple statements end with a semicolon, while compound statements are sequences of simple statements enclosed in braces, { and }. Statements can involve variable declarations, evaluation of an expression and storage of the result to a variable through the use of the assignment operator =, or function calls. The various types of statements are explained in the sections that follow.

A.9.1 Comments

Comments are blocks of text that are ignored by the compiler but are useful for anyone who is reading the code. They are used to document the behaviour of code and describe design decisions. Double slashes, `//`, indicate the start of a comment that extends to the end of the current line. Comments can also be enclosed between `/*` and `*/`. Such comments may span multiple lines.

A.9.2 Expressions and Operators

Expressions can involve a variety of operators. The assignment operator = is used to save the result of evaluating its right hand side to the variable on the left hand side, as in `x = y + 5`.

The arithmetic operators are `+`, `-`, `*`, `/`, `%`, for addition, subtraction, multiplication, division, and remainder upon division, respectively, as well as the increment (`++`) and decrement (`--`) operators. These latter two operators have different effects when they appear before or after the variable they are applied to. When appearing before a variable, as in `b = ++a` or `b = --a`, they return the value in the variable after the operation is performed on it. In other words, `b = ++a` is equivalent to `a = a+1; b = a;` and `b = --a` is equivalent to `a = a-1; b = a;`. In contrast, when the operators appear after the variable, they return the value of that variable before it is modified. In this case, `b = a++` is equivalent to `b = a; a = a+1;` and `b = a--` is equivalent to `b = a; a = a-1;`. The pre- and post-increment/decrement operators are a common source of confusion, and code should be written so that the task it performs is clear. These operators can also be used without assignment, for example `a++;` by itself is equivalent to `a = a+1`.

Relational operators are used to compare values, and they are `==` (equal), `!=` (not equal), `<` (less than), `>` (greater than), `<=` (less than or equal), `>=` (greater than or equal). The difference between the assignment and equality operators is particularly important because assignment can be used in a conditional expression and returns the value that was assigned.

The logical operators are `||` (or), `&&` (and), and `!` (not). The last of these is used before the expression it modifies, for example `!(a || b)` is logically equivalent to `!a && !b`. Readers may consult standard references about the precedence rules for these operators. Parentheses can be used to specify the order of subexpression evaluation.

A.9.3 Data Types

The type of data stored in each variable must be explicitly stated before the variable is used in another statement. The names of all variables and functions are case sensitive. The main data types used in the code in this chapter are `int`, `unsigned int`, and `double`, which on the architectures we use correspond to a 32 bit signed integer (positive and negative values allowed), 32 bit unsigned (non-negative) integer, and 64 bit (double precision) floating point value, respectively. Text characters (or small numbers) are stored in `char` variables, which occupy one byte (8 bits). In special circumstances the unsigned integer type `size_t` is used, for example, to store the sizes of memory regions in bytes. On common 64 bit x86 architectures, `size_t` is a 64 bit unsigned integer. Variables that store text are arrays of `chars` (see Sect. A.9.8).

The statements `int i;` and `double d;` declare integer variables `i` and `d` to be an `int` and a `double`, respectively. Variables should not be used until they are given an initial value. Variables can be initialised when they are declared: `int x = 5;` is effectively shorthand for `int x; x = 5;`.

Variables declared with the `const` keywords are read-only variables. After they are initialised, they cannot be modified. Compilation of `const int five = 5; five = 6;` will stop with an error message.

A.9.4 Composite Data Types

Structures are composite data types that group several variables together. The components of a structure can be used in expressions or modified by writing the name of the structure variable followed by a dot and then the name of the variable within the structure. For example, to describe a vector structure containing two variables `x` and `y` we write:

```
struct vect {
   double x; double y;
};
```

We can then create one of these vectors, initialise it, and compute its norm:

```
vect v;
v.x = 5.0;
v.y = 0.1;
double norm = sqrt(v.x*v.x + v.y*v.y);
```

A.9.5 *Variable Scope*

Variables declared within a block of code or a function, i.e. between { and }, can only be used inside that block/function. The memory for these variables is automatically managed: space is set aside for the variables before entering the block/function and it is released at the end.

Global variables are declared outside of any function and can be used anywhere in the code (after they have been declared).

A.9.6 *Pointers*

An important feature of C and C++ is pointer variables. These are special variables that "refer" or "point" to another variable rather than holding a value. Such variables instead store the location of the variable they "point" to, i.e. their address in the system's memory.

The syntax for declaring a pointer is `type *v`, which declares that `v` is a variable that contains the location in memory of a variable of type `type`. Supposing we have declared `int *i;`, it is incorrect to write `i = 5;` because one would be using an integer value (5) where a valid memory address is needed (one that the program is permitted to access). To use `i` correctly, we must ensure that `i` points to a location in memory that has been reserved for an `int`. This can be done by explicitly allocating memory (see Sect. A.9.7) for an integer or by assigning the pointer the address of an integer variable.

To assign a pointer variable the address of another variable, we use the "address of" operator `&`. For example, we can write `int a = 10; int *i = &a;` to create a pointer `i` that refers to the variable `a`. We can then either directly change `a`, using `a = 2;`, or change it through the pointer `i` by using `*i = 2;`. Here, `*` is the pointer dereferencing operator, which indicates that we want to work with the value the pointer refers to rather than the location of that value in the computer's memory. Informally, we can say that `i` is the pointer, and `*i` is the value that the pointer refers to.

Pointer Arithmetic

When pointers refer to elements in an array (cf. Sect. A.9.8), arithmetic expressions involving the pointer variable can be used to access other elements in that array. For example, when `p` points to a `double`, `p+1` points to the next double, and `p-1` points to the previous double. Parentheses are essential for pointer arithmetic: `*(p+1)` refers to the contents of the double after the one at `p`, while `*p+1` is the result of adding 1 to the value pointed to by `p`.

Pointers must not be used to refer to memory that has not been reserved for use by the program (Sect. A.9.7), i.e. that is outside the bounds of an allocated array (Sect. A.9.8 and Sect. A.9.7). In such cases, behaviour is undefined and can cause the operating system to terminate the program.

A.9.7 Dynamic Memory Allocation

When the amount of memory that needs to be reserved for a variable is not known at compilation time or the required amount of memory is too large to reside in the space that is managed automatically (called the stack), programmers need to explicitly reserve regions of memory in what is called the "heap" for these variables. The allocation of memory for variables during the execution of a program is called dynamic memory allocation.

In C, the function `void* malloc(size_t size)` requests `size` bytes of memory and returns the address of the first reserved byte or a null pointer (value of zero) if the request could not be satisfied due to insufficient memory being available. Since the return type is a pointer to `void` it must be converted to a pointer to the correct type of variable. For example, the statement `double *data = (double*) malloc(100*sizeof(double))` requests space for 100 `doubles`. Note that `malloc` knows nothing about the type of variable being allocated: the `size` specified in `malloc` is the number of bytes required, not the number of variables of the desired type. We therefore use the `sizeof` operator to determine the number of bytes that a `double` occupies, since the number of bytes used for different variable types can vary across platforms.

The memory reserved by `malloc` is not initialised: it contains whatever was held at those memory addresses previously. This memory remains reserved until a later call to `void free(void *ptr)`. This function releases the memory, allowing it to be reserved by subsequent calls to `malloc`. Repeated use of `malloc` without matching calls to `free` will cause a program to use more and more memory over time. This is called a *memory leak* and must be avoided because it can interfere with other programs running on a system and will eventually lead to a call to `malloc` failing due to a lack of available memory.

In C++, the operators `new`, `delete`, and `delete[]` are used to allocate memory, free individual variables, and free arrays of variables, respectively. These operators are useful when working with the object oriented features in C++. The C functions `malloc` and `free` may still be used in C++ and they are sufficient for the simple data types we employ in the code.

A.9.8 *Arrays*

Many algorithms use sets of data that are conveniently stored in arrays. Arrays are sequences of variables of the same type that occupy contiguous blocks of memory. The values in arrays are accessed through their index (location) within the array. The syntax for declaring an array is

```
type varname[N];
```

where `N` is an integer constant that specifies the number of elements of type `type` that are available in the array with name `varname`. When the initial content of the array is known, the array can be initialised when it is declared, for example as

```
int integers[] = {1,2,3,4,5};
```

for an array of five integers. In this case, the number of elements in the array does not need to be specified and is inferred from the initialising list of values. If initial values are not provided, no assumptions can be made about the contents of the array; it does not contain zeroes by default, and its initial contents must be specified in subsequent code.

Elements of an array can be read or modified by using the array subscript operator that consists of an integer expression in between brackets. For example, the third element of the `integers` array could be modified by writing `integers[2] = 10;`. The syntax is the same when an element of an array is used in an expression, such as `x = integers[2]+5;`, which adds 5 to the third element in `integers` and stores the result to a variable `x`. Note that the first element of an array has index 0, and the index of the last element in an array with N elements is $N-1$. Programmers must ensure that index expressions are bounded between these limits; it is an error to access memory outside the limits, and can cause the operating system to terminate the program.

Dynamically allocated arrays are used for the large data sets that store the populations for the LBM simulations, as described in Sect. 13.3.1. When a pointer is known to refer to a section of memory that is occupied by an array, the array subscript operator can be used with the pointer. For example, if `double *p` points to the start of an array of 10 `doubles`, one may use `p[0]`, `p[1]`, up to `p[9]` to read or modify the elements of this array. Array indexing expressions with pointers are equivalent to applying an offset to the pointer, for example `p[5]` is equivalent to `*(p+5)`.

Text is stored and manipulated using arrays of `chars`. Such variables can be initialised using text entered in between double quotes, for example `"sample text"`. The first element of the resulting array is the letter `'s'`. Backslashes, \, followed by one letter can be used to include special characters in literal text: among others, `\n` is replaced with a new line character, `\t` represents a tab character, and `\\` is replaced with a single backslash.

A.9.9 If Statement

Several special statements are used to selectively execute and repeat sequences of statements, the first of which is an `if` statement:

```
if( conditional_expression )
{
  // statements executed if conditional_expression is true
}
else
{
  // statements executed if conditional_expression is false
}
```

If the `conditional_expression` evaluates to a logical value that is considered `true`, the first block of statements (enclosed between { and }) is executed. Otherwise the block following `else` is executed. When blocks contain only a single statement, the braces are often omitted. Such unneeded braces are usually omitted when a chain of `if` statements is used to test which one of several conditions holds. For example, it is common to write

```
if( case_1 )
{
  // statements for case 1
}
else if( case_2 )
{
  // statements for case 2
}
else if( case_3 )
{
  // statements for case 3
}
else
{
  // for when none of the previous cases hold
}
```

This is much easier to read than code that includes additional braces around each nested `if` statement.

A.9.10 While Loop

A `while` loop is written as:

```
while( conditional_expression )
{
  // statements executed while conditional_expression is true
}
```

When this loop is executed, the conditional expression is evaluated and the body is executed if the result is `true`. Evaluation of the conditional and execution of the inner block continue indefinitely until the conditional evaluates as `false`. When the conditional is `false`, execution continues with the statements that follow after the `while` loop.

A.9.11 For Loop

The second common type of loop is a `for` loop that is commonly used when an index variable is needed to keep track of the iterations of the loop. This loop has the structure

```
for(init; condition; post_loop)
{
  // body of for loop
}
```

and it is effectively equivalent to the code

```
{
  init;
  while(condition)
  {
    // body of for loop

    post_loop;
  }
}
```

This shows that a `for` loop consists of four elements: an initialisation statement `init` that is executed before the loop starts, a conditional expression `condition`, a body that is repeated as long as the conditional expression is `true`, and an update statement `post_loop` that is executed after each repetition of the loop body.

`for` loops are useful when repeating a task a particular number of times and using the iteration number within the body. For example, one might compute a sum as follows:

```
double sum = 0.0;
for(int n = 1; n < 5; ++n)
{
  sum = sum + 1.0/n;
}
```

The initialisation statement declares an integer variable `n` and initialises it to 0. The update statement increments this integer, and the conditional expression allows terms to be added to the sum as long as `n` is less than 5 (i.e. up to and including 4). Variables declared in the initialisation statement can only be used within the `for` loop.

A.9.12 Functions

Functions are defined as follows:

```
return_type function_name(type1 param1, type2 param2)
{
  // body of function
  return return_value;
}
```

This creates a function with the name `function_name` whose body consists of a sequence of statements that uses the data supplied in the parameters to generate an output. The output value is `return_value`, a variable of type `return_type`.

The parameters of the function (`param1` and `param2` or more as needed) are variables that can only be used within the body of the function. The result (or output) of the function is specified with a `return` statement that ends execution of the function, allowing execution to continue in whichever function called this function. The use of `void` as the return type indicates that the function does not return a value. In this case, `return_value` is omitted from the `return` statement, which then becomes `return;`. If no `return;` statement is used in a `void` function, the function automatically returns when the end of the its body is reached.

Functions are invoked (or called) using their name and a list of the values (called arguments) to be used for each parameter. The values of the arguments are copied into temporary variables that are then used within the function. Functions that return a value can be used within expressions, such as when using mathematical functions: `y = sin(2*x)+1.5;`. Functions that do not return a value (`void` functions) are used as a statement by themselves, for example `perform_task(value1,value2);`. Non-`void` functions can also be used in this way when their return value is not needed, such as `finish_task();` instead of `status = finish_task();` when `status` is not used subsequently.

When function parameters are pointers, the function body can modify the data they point to unless a `const` keyword is used to disallow modification.

A.9.13 Screen and File Output

The code accompanying this book uses the standard C functions for displaying text and writing data to files. The complete details about the functions mentioned below can be found in many references about the C language.

The `void printf(const char* format_string,...)` function is used to display text and convert other variables, such as integers and floating point numbers, to text. This conversion is performed by scanning `format_string` for special character sequences, called format specifiers, that start with a percent sign, `%`. The ellipsis in the function definition indicates a list of variables that is matched (in order) with each format specifier in `format_string`. The characters after the percent sign in the format specifiers indicate how the variables in the variable list should be interpreted and displayed as text. For example, `%d` is used to show signed integers in decimal notation, and `%e`, `%f`, or `%g` are used for displaying floating point numbers in different ways. Format specifiers can also include numbers and other special characters to further describe how the variables will be represented as text.

Binary data is written to files using the `fopen`, `fwrite`, and `fclose` functions as demonstrated in Sect. 13.3.3.

A.9.14 Header Files

When a source code file is compiled, the compiler does not need to know all the details about the functions and variables that are used in that file. The compiler only needs to know enough to set up function calls, allocate memory correctly, and check whether expressions involving the variables/function are permitted. The compiler assumes that the details of what functions do will be supplied later (during linking; see Sect. A.9.15). The declarations of functions (their names, parameters, and return types[2]) and their definitions (what the function does) can therefore be separated into different files. This helps organise the source code for long programs.

Files that contain the declarations of functions and variables that are used in other files are called header files, and their names end in the extension `.h`. Source files that make use of the functions and variables declared in header files must indicate that these files should be loaded before continuing with compilation. This is done with the preprocessor directive[3] `#include`. This directive has two forms: `#include <filename>` and `#include "filename"`. The first form is used to load header files for standard functions, such as `stdio.h` and `math.h`, and other libraries installed in system-specific directories. The second form is used to load header files that are stored together with the source code. The preprocessor searches for these header files in the directory that contains the source file being compiled.

A.9.15 Compilation and Linking

Compilers are programs that translate source code into binary files that contain the instructions that a processor will perform when running the program. The process of converting source code into machine instructions is called compilation. Compilation

[2]The declarations of functions are effectively the definitions of the functions with the body removed and replaced with a semicolon to indicate the end of the declaration.

[3]This is a special statement that is handled by the preprocessor, a program that performs several pre-compilation tasks that include inserting the contents of any files specified with `#include` into the source code that is then passed along to the compiler.

of a source code file produces an object file with machine instructions together with information about the functions and data structures it contains. Object files do not necessarily contain all the functions and data structures needed to form a complete program. A process called linking combines all the object files that are needed to generate a particular program. Linking creates the binary executable file that can then be run by an operating system.

In the GNU Compiler Collection, the compiler for C code is `gcc` and the compiler for C++ code is `g++`. To use `g++` to compile code for this book, a sample command is

```
g++ -c -O3 source1.cpp -o source1.o
```

This compiles the source code file `source1.cpp` with level 3 optimisation to generate the object file `source1.o`. The `-c` option indicates that we are only compiling the source file into an object file, linking should not yet be performed, and the output is not a full program.

To link several object files (and optionally first compile a source code file as well), the command is:

```
g++ -O3 source1.o source2.o main.cpp -o program
```

This compiles `main.cpp` and links the result with `source1.o` and `source2.o` to generate the program named `program`. The same command `g++` is used for both compilation and linking here. `g++` invokes other programs to carry out different steps in the compilation process, such as `ld`, which is the "linker" used for linking. `g++` determines what task to perform from the options that are used (such as `-c`) and the extensions of the files that are specified (`.cpp` or `.cc` for C++ source code and `.o` for object files).

A shortcut for quickly compiling and linking several source files directly to an executable is available:

```
g++ -O3 source1.cpp source2.cpp main.cpp -o program
```

This generates the executable file `program` from the source code files `source1.cpp`, `source2.cpp`, and `main.cpp`. On Windows, the file name of the output executable requires the extension `.exe`. After compilation, one can run the generated program on the command line by using the command `./program` on Unix systems and `program.exe` in Windows.

References

1. P.A. Thompson, *Compressible-Fluid Dynamics* (McGraw-Hill, New York, 1972)
2. L.E. Kinsler, A.R. Frey, A.B. Coppens, J.V. Sanders, *Fundamentals of Acoustics*, 4th edn. (Wiley, New York, 2000)
3. P.J. Dellar, Phys. Rev. E **64**(3) (2001)
4. P.J. Dellar, J. Comput. Phys. **259**, 270 (2014)
5. E.M. Viggen, The lattice Boltzmann method: Fundamentals and acoustics. Ph.D. thesis, Norwegian University of Science and Technology (NTNU), Trondheim (2014)

6. T. Shao, T. Chen, R. Frank, Math. Comp. **18**, 598 (1964)
7. M. Abramowitz, I. Stegun, *Handbook of Mathematical Functions with Formulas, Graphs, and Mathematical Tables* (U.S. Government Printing Office, Washington, D.C., 1964)
8. X. Shan, X.F. Yuan, H. Chen, J. Fluid Mech. **550**, 413 (2006)
9. W.P. Yudistiawan, S.K. Kwak, D.V. Patil, S. Ansumali, Phys. Rev. E **82**(4), 046701 (2010)
10. X. He, L.S. Luo, Phys. Rev. E **56**(6), 6811 (1997)
11. S. Ubertini, P. Asinari, S. Succi, Phys. Rev. E **81**(1), 016311 (2010)
12. D. d'Humières, I. Ginzburg, M. Krafczyk, P. Lallemand, L.S. Luo, Phil. Trans. R. Soc. Lond. A **360**, 437 (2002)
13. I. Ginzburg, F. Verhaeghe, D. d'Humières, Commun. Comput. Phys. **3**, 427 (2008)
14. R. Rubinstein, L.S. Luo, Phys. Rev. E **77**(036709), 1 (2008)
15. K. Suga, Y. Kuwata, K. Takashima, R. Chikasue, Comput. Math. Appl. **69**(6), 518 (2015)
16. M. Geier, A. Greiner, J. Korvink, Phys. Rev. E **73**(066705), 1 (2006)
17. K. Premnath, S. Banerjee, J. Stat. Phys. **143**, 747 (2011)

Index

T. Krüger et al., *The Lattice Boltzmann Method*, Graduate Texts in Physics,
DOI 10.1007/978-3-319-44649-3

Printed in the United States
By Bookmasters